Handbook of Plant Food Phytochemicals

Handbook of Plant Food Phytochemicals

Sources, Stability and Extraction

Edited by

B.K. Tiwari
Food and Consumer Technology Department
Hollings Faculty
Manchester Metropolitan University
Old Hall Lane
Manchester
UK

Nigel P. Brunton
School of Agriculture and Food Science
University College Dublin
Dublin
Ireland

Charles S. Brennan
Faculty of Agriculture and Life Sciences
Lincoln University
Lincoln
Canterbury
New Zealand

A John Wiley & Sons, Ltd., Publication

This edition first published 2013 © 2013 by John Wiley & Sons, Ltd

Wiley-Blackwell is an imprint of John Wiley & Sons, formed by the merger of Wiley's global Scientific, Technical and Medical business with Blackwell Publishing.

Registered Office
John Wiley & Sons, Ltd., The Atrium, Southern Gate, Chichester, West Sussex, PO19 8SQ, UK

Editorial Offices
9600 Garsington Road, Oxford, OX4 2DQ, UK
The Atrium, Southern Gate, Chichester, West Sussex, PO19 8SQ, UK
2121 State Avenue, Ames, Iowa 50014-8300, USA

For details of our global editorial offices, for customer services and for information about how to apply for permission to reuse the copyright material in this book please see our website at www.wiley.com/wiley-blackwell.

The right of the authors to be identified as the authors of this work has been asserted in accordance with the UK Copyright, Designs and Patents Act 1988.

All rights reserved. No part of this publication may be reproduced, stored in a retrieval system, or transmitted, in any form or by any means, electronic, mechanical, photocopying, recording or otherwise, except as permitted by the UK Copyright, Designs and Patents Act 1988, without the prior permission of the publisher.

Designations used by companies to distinguish their products are often claimed as trademarks. All brand names and product names used in this book are trade names, service marks, trademarks or registered trademarks of their respective owners. The publisher is not associated with any product or vendor mentioned in this book.

Limit of Liability/Disclaimer of Warranty: While the publisher and author(s) have used their best efforts in preparing this book, they make no representations or warranties with respect to the accuracy or completeness of the contents of this book and specifically disclaim any implied warranties of merchantability or fitness for a particular purpose. It is sold on the understanding that the publisher is not engaged in rendering professional services and neither the publisher nor the author shall be liable for damages arising herefrom. If professional advice or other expert assistance is required, the services of a competent professional should be sought.

Library of Congress Cataloging-in-Publication Data

Handbook of plant food phytochemicals : sources, stability and extraction / edited by Brijesh Tiwari, Nigel Brunton, Charles S. Brennan.
    p. cm.
  Includes bibliographical references and index.
  ISBN 978-1-4443-3810-2 (hardback : alk. paper) – ISBN 978-1-118-46467-0 (epdf) –
ISBN 978-1-118-46469-4 (emobi) – ISBN 978-1-118-46468-7 (epub) – ISBN 978-1-118-46471-7 (obook)
1. Phytochemicals. 2. Plants–Composition. 3. Food–Composition. 4. Food industry and trade.
I. Tiwari, Brijesh K. II. Brunton, Nigel. III. Brennan, Charles S.
  QK861.H34 2012
  580–dc23
                                2012024779

A catalogue record for this book is available from the British Library.

Wiley also publishes its books in a variety of electronic formats. Some content that appears in print may not be available in electronic books.

Cover image credits, left to right: © iStockphoto.com/KevinDyer; © iStockphoto.com/aluxum;
© iStockphoto.com/FotografiaBasica
Cover design by Meaden Creative

Set in 10/12pt Times by SPi Publisher Services, Pondicherry, India
Printed and bound in Malaysia by Vivar Printing Sdn Bhd

1 2013

Contents

Contributor list   xiii

1 Plant food phytochemicals   1
B.K. Tiwari, Nigel P. Brunton and Charles S. Brennan

   1.1 Importance of phytochemicals   1
   1.2 Book objective   2
   1.3 Book structure   2

PART I   CHEMISTRY AND HEALTH   5

2 Chemistry and classification of phytochemicals   7
Rocio Campos-Vega and B. Dave Oomah

   2.1 Introduction   7
   2.2 Classification of phytochemicals   8
      2.2.1 Terpenes   8
      2.2.2 Polyphenols   13
      2.2.3 Carotenoids   15
      2.2.4 Glucosinolates   15
      2.2.5 Dietary fiber (non starch polysaccharides)   16
      2.2.6 Lectins   17
      2.2.7 Other phytochemicals   18
   2.3 Chemical properties of phytochemicals   21
      2.3.1 Terpenes   21
      2.3.2 Polyphenols   22
      2.3.3 Carotenoids   23
      2.3.4 Glucosinolates   24
      2.3.5 Dietary fiber (non starch polysaccharides)   26
      2.3.6 Lectins   26
      2.3.7 Other phytochemicals   28
   2.4 Biochemical pathways of important phytochemicals   34
      2.4.1 Shikimate pathway   34
      2.4.2 Isoprenoid pathway   37
      2.4.3 Polyketide pathway   38
      2.4.4 Secondary transformation   40
      2.4.5 Glucosinolate biosynthesis   40
   References   41

3 Phytochemicals and health   49
Ian T. Johnson

   3.1 Introduction   49

3.2		Bioavailability of phytochemicals	50
	3.2.1	Terpenes	51
	3.2.2	Polyphenols	52
	3.2.3	Carotenoids	53
	3.2.4	Glucosinolates	54
	3.2.5	Lectins	55
3.3		Phytochemicals and their health-promoting effects	55
	3.3.1	Phytochemicals as antioxidants	56
	3.3.2	Blocking and suppressing the growth of tumours	59
	3.3.3	Modifying cardiovascular physiology	62
3.4		General conclusions	63
		References	64

4 Pharmacology of phytochemicals — 68
José M. Matés

4.1	Introduction	68
4.2	Medicinal properties of phytochemicals	69
	4.2.1 Therapeutic use of antioxidants	73
	4.2.2 Phytochemicals as therapeutic agents	75
4.3	Phytochemicals and disease prevention	78
	4.3.1 Pharmacologic effects of phytochemicals	80
4.4	Phytochemicals and cardiovascular disease	82
4.5	Phytochemicals and cancer	88
4.6	Summary and conclusions	95
	References	96

PART II SOURCES OF PHYTOCHEMICALS — 105

5 Fruit and vegetables — 107
Uma Tiwari and Enda Cummins

5.1	Introduction	107
5.2	Polyphenols	107
5.3	Carotenoids	113
5.4	Glucosinolates	117
	5.4.1 Variations in glucosinolates	119
5.5	Glycoalkaloids	120
5.6	Polyacetylenes	121
5.7	Sesquiterpene lactones	123
5.8	Coumarins	124
5.9	Terpenoids	125
5.10	Betalains	125
5.11	Vitamin E or tocols content in fruit and vegetables	126
5.12	Conclusions	129
	References	129

6 Food grains — 138
Sanaa Ragaee, Tamer Gamel, Koushik Seethraman, and El-Sayed M. Abdel-Aal

| 6.1 | Introduction | 138 |

	6.2	Phytochemicals in cereal grains	139
		6.2.1 Dietary fiber	139
		6.2.2 Phenolic compounds	141
		6.2.3 Other phytochemicals	143
	6.3	Phytochemicals in legume grains	144
		6.3.1 Dietary fiber	144
		6.3.2 Phenolic acids	145
		6.3.3 Isoflavones	146
		6.3.4 Saponins	146
		6.3.5 Anthocyanins	147
		6.3.6 Lignans	148
		6.3.7 Other phytochemicals	148
	6.4	Stability of phytochemicals during processing	149
	6.5	Food applications and impact on health	152
	6.6	Cereal-based functional foods	152
	6.7	Legume-based functional foods	153
	References		154
7	**Plantation crops and tree nuts: composition, phytochemicals and health benefits**		**163**
	Narpinder Singh and Amritpal Kaur		
	7.1	Introduction	163
	7.2	Composition	165
	7.3	Phytochemicals content	167
	7.4	Health benefits	174
	References		175
8	**Food processing by-products**		**180**
	Anil Kumar Anal		
	8.1	Introduction	180
	8.2	Phytochemicals from food by-products	181
		8.2.1 Biowaste from tropical fruit and vegetables	181
		8.2.2 Citrus peels and seeds	181
		8.2.3 Mango peels and kernels	182
		8.2.4 Passion fruit seed and rind	183
		8.2.5 Pomegranate peels, rinds and seeds	184
		8.2.6 Mangosteen rind and seeds	184
	8.3	By-products from fruit and vegetables	187
		8.3.1 Apple pomace	187
		8.3.2 By-products from grapes	187
		8.3.3 Banana peels	188
		8.3.4 Tomato	188
		8.3.5 Carrot	188
		8.3.6 Mulberry leaves	189
	8.4	Tuber crops and cereals	189
		8.4.1 Cassava	189
		8.4.2 Defatted rice bran	189
	8.5	Extraction of bioactive compounds from plant food by-products	190

	8.6	Future trends	190
	References		192

PART III   IMPACT OF PROCESSING ON PHYTOCHEMICALS — 199

9 On farm and fresh produce management — 201
Kim Reilly

9.1	Introduction		201
9.2	Pre-harvest factors affecting phytochemical content		202
	9.2.1	Tissue type and developmental stage	208
	9.2.2	Fertilizer application – nitrogen, phosphorus, potassium, sulphur and selenium	210
	9.2.3	Seasonal and environmental effects – light and temperature	212
	9.2.4	Biotic and abiotic stress	214
	9.2.5	Means of production – organic and conventional agriculture	216
	9.2.6	Other factors	217
9.3	Harvest and post-harvest management practices		218
	9.3.1	Harvest and post-harvest management of onion	218
	9.3.2	Harvest and post-harvest management of broccoli	220
	9.3.3	Harvest and post-harvest management of carrot	221
9.4	Future prospects		222
	9.4.1	Growing bio-fortified crops – optimized agronomic and post-harvest practices	222
	9.4.2	Edible sprouts	222
	9.4.3	Variety screening and plant breeding for bio-fortified crops	223
	9.4.4	Novel uses for crops and crop wastes	224
References			225

10 Minimal processing of leafy vegetables — 235
Rod Jones and Bruce Tomkins

10.1	Introduction	235
10.2	Minimally processed products	236
10.3	Cutting and shredding	237
10.4	Wounding physiology	238
10.5	Browning in lettuce leaves	240
10.6	Refrigerated storage	241
10.7	Modified atmosphere storage	242
10.8	Conclusions	243
References		244

11 Thermal processing — 247
Nigel P. Brunton

11.1	Introduction	247
11.2	Blanching	248
11.3	Sous vide processing	250
11.4	Pasteurisation	251

		11.5 Sterilisation	254
		11.6 Frying	255
		11.7 Conclusion	257
		References	257

12 Effect of novel thermal processing on phytochemicals — 260
Bhupinder Kaur, Fazilah Ariffin, Rajeev Bhat, and Alias A. Karim

- 12.1 Introduction — 260
- 12.2 An overview of different processing methods for fruits and vegetables — 261
- 12.3 Novel thermal processing methods — 261
- 12.4 Effect of novel processing methods on phytochemicals — 264
  - 12.4.1 Ohmic heating — 265
  - 12.4.2 Microwave heating — 266
  - 12.4.3 Radio frequency — 268
- 12.5 Challenges and prospects/future outlook — 268
- 12.6 Conclusion — 269
- References — 269

13 Non thermal processing — 273
B.K. Tiwari, PJ Cullen, Charles S. Brennan and Colm P. O'Donnell

- 13.1 Introduction — 273
- 13.2 Irradiation — 273
  - 13.2.1 Ionising radiation — 274
  - 13.2.2 Non ionising radiation — 274
- 13.3 High pressure processing — 281
- 13.4 Pulsed electric field — 284
- 13.5 Ozone processing — 286
- 13.6 Ultrasound processing — 289
- 13.7 Supercritical carbon dioxide — 291
- 13.8 Conclusions — 292
- References — 293

PART IV STABILITY OF PHYTOCHEMICALS — 301

14 Stability of phytochemicals during grain processing — 303
Laura Alvarez-Jubete and Uma Tiwari

- 14.1 Introduction — 303
- 14.2 Germination — 304
- 14.3 Milling — 307
- 14.4 Fermentation — 312
- 14.5 Baking — 315
- 14.6 Roasting — 323
- 14.7 Extrusion cooking — 324
- 14.8 Parboiling — 327
- 14.9 Conclusions — 327
- References — 327

15 Factors affecting phytochemical stability — 332
Jun Yang, Xiangjiu He, and Dongjun Zhao

- 15.1 Introduction — 332
- 15.2 Effect of pH — 335
- 15.3 Concentration — 337
- 15.4 Processing — 338
  - 15.4.1 Processing temperature — 338
  - 15.4.2 Processing type — 341
- 15.5 Enzymes — 346
- 15.6 Structure — 349
- 15.7 Copigments — 350
- 15.8 Matrix — 353
  - 15.8.1 Presence of SO_2 — 353
  - 15.8.2 Presence of ascorbic acids and other organic acids — 354
  - 15.8.3 Presence of metallic ions — 355
  - 15.8.4 Others — 356
- 15.9 Storage conditions — 357
  - 15.9.1 Light — 357
  - 15.9.2 Temperature — 358
  - 15.9.3 Relative humidity (RH) — 360
  - 15.9.4 Water activity (a_w) — 361
  - 15.9.5 Atmosphere — 361
- 15.10 Conclusion — 363
- References — 364

16 Stability of phytochemicals at the point of sale — 375
Pradeep Singh Negi

- 16.1 Introduction — 375
- 16.2 Stability of phytochemicals during storage — 375
  - 16.2.1 Effect of water activity — 376
  - 16.2.2 Effect of temperature — 376
  - 16.2.3 Effect of light and oxidation — 379
  - 16.2.4 Effect of pH — 381
- 16.3 Food application and stability of phytochemicals — 381
- 16.4 Edible coatings for enhancement of phytochemical stability — 382
- 16.5 Modified atmosphere storage for enhanced phytochemical stability — 383
- 16.6 Bioactive packaging and micro encapsulation for enhanced phytochemical stability — 384
- 16.7 Conclusions — 387
- References — 387

PART V ANALYSIS AND APPLICATION — 397

17 Conventional extraction techniques for phytochemicals — 399
Niamh Harbourne, Eunice Marete, Jean Christophe Jacquier and Dolores O'Riordan

- 17.1 Introduction — 399
- 17.2 Theory and principles of extraction — 399

		17.2.1	Conventional extraction methods	400
		17.2.2	Factors affecting extraction methods	401
		17.2.3	Limitations of extraction techniques	404
	17.3	Examples of conventional techniques		405
		17.3.1	Roots	405
		17.3.2	Leaves and stems	405
		17.3.3	Flowers	407
		17.3.4	Fruits	407
	17.4	Conclusion		409
	References			409
18	**Novel extraction techniques for phytochemicals**			**412**
	Hilde H. Wijngaard, Olivera Trifunovic and Peter Bongers			
	18.1	Introduction		412
	18.2	Pressurised solvents		413
		18.2.1	Supercritical fluid extraction	413
		18.2.2	Pressurised liquid extraction (PLE)	419
	18.3	Enzyme assisted extraction		421
	18.4	Non-thermal processing assisted extraction		423
		18.4.1	Ultrasound	423
		18.4.2	Pulsed electric fields	424
	18.5	Challenges and future of novel extraction techniques		426
	References			428
19	**Analytical techniques for phytochemicals**			**434**
	Rong Tsao and Hongyan Li			
	19.1	Introduction		434
	19.2	Sample preparation		436
		19.2.1	Extraction	436
		19.2.2	Sample clean-up	438
	19.3	Non-chromatographic spectrophotometric methods		439
		19.3.1	Total phenolic content (TPC)	440
		19.3.2	Total flavonoid content (TFC)	440
		19.3.3	Total anthocyanin content (TAC)	441
		19.3.4	Total carotenoid content (TCC)	441
		19.3.5	Methods based on fluorescence	441
		19.3.6	Colorimetric methods for other phytochemicals	442
	19.4	Chromatographic methods		442
		19.4.1	Conventional chromatographic methods	442
		19.4.2	Instrumental chromatographic methods	443
	References			447
20	**Antioxidant activity of phytochemicals**			**452**
	Ankit Patras, Yvonne V. Yuan, Helena Soares Costa and			
	Ana Sanches-Silva			
	20.1	Introduction		452
	20.2	Measurement of antioxidant activity		453
		20.2.1	Assays involving a biological substrate	453

	20.2.2	Assays involving a non-biological substrate	454
	20.2.3	Ferrous oxidation–xylenol orange (FOX) assay	455
	20.2.4	Ferric thiocyanate (FTC) assay	455
	20.2.5	Hydroxyl radical scavenging deoxyribose assay	456
	20.2.6	1,1-diphenyl-2-picrylhydrazyl (DPPH•) stable free radical scavenging assay	456
	20.2.7	Azo dyes as sources of stable free radicals in antioxidant assays	457
	20.2.8	Oxygen radical absorbance capacity (ORAC) assay	458
	20.2.9	Total radical-trapping antioxidant parameter (TRAP) assay	459
	20.2.10	ABTS•+ radical cation scavenging activity	460
	20.2.11	Ferric reducing ability of plasma (FRAP) assay	460
	20.2.12	Inhibition of linoleic acid oxidation as a measure of antioxidant activity	461
	20.2.13	Other assays – methods based on the chemiluminescence (CL) of luminol	462
	20.2.14	Comparison of various methods for determining antioxidant activity: general perspectives	462
	20.2.15	Discrepancies over antioxidant measurement	463
20.3	Concluding remarks		465
References			466

21 Industrial applications of phytochemicals 473
Juan Valverde

21.1	Introduction		473
21.2	Phytochemicals as food additives		474
	21.2.1	Flavourings	475
	21.2.2	Sweeteners and sugar substitutes	476
	21.2.3	Colouring substances	477
	21.2.4	Antimicrobial agents/essential oils	478
	21.2.5	Antioxidants	480
21.3	Stabilisation of fats, frying oils and fried products		481
21.4	Stabilisation and development of other food products		488
	21.4.1	Anti-browning effect of phytochemicals in foods	488
	21.4.2	Colour Stabilisation in meat products	490
	21.4.3	Antimicrobials to extends shelf life	491
21.5	Nutracetical applications		492
	21.5.1	Phytosterol and phytostanol enriched foods	492
	21.5.2	Resveratrol enriched drinks and beverages	492
	21.5.3	Isoflavone enriched dairy-like products	493
	21.5.4	β-glucans	493
	21.5.5	Flavonoids	494
21.6	Miscellaneous industrial applications		494
	21.6.1	Cosmetic applications	494
	21.6.2	Bio-pesticides	495
References			495

Index 502

Contributor list

Editors

B.K. Tiwari
Food and Consumer Technology,
Manchester Metropolitan University,
Manchester, UK

Nigel P. Brunton
School of Agriculture and Food Science,
University College Dublin,
Dublin, Ireland

Charles S. Brennan
Faculty of Agriculture and Life Sciences,
Lincoln University, Lincoln,
Canterbury, New Zealand

Contributors

El-Sayed M. Abdel-Aal
Guelph Food Research Centre,
Agriculture and Agri-Food Canada,
Guelph, Ontario, Canada

Laura Alvarez-Jubete
School of Food Science and
Environmental Health,
Dublin Institute of Technology,
Dublin, Ireland

Anil Kumar Anal
Food Engineering and Bioprocess
Technology,
Asian Institute of Technology,
Klongluang, Thailand

Fazilah Ariffin
Food Biopolymer Research Group,
Food Technology Division,
School of Industrial Technology,
Universiti Sains Malaysia,
Penang, Malaysia

Rajeev Bhat
Food Biopolymer Research Group,
Food Technology Division,
School of Industrial Technology,
University Sains Malaysia,
Penang, Malaysia

Peter Bongers*
Structured Materials and
Process Science,
Unilever Research and Development
Vlaardingen,
The Netherlands

Charles S. Brennan
Faculty of Agriculture and Life Sciences,
Lincoln University, Lincoln,
Canterbury, New Zealand

Nigel P. Brunton
School of Agriculture and Food Science,
University College Dublin,
Dublin, Ireland

Rocio Campos-Vega
Kellogg Company Km,
Campo Militar,
Querétaro, México

Helena Soares Costa
National Institute of Health Dr Ricardo Jorge, Food and Nutrition Department,
Lisbon, Portugal

PJ Cullen
School of Food Science and Environmental Health,
Dublin Institute of Technology,
Dublin, Ireland

Enda Cummins
UCD School of Biosystems Engineering,
Agriculture and Food Science Centre,
Dublin, Ireland

Tamer Gamel
Guelph Food Research Centre,
Agriculture and Agri-Food Canada,
Guelph, Ontario, Canada

Niamh Harbourne
Food and Nutritional Sciences,
University of Reading,
Reading, UK

Xiangjiu He
School of Pharmaceutical Sciences,
Wuhan University,
Wuhan, Hubei, China

Jean Christophe Jacquier
School of Agriculture and Food Science
University College Dublin,
Dublin, Ireland

Ian T. Johnson
Institute of Food Research,
Norwich Research Park, Colney,
Norwich, UK

Rod Jones
Department of Primary Industries,
Victoria, Australia

Alias A. Karim
Food Biopolymer Research Group,
Food Technology Division,
School of Industrial Technology,
Universiti Sains Malaysia,
Penang, Malaysia

Amritpal Kaur
Department of Food Science and Technology,
Guru Nanak Dev University,
Amritsar, India

Bhupinder Kaur
Food Biopolymer Research Group,
Food Technology Division,
School of Industrial Technology,
Universiti Sains Malaysia,
Penang, Malaysia

Hongyan Li
Guelph Food Research Centre,
Agriculture and Agri-Food Canada,
Guelph, Ontario, Canada

Eunice Marete
School of Agriculture and Food Science
University College Dublin,
Dublin, Ireland

José M. Matés
Department of Molecular Biology and Biochemistry,
Faculty of Sciences, Campus de Teatinos,
University of Málaga, Spain

Pradeep Singh Negi
Human Resource Development,
Central Food Technological Research Institute (CSIR),
Mysore, India

Colm P. O'Donnell
UCD School of Biosystems Engineering,
University College Dublin
Belfield, Dublin, Ireland

B. Dave Oomah
Pacific Agri-Food Research Centre,
Agriculture and Agri-Food Canada,
Summerland,
British Columbia, Canada

Dolores O'Riordan
School of Agriculture and Food Science
University College Dublin,
Dublin, Ireland

Ankit Patras
Department of Food Science,
University of Guelph, Guelph
Ontario, Canada

Sanaa Ragaee
Department of Food Science,
University of Guelph, Guelph,
Ontario, Canada

Kim Reilly
Horticulture Development Unit,
Teagasc, Kinsealy Research Centre,
Dublin, Ireland

Ana Sanches-Silva
National Institute of Health Dr Ricardo
Jorge, Food and Nutrition Department,
Lisbon, Portugal

Koushik Seethraman
Department of Food Science,
University of Guelph,
Guelph, Ontario, Canada

Narpinder Singh
Department of Food Science and
Technology,
Guru Nanak Dev University,
Amritsar, India

B.K. Tiwari
Food and Consumer Technology,
Manchester Metropolitan University,
Manchester, UK

Uma Tiwari
UCD School of Biosystems Engineering,
Agriculture and Food Science Centre,
Dublin, Ireland

Bruce Tomkins
Department of Primary Industries,
Victoria, Ferntree Gully,
DC, Australia

Olivera Trifunovic
Structured Materials and Process Science,
Unilever Research and Development
Vlaardingen,
The Netherlands

Rong Tsao
Guelph Food Research Centre,
Agriculture and Agri-Food Canada,
Guelph, Ontario, Canada

Juan Valverde
Teagasc Food Research Centre Ashtown,
Dublin, Ireland

Hilde H. Wijngaard
Dutch Separation Technology Institute,
Amersfoort, The Netherlands

Jun Yang
Frito-Lay North America R&D,
PepsiCo Inc.,
Plano, TX, USA

Yvonne V. Yuan
School of Nutrition,
Ryerson University,
Toronto, Ontario, Canada

Dongjun Zhao
Department of Food Science,
Cornell University,
Ithaca, NY, USA

1 Plant food phytochemicals

B.K. Tiwari, Nigel P. Brunton and Charles S. Brennan

1.1 Importance of phytochemicals

Type the word 'phytochemical' into any online search engine and it will return literally thousands of hits. This is a reflection of the role plant derived chemicals have played in medicine and other areas since humans have looked to nature to provide cures for various ailments and diseases. While it is often stated, it is worth repeating that the evolution of modern medicine derived from applying scientific principles to herbalism and to this day plants derived compounds provide the skeletons for constructing molecules with the abilities to cure many diseases. In recent times applications of phytochemicals have extended into other areas especially nutraceuticals and functional foods. The focus here is not on curing existing conditions but delaying the onset of new ones and it is not surprising to note that plant foods and plant derived components make up the vast majority of compounds with European Food Safety Authority validated Article 13.1 health claims. Whilst there has been a renewed interest in the use of medicinal plants to treat diseases in recent times and the use of phytochemicals as pharmaceuticals is covered in the present book, this is not the core theme of the book. Given that plant foods are still a major component of most diets worldwide the greatest significance of phytochemicals derives from their role in human diets and health. In fact it is only in relatively recent times that due recognition has been given to the importance of phytochemicals in maintaining health. This has driven a huge volume of work on the subject ranging from unravelling mechanisms of biological significance to discovery and stability studies.

An overview of the health benefits of phytochemicals is essential as many phytochemicals have been reported to illicit both positive and negative biological effects. In recent times some evidence for the role of specific plant food phytochemicals in protecting against the onset of diseases such as cancers and heart diseases has been put forward. Most researchers in this field will however agree that in most cases more evidence is needed to prove the case for the ability of phytochemicals to delay the onset of these diseases. Nevertheless the increasing awareness of consumers of the link between diet and health has exponentially increased the number of scientific studies into the biological effects of these substances.

1.2 Book objective

The overarching objective, therefore, of the *Handbook of Plant Food Phytochemicals* is to provide a bird's eye view of the occurrence, significance and factors affecting phytochemicals in plant foods. A key of objective of the handbook is to critically evaluate some of these with a particular emphasis on evidence for or against quantifiable beneficial health effects being imparted via a reduction in disease risk through the consumption of foods rich in phytochemicals.

1.3 Book structure

The book is divided into five parts. Part I deals with the health benefits and chemistry of phytochemicals, Part II summarises phytochemicals in various food types, Parts III and IV deal with a variety of factors that can affect phytochemical content and stability and Part V deals with a range of analytical techniques and applications of phytochemicals. The subject of the biological activity of phytochemicals is approached both from a disease risk reduction perspective in Chapter 3 and from a more traditional pharmacological viewpoint in Chapter 4. Together these chapters are intended to give the reader a sound basis for understanding the biological significance of these substances and to contextualise their roles either as a medicinal plant or as a nutraceutical/functional food. Key to understanding both the stability and biological role of phytochemicals is a sound knowledge of their chemistry and biochemical origin. This often neglected topic is covered in detail here along with an overview of the classification of these compounds. This reflects the ambition of the book to serve as a reference text for students in the field and is intended to provide a basis for understanding some of the complex subjects covered in earlier chapters.

The chemical diversity and number of plant food phytochemicals with reported abilities to protect against diseases numbers in the many thousands. Therefore, to cover all these substances in detail would be impossible. However, myself and my fellow editors felt that providing readers with a reference manuscript for plant food phytochemicals and a basic understanding of the types of phytochemicals in plant foods was essential. Part II of the handbook covers this subject matter by giving an overview of the phytochemicals present in four food categories – fruit and vegetables, food grains, natural products and tree nuts and food processing by-products. Fruits and vegetables are perhaps the best recognised source of phytochemicals and this is reflected in the depth and volume of literature on this food type. Chapter 5 summarises information on major phytochemicals groups in fruits and vegetables as well as some of the more obscure and recently emerged groups. From a consumption perspective food grains form a huge proportion of most diets worldwide – however, due recognition of grains as sources of phytochemicals has only emerged relatively recently. Chapter 6 summarises the phytochemical composition of both cereals and legumes and underlines the importance of this food group as a source of phytochemicals in human diets. Early humans were of course hunter gatherers and nuts would have been important of their diets. It is therefore perhaps not surprising that tree nuts and other natural products have been shown to contain a range of phytochemicals with the potential to deliver benefits beyond basic nutrition. The importance of tree nuts as sources of these compounds is hence covered in detail in Chapter 7 along with related food types such as plantation products. Whilst a core objective of the handbook is to cover the breadth of subject matter in phytochemicals from plant foods this is not merely an academic exercise. Phytochemicals have real commercial uses and this is given due recognition in Chapters 7 and 8 where an overview of the application of phytochemicals derived from foods grains and trees is given. In fact throughout the handbook authors provide detailed information and examples of real applications of plant food derived phytochemicals with a view to underlining the commercial importance of these compounds. Food processing by-products do not of course constitute a food group – however, they have become hugely important sources of phytochemicals in recent times and Chapter 8 is dedicated to revealing the potential of food processing by-products as sources of phytochemicals with real commercial potential. Recovering value from by-products is of course hugely significant to food processors as they seek to maximise the value of a resource that hitherto was considered a waste. This also reflects the drive to identify more sustainable food processing practices and increasing pressures from regulators to reduce waste.

As with most other foods, plant foods are often not consumed in their native form. Therefore, investigators have long been interested in developing an understanding of how processing effects phytochemical composition with view to maximising their potential health promoting properties. Today's consumers are demanding foods that are healthy, convenient and appetising. The drive for healthy foods has fuelled interest in the effect of processing on the level of components responsible for imparting this benefit, especially phytochemicals. Therefore, much work has been devoted to assessing the effect of processing and storage on levels of potentially important phytochemicals in foods. In addition, a number of novel thermal and non-thermal technologies designed to achieve microbial safety, while minimising the effects on its nutritional and quality attributes, have recently become available. Minimising changes in phytochemicals during processing is a considerable challenge for food processors and technologists. Thus, there is a requirement for detailed industrially relevant information concerning phytochemicals and their application in food products. In addition, industrial adoption of novel processing techniques is in its infancy. Applications of new and innovative technologies and resulting effects on those food products either individually or in combination are always of great interest to academic, industrial, nutrition and health professionals. Part III gives an oversight as to how processing affects phytochemicals in plant foods. This is an area that has received huge attention recently and this has reflected the number of chapters dedicated to it in the handbook. This part of the handbook also summarises and evaluates an area that is often neglected when in the phytochemicals arena but can have profound impact on final phytochemical content, namely on farm and fresh produce management. Given the investment and scale of research required to carry out replicated field trials elucidating the impact of pre-harvest factors, such as fertiliser application, light, temperature, biotic and abiotic stress, this area has perhaps been the most challenging of any of the 'farm to fork' factors involved in determining the phytochemical content of plant foods. Indeed assessing the relative effects of intensive and organic farming practices is a highly controversial area but one that consumers appear to take an active interest in given the premium demand for organically produced plant foods. Post-harvest management pertains to the period between harvesting of the plant food and its arrival at the processing plant. This covers many operations including mechanical harvesting, storage and transport. Unsurprisingly many of these operations constitute a stress to the still respiring plant food and thus can activate or deactivate pathways leading to the synthesis of phytochemicals. Ready to eat fruit and vegetables are a relatively recent phenomenon on supermarket shelves. Their emergence is a reflection of consumers' busy lifestyles and the need to provide healthy and convenient solutions for time poor customers who desire a healthy diet. Products of this nature are often referred to as minimally processed and are subjected to a variety of operations ranging from peeling and cutting to washing. Unlike plant foods, which have been subjected to heat processing, minimally processed products remain viable, albeit in many cases in a wounded state. Therefore a wide variety of responses to minimal processing have been reported and these are summarised and evaluated in Chapter 10, with a particular emphasis on salad mixes. A huge spectrum of full processing techniques is available to food processors nowadays. These range from severe (canning) to mild (*sous-vide* processing) to non-thermal examples such as high pressure processing, ultrasound and irradiation. Not surprisingly these can have a range of effects on phytochemical content and Chapters 11, 12 and 13 summarise the work done to date on these processes. Grains and pulses undergo a distinctly different processing route to other plant foods involving germination, milling, fermentation and finally baking. Therefore we have dedicated a standalone chapter to food grains, which reviews reports on the grain processing techniques on the content of phytochemicals. Finally, in tune with the farm to fork approach adopted by the handbook, the last chapter in Part III reviews the stability of foods containing phytochemicals during storage after processing. Like most chemical constituents the nature of the matrix they are contained in has a profound effect on their stability. Therefore, in Chapter 15 the stability of phytochemicals with different properties such as low moisture contents, ethnic foods and of course traditional foods is reviewed.

The final part of the book deals with perhaps the first question a researcher must ask him/herself when entering the field namely how do we extract these compounds and how do we measure them. The chapter on extraction is particularly relevant as this is an important consideration not only when analysing these compounds but also when preparing to include them as an ingredient in another food. Phytochemical analysis techniques are advancing at an exponential rate and therefore a chapter reviewing the state of the art in this discipline was one of the first we put on paper when deciding on the content of the book. Finally, the reason we have dedicated a book to the subject of phytochemicals in plant foods is because they have very real applications in industry and everyday life. The final chapter of the handbook drives this point home by providing real examples of industrial uses for phytochemicals ranging from maintaining stability in oxidatively labile foods to enhancing the health promoting properties of others. To conclude we hope you find the proceeding chapters to be informative, clear, concise and that they provide a clear thinking perspective on a subject matter that has benefitted mankind from many perspectives and will no doubt continue to do so into the future.

Part I
Chemistry and Health

2 Chemistry and classification of phytochemicals

Rocio Campos-Vega[1] and B. Dave Oomah[2]

[1] Kellogg Company Km. Querétaro, Qro. México
[2] Pacific Agri-Food Research Centre, Agriculture and Agri-Food Canada, Summerland, British Columbia, Canada

2.1 Introduction

The word 'biodiversity' is on nearly everyone's lips these days, but 'chemodiversity' is just as much a characteristic of life on Earth as biodiversity. Living organisms produce several thousands of different structures of low-molecular-weight organic compounds. Many of these have no apparent function in the basic processes of growth and development, and have been historically referred to as natural products or secondary metabolites. The importance of natural products in medicine, agriculture and industry has led to numerous studies on the synthesis, biosynthesis and biological activities of these substances. Yet we still know comparatively little about their actual roles in nature.

Clearly such research has been stimulated by scientific curiosity in the substances and mechanisms involved in the protective effects of fruits and vegetables. Dietary phytonutrients appear to lower the risk of cancer and cardiovascular disease. Studies on the mechanisms of chemoprotection have focused on the biological activity of plant-based phenols and polyphenols, flavonoids, isoflavones, terpenes, and glucosinolates. However, most, if not all, of these bioactive compounds are bitter, acrid, or astringent and therefore aversive to the consumer. Some have long been viewed as plant-based toxins. The analysis of phytochemicals is complicated due to the wide variation even within the same group of compounds, and the metabolic degradation or transformation that may occur during crushing or processing of plants (e.g. for *Allium* and *Brassica* compounds), thus increasing the complexity of the mixture. Many phytochemical analyses require mass spectroscopy and therefore are time-consuming and expensive. Furthermore, some compounds tend to bind to macromolecules, making quantitative extraction difficult. Furthermore, many plant food phytochemicals that are poorly absorbed by humans usually undergo metabolism and rapid excretion. It is clear from *in vitro* and animal data that the actions of some phytochemicals are likely to be achieved only at doses much higher than those present in edible plant foods. Thus, extraction or synthesis of the active ingredient is essential if they are to be of prophylactic or therapeutic value in human subjects.

Handbook of Plant Food Phytochemicals: Sources, Stability and Extraction, First Edition.
Edited by B.K. Tiwari, Nigel P. Brunton and Charles S. Brennan.
© 2013 John Wiley & Sons, Ltd. Published 2013 by John Wiley & Sons, Ltd.

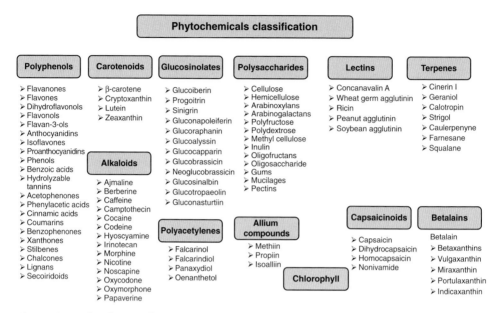

Figure 2.1 Classification of phytochemicals.

2.2 Classification of phytochemicals

Many phytochemicals have a range of different biochemical and physiological effects, isoflavonoids, for example have antioxidant and anti-oestrogenic activities. These activities may require different plasma or tissue concentrations for optimum effects. A diagram illustrating the classification of the phytochemicals covered in this chapter is shown in Figure 2.1.

In addition, plants contain mixtures of phytochemicals (Table 2.1), with considerable opportunity for interaction (Rowland et al., 1999). Plant secondary metabolites are an enormously variable group of phytochemicals in terms of their number, structural heterogeneity, and distribution.

A summary of the main groups of bioactive chemicals in edible plants, their sources, and their biological activities is presented in Table 2.2 (Rowland et al., 1999).

2.2.1 Terpenes

The term terpenes originates from turpentine (*balsamum terebinthinae*). Turpentine, the so-called "resin of pine trees", is the viscous pleasantly smelling balsam that flows upon cutting or carving the bark and the new wood of several pine tree species (*Pinaceae*). Turpentine contains the "resin acids" and some hydrocarbons, which were originally referred to as terpenes. Traditionally, all natural compounds built up from isoprene subunits and, for the most part, originating from plants are denoted as terpenes (Breitmaier, 2006).

All living organisms manufacture terpenes for certain essential physiological functions and therefore have the potential to produce terpene natural products. Given the many ways in which the basic C_5 units can be combined together and the different selection pressures under which organisms have evolved, it is not surprising to observe the enormous number and

Table 2.1 Phytochemical content of some edible plants (modified from Caragay, 1992; Rowland et al., 1999)

Plants	Flavo-noids	Isofla-vonoids	Lig-nans	Organo-sulphides	Glucosi-nolates	Phenolic acids	Oligo-saccharides	Terpe-nes	NSP	Alka-loids	Poly-acetylenes	Chloro-phyll	Capsaici-noids	Beta-lains	Carote-noids
Soybeans	✓	✓				✓			✓						
Cereals	✓		✓			✓			✓					✓	
Garlic and onions				✓		✓	✓								
Cruciferae	✓			✓	✓	✓	✓	✓	✓						
Solanacae	✓					✓		✓	✓						
Umbeliderae	✓					✓		✓	✓						
Citrus fruits	✓					✓		✓	✓						
Green tea	✓					✓				✓					
Legumes		✓							✓						
Blueberry	✓		✓												
Grapes						✓									
Tomato															✓
Carrots											✓				
Pepper												✓	✓		
Beets													✓	✓	
Amaranthus caudatus									✓					✓	
Flaxseed	✓		✓			✓									

Table 2.2 Sources and biological activities of phytochemicals (adapted from Rowland et al., 1999)

Group	Examples	Main food sources	Activity and functional marker
Fiber and related compounds	NSP Soluble (e.g. pectins, gums) Insoluble (celluloses) Resistant starch, retrograded starch Phytate Oligosaccharides	Fruit (apples, citrus), oats, soybean, algae Cereals (wheat, rye), vegetables High-amylose starches, processed starches, whole grains, and seeds Cereals, grains, soybeans Chicory, soybeans, artichokes, onion	Lowers serum cholesterol Prevents colon and breast cancer, diverticular disease Alleviates constipation Increases butyrate in faeces Prevents colon cancer Binds minerals. Prevents Colon cancer Modifies gut flora, modulates lipid metabolism, Cancer prevention?
Flavonoids	Flavonols: quercetin, kaempferol Flavanones: tangeritin, naringenin, hesperitin Flavanols: catechins, epicatechins	Vegetables (onion, lettuce, tomatoes, peppers) wine, tea Citrus fruits Green tea	Antioxidants, modulate phase 1 enzymes, inhibit protein kinase C. Prevent cancer protect CVD? Modulate immune response?
Tea polyphenols Derived tannins Isoflavonoids	Catechins, epicatechins Theaflavins, thearubigens Daidzein, genistein	Green tea Black tea, red wine, roasted coffee Soybean products	Antioxidants prevent CHD? Anti-oestrogenic effects, effects on serum lipids, prevent breast and prostate cancers
Lignans	Secoisolariciresinol, matairesinol	Rye bran, flaxseed, berries, nuts	Antioxidant and anti-oestrogenic effects Prevent colon and prostate cancer?
Glucosinolates Isothiocynates	Glucobrassicin, indole-3-carbinol Allylisothiocynates, indoles, sulforaphane	Cruciferous vegetables (broccoli, cabbage, Brussel sprouts, watercress, mustard)	Induces phase 2 enzymes Cancer prevention?
Simple phenols Phenolic acids, condensed phenols	p-Cresol, ethyl phenol, hydroquinone Gallic acid, tannins, ellagic acid	Raspberry, cocoa beans, green tea, black tea, strawberries	Antioxidants

Monoterpenes	D-Limonene, D-carvone, perillyl alcohol	Citrus fruits, cherries, herbs	Induce Phase I and Phase II enzymes Anti-tumour activity
Hydroxycinnamic acid	Caffeic, ferulic, chlorogenic acids, curcumin	Apples, pears, coffee, mustard, curry	Inhibit nitrosation by trapping nitrite, nucleophiles, antioxidants
Phytosterols	β-Sitosterol, campesterol, stigmasterol	Vegetable oils (soybean, rape seed, maize, sunflower)	Lower serum cholesterol
Alkaloids	Caffeine, Codeine, Noscopine, Quinidine	Berberis vulgaris, Cinchona ledgeriana	Anticancer agents, glycosidase inhibitors, Analgesic
Polyacetylenes	Falcarindiol, Falcarinol, Crepenycic, Steariolic, Teriric acids	Carrots	Anti cancer properties
Chlorophyll	Chlorophyll	Plants, algae and cyanobacteria	Antioxidant
Betalains	Vulgaxanthin, Miraxanthin, Portulaxanthin, Indicaxanthin	Plants: amaranth, cactus fruits	Antioxidant
Organosulphides (allium compounds)	Diallyl sulphide, allyl methyl sulphide, S-allylcysteine	Garlic, onions, leeks	Induce Phase II enzyme, affects serum lipids and platelet aggregation Prevent cancer

12 Handbook of Plant Food Phytochemicals

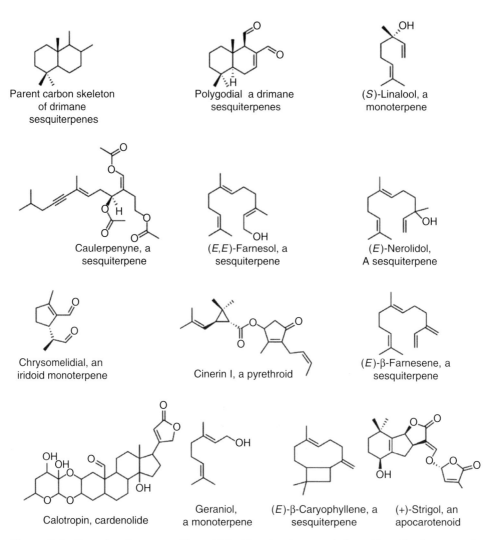

Figure 2.2 Examples of terpenes with established functions in nature (adapted from Gershenzon and Dudareva, 2007).

diversity of structures elaborated (Gershenzon and Dudareva, 2007). Terpenes (also known as terpenoids or isoprenoids) are the largest group of natural products comprising approximately 36 000 terpene structures (Buckingham, 2007), but very few have been investigated from a functional perspective (Figure 2.2).

The classification of terpenoids is based on the number of isoprenoid units present in their structure. The largest categories consist of compounds with two (monoterpenes), three (sesquiterpenes), four (diterpenes), five (sesterterpenes), six (triterpenes), and eight (tetraterpenes) isoprenoid units (see Figure 2.3) (Ashour et al., 2010).

Terpenoids have well-established roles in almost all basic plant processes, including growth, development, reproduction, and defence (Wink and van Wyk, 2008). Gibberellins, a large group of diterpene plant hormones involved in the control of seed germination, stem elongation, and flower induction (Thomas et al., 2005) are among the best-known lower

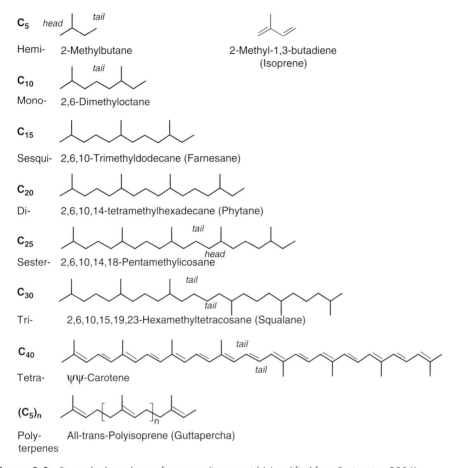

Figure 2.3 Parent hydrocarbons of terpenes (isoprenoids) (modified from Breitmaier, 2006).

(C5–C20) terpenes. Another terpenoid hormone, abscisic acid (ABA), is not properly considered a lower terpenoid, since it is formed from the oxidative cleavage of a C40 carotenoid precursor (Schwartz et al., 1997).

2.2.2 Polyphenols

Polyphenols, secondary plant metabolites are the most abundant antioxidants in human diets. These compounds are designed with an aromatic ring carrying one or more hydroxyl moieties. Several classes can be considered according to the number of phenol rings and to the structural elements that bind these rings. In this context, two main groups of polyphenols, termed flavonoids and nonflavonoids, have been traditionally adopted. As seen in Figures 2.4 and 2.5, the flavonoid group comprises compounds with a C6-C3-C6 structure: flavanones, flavones, dihydroflavonols, flavonols, flavan-3-ols, anthocyanidins, isoflavones, and proanthocyanidins. The nonflavonoids group is classified according to the number of carbons and comprises the following subgroups: simple phenols, benzoic acids, hydrolyzable tannins, acetophenones and phenylacetic acids, cinnamic acids, coumarins, benzophenones, xanthones, stilbenes, chalcones, lignans, and secoiridoids (Andrés-Lacueva et al., 2010).

14 Handbook of Plant Food Phytochemicals

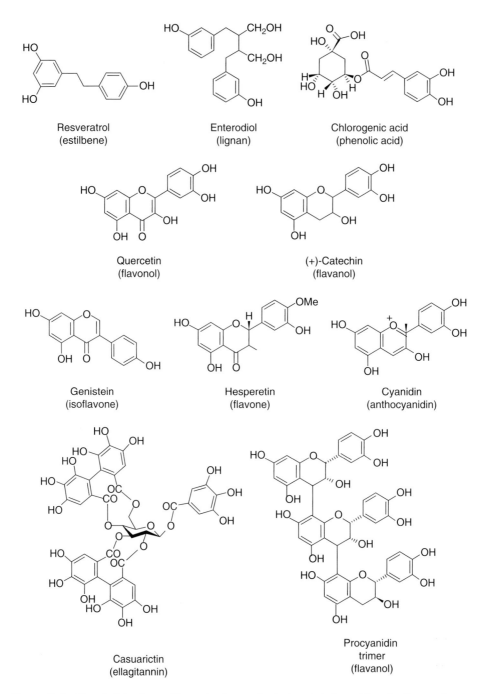

Figure 2.4 Chemical structures of the main classes of polyphenols (adapted from Scalbert and Williamson, 2000).

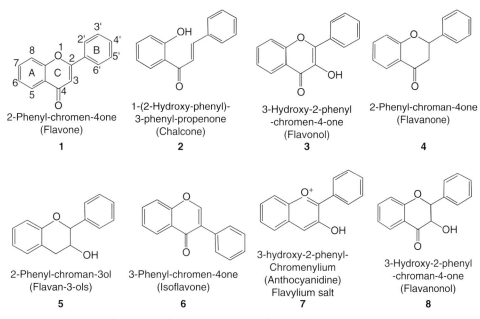

Figure 2.5 Chemical structures of some representative flavonoids (adapted from Tapas et al., 2008).

2.2.3 Carotenoids

Carotenoids are fat-soluble natural pigments with antioxidant properties (Krinsky and Yeum, 2003), with various other additional physiological functions, such as immunostimulation (McGraw and Ardia, 2003). The more than 600 known carotenoids are generally classified as xanthophylls (containing oxygen) or carotenes (purely hydrocarbons with no oxygen). Carotenoids in general absorb blue light and serve two key roles in plants and algae: they absorb light energy for use in photosynthesis, and protect chlorophyll from photodamage (Armstrong and Hearst, 1996). In humans, four carotenoids (α-, β-, and γ-carotene, and β-cryptoxanthin) have vitamin A activity (i.e. can be converted to retinal), and these and other carotenoids can also act as antioxidants (Figure 2.6). In the eye, certain other carotenoids (lutein and zeaxanthin) apparently act directly to absorb damaging blue and near-ultraviolet light, in order to protect the macula lutea. People consuming diets rich in carotenoids from natural foods, such as fruits and vegetables, are healthier and have lower mortality from a number of chronic illnesses (Diplock et al., 1998).

2.2.4 Glucosinolates

Glucosinolates (GLS), a group of plant thioglucosides found among several vegetables (Larsen, 1981), are a class of organic compounds containing sulfur and nitrogen and are derived from glucose and an amino acid (Anastas and Warner, 1998). Over 100 different GLS have been characterized since the first crystalline glucosinolate, sinalbin, was isolated from the seeds of white mustard in 1831. GLS occur mainly in the order Capparales, principally in the Cruciferae, Resedaceae, and Capparidaceae families, although their presence in other families has also been reported (Larsen, 1981). Some economically important GLS containing plants are white mustard, brown mustard, radish, horse radish,

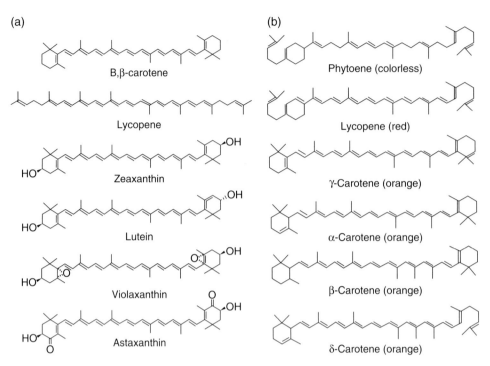

Figure 2.6 Some examples of carotenoids ((a) adapted from Sliwka et al., 2010; (b) adapted from Yahia et al., 2010)).

cress, kohlrabi, cabbages (red, white, and savoy), brussel sprouts, cauliflower, broccoli, kale, turnip, swede, and rapeseed (Fenwick et al., 1989).

GLS hydrolysis and metabolic products have proven chemoprotective properties against chemical carcinogens by blocking the initiation of tumours in various tissues, for example, liver, colon, mammary gland, and pancreas. They exhibit their effect by inducing Phase I and II enzymes, inhibiting the enzyme activation, modifying the steroid hormone metabolism and protecting against oxidative damage. GLS facilitate detoxificiation of carcinogens by inducing Phase I and Phase II enzymes. Some enzymes of Phase I reaction that activate the carcinogens, are selectively inhibited by glucosinolate metabolites (Das et al., 2000).

2.2.5 Dietary fiber (non starch polysaccharides)

Dietary fiber is the edible parts or analogous carbohydrates resistant to digestion and absorption in the small intestine with complete or partial fermentation in the large intestine. Dietary fiber includes polysaccharides, oligosaccharides, lignin, and associated plant substances. Dietary fibers promote beneficial physiological effects including laxation, and/or blood cholesterol attenuation, and/or blood glucose attenuation (AACC, 2001) (Table 2.3). Dietary fibers are polymers of monosaccharides joined through glycosidic linkages and are defined and classified in terms of the following structural considerations: (a) identity of the monosaccharides present; (b) monosaccharide ring forms (six-membered pyranose or five-membered furanose); (c) positions of the glycosidic linkages; (d) configurations (a or b) of the glycosidic linkages; (e) sequence of monosaccharide residues in the chain, and (f) presence or absence of non-carbohydrate substituents. Monosaccharides commonly present in cereal

Table 2.3 Constituents of dietary fiber according to the definition of the American Association of Cereal Chemists (adapted from Jones, 2000)

Non starch polysaccharides (NSP) and resistant
Cellulose
Hemicellulose
Arabinoxylans
Arabinogalactans
Polyfructose
Inulin
Oligofructans
Galacto-oligosaccharides
Gums
Mucilages
Pectins
Analogous carbohydrates
Indigestible dextrins
Resistant maltodextrins (from maize and others sources)
Resistant potato dextrins
Synthesized carbohydrate compounds
Polydextrose
Methyl cellulose
Hydroxypropylmethyl cellulose
Indigestible ("resistant") starches
Lignin substances associated with the NSP and lignin complex in
Plants
Waxes
Phytate
Cutin
Saponins
Suberin
Tannins

cell walls are: (a) *hexoses* – D-glucose, D-galactose, D-mannose; (b) *pentoses* – L-arabinose, D-xylose; and (c) *acidic sugars* – D-galacturonic acid, D-glucuronic acid and its 4-O-methyl ether (Choct, 1997).

According to Cummings (1997), the health benefits of DF do not provide a distinct disease-related characteristic that can be exclusively associated with it. Constipation comes closest to fulfilling such a criterion and it is clear that some functional and physiological effects have been demonstrated with some specific fibers: (a) faecal bulking or stool output (ispaghula, xanthan gum, and wheat bran); (b) lowering of postprandial blood glucose response (highly viscous guar gum or β-glucans); (c) lowering of plasma (LDL-) cholesterol (highly viscous guar gum, β-glucans or oat bran, pectins, psyllium). Other effects have not yet been demonstrated in human subjects, such as colonic health effects related to fermentation products, although a substantial body of evidence is available from *in vitro* or animal models (Champ *et al.*, 2003 and references therein).

2.2.6 Lectins

Lectins (from *lectus*, the past participle of *legere*, to select or choose) are defined as carbohydrate binding proteins other than enzymes or antibodies and exist in most living

Table 2.4 Examples of lectins, the families to which they belong and their glycan ligand specificities (modified from Ambrosi et al., 2005 and references therein)

Lectin name	Family	Glycan ligands
Plant lectins		
Concanavalin A (Con A; jack bean)	Leguminosae	Man/Glc
Wheat germ agglutinin (WGA; wheat)	Gramineae	(GlcNAc)1–3, Neu5Ac
Ricin (castor bean)	Euphorbiaceae	Gal
Phaseolus vulgaris (PHA; French bean)	Leguminosae	Unknown
Peanut agglutinin (PNA; peanut)	Leguminosae	Gal, Galb3GalNAca (T-antigen)
Soybean agglutinin (SBA; soybean)	Leguminosae	Gal/GalNAc
Pisum sativum (PSA; pea)	Leguminosae	Man/Glc
Lens culinaris (LCA; lentil)	Leguminosae	Man/Glc
Galanthus nivalus (GNA; snowdrop)	Amaryllidaceae	Man
Dolichos bifloris (DBA; horse gram)	Leguminosae	GalNAca3GalNAc, GalNAc
Solanum tuberosum (STA; potato)	Solanaceae	(GlcNAc)n

organisms, ranging from viruses and bacteria to plants and animals. Some examples are given in Table 2.4. Their involvement in diverse biological processes in many species, such as clearance of glycoproteins from the circulatory system, adhesion of infectious agents to host cells, recruitment of leukocytes to inflammatory sites, cell interactions in the immune system, in malignancy and metastasis, has been shown (Ambrosi et al., 2005 and references therein).

2.2.7 Other phytochemicals

2.2.7.1 Alkaloids

The term "alkaloid" was coined by the German pharmacist Carl Friedrich Wilhelm Meissner in 1819 to refer to plant natural products (the only organic compounds known at that time) showing basic properties similar to those of the inorganic alkalis (Friedrich and Von, 1998) The ending "-oid" (from the Greek *eidv*, appear) is still used today to suggest similarity of structure or activity, as is evident in names of more modern vintage such as terpenoid, peptoid, or vanilloid (Hesse, 2002).

Among the secondary metabolites that are produced by plants, alkaloids figure as a very prominent class of defense compounds. Over 21 000 alkaloids have been identified, which thus constitute the largest group among the nitrogen-containing secondary metabolites (besides 700 nonprotein amino acids, 100 amines, 60 cyanogenic glycosides, 100 glucosinolates, and 150 alkylamides) (Roberts and Wink, 1998; Wink, 1993). An alkaloid never occurs alone; alkaloids are usually present as a mixture of a few major and several minor alkaloids of a particular biosynthetic unit, which differ in functional groups (Wink, 2005).

2.2.7.2 Polyacetylenes

Polyacetylenes are examples of bioactive secondary metabolites that were previously considered undesirable in plant foods due to their toxicity (Czepa and Hofmann, 2004) (Figure 2.7). However, a low daily intake of these "toxins" may be an important factor in the search for an explanation of the beneficial effects of fruit and vegetables on human

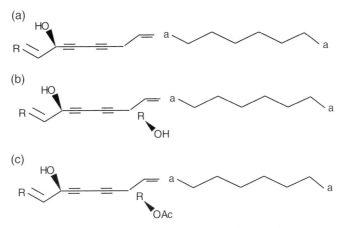

Figure 2.7 Polyacetylenes structure (a) Falcarinol (FaOH), (b) Falcarindiol (FaDOH), (c) Falcarindiol 3-acetate (FaDOAc).

health. For example, polyacetylenes isolated from carrots have been found to be highly cytotoxic against numerous cancer cell lines. Over 1400 different polyacetylenes and related compounds have been isolated from higher plants.

Aliphatic C17-polyacetylenes of the falcarinol type such as falcarinol and falcarindiol (Figure 2.7), are widely distributed in the Apiaceae and Araliaceae (Bohlmann *et al.*, 1973; Hansen and Boll, 1986), and consequently nearly all polyacetylenes found in the utilized/edible parts of food plants of the Apiaceae, such as carrot, celeriac, parsnip, and parsley are of the falcarinol-type. Falcarinol, a polyacetylene with anti-cancer properties, is commonly found in the Apiaceae, Araliaceae, and Asteraceae plant families (Zidorn *et al.*, 2005). Other polyacetylenes had been reported from other plants like *Centella asiatica*, *Bidens pilosa* (*Cytopiloyne*), *Panax quinquefolium* L. (American ginseng), and *Dendranthema zawadskii* (Dendrazawaynes A and B), among others.

2.2.7.3 Allium *compounds*

Early investigators identified volatile odour principles in garlic oils – however, these compounds were only generated during tissue damage and preparation. Indeed, the vegetative tissues of *Allium* species are usually odour-free, and it is this observation that led to the hypothesis that the generation of volatile compounds from *Allium* species arose from non-volatile precursor substances. It was in the laboratory of Stroll and Seebrook in 1948 that the first stable precursor compound, (+)-*S*-allyl-L-cysteine sulfoxide (ACSO), commonly known as alliin, was identified; it makes garlic unique sulfur-containing molecules among vegetables (Stoll and Seebeck, 1947). Alliin is the parental sulfur compound that is responsible for the majority of the odorous volatiles produced from crushed or cut garlic. Three additional sulfoxides present in the tissues of onions were later identified in the laboratory of Virtanen and Matikkala, these being (+)-*S*-methyl-L-cysteine sulfoxide (methiin; MCSO), (+)-*S*-propyl-L-cysteine sulfoxide (propiin; PCSO), and (+)-*S*-*trans*-1-propenyl-L-cysteine sulfoxide or isoalliin (TPCSO). Isoalliin is the major sulfoxide present within intact onion tissues and is the source of the *A. cepa* lachrymatory factor (Virtanen and Matikkala, 1959). With regards to chemical distribution, (+)-*S*-methyl-L-cysteine sulfoxide is by far the most

Table 2.5 S-Alk(en)yl cysteine in *Allium* spp (modified from Rose et al., 2005)

Common name	Chemical name	Chemical structure
Methiin	S-Methyl-L-cysteine sulfoxide	H₃C–S(=O)–CH₂–CH(NH₂)–COOH
Aliin	S-Allyl-L.cysteine sulfoxide	H₂C=CH–CH₂–S(=O)–CH₂–CH(NH₂)–COOH
Propiin	S-Propyl-L-cysteine sulfoxide	H₃C–CH₂–CH₂–S(=O)–CH₂–CH(NH₂)–COOH
Isoalliin	S-Propenyl-L-cysteine sulfoxide	H₃C–CH=CH–S(=O)–CH₂–CH(NH₂)–COOH
Ethiin	S-Ethyl-L-cysteine sulfoxide	H₃C–CH₂–S(=O)–CH₂–CH(NH₂)–COOH
Butiin	S-n-Butyl-L-cysteine sulfoxide	H₃C–CH₂–CH₂–CH₂–S(=O)–CH₂–CH(NH₂)–COOH

ubiquitous, being found in varying amounts in the intact tissues of *A. sativum*, *A. cepa*, *A. porrum*, and *A. ursinum* L (Table 2.5).

Upon hydrolysis and oxidation, oil-soluble allyl compounds, which normally account for 0.2–0.5% of garlic extracts, such as diallyl sulfide (DAS), 5 diallyl disulfide (DADS), diallyl trisulfide (DATS), and other allyl polysulfides (2), are generated. Alternatively, it can be slowly converted into watersoluble allyl compounds, such as *S*-allyl-cysteine and *S*-allylmercaptocysteine (SAMC) (Filomeni *et al.*, 2008 and references there in).

2.2.7.4 Chlorophyll

Chlorophyll (also chlorophyl) is a green pigment found in almost all plants, algae, and cyanobacteria. Its name is derived from the Greek words *chloros* ("green") and *phyllon* ("leaf"). Chlorophyll is an extremely important biomolecule, critical in photosynthesis, which allows plants to obtain energy from light. Chlorophyll absorbs light most strongly in the blue portion of the electromagnetic spectrum, followed by the red portion. However, it is a poor absorber of green and near-green portions of the spectrum; hence the green color of chlorophyll-containing tissues (Speer, 1997). Chlorophyll was first isolated by Joseph Bienaimé Caventou and Pierre Joseph Pelletier in 1817 (Pelletier and Caventou, 1951).

In pepper, unripe fruit colors can vary from ivory, green, or yellow. The green color derives from accumulation of chlorophyll in the chloroplast while ivory indicates chlorophyll degradation as the fruit ripens (Wang *et al.*, 2005). The persistent presence of chlorophyll in fruit ripening to accumulate other pigments like carotenoids or anthocyanins produces brown or black mature fruit colors. Chlorophyll in black pepper fruit is 14-fold higher compared to violet fruit (Lightbourn *et al.*, 2008).

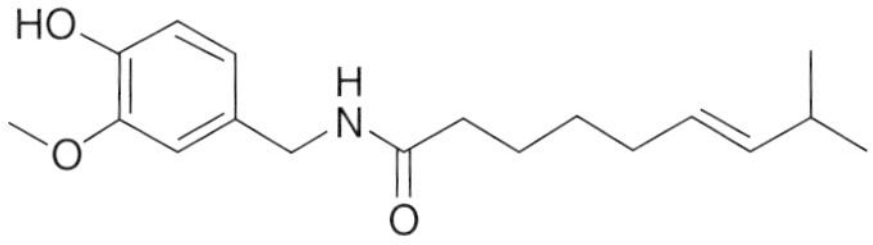

Figure 2.8 Capsaicin.

2.2.7.5 Betalains

The name "betalain" comes from the Latin name of the common beet (*Beta vulgaris*), from which betalains were first extracted. The deep red color of beets, bougainvillea, amaranth, and many cacti results from the presence of pigments. Betalains are a class of red and yellow indole-derived pigments found in plants of the Caryophyllales, where they replace anthocyanin pigments, as well as some higher order fungi (Strack and Schliemann, 2003).

There are two categories of betalains: Betacyanins include the reddish to violet betalain pigments and betaxanthins that are those betalain pigments that appear yellow to orange. Among the betaxanthins present in plants include vulgaxanthin, miraxanthin and portulaxanthin, and indicaxanthin (Salisbury et al., 1991).

The few edible known sources of betalains are red and yellow beetroot (*Beta vulgaris* L. ssp. vulgaris), coloured Swiss chard (*Beta vulgaris* L. ssp. cicla), grain or leafy amaranth (*Amaranthus* sp.), and cactus fruits, such as those of Opuntia and Hylocereus genera (Azeredo, 2009 and references there in).

2.2.7.6 Capsaicinoids

The nitrogenous compounds produced in pepper fruit, which cause a burning sensation, are called capsaicinoids. Capsaicinoids are purported to have antimicrobial effects for food preservation (Billing and Sherman, 1998), and their most medically relevant use is as an analgesic (Winter et al., 1995). Capsaicinoids have been used successfully to treat a wide range of painful conditions including arthritis, cluster headaches, and neuropathic pain. The analgesic action of the capsaicinoids has been described as dose dependent, and specific for polymodal nociceptors. The gene for the capsaicinoid receptor has been cloned (TRPV1) and the receptor transduces multiple pain-producing stimuli (Caterina et al., 1997; Tominaga et al., 1998). Capsaicin (*trans*-8-*N*-vamllyl-6-nonenamide) is an acrid, volatile alkaloid responsible for hotness in peppers (Figure 2.8).

2.3 Chemical properties of phytochemicals

2.3.1 Terpenes

The basic structure of terpenes follows a general principle: *2-Methylbutane* residues, less precisely but usually also referred to as *isoprene* units, (C_5) n, build up the carbon skeleton of terpenes; this is the *isoprene* rule 1 formulated by Ruzicka (1953) (Figures 2.2 and 2.3). The isopropyl part of 2-methylbutane is defined as the *head*, and the ethyl residue as the *tail* (Breitmaier, 2006). In nature, terpenes occur predominantly as hydrocarbons, alcohols and their glycosides, ethers, aldehydes, ketones, carboxylic acids, and esters (Breitmaier, 2006).

Several important groups of plant compounds, including cytokinins, chlorophylls, and the quinone-based electron carriers (the plastoquinones and ubiquinones), have terpenoid side chains attached to a non-terpenoid nucleus. These side chains facilitate anchoring to or movement within membranes. In plants, prenylated proteins may be involved in the control of the cell cycle (Qian *et al.*, 1996; Crowell, 2000), nutrient allocation (Zhou *et al.*, 1997), and abscisic acid signal transduction (Clark *et al.*, 2001).

The most abundant hydrocarbon emitted by plants is the hemiterpene (C5) isoprene, 2-methyl-1,3-butadiene. Emitted from many taxa, especially woody species, isoprene has a major impact on the redox balance of the atmosphere, affecting ozone, carbon monoxide, and methane levels (Lerdau *et al.*, 1997). The release of isoprene from plants is strongly influenced by light and temperature, with the greatest release rates typically occurring under conditions of high light and high temperature (Lichtenthaler, 2007). Although the direct function of isoprene in plants themselves has been a mystery for many years, there are now indications that it may serve to prevent cellular damage at high temperatures, perhaps by reacting with free radicals to stabilize membrane components (Sasaki *et al.*, 2007).

2.3.2 Polyphenols

Simple phenols (C_6), the simplest group, are formed with an aromatic ring substituted by an alcohol in one or more positions as they may have some substituent groups, such as alcoholic chains, in their structure (Andrés-Lacueva *et al.*, 2010). *Phenolic acids* (C_6-C_1) with the same structure as simple phenols are hydroxylated derivatives of benzoic and cinnamic acids (Herrmann, 1989; Shahidi and Naczk, 1995). They act as cell wall support materials (Wallace and Fry, 1994) and as colourful attractants for birds and insects helping seed dispersal and pollination (Harborne, 1994). *Hydrolyzable tannins* are mainly glucose esters of gallic acid. Two types are known: the gallotannins, which yield only gallic acid upon hydrolysis, and the ellagitannins, which produce ellagic acid as the common degradation product (see Figure 2.4) (Andrés-Lacueva *et al.*, 2010).

Acetophenones are aromatic ketones, and *phenylacetic acids* have a chain of acetic acid linked to benzene. Both have a C6-C2 structure. *Hydroxycinnamic acids* are included in the phenylpropanoid group (C6-C3). They are formed with an aromatic ring and a three-carbon chain. There are four basic structures: the coumaric, caffeic, ferulic, and sinapic acids. In nature, they are usually associated with other compounds such as chlorogenic acid, which is the link between caffeic and quinic acids (Andrés-Lacueva *et al.*, 2010). *Coumarins* belong to the benzopyrone group of compounds, all of which consist of a benzene ring joined to a pyrone. They may also be found in nature, in combination with sugars, as glycosides. They can be categorized as simple furanocoumarins, pyranocoumarins, and coumarins substituted in the pyrone ring (Murray *et al.*, 1982). *Benzophenones* and *xanthones* have the C6-C1-C6 structure. The basic structure of benzophenone is a diphenyl ketone, and that of xanthone is a 10-oxy-10*H*-9- oxaanthracene. More than 500 xanthones are currently known to exist in nature, and approximately 50 of them are found in the mangosteen with prenyl substituents (Andrés-Lacueva *et al.*, 2010). *Stilbenes* have a 1,2-diphenylethylene as their basic structure (C6-C2-C6). Resveratrol, the most widely known compound, contains three hydroxyl groups in the basic structure and is called 3,4,5-trihydroxystilbene. Stilbenes are present in plants as *cis* or *trans* isomers. *Trans* forms can be isomerized to *cis* forms by UV radiation (Lamuela-Raventós *et al.*, 1994). *Lignans* in the strict sense are phenylpropanoid dimers linked by a C-C bond between carbons 8 and 8'prime' in the side chain; they can be divided into several subgroups, depending on other linkages and substitution patterns introduced into the original hydroxycinnamyl alcohol dimmer. More than 55 plant families contain lignans, mainly gymnosperms and dicotyledonous angiosperms (Dewick, 1989).

Flavonoids constitute one of the most ubiquitous groups of all plant phenolics. So far, over 8000 varieties of flavonoids have been identified (De Groot and Raven, 1998). In plants, flavonoids are usually glycosylated mainly with glucose or rhamnose, but they can also be linked with galactose, arabinose, xylose, glucuronic acid, or other sugars (Vallejo *et al.*, 2004). All flavonoids contain 15 carbon atoms in their basic nucleus: two six-membered rings linked with a three-carbon unit, which may or may not be parts of a third ring (Middleton, 1984). The rings are labeled A, B, and C (see Figure 2.5). The individual carbon atoms are based on a numbering system that uses ordinary numerals for the A and C and "primed" numerals for B-ring (1). Primed modified numbering system is not used for chalcones (2) and the isoflavones derivatives (6): the pterocarpans and the rotenoids. The different ways to close this ring associated with the different oxidation degrees of ring A provide the various classes of flavonoids. The six-membered ring condensed with the benzene ring is either a α-pyrone (flavones (1) flavonols (3) or its dihydroderivative (flavanones (4) and flavan-3-ols (5)). The position of the benzenoid substituent divides the flavonoids into two classes: flavonoids (1) (2-position) and isoflavonoids (6) (3-position). Most flavonoids occur naturally associated with sugar in conjugated form and, within any one class, may be characterized as monoglycosidic, diglycosidic, etc. The glycosidic linkage is normally located at position 3 or 7 and the carbohydrate unit can be L-rhamnose, D-glucose, glucorhamnose, galactose, or arabinose (Tapas *et al.*, 2008 and references therein).

2.3.3 Carotenoids

Carotenoids consist of 40 carbon atoms (tetraterpenes) with conjugated double bonds. They consist of eight isoprenoid units joined in such a manner that the arrangement of isoprenoid units is reversed at the center of the molecule so that the two central methyl groups are in a 1,6-position and the remaining nonterminal methyl groups are in a 1,5-position relationship. They can be acyclic or cyclic (mono- or bi-, alicyclic or aryl) (see Figure 2.6) (Yahia and Ornelas-Paz, 2010).

Carotenoids are often used in visual displays through deposition in skin or feathers. Given these multiple uses that all require substantial amounts of carotenoids for normal functioning, carotenoids have been suggested to be in limited supply for reproduction, health related functions, or the expression of sexual coloration. For example, it has been suggested that carotenoids may limit vital functions, such as scavenging of free radicals, eliminating peroxides, and enhancing immune function (production of lymphocytes, enhancement of phagocytic ability of neutrophils and macrophages, production of tumor immunity), in which they have been shown to be involved (Møller *et al.*, 2000).

In nature, carotenoids exist as only two varieties: (1) unelaborated hydrocarbons, or (2) with functional groups, these are always attached via oxygen to the carotenoid skeleton. Carotenoids with heteroatoms other than oxygen have not yet been discovered in nature, but have been synthesized (Pfander, 1976). Hydrocarbon carotenoids generally form colored monomolecular solutions in nonpolar organic solvents, whereas and typically remain colorless in water. At extremely low carotenoid concentration, water unexpectedly exhibits an orange tint (the highly unsaturated polyene chain acting as a hydrophilic component). Strangely enough, the first carotenoid aggregates in water were obtained from β,β-carotene (von Euler *et al.*, 1931). Its well-known hydrophobicity did not prevent other studies with β,β-carotene, and lycopene, an acyclic carotenoid hydrocarbon (Song and Moore, 1974; Bystritskaya and Karpukhin, 1975; Mortensen *et al.*, 1997; Lindig and Rodgers, 1981). The many natural carotenols and carotenones (zeaxanthin, lutein, viloxanthin, astaxanthin) are undoubtedly more suited for aggregation studies in water (Mori, 2001; Zsila *et al.*, 2001; Billsten *et al.*, 2005; Köpsel *et al.*, 2005).

The overwhelming majority of the ~750 known naturally occurring carotenoids are hydrophobic (Britton *et al.*, 2004). It is therefore a striking paradox that the most utilized carotenoid since antiquity is extremely water-soluble: crocin has no saturation point in water. Crocin illustrates the typical surfactant structure, the hydrophobic polyene chain linked to two hydrophilic sugars; it is surface active, and the molecules associate to small oligomers at high concentration. The surface and aggregation properties of crocin have only recently been determined (Nalum-Naess *et al.*, 2006). Meanwhile, other natural sugar carotenoids have been isolated and characterized, however, the low occurrence and abundance of these "red sugar derivatives" prevents practical applications (Dembitsky, 2005). Another group of naturally occurring carotenoids – sulfates – are considerably less hydrophilic; the first characterized compound was bastaxanthin sulphate (Hertzberg *et al.*, 1983). A proposed application of carotenoid sulfates as feed/flesh colorants for cultured fish requires the additional help of an organic solvent for good outcomes (Yokyoyama and Shizusato, 1997). The "strange" appearance of the first recorded carotenoid sulfate visible spectrum in water was not immediately recognized as a sign of *H*-aggregation (Hertzberg and Liaaen-Jensen, 1985). The aggregation of a carotenoid sulfate was later observed as a negative outcome (Oliveros *et al.*, 1994). Norbixin is the other carotenoid utilized since ancient times; it is reported to be water-soluble up to 5%. Recent measurements could not confirm solubility; only negligible dispersibility was observed (Breukers *et al.*, 2009).

In the modern age, in addition to crocin and norbixin, several carotenoids have become extremely important commercially. These include, in particular, astaxanthin (fish, swine, and poultry feed, and recently human nutritional supplements); lutein and zeaxanthin (animal feed and poultry egg production, human nutritional supplements); and lycopene (human nutritional supplements). The inherent lipophilicity of these compounds has limited their potential applications as hydrophilic additives without significant formulation efforts; in the diet, the lipid content of the meal increases the absorption of these nutrients, however, parenteral administration to potentially effective therapeutic levels requires separate formulation that is sometimes ineffective or toxic (Lockwood *et al.*, 2003).

2.3.4 Glucosinolates

Glucosinolates are amino acid-derived secondary plant metabolites found exclusively in cruciferous plants. The majority of cultivated plants that contain glucosinolates belong to the family of *Brassicaceae* such as brussel sprouts, cabbage, broccoli, and cauliflower. These are the major source of glucosinolates in the human diet – about 120 different glucosinolates have been characterized. Glucosinolates and their breakdown products are of particular interest because of their nutritive and antinutritional properties, their potential adverse effects on health, their anticarcinogenic properties, and finally the characteristic flavour and odour they give to many vegetables (Verkerk and Dekker, 2008).

The majority of cultivated plants that contain glucosinolates belong to the family of *Brassicaceae*. Mustard seed, used as a seasoning, is derived from *B. nigra*, *B. juncea* (L.) Coss, and *B. hirta* species. Vegetable crops include cabbage, cauliflower, broccoli, brussel sprouts, and turnip of the *B. oleracea* L., *B. rapa* L., *B. campestris* L., and *B. napus* L. species. Kale of the *B. oleracea* species is used for forage, pasture, and silage. *Brassica* vegetables such as brussel sprouts, cabbage, broccoli, and cauliflower are the major source of glucosinolates

Table 2.6 Glucosinolates commonly found in *Brassicca* vegetables (adapted from Verkerk and Dekker, 2008)

Trivial name	Chemical name (side chain R)
Aliphatic glucosinolates	
Glucoiberin	3-Methylsulfinylpropyl
Progoitrin	2-Hydroxy-3-butenyl
Sinigrin	2-Propenyl
Gluconapoleiferin	2-Hydroxy-4-pentenyl
Glucoraphanin	4-Methylsulfinylbutyl
Glucoalyssin	5-Methylsulfinylpentyl
Glucocapparin	Methyl
Glucobrassicanapin	4-Pentenyl
Glucocheirolin	3-Methylsulfonylpropyl
Glucoiberverin	3-Methylthiopropyl
Gluconapin	3-Butenyl
Indole glucosinolates	
4-Hydroxyglucobrassicin	4-Hydroxy-3-indolylmethyl
Glucobrassicin	3-Indolylmethyl
4-Methoxyglucobrassicin	4-Methoxy-3-indolylmethyl
Neoglucobrassicin	1-Methoxy-3-indolylmethyl
Aromatic glucosinolates	
Glucosinalbin	*p*-Hydroxybenzyl
Glucotropaeolin	Benzyl
Gluconasturtiin	2-Phenethyl

in the human diet. They are frequently consumed by humans from Western and Eastern cultures (McNaughton and Marks, 2003). In the Netherlands, the average consumption of these vegetables is more than 36 g *Brassica* per person per day (Godeschalk, 1987). The typical flavor of *Brassica* vegetables is largely due to glucosinolate-derived volatiles. The versatility of these compounds is also demonstrated by the fact that glucosinolates are quite toxic to some insects and therefore could be included as one of many natural pesticides. However, a small number of insects, such as the cabbage aphids, use glucosinolates to locate their favorite plants as feed and to find a suitable environment to deposit their eggs (Barker *et al.*, 2006). Furthermore, glucosinolates show antifungal and antibacterial properties (Fahey *et al.*, 2001).

Only a limited number of glucosinolates have been investigated thoroughly although there are about 120 different ones currently characterized. A considerable amount of data on levels of total and individual glucosinolates are now available. The levels of total glucosinolates in plants may depend on variety, cultivation conditions, climate, and agronomic practice, while the levels within a particular plant vary between the parts of the plant. Generally the same glucosinolates occur in a particular sub-species regardless of genetic origin, and in most species only between one and four glucosinolates are found in relatively high concentrations (Table 2.6). Glucosinolates are chemically stable and biologically inactive when separated within sub-cellular compartments throughout the plant. However, tissue damage caused by pests, harvesting, food processing, or chewing initiates contact with the endogenous enzyme myrosinase in the presence of water leading to hydrolysis releasing a broad range of biologically active products such as isothiocyanates (ITCs), organic cyanides, oxazolidinethiones, and ionic thiocyanate.

Glucosinolate breakdown products exert a variety of toxic and antinutritional effects in higher animals amongst which the adverse effects on thyroid metabolism are the most thoroughly studied (Tripathi and Mishra, 2007). Tiedink *et al.* (1990, 1991) investigated the role of indole compounds and glucosinolates in the formation of *N*-nitroso compounds in vegetables. These studies revealed that the indole compounds present in *Brassica* vegetables can be nitrosated and thereby become mutagenic. However, the nitrosated products are stable only in the presence of large amounts of free nitrite.

2.3.5 Dietary fiber (non starch polysaccharides)

Polysaccharides are widespread biopolymers, which quantitatively represent the most important group of nutrients in botanical feed. Carbohydrates constitute a diverse nutrient category ranging from sugars easily digested by monogastric animals in the small intestine to dietary fiber fermented by microbes in the large intestine.

The structure of the plant cell wall influences the physical and chemical properties of the individual NSP and these vary considerably between different polymers and different molecular weights of the same polymer (Choct, 1997). Another factor that differentiates the physical properties among polysaccharides is the way the monomer units of polysaccharides are linked together (Moms, 1992). Different sugars linked together in the same way often give polysaccharides with very similar physical properties.

On the other hand, despite being built up from the same monomer units, polysaccharides can have different physical properties when the monomer units are linked together in different ways. The physiological effects of NSP on digestion and absorption of nutrients in human and monogastric animals have been attributed to its physicochemical properties. The main physicochemical properties of NSP that are of nutritional significance include: (a) hydration; (b) viscosity; (c) cation exchange capacity; and (d) organic compound absorptive properties. The hydration properties of NSP influence its water holding and binding capacity (Bach Knudsen, 2001). These depend on the physicochemical structure of the molecule and its ability to incorporate water within the molecular matrix. The viscosity properties of the NSP depend on its molecular weight or size (linear or branched), ionically charged groups, the surrounding structures, and the concentration of NSP (Smits and Annison, 1996). The cation exchange capacity is formed because the three-dimensional structure of the NSP molecule allows a chelation of ions to occur. The organic compound absorptive properties of NSP are due to its capacity to bind small molecules by both hydrophobic and hydrophilic bond interactions.

2.3.6 Lectins

Although it seems apparent now that Weir Mitchell had already observed lectin activity in rattle snake venom before (Kilpatrick, 2002) it wasn't until at least six years later, when Stillmark reported the dramatic action of ricin on red blood cells and then Helin followed it up by a similar report on abrin, that agglutinins caught the attention of the medical community. Reports of hemagglutinins from various sources were quick to follow. Besides plants, agglutinins were discovered in fungi, bacteria, viruses, invertebrates, and vertebrates. Although this early period established, beyond any doubt, the proteinaceous nature of lectins and their cell-agglutination and precipitation capabilities, lectin research thereafter was beset with problems and difficulties for the next quarter of a century. Studies, by Sugishita, Jonsson, Boyd, and Renkonen, provided the proverbial "shot in the arm" for research on lectins by identifying lectins as cell-recognition molecules that could have practical applications (Kocoureck, 1986). Reports of blood-group specificity, mitogenicity, and tumor cell-binding of lectins followed almost immediately.

The number of known properties and possible applications of lectins grew rapidly. Concanavalin A (Con A), a lectin from jack beans, became the first lectin to be crystallized and then extensively characterized by Sumner and Howell (1936) who also showed for the first time that sucrose could inhibit its agglutination activity. Two other major discoveries set the tone of the research that was to follow. Funatsu and his collaborators isolated the first non-toxic lectin from *Ricinus communis*, shattering the prevalent notion at that time that lectins were necessarily toxic proteins (Ishiguro *et al.*, 1964). Secondly, it was shown that several of these lectins, such as that from soybean, were glycoproteins (Lis and Sharon, 1973).

On the other hand, the effect of plant lectins on different cell types had already set the agenda for early research on them, leading to an extensive search for lectins in plant extracts and identification of a large number of lectins with practical applications. Such an objective did not require identification of the biological function of the protein *per se*. Indeed, in several cases where biological functions have been hypothesized or proven, the effect of the plant lectins on microbial or animal cells has provided clues to their putative function *in vivo*. Research on the endogenous roles of plant lectins has therefore been a late starter, although some progress has been made in this direction. Despite this, interest in studying plant lectins has been sustained, owing to the fact that their natural abundance makes their applications in a large number of areas much more feasible (Komath *et al.*, 2006 and references therein).

Glycosylation is the key step in a number of processes at the cellular level. Cell-surface oligosaccharides get altered in various kinds of pathological conditions including malignant transformations. With developments in the closely-related field of glycobiology, it has now become evident that oligosaccharide-mediated recognition plays a very important role in various biological processes such as fertilization, immune defence, viral replication, parasitic infection, cell–matrix interaction, cell–cell adhesion, and enzymatic activity. Lectins have been implicated in most, if not all of these recognition events. The strict selectivity that this kind of recognition requires imposes a stringent geometry upon both the ligand and the corresponding receptor, thus conferring unique sugar-specificities upon lectins (Sharon and Lis, 2004).

Carbohydrates can interact with lectins *via* hydrogen bonds, metal coordination bonds and van der Waal interaction and hydrophobic interaction. Selectivity results from specific hydrogen bonding and/or metal coordination bonds with key hydroxyls of the carbohydrates, which can act as both acceptors and donors of hydrogen bonds. Water molecules often act as bridges in these interactions. The hydroxyl at the C4 position, in particular, seems to be a decisive player in these events. Steric exclusions minimize unwanted recognition, further fine-tuning the saccharide specificity of the lectin. Subsite binding and subunit multivalency, where possible, increase the binding selectivity manifold (Rinni, 1995). In subsite binding, the primary binding site appears critical for carbohydrate recognition, but secondary binding sites contribute to enhanced affinity of the lectin towards specific oligosaccharides. For example, the legume lectins *Lathyrus ochrus* isolectin II (LOL II) and Con A are both Man/Glc specific lectins, but their oligosaccharide preferences are very different. LOL II has several-fold higher affinity for oligosaccharides that have additional a (1–6)-linked fucose residues, while Con A does not (Rinni, 1995; Weis and Drickamer, 1996). In subunit multivalency, several subunits of the same lectin contribute to the binding by recognizing different extensions of the carbohydrate or different chains of a branched oligosaccharide. This kind of binding is exhibited, among others, by the asialoglycoprotein receptor, the mannose binding protein (MBP) from the serum, the chicken hepatic lectin, and the cholera toxin (Drickamer, 1997; Elgavish and Shaanan, 1997). It appears that the monosaccharide specificity of a lectin, although useful, need not necessarily tell the complete story. It has become evident in numerous cases, particularly of lectins with proven or putative biological functions, that multivalency of the receptor is a prerequisite for recognition. Thus, the MBP, for example, binds to monomeric mannose units and simply releases them but, when it binds to the oligomannosides on a pathogen that has the same spacing as the trimers of MBP, it triggers a biological response that results in complement fixation (Komath *et al.*, 2006 and references therein).

2.3.7 Other phytochemicals

2.3.7.1 Alkaloids

Plant alkaloids are important privileged compounds with many pharmacological activities (Beghyn *et al.*, 2008; Facchini and Luca, 2008). In fact, alkaloid-containing plants have been recognized and exploited since ancient human civilization, from the utilization of *Conium maculatum* (hemlock) extract containing neurotoxin alkaloids to poison Socrates, to the use of coffee and tea as mild stimulants (Kutchan, 1995). Today, numerous alkaloids are pharmacologically well characterized and are used as clinical drugs, ranging from cancer chemotherapeutics to analgesic agents (Table 2.7).

Table 2.7 Alakloids with pharmaceutical applications (adapted from Leonard *et al.*, 2009)

Alkaloid	Plant species
Ajmaline	*Rauwolfia sellowii*
Berberine	*Berberis vulgaris*
Caffeine	*Cofee arabica*
Camptothecin	*Camptotheca acuminata*
Cocaine	*Erythroxylon coca*
Codeine	*Papaver somniferum*
Hyoscyamine	*Hyoscyamus muticus*
Irinotecan	–
Morphine	*Papaver somniverum*
Nicotine	*Nicotiana tabacum*
Noscapine	*Papaver somniverum*
Oxycodone	–
Oxymorphone	–
Papaverine	*Papaver somniverum*
Quinidine	*Cinchona ledgeriana*
Quinine	*Cinchona ledgeriana*
Reserpine	*Rauwolfia nitida*
Sanguinarine	*Sanguinaria canadiensis*
Scopolamine	*Hyoscyamus muticus*
Strychnine	*Strychnos nux-vomica*
Topotecan	–
Vinblastine	*Catharanthus roseus*
Vincristine	*Catharanthus roseus*
Vindesine	–
Vinflunine	–
Vinorelbine	–
Yohimbine	*Pausinystalia yohimbe*

The clinical value of vinca alkaloids, for example, isolated from the Madagascar periwinkle, *Catharantus roseus* G. Don., was clearly identified as early as 1965 and so this class of compounds has been used as anti-cancer agents for over 40 years and represents a true lead compound for drug development (Sipiora *et al.*, 2000).

A number of other alkaloids are known to have a bitter taste, and the response of cultured rat trigeminal ganglion neurons to bitter tastants has been studied (Liu and Simon, 1998). The authors investigated the responses of rat chorda tympani and glossopharnygeal neurons to a variety of bitter-tasting alkaloids. Of the 89 neurons tested, 34% responded to 1mM nicotine, 7% to 1mM caffeine, 5% to 1mM denatonium benzoate, 22% to 1mM quinine hydrochloride, 18% to 1mM strychnine, and 55% to 1mM capsaicin. These data suggest that neurons from the trigeminal ganglion respond to the same bitter-tasting chemical stimuli as do taste receptor cells and are likely to contribute information sent to the higher central nervous system regarding the perception of bitter/irritating chemical stimuli.

Many alkaloids mimicking the structures of monosaccharides or oligosaccharides have been isolated from plants and microorganisms. Such alkaloids are easily soluble in water because of their polyhydroxylated structures and inhibit glycosidases because of a structural resemblance to the sugar moiety of the natural substrate. Glycosidases are involved in a wide range of important biological processes, such as intestinal digestion, post-translational processing of the sugar chain of glycoproteins, quality-control systems in the endoplasmic reticulum (ER) and ER associated degradation mechanism, and the lysosomal catabolism of glycoconjugates. Inhibition of these glycosidases can have profound effects on carbohydrate catabolism in the intestines, on the maturation, transport, and secretion of glycoproteins, and can alter cell–cell or cell–virus recognition processes. The realization that glycosidase inhibitors have enormous therapeutic potential in many diseases such as diabetes, viral infection, and lysosomal storage disorders has led to increasing interest in and demand for them (Asako, 2008). Acarbose, a potent inhibitor of intestinal sucrase, was effective in carbohydrate loading tests in rats and healthy volunteers, reducing postprandial blood glucose and increasing insulin secretion (Puls *et al.*, 1977).

A possible way to suppress hepatic glucose production and lower blood glucose in type 2 diabetes patients may be through inhibition of hepatic glycogen phosphorylase. Fosgerau *et al.* (2000) reported that in enzyme assays 1,4-dideoxy-1,4-imino-d-arabinitol (DAB, alkaloid isolated from the leaves of *Morus bombycis* in Japan) is a potent inhibitor of hepatic glycogen phosphorylase. Furthermore, in primary rat hepatocytes, DAB was shown to be the most potent inhibitor of basal and glucagon-stimulated glycogenolysis ever reported (Andersen *et al.*, 1999).

2.3.7.2 Polyacetylenes

Acetylenic natural products include all compounds with a carbon-carbon triple bond or alkynyl functional group. While not always technically accurate, the term "polyacetylenes" is often used interchangeably to describe this class of natural products, although they are not polymers and many precursors and metabolites contain only a single acetylenic bond. These compounds tend to be unstable, succumbing either to oxidative, photolytic, or pH-dependent decomposition, which originally provided substantial challenges for their isolation and characterization (Minto and Blacklock, 2008). The earliest isolated alkyne-bearing natural product was dehydromatricaria ester, which was isolated, but not fully characterized, in 1826. No compound was characterized as being acetylenic until 1892 (tariric acid, 5T)

(Arnaud, 1892, 1902), after which only a handful of compounds were isolated before 1952. A lecture by N. A. Sörensen to the Royal Chemical Society in Glasgow describes the early history of polyacetylenic natural product chemistry (Sörensen, 1961).

A diacetylenic 1,6-dioxaspiro[4.5]decane gymnasterkoreayne G (23A) was isolated from the aerial parts of *Matricaria aurea* (Asteraceae) and, along with four known gymnasterkoreaynes, was active in a transcription factor inhibitory screen of gymnasterkoraiensis leaf extract. Elevated levels of the NFAT transcription factor have been linked to autoimmune responses and inflammation. While 23B showed lower NFAT inhibition than the threo-diol-containing gymnasterkoreayne E (23C), the differential activities of 23B, 23C, and the epoxydiyne gymnasterkoreayne B (23D) illuminate the importance of the stereochemical arrangement of the oxygen functionalities in maximizing this inhibitory effect. Several of the gymnasterkoreaynes A-F exhibit anti-cancer activity. The gymnasterkoreaynes are found with polyacetylenes 23F and 23E, the latter being their likely direct precursor. Three new diacetylenic spiroketals (23G-I) were isolated from *Plagius flosculosus* and examined for cytotoxicity. They were found to be less active against Jurat T and HL-60 leukemia cells than known compounds that contained two unsaturated rings. Reduced sensitivity of Bcl-2-overexpressing cells to these natural products suggested a mechanism of action involving the mitochondrial apoptotic pathway (Minto and Blacklock, 2008 and references therein).

Cytopiloyne, was identified from the *Bidens pilosa* extract using *ex vivo* T cell differentiation assays based on a bioactivity-guided fractionation and isolation procedure. Its structure was elucidated as 2β-d-glucopyranosyloxy-1-hydroxytrideca-5,7,9,11-tetrayne by various spectroscopic methods. Functional studies showed that cytopiloyne was able to inhibit the differentiation of naïve T helper (Th0) cells into type I T helper (Th1) cells but to promote the differentiation of Th0 cells into type II T helper (Th2) cell (Chiang *et al.*, 2007). It has also been demonstrated that polyacetylene aglycones of *B. pilosa*, namely 1,2-dihydroxytrideca-5,7,9,11-tetrayne and 1,3- dihydroxy-6(E)-tetradecene-8,10,12-triyne, exhibit significant and potent anti-angiogenic activities. The ability of both compounds to block angiogenesis is possibly in part through induction of p27(Kip1) and regulation of other cell cycle mediators including p21(Cip1) and cyclin E (Wu *et al.*, 2004).

Slight variations in polyacetylene structure result in extreme variations in biological activities. Low toxicity cicutol (46A), 4B, and 4C can be contrasted with lethal K+-current blocker cicutoxin from water hemlock (46B) (Straub *et al.*, 1996). The toxicity of 46B was found to have three structural requirements: an allylic alcohol, a long-conjugated (E)-polyene, and a terminal hydroxy group (Uwai *et al.*, 2000).

2.3.7.3 Allium compounds

The ACSOs are found in the cytoplasm of onion cells, physically separated from alliinase. When the tissues of any *allium* are disrupted, the enzyme alliinase hydrolyses the flavor precursors. The result is a wide range of reactive organosulphur compounds with characteristic flavor and striking bioactivity. The first products of the reaction between alliinase and the flavor precursors are the highly reactive sulphenic acids. In garlic, the 2-propene sulphenic acid condenses to form the thiosulphinate allicin (allyl-2-propenethiosulphinate), which gives it its characteristic flavor. In aged extracts of garlic, allicin can disproportionate (react with itself) to form the sulphides, thiosulphonates, and the trisulphur compound called ajoene. Ajoene has notable antithrombitic activity (Randle and Lancaster, 2002).

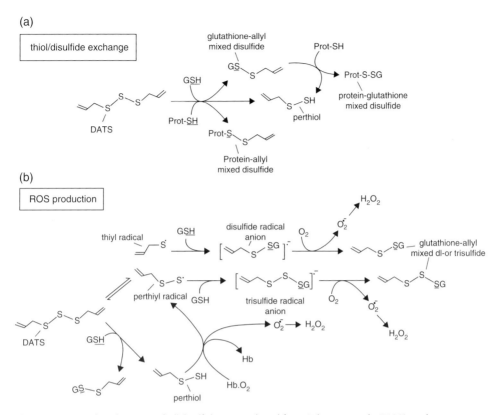

Figure 2.9 Redox chemistry of allyl sulfides Reproduced from (Filomeni et al., 2008), with permission from the American Society for Nutrition.

The chemical structure and reactivity of allyl compounds rather favor a pro-oxidant activity. In fact, oil-soluble allyl compounds are the main source of disulfides and polysulfides and, due to the high intracellular abundance of reduced glutathione (GSH) and protein thiols, they can mediate thiol/disulfide exchange by determining decrease of GSH and thiolation of reactive cysteine residues on proteins (Figure 2.9(a)). Whereas the former reaction induces oxidative unbalance, the latter yields reversible alterations of protein function, as demonstrated for the nonselective cation channel transient receptor potential-A1 of sensory nerve endings upon treatment with DADS, which underlies its pungent effects. Allyl disulfides and polysulfides can also produce reactive oxygen species (ROS) directly by reactions relying upon the homolytic cleavage of disulfide bond. This leads to the formation of allyl-(per)thiyl radicals, which can rapidly react with GSH, thus forming disulfide or polysulfide radical anions and reducing oxygen to produce ROS. Superoxide and hydrogen peroxide can be produced also as by-products of the reaction between perthiol and oxygen (e.g. O_2 bound to hemoglobin; Figure 2.9(b)) (Filomeni et al., 2008).

2.3.7.4 Chlorophyll

Chlorophyll is a chlorin pigment, which is structurally similar to and produced through the same metabolic pathway as other porphyrin pigments such as heme. At the center of the chlorin ring is a magnesium ion. For the structures depicted in this chapter, some of the ligands attached to the Mg^{2+} center are omitted for clarity. The chlorin ring can have several different

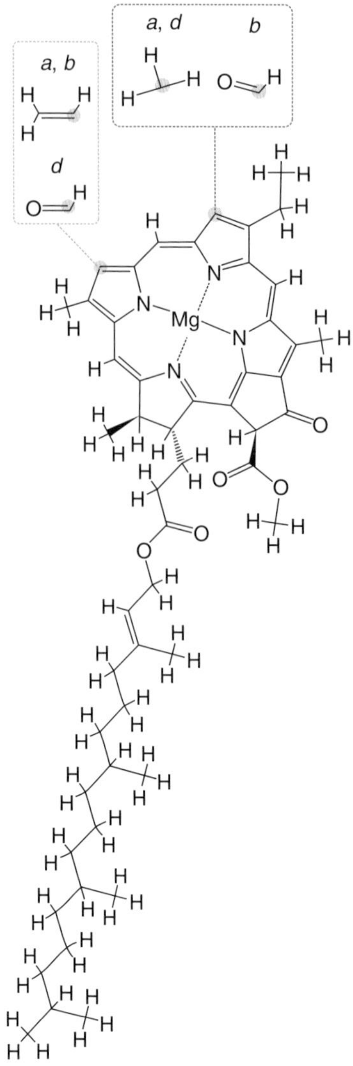

Figure 2.10 Chlorophyll structure.

side chains, usually including a long phytol chain. There are a few different forms that occur naturally, but the most widely distributed form in terrestrial plants is chlorophyll *a* (Figure 2.10). The general structure of chlorophyll *a* was elucidated by Hans Fischer in 1940, and by 1960, when most of the stereochemistry of chlorophyll *a* was known, Robert Burns Woodward published a total synthesis of the molecule as then known (Woodward et al., 1960). In 1967, the last remaining stereochemical elucidation was completed by Ian Fleming (Fleming, 1967) and in 1990 Woodward and co-authors published an updated synthesis (Woodward et al., 1990).

2.3.7.5 Betalains

Betalains are water-soluble nitrogen-containing pigments, which are synthesized from the amino acid tyrosine into two structural groups: the red-violet betacyanins and the

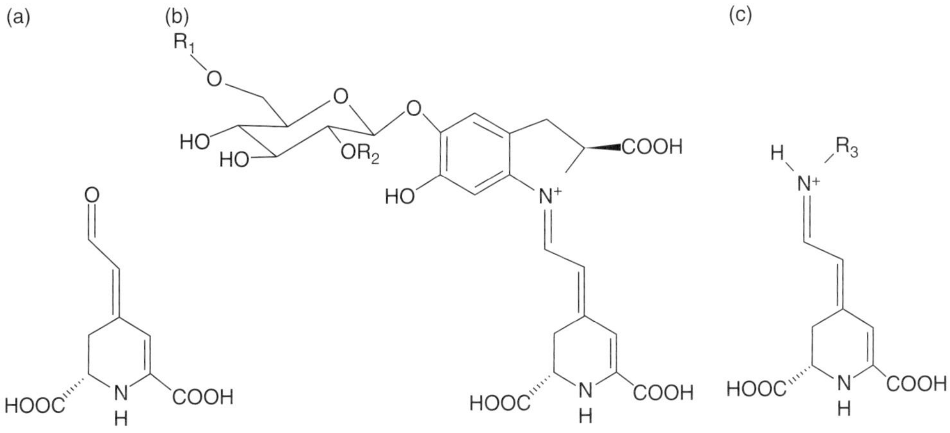

Figure 2.11 General structures of betalamic acid (a), betacyanins (b), and betaxanthins (c). Betanin: R1 = R2 = H. R3 = amine or amino acid group (Strack et al., 2003).

yellow-orange betaxanthins. Betalamic acid, presented in Figure 2.11(a), is the chromophore common to all betalain pigments. The nature of the betalamic acid addition residue determines the pigment classification as betacyanin or betaxanthin (Figures 2.11(b) and 2.11(c)), respectively (Azeredo, 2009). They are not related chemically to the anthocyanins and are not even flavonoids (Raven et al., 2004). Each betalain is a glycoside, and consists of a sugar and a colored portion. Their synthesis is promoted by light (Salisbury and Cleon, 1991).

In natural plant, betalains play important roles in physiology, optical attraction for pollination, and seed dispersal (Piattelli, 1981). They also function as reactive oxygen species (ROS) scavengers, protect plants from damages caused by wounding and bacterial infiltration as seen in red beet (*Beta vulgaris* subsp. *vulgaris*) (Sepúlveda-Jiménez et al., 2004), and function as UV-protecter in ice plant (*Mesembryanthemum crystallinum*) (Vogt et al., 1999).

2.3.7.6 Capsaicinoids

Capsaicinoids all share a common aromatic moiety, the vanillylamine, and differ in the length and degree of unsaturation of the fatty acid side chain (Bennett and Kirby, 1968; Leete and Louden, 1968). The perception of burn from these individual capsaicinoids will also vary slightly; capsaicin (Figure 2.9) and dihydrocapsaicin are the hottest and deliver their bite everywhere from the mid-tongue and palate to down in the throat (Krajewska and Powers, 1998).

Capsaicinoids start to accumulate 20 days post anthesis and synthesis usually persists through fruit development. The site of synthesis and accumulation of the capsaicinoids is the epidermal cells of the placenta in the fruit. Ultimately, the capsaicinoids are secreted extracellularly into receptacles between the cuticle layer and the epidermal layer of the placenta. These receptacles of accumulated capsaicinoids are macroscopically visible as pale yellow to orange droplets or blisters on the placenta of many chile types are odorless and tasteless (Guzman et al., 2010 and references therein).

While it is used as an ingredient in pepper sprays, capsaicin and its dihydro derivatives all exhibit anti-inflammatory properties (Sancho et al., 2002). Kim et al. (2003) examined the anti-inflammatory mechanism of capsaicin on the production of inflammatory molecules in liposaccharides (LPS)-stimulated murine peritoneal macrophages. Capsaicin suppressed PGE2 production by inhibiting *COX-2* enzyme and inducible nitric-oxide synthase (iNOS) expression in a dose-dependent manner. Lee et al. (2000) showed capsaicin induced apoptosis in A172 human glioblastoma cells in a time and dose-dependent manner. The mechanism whereby capsaicin induced apoptosis may involve reduction of the basal generation of ROS.

2.4 Biochemical pathways of important phytochemicals

In plants three pathways: shikimate, isoprenoid, and polyketide are particularly the source of most secondary metabolites. After the formation of the major basic skeletons, further modifications result in plant species specific compounds. The "decorations" concern, for example hydroxy, methoxy, aldehyde, carboxyl groups, and substituents adding further carbon atoms to the molecule, such as prenyl-, malonyl-, and glucosyl-moieties. Moreover, various oxidative reactions may result in loss of certain fragments of the molecule or rearrangements leading to new skeletons (Verpoorte and Alfermann, 2000).

2.4.1 Shikimate pathway

The shikimate pathway is the major source of aromatic compounds (Bentley, 1990; Haslam, 1993; Herrmann, 1995; Schmidt and Amrhein, 1995). It is found in microorganisms and plants, but not in mammals. The main trunk of the shikimate pathway consists of reactions catalyzed by seven enzymes. The best studied of these are the penultimate enzyme, the 5-enol-pyruvoyl shikimate-3-P synthase, the primary target site for the herbicide glyphosate, and the first enzyme, DAHP synthase, controls carbon flow into the shikimate pathway. DAHP synthase catalyzes the condensation of phosphoenolpyruvate (PEP) and erythrose-4-P to yield DAHP and Pi. Even though the enzyme was discovered in *Escherichia* coli more than three decades ago and has been purified to electrophoretic homogeneity from a number of sources, the fine structure of DAHP, the product of the enzyme-catalyzed reaction, was not described until many years later (Garner and Herrmann, 1984).

The pathway starts with the condensation of D-erythrose 4-phosphate and phosphoenolpyruvate. In a series of reactions a cyclic compound, 3-dehydroquinate, is obtained. In two further steps this yields shikimate, which after phosphorylation is coupled by the enzyme EPSP synthase with phosphoenolpyruvate to give 5-enolpyruvylshikimate-3-phosphate (EPSP). This enzyme is the target for glyphosate, the herbicide. Dephosporylation of EPSP eventually results in chorismate, from where the pathway diverges into two major branches, leading to respectively phenylalanine/tyrosine and tryptophan. In terms of carbon fluxes some minor branches lead to isochorismate, 4-hydroxybenzoic acid, and 4-aminobenzoic acid, from which series of different secondary metabolites are derived. All branches lead to products necessary for primary metabolism and primary functions in cells, but also secondary metabolite pathways are derived from these branches. From an early intermediate of the shikimate pathway (3-dehydroshikimate) gallic acid derivatives are formed (Figure 2.12) (Verpoorte and Alfermann, 2000).

The majority of shikimate-derived natural products are formed from the end products of the shikimate pathway, i.e. the aromatic amino acids. Of these, phenylalanine in particular gives rise to a tremendous variety of different phenylpropanoid compounds, e.g. the flavonoids or the lignans (van Sumere and Lea, 1985; Haslam, 1993), although all three aromatic amino acids along with anthranilic acid are the precursors of numerous alkaloids (Southon and Buckingham, 1989; Haslam, 1993). A smaller number of natural products are derived from variants of the shikimate pathway, which branch off at different points along the main metabolic sequence. It not only generates end products that serve as the starting materials for the biosynthesis of countless natural products but, as a very complex metabolic pathway, it also provides ample opportunities for "derailments" along the pathway, which, through often very intriguing chemistry, lead to additional unique secondary metabolites line phenazines and esmeraldins, as has been reviewed by Floss (1997).

The classification based on biosynthetic origin has as major examples the terpenoids, phenylpropanoids, and polyketides with terpenoids as the largest group. These compounds are all derived from the isoprenoid biosynthetic pathway, which uses a C5 building block to build up C10 (monoterpenes), C15 (sesquiterpenes), C20 (diterpenes), C30 (steroids and triterpenes), and C40 (carotenoids) compounds. In the other two groups a few basic building blocks phenylalanine/tyrosine (C9) and acetate (C2), are used to assemble a basic skeleton from which respectively the phenylpropanoids and polyketides are derived. In Figure 2.12 some major group of secondary metabolites derived from the terpenoid and phenylpropanoid pathways in plants are summarized. These two pathways are most important for secondary metabolite formation in plants; the polyketide pathway is particularly well-developed in microorganisms.

The phenylpropanoid pathway is one of the most important metabolic pathways in plants in terms of carbon flux (Bentley, 1990; Haslam, 1993; Strack, 1997). In a cell more than 20% of the total metabolism can go through this pathway, the enzyme chorismate mutase is an important regulatory point. This pathway leads to, among others, lignin, lignans, flavonoids, and anthocyanins mediated by phenylalanine ammonia lyase (PAL), which converts phenylalanine into *trans-cinnamic* acid by a non-oxidative deamination (Verpoorte and Alfermann, 2000).

Cinnamate and its hydroxy-derivatives are also the precursor for a broad range of other phenolics such as coumarins, formed by lactonization after introduction of an ortho hydroxy group in cinnamate, and benzoic acid derivatives such as salicylic acid by cleavage at the double bond in the side chain of cinnamate. Conversion of the carboxylic group in the (hydroxy) cinnamates to an alcohol yields the building blocks for lignin and the lignans (Dawson et al., 1989). The two major classes of alkaloids, the isoquinoline and the indole alkaloids, are derived from the aromatic amino acids. The isoquinoline alkaloids are formed from dopamine, which is condensed with 4-hydroxyphenyl acetaldehyde (both formed from tyrosine), yielding the benzylisoquinoline tiorcoclaurine. This compound in a series of steps is converted into reticuline, the precursor for numerous isoquinoline alkaloids such as morphine, sanguinarine, and berberine (Hashimoto and Yamada, 1994; Kutchan, 1995). Other types of phenolic compounds are derived from other branches of the chorismate pathway (Figure 2.12). For example, the isochorismate branch leads to anthraquinones (e.g. in some Rubiaceae plants). Naphtoquinones are derived from 4-hydroxybenzoic acid.

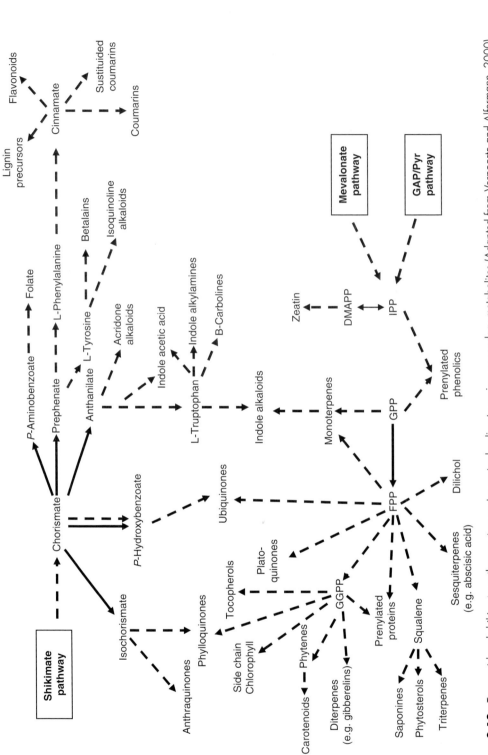

Figure 2.12 Terpenoid and shikimate pathways, two major routes leading to various secondary metabolites (Adapted from Verpoorte and Alfermann, 2000). GGP, geranylgeranyl pyrophosphate; FPP, farnesyl pyrophosphate; IPP, isopentenyl pyrophosphate; DMAPP, dimethylallyl pyrophosphate; GGPP, geranylgeranyl diphosphate synthase; GAP, glyceraldehyde-3-phosphate; Pyr, pyruvate.

2.4.2 Isoprenoid pathway

The other important pathway in plants is that of the terpenoids, also known as isoprenoid pathway (Nes et al., 1992; McGarvey and Croteau, 1995; Torsell, 1997). Terpenoids include more than one third of all known secondary metabolites (Figure 2.13). Moreover, the C5-building block is also incorporated in many other skeletons, e.g. in anthraquinones, naphtoquinones, cannabinoids, furanocoumarines, and terpenoid indole alkaloids. In the "decoration" type of reactions in various types of secondary metabolites C5-units are attached to the basic skeleton, e.g. hop bitter acids, flavonoids, and isoflavonoids (Tahara and Ibrahim, 1995; Barron and Ibrahim, 1996).

The biosynthetic pathway to terpenoids (Figure 2.13) is conveniently treated as comprising four stages, the first of which involves the formation of isopentenyl diphosphate (IPP), the biological C5 isoprene unit. Plants synthesize IPP and its allylic isomer, dimethylallyl

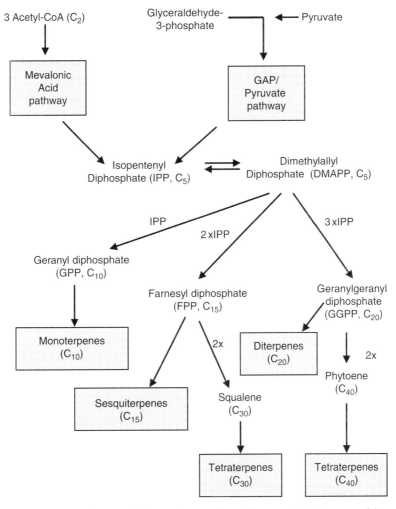

Figure 2.13 Overview of terpenoids buisynthesis in plants, showing the basic stages of this process and major groups of ebd products. CoA, coenzyme A; GAP, glyceraldehyde-3-phosphate (adapted from Ashour and Gershenzon, 2010).

diphosphate (DMAPP), by one of two routes: the well-known mevalonic acid pathway, or the newly discovered methylerythritol phosphate (MEP) pathway. In the second stage, the basic C5 units condense to generate three larger prenyl diphosphates, geranyl diphosphate (GPP, C10), farnesyl diphosphate (FPP, C15), and geranylgeranyl diphosphate (GGPP, C20). In the third stage, the C10–C20 diphosphates undergo a wide range of cyclizations and rearrangements to produce the parent carbon skeletons of each terpene class. GPP is converted to the monoterpenes, FPP is converted to the sesquiterpenes, and GGPP is converted to the diterpenes. FPP and GGPP can also dimerize in a head-to-head fashion to form the precursors of the C30 and the C40 terpenoids, respectively. The fourth and final stage encompasses a variety of oxidations, reductions isomerizations, conjugations, and other transformations by which the parent skeletons of each terpene class are converted to thousands of distinct terpene metabolites (Ashour et al., 2010).

The most exciting advance in the field of plant terpenoid biosynthesis is the discovery of a second route for making the basic C5 building block of terpenes completely distinct from the mevalonate pathway (Lichtenthaler, 2000). This route, which starts from glyceraldehyde phosphate and pyruvate (Figure 2.13), has also been detected in bacteria and other microorganisms. Different studies have demonstrated that an assortment of terpenoids from angiosperms, gymnosperms, and bryophytes, including monoterpenes (Eisenreich et al., 1997), diterpenes (Knoss et al., 1997; Jennewein and Croteau, 2001), carotenoids (Lichtenthaler et al., 1997), and the side chains of chlorophyll (phytol) and quinones (Lichtenthaler et al., 1997) are formed in a non-mevalonate fashion, while the labeling of sesquiterpenes and sterols was consistent with their origin from the mevalonate pathway (Lichtenthaler et al., 1997).

The non-mevalonate route to terpenoids appears to be localized in the plastids. In plant cells, terpenoids are manufactured both in the plastids and the cytosol (Gray, 1987; Kleinig, 1989). As a general rule, the plastids produce monoterpenes, diterpenes, phytol, carotenoids, and the side chains of plastoquinone and α-tocopherol, while the cytosol/ER compartment produces sesquiterpenes, sterols, and dolichols. In the studies discussed here, nearly all of the terpenoids labeled by deoxyxylulose (Sagner et al., 1998; Eisenreich et al., 2001) and 2-C-methyl erythritol feeding (Duvold et al., 1997) or showing 13C-patterns indicative of a non-mevalonate origin (Cartayrade et al., 1994; Lichtenthaler et al., 1997; Eisenreich et al., 2001) are thought to be plastid derived. A strict division between the mevalonate and non-mevalonate pathways may not always exist for a given end product. The biosynthesis of certain terpenoids appears to involve the participation of both routes (Nabeta et al., 1995; Piel et al., 1998).

2.4.3 Polyketide pathway

The poyketide pathway plays an important role in primary metabolism in the biosynthesis of fatty acids. The fatty acids are the basis for various secondary metabolites, but the polyketide pathway also directly leads to secondary metabolites, particularly in microorganisms, but also in plants (Luckner, 1990; Borejsza-Wysocki and Hrazdina, 1996; Torsell, 1997). The C2 polyketide building block acetyl- CoA ester is first converted into the more reactive malonyl-CoA ester. This compound is then used in various reactions, also at the start of the mevalonate pathway (Figure 2.14). In the fatty acid biosynthesis, acetyl-CoA is the starter molecule, bound to a thiol group in the fatty acid synthetase enzyme complex, the malonyl-CoA is bound to another vicinal thiol group in the acyl carrier protein (ACP) and is subsequently condensed with the acetyl group. The acetoacetyl-ACP

Chemistry and classification of phytochemicals 39

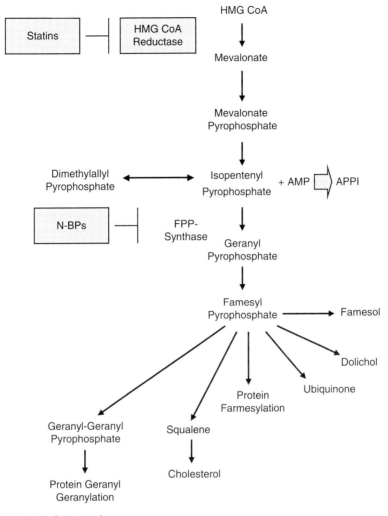

Figure 2.14 Mevalonate pathway.

is reduced to give the butyryl-ACP, which reacts with a further malonyl-CoA, thus in a series of reactions the fatty acids are built up. From the fatty acid pathway various secondary metabolites, such as alkanes, acetogenins, and jasmonates are formed (Verpoorte and Alfermann, 2000).

The malonyl-CoA is also part of the flavonoid biosynthesis, coumaryl-CoA is condensed with three molecules of malonyl-CoA, after which ring closure yields naringenin. The enzyme stilbene synthase results in the formation of stilbenes such as resveratrol using the same substrates. The condensation of coumaryl-CoA with one malonyl-CoA leads to benzalacetones (Borejsza-Wysocki and Hrazdina, 1996). Other examples of plant secondary metabolites derived from the polyketide pathway are 6-methylsalicylic acid and coniine (four C2 units), plumbagin (six C2 units), and anthraquinones (eight C2 units) (Figure 2.15). However, the anthraquinones are, in some plant families, derived from the chorismate pathway (Verpoorte and Alfermann, 2000).

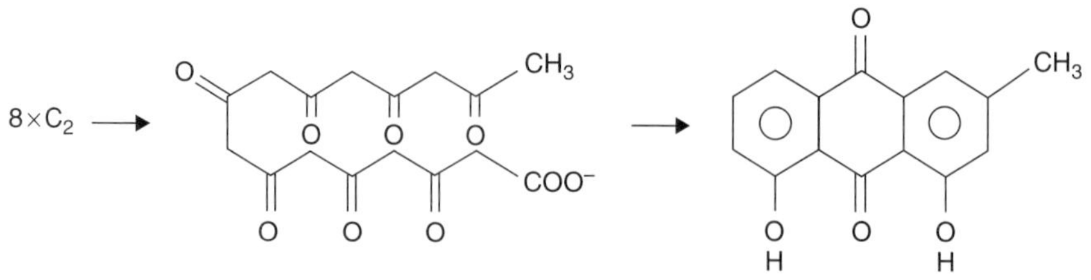

Figure 2.15 Polyketide biosynthetic pathway leading to anthraquinones (from Verpoorte and Alfermann, 2000).

2.4.4 Secondary transformation

The cyclic terpenes formed initially are subject to an assortment of further enzymatic modifications, including oxidations, reductions, isomerizations, and conjugations, to produce the wide array of terpenoid end products found in plants. Unfortunately, few of these conversions have been well studied, and there is little evidence from most of the biosynthetic routes proposed, except in the case of the gibberellin pathway (Yamaguchi, 2008). Many of the secondary transformations belong to a series of well-known reaction types that are not restricted to terpenoid biosynthesis (Mihaliak et al., 1993). The enzymes involved are often cytochrome P450 enzymes, dioxygenases, and peroxidases.

2.4.5 Glucosinolate biosynthesis

The pathway of glucosinolate biosynthesis has been studied since the 1960s and the identity of many intermediates, enzymes, and genes involved is now known. The biosynthesis of glucosinolates was recently reviewed extensively by Halkier and Gershenzon (2006).

Kjaer and Conti (1954) suggested that amino acids may be natural precursors of the aglycone moiety of glucosinolates based on the similarities between the carbon skeletons of some amino acids and the glucosinolates. This hypothesis was confirmed by studies of the different biosynthetic stages. Most of these studies involved the administration of variously labeled compounds (3H, 14C, 15N, or 35S) to plants and the assessment of their relative efficiencies as precursors on the basis of the extent of incorporation of isotope into the glucosinolate. The classification of glucosinolates, as shown in Table 2.6, depends on the amino acid from which they are derived; aliphatic glucosinolates are derived from alanine, leucine, isoleucine, methionine, or valine; aromatic glucosinolates are derived from phenylalanine or tyrosine; and indole glucosinolates are derived from tryptophane (Sørensen, 1990).

The biosynthesis of glucosinolates from amino acids can be divided into three separate steps. The first step is the chain elongation of aliphatic and aromatic amino acids by inserting methylene groups into their side chains. Second, the metabolic modification of the amino acids (or chain-extended derivatives of amino acids) takes place via an aldoxime intermediate. The same modifications also occur in the biosynthetic route of cyanogenic glycosides. However, the co-occurrence of glucosinolates and cyanogenic glycosides in the same plant is very rare (an example is *Carica papaya*). The biosynthesis of the cyanogenic glycosides has been elucidated in more detail by Halkier and Lindberg-Møller (1991) and by Koch et al. (1992). Third, following the formation of the aldoxime, the glucosinolate is formed by various secondary transformations such as S-insertion, glucosylation, and

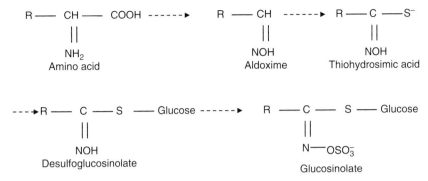

Figure 2.16 The simplified biosynthesis of the glucosinolate core structure.

sulfation. Further modification of the side chain can occur in the formed glucosinolate by, for example, oxidation and/or elimination reactions (Figure 2.16).

References

Ambrosi, M., Cameron, N.R., and Davis, B.G. (2005) Lectins: tools for the molecular understanding of the glycocode. *Organic and Biomolecular*, 3(9), 1593–1608.

American Association of Cereal Chemists (2001) The definition of dietary fiber. Report of the dietary fiber definition committee to the board of directors of the American Association of Cereal Chemists. *Cereal Foods World*, 46(3), 112–126.

Anastas, P.T. and Warner, G.C. (1998) *Green Chemistry: Theory and Practice*, Oxford University Press, New York.

Andersen, B., Rassov, A., Westergaard, N., and Lundgren, K. (1999) Inhibition of glycogenolysis in primary rat hepatocytes by 1,4-dideoxy-1,4-imino-D-arabinitol. *Biochemical Journal*, 342, 545–550.

Andrés-Lacueva, C., Medina-Remon, A., Llorach, R., Urpi-Sarda, M., Khan, N. Chiva-Blanch, G., Zamora-Ros, R., Rotches-Ribalta, M., and Lamuela-Raventós, R.M. (2010) Phenolic compounds: chemistry and occurrence in fruits and vegetables. In Laura A. de la Rosa, Emilio Alvarez-Parrilla and Gustavo A. Gonzalez-Aguilar (eds), *Fruit and Vegetable Phytochemicals: Chemistry, nutritional value and stability*. Wiley-Blackwell, pp. 53–87.

Armstrong, G.A. and Hearst, J.E. (1996) Carotenoids 2: Genetics and molecular biology of carotenoid pigment biosynthesis. *Journal of the Federation of American Societies for Experimental Biology*, 10(2), 228–37.

Arnaud, A. (1892) Sur un nouvel acide gras non saturé de la série CnH2n−4O2. *Bulletin de la Société Chimique de France*, pp. 233–234.

Arnaud, M.A. (1902) Sur la constitution de l'acide taririque. *Bulletin de la Société Chimique de France*, pp. 489–496.

Asano, N. (2008) Glycosidase-Inhibiting alkaloids: isolation, structure, and application. In E. Fattorusso and O. Taglialatela-Scafati (eds) *Modern Alkaloids: Structure, isolation, synthesis and biology*, Wiley-VCH Verlag GmbH & Co. KGaA, Weinheim.

Ashour, M., Wink M., and Gershenzon, J. (2010) Biochemistry of terpenoids: monoterpenes, sesquiterpenes and diterpenes. In Wink M. (ed.) *Biochemistry of Plant Secondary Metabolism, Annual Plant Reviews*, Second Edition, Wiley-Blackwell, pp. 258–303.

Azeredo, H.M.C. (2009) Betalains: properties, sources, applications, and stability – a review. *International Journal of Food Science and Technology*, 44, 2365–2376.

Bach Knudsen, K.E. (2001) The nutritional significance of "dietary fiber" analysis. *Animal Feed Science and Technology*, 90, 3–20.

Baranska, M. and Schulz, H. (2005) Spatial tissue distribution of polyacetylenes in carrot root. *Analyst*, 130, 855–859.

Barker, A.M., Molotsane, R., Müller, C., Schaffner, U., and Stadler, E. (2006) Chemosensory and behavioural responses of the sawfly, Athalia rosae, to glucosinolates and isothiocyanates. *Chemoecology*, 16(4), 209–218.

Barron, D. and Ibrahim, R.K. (1996) Isoprenylated flavonoids a survey. *Phytochemistry*, 43(5), 921–982.

Beghyn, T., Deprez-Poulain, R., Willand, N., Folleas, B. and Deprez, B. (2008) Natural compounds: leads or ideas? Bioinspired molecules for drug discovery. *Chemical Biology and Drug Design*, 72, 3–15.

Bennett, D.J. and Kirby, G.W. (1968) Constitution and biosynthesis of capsaicin. *Journal of the Chemical Society Communications*, 442–446.

Bentley, R. (1990) The shikimate pathway – A metabolic tree with many branche. *Critical Reviews in Biochemical and Molecular Biology*, 25(5), 307–334.

Billing, J. and Sherman, P.W. (1998) Antimicrobial functions of spices: why some like it hot. *Quarterly Review in Biology*, 73, 3–49.

Billsten, H.H., Sundström, V., and Polivka, T. (2005) Self-assembled aggregates of the carotenoid zeaxanthin: Time resolved study of excited states. *Journal of Physical Chemistry A*, 109(8), 1521–1529.

Bohlmann, F., Burkhardt, T., and Zdero, C. (1973) *Naturally Occurring Acetylenes*, Academic Press, London.

Borejsza-Wysocki, W. and Hrazdina, G. (1996) Aromatic polyketide synthases. *Plant Physiology*, 110, 791–799.

Breitmaier, E. (2006) Terpenes: Importance, General Structure, and Biosynthesis. Terpenes: Flavors, Fragrances, Pharmaca, Pheromones, Wiley-VCH Verlag GmbH & Co. KGaA, pp. 1–9.

Breukers, S., Opstad, C.L., Sliwka, H.R., and Partali, V. Hydrophilic carotenoids: Surface properties and aggregation behaviour of the potassium salt of the highly unsaturated diacid norbixin. *Helvetica Chimica Acta*, 92, (2009), 1741–1747.

Britton, G., Liaaen-Jensen and Pfander, H. (eds) (2007) *Carotenoids – Handbook*. Compiled First Edition, Birkhauser Basel book, by A.Z. Mercadante e E.S. Egeland.

Buckingham (2007) Biochemistry of plant secondary metabolism. In *Dictionary of Natural Products* Wink, M. (ed.), Chapman and Hall/CRC, London, volume 2.

Bystritskaya, E.V. and Karpukhin, O.N. (1975) Effect of the aggregate state of a medium on the quenching of singlet oxygen. *Doklady Akademii Nauk SSSR*, 221, 1100–1103.

Caragay, A.B. (1992) Cancer-preventive foods and ingredients. *Food Technology*, 46, 65–68.

Cartayrade, A., Schwarz, M., Jaun, B., and Arigoni, D. (1994) Detection of two independent mechanistic pathways for the early steps of isoprenoid biosynthesis. In *Ginkgo Biloba Second Symposium of the European Network on Plant Terpenoids*, Strasbourg/Bischenberg, January 23–27. Abstract, p. 1.

Caterina, M.J., Schumacher, M.A., Tominaga, M., Rosen, T.A., Levine, J.D., and Julius, D. (1997) The capsaicin receptor: a heat-activated ion channel in pain pathway. *Nature*, 389, 816–824.

Champ, M., Langkilde, A.M., Brouns, F., Kettlitz B., and Le Bail Collet Y. (2003) Advances in dietary fiber characterisation. Definition of dietary fiber, physiological relevance, health benefits and analytical aspects. *Nutrition Research Reviews*, 16, 71–82.

Chiang, Y.M., Chang, C.L.T., Chang, S.L., Yang, W.C., and Shyur, L.F. (2007) Cytopiloyne, a novel polyacetylenic glucoside from Bidens pilosa, functions as a T helper cell modulator. *Journal of Ethnopharmacology*, 110(3), 532–538.

Choct, M. (1997) Feed non-starch polysaccharides: chemical structures and nutritional significance. *Feed Milling International*, 191, 13–26.

Clark, G.B., Thompson, G., and Roux, S.J. (2001) Signal transduction mechanisms in plants: an overview. *Current Science*, 80, 170–177.

Crowell, D.N. (2000) Functional implications of protein isoprenylation in plants. *Progress Lipid Research*, 39(5), 393–408.

Cummings, J.H. (1997) The large intestine in nutrition and disease. In *Danone Chair Monograph*, Institute Danone, Bruxelles, pp. 103–110.

Czepa, A. and Hofmann, T. (2004) Quantitative studies and sensory analyses on the influence of cultivar, spatial tissue distribution, and industrial processing on the bitter off-taste of carrots (Daucus carota L.) and carrot products. *Journal of Agricultural and Food Chemistry*, 52, 14, 4508–4514.

Das, S., Tyagi, A.K., and Kaur H. (2000) Cancer modulation by glucosinolates. *Current Science*, 79(12), 25.

Dawson, G.W., Hallahan, D.L., Mudd, A. (1989) Secondary plant metabolites as targets for genetic modification of crop plants for pest resistance. *Pesticide Science*, 27, 191–201.

De-Groot, H. and Raven, U. (1998) Tissue injury by reactive oxygen species and the protective effects of flavonoids. *Fundamental Clinical Pharmacology*, 12, 249–255.

Dembitsky, V.M. (2005) Astonishing diversity of natural surfactants: 3. Carotenoid glycosides and isoprenoid glycolipids. *Lipids*, 40(6), 535–557.

Dewick, P.M. (1989) Biosynthesis of lignans, In *Studies in Natural Products Chemistry*. Edited by Atte-ur-Rahman. Structure Elucidation. Elsevier, Amsterdam, 5, pp. 459–503.

Diplock, A.T., Charleux, J.L., Crozier-Willi, G., Kok, F.J., Rice-Evans, C., Roberfroid, M., Stahl, W., Vina-Ribes, J., (1998) Functional food science and defense against reactive oxidative species. *British Journal of Nutrition*, 1, S77–S112.

Drickamer, K. (1997) Making a fitting choice: common aspects of sugar-binding sites in plant and animal lectins. *Structure*, 5, 465–468.

Duvold, T., Cali, P., Bravo, J.M., and Rohmer, M. (1997) Incorporation of 2-C-methyl-D-erythritol, a putative isoprenoid precursor in the mevalonate-independent pathway, into ubiquinone and menaquinone of *Escherichia* coli. *Tetrahedron Letters*, 38, 6181–6184.

Eisenreich, W., Rohdich, F., and Bacher, A. (2001) Deoxyxylulose phosphate pathway to terpenoids. *Trends in Plant Science*, 6, 78–84.

Eisenreich, W., Sagner, S., Zenk, M.H., and Bacher, A. (1997) Monoterpenoid essential oils are not of mevalonoid origin. *Tetrahedron Letters*, 38, 3889–3892.

Elgavish, S. and Shaanan, B. (1997) Lectin-carbohydrate interactions: different folds, common recognition principles. *Trends in Biochemical Science*, 22, 462–467.

Facchini, P.J. and De Luca, V. (2008) Opium poppy and Madagascar periwinkle: model non-model systems to investigate alkaloid biosynthesis in plants. *Plant Journal*, 54, 763–784.

Fahey, J.W., Zalcmann, A.T., and Talalay, P. (2001) The chemical diversity and distribution of glucosinolates and isothiocyanates among plants. *Phytochemistry*, 56(1), 5–51.

Fenwick, G.R., Heaney R.K., and Mawson, R., (1989) Glucosinolates. In Cheeke, P.R. (ed.) *Toxicants of Plant Origin, vol 2. Glycosides*. Boca Raton, FL, CRC Press, p. 277.

Filomeni, G., Rotilio, G., and Ciriolo, G.M.R. (2008) Molecular transduction mechanisms of the redox network underlying the antiproliferative effects of allyl compounds from garlic. *Journal of Nutrition*, 138, 2053–2057.

Fleming, I. (1967) Absolute configuration and the structure of chlorophyll. *Nature*, 216, 151–152.

Floss, H.G. (1997) Natural products derived from unusual variants of the shikimate pathway. *Natural Product Reports*, pp. 433–452.

Fosgerau, K., Westergaard, N., Quistorff, B., Grunner, N., Kristiansen, M., and Lundgren, K. (2000) Kinetic and functional characterization of 1,4-dideoxy-1, 4-imino-d-arabinitol: a potent inhibitor of glycogen phosphorylase with anti-hyperglyceamic effect in ob/ob mice. *Archives of Biochemistry and Biophysics*, 380, 274–284.

Friedrich, C. and Von Domarus, C. (1998) Carl Friedrich Wilhelm Meissner (1792–1853) – pharmacist and alkaloid researcher. *Pharmazie*, 53, 67–73.

Garner, C.C. and Herrmann, K.M. (1984) Structural analysis of 3-deoxyo- arabino-heptulosonate 7-phosphate by 'H- and natural-abundance I3C-NMR spectroscopy. *Carbohydrate Research*, 132, 317–322.

Gershenzon, J. and Dudareva, N. (2007) The function of terpene natural products in the natural world. *Nature Chemical Biology*, 3, 408–414.

Godeschalk, F.E. (1987) Consumptie van voedingsmiddelen in Nederland in 1984 en 1985. *Periodieke Rapportage*,volume 64, LEI, The Hague.

Gray, J.C. (1987) Control of isoprenoid biosynthesis in higher plants. *Advances in Botanical Research*, 14, 25–91.

Guzman, I., Bosland, P.W., and O'Connell, M.A. (2010) Heat, color, and flavor compounds in *Capsicum* fruit 2010. In David R. Gang, *The Biological Activity of Phytochemicals*, Springer-Verlag, London, pp. 109–126.

Halkier, B.A. and Gershenzon, J. (2006) Biology and biochemistry of glucosinolates. *Annual Review of Plant Biology*, 57, 303–333.

Halkier, B.A. and Lindberg-Moller, B. (1991) Involvement of cytochrome P-450 in the biosynthesis of dhurrin in Sorghum bicolor (L.) Moench. *Plant Physiology*, 96(1), 10–17.

Hansen, L. and Boll, P.M. (1986) Polyacetylenes in Araliaceae: Their chemistry, biosynthesis and biological significance. *Phytochemistry*, 25, 285–293.

Harborne, J.B. (1995) *The Flavonoids: Advances in Research Since 1986*. Chapman & Hall, London, 72(3), p. A73.

Hashimoto, T. and Yamada, Y. (1994) Alkaloid biogenesis: molecular aspects. *Annual Review Plant Physiology, Plant Molecular Biology*, 45, 257–285.

Haslam, E. (1993) *Shikimic Acid: Metabolism and Metabolites*. John Wiley & Sons, Ltd., Chichester.

Haworth, R.D. (1942) The chemistry of the lignan group of natural products. *Journal of the Chemistry Society*, 0, 448–456 (doi:10.1039/JR9420000448).

Herrmann, K. (1989) Occurrence and content of hydroxycinnamic and hydroxybenzoic acid compounds in food. *Critical Reviews in Food Science Nutrition*, 28(4), 315–347.

Herrmann, K.M. (1995) The shikimate pathway: early steps in the biosynthesis of aromatic compounds. *Plant Cell*, 7, 907–919.

Hertzberg, S. and Liaaen-Jensen, S. (1985) Carotenoid sulfates 4. Syntheses and properties of carotenoid sulfates. *Acta Chemica Scandinavica Series B—Organic Chemistry and Biochemistry*, 39, 629–638.

Hertzberg, S., Ramdahl, T., Johansen, J.E., and Liaaen-Jensen, S. (1983) Carotenoid sulfates 2. Structural elucidation of bastaxanthin. *Acta Chemica Scandinavica*, B37, 267–280.

Hesse, M. (2002) *Alkaloids. Nature's Curse or Blessing?* Wiley-VCH, Weinheim, pp. 1–5.

Ishiguro, M., Takahashi, T., Hayashi, K., and Funatsu, M. (1964) Biochemical studies of ricin. I. Purification of ricin. *Journal of Biochemistry*, 56, 325.

Jennewein, S. and Croteau, R. (2001) Taxol: biosynthesis, molecular genetics, and biotechnological applications. *Applcation Microbiology Biotechnology*, 57, 13–9.

Jones, J.M. (2000) Update on defining dietary fiber. *Cereal Foods World*, 45, 219–220.

Kilpatrick, D.C. (2002) Animal lectins: an historical introduction and overview. *Biochimica Biophysica Acta*, 1572, 187.

Kim, C.S., Kawada, T., Kim, B.S., Han, I.S., Choe, S.Y., Krata. T., and Yu, R. (2003) Capsaicin exhibits antiinflammatory property by inhibiting IkB-a degradation in LPS-stimulated pertiotoneal macrophages. *Cellular Signalling*, 15, 299–306.

Kjaer, A. and Conti, J. (1954) Isothiocyanates, VII: A convenient synthesis of erysoline. *Acta Chemica Scandinavica*, 8, 295–298.

Kleinig, H. (1989) The role of plastids in isoprenoid biosynthesis. *Annual Review Plant Physiology and Plant Molecular Biology*, 40, 39–59.

Knoss, W., Reuter, B., and Zapp, J. (1997) Biosynthesis of the labdane diterpene marrubiin in Marrubium vulgare via a non-mevalonate pathway. *Biochemical Journal*, 326(2), 449–54.

Koch, B., Nielson, V.S., and Halkier, B.A. (1992) Biosynthesis of cyanogenic glucosides in seedlings of cassava. *Archives of Biochemistry and Biophysics*, 292, 141–150.

Kocoureck, J. *The Lectins: Properties, Functions and Applications in Biology and Medicine*, ed. I. E. Liener, N. Sharon and I. J. Goldstein, Academic Press, New York, (1986).

Komath, S.S., Kavitha, M., and Swamy, M.J. (2006) Beyond carbohydrate binding: new directions in plant lectin research. *Organic and Biomolecular Chemistry*, 4, 973–988.

Köpsel, C., Möltgen, H., Schuch, H., Auweter, H., Kleinermanns, K., Martin, H.D., and Bettermann, H. (2005) Structure investigations on assembled astaxanthin molecules. *Journal of Molecular Structure*, 750, 109–115.

Krajewska, A. and Powers J. (1988) Sensory properties of naturally occurring capsaicinoids. *Journal of Food Science*, 53, 902–905.

Krinsky, N.I. and Yeum, K.J. (2003) Carotenoid-radical interactions. *Biochemical and Biophysical Research Communications*, 305, 754–760.

Kutchan, T.M. (1995) Alkaloid biosynthesis – the basis for metabolic engineering of medicinal plants. *Plant Cell*, 7, 1059–1070.

Lamuela-Raventós, R.M. and Waterhouse, A.L. (1994) A direct HPLC separation of wine phenolics. *American Journal of Enology and Viticulture*, 45, 1–5.

Larsen, P.O. Glucosinolates. In *The Biochemistry of Plants* Stumpf, P.K. and Conn, E.E. (eds) Academic Press, London, 7, (1981), 501.

Lee, Y.S., Nam, D.H., and Kim, J.A. (2000) Induction of apoptosis by capsaicin in A172 human glioblastoma cells. *Cancer Letters*, 161, 121–130.

Leete, E. and Louden, M. (1968) Biosynthesis of capsaicin and dihydrocapsaicin in *Capsicum frutescens*. *Journal of the American Chemical Society*, 90, 6837–6841.

Leonard, E., Runguphan, W., O'Connor, S., and Prather, J.K. (2009) Opportunities in metabolic engineering to facilitate scalable alkaloid production. *Nature Chemical Biology*, 5, 292–300.

Lerdau, M., Guenther, A., and Monson, R. (1997) Plant production and emission of volatile organic compounds. *Bioscience*, 47, 373–383.

Lichtenthaler, H.K. (2000) Non-mevalonate isoprenoid biosynthesis: enzymes, genes and inhibitors. *Biochemical Society Transactions*, 28, 785–789.

Lichtenthaler, H.K. (2007) Biosynthesis, accumulation and emission of carotenoids, alpha-tocopherol, plastoquinone, and isoprene in leaves under high photosynthetic irradiance. *Photosynthesis Research*, 92, 163–179.

Lichtenthaler, H.K., Schwender, J., Disch, A., and Rohmer, M. (1997) Biosynthesis of isoprenoids in higher plant chloroplasts proceeds via a mevalonate-independent pathway. *FEBS Letters*, 400, 271–274.

Lightbourn, G.J., Griesbach, R.J., Novotny, J.A., Clevidence, B.A., Rao, D.D., and Stommel, J.R. (2008) Effects of anthocyanin and carotenoid combinations on foliage and immature fruit color of *Capsicum annuum* L. *J HERED (Journal of Heredity)* 99, 105–111.

Lindig, B.A. and Rodgers, M.A.J. (1981) Rate parameters for the quenching of singlet oxygen by water-soluble and lipid-soluble substrates in aqueous and micellar systems. *Photochemistry and Photobiology*, 33, 627–634.

Lis, H. and Sharon, N. (1973) The biochemistry of plant lectins (phytohemagglutinins). *Annual Reviews on Biochemistry*, 42(0), 541–574.

Liu, L. and Simon, S.A. (1998) Response or cultured rat trigerminal ganglion neurons to bitter tastants. *Chemical Senses*, 23, 125–130.

Lockwood, S.F., O'Malley, S., and Mosher, G.L. (2003) Improved aqueous solubility of crystalline astaxanthin (3,3'-dihydroxy-β,β-carotene-4,4'-dione) by Captisol (sulfobutyl ether β-cyclodextrin). *Journal of Pharmaceutical Sciences*, 92, 922–926.

Luckner, M. (1990) *Secondary Metabolism in Microorganisms, Plants and Animals*, Springer-Verlag, Berlin. Mc Naughton, S.A. and Marks, G.C. (2003) Development of a food composition database for the estimation of dietary intakes of glucosinolates, the biologically active constituents of cruciferous vegetables. *British Journal of Nutrition*, 90, 687–697.

McGarvey, D.J. and Croteau, R. (1995) Terpenoid metabolism. *Plant Cell*, 7, 1015–1026.

McGraw, K.J. and Ardia, D.R. (2003) Carotenoids, immunocompetence, and the information content of sexual colors: an experimental test. *American Naturalist*, 162, 704–712.

Middleton, E. (1984). The flavonoids. *Trends in Pharmacology Science*, 5, 335–338.

Mihaliak, C.A., Karp, F., and Croteau, R. (1993) Cytochrome P450 terpene hydroxylases., In P.J. Lea (ed.) *Enzymes of Secondary Metabolism*, Academic Press, London, pp. 261–279.

Minto, R.E. and Blacklock, B.J. (2008) Biosynthesis and function of polyacetylenes and allied natural products. *Progress in Lipid Research*, 47(4), 233–306.

Møller, A.P. and Jennions, M.D. (2000) Testing and adjusting for publication bias. *Trends in Ecology and Evolution*, 16, 580–586.

Moms, E.R. Physico-chemical properties of food polysaccharides. In *Dietary Fibre: A Component of Food, Nutritional Function in Health and Disease*. T. F. Schweizer and C. A. Edwards (eds). Springer-Verlag, London, (1992), pp. 41–56.

Mori, Y. (2001) Introductory studies on the growth and characterization of carotenoid solids: An approach to carotenoid solid engineering. *Journal of Raman Spectroscopy*, 32, 543–550.

Mortensen, A., Skibsted, L.H., Sampson, J., Rice-Evans, C., and Everett, S.A. (1997) Comparative mechanisms and rates of free radical scavenging by carotenoid antioxidants. *FEBS Letters*, 418, 91–97.

Murray, R., Mendez, J., and Brown, S. (1982) *The Natural Coumarins: Occurrence, Chemistry and Biochemistry*. Chichester, UK, John Wiley & Sons, Ltd.

Nabeta, K., Ishikawa, T., Kawae, T., and Okuyama, H. (1995) Biosynthesis of heteroscyphic acid in cell cultures of Heteroscyphus planus: nonequivalent labelling of C-5 units in diterpene biosynthesis. *Journal of the Chemical Society Series, Chemical Communications*, 6, 681–682.

Nalum-Naess, S., Elgsaeter, A., Foss, B.J., Li, B.J., Sliwka, H.R., Partali, V., Melo, T.B., and Naqvi, K.R. (2006) Hydrophilic carotenoids: Surface properties and aggregation of crocin as a biosurfactant. *Helvetica Chimica Acta*, 89, 45–53.

Nes, W.D., Parish, E.J., and Trazskos, J.M. (1992) *Regulation of Isopentenoid Metabolism*. American Chemical Society, Washington, DC, ACS Symposium Series no. 497.

Oliveros, E., Braun, A.M., Aminiansaghafi, T., and Sliwka, H.R. (1994) Quenching of singlet oxygen (1Δg) by carotenoid derivatives—Kinetic analysis by near-infrared luminescence. *New Journal of Chemistry*, 18, 535–539.

Pelletier, J. and Caventou, J. (1951) Editors' outlook. *Journal of Chemical Education*, 28 (9), p. 454.

Pfander, H. (1979) Synthesis of carotenoid glycosylesters and other carotenoids. *Pure and Applied Chemistry*, 51, 565–580.

Piattelli, M. (1981) The betalains: structure, biosynthesis, and chemical taxonomy. In Conn EE (eds.) The Biochemistry of Plants, Academic Press, New York, pp. 557–575.

Piel, J., Donath, J., Bandemer, K., and Boland, W. (1998) Mevalonate-independent biosynthesis of terpenoid volatiles in plants: induced and constitutive emission of volatiles. *Angewadte Chemie*, 37, 2478–2481.

Puls, W., Keup, U., Krause, H.P., Thomas, G., and Hoffmeister, F. (1977) Glucosidase inhibition. A new approach to the treatment of diabetes, obesity and hyperlipoproteinaemia. *Naturwissenschaften*, 64, 536–537.

Qian, D.Q., Zhou, D.F., Ju, R., Cramer, C.L., and Yang, Z.B. (1996) Protein farnesyl transferase in plants: molecular characterization and involvement in cell cycle control. *Plant Cell*, 8, 2381–2394.

Randle, W.M. and Lancaster J.E. (2002) Sulphur Compounds in *Allium*s in Relation to Flavour Quality. In H.D. Rabinowitch (ed.) *Allium Crop Science: Recent Advances*. CAB International, Wallingford Oxon OX10 8DE, UK. p. 330–350.

Raven, P.H., Ray F. Evert, and Susan E.E. (2004) *Biology of Plants* (7th edition), W. H. Freeman and Company, New York. p. 465.

Rinni, J. M. (1995) Lectin structure. *Annual Review of Biophysics and Biomolecular Structure*, 24, 551–557.

Roberts, M.F. and Wink, M. (1998) *Alkaloids – Biochemistry, Ecological Functions and Medical Applications*, Plenum Press, New York.

Rose, P., Whiteman, M., Moore, P.K., and Zhu, Z.Y. (2005) Bioactive S-alk(en)yl cysteine sulfoxide metabolites in the genus *Allium*: the chemistry of potential therapeutic agents. *Natural Product Reports*, 22, 351–368.

Rowland, I. (1999) Optimal nutrition: fibre and phytochemicals. *Proceedings of the Nutrition Society*, 58, 415–419.

Ruzicka, L. (1953) The isoprene rule and the biogenesis of terpenic compounds. *Experientia*, 9, 357.

Sagner, S., Latzel, C., Eisenreich, W., Bacher, A., and Zenk, M.H. (1998) Differential incorporation of 1-deoxy-D-xylulose into monoterpenes and carotenoids in higher plants. *Chemical Communications*, 2, 221–222.

Salisbury, F.B. and Cleon, W. R. (1991) *Plant Physiology* (4th edition), Wadsworth Publishing, Belmont, California pp. 325–326.

Sancho, R., Lucena, C., Machio, A., Caldzado, M.A., Blanco-Molina, M., Minassi, A., Appendino, G., and Munoz, E. (2002) Immunosuppresive activity of capsaicinoids: Capsiate derived from sweet peppers inhibits NF-κB activation and is a potent anti-inflammatory compound *in vivo*. *European Journal of Immunology*, 32, 1753–1763.

Sasaki, K., Saito, T., Lämsä, M., Oksman-Caldentey, K.M., Suzuki, M., and Ohyama, K. (2007) Plants utilize isoprene emission as a thermotolerance mechanism. *Plant Cell Physiology*, 48, 1254–1262.

Scalbert, A. and Williamson, G. (2000) Dietary intake and bioavailability of polyphenols. *Journal of Nutrition*, 130, 2073S–2085S.

Schmidt, J. and Amrhein, N. (1995) Molecular organization of the shikimate pathway in higher plants. *Phytochemistry*, 39, 737–749.

Schwartz, S.H., Tan, B.C., Gage, D.A., Zeevaart, J.A.D., and McCarty, D.R. (1997) Specific oxidative cleavage of carotenoids by vp14 of maize. *Science*, 276, 1872–1874.

Sepúlveda-Jiménez, G., Rueda-Benítez, P., Porta, H., and Rocha-Sosa, M. (2004) Betacyanin synthesis in red beet (*Beta vulgaris*) leaves induced by wounding and bacterial infiltration is preceded by an oxidative burst. *Physiological and Molecular Plant Pathology*, 64, 25–133.

Shahidi, F. and Naczk, M. (1995) *Food Phenolics. Sources, Chemistry, Effects, Applications*, Technomic Publishing Company, Inc. Lancaster, USA.

Sharon, N. and Lis, H. (2004) History of lectins: from hemagglutinins to biological recognition molecules. *Glycobiology*, 14, 53R–62R.

Sipiora, M.L., Murtaugh, M.A., Gregoire, M.B., and Duffy, V.B. (2000) Bitter taste perception and severe vomiting in pregnancy. *Physiology and Behavior*, 69, 259–267.

Sliwka, H.R. (2010) Conformation and circular dichroism of β,β-carotene derivatives with nitrogen-, sulfur-, and selenium-containing substituents. *Helvetica Chimica Acta*, 82, 161–169.

Smits, C.H.M. and Annison, G. (1996) Non-starch polysaccharides in broiler nutrition towards a physiologically valid approach to their determination. *World's Poultry Science Journal*, 52, 203–221.

Song, P.S. and Moore, T.A. (1974) On the photoreceptor pigment for phototropism and phototaxis: Is a carotenoid the most likely candidate? *Photochemistry and Photobiology*, 19, 435–441.

Sørensen, H. (1990) Glucosinolates: Structure properties function In F. Shahdi (ed.) *Canola and Rapeseed*. Van Nostrand, New York, pp. 149–172.

Sörensen, N.A (1961). Some naturally occurring acetylenic compounds. *Proccedings of the Chemical Society*, pp. 98–110.

Southon, W. and Buckingham, J. (1989) *Dictionary of Alkaloids*, Chapman & Hall, London.

Speer, B.R. (1997) "Photosynthetic Pigments". UCMP Glossary (online), University of California Museum of Paleontology.

Stoll, A. and Seebeck, E. (1947) Über Allin, die genuine muttersubstanz des knoblauchöls. *Experentia*, 3, 114–115.

Strack, D. (1977) Phenolic Metabolism. In P.M. Dey and J.B. Harborne (eds) *Plant Biochemistry*, Academic Press, San Diego, pp. 387–416.

Strack, D., Vogt, T., and Schliemann, W. (2003) Recent advances in betalain research. *Phytochemistry*, 62(3), 247–269.
Strauß, U., Wittstock, U., Schubert, R., Teuscher, E., Jung, S., and Mix, E. (1996) Cicutoxin from Cicuta virosa – a new and potent potassium channel blocker in T lymphocytes. *Biochemical and Biophysical Research Communications*, 219, 332–336.
Tahara, S. and Ibrahim, R.K. (1995) Prenylated isoflavonoids a survey. *Phytochemistry*, 38, 1073–1094.
Tapas, A.R., Sakarkar, D.M., and Kakde R.B. (2008) Flavonoids as nutraceuticals. *Tropical Journal of Pharmaceutical Research*, 7, 1089–1099.
Thomas, S.G., Rieu, I., and Steber, C.M. (2005) Gibberellin metabolism and signaling. *Vitamins & Hormones*, 72, 289–338.
Tiedink, H.G.M., Hissink, A.M., Lodema, S.M., van Broekhoven, L.W., and Jongen, W.M.F. (1990) Several known indole compounds are not important precursors of direct mutagenic N-nitroso compounds in green cabbage. *Mutation Research*, 232, 199–207.
Tiedink, H.G.M., Malingre, C.E., van Broekhoven, L.W., Jongen, W.M.F., Lewis, J., and Fenwick, G.R. (1991) The role of glucosinolates in the formation of N-nitroso compounds. *Journal of Agricultural Food Chemistry*, 39, 922–926.
Tominaga, M., Caterina, M.J., Malmberg, A.B., Rosen, T.A., Gilbert, H., Skinner, K., Raumann, B.E., Basbaum, A.I., and Julius, D. (1998) The cloned capsaicin receptor integrates multiple painproducing stimuli. *Neuron*, 21, 531–543.
Torsell, K.B.G. (1997) *Natural Products Chemistry*, Apotekarsocieteten, Stockholm.
Tripathi, M.K. and Mishra, A.S. (2007) Glucosinolates in animal nutrition: A review. *Animal Feed Science and Technology*, 132, 1–27.
Uwai, K., Ohashi, K., Takaya, Y., Ohta, T., Tadano, T., Kisara, K., Shibusawa, K., Sakakibara, R., and Oshima, Y. (2000) Exploring the structural basis of neurotoxicity in C17-polyacetylenes isolated from water hemlock. *Journal of Medicinal Chemistry*, 43, 4508–4515.
Vallejo, F., Tomás-Barberán, F.A., and Ferreres, F. (2004) Characterisation of flavonols in broccoli (*Brassica oleracea* L. var. italica) by liquid chromatography-UV diode-array detection-electrospray ionisation mass spectrometry. *Journal of Chromatography A*, 1054(1–2), 181–193.
Van Sumere, C.F. and Lea, P. J. (1985) *The Biochemistry of Plant Phenolics*, Clarendon Press, Oxford.
Verkerk, R. and Dekker, M. (2008) Glucosinolates. In John Gilbert and Hamide Z. Senyuva (eds) *Bioactive Compounds in Foods*. Oxford, Blackwell Publishing Ltd., pp. 31–47.
Verpoorte, R. and Alfermann, A.W. (2000) Secondary metabolismo. In R. Verpoorte, *Metabolic Engineering of Plant Secondary Metabolism*, Springer-Verlag, London, pp. 1–29.
Virtanen, A.I. and Matikkala, E.J. (1959) The isolation of S- mcthylcysteinesulfoxide and Sn-propylcysteinesulfoxide from onion (*Allium* ccpa) and the antibiotic activity of crushed onion. *Acta Chemica Scandinavica*, 13, 1898–1900.
Vogt, T., Ibdah, M., Schmidt, J., Wray, V., Nimtz, M., and Strak, D. (1999) Light-induced betacyanin and flavonol accumulation in bladder cells of *Mesembryanthemum crystallinum*. *Phytochemistry*, 52, 583–592.
Von Euler, H., Hellström, H., and Klussmann, E. (1931) Carotenoiden (Cartenoids). In Arkiv (ed.) *Physikalisch, Chemische Beobachtungen und Messungen* (Physical, Chemical, and Biological Functions and Properties). Mineralogi och Geologi, pp. 1–4.
Wallace, G. and Fry, S.C. (1994) Phenolic components of the plant cell wall. *International Review of Cytology*, 151, 229–267.
Wang, H.C., Huang, X.M., Hu, G.B., Yang, Z., and Huang, H.B. (2005) A comparative study of chlorophyll loss and its related mechanism during fruit maturation in the pericarp of fast- and slow-degreening litchi pericarp. *Scientia Horticulturae*, 106, 247–257.
Weis, W.I. and Drickamer, K. (1996) Structural basis of lectin-carbohydrate recognition. *Annual Reviews of Biochemistry*, 65, 441–473.
Wink, M. (1993) Allelochemical properties and the raison d'etre of alkaloids. In Cordell G. (ed.), *The Alkaloids: Chemistry and Biology*, vol. 43 Academic Press, San Diego, pp. 1–118.
Wink, M. (2005) Zeitschr Phytother, teitschrift für. *Phytotherapie*, 26, 271–274.
Wink, M. and Van Wyk, B.E. (2008) *Mind-altering and Poisonous Plants of the World*. Timber, Portland, OR.
Winter, J., Bevan, S., and Campbell, E.A. (1995) Capsaicin and pain mechanisms. *British Journal of Anaesthesia*, 75, 157–116.
Woodward, R.B., Ayer, W.A., and Beaton, J.M, (1960) The total synthesis of chlorophyll. *Journal of the American Chemical Society*, 82(14), 3800–3802.

Woodward, R.B., Ayer, W.A., Beaton, J.M., Bickelhaupt, F., Bonnett, R., Buchschacher, P., Closs, G.L., Dutler, H., Hannah, Hauck, F.P., Ito, S., Langemann, A., Le Goff, E., Leimgruber, W., Lwowski, W., Sauer, J., Valenta, Z., and Volz, H. (1990). The total synthesis of chlorophyll a. *Tetrahedron*, 46(22), 7599–7659.

Wu, L.W., Chiang, Y.M., Chuang, H.C., Wang, S.Y., Yang, G.W., Chen, Y.H., Lai, L.Y., and Shyur, L.F. (2004) Polyacetylenes function as anti-angiogenic agents. *Pharmaceutical Research*, 21, 11.

Yahia, E.M. and Ornelas-Paz J.J. (2010) Chemistry, stability, and biological actions of carotenoids. In Laura A. la de Rosa, Emilio Alvarez-Parrilla, Gustavo A. Gonzalez-Aguilar, *Fruit and Vegetable Phytochemiclas: Chemistry, Nutritional Value and Stability*, (first edition) Oxford, Wiley-Blackwell, pp. 177–222.

Yamaguchi, S. (2008) Gibberellin metabolism and its regulation. Annual. Review. *Plant Biology*, 59, 225–251.

Yokyoyama, A. and Shizusato, Y. (1997) Carotenoid sulfate and its production, Kaiyo *Biotechnology*, JP 9084591.

Zhou, D.F., Qian, D.Q., Cramer, C.L., and Yang, Z.B. (1997) Developmental and environmental regulation of tissue- and cell-specific expression for a pea protein farnesyltransferase gene in transgenic plants. *Plant Journal*, 12, 921–930.

Zidorn, C., Johrer, K., Ganzera, M., Schubert, B., Sigmund, E.M., Mader, J., Greil, R., Ellmerer, E.P., and Stuppner, H. (2005) Polyacetylenes from the Apiaceae vegetables carrot, celery, fennel, parsley, and parnsnip and their cytotoxic activities. *Journal of Agricultural and Food Chemistry*, 53, 2518–2523.

Zsila, F., Deli, J., Bikadi, Z., and Simonyi, M. (2001) Supramolecular assemblies of carotenoids. *Chirality*, 13, 739–744.

3 Phytochemicals and health

Ian T. Johnson

Institute of Food Research, Norwich Research Park, Colney, Norwich, UK

3.1 Introduction

Although the term *phytochemical* could be applied to any chemical constituent of plants, the term is used in this chapter to describe biologically active organic substances found in plants used by humans as food, which may be beneficial for health, but for which no specific human deficiency disorder has been identified. Thus nutrients are excluded from the discussion by definition, as are, for practical reasons, the carbohydrate polymers comprising *dietary fibre*. In general, phytochemicals are *secondary plant metabolites*; that is, substances synthesised by plant cells, but which serve some function beyond the primary needs of the cell, and contribute to the survival of the whole plant as a functional organism. Some phytochemicals confer colour or scent, others act as signalling molecules, either within the plant itself, or in interactions with other organisms, and many are believed to function as natural pesticides. Some of these substances are pharmacologically active, whilst others are either profoundly unpalatable or highly toxic. Obviously these properties exclude many classes of secondary plant metabolite from the human food chain, but thousands of food-borne phytochemicals are consumed in significant quantities, even in Western economies that cultivate and consume only a relatively small number of plant varieties as food.

Fruits, vegetables and cereals have long been recognised as important sources of vitamins and mineral micronutrients, but interest in the potentially beneficial effects of phytochemicals on human health began with epidemiological studies showing protective effects of plant foods against several of the chronic diseases of old age, and particularly against cancer. One of the most influential studies was the review of Block and colleagues (Block *et al.*, 1992), who collated and summarised the published evidence for a relationship between fruit and vegetable consumption and the risk of cancer at many sites. Overall they observed strong evidence for a protective effect against a range of different cancers in many populations, and concluded that on average, individuals in the lowest population quartile for fruit and vegetable intake experienced about twice the risk of cancer compared to those in the highest quartile. Steinmetz and Potter (Steinmetz *et al.*, 1991) came to similar conclusions, as did

Handbook of Plant Food Phytochemicals: Sources, Stability and Extraction, First Edition.
Edited by B.K. Tiwari, Nigel P. Brunton and Charles S. Brennan.
© 2013 John Wiley & Sons, Ltd. Published 2013 by John Wiley & Sons, Ltd.

the World Cancer Research Fund in its 1997 report on Food, Nutrition and the Prevention of Cancer, which described 'convincing' evidence for protective effects of fruits and vegetables against cancers of the upper aerophagic tract, stomach and lung, and of vegetables against cancers of the colon and rectum (World Cancer Research Fund, 1997). This rising tide of evidence, coupled with improved analytical procedures and growing interest in the interactions between plant secondary metabolites and mammalian cells stimulated interest in the possibility that plants might confer health benefits beyond those that could be attributed to their nutrient content alone, and triggered a surge in research on the biological properties of phytochemicals (Johnson et al., 1994).

Clearly, if phytochemicals are to be of benefit to human health, they must reach their target tissues in physiologically significant quantities. Some secondary plant metabolites may act entirely within the lumen of the alimentary tract, perhaps by functioning as quenching agents for free radicals, or by interacting directly with gut epithelial cells, without ever crossing the intestine and entering the blood stream. This type of localised activity could perhaps account for protective effects of some fruits, vegetables or functional foods against digestive diseases such as gastric or colorectal cancer, but if phytochemicals play a larger role in human health, they must first cross the gut and enter the circulation in active forms. The complex issue of *bioavailability* is discussed in the next section of this chapter. Assuming that the necessary active concentrations are achieved, the next question is how do these various nonessential but nevertheless beneficial substances act to preserve human health? During the early stages of research on phytochemicals it was assumed that since so many could act as antioxidants *in vitro*, this would prove to be their most important role in human metabolism – indeed the terms *phytochemical* and *antioxidant* remain almost synonymous in the minds of consumers and some commercial marketing departments. However over the last decade or so it has become clear from studies *in vitro* and with animal models that phytochemicals interact with mammalian physiology and metabolism in many unexpected ways that might benefit human health, whilst in parallel with these developments, the true significance of phytochemicals as antioxidants has had to be re-evaluated. Sections 3 and 4 of this chapter will provide a critical discussion of these issues.

3.2 Bioavailability of phytochemicals

One very important characteristic of phytochemicals that distinguishes them from organic micronutrients is the lack of any evidence for specialised adaptations of the human body that might serve to maximise their absorption and delivery to the tissues. Indeed many phytochemicals are transferred only sparingly across the intestinal mucosa, and those compounds that do cross the intestinal surface in significant quantities tend to be rapidly metabolised by the Phase II enzymes, which convert potentially toxic molecules to water soluble conjugates. Much of this metabolism occurs in the gut mucosa, and a large fraction of the products are actively secreted back into the gut lumen (Petri et al., 2003), there to be either metabolised further by the gut microflora, or lost in the faeces. In many cases, any unmetabolised compounds that do enter the circulation undergo metabolic conversions during their first pass through the liver, so that it is the modified forms that reach the target tissues, not the native compound found in the plant. Unfortunately, much of the evidence linking phytochemicals to the health benefits of fruit and vegetables has come from *in vitro* research, in which isolated cells and tissue preparations have been exposed to unrealistically high concentrations of pure, unmetabolised compounds. In this section, the current state of knowledge with regard to the bioavailability to humans of the main classes of phytochemicals will be briefly reviewed.

3.2.1 Terpenes

The terpenes form a large class of organic compounds based upon the isoprene unit, which has the molecular formula C_5H_8. All terpenes have the general formula $(C_5H_8)_n$, but their isoprene constituents may be present as linear chains, or as a combination of both rings and chains. Chemically modified terpenes are very common in nature, and are termed terpenoids or isoprenoids. Terpenes and their derivatives occur widely in the plant kingdom, often as components of resins and essential oils. They are often coloured or pungently scented, and they enter the human food chain as constituents of citrus fruits, or as aromatic food ingredients, such as ginger, cinnamon and cloves. The carotenoids are a particular class of terpenoids, based on eight isoprene units, which will be considered separately below.

Because of their highly lipophylic behaviour, terpenes and their derivatives are likely to cross biological membranes readily by passive diffusion. However their solubility in the aqueous phase of the gut lumen will be low; and their bioavailability will probably depend upon emulsification and partioning into the micellar phase during gastrointestinal lipid digestion. Apart from the carotenoids, discussed in section 3.2.3, studies on the intestinal transport of terpenoids have been relatively few in number. One important exception however is the compound d-limonene, which is a monoterpene ($C_{10}H_{16}$) based on two isoprene units containing a single ring. Citrus peel oils contain about 90% d-limonene and significant quantities are present in conventional citrus juices. Average intakes have been estimated to be about 0.27 mg/kg body weight per day in the USA, but may range up to 1 mg/kg per day in heavy consumers of citrus juices (FAEM Association, 1991). D-Limonene has attracted considerable attention because of its anticarcinogenic effects in rodent models of skin, stomach and mammary cancer. As with many other phytochemicals, the native food-borne compound does not appear in high concentrations in plasma, but the major metabolite perillic acid has been shown to be biologically active. Chow et al. (2002) have argued that citrus juices containing a high level of peel are important constituents of Mediterranean diets and that heavy consumption of such juices may contribute to low levels of certain cancers in countries such as Spain and Southern Italy.

Crowell et al. (1994) determined the plasma concentrations of d-limonene, perillic acid, dihydroperillic acid and limonene-1,2-diol in human volunteers after administration of d-limonene (100 mg/kg body weight) in the form of orange oil incorporated into a food product. Only traces of unmetabolised d-limonene were detected in plasma, but average concentrations of perillic acid, dihydroperillic acid and limonene-1,2-diol reached 35, 33 and 16 micromolar respectively. The pharmacokinetics of perillic acid after consumption of what was described as a Mediterranean-style lemonade made from whole lemons and containing up to 596 mg of d-limonene per 40 oz dose were investigated by Chow et al. (2002). The concentration of perillic acid peaked at one hour after consumption, indicating rapid absorption of d-limonene in the proximal gastrointestinal tract, and ranged between 4.5 and 14.0 microM. Subsequent work has shown that d-limonene consumed in this way is deposited to a significant extent in human adipose tissue (Miller et al., 2010). Evidently d-limonene, and presumably many other terpenoids with similar physical properties, are absorbed and metabolised to a significant extent from commonly consumed foods, but the biological significance of this for human beings remains largely unexplored.

3.2.2 Polyphenols

Much of the complexity of the problems associated with the bioavailability of phytochemicals in general can be judged from the growing literature on the absorption and metabolism of food-borne polyphenols. Amongst this huge group of food-borne substances, the anthocyanins and flavanols have received particular attention. Anthocyanins are a large group of phenolic compounds found abundantly in fruit juices, berries of various types, and wine. Manach et al. (2005) reviewed published data on the absorption and metabolism of anthocyanins in humans, and concluded that from most food sources, only a very small fraction is absorbed, that the small amount of absorption that does occur takes place very rapidly in the stomach and upper intestine, and that excretion of the absorbed fraction is rapid and efficient. In most human bioavailability studies, the administered doses were in the range of hundreds of milligrams, and resulted in peak plasma concentrations in the 10–50 nmol/L range. The average bioavailability of the anthocyanins has been reported in many studies to be less than 1%. Unlike other polyphenols, unmetabolised anthocyanin glycosides are often detected in blood and urine, but there is evidence that anthocyanin glucuronides and sulphates are unstable in urine (Felgines et al., 2003) and that as a consequence their abundance, and hence the total absorption and excretion of the anthocyanins, may have been underestimated in many studies that did not allow for this.

Like the anthocyanins, flavonols are also present in plants as a mixture of water soluble glycosides, and this is also the form in which they are released into the alimentary tract during digestion. They too are commonly found in fruits and vegetables used for human consumption, although they tend to be present in the diet at lower concentrations than the anthocyanins. One of the first mechanistic studies on the absorption of flavonols in humans was conducted by Hollman and colleagues (Hollman et al., 1995), using volunteers who had previously undergone surgery for the removal of a diseased large intestine, and whose small intestine emptied via a permanent orifice at the abdominal surface (ileostomists). Because the digestive residues from the small intestine can be collected and analysed, such patients are often used to study the digestion of food constituents before they are exposed to the intense metabolic activity of the colonic microflora. Hollman et al. measured the disappearance of quercetin glycosides from test-meals of fried onion during their passage through small intestinal lumen, and compared it with the disappearance of pure quercetin aglycone. Their study showed that the absorption of the quercetin glucosides found in food was more efficient than the absorption of quercetin aglycone, and the absorption of the rhamnoglycoside, rutin was even less efficient. It was argued that this was evidence for absorption of the intact glucosides via the specialised glucose transport channels of the small intestinal epithelial cells. This study generated considerable interest and stimulated further research using in vitro systems, animal models and human volunteers, both to test the hypothesis, and to elucidate the metabolic fate of quercetin and other polyphenols in humans. As a result it is now well established that although flavonol glucosides do interact with the glucose transporters of intestinal epithelial cells, their effect is mainly to act as weak inhibitors of glucose absorption (Gee et al., 2000). In practice, most quercetin glycosides are readily hydrolysed by the digestive enzyme lactase phlorizin hydrolase (LPH), which is localised at the epithelial surface. A small fraction of one quercetin glucoside commonly found in foods, quercetin-4'-glucoside, may remain intact long enough to cross the epithelium via the glucose transporter, but the similarly abundant compound quercetin-3-glucoside appears to be absorbed entirely by the passage of free quercetin following hydrolysis by LPH (Day et al., 2003). In any case, unmodified flavonoid glycosides do not reach the human circulation. Only the metabolised flavonoids (e.g. glucuronides, sulphates) are found in the blood, and it is these compounds that must be studied in order to fully define and understand the physiological effects of flavonoids in the human body. The absorption of quercetin is somewhat slower than that of the anthocyanins, as is its excretion in urine. Prolonged dietary supplementation with quercetin can lead to plasma concentrations in the 1–2 µmol/L range (Conquer et al., 1998).

Another aspect of polyphenol metabolism that is poorly understood, but should not be neglected, is the large proportion of the ingested dose that remains in the gut lumen, but which is broken down to simpler, often more readily absorbable compounds, by the gut microflora (Forester et al., 2009). Bacterial metabolism of polyphenols includes ring-fission, and leads to a complex range of metabolites including aldehydes and phenolic acids. Many of these compounds are taken up into the circulation by passive absorption across the colon, and may also exert local anti-inflammatory activity in the gut lumen, which could be important for the maintenance of mucosal homeostasis and health (Larrosa et al., 2009).

3.2.3 Carotenoids

Carotenoids are terpenoids containing forty carbon atoms, and are found throughout the plant kingdom, mainly as components of chloroplasts, where they occur as pigments in close association with the photosynthetic apparatus. The two main classes of carotenoids are the carotenes, which contain no oxygen atoms, and the xanthophylls, which do. There are about 600 known carotenoids in nature, but relatively few are thought to be of nutritional significance for humans. The provitamin A carotenoids (beta-carotene, alpha-carotene, gamma-carotene and beta-cryptoxanthin) are important because they are converted in the human intestinal mucosa to vitamin A. Beta-carotene and other carotenoids are potent antioxidants, and certain compounds, including the xanthophyll lutein, accumulate in the macula lutea of the human eye and the corpus luteum of the ovaries, where they are thought to plan an important protective role against free-radical mediated damage.

Because of their well established nutritional role in vitamin A metabolism, and their putative function as phytochemicals in their own right, the bioavailability of carotenoids has received much attention. Being both hydrophobic and tightly bound within robust intracellular structures, the bioavailability of carotenoids depends upon their physical release from the plant tissue, and incorporation into a suitable lipid phase, either during food processing or in the intestinal lumen during digestion. The details of this initial stage, termed *bioaccessibility*, vary markedly between different food sources. The release of carotenoids from the cells of fruit and vegetable tissues is greatly facilitated by thermal processing, but also exposes the molecules to the possibility of chemical degradation. For example, lycopene is released from tomatoes by thermal processing, but becomes susceptible to *cis*-isomerisation, which may modify its biological activity (Schierle et al., 1997).

Having been released into the gut lumen as an emulsion, the absorption of carotenoids occurs via the mixed micelle phase formed in the presence of bile salts during lipid absorption. The presence of adequate quantities of lipid in the digesta is thus an essential prerequisite for uptake of carotenoids, and their bioavailability depends on the contemporaneous intake of dietary fat. This is an excellent example of the extent to which the micronutrient or phytochemical content of the diet may be an inadequate predictor of its biological effects if the issue of bioavailability is ignored. In an interesting study, Unlu et al. (2005) explored the effects of the lipid-rich fruit avocado, or extracted avocado oil, on the bioavailability to humans of carotenoids from salsa or salads. The addition of 150 g of avocado to salsa enhanced the area under the plasma concentration vs time curve (AUC) for lycopene and beta-carotene by 4.4-fold and 2.6-fold respectively. Similarly 150 g of avocado or 24 g of avocado oil added to salad increased the uptake of alpha-carotene, beta-carotene and lutein by 7.2-, 15.3- and 5.1-fold respectively.

A full understanding of bioavailability implies a description of the delivery of the substance under investigation to target tissues. In the case of the carotenoids this is made more complex by their hydrophobicity, which ensures that they are transferred to the circulation with the chylomicrons, and transported as components of the plasma lipoproteins. About 80% of plasma beta-carotene and lycopene are transported by low density lipoproteins (LDL) but lutein and zeaxanthin also occur at significant levels in high density lipoproteins (HDL). Lipoprotein metabolism varies between individuals to an extent that can significantly modify the apparent concentrations of carotenoids in plasma. Because of this, Faulks and Southon have cautioned against the assessment of carotenoids bioavailability without also taking into account such between-subject differences (Faulks et al., 2005).

3.2.4 Glucosinolates

The glucosinolates are another complex group of biologically active compounds, which occur in cruciferous plants, and enter the human food chain in *Brassica* vegetables such as cabbages, broccoli and brussel sprouts, and in cruciferous salad vegetables including mustard greens, rocket and radishes (Mithen et al., 2000). All glucosinolates contain a common sulphur group, linked to a variable side chain, and a glucose molecule. They are stable, water soluble glycosides, sequestered within the plant tissue until acted upon by endogenous hydrolytic enzyme, myrosinase, which is released by mechanical disruption of the tissue. Hydrolysis liberates glucose and an unstable intermediate which undergoes further reactions to release a variety of products, the most important of which are the isothiocyanates. These pungent compounds impart flavour and aroma to cruciferous vegetables and herbs. They are released from raw plant tissue during food preparation, or by chewing and digestion, and they are absorbed passively across the intestinal surface. Like flavonoids, they are rapidly metabolised both in the gut epithelial cells and in the liver. Petri et al. (2003) used intraluminal tubes to infuse liquidised broccoli containing the isothiocyanate sulforaphane into the intact human intestine, and to recover the luminal contents for analysis. This study showed that most of the absorbed sulforaphane was metabolised to glucuronides and sulphates, and a large proportion was re-secreted into the gut. Nevertheless a much larger fraction of an ingested dose of isothiocyanates is absorbed and metabolised than is the case for the flavonoids, and low concentrations of intact isothiocyanates can be detected in human plasma (Verkerk et al., 2009). Interestingly, the concentration of sulforaphane metabolites in the urine of human volunteers after consumption of a test-meal of broccoli depends on their genetic status in relation to one of the major classes of Phase II enzymes, glutathione-S transferase (GST), which varies markedly in activity between individuals because of polymorphisms in the genes coding for the various components of its super-family (Gasper et al., 2005). Variations in the expression of the various sub-types of GST may influence the availability of isothiocyanate metabolites to the tissues, and seem to determine the degree to which humans benefit from the anticarcinogenic effects of *Brassica* vegetables (London et al., 2000).

3.2.5 Lectins

The lectins, unlike the other main classes of phytochemicals reviewed here, are proteins, of diverse structure and high molecular weight. They occur in the human food chain mainly as plant constituents, but they are found in the animal kingdom as well. Their defining characteristic is their capacity to bind specifically to carbohydrates, and most notably to the carbohydrate moieties of glycoproteins or glycolipids that occur as constituents of cell membranes. It is this property that accounts for their frequent role in mechanisms involving specific bio-recognition phenomena, and for their laboratory use in cellular agglutination reactions. Plant lectins may also have evolved as natural pesticides; many act as anti-nutritional factors and can be toxic to humans (Vasconcelos *et al.*, 2004).

Lectins are generally very resistant to digestion in the gut, and their high molecular weight makes them poor candidates for intestinal absorption. They do however frequently show a strong tendency to interact with glycoconjugation sites on the mucosal surfaces of the intestine, and this is thought to account for many of their well documented biological activities in the gastrointestinal tract. In animals, these effects include stimulation of intestinal epithelial cell proliferation to higher than normal levels, an effect which has been reported to occur in humans (Ryder *et al.*, 1998). The lectin (phytohemagglutinin) derived from uncooked beans (*Phaseolus vulgaris*) causes aberrant growth and precocious maturation of the gastrointestinal tract in suckling rats. Linderoth *et al.* (2006) showed that the effects on the gut mucosa occurred when the lectin was given by direct introduction into the alimentary tract (enteral exposure), but not when it was given by sub-cutaneous injection. However subcutaneous exposure did lead to effects on systemic organs not seen after enteral exposure. These results suggest that this lectin is highly biologically active within the gut lumen but is unlikely to be absorbed and become available to other organs via the circulation. One other possible route of delivery of biologically active lectins to sub-epithelial tissues in the gut is via uptake and translocation by intestinal M cells, which are known to sample intraluminal proteins and present them to the lymphoid cells of the gastrointestinal immune system. Transport of the mistletoe lectin (Viscum album L, var. coloratum agglutinin) through this pathway has been demonstrated using an *in vitro* model of the intestinal mucosa (Lyu *et al.*, 2008) but it is not clear whether this mechanism is of biological importance to human consuming lectins from conventional food sources.

3.3 Phytochemicals and their health-promoting effects

The very strong evidence for protective effects of plant foods against cancer and cardiovascular disease that began to appear in the early 1990s prompted a very significant burst of research activity on the biological effects of phytochemicals in *in vitro* systems, animal models and humans, but in 2003 the authors of a report published by the International Agency for Research on Cancer (IARC, 2003) came to somewhat more cautious conclusions about the benefits of fruit and vegetable consumption than previous authors (Block *et al.*, 1992), Their findings were that: 'There is *limited evidence* for a cancer-preventive effect of consumption of fruit and of vegetables for cancers of the mouth and pharynx, oesophagus, stomach colon-rectum, larynx, lung, ovary (vegetables only), bladder (fruit only) and kidney. There is *inadequate* evidence for a cancer-preventive effect of consumption of fruit and vegetables for all other sites'. This general trend towards a more conservative assessment of the protective effects of fruits and vegetables against cancer has continued. The most recent report (Boffetta et al., 2010) on fruits and vegetables from the very large European Prospective Investigation into Cancer and Nutrition (EPIC), concluded that a statistically significant protective effect was detectable, but for men and women combined it amounted to only around 10% reduction in overall risk of cancer for the highest quintile of fruit and vegetable consumption (>647 g/day) compared to the lowest quintile (0–226 g/day).

The relationship between fruit and vegetable consumption and risk of coronary heart disease has been much less intensely studied than that for cancer, but recent research suggests a somewhat similar level of protection. For example, Dauchet et al. (2006) conducted a meta-analysis of nine cohort studies and calculated that across the entire population of 91 000 men and 130 000 women, the risk of coronary heart disease was decreased by 4% for each extra portion of fruits and vegetables consumed per day, and by about 7% for each additional portion of fruit. Against this background, some of the most important lines of evidence for the protective mechanisms of particular groups of phytochemicals against cancer and some forms of cardiovascular disease are discussed in the remainder of this section.

3.3.1 Phytochemicals as antioxidants

Free radicals are highly reactive, short-lived species generated by a variety of biological mechanisms, including inflammation (Hussain et al., 2003), or as a side effect of the reactions occurring during normal oxidative metabolism (Poulsen et al., 2000). During their short lifespan they readily interact with macromolecules, including lipids, proteins and nucleic acids, damaging their structures and often modifying their functionality. Mammalian cells have evolved a complex arsenal of antioxidant mechanisms to defend their constituent macromolecules from free-radical mediated damage but, even so, the steady-state level of oxidative DNA adduct formation caused by free radicals such as the hydroxyl radical ($\cdot$OH) released from H_2O_2 in the presence of iron, nitric oxide (NO$\cdot$) and peroxynitrite (ONOO) has been estimated to be about 66 000 adducts per cell (Helbock et al., 1998). The cumulative effects of such damage include mutations resulting from faulty DNA repair, and double strand-breaks (Bjelland et al., 2003). Free-radical reactions can also cause oxidative damage to proteins such as p53 that are involved in the regulation of cellular proliferation and apoptosis, and can thereby contribute directly to tumour promotion (Hofseth et al., 2003). High levels of free-radical production also occur in vascular tissues during the development of disease. Superoxide reacts with NO, forming peroxynitrite, and impairing NO-mediated processes essential to the maintenance of vascular health.

Fruits and vegetables are rich in both antioxidant nutrients such as ascorbate, which is a powerful electron donor, and which therefore acts as a reducing agent in a range of free-radical and other reactions. The donation of an electron by ascorbate, in reactions with O_2^- and OH$\cdot$, gives as a product the radical semidehydroascorbate, which is only weakly reactive. Vitamin E (α-tocopherol) is an important lipid-soluble antioxidant nutrient, that tends to accumulate in cell membranes, and which acts by reacting with lipid free-radicals, blocking peroxidation chain reactions, and thus protecting cell membranes from oxidative damage. Many naturally occurring polyphenols act as chain-breaking antioxidants in a similar way to vitamin E.

The fact that polyphenols and other secondary plant metabolites exhibit strong antioxidant activity in vitro led to the hypothesis that many of the putative protective effects of fruits and vegetables against cardiovascular disease and cancer are a direct consequence of strengthened antioxidant defences. Much of the experimental work underpinning this hypothesis was based on the use *in vitro* systems, but there have also been many attempts to demonstrate direct benefits of dietary antioxidant supplementation in human volunteers, using antioxidant activity in plasma or target tissues, or changes in the production of end-products of oxidative damage, as biomarkers. One widely used technique for the investigation of antioxidant effects in biological systems is the oxygen radical absorbance capacity assay (ORAC), which works by measuring the effect of some biological sample on a standard free-radical mediated reaction between *R*-phycoerythrin and a peroxyl radical generator, 2,2´-azobis(2-amidinopropane) dihydrochloride (AAPH). The synthetic, water-soluble antioxidant Trolox® is often used as a standard, so that the antioxidant activity of the biological system under investigation can be expressed in Trolox equivalents. Cao *et al.* (1998a) used the ORAC assay to explore the effects of fruit and vegetable consumption on the antioxidant capacity of plasma in a group of healthy non-smoking volunteers. At the outset of the study, the baseline antioxidant capacity of their plasma was positively correlated with their fruit and vegetable intake as estimated from a food-frequency questionnaire. The subjects then entered a metabolic laboratory, where they all consumed one or other of two controlled diets consisting of ten servings of fruit and vegetables per day for 15 days, or a similar diet with two additional servings of broccoli, with a washout period of six weeks between experiments. All subjects showed a significant increase in the antioxidant capacity of the plasma in response to both experimental diets. These effects were associated with an increase in α-tocopherol (vitamin E) in the plasma, but it was shown that the increased antioxidant capacity could not be accounted for by antioxidant nutrients alone. The authors therefore proposed that phytochemicals, including flavonoids, were the probable cause of the observed effects. In a separate study from the same laboratory the acute effects of strawberries, spinach, red wine and vitamin C were evaluated in elderly women (Cao *et al.*, 1998b). As in the previous study, the theoretical effects of other sources of antioxidant nutrient activities were accounted for, and shown not to fully explain the observed increases in antioxidant activity. The authors concluded that much of the excess antioxidant capacity was due to absorption of food-borne polyphenolic phytochemicals, but this conclusion was not directly verified.

Other studies have confirmed that dietary intervention with flavonoid-rich berries and other fruits leads to a significant increase in the antioxidant activity in human plasma (Pedersen *et al.*, 2000), but the causal relationship between this effect and reductions in the risk of disease remains largely hypothetical. Furthermore, the precise reasons for the observed changes in plasma antioxidant capacity in response to dietary intervention often remain ambiguous. Much of the work in this field has been based on the assumption that the antioxidant effects of fruits and vegetables can be ascribed largely to their phytochemical content, but the relevance of the antioxidant effects observed *in vitro* to clinical findings has been challenged by Lotito and colleagues, who argued that the rise in antioxidant capacity following fruit and vegetable consumption is often caused by an increase in plasma urate levels (Lotito *et al.*, 2004). Uric acid accumulates in human plasma as an end-product of purine metabolism, and can reach concentrations close to 0.5 mM/L. Ames (1981) showed that uric acid is a powerful antioxidant and argued that it accounts for most of the antioxidant capacity of human plasma. Given the low bioavailability of flavonoids, which seldom reach concentrations in the micromolar range in human plasma, it seems highly unlikely that they can make a major contribution to antioxidant activity when consumed from conventional fruits and vegetables. Furthermore, it has been shown that a post-prandial increase in urate levels occurs in response to the metabolism of fructose via fructo-kinase mediated production of fructose 1-phosphate, which enables the rate of adenosine monophosphate degradation to urate to rise (Lotito *et al.*, 2004). This transient rise in plasma urate levels may be the primary cause of increased postprandial antioxidant capacity after ingestion of apples and other fructose-rich foods.

Another approach to testing the antioxidant hypothesis is to conduct dietary interventions with foods rich in antioxidant phytochemicals and then to search for evidence of a reduction in free-radical mediated damage to macromolecules. Actual disease endpoints are extremely difficult to study under controlled experimental conditions, but biomarkers have been used in this way to study the effects of dietary intervention on oxidative damage in human trials. Several different markers of oxidative damage to lipids, proteins and DNA have been employed for this purpose. One very widely used measure of lipid peroxidation is the level of thiobarbituric acid-reactive substances (TBARS) present either in plasma or in low density lipoproteins obtained from the blood (Wade et al., 1989). In a small study with five subjects, Young et al. (1999) explored the effects of three daily doses of blackberry and apple juice (750, 1000 and 1500 ml) consumed for one week, on markers of lipid and protein peroxidation. Total plasma TBARS were reduced following the intervention with 1500 ml of juice but plasma 2-amino-adipic semialdehyde residues increased with time and dose, suggesting an unexpected pro-oxidant effect of the juice on plasma proteins. Bub et al. (2000) used a similar approach to measure changes in lipid peroxidation in 23 healthy male subjects after a period of dietary antioxidant depletion, and after intervention periods with 330 ml tomato juice, 330 ml carrot juice and finally with 10 g of spinach powder. Consumption of tomato juice reduced plasma TBARS by 12% ($P<0.05$) and lipoprotein oxidisability as measured by an increased lag time by 18% ($P<0.05$). However carrot juice and spinach powder had no effect on lipid peroxidation, and antioxidant status did not change during any of the study periods. In contrast van den Bergh et al. undertook a randomised placebo-controlled cross-over trial, lasting three weeks with a two-week washout period between treatments, in a group of 22 male smokers with a relatively low vegetable and fruit intake (van den Berg et al., 2001). During the treatment phase the subjects consumed a vegetable burger and fruit drink, and showed increased plasma levels of vitamin C, carotenoids and total antioxidant capacity. However there were no effects on any marker of oxidative damage to lipids, proteins or DNA, or on other biomarkers of oxidative stress.

A group of Dutch and Scandinavian collaborators undertook a large and very thorough human intervention study to explore in some depth the effects of prolonged dietary supplementation with fruits and vegetables on antioxidant status and other aspects of metabolism in humans. For 25 days, a group of 43 healthy volunteers consumed either 600 g of fruits and vegetables per day, an equivalent quantity of vitamins and minerals or a placebo (Dragsted et al., 2004). The so-called '6-a-day' study was designed to explore both the direct antioxidant effects of prolonged fruit and vegetable consumption, and the induction of enzymes involved in the metabolism, conjugation and excretion of potentially toxic substances. The use of a positive control group consisting of subjects receiving micronutrient supplements whilst consuming an essentially fruit- and vegetable-free control diet also enabled the researchers to deduce what proportion of any physiological response to the fruit and vegetables supplementation could be ascribed to phytochemicals. In practice however, despite the high levels of supplementation with fruits and vegetables and the assessment of a variety of sophisticated biomarkers, few important biological effects were observed. None of the markers of plasma antioxidant capacity that were measured showed any statistically significant response to dietary intervention. There was some evidence of an increased resistance of plasma lipoproteins to oxidation, but also an increase in protein carbonyl formation at lysine residues, which is indicative of increased protein oxidation. Interestingly, the latter effect was attributed to a pro-oxidant effect of ascorbate, which is known to occur under certain conditions. This finding emphasises the complexity of free-radical biology in living systems other than simple *in vitro* models. In another paper from the same study, it was reported that neither the prolonged period of fruit and vegetable depletion, nor the supplementation with either fruits and vegetables or micronutrients had any significant effects on the levels of oxidative damage to DNA (Moller *et al.*, 2003). The authors concluded that the inherent antioxidant defence systems of these healthy human subjects were sufficient to protect their circulating mononuclear cells from oxidative damage.

3.3.2 Blocking and suppressing the growth of tumours

The development of cancer is a prolonged, multi-stage process, involving a progressive series of molecular events, beginning with damage to DNA in a single dividing cell. Cells that have undergone the first step of *initiation* and continue to divide and multiply, are increasingly vulnerable to further mutations, leading to an increasingly abnormal phenotype that gradually acquires the ability to migrate to other tissues and establish secondary tumours. Some of the earliest studies on the ability of natural food-borne chemicals to inhibit the development of cancer were conducted by Wattenberg, who observed that anticarcinogenic chemicals could be defined as either *blocking agents*, which act immediately before or during the initiation of carcinogenesis by chemical carcinogens, or as *suppressing agents*, which act at later stages of promotion and progression (Wattenberg *et al.*, 1985). Blocking agents are drugs or phytochemicals that prevent the initial damage to DNA by chemical carcinogens, either by inhibiting their activation from procarcinogens or by enhancing their detoxification and excretion. These effects occur primarily through changes in the activity of the Phase II metabolic enzymes mentioned earlier in the context of bio-availability. Phase II enzymes act downstream from Phase I metabolism, which is mainly due to the cytochrome p450 enzymes that orchestrate the oxidation, reduction and hydrolysis of environmental chemicals such as drugs, toxins and carcinogens. The products of Phase I metabolism are often highly reactive genotoxic intermediates that form substrates for Phase II enzymes such as glutathione *S*-transferase (GST), NAD:quinone reductase and γ-glutamylcysteine synthetase. Phase II catalyses the formation of less reactive, water-soluble conjugates that are readily excreted via the kidneys or in bile. Certain phytochemicals induce the transcription of genes expressing Phase I and II enzymes, and the most effective are those that selectively induce Phase II enzymes, without simultaneously inducing activation of carcinogens via increased Phase I activity (Prochaska *et al.*, 1988). Several groups of phytochemicals have now been identified as potent inducers of Phase II enzymes; two of the most actively investigated are the flavanols, including epigallocatechin gallate (EGCG), which is the principal biologically active component of green tea (Chou *et al.*, 2000), and the isothiocyanate sulforaphane derived from broccoli (Talalay *et al.*, 2001).

A very substantial amount of research to elucidate the mechanisms of action of anticarcinogenic phytochemicals has been done using cultured tumour cells *in vitro*, but much of this work is also supported by studies with experimental animals. For example, the compound indole-3-carbinol obtained from *Brassica* vegetables (Morse *et al.*, 1990) and the isothiocyanate phenethyl isothiocyanate (PEITC) from watercress (Hecht, 1996) have been shown to modify the metabolism of the tobacco smoke carcinogen, 4-(methylnitrosamino)-1-(3-pyridyl)-1-butanone (NNK) and inhibit the development of lung tumours in rats. In the case of NNK, the shunting of NNK metabolism away from the lung leads to increased metabolism in the liver, and higher urinary excretion of NNK metabolites. In some cases it has been possible to confirm the existence of such anticarcinogenic activity in studies with human volunteers. Thus smokers who consumed 170 g of watercress (*Rorippa nasturtium-aquaticum*) per day for three days showed increased urinary excretion of two NNK metabolites, 4-(methylnitrosamino)-1-(3-pyridyl)-1-butanol (NNAL) and (4-methylnitrosamino)-1-(3-pyridyl)but-1-yl)-beta-omega-D-glucosiduronic acid (Hecht, 1995). Overall, the evidence from both experimental and epidemiological studies (London et al., 2000) is consistent with the hypothesis that glucosinolate breakdown products modulate Phase II metabolism of tobacco smoke carcinogens in humans, and help prevent lung cancer, at least in some genetically distinct sub-groups in the population. The significance of these effects in relation to public health remain to be fully established, but the evidence has been strong enough to encourage the development of *Brassica* varieties rich in glucosinolates for human consumption (Mithen et al., 2003).

Prolonged exposure to carcinogens such as those present in cigarette smoke inevitably leads to an accumulation of genetic damage, and to further molecular events favouring the development of cancer. These include the appearance of further mutations and other genetic abnormalities that cause progressively abnormal gene expression. This so-called *promotion* stage of cancer development is characterised by poorly regulated cell proliferation and differentiation, and a reduced tendency for damaged cells to undergo programmed 'suicide' (apoptosis). Eventually the surviving cells acquire the full cancer phenotype, but there are several biologically plausible mechanisms whereby phytochemicals may delay or interrupt this process and thereby lead to tumour suppression. For example it is increasingly recognised that inflammation is a risk-factor for certain cancers (Balkwill et al., 2001) and there is strong evidence that prolonged use of aspirin and other anti-inflammatory drugs reduce the risk of cancers of the colon and other sites (Chan et al., 2005). These various lines of evidence have focused attention on the molecular mechanisms of inflammation, on the pathways through which they may promote cancer and on the phytochemicals that may be used to inhibit them.

One key factor in the activation of inflammatory processes in human disease is nuclear transcription factor κB (NF-κB). In its inactive form NF-κB resides in the cytoplasm as a complex with its main regulatory protein IκB. The activation pathway for NF-κB involves phosphorylation of IκB by the enzyme IκB kinase (IKK), which marks it for destruction by proteolytic enzymes. This step frees NF-κB to translocate to the nucleus, where it binds to a specialised sequence motif in the nuclear DNA, and functions as a transcription factor favouring the expression of at least 200 genes involved in the regulation of inflammation, cell proliferation, differentiation and apoptosis. There is strong evidence that the chronic, abnormal up-regulation of NF-κB is a key factor in the promotion and growth of many tumours (Karin et al., 2002). A variety of secondary plant metabolites (resveratrol, curcumin, limonene, glycyrrhizin, gingerol, indole-3-carbinol, genistein, apigenin) have been shown to inhibit NF-κB activity at various stages in its regulatory pathway. To take one example, curcumin, which is an established anticarcinogenic plant metabolite found in the spice cumin (*Cuminum cyminum*), suppresses TNF-induced activation of IKK (Singh et al., 1995). In contrast, caffeic acid phenethyl ester has been shown to prevent the binding of NF-κB to its target DNA sequence (Natarajan et al., 1996).

The enzyme cyclooxygenase (prostaglandin H synthase) exists as two distinct isoforms; COX-1, which is expressed, in normal healthy tissues, produces prostaglandins essential to platelet aggregation and gastric mucosal integrity, whereas *COX-2* produces prostaglandins involved in inflammatory processes. The downstream effects of NF-κB include increased expression of *COX-2*, and so inhibition of NF-κB can inhibit inflammation by this route. Other phytochemicals also act as naturally occurring COX inhibitors, amongst which perhaps the earliest and best known example is salicylate, which was originally isolated from the willow tree (*Salix alba*). Both *COX-1* and *COX-2* are inhibited by aspirin, an acetylated derivative of salicylate. Salicylates have been shown to irreversibly inhibit the COX enzymes by selectively acetylating the hydroxyl group of a single serine residue, and also to suppress NF-κB, by inhibiting IKK kinase activity (Yin *et al.*, 1998). Many flavonoids are also *COX-2* enzyme inhibitors, and some (apigenin, chrysin and kaempferol) can suppress *COX-2* transcription by mechanisms including activation of the peroxisome proliferator-activated receptor (PPAR) gamma transcription factor (Liang *et al.*, 2001) and inhibition of NF-κB expression (Liang *et al.*, 1999). As noted previously, flavonoids are extensively metabolised during and after absorption, but *in vitro* studies have established that *COX-2* transcription is inhibited by flavonoid metabolites found in human plasma, including quercetin 3-glucuronide, quercetin 3'-sulphate and 3' methylquercetin 3-glucuronide (O'Leary *et al.*, 2004).

At a later stage in cancer promotion, anticarcinogenic phytochemicals may act directly on tumour growth by inhibiting cell proliferation (mitosis), or favouring cell death (apoptosis). The Wnt proteins are extracellular signalling molecules involved in the regulation of cell proliferation via the β-catenin signal pathway. They play an important role in gut formation during embryogenesis, and they contribute to the maintenance of normal gut morphology in the adult. About nineteen Wnt genes are known to code for cysteine-rich Wnt glycoproteins that are released into the extracellular environment. Their function is to regulate signalling by the cytoplsmic protein β-catenin in target cells, by interacting with the membrane receptors Frizzled and LRP. β-catenin regulates many aspects of cellular organisation, including cytoskeletal structure, cell proliferation and apoptosis (Wikramanayake *et al.*, 2003), and it is itself tightly regulated by a sequence of interactions with other proteins. It is present in the cytoplasm as a complex with the adenomatous polyposis coli protein (APC) and the scaffolding protein Axin. This complex then associates with casein kinase I (CKI), which phosphorylates the N terminus of β-catenin, and glycogen synthase kinase 3β, an enzyme that phosphorylates other β-catenin residues. The phosphorylated β-catenin molecule is then marked for degradation, which tightly regulates the levels of β-catenin in the cytoplasm. In normal cells there is a relatively large and stable pool of inactive β-catenin associated with the cytoskeletal protein cadherin, and a small labile pool in the cytoplasm. However in cancer cells the degradation pathway is often suppressed and the balance is altered in favour of the labile cytoplasmic pool (Gregorieff *et al.*, 2005). Active β-catenin then migrates to the nucleus, where it activates transcription factors regulating *COX-2*, and many other genes linked to cell proliferation.

It is well established that synthetic *COX-2* inhibitors suppress β-catenin mediated gene transcription in colorectal carcinoma cells, and a number of phytochemicals, including quercetin (Park *et al.*, 2005), also interact with the β-catenin pathway *in vitro* (Jaiswal *et al.*, 2002). It has also been shown that both green tea, and its active flavanoid constituent epigallocatechin gallate (EGCG), suppressed nuclear β-catenin activity in kidney tumour cells *in vitro* (Dashwood *et al.*, 2002). Furthermore, in the *APC^{min}* mouse, which is a widely used animal model of colorectal cancer, treatment with green tea and the *COX-2* inhibitor sulindac both suppressed the growth of tumours (Orner *et al.*, 2003).

One of the most important characteristics of a tumour cell is its ability to evade the induction of programmed cell death, which is a normal response to the many genetic abnormalities that typify cancer. In principle, the enhancement of apoptosis could eliminate genetically damaged cells from a tissue, or tip the balance of cell proliferation in a tumour towards regression rather than growth (Johnson, 2001). Several classes of phytochemicals, including organolsulphur compounds from garlic (*Allium sativum*) and isothiocyanates from cruciferous plants have been shown to induce apoptosis *in vitro*. It has been mentioned that glucosinolate breakdown products act as powerful inducers of Phase II enzymes and modulate carcinogen metabolism, both *in vitro* and *in vivo*, but this may not be their only mode of action. There is also much evidence that sulforaphane and other isothiocyanates can block mitosis and initiate apoptosis in a variety of epithelial cell lines and tissues. Using an animal model, Smith *et al.* showed that both oral administration of the pure glucosinolate sinigrin (Smith *et al.*, 1998), which is the precursor of allyl isothiocyanate, and consumption of diets rich in raw brussel sprouts (*Brassica oleracea* var. *gemmifera*) rich in sinigrin (Smith *et al.*, 2003), both caused an amplification of the apoptotic response induced in rat colorectal crypts 48 h after exposure to the chemical carcinogen 1, 2 dimethylhydrazine (DMH).

3.3.3 Modifying cardiovascular physiology

Like cancer, cardiovascular disease, which includes both coronary heart disease and stroke, is a major cause of both death and long-term morbidity in the developed world, and a similar amount of effort has been devoted toward understanding its causes at the cellular and molecular level. All the major and minor blood vessels, including the capillaries, are lined by squamous epithelial cells, which collectively comprise the *endothelium*. Endothelial cells play a crucial role in the maintenance of normal vascular physiology through their surface properties, their barrier functions and their importance in the regulation of vasodilation. Disruption of these physiological mechanisms, coupled with the onset of endothelial inflammation, is important in the development of cardiovascular disease, and there is much interest in the possible role of phytochemicals in their maintenance. Interest in the possibility that phytochemicals may help to prevent heart disease and stroke began with epidemiological data showing reduced risk of disease in heavy consumers of fruits and vegetables, but recently the attention of both scientists and food manufacturers has become more focussed on a few rich sources of dietary polyphenols, including grapes and wine, tea and cocoa products (Ghosh *et al.*, 2009).

Endothelium-dependent vasodilation is regulated primarily through the signalling molecule, nitric oxide (NO), a short-lived diffusible gas that readily crosses cell membranes. The maintenance of optimal levels of NO within the vascular endothelial tissues is essential to vascular health because of its role as a smooth muscle relaxant and platelet aggregation antagonist, and its inhibitory activity against NF-κB dependent expression of cytokines and inflammatory factors that mediate the formation of atherosclerotic plaque. NO levels are controlled largely by the activity of endothelial nitric oxide synthase (eNOS), which catalyses oxidation of the guanidine group of L-arginine, releasing NO and L-citrulline. Interest in the role of phytochemicals as modulators of eNOS activity began with observations showing that treatment of vascular tissue with red wine or with polyphenols derived from red wine, caused NO-mediated dilation of isolated blood vessels *in vitro* (Fitzpatrick *et al.*, 1993). These *in vitro* effects have since been shown to be due to the induction by red wine polyphenols of eNOS activity in endothelial cells, leading to a sustained increase in production of NO (Leikert *et al.*, 2002).

The standard technique for the investigation of endothelial function in humans is the measurement of flow-mediated dilatation (FMD). FMD occurs when increased flow within vessels is detected by the endothelial cells, which respond by releasing dilator factors, the most important of which is probably NO. The effect can be measured non-invasively in humans by using an inflatable cuff to regulate the flow of blood into the vessels of the forearm. This technique has been widely used to study the effects of phytochemical supplements and other nutritional interventions. In one study, healthy male volunteers received a high-fat diet, which led as expected to a reduction in FMD, but the adverse effects were prevented by simultaneous daily consumption of 240 ml of red wine for 30 days (Cuevas *et al.*, 2000). However in another study, red wine consumption failed to improve FMD in type II diabetics, although it did have the useful benefit of improving insulin sensitivity (Napoli *et al.*, 2005).

Cocoa powder, which is prepared from pods of the cocoa tree *Theobroma cacao*, is amongst the most promising sources of biologically active flavonoids, principally oligomeric procyanidins, currently available to the food industry. In a double-blinded controlled intervention trial, Fisher *et al.* (2003) administered approximately 821 mg of flavanols/day, containing (-)-epicatechin and (+)-catechin, as well as oligomeric procyanidins, and measured changes in peripheral vasodilation. The cocoa supplementation induced significant vasodilation, which was reversed by infusion of a nitric oxide synthase inhibitor. In another controlled trial with chocolate, it was shown that consumption of moderate daily quantities (46 g) of dark chocolate, rich in flavonoids, led to a measurable increase in plasma concentrations of epicatechin, to more than 200 nM/L, compared to around 18 nM/L in the control group, and increased FMD significantly compared to controls (Engler *et al.*, 2004). No reduction in blood pressure was observed by Engler *et al.*, but reductions in blood pressure after supplementation with dark chocolate (Grassi *et al.*, 2005) or with relatively high doses of cocoa (Davison *et al.*, 2010) have been reported. It is interesting to note that although chocolate is a high calorie product containing relatively high levels of sucrose and fat, epidemiological evidence shows an inverse correlation between chocolate consumption and coronary heart disease in the USA, even after correction for other variables (Djousse *et al.*, 2010). Cross-sectional population studies cannot establish causal mechanisms, but they do provide evidence in support of hypotheses derived from mechanistic studies.

Tea, in both its black and green forms, is a widely consumed beverage and one of the major sources of biologically active flavonoids in the human diet. It also has the advantage that tea drinking is not associated with any significant risk of over-consumption of either alcohol or energy. As with red wine and cocoa, epidemiological studies do suggest an inverse relationship between both black and green tea consumption and the risk of coronary heart disease (Hodgson *et al.*, 2010), though as is usually the case it has often been difficult to completely separate the effects of tea from those of confounding factors. A recent dose-response study provides some evidence for effects of black tea consumption on blood pressure in humans, which if confirmed could prove to be of considerable significance for public health (Grassi *et al.*, 2009).

3.4 General conclusions

Whilst it is probably true to say that the evidence for protective effects of diets rich in fruits and vegetables against chronic disease has tended to become less impressive with the passage of time, our understanding of the biological effects of their constituent phytochemical has grown at a near exponential rate. Clearly the so-called *antioxidant hypothesis* for the protective effects of fruits and vegetables remains, at best, unproven. There is little doubt that plant foods are rich in antioxidant constituents, but their poor bioavailability probably limits their effectiveness as regulators of antioxidant damage in humans. Even where there is evidence that consumption of high levels of fruits and vegetables modifies some biomarkers of antioxidant capacity and redox status, the active constituents of these dietary supplements may not be phytochemicals, and it is far from clear that the health of western consumers eating a normal diet is indeed compromised by a shortage of antioxidant nutrients. At present then, the lack of consistency of evidence across the field makes it difficult to reach a definitive conclusion about the real significance of antioxidant phytochemicals for human health. Nevertheless the last two decades have provided an abundance of new evidence for other potentially important protective mechanisms operating at the cellular and organ levels, and research on all aspects of phytochemicals and their physiological and biochemical effects continues apace. This growing evidence-base has stimulated interest in the broad concept of chemoprevention, focussed attention on particular fruits and vegetables rich in the most active compounds, and encouraged a more mechanistic approach to the epidemiology of diet and disease. It seems likely that new plant varieties and novel products based on these advances will emerge and become commercially viable in the very near future.

References

Ames, B.N. (1981) Uric acid provides an antioxidant defence in humans against oxidant and radical caused ageing and cancer: a hypothesis. *Proceedings of the National Acedemy of Science USA*, 78, 6858.

Balkwill, F. and Mantovani, A. (2001) Inflammation and cancer: back to Virchow? *Lancet*, 357, 539–545.

Bjelland, S. and Seeberg, E. (2003) Mutagenicity, toxicity and repair of DNA base damage induced by oxidation. *Mutation Research*, 531, 37–80.

Block, G., Patterson, B. and Subar, A. (1992) Fruit, vegetables, and cancer prevention: a review of the epidemiological evidence. *Nutrition and Cancer*, 18, 1–29.

Boffetta, P., Couto, E., Wichmann, J., Ferrari, P., Trichopoulos, D., et al. (2010) Fruit and vegetable intake and overall cancer risk in the European Prospective Investigation into Cancer and Nutrition (EPIC). *Journal of the National Cancer Institute*, 102, 529–537.

Bub, A., Watzl, B., Abrahamse, L., Delincee, H., Adam, S., et al. (2000) Moderate intervention with carotenoid-rich vegetable products reduces lipid peroxidation in men. *Journal of Nutrition*, 130, 2200–2206.

Cao, G., Booth, S.L., Sadowski, J.A. and Prior, R.L. (1998a) Increases in human plasma antioxidant capacity after consumption of controlled diets high in fruit and vegetables. *American Journal of Clinical Nutrition*, 68, 1081–1087.

Cao, G., Russell, R.M., Lischner, N. and Prior, R.L. (1998b) Serum antioxidant capacity is increased by consumption of strawberries, spinach, red wine or vitamin C in elderly women. *Journal of Nutrition*, 128, 2383–2390.

Chan, A.T., Giovannucci, E.L., Meyerhardt, J.A., Schernhammer, E.S., Curhan, G.C., et al. (2005) Long-term use of aspirin and nonsteroidal anti-inflammatory drugs and risk of colorectal cancer. *Journal of the American Medical Association*, 294, 914–923.

Chou, F.P., Chu, Y.C., Hsu, J.D., Chiang, H.C. and Wang, C.J. (2000) Specific induction of glutathione S-transferase GSTM2 subunit expression by epigallocatechin gallate in rat liver. *Biochemical Pharmacology*, 60, 643–650.

Chow, H.H., Salazar, D. and Hakim, I.A. (2002) Pharmacokinetics of perillic acid in humans after a single dose administration of a citrus preparation rich in d-limonene content. *Cancer Epidemiology, Biomarkers and Prevention*, 11, 1472–1476.

Conquer, J.A., Maiani, G., Azzini, E., Raguzzini, A. and Holub, B.J. (1998) Supplementation with quercetin markedly increases plasma quercetin concentration without effect on selected risk factors for heart disease in healthy subjects. *Journal of Nutrition*, 128, 593–597.

Crowell, P.L., Elson, C.E., Bailey, H.H., Elegbede, A., Haag, J.D., et al. (1994) Human metabolism of the experimental cancer therapeutic agent d-limonene. *Cancer Chemotherapy and Pharmacology*, 35, 31–37.

Cuevas, A.M., Guasch, V., Castillo, O., Irribarra, V., Mizon, C., et al. (2000) A high-fat diet induces and red wine counteracts endothelial dysfunction in human volunteers. *Lipids*, 35, 143–148.

Dashwood, W.M., Orner, G.A. and Dashwood, R.H. (2002) Inhibition of beta-catenin/Tcf activity by white tea, green tea, and epigallocatechin-3-gallate (EGCG): minor contribution of H_2O_2 at physiologically relevant EGCG concentrations. *Biochemical Biophysical Research Communications*, 296, 584–588.

Dauchet, L., Amouyel, P., Hercberg, S. and Dallongeville, J. (2006) Fruit and vegetable consumption and risk of coronary heart disease: a meta-analysis of cohort studies. *Journal of Nutrition*, 136, 2588–2593.

Davison, K., Berry, N.M., Misan, G., Coates, A.M., Buckley, J.D., et al. (2010) Dose-related effects of flavanol-rich cocoa on blood pressure. *Journal of Human Hypertension*, 24, 568–576.

Day, A.J., Gee, J.M., DuPont, M.S., Johnson, I.T. and Williamson, G. (2003) Absorption of quercetin-3-glucoside and quercetin-4'-glucoside in the rat small intestine: the role of lactase phlorizin hydrolase and the sodium-dependent glucose transporter. *Biochemical Pharmacology*, 65, 1199–1206.

Djousse, L., Hopkins, P.N., North, K.E., Pankow, J.S., Arnett, D.K., et al. (2010) Chocolate consumption is inversely associated with prevalent coronary heart disease: The National Heart, Lung, and Blood Institute Family Heart Study. *Clinical Nutrition*, 30(2), 182–187.

Dragsted, L.O., Pedersen, A., Hermetter, A., Basu, S., Hansen, M., et al. (2004) The 6-a-day study: effects of fruit and vegetables on markers of oxidative stress and antioxidative defense in healthy nonsmokers. *American Journal of Clinical Nutrition*, 79, 1060–1072.

Engler, M.B., Engler, M.M., Chen, C.Y., Malloy, M.J., Browne, A., et al. (2004) Flavonoid-rich dark chocolate improves endothelial function and increases plasma epicatechin concentrations in healthy adults. *Journal of the American College of Nutrition*, 23, 197–204.

FAEM Association (1991) D-Limonene Monograph. Flavor and Extract Manufacturers' Association, Washington DC.

Faulks, R.M. and Southon, S. (2005) Challenges to understanding and measuring carotenoid bioavailability. *Biochimica et Biophysica Acta*, 1740, 95–100.

Felgines, C., Talavera, S., Gonthier, M.P., Texier, O., Scalbert, A., et al. (2003) Strawberry anthocyanins are recovered in urine as glucuro- and sulfoconjugates in humans. *Journal of Nutrition*, 133, 1296–1301.

Fisher, N.D., Hughes, M., Gerhard-Herman, M. and Hollenberg, N.K. (2003) Flavanol-rich cocoa induces nitric-oxide-dependent vasodilation in healthy humans. *Journal of Hypertension*, 21, 2281–2286.

Fitzpatrick, D.F., Hirschfield, S.L. and Coffey, R.G. (1993) Endothelium-dependent vasorelaxing activity of wine and other grape products. *American Journal of Physiology*, 265, H774–778.

Forester, S.C. and Waterhouse, A.L. (2009) Metabolites are key to understanding health effects of wine polyphenolics. *Journal of Nutrition*, 139, 1824S–1831S.

Gasper, A.V., Al-Janobi, A., Smith, J.A., Bacon, J.R., Fortun, P., et al. (2005) Glutathione S-transferase M1 polymorphism and metabolism of sulforaphane from standard and high-glucosinolate broccoli. *American Journal of Clinical Nutrition*, 82, 1283–1291.

Gee, J.M., DuPont, M.S., Day, A.J., Plumb, G.W., Williamson, G., et al. (2000) Intestinal transport of quercetin glycosides in rats involves both deglycosylation and interaction with the hexose transport pathway. *Journal of Nutrition*, 130, 2765–2771.

Ghosh, D. and Scheepens, A. (2009) Vascular action of polyphenols. *Molecular Nutrition and Food Research*, 53, 322–331.

Grassi, D., Mulder, T.P., Draijer, R., Desideri, G., Molhuizen, H.O., et al. (2009) Black tea consumption dose-dependently improves flow-mediated dilation in healthy males. *Journal of Hypertension*, 27, 774–781.

Grassi, D., Necozione, S., Lippi, C., Croce, G., Valeri, L., et al. (2005) Cocoa reduces blood pressure and insulin resistance and improves endothelium-dependent vasodilation in hypertensives. *Hypertension*, 46, 398–405.

Gregorieff, A. and Clevers, H. (2005) Wnt signaling in the intestinal epithelium: from endoderm to cancer. *Genes and Development*, 19, 877–890.

Hecht, S.S. (1995) Chemoprevention by isothiocyanates. *Journal of Cellular Biochemistry*, Supplement, 22, 195–209.

Hecht, S.S. (1996) Chemoprevention of lung cancer by isothiocyanates. *Advances in Experimental Medicine and Biology*, 401, 1–11.

Helbock, H.J., Beckman, K.B., Shigenaga, M.K., Walter, P.B., Woodall, A.A., et al. (1998) DNA oxidation matters: the HPLC-electrochemical detection assay of 8-oxo-deoxyguanosine and 8-oxo-guanine. *Proceedings of the Natlional Academy of Sciences USA*, 95, 288–293.

Hodgson, J.M. and Croft, K.D. (2010) Tea flavonoids and cardiovascular health. *Molecular Aspects of Medicine*, 31(6), 495–502.

Hofseth, L.J., Saito, S., Hussain, S.P., Espey, M.G., Miranda, K.M., et al. (2003) Nitric oxide-induced cellular stress and p53 activation in chronic inflammation. *Proceedings of the Natlional Academy of Sciences USA*, 100, 143–148.

Hollman, P.C., de Vries, J.H., van Leeuwen, S.D., Mengelers, M.J. and Katan, M.B. (1995) Absorption of dietary quercetin glycosides and quercetin in healthy ileostomy volunteers. *American Journal of Clinical Nutrition*, 62, 1276–1282.

Hussain, S.P., Hofseth, L.J. and Harris, C.C. (2003) Radical causes of cancer. *Nature Reviews Cancer*, 3, 276–285.

IARC (2003) *Fruit and Vegetables*. IARC Press, Lyon.

Jaiswal, A.S., Marlow, B.P., Gupta, N. and Narayan, S. (2002) Beta-catenin-mediated transactivation and cell-cell adhesion pathways are important in curcumin (diferuylmethane)-induced growth arrest and apoptosis in colon cancer cells. *Oncogene*, 21, 8414–8427.

Johnson, I., Williamson, G. and Musk, S. (1994) Anticarcinogenic factors in plant foods: a new class of nutrients? *Nutrition Research Reviews*, 7, 175–204.

Johnson, I.T. (2001) Mechanisms and possible anticarcinogenic effects of diet related apoptosis in colorectal mucosa. *Nutrition Research Reviews*, 14, 229–256.

Karin, M., Cao, Y., Greten, F.R. and Li, Z.W. (2002) NF-kappaB in cancer: from innocent bystander to major culprit. *Nature Reviews Cancer*, 2, 301–310.

Larrosa, M., Luceri, C., Vivoli, E., Pagliuca, C., Lodovici, M., et al. (2009) Polyphenol metabolites from colonic microbiota exert anti-inflammatory activity on different inflammation models. *Molecular Nutrition and Food Research*, 53, 1044–1054.

Leikert, J.F., Rathel, T.R., Wohlfart, P., Cheynier, V., Vollmar, A.M., et al. (2002) Red wine polyphenols enhance endothelial nitric oxide synthase expression and subsequent nitric oxide release from endothelial cells. *Circulation*, 106, 1614–1617.

Liang, Y.C., Huang, Y.T., Tsai, S.H., Lin-Shiau, S.Y., Chen, C.F., et al. (1999) Suppression of inducible cyclooxygenase and inducible nitric oxide synthase by apigenin and related flavonoids in mouse macrophages. *Carcinogenesis*, 20, 1945–1952.

Liang, Y.C., Tsai, S.H., Tsai, D.C., Lin-Shiau, S.Y. and Lin, J.K. (2001) Suppression of inducible cyclooxygenase and nitric oxide synthase through activation of peroxisome proliferator-activated receptor-gamma by flavonoids in mouse macrophages. *FEBS Letters*, 496, 12–18.

Linderoth, A., Prykhod'ko, O., Pierzynowski, S.G. and Westrom, B.R. (2006) Enterally but not parenterally administered Phaseolus vulgaris lectin induces growth and precocious maturation of the gut in suckling rats. *Biology of the Neonate*, 89, 60–68.

London, S.J., Yuan, J.M., Chung, F.L., Gao, Y.T., Coetzee, G.A., et al. (2000) Isothiocyanates, glutathione S-transferase M1 and T1 polymorphisms, and lung-cancer risk: a prospective study of men in Shanghai, China. *The Lancet*, 356, 724–729.

Lotito, S.B. and Frei, B. (2004) The increase in human plasma antioxidant capacity after apple consumption is due to the metabolic effect of fructose on urate, not apple-derived antioxidant flavonoids. *Free Radical Biology and Medicine*, 37, 251–258.

Lyu, S.Y. and Park, W.B. (2008) Transport of mistletoe lectin by M cells in human intestinal follicle-associated epithelium (FAE) in vitro. *Archives of Pharmacal Research*, 31, 1613–1621.

Manach, C., Williamson, G., Morand, C., Scalbert, A. and Remesy, C. (2005) Bioavailability and bioefficacy of polyphenols in humans. I. Review of 97 bioavailability studies. *American Journal of Clinical Nutrition*, 81, 230S–242S.

Miller, J.A., Hakim, I.A., Chew, W., Thompson, P., Thomson, C.A., et al. (2010) Adipose tissue accumulation of d-limonene with the consumption of a lemonade preparation rich in d-limonene content. *Nutrition and Cancer*, 62, 783–788.

Mithen, R., Faulkner, K., Magrath, R., Rose, P., Williamson, G., et al. (2003) Development of isothiocyanate-enriched broccoli, and its enhanced ability to induce phase 2 detoxification enzymes in mammalian cells. *Theoretical and Applied Genetics*, 106, 727–734.

Mithen, R.F., Dekker, M., Verkerk, R., Rabot, S. and Johnson, I.T. (2000) The nutritional significance, biosynthesis and bioavailability of glucosinolates in human foods. *Journal of the Science of Food and Agriculture*, 80, 967–984.

Moller, P., Vogel, U., Pedersen, A., Dragsted, L.O., Sandstrom, B., et al. (2003) No effect of 600 grams fruit and vegetables per day on oxidative DNA damage and repair in healthy nonsmokers. *Cancer Epidemiology, Biomarkers and Prevention*, 12, 1016–1022.

Morse, M., LaGreca Amin, S. and Chung, F. (1990) Effects of Indole-3-Carbinol on Lung Tumorigenesis and DNA Methylation Induced by 4-(Methylnitrosamino)-1-(3-Pyridyl)-1-butanone (NNK), and on the Metabolism and Disposition of NNK in A/J Mice1. *Cancer Research*, 50(9), 2613–1617.

Napoli, R., Cozzolino, D., Guardasole, V., Angelini, V., Zarra, E., et al. (2005) Red wine consumption improves insulin resistance but not endothelial function in type 2 diabetic patients. *Metabolism*, 54, 306–313.

Natarajan, K., Singh, S., Burke, T.R., Jr., Grunberger, D. and Aggarwal, B.B. (1996) Caffeic acid phenethyl ester is a potent and specific inhibitor of activation of nuclear transcription factor NF-kappa B. *Proceedings of the Natlional Academy of Sciences USA*, 93, 9090–9095.

O'Leary, K.A., de Pascual-Tereasa, S., Needs, P.W., Bao, Y.P., O'Brien, N.M., et al. (2004) Effect of flavonoids and vitamin E on cyclooxygenase-2 (*COX-2*) transcription. *Mutation Research*, 551, 245–254.

Orner, G.A., Dashwood, W.M., Blum, C.A., Diaz, G.D., Li, Q., et al. (2003) Suppression of tumorigenesis in the Apc(min) mouse: down-regulation of beta-catenin signaling by a combination of tea plus sulindac. *Carcinogenesis*, 24, 263–267.

Park, C.H., Chang, J.Y., Hahm, E.R., Park, S., Kim, H.K., et al. (2005) Quercetin, a potent inhibitor against beta-catenin/Tcf signaling in SW480 colon cancer cells. *Biochemical and Biophysical Research Communications*, 328, 227–234.

Pedersen, C.B., Kyle, J., Jenkinson, A.M., Gardner, P.T., McPhail, D.B., et al. (2000) Effects of blueberry and cranberry juice consumption on the plasma antioxidant capacity of healthy female volunteers. *European Journa; of Clinical Nutrition*, 54, 405–408.

Petri, N., Tannergren, C., Holst, B., Mellon, F.A., Bao, Y., et al. (2003) Absorption/metabolism of sulforaphane and quercetin, and regulation of phase II enzymes, in human jejunum in vivo. *Drug Metababolism and Disposition*, 31, 805–813.

Poulsen, H.E., Jensen, B.R., Weimann, A., Jensen, S.A., Sorensen, M., et al. (2000) Antioxidants, DNA damage and gene expression. *Free Radical Research*, 33 Suppl, S33–39.

Prochaska, H.J. and Talalay, P. (1988) Regulatory mechanisms of monofunctional and bifunctional anticarcinogenic enzyme inducers in murine liver. *Cancer Research*, 48, 4776–4782.

Ryder, S.D., Jacyna, M.R., Levi, A.J., Rizzi, P.M. and Rhodes, J.M. (1998) Peanut ingestion increases rectal proliferation in individuals with mucosal expression of peanut lectin receptor. *Gastroenterology*, 114, 44–49.

Schierle, J., Bretzel, W., Buhler, I., Faccin, N., Hess, N., et al. (1997) Content and isomeric ratio of lycopene in food and human plasma. *Food Chemistry*, 59, 459–465.

Singh, S. and Aggarwal, B.B. (1995) Activation of transcription factor NF-kappa B is suppressed by curcumin (diferuloylmethane). *Journal of Biological Chemistry*, 270, 24995–25000.

Smith, T.K., Lund, E.K. and Johnson, I.T. (1998) Inhibition of dimethylhydrazine-induced aberrant crypt foci and induction of apoptosis in rat colon following oral administration of the glucosinolate sinigrin. *Carcinogenesis*, 19, 267–273.

Smith, T.K., Mithen, R. and Johnson, I.T. (2003) Effects of Brassica vegetable juice on the induction of apoptosis and aberrant crypt foci in rat colonic mucosal crypts in vivo. *Carcinogenesis*, 24, 491–495.

Steinmetz, K.A. and Potter, J.D. (1991) Vegetables, fruit, and cancer. I. Epidemiology. *Cancer Causes and Control*, 2, 325–357.

Talalay, P. and Fahey, J.W. (2001) Phytochemicals from cruciferous plants protect against cancer by modulating carcinogen metabolism. *Journal of Nutrition*, 131, 3027S–3033S.

Unlu, N.Z., Bohn, T., Clinton, S.K. and Schwartz, S.J. (2005) Carotenoid absorption from salad and salsa by humans is enhanced by the addition of avocado or avocado oil. *Journal of Nutrition*, 135, 431–436.

van den Berg, R., van Vliet, T., Broekmans, W.M., Cnubben, N.H., Vaes, W.H., et al. (2001) A vegetable/fruit concentrate with high antioxidant capacity has no effect on biomarkers of antioxidant status in male smokers. *Journal of Nutrition*, 131, 1714–1722.

Vasconcelos, I.M. and Oliveira, J.T. (2004) Antinutritional properties of plant lectins. *Toxicon*, 44, 385–403.

Verkerk, R., Schreiner, M., Krumbein, A., Ciska, E., Holst, B., et al. (2009) Glucosinolates in Brassica vegetables: the influence of the food supply chain on intake, bioavailability and human health. *Molecular Nutrition and Food Research*, 53 Suppl 2, S219.

Wade, C.R. and van Rij, A.M. (1989) Plasma malondialdehyde, lipid peroxides, and the thiobarbituric acid reaction. *Clinical Chemistry*, 35, 336.

Wattenberg, L.W., Hanley, A.B., Barany, G., Sparnins, V.L., Lam, L.K., et al. (1985) Inhibition of carcinogenesis by some minor dietary constituents. *Princess Takamatsu Symposium*, 16, 193–203.

Wikramanayake, A.H., Hong, M., Lee, P.N., Pang, K., Byrum, C.A., et al. (2003) An ancient role for nuclear beta-catenin in the evolution of axial polarity and germ layer segregation. *Nature*, 426, 446–450.

World Cancer Research Fund, W.C.R. (1997) *Food, Nutrition and the Prevention of Cancer: a Global Perspective*, American Institute for Cancer Research, Washington DC, pp. 216–251.

Yin, M.J., Yamamoto, Y. and Gaynor, R.B. (1998) The anti-inflammatory agents aspirin and salicylate inhibit the activity of I(kappa)B kinase-beta. *Nature*, 396, 77–80.

Young, J.F., Nielsen, S.E., Haraldsdottir, J., Daneshvar, B., Lauridsen, S.T., et al. (1999) Effect of fruit juice intake on urinary quercetin excretion and biomarkers of antioxidative status. *American Journal of Clinical Nutrition*, 69, 87–94.

4 Pharmacology of phytochemicals

José M. Matés

Department of Molecular Biology and Biochemistry, Faculty of Sciences, Campus de Teatinos, University of Málaga, Málaga, Spain

Abbreviations: AP-1: activator protein-1, CaP: pancreas cancer, CFRs: cyclic reductions in coronary flow, COX: cyclo-oxygenase, CVD: cardiovascular disease, EGCG: epigallocatechin gallate, γ-GCS: gamma-glutamylcysteine synthetase, GSH: glutathione, GST: glutathione S-transferases, HCC: hepatocellular carcinoma, HDL-C: high-density lipoprotein-cholesterol, I3C: indole-3-carbinol, LDL-C: low-density lipoprotein-cholesterol, MAPK: mitogen-activated protein kinases, NF-κB: nuclear factor-kappaB, NO: nitric oxide, NOS: nitric oxide synthase, NQO1: NAD(P)H:quinone oxidoreductase, PK: protein kinase, ROS: reactive oxygen species, RSV: resveratrol, SOD: superoxide dismutase, TG: triglycerides, UV: ultraviolet, VEGF: vascular endothelial growth factor.

4.1 Introduction

Since early in the history of medicine, an association between phytochemicals and disease has persisted. Galen of Pegamon (129–199 AD), a Greek physician and follower of Hippocrates' teachings was said to have prescribed various foods, including peeled barley, and various vegetables for the treatment of cancer. The beneficial effects of fruits and vegetables have been attributed to, among other things, the high content of bioactive compounds that are non-nutrient constituents commonly present in food (Siddiqui *et al.*, 2009). Natural dietary components, obtained from several fruits, vegetables, nuts and spices have drawn a considerable amount of attention due to their demonstrated ability to partially prevent cardiovascular disease (CVD) and suppress carcinogenesis in animal models, or delay cancer formation in humans. It has been ascribed in part to antioxidants in plant bionutrients inactivating reactive oxygen species (ROS) involved in initiation or progression of these diseases (Duthie *et al.*, 2006). It is estimated that approximately 8000 phytochemicals are present in whole foods, and there are quite possibly many more (Liu, 2004). These compounds with much more complex scope, interaction, and magnitude may act on different targets with different mechanisms of action. Over the centuries, no fewer than 3000 plant species have been used for chemotherapy and chemoprevention. The World Health

Handbook of Plant Food Phytochemicals: Sources, Stability and Extraction, First Edition.
Edited by B.K. Tiwari, Nigel P. Brunton and Charles S. Brennan.
© 2013 John Wiley & Sons, Ltd. Published 2013 by John Wiley & Sons, Ltd.

Organization (WHO) has estimated that approximately 80% of the population in some Asian and African countries still depends on complementary and traditional medicine for the prevention and treatment of diseases, most of which involves the use of plant extracts (Naczk and Shahidi, 2006).

Most natural products can be classified into three major groups: nitrogen-containing compounds, terpenoids and phenolic compounds. The major class of nitrogen-containing compounds is represented by alkaloids, synthesised principally from aspartic acid, tryptophan, arginine and tyrosine. More than 10 000 different alkaloids have been discovered in species from over 300 plant families. These compounds protect plants from a variety of herbivorous animals, and many possess pharmacologically important activities (Zenk and Juenger, 2007). Terpenoids (also called terpenes) are a large and diverse class of naturally occurring organic chemicals derived from five-carbonisopreneunits assembled and modified in thousands of ways. These chemically-different compounds are grouped in a unique class, since they are all derived from acetyl-CoA. Many plant terpenoids are toxins and feeding deterrents to herbivores or are attractants of various sorts. Plant terpenoids are used extensively for their aromatic qualities. They play a role in traditional herbal remedies and are under investigation for antibacterial, anti-neoplastic and other pharmaceutical functions. Phenolic compounds are widely distributed in the plant kingdom. Plant tissues may contain up to several grams per kilogram. Flavonoids are the most abundant, commonly known for their antioxidant activity and for their use in human diet, due to their widespread distribution, and their relatively low toxicity, compared to other active plant compounds, i.e. alkaloids (Le Marchand, 2002). Among plant chemopreventive agents we will highlight indoles, catechins, vitamins, isoflavonoids (silymarin) and phenols (resveratrol, and curcumin) (Gerhäuser et al., 2003).

Many phytochemicals of differing chemical structure have medicinal properties. They activate cytoprotective enzymes and inhibit DNA damage to block initiation in healthy cells, or modulate cell signalling to eliminate unhealthy cells at later stages in the carcinogenic process. *In vitro* results for several well-studied compounds indicate that each can affect many aspects of cell biochemistry (Manson et al., 2007). Many phytochemicals are poorly bioavailable and evidence suggests that combinations may be more effective than single agents. There may also be advantages in combining them with chemo- or radio-therapies (Manson et al., 2007). In this chapter we deal with the use and efficacy of phytochemicals as pharmaceuticals, i.e. as chemicals intended for the cure and treatment of disease. In addition, we will outline phytochemicals that have progressed to be used as therapeutic drugs. Finally, we will analalyse some active component of phytochemicals that has served as the basis for drug development.

4.2 Medicinal properties of phytochemicals

Nowadays, more than 600 functional non-nutrient food factors in vegetables and fruits are considered to be effective for health promotion and medicinal properties (Table 4.1). The optimal intake of various phytochemicals per capita has been calculated as more than 10 micromole per day; such as catechin, isoflavones, isothiocyanate, ferulic acid, quercetin, cinnamic acid and chlorogenic acid (Watanabe et al., 2004). Epidemiological studies find that whole grain intake is protective against cancer, CVD, diabetes and obesity. Whole grains are rich in nutrients and antioxidant phytochemicals with known health benefits. Published whole grain feeding studies report improvements in biomarkers with whole grain

Table 4.1 Medicinal properties of dietary foods and their phytochemicals

	Cereal grains	Berries and apples	Grapes, pulses, and nuts	Spices, allium vegetables, and herbs
Phytochemical	Lignans Tocotrienols Phenolic compounds Phytic acid Sphingolipids Phytosterols Tannins Vitamins B Vitamin E	Anthocyanins Hydroxycinnamic acids Dihydrochalcones Flavan-3-ols Procyanidins Pterostilbene Vitamin C	Resveratrol Anthocyanins Catechins Flavonols Procyanidins Genistein	Curcumin Carotenoids Hydroxycinnamic acids 6-Gingerol Diallyldisulfide Allyl sulfides Eugenol Quercetin Catechins
Dietary food	Wheat Rice Maize Oats Rye Barley Triticale Sorghum Millet	Blackberries Black rasberries Red rasberries Blueberries Cranberries Strawberries Apple	Grape Red wine Lentils Chickpeas Beans Soy Peanuts Pine nuts Walnuts	Curry Saffron Cinnamon Clove Ginger Garlic Onion Basil Tea
Disease	Cancer CVD Stroke Hypertension Obesity Metabolic syndrome Type 2 diabetes	Cancer CVD Neurodegenerative diseases Obesity	Cancer CVD Neurodegenerative diseases	Cancer CVD Neurodegenerative diseases
Mechanism	Antioxidant Hormones Immune system Insulin LDL TG Cholesterol	Antioxidant Brain ischemia Immune system LDL TG Cholesterol	Antioxidant Brain ischemia Immune system Anti-inflammatory Anti-aggregatory Anti-apoptosis	Antioxidant Brain atrophy Immune system Angiogenesis HDL LDL TG Cholesterol

consumption, such as blood-lipid improvement, and antioxidant protection (Slavin *et al.*, 2004). The major cereal grains include wheat, rice and maize, with others as minor grains (Table 4.1). Buckwheat, wild rice and amaranth are not botanically true grains but are typically associated with the grain family due to their similar composition (Slavin *et al.*, 2003). Components in whole grains associated with improved health status include dietary fibre, starch, unsaturated fatty acids, minerals, phytochemicals and enzyme inhibitors (Table 4.1). In the grain-refining process the bran is removed, resulting in the loss of dietary fibre, vitamins, minerals, lignans, phyto-oestrogens, phenolic compounds and phytic acid (Slavin *et al.*, 2004). Antinutrients found in grains include digestive enzyme (protease and amylase) inhibitors, phytic acid, haemagglutinins and phenolics and tannins. Protease inhibitors, phytic acid, phenolics and saponins have been shown to reduce the risk of cancer of the colon and breast in animals. Phytic acid, lectins, phenolics, amylase inhibitors and saponins have also been shown to lower plasma glucose, insulin and/or plasma cholesterol and triacylglycerols (Slavin *et al.*, 2003). Phytic acid forms chelates with various metals, suppressing damaging Fe-catalysed redox reactions (Slavin *et al.*, 2004). Hormonally active compounds called lignans may protect against hormonally mediated diseases (Adlercreutz *et al.*, 1997). Lignans are compounds processing a 2,3-dibenzylbutane structure and exist as minor constituents of many plants where they form the building blocks for the formation of lignin in the plant cell wall. The plant lignans secoisolariciresinol and matairesinol are converted by human gut bacteria to the mammalian lignans enterolactone and enterodiol (Slavin *et al.*, 2003). Plant sterols and stanols are found in oilseeds, grains, nuts and legumes. These compounds are known to reduce serum cholesterol (Yankah and Jones, 2001). It is believed that phytosterols inhibit dietary and biliary cholesterol absorption from the small intestine. Phytosterols displace cholesterol from micelles, which reduces cholesterol absorption and increases its excretion (Hallikainen *et al.*, 2000).

Whole grain intake is associated with reduced risk of chronic disease. Specifically, there is a decreased risk of obesity, coronary heart disease, hypertension, stroke, metabolic syndrome, type 2 diabetes and some cancers observed among the highest whole grain eaters compared with those eating little or no whole grains (Jones *et al.*, 2008). Additional epidemiological studies have associated consumption of whole grains and whole grain products with reduced incidence of chronic diseases such as CVD, diabetes and cancer (Adom *et al.*, 2005). The health beneficial phytochemicals of wheat are distributed as free, soluble-conjugated and bound forms in the endosperm, germ and bran fractions of whole grain (Adom and Liu, 2002). Health benefits of grains have been attributed in part to the unique phytochemical content and distribution of grains. Grain phytochemicals also include derivatives of benzoic and cinnamic acids, anthocyanidins, quinones, flavonols, chalcones, amino phenolics compounds, tocopherols and carotenoids (Adom *et al.*, 2005). Grain phytochemicals exert their health benefits through multifactorial physiologic mechanisms, including antioxidant activity, mediation of hormones, enhancement of the immune system and facilitation of substance transit through the digestive tract, butyric acid production in the colon and absorption and/or dilution of substances in the gut (Adom and Liu, 2002). The bran/germ fraction of whole wheat may therefore impart greater health benefits when consumed as part of a diet and thus help reduce the risk of chronic diseases (Thompson *et al.*, 1994). Nonetheless, the endosperm fraction also makes some significant contributions to the overall health benefits as outlined here(Adom *et al.*, 2005).

Phenolic compounds fall into two major categories: phenolic acids and flavonoids. The phenolic acids are benzoic or cinnamic acid derivatives, whereas the flavonoids are largely tannins and anthocyanins (Dykes *et al.*, 2006). In comparison with sorghum, other cereal brans examined, such as oat, rice and wheat, had low phenolic contents and low antioxidant potential (Farrar et al., 2008). There are hundreds of phytochemical components in soybeans and soy-based foods. In recent years, accumulating evidence has suggested that the isoflavones or soy proteins stripped of phytochemicals only reflect certain aspects of health effects associated with soy consumption. Other phytochemicals, either alone or in combination with isoflavones or soy protein, may be involved in the health effects of soy (Kang et al., 2010). Polyphenols comprise a wide variety of compounds, divided into several classes (i.e. hydroxybenzoic acids, hydroxycinnamic acids, anthocyanins, proanthocyanindins, flavonols, flavones, flavanols, flavanones, isoflavones, stilbenes and lignans), that occur in fruits and vegetables, wine and tea, chocolate and other cocoa products (Manach et al., 2004). Epidemiological studies showed that increased intake of polyphenols were associated with reduced risk of CVD, cancer and neurodegenerative disorders. Several polyphenols have been demonstrated to have clear antioxidant properties in vitro as they can act as chain breakers or radical scavengers depending on their chemical structures, which also influence their antioxidant power (Rice-Evans, 2001). A hierarchy has been established for the different polyphenolic compounds within each class on the basis of their capability to protect lipids, proteins or DNA against oxidative injury (Heijnen et al., 2002). This concept, however, appears now to be a simplistic way to conceive their activity (Masella et al., 2005). First of all, pro-oxidant effects of polyphenols have also been described to have opposite effects on basic cell physiological processes (Elbling et al., 2005): for example, if as antioxidants they improve cell survival, as pro-oxidants they may indeed induce apoptosis, cell death and block cell proliferation (Lambert et al., 2005). It should be noted that intracellular redox status, which is influenced by antioxidants, can regulate different transcription factors, which in turn regulate various cell activities (Kwon et al., 2003).

Recent advances have been made in our scientific understanding of how berries promote human health and prevent chronic illnesses (Table 4.1). Berry bioactives encompass a wide diversity of phytochemicals ranging from fat-soluble/lipophilic to water-soluble/hydrophilic compounds (Seeram et al., 2010). Long-term feeding of blueberries to rats hindered and even reversed the onset of age-related neurologic dysfunctions, such as a decline in neuronal signal transduction, and cognitive, behavioral and motor deficits. In addition, Stoner and coworkers showed that supplementation with black raspberries in the diet reduced the multiplicity and incidence of esophageal tumours in N-nitrosomethylbenzylamine-treated rats (Stoner et al., 1999).

The tocopherols (α-, β-, γ- and δ-tocopherol) and resveratrol (RSV) are phytochemicals with alleged beneficial effects against atherosclerosis, vascular diseases and different cancers (Bishayee et al., 2010). Although they both can act as antioxidants, they also modulate signal transduction and gene expression by non-antioxidant mechanisms (Reiter et al., 2007). Apples are widely and commonly consumed and are one of the main contributors of phytochemicals in the human diet (Table 4.1), making them the largest source of dietary phenolics (Yang and Liu, 2009). It is believed that chemotherapeutic combination approaches have been used to reduce drug toxicity, to delay the development of cancer cells, and to reach a greater effect than with one active drug alone. Antioxidant synergism has been observed with different compounds such as vitamins E and C, vitamin E and β-carotene, catechin and malvidin 3-glucoside, flavonoids and urate, and tea polyphenols and vitamin E (Yang and Liu, 2009). The phytochemicals in fruits may act independently or in combination as anti-cancer agents. The additive and synergistic effects of phytochemicals in fruits may be responsible for their potent anti-cancer activities, and the benefit of a diet rich in fruits is attributed to the complex mixture of phytochemicals present in whole foods.

Among spices, saffron displayed the highest antioxidant capacity, whereas among dried fruits, prunes exhibited the highest value. Among cereal products, whole meal buckwheat and wheat bran had the greatest total antioxidant capacity. Among pulses and nuts, broad beans, lentils and walnuts had the highest antioxidant capacity, whereas chickpeas, pine nuts and peanuts were less effective. The contribution of bound phytochemicals to the overall antioxidant capacity was relevant in cereals as well as in nuts and pulses (Pellegrini *et al.*, 2006). Of note, the polyphenol curcumin (1,7-bis(4-hydroxy-3-methoxyphenyl)-1,6-heptadiene-3,5-dione), a natural yellow pigment extracted from the rhizome of the turmeric plant *Curcuma longa* (von Metzler *et al.*, 2009), was shown to inhibit the activation of the transcriptional factor nuclear factor-kappaB (NF-κB), and to inhibit the proteasome-ubiquitin system (Bharti *et al.*, 2004). Curcumin modifies the invasive potential of breast cancer cells (Squires *et al.*, 2003). Another polyphenol, (−)-epigallocatechin-3-gallate (EGCG), was found to inhibit neovascularisation in the chick chorioallantoic membrane assay and when given in drinking water could significantly suppress VEGF (vascular endothelial growth factor)-induced corneal neovascularisation (Manson *et al.*, 2007). Such results suggest that EGCG may be a useful inhibitor of angiogenesis *in vivo* (Gilbert and Liu, 2010). A number of phytochemicals also affect expression of cadherins, catenins and matrix metalloproteinases (Manson *et al.*, 2007). Also of increasing importance is the investigation of combinations of phytochemicals or their use in conjunction with other therapies, to increase efficacy or decrease unwanted side effects (Howells *et al.*, 2007). It has been shown in breast cell lines that indole-3-carbinol (I3C) exhibits enhanced efficacy in combination with Src or EGFR kinase inhibitors, and *in vivo* I3C prevented the hepatotoxicity of trabectidin (ET743), an experimental antitumour drug with promising activity in sarcoma, breast and ovarian carcinomas, without compromising antitumour efficacy (Manson *et al.*, 2007). Curcumin enhances the efficacy of oxaliplatin in both p53-^{positive} and p53^{mutant} colon cancer cells (Howells *et al.*, 2007). However, caution is required, since it has also been reported to compromise the efficacy of some chemotherapeutic drugs in human breast cancer models (Somasundaram *et al.*, 2002). Additionally, using a human osteoclast system, curcumin abrogated both osteoclast differentiation and bone resorbing activities, preventing the IκB phosphorylation (von Metzler, 2009).

4.2.1 Therapeutic use of antioxidants

Antioxidants work effectively as disease preventing species. The three major types of ROS are superoxide anion radical ($O_2^{\bullet-}$), constitutively present in cells because of leakage from the respiratory chain in mitochondria, hydrogen peroxide (H_2O_2), resulting from the dismutation of $O_2^{\bullet-}$ or directly from the action of oxidase enzymes, and hydroxyl radical ($^\bullet OH$), a highly reactive species that can modify purine and pyrimidine bases and cause strand breaks that result in oxidatively damaged DNA (Matés *et al.*, 2010). Free radical compounds result from normal metabolic activity as well as from the diet and environment (Matés *et al.*, 2002), contributing to general inflammatory response and tissue damage (Matés *et al.*, 2008). Antioxidants protect DNA from oxidative damage and mutation, leading to cancer (Matés *et al.*, 2006). Antioxidants are considered as the most promising chemopreventive agents against various human cancers (Matés, 2000). However, some antioxidants play paradoxical roles, acting as double-edged swords. A primary property of effective and acceptable chemopreventive agents should be freedom from toxic effects in population (Kawanishi *et al.*, 2005). In spite of identification, use of effective cancer chemopreventive agents has become an important issue in public health-related research; miscarriage of the intervention by some antioxidants makes necessary the evaluation of safety before recommending use of antioxidant supplements for chemoprevention (Calabrese et al., 2010).

A number of epidemiological studies initially indicated utility of antioxidants in disease prevention, particularly for CVD and cancer. Regardless, recent conflicting results from intervention trials have identified negative consequences associated with antioxidant supplement use and a presumed reduction in ROS (Seifried et al., 2007). This apparent conundrum of antioxidant effects on ROS has recently been examined in light of molecular evidence for the role(s) of ROS in development and progression of cancer and CVD, especially since there has been a flurry of studies potentially linking some antioxidants with increased mortality and CVD (Seifried et al., 2006). Alternatively, compounds in a plant-based diet may increase the capacity of endogenous antioxidant defenses and modulate the cellular redox state. Changes in the cellular redox state, conveying physiologic stimuli through regulation of signaling pathways, may have wide-ranging consequences for cellular growth and differentiation (Haddad et al., 2002). In addition, it has been well documented that phytochemicals modulate protein kinase (PK) activities, serve as ligands for transcription factors and modulate protease activities (Moskaug, 2005).

Polyphenols are among the most abundant phytochemicals in human food items and, of these, flavonoids are probably the most deeply studied. Low concentrations of flavonoids stimulated transcription of a critical gene for glutathione (GSH) synthesis in cells. Both onion extracts and pure flavonoids transactivated human gamma-glutamylcysteine synthetase (γ-GCS) through antioxidant response elements in the promoter in both COS-1 cells and HepG2 cells, with quercetin being the most potent flavonoid. Structurally similar flavonoids were not as potent; myricetin, with only one hydroxyl group more than quercetin, was inactive, which emphasises the apparent specificity of human γ-GCS induction (Myhrstad et al., 2002). In vivo feeding experiments with polyphenol-rich diets revealed large differences in human γ-GCS promoter activity responses among individual animals. Some animals responded and some did not. One possible explanation for this phenomenon may be related to differences in bacterial populations in the gut microbial flora influencing the extent of enzymatic hydrolysis of polyphenol conjugates (Scalbert et al., 2000). On the other hand, flavonoid antioxidant scavenging of free radicals often involves formation of a radical of the flavonoid itself. Quercetin is oxidised to a quinone when serving as an antioxidant, and Boots et al. (2003) showed that such quinones react with thiols. Therefore, it could be speculated that free radical-oxidised quercetin reacts with thiols in Keap1, the key regulatory protein in transcriptional regulation of antioxidant-responsive genes through Nrf2. Quercetin and myricetin are known to auto-oxidise at physiologic pH, and subsequent reduction of glutathione concentrations can possibly explain transcriptional up-regulation of both γ-GCS subunits (Tian et al., 1997).

Although the redox potentials of most flavonoid radicals are lower than those of $O_2^{\bullet-}$ and peroxyl radicals ($ROO^{\bullet-}$) (Moskaug, 2005), the effectiveness of the radicals in generating lipid peroxidation, DNA adducts and mutations may still be significant in disease development (Skibola et al., 2000). Also of concern is the observation that some flavonoids inhibit enzymes (such as topoisomerases) involved in DNA structure and replication, and it has been suggested that high intake of flavonoids predisposes subjects to the development of certain childhood leukemias (Strick et al., 2000). Flavonoid supplementation as a general recommendation to increase cellular GSH concentrations may also be troublesome, because glutathione has a major role in overall redox regulation of cell functions and is not suitable as a therapeutic target for substances that alter cellular concentrations by orders of magnitudes (Moskaug, 2005). Interesting results add modulation of intracellular GSH concentrations to the list of possible disease-preventing effects of polyphenols, with the implication that they modulate GSH-dependent cellular processes, such as detoxification of xenobiotics, glutathionylation of proteins and regulation of redox switching of protein functions in major cellular processes (Carlsen *et al.*, 2003).

Recent findings suggest that several heavily studied phytochemicals exhibit biphasic dose responses on cells with low doses activating signaling pathways that result in increased expression of genes encoding cytoprotective proteins including antioxidant enzymes, protein chaperones, growth factors and mitochondrial proteins. Examples include the transcription factor Nrf2, which binds the antioxidant response element (ARE) upstream of genes encoding cytoprotective antioxidant enzymes and Phase II proteins (Mattson, 2008). The latter pathway is activated by curcumin, sulforaphane (present in broccoli) and allicin (present in garlic). Other phytochemicals may activate the sirtuin-FOXO pathway resulting in increased expression of antioxidant enzymes and cell survival-promoting proteins; RSV has been shown to activate this pathway (Frescas *et al.*, 2005). Ingestion of other phytochemicals may activate the hormetic transcription factors NF-κB and cAMP response element-binding (CREB) resulting in the induction of genes encoding growth factors and anti-apoptotic proteins (Mattson *et al.*, 2006). Allicin and capsaicin activate transient receptor potential ion channels, and RSV activates sirtuin-1 (Mattson, 2008). Isothiocyanates present at high levels in broccoli and watercress induced the expression of cytoprotective Phase IIproteins in liver, intestinal and stomach cells (McWalter *et al.*, 2004); the curry spice curcumin has been reported to induce adaptive stress response genes and protect cells in animal models of cataract formation, pulmonary toxicity, multiple sclerosis and Alzheimer's disease (Mattson, 2008); and RSV can activate stress response pathways and protect cells in models of myocardial infarction and stroke (Baur and Sinclair, 2006). Several epidemiological studies have shown beneficial effects of green tea in cancer and CVD (Kuriyama, 2008). Also, it has been demonstrated that coffee drinking may reduce the risk of liver cancer (Xu *et al.*, 2009). Furthermore, dietary supplementation rich in polyphenols such as blueberries and apple juice showed neuroprotection for focal brain ischemia and Alzheimer's disease (Ortiz and Shea, 2004). Individual grape compounds (Table 4.1) contain antioxidative, anti-inflammatory, antiapoptosis, antivirus, antiallergy, platelet antiaggregatory and/or anticarcinogenic properties (Aggarwal and Shishodia, 2006). Grape polyphenols reduced macrophage atherogenicity in mice, ameliorated cerebral ischemia-induced neuronal death in gerbils and exhibited a cardioprotective effect in humans (Zern *et al.*, 2005). Freeze-dried grape powder contains anthocyanidins, catechin, epicatechin, quercetin, RSV and kaempferol (Xu *et al.*, 2009). Anthocyanidin demonstrates cytotoxic effects in human breast, lung and gastric adenocarcinoma cells (Xu *et al.*, 2007).

4.2.2 Phytochemicals as therapeutic agents

Herbal medicine has clearly recognisable therapeutic effects. The results obtained support prior observations and future pharmacologic uses concerning a huge amount of species (Table 4.2). We can outline tannins from *Acacia catechu* for respiratory diseases, flavonoids from *Aesculus indica* for joint pain, glycosides from *Azadirachta indica* as antipyretic, saponins, tannins, alkaloids and cardiac glycosides from *Euphorbia hirta* for asthma, alkaloids from *Taxus wallichiana* for anti-cancer theraphy and lignan glucosides from *Tinospora sinensis* for diabetes. Immunostimulant, antibacterial, analgesic and antiprotozoal characteristics of *Andrographis paniculata* extract have also been demonstrated. Crude root extract of *Podophyllum hexandrum* (Berberidaceae) was used as hepato-protective (Kunwar *et al.*,

Table 4.2 Therapeutic properties of some exotic plants and their phytochemicals

Plant/herb	Phytochemical	Symptom/disease	Use
Acacia catechu	Quercetin	Hepato-protective	Crude extracts
	Epicatechin	Hypoglycaemic	Wood tea
	Tannin	Cold	
	Cyanidanol	Cough	
Aconitum spicatum	Norditerpenoids alkaloids	Analgesic	Crude extracts
Aesculus indica	Flavonoids	Analgesic	Crude extracts
Andrographis paniculata	Flavonoids	Fever	Crude extracts
Andrographis paniculata	Flavonoids	Analgesical	Crude extracts
	Diterpenoids	Antibiotic	
Anisomeles indica	Diterpenoid	Urinary affections	Crude extracts
Anoectochilus formosanus	Glucosides	Wound healing	Crude extract
Azadirachta indica	Triterpenes	Fever	Leaf extracts
Cannabis sativa	Lectins	Control bleeding	Leaf extract
Dissotis rotundifolia	Alkaloids	Diarrhea	Leaves decoction
	Saponins	Rheumatism	Crude extracts
	Cardiac glycosides	Dysentery	
Epimedium grandiflorum	Epimedin a, b and c	Sexual dysfunction	Crude extract
Epimedium sagittatum	Icariin	Osteoarthritis	
	Flavonol glycosides	Osteoporosis	
Euphorbia hirta	Flavonoids	Asthma	Leaf extracts
Ficus religiosa	Ellagic acid	Diabetes	Crude extract
	Flavonoids	Alzheimer disease	Bark extract
Juniperus virginiana	Lignans	Cancer	Leaves extract
	Flavonoids	Rheumatoid arthritis	
Lepidium sativum	Imidazole	Antihypertensive	Seeds extracts
	Lepidine	Diuretic	
	Semilepidinoside a and b	Antiasthmatic Anti-inflammatory	
	Phenolic compounds	Hypothermic	
	Fatty acids	Analgesic	
	Ascorbic acid	Coagulant	
	β-carotene	hypoglycaemic	
	Glucosinolates		
Parmelia sulcata	Polyphenols	Antibacterial	Crude extracts
	Ascorbic acid	Antifungal	
	β-carotene	Antiviral	
	Anthocyanins	Immunomodulator	
	Flavonoids		
Podophyllum hexandrum	Lignans	Rheumatism	Root extract
	Flavonoids	Mental disorders	Rhizome extract
Prosopis africana	Tannins	Anti-inflammatory	Stem bark extract
	Sapononins	Antimicrobial	Leaves extract
Scutellaria discolor	Flavonoids	Antiviral	Root extract
	Glucosides	Rheumatism	Stem extract
Skimmia anquetilia	Linalool	Headache	Leaf extract
Taxus wallichiana	Lignans	Tumor control	Leaf extracts
Tinospora sinensis	Tannins	Diabetes	Stem extracts
Vanda roxburghii	Glucosides	Anti-inflammatory	Root extract
Vanda tessellate	Alkaloids	Sexual dysfunction	Crude extract
Vitex negundo	Glycosides	Asthma	Leaf extract
	Camphene	Antibiotic	Bark extract
	Flavonoids	Cancer	

2010). The hepato-protective and hypoglycemic properties of *Acacia catechu* could be attributed to the quercetin and epicatechin respectively. Lectins of *Cannabis sativa* possess haema-gluttinating properties that corroborate the indigenous use of the leaf extract to control bleeding. Vegetale oil such as α-pinene obtained from crude leaf extract of *Vitex negundo* is recommended as antitussive and anti-asthma, antibacterial, antifungal, hypoglycemic, anti-cancer, acne control and inhibitor of edema to tracheal contraction (Kunwar *et al.*, 2010). Linalool also possesses an anxiolytic effect, and this effect probably substantiates the folk uses of *Skimmia anquetilia* leaves as medicine for headache (Kunwar *et al.*, 2010).

The results of pharmacological studies in *Ficus religiosa* will further expand the existing therapeutic activity of tannins, saponins, flavonoids, steroids and cardiac glycosides, and provide convincing support to its future clinical use in modern medicine (Singh *et al.*, 2011). *Lepidium sativum* has been studied for its medicinal use in many diseases (Table 4.2), including as a bone fracture healing agent (Najeeb-Ur-Rehman *et al.*, 2011). Orally, *Epimedium* has traditionally been used to treat impotence, involuntary ejaculation, weak backs and knees, postmenopausal bone loss, arthralgia, mental and physical fatigue, memory loss, hypertension, coronary heart disease, bronchitis, chronic hepatitis, HIV/AIDS, polio, chronic leukopenia and viral myocarditis. It is also used to arouse sexual desire. In clinics, *Epimedium* is used to treat osteoporosis, climacteric period syndrome, breast lumps, hyperpiesia and coronary heart disease (Ma *et al.*, 2011). Taking into account their therapeutic efficiency and economical considerations, the total flavonoids and/or active ingredients might be developed into new drugs for the treatment of various diseases, especially sexual dysfunction, osteoporosis and immunity-related diseases (Ma *et al.*, 2011).

Dissotis rotundifolia is used mainly for the treatment of rheumatism and painful swellings. The leaves decoction is used to relieve stomach ache, diarrhoea, dysentery, cough, prevent miscarriage/abortion, conjunctivitis, circulatory problems and venereal diseases (Abere *et al.*, 2010). Extracts of *D. rotundifolia* have been found to possess antimicrobial and antispasmodic activities, which makes it a good candidate for further works in diarrhoea management (Abere *et al.*, 2010). Lignans podophyllotoxin, deoxypodophyllotoxin, demethylpodophyllotoxin and podophyllotoxone are four therapeutically potent anti-cancer secondary metabolites found in *Juniperus* and *Podophyllum* species collected from natural populations in Himalayan environments and the botanical gardens of Rombergpark and Haltern (Germany). *Juniperus virginiana* has been used for treatmwent of genital warts, psoriasis and multiple sclerosis. *Podophyllum hexandrum* has been used to treat constipation, cold, fever and septic wounds (Kusari *et al.*, 2011). *Vanda tessellata* is a potent aphrodisiac and fertility booster in mice. These results could be extrapolated to humans and preliminary tests could be done to see if researchers develop another drug like Viagra™.

A series of experiments were conducted on *Anoectochilus formosanus*, and accentuated the possibility of its commercial application for healing of different diseases. Wound healing properties of the extract of *Vanda roxburghii* is investigated, as it is reported in Ayurveda as a strong candidate of medicinal plant used in anti-inflammatory, antiarthritic, treatment of otitis externa and sciatica (Hossain, 2011). The combinations of sulfamethoxazole plus protocatechuic acid, sulfamethoxazole plus ellagic acid, sulfamethoxazole plus gallic acid and tetracycline plus gallic acid show synergistic mode of interaction. The identified synergistic combinations can be of potent therapeutic value against *P. aeruginosa* infections. These findings have potential implications in delaying the development of resistance as the antibacterial effect is achieved with lower concentrations of both drugs (antibiotics and phytochemicals). The present study clearly highlights the low toxic potential of phytochemicals as antibacterial compounds and makes suggestions on the possibility of use of the above shown synergistic drug and herb combinations for combating infections caused by this pathogen (Jayaraman et al., 2010). A large number of plant-derived triterpenoids are known to exhibit cytotoxicity against a variety of tumour cells as well as anti-cancer efficacy in preclinical animal models. Numerous triterpenoids have been synthesised by structural modification of natural compounds. Some of these analogues are considered to be the most potent anti-inflammatory and anticarcinogenic triterpenoids known in the prevention and therapy of human breast cancer (Bishayee et al., 2011).

4.3 Phytochemicals and disease prevention

Since ancient times, natural products, herbs and spices have been used for preventing several diseases. The term chemoprevention was coined in the late 1970s and referred to the prevention of cancer by selective use of phytochemicals or their analogues. The concept of using naturally derived chemicals as potential chemopreventive agents has advanced the field dramatically. Throughout the years, a vast number of chemopreventive agents present in natural products have been evaluated using various experimental models. A number of them have progressed to early clinical trials. More recently, the focus has been directed towards molecular targeting of chemopreventive agents to identify mechanism(s) of action of these newly discovered bioactive compounds. Moreover, it has been recognised that single agents may not always be sufficient to provide chemopreventive efficacy and, therefore, the new concept of combination chemoprevention by multiple agents or by the consumption of whole food has become an increasingly attractive area of study (Mehta et al., 2010).

Preventing many chronic diseases requires healthy dietary habits. Achieving a better balance of grain-based foods through the inclusion of whole grains is one scientifically supported dietary recommendation. Epidemiological and other types of research continue to document health benefits for diverse populations who have adequate intakes of both folic acid-fortified grain foods and whole grains. Folic acid fortification of grains is associated with reduced incidence of neural tube and other birth defects and may be related to decreased risk of other chronic disease and may contribute to specific health-maintaining and disease-preventing mechanisms (Jones et al., 2008). Despite the high levels of polyphenolic phytochemicals in grain cereals and their position as a major food staple, there has been a lack of research on the effects on both animal and human health and disease prevention. Cereal brans with a high phenolic content and high antioxidant properties inhibited protein glycation mediated by the reducing sugar fructose. These results suggest that certain varieties of cereal bran may affect critical biological processes that are important in diabetes and insulin resistance (Farrar et al., 2008). The consistent consumption of foods that contain significant levels of phytochemicals and dietary fibre correlates with tangible disease prevention. For example, whole grain comsumption is known to help in reducing the incidence of heart disease, metabolic syndrome, neuropathy, diabetes and other chronic diseases, partly due to components in cereal brans, especially dietary fibre and phytochemicals (Awika et al., 2005).

The combination of phytochemicals with relatively broad specificity on enzymes involved in signal transduction and gene expression may increase their activity in disease prevention by modulating several different molecular targets (Reiter et al., 2007). Moreover, with further development of nutrigenomics, on the basis of a simple gene test, physicians can personalise food medicine, which makes it possible for patients to control their weight, optimise their health and reduce the risk of cancer, diabetes and liver diseases (Xu et al., 2009).

Dietary phytochemicals have the potential to moderate deregulated signalling or reinstate checkpoint pathways and apoptosis in damaged cells, while having minimal impact on healthy cells. These are ideal characteristics for chemopreventive and combination anti-cancer strategies, warranting substantial research effort into harnessing the biological activities of these agents in disease prevention and treatment (Manson *et al.*, 2007). The absorption, metabolism, distribution and excretion profile of bioactive compounds is essential to assess the full potential of promising chemopreventive agents and may help guide in the design of novel synthetic analogues (Siddiqui *et al.*, 2009). In order to optimise the chances of success in cancer chemoprevention trials, the ability to identify those individuals most likely to benefit is clearly important. In the case of primary prevention to inhibit the earliest stages of tumour development, selection has traditionally been based on known environmental and lifestyle risk factors, genetic predisposition and family history (Tsao *et al.*, 2004). Secondary prevention is appropriate for those who have already developed pre-malignant lesions, such as intraepithelial neoplasia or intestinal polyps, the progress of which can be monitored in response to chemopreventive treatments (Manson *et al.*, 2007). Several dietary compounds, including indoles and polyphenols, have shown promise in this respect, with regression of respiratory papillomatosis, cervical, vulvar and prostate intraepithelial neoplasia and oral leukoplakia (Thomasset *et al.*, 2007). A third strategy is tertiary prevention, which focuses on patients who have been successfully treated for a primary tumour, in order to inhibit development of second primary tumours. Greatest success to date in this respect has resulted from the use of drugs such as tamoxifen and its analogues for breast cancer, and retinoids for skin, head and neck and liver cancer. If phytochemicals have a role at this stage, it is most likely to be as part of a combined therapy (Manson *et al.*, 2007).

Modulation of detoxification enzymes is a main mechanism by which diet may influence risk of cancer and other diseases. However, genetic differences in taste preference, food tolerance, nutrient absorption, and metabolism and response of target tissues all potentially influence the effect of diet on disease risk. Thus, disease prevention at the individual and population level needs to be evaluated in the context of the totality of genetic background and exposures to both causative agents and chemopreventive compounds. Polymorphisms in the detoxification enzymes that alter protein expression and/or function can modify risk in individuals exposed to the relevant substrates. Genotypes associated with more favourable handling of carcinogens may be associated with less favourable handling of phytochemicals. For example, glutathione S-transferases (GST) detoxify polycyclic aromatic hydrocarbons present in grilled meats. GSTs also conjugate isothiocyanates, the chemopreventive compounds found in cruciferous vegetables. Polymorphisms in the GSTM1 and GSTT1 genes result in complete lack of GSTM1-1 and GSTT1-1 proteins, respectively. In some observational studies of cancer, cruciferous vegetable intake confers greater protection in individuals with these polymorphisms. A recent study of sulforaphane pharmacokinetics suggests that lack of the GSTM1 enzyme is associated with more rapid excretion of sulforaphane. Many phytochemicals are also conjugated with glucuronide and sulfate moieties, and are excreted in urine and bile. Polymorphisms in UDP-glucuronosyltransferases and sulfotransferases may contribute to the variability in phytochemical clearance and efficacy (Lampe, 2007).

The cancer chemopreventive activity of cruciferous vegetables such as cabbage, watercress and broccoli, *Allium* vegetables such as garlic and onion, green tea, citrus fruits, tomatoes, berries, ginger and ginseng, as well as some medicinal plants have been discussed. Several compounds, such as brassinin (from cruciferous vegetables like Chinese cabbage), sulforaphane (from broccoli) and its analogue sulforamate, withanolides (from tomatillos), and resveratrol (from grapes and peanuts among other foods), are in preclinical or clinical trials for cancer chemoprevention. Phytochemicals of these types have great potential in the fight against human cancer (Park and Pezzuto, 2002). Of particular significance, Indian habitual diets, which are based predominantly on plant foods like cereals, pulses, oils and spices, are all good sources of phytochemicals, particularly dietary fibre, vitamin E, carotenoids and phenolic compounds (Rao, 2003). According to the recent pharmacological findings, garlic is preventive rather than therapeutic. Epidemiological studies in China, Italy and the USA showed the inverse relationship between stomach and colon cancer incidences and dietary garlic intake. Anti-carcinogenic activities of garlic and its constituents including sulfides and S-allyl cysteine, have been demonstrated using several animal models. Garlic preparations has been also shown to lower serum cholesterol and triglyceride (TG) levels, which are major risk factors of CVD, through inhibition of their bio-synthesis in the liver, and to inhibit oxidation of low density lipoprotein (LDL). Furthermore, *in vitro* and *in vivo* studies have revealed that aged garlic extract stimulated immune functions, such as proliferation of lymphocyte, cytokine release and phagocytosis (Table 4.1). More recently, aged garlic extract has been demonstrated to prolong life span of senescence accelerated mice and prevent brain atrophy (Sumiyoshi, 1997). Besides, glucosinolates and eugenol (4-allyl-1-hydroxy-2-methylbenzene) are phytochemicals with cytochrome P-450 inducing activity. They have shown cholesterolemic effects in humans, increasing plasma high-density lipoprotein (HDL) concentrations (Hassel, 1998).

Functional foods are foods similar in appearance to a conventional food, consumed as part of the usual diet, with demonstrated physiological benefits, and/or to reduce the risk of chronic disease beyond basic nutritional functions (Hasler *et al.*, 2004). Broccoli, carrots or tomatoes would be considered functional foods because they are rich in such physiologically active components as sulforaphane, β-carotene and lycopene, respectively (Sloan *et al.*, 2002). Finally, there exists a growing selection of functional food components marketed under the umbrella of dietary supplements (Hasler *et al.*, 2004). This category also includes a large number of herbal-enriched products that make a variety of structure/function claims. Examples include cereal fortified with ginkgo biloba, which is marketed as reducing symptoms of dementia, or juices with echinacea, which are marketed for boosting the immune system (Ernst and Pittler, 1999). Pharmaceutical companies have isolated many food components into supplement form to achieve disease prevention. These compounds include diallylsulfides (garlic), isoflavones (soy), anthocyanin (bilberry extract) and glycyrrhizin (licorice) (Fletcher and Fairfield, 2002).

4.3.1 Pharmacologic effects of phytochemicals

Some alkaloids (aconitine, anisodamine, berberine, charantine, leurosine) show antidiabetic effects. *Acacia catechu* wood tea works as an expectorant. Additionally, the tannin and cyanidanol of the plant impart astringent activity, which helps to alleviate diarrhoea (Kunwar *et al.*, 2010). Usnic acid and vulpunic acid of lichens are mitotic regulators and own antibiotic properties. *Parmelia sulcata* lichen manifests antibacterial and antifungal activities. Further pharmacological evaluation of the extracts of those species that reveal weak pharmacological validities are needed before they can be used as therapeutic potentials. The compounds that contribute to the antioxidative properties are polyphenols, vitamin C, β carotene, anthocyanins and flavonoids. Ellagic acid of *Fragaria nubicola* is also responsible for antioxidant activity. Wogonin of *Scutellaria discolor* is considered as a most potent antiviral and anxiolytic compound. Plant root extract is also useful for rheumatism (Kunwar *et al.*, 2010). Fresh plant materials, crude extracts and isolated components of *Ficus religiosa* showed a wide spectrum of *in vitro* and *in vivo* pharmacological activities like, antidiabetic, cognitive enhancer, wound healing [in combination with other herbs like *Ageratum conyzoides* (root), *C. longa* (rhizome), *Ficus religiosa* (stem-bark) and *Tamarindus indica* (leaf)], anticonvulsant (modulation of glutamatergic and/or GABAergic functions), anti-inflammatory, anti-infectious diseases, hypolipidemic, antioxidant, immunomodulatory, parasympathetic, anti-tumour and hypotensive (Singh *et al.*, 2011). *Lepidium sativum* Linn. is commonly known as 'Common cress', 'Garden cress' or 'Halim'. Its seeds are popularly used as gastrointestinal stimulant, laxative, gastroprotective and digestive aid. In addition, the plant has been reported to have other properties, such as antibacterial, anti-asthmatic, diuretic, aphrodisiac and abortifacient. The plant has been reported to contain alkaloids (imidazole, lepidine, semilepidinoside A and B), β-carotenes, ascorbic, linoleic, oleic, palmitic and stearic acids, cucurbitacins and cardenolides. Moreover, a few phenolic constituents, such as sinapic acid and sinapin, were isolated from its seed extract (Najeeb-Ur-Rehman *et al.*, 2011).

Epimedium, is a genus of about 52 species in the family Berberidaceae. Modern pharmacology studies and clinical practice demonstrated that Epimedium and its active compounds possess wide pharmacological actions, especially in hormone regulation, anti-osteoporosis, immunological function modulation, anti-oxidation and anti-tumour, anti-aging, anti-atherosclerosis and anti-depressant activities. Currently, effective monomeric compounds or active parts have been screened for pharmacological activity from *Epimedium in vivo* and *in vitro*. *Epimedium* pharmacological actions have attracted extensive attention (Ma *et al.*, 2011). The major active constituents of *Herba Epimedii* are flavonoids, and among them epimedin A, B, C and icariin are considered major bioactive components that make up more than 52% of the total flavonoids in *Herba Epimedii*. A double-blind clinical trial relating to the effect of *Epimedium* Herbal Complex Supplement on sexual satisfaction in healthy men was compared with Viagra™ (Ma *et al.*, 2011). Berberine, a traditional plant alkaloid, is used in Ayurvedic and Chinese medicine for its antimicrobial and antiprotozoal properties. Interestingly, current clinical research on berberine has revealed its various pharmacological properties and multi-spectrum therapeutic applications, including diabetes, cancer, depression, hypertension and hypercholesterolemia (Vuddanda *et al.*, 2010). *Dissotis rotundifolia* revealed the presence of alkaloids, saponins and cardiac glycosides. The pharmacological effectiveness of glycosides is dependent on the aglycones, but the sugars render the compounds more soluble and increase the power of fixation of the glycosides. On the basis of the overall results from several investigations, the use of *D. rotundifolia* in the treatment of diarrhoea, veneral diseases, dysentery and relief of stomach ache is justified (Abere *et al.*, 2010).

Orchid phytochemicals are generally categorised as alkaloids, flavonoids, carotenoids, anthocyanins and sterols. A few studies have been conducted on animal bodies, such as mice, rabbits, frogs and guinea pigs, which created optimism that life saving phytochemicals, like Taxol, Vinblastine or Quinine, will be proved. Organic compounds, called stilbenoids, inhibited aortic contractions provoked by noradrenaline and caused vasodialation, the relaxation and widening of blood vessels in the body. Again, the implications of these chemicals for usage in human models may be promising in cardiology, pending further examination (Hossain, 2011). In a recent study, the extract of the stem bark of Prosopis africana was evaluated for analgesic and anti-inflammatory activities in rats, comparable to that of piroxicam – the standard agent used. The preliminary phytochemical screening revealed the presence of flavonoids, saponins, carbohydrates, cardiac glycosides, tannins, terpenes and alkaloids (Ayanwuyi *et al.*, 2010). *Pseudomonas aeruginosa* is a major nosocomial pathogen, particularly dangerous to cystic fibrosis patients and populations with weak immune system. In a recent study the *in vitro* activities of seven antibiotics (ciprofloxacin, ceftazidime, tetracycline, trimethoprim, sulfamethoxazole, polymyxin B and piperacillin) and six phytochemicals (protocatechuic acid, gallic acid, ellagic acid, rutin, berberine and myricetin) against five *P. aeruginosa* isolates, alone and in combination, have been evaluated (Jayaraman *et al.*, 2010).

4.4 Phytochemicals and cardiovascular disease

The heart is an aerobic organ, and most of the energy required for the contraction and maintenance of ion gradients comes from oxidative phosphorylation, generating a large amount of ROS (Matés *et al.*, 2009b). Therefore, a great deal of attention has focused on the naturally occurring antioxidant phytochemicals as potential therapy for CVD. Until 500 generations ago, all humans consumed only wild and unprocessed food foraged and hunted from their environment. These circumstances provided a diet high in lean protein, polyunsaturated fats (especially omega-3 fatty acids), monounsaturated fats, fibre, vitamins, minerals, antioxidants and other beneficial phytochemicals. Historical and anthropological studies show hunter-gatherers generally to be healthy, fit and largely free of the degenerative CVD common in modern societies (O'Keefe and Cordain, 2004). CVD is the number one cause of death and disability of both men and women in the USA with a high impact on human health and community social costs (Anderson, 2002). Many compounds in grains, including antioxidants, phytic acid, lectins, phenolic compounds, amylase inhibitors and saponins, have been shown to alter risk factors for CVD. It is probable that the combination of compounds in grains, rather than any one component, explains their protective effects in CVD (Slavin *et al.*, 2004). The phytochemical-rich diet included dried fruits, nuts, tea, whole grain products, fresh fruits and vegetables. The whole food diets significantly lowered serum cholesterol and LDL-cholesterol (LDL-C) and decreased measures of antioxidant defence, all biomarkers of decreased risk of chronic disease (Slavin *et al.*, 2003). Refined diets that do not include whole grains were associated with higher serum cholesterol levels (Slavin *et al.*, 2004). Recent studies find that serum enterolactone is associated with reduced CVD-related and all-cause death (Slavin *et al.*, 2003). As already stated, whole grains are rich in compounds such as tocotrienols (a form of vitamin E) and other plant sterols (i.e. β-sitosterol), and short-chain fatty acids (i.e. acetate, butyrate and propionate), which can lower cholesterol (Slavin *et al.*, 2004). Oxidative stress induced by ROS plays an important role in the aetiology of CVD. In particular, the LDL-oxidisation has a key role in the pathogenesis of atherosclerosis and cardiovascular heart diseases through the initiation of plaque formation process. Dietary phytochemical products such antioxidant vitamins (A, C and E) and bioactive food components (α- and β-carotene) have shown an antioxidant effect in reducing both oxidative marker stress and LDL-oxidisation process. Lycopene, an oxygenated carotenoid with great antioxidant properties, has shown both in epidemiological studies and supplementation human trials a reduction of cardiovascular risk (Riccioni *et al.*, 2008).

Epidemiological studies suggest that diets rich in polyphenols may be associated with reduced incidence of cardiovascular disorders (mainly coronary heart disease and myocardial infarction). Current evidence suggests that polyphenols, acting at the molecular level, improve endothelial function and inhibit platelet aggregation. In view of their antithrombotic, anti-inflammatory, and anti-aggregative properties, these compounds may play a role in the prevention and treatment of CVD. The antioxidant activity of several polyphenols positively correlated with the presence of a catechol ring in their molecular structure. Catechin,

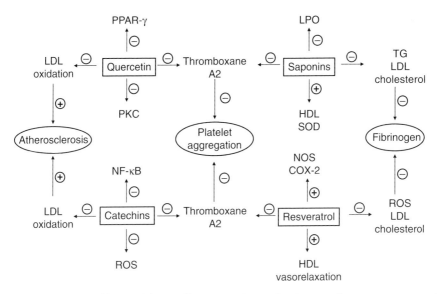

Figure 4.1 Summary of key modulatory effects on cardiovascular diseases of quercetin, resveratrol, saponins and catechins.
COX: cyclo-oxygenase, HDL: high-density lipoprotein, LDL: low-density lipoprotein, LPO: lipid peroxidation, NF-κB: nuclear factor kappaB, NOS: nitric oxide synthase, PKC: protein kinase C, PPAR-γ: proliferator-activated receptor gamma, ROS: reactive oxygen species, SOD: superoxide dismutase, TG: triglycerides.

3,4-dihydroxycinnamic acid, 3,4-dihydroxyhydrocinnamic acid, 3,4-dihydroxyphenylacetic acid, feluric acid, gallic acid and quercetin presented significant antioxidant capacity at concentrations commensurate with human plasma. The anti-atherosclerotic action of polyphenols is based on the removal of already formed ROS from the blood and on the inhibition of enzymes generating ROS, such as lipoxygenase, cyclo-oxygenase (COX), xanthine oxidase and NADPH oxidase by polyphenols. Polyphenols, particularly quercetin, are able to chelate pro-oxidant metals (mainly iron). Catechins and procyanidins block the enzymatic production of ROS, while quercetin displays a similar action of inhibiting the PKC-dependent NADPH oxidase (Figure 4.1). Moreover, polyphenols, as scavengers of the $O_2^{\bullet-}$, limits its reaction with nitric oxide (NO), as a result of which highly toxic ONOO⁻ is formed. Quercetin inhibits platelet reactivity through blocking collagen receptor (GPVI)-dependent activation. Molecular studies have also demonstrated an antagonistic action of flavones (apigenin) and isoflavones (genistein) on thromboxane A receptors. Flavonols and their derivatives, procyanidins, inhibit the expression of endothelial adhesion molecules VCAM-1, ICAM-1 and E-selectin and the activation of leukocytes and thus prevent the formation of platelet-leukocyte aggregates. Quercetin also reduces the activation of peroxisome proliferator-activated receptor gamma (PPAR-γ) (Michalska et al., 2010). Regarding this, Hayek et al. (1997) observed reduced susceptibility to LDL oxidation and an attenuation of the development of atherosclerotic lesions in the aortic arch in mice fed red wine or quercetin and, to a lesser extent, in mice fed catechins.

Oxidation of LDL plays a crucial role in the initiation mechanism of atherosclerosis (Berliner et al., 1996). An epidemiologic study indicated that European populations with higher plasma concentrations of natural antioxidants, ascorbic acid and α-tocopherol have a lower incidence of coronary heart disease (Miura et al., 2001). Several epidemiologic investigations indicated that flavonoid intake is inversely associated with the mortality of coronary heart disease (Hertog et al., 1997). Apart from lowering cardiovascular risk factors associated with diabetes, phytosterols (β-sitosterol in particular) have been shown directly lowering fasting blood glucose levels by cortisol inhibition. Among the most abundant compounds from *Aloe ferox* leaf are protocatechuic, p-coumaric, p-toluic, benzoic, hydroxyphenylacetic, xanthine and β-sitosterol. Due to the occurrence of the polyphenols, phytosterols and perhaps the indoles present, *A. ferox* leaf gel may show promise in alleviating or preventing the symptoms associated with CVD (Loots et al., 2007). On the other hand, regular consumption of fruit *Euphoria longana Lam.* (longan), whose major components were identified as gallic acid, corilagin and ellagic acid, is associated with a lower risk of CVD (Rangkadilok et al., 2005). In addition, Scottish Heart Health Study and other studies all indicated an inverse correlation between black tea consumption and the risk of coronary heart disease (Miura et al., 2001). Green tea leaves (*Camellia sinensis*) contain antioxidative tea catechins consisting of various flavan 3-ols as follows: (+)-catechin, (–)-epicatechin, (–)-epicatechin gallate, (–)-epigallocatechin and (–)-epigallocatechin-3-gallate (EGCG). Green tea catechins exert potent inhibitory effects on Cu^{2+}-mediated oxidative modification of LDL *in vitro*. Daily consumption of catechins prevents the development of atherosclerosis (Figure 4.1). Green tea, and its principal component (EGCG), exerts a much stronger antioxidative effect than teaflavin and tealubidin, the major components of black tea. Several lines of investigation demonstrated that catechins have potent scavenging effects on $O_2^{\bullet-}$ and $^{\bullet}OH$ (Miura et al., 2001). Lin and Lin (1997) reported that EGCG blocks the induction of nitric oxide synthase (NOS) by down-regulating lipopolysaccharide-induced activity of the transcription factor, NF-κB, which is a pleiotropic mediator for induction of genes, including genes relevant to atherogenesis.

Accumulating evidence has suggested that the isoflavones or soy protein only reflected certain aspects of health effects associated with soy consumption. For example, using primate models of atherosclerosis, the intact soy protein has been shown to be effective in lowering cholesterol. Studies have shown that non-isoflavone compounds, such as soyasaponins, phytic acid or plant sterols, display a wide range of bioactivities, including cardiovascular protective effects (Kang et al., 2010). Soyasaponins showed their cardiovascular protective effects through several different mechanisms (Oakenfull and Sidhu, 1990). In animal models, soyasaponins were found to significantly reduce the serum total cholesterol, LDL-C and TG concentrations and to increase the HDL-cholesterol (HDL-C) levels (Xiao et al., 2005). 24-Methylenecycloartanol, in combination with soysterol, greatly reduced plasma cholesterol and enhanced cholesterol excretion in rats (Kang et al., 2010). Total soyasaponins prevented the decrease of blood platelets and fibrinogen, and the increase of fibrin degradation products in the disseminated intravascular coagulation (Figure 4.1). *In vitro* experiments, total soyasaponins, soyasaponins I, II, A_1, and A_2 inhibited the conversion of fibrinogen to fibrin (Kang et al., 2010). Total soyasaponins decreased elevated blood sugar and lipid peroxidation levels and increased the decreased levels of superoxide dismutase (SOD) in diabetic rats (Wang et al., 1993). Phytosterols (not restricted to soy-based sources), which have long been known to reduce intestinal cholesterol absorption, lead to decreased blood LDL-C levels and lower CVD risk (Kang et al., 2010).

There has been great deal of focus on the naturally occurring antispasmodic phytochemicals as potential therapy for CVD. Diterpenes exert several biological activities such as anti-inflammatory action, antimicrobial and antispasmodic activities. Several diterpenes have been shown to have pronounced cardiovascular effects, for example, grayanotoxin I produces positive inotropic responses, forskolin is a well-known activator of adenylate cyclase, eleganolone and 14-deoxyandrographolide exhibit vasorelaxant properties and marrubenol inhibits smooth muscle contraction by blocking L-type calcium channels. In the last few years, the biological activity of kaurane and pimarane-type diterpenes, which are the main secondary metabolites isolated from the roots of *Viguiera robusta* and *V. arenaria*, respectively, has been investigated. These diterpenoids exhibit vasorelaxant action and inhibit the vascular contractility mainly by blocking extracellular Ca^{2+} influx. Moreover, kaurane and pimarane-type diterpenes decreased mean arterial blood pressure in normotensive rats. Diterpenes likely fulfil the definition of a pharmacological preconditioning class of compounds and give hope for the therapeutic use in CVD (Tirapelli *et al.*, 2008). The aqueous extracts from three popular Thai dietary and herbal plants, *Cratoxylum formosum*, *Syzygium gratum* and *Limnophila aromatica* possessed high free radical scavenging and antioxidant activities. Vascular responsiveness to bradykinin, acetylcholine and phenylephrine in phenylhydrazine-control rats was markedly impaired. Moreover, the plant extracts prevented loss of blood GSH and suppressed formation of plasma malondialdehyde, plasma NO metabolites and blood $O_2^{\bullet-}$. It was concluded that the plant extracts possess antioxidants and have potential roles in protection of vascular dysfunction (Kukongviriyapan *et al.*, 2007).

Recent studies have highlighted the role of dietary fibre, particularly water-soluble varieties, in decreasing the risk of CVD. Several types of soluble fibre, including psyllium, β-glucan, pectin and guar gum, have been shown to decrease LDL-C in well-controlled intervention studies, whereas the soluble fibre content of legumes and vegetables has also been shown to decrease LDL-C (Bazzano, 2008). Surprisingly, the consumption of insoluble fibre from whole grains, though metabolically inert, has been associated with a reduction in the risk of developing coronary heart disease in epidemiological studies. The likely reason is that whole grains, like nuts, legumes and other edible seeds, contain many bioactive phytochemicals and various antioxidants. After cereals, nuts are the vegetable foods that are richest in fibre, which may partly explain their benefit on the lipid profile and cardiovascular health (Salas-Salvadó *et al.*, 2006). On the other hand, men who had low concentrations of β-caroteno and vitamin C in their blood had a significantly increased risk of subsequent ischemic and coronary heart disease, suggesting that carotenoid-containing diets are protective against CVD. In conclusion, the consumption of β-carotene-rich foods has been associated consistently with a decreased risk of CVD. In contrast, supplementation with β-carotene in major intervention trials generally has failed to reduce the incidence of CVD (Mayne, 1996).

Stilbenes have been shown to protect lipoproteins from oxidative damage and to have chemopreventive activity (Vitrac *et al.*, 2005). Resveratrol (3,4′,5-trihydroxy-trans-stilbene), or (E)-5-(p-hydroxystyryl)resorcinol, is a naturally occurring polyphenolic compound abundant in grapes, peanuts, red wine, pines and other leguminosae family plants in response to injury, ultraviolet irradiation and fungal attack (Matés *et al.*, 2009a). Epidemiological evidence has shown that CVD is less prevalent in the French population than expected in light of their saturated fat intake and serum cholesterol concentrations (Zenebe *et al.*, 2001). The protective effect of moderate consumption (two to three units) of red wine on the risk of CVD morbidity and mortality, however, has been consistently shown in many epidemiological studies (Zenebe and Pechanova, 2002). Phenolic compounds and especially a group of flavonoids seem to be responsible for the majority of protective effects of red wine on CVD, particularly their antithrombic, antioxidant, anti-ischemic, vasorelaxant and antihypertensive properties (Zenebe et al., 2001). Polyphenols have been shown to be able to modulate the process of thrombosis in several systems. Fuster et al. (1992) reported that a reduced rate of development of atherosclerosis and coronary artery disease caused by daily intake of flavonoids was based mainly on the possibility of flavonoids to inhibit acute thrombus formation. One of the most recognised and widely studied compounds is RSV, a phytoalexin member of a family of polyphenols called viniferins. Although RSV was first isolated in 1940 from the roots of white hellebore (*Veratrum grandiflorum*), the importance of RSV was recognised only after the widely publicised historic French paradox associated with drinking of red wine. Both epidemiological and experimental studies have revealed that drinking wine, particularly red wine, in moderation protects cardiovascular health. A growing body of evidence supports the role of RSV as evidence based cardiovascular medicine. RSV protects the cardiovascular system in multidimensional ways. The most important point about RSV is that, at a very low concentration, it inhibits apoptotic cell death, thereby providing protection from various diseases including myocardial ischemic reperfusion injury, atherosclerosis and ventricular arrhythmias (Figure 4.1). Both in acute and in chronic models, RSV-mediated cardioprotection is achieved through the preconditioning effect, rather than direct effect as found in conventional medicine. The same RSV when used in higher doses facilitates apoptotic cell death and behaves as a chemopreventive alternative. RSV likely fulfils the definition of a pharmacological preconditioning compound and gives hope for the therapeutic promise of alternative medicine (Das and Das, 2007).

Evidence indicates that some polyphenols modulate specific pathways regulating the expression and activation of genes involved in the control of the cardiovascular system (Zenebe et al., 2001). It is possible that red wine polyphenols decrease degradation of basal levels of NO, preventing its destruction by superoxides, or stimulate NO synthase in endothelial cells. It is conceivable that both mechanisms are active *in vivo* (Andriambeloson et al., 1997). Red wine polyphenols reduced the level of thromboxane A_2 similarly to acetylsalicylic acid. Polyphenols, in contrast to acetylsalicylic acid, had a shorter-term effect on coronary blood flow but interfered with glycoprotein receptors on endothelial cells. Several polyphenols have been also shown to interfere with several enzyme systems critically involved in cellular responses, such as tyrosine and serine-threonine PKs, phospholipases and COXs (Middleton et al., 2000). Adhesion of platelets to the subendothelial matrix, after vessel damage, is a triggering mechanism of thrombus formation, and thus platelet inhibition by red wine may partially explain the prevention of thrombus growth (Zenebe et al., 2001). In humans, Pace-Asciak et al. (1995) showed that polyphenolic compounds from red wine, especially quercetin, catechin and RSV, inhibited the synthesis of thromboxane in platelets and of leukotriene in neutrophils. In their experiments, RSV and quercetin exhibited a dose-dependent inhibition of thromboxane-induced and ADP-induced platelet aggregation, while epicatechin, α-tocopherol and butylated hydroxytoluene were inactive. RSV also inhibited synthesis of thromboxane B_2 and hydroxyheptadecatrienoate, and slightly inhibited synthesis of 12-hydroxyeicosatetraenoate. Alcohol-free red wine only inhibited the synthesis of thromboxane B_2. Interestingly, cyclic reductions in coronary flow (CFRs) were eliminated by red wine and grape juice when given intravenously or intragastrically; however, a 2.5-fold greater amount of grape juice than red wine was needed for the elimination of CFRs. In the case of white wine, the elimination of CFRs was not significant (Zenebe et al., 2001). Quercetin and rutin were also found to eliminate CFRs in the same model. Measurement of quercetin, rutin and RSV content of red wine, white wine and grape juice indicated that flavonoid content was several-fold higher in red wine and grape juice than in white wine (Wollny et al., 1999). Red wine consumption was also found to increase plasma HDL concentrations characterised by their antiatherogenic effects (Zenebe et al., 2001). All the mechanisms by which red wine polyphenols exert their antiatherogenic effect appear to be crucial in the prevention and treatment of CVD. Sato et al. (2000) found that an ethanol-free red wine extract as well as RSV protected the heart from detrimental effects of ischemia-reperfusion injury, as seen by improved postischemic ventricular function and reduced myocardial infarction. Both the red wine extracts and RSV reduced oxidative stress in the heart, as indicated by decreasing malondyaldehyde formation. A reduction effect of several flavonoids on acute regional myocardial ischemia in isolated rabbit hearts was also reported (Zenebe et al., 2001). Ning et al. (1993) showed that flavone administration improved functional recovery in the reperfused heart after a bout of global ischemia. The effect of flavone on postischemic recovery was proposed to be caused by its stimulation of the cytochrome P450 system. Quercetin was reported to exert a protective effect by preventing the decrease in the xanthine dehydrogenase to oxidase ratio observed during ischemia-reperfusion in rats (Zenebe et al., 2001). The protective effects of flavonoids in cardiac ischemia are also associated with their ability to inhibit mast cell secretion, which may be involved in cardiovascular inflammation, at present considered one of the key factors in coronary artery disease (Ridker et al., 1998).

Polyphenolic compounds have also the ability to relax precontracted smooth muscle of aortic rings with intact endothelium; moreover, some of them are able to relax endothelium-denuded arteries (Andriambeloson et al., 1997). Because red wine polyphenols consist of hydroxycinnamic acid, proanthocyanidins, anthocyanins, flavanes and flavonols, the question of which substance(s) may be responsible for increased NO synthesis had to be addressed (Zenebe et al., 2001). From anthocyanin-enriched wine extracts, aglycone-, monoglycoside- and diglycoside-enriched fractions induced endothelium-dependent vasorelaxation, similar to that elicited by the original red wine polyphenolic extract. The representative derivatives of phenolic acid (benzoic, vanillic and gallic acid), hydroxycinnamic acid (p-coumaric and caffeic acid), flavanols (catechine and epicatechine) and the higher polymerenriched fraction of condensed tannins failed to induce endothelium-dependent vasorelaxation (Stoclet et al., 2000). Mechanisms implicated in the vasorelaxant effects of flavonoids may also include inhibition of cyclic nucleotide phosphodiesterases and activation of Ca^{2+}- activated K^+ channels (Zenebe et al., 2001). Both an increase in NO synthase activity and a decrease in phosphodiesterase activity may lead to increased cyclic guanosine monophosphate (cGMP) concentration, resulting in vasorelaxation and inhibition of platelet aggregation. The ability of polyphenolic compounds to activate the NO-cGMP system seems to be associated also with their antihypertensive effect. Mizutani et al. (1999) reported that in vivo administration of an extract of polyphenolic compounds from wine attenuated elevated blood pressure in spontaneously hypertensive rats, and Hara (1992) found that in vivo administration of an extract of polyphenolic compounds from tea reduced blood pressure and decreased risk of stroke in susceptible rats. Improved biomechanical properties of aorta, lowering of cholesterol concentrations and inhibition of LDL oxidation were suggested as the mechanisms responsible for blood pressure reduction (Zenebe et al., 2001). This hemodynamic effect of red wine polyphenolic compounds was associated with augmented endothelium-dependent relaxation and a modest induction of gene expression of inducible NO synthase and *COX-2* within the arterial wall, which together maintained unchanged agonist-induced contractility (Diebolt et al., 2001).

4.5 Phytochemicals and cancer

Cancer induction, growth and progression are multi-step events and numerous studies have demonstrated that various dietary agents interfere with these stages of cancer. Fruits and vegetables represent an untapped reservoir of various nutritive and nonnutritive phytochemicals with potential cancer chemopreventive activity. Overall, completed studies from various scientific groups conclude that a large number of phytochemicals are excellent sources of various anti-cancer agents and their regular consumption should thus be beneficial to the general population (Kaur *et al.*, 2009). Evidence has shown that dietary polyphenolic compounds including anthocyanidins from berries, catechins from green tea, curcumin from turmeric, genistein from soy, lycopene from tomatoes and quercetin from red onions and apples are phytochemicals with significant anti-cancer properties (Bishayee *et al.*, 2010). Identification of dietary phytochemicals that target key molecules regulating apoptosis, invasion and angiogenesis has become a major focus of cancer chemoprevention in recent years (Yang *et al.*, 2009). Many compounds have also shown a high efficacy on tumour angiogenesis (Jeong *et al.*, 2011), i.e. polyphenolic compounds (flavonols, flavones, flavanols, isoflavones, phenolic acids), non-flavonoids polyphenols (stilbenes and pterostilbenes), terpenoids (terpenes, sesquiterpenes) and indoles (sulforaphane) (Table 4.3) (El-Najjar *et al.*, 2010).

In recent years, the effects of phytochemicals on cell transformation and suppression of transformed cells during the different phases of carcinogenesis have been a topic of interest to many laboratories (Kang, 2010). Among specific groups, *Allium* vegetables and garlic were considered to offer probable protection against stomach and colorectal cancers, respectively. Other groups, i.e. cruciferous vegetables (source of isothiocyanates and indoles) or tea (source of polyphenols), received little attention in this report. Overall, the link between diet and health seems to be much more complicated than previously anticipated. Further investigations into the potential of phytochemicals, which take account of modifying factors, a potential threshold effect and cancer subgroups, are essential to establish their effective use in chemoprevention (Moiseeva and Manson, 2009). Where phytochemicals (I3C, tea polyphenols and curcumin) have been investigated in extended trials, they have been associated with very few side effects (Rosen and Brison, 2004). In this regard, I3C has shown great promise as a chemopreventive agent for several types of cancer, yet enthusiasm for this compound has been somewhat diminished due to its unstable characteristics upon exposure

Table 4.3 Phytochemicals showing a high efficacy on tumour angiogenesis

Flavonoids polyphenolics	Non-flavonoids polyphenols	Other polyphenolic compounds	Terpenoids	Coumarins	Miscellaneous
quercetin	resveratrol	epigallocatechin gallate	campesterol	Decursin	sulforaphane
apigenin	curcumin	genistein	celastrol	decursinol angelate	11,11′-dideoxyverticillin
morelloflavone		gallic acid ellagic acid 1,2,3,4,6-penta-O-galloyl-β-D-glucose			erianin pedicularioside G thymoquinone

to acids in the stomach (Mehta *et al.*, 2010). Resveratrol also merits further clinical evaluation as a potential colorectal cancer chemopreventive agent. Recent results suggest that daily doses of resveratrol produce levels in the human gastrointestinal tract of an order of magnitude sufficient to elicit anticarcinogenic effects (Patel *et al.*, 2010).

The lower rates of several chronic diseases in Asia, including certain types of cancer, have been partly attributed to consumption of large quantities of soy foods (Kang, 2010). Genistein from soy has demonstrated breast and prostate cancer preventive activities (Mage and Rowland, 2004). Conversely, the tumour-promoting effects of high doses of genistein have been confirmed by the USA National Toxicology Program. Moreover, combining EGCG and genistein in the diet enhanced intestinal tumourigenesis (Moiseeva and Manson, 2009). Increasing evidence suggests the potential toxicity of some dietary phytochemicals. For instance, overdose of flavonoids could increase the risk of leukemia in offspring (Ross *et al.*, 1994). It was also reported that EGCG can reduce cell viability, which was associated with increased production of ROS and depletion of GSH. Therefore, it is required to assess the adverse effects of certain diet-derived compounds (Yang *et al.*, 2010). The purported benefits of healthy dietary agents are challenged by the uncertain results regarding the lowering of cancer risk that were obtained in large-scale intervention studies using specific single dietary ingredients at supraphysiological doses (Hsieh and Wu, 2009). The concept of functional synergy was tested by investigating the combination of EGCG and genistein, derived from tea and soy products commonly found in a traditional Asian diet, with quercetin, present in abundance in fruits and vegetables, for efficacy against CaP (Conte *et al.*, 2004). Each chosen agent reportedly has shown anti-CaP activities, with overlapping and distinct molecular actions and targets. For example, EGCG acts at G_1/S whereas genistein affects the G_2/M checkpoint of the cell cycle (Hsieh and Wu, 2009). EGCG exerts epigenetic control by inhibiting DNA methyl-transferases (Fang *et al.*, 2003). Synergy was observed between EGCG, genistein and quercetin in regards to the control of androgen receptor, p53 and NAD(P)H:quinone oxidoreductase (NQO1) (Hsieh and Wu, 2009). As stated before, a combination of agents is more effective than any single constituent in achieving chemopreventive effects (Nakamura *et al.*, 2009). For this reason, studies on synergistic effects of different phytochemicals might contribute to the chemopreventive strategies against malignant tumours.

Genistein is a soy-derived isoflavone with multiple biochemical effects, including the alteration of cell cycle-regulatory kinase activities (Banerjee *et al.*, 2008). Previous studies indicated that genistein enhanced the induction of apoptosis by chemotherapeutic agents, and increased radiosensitivity in several cancer cell lines (Sarkar and Li, 2006). Genistein is also known as an estrogen receptor agonist and it can antagonise the proliferation of breast cancer cells by estradiol (Figure 4.2). I3C and genistein synergistically induces apoptosis in human colon cancer HT-29 cells by inhibiting Akt phosphorylation and progression of autophagy (Nakamura *et al.*, 2009). It is believed that phenolics can exert their effects on the different signaling pathways such as mitogen-activated protein kinases (MAPK), activator protein-1 (AP-1) or NF-κB either separately or sequentially, as well as possibly interacting between/among these pathways, which can offer complementary and overlapping mechanisms of action. Bioactive compounds can offer additive or synergistic interaction through different biochemical targets (Yang and Liu, 2009). For example, quercetin could enhance the action of carboxyamidotriazole in human breast carcinoma MDA-MB-435 cells (Liu, 2004). Other anti-cancer polyphenols are found in tea, in particular catechins. In green tea, EGCG, (−)-epicatechin-3-gallate and (−)-epicatechin are the major compounds. EGCG reduces the growth of gastric cancer (Zhu *et al.*, 2007) and

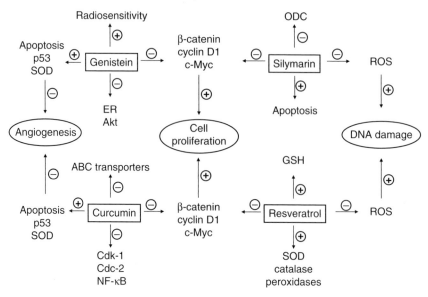

Figure 4.2 Summary of key modulatory effects on cancer of genistein, silymarin, curcumin and resveratrol
Akt: protein kinase B, Cdc: cell division cycle, Cdk: cyclin-dependent kinases, COX: cyclo-oxygenase, GSH: glutathione, ER: estrogen receptor, NF-κB: nuclear factor kappaB, ODC: ornithine decarboxylase, ROS: reactive oxygen species, SOD: superoxide dismutase.

inhibits the progression of human pancreatic cancer cells, inducing apoptosis. Several observations have indicated that a number of anti-neoplastic effects could be found among several well-studied anthraquinones, including the phenolic compounds emodin, aloe-emodin and rhein (Lentini et al., 2010).

Ginger rhizome (*Zingiber officinale*) is consumed worldwide as a spice and herbal medicine. It contains pungent phenolic substances collectively known as gingerols (Lee et al., 2008). One of the gingerols, 6-gingerol, was found to be a major pharmacologically active component of ginger. It has anti-inflammatory, antioxidant and anti-cancer activities (Kim et al., 2005). Genotoxic effects of 6-gingerol in HepG2 cells can be used as a suitable system for the prediction of toxicity, carcinogenicity and cell genotoxicity in humans (Yang et al., 2010). A decrease of GSH was observed in HepG2 cells exposed to 6-gingerol, which indicates GSH, as a main intracellular antioxidant, plays a vital role in defence against genotoxic effects induced by 6-gingerol. As a natural product of ginger, 6-gingerol is recommended for prevention of cancer and other diseases (Kim et al., 2005). 6-Gingerol is anti-mutagenic as well as mutagenic depending on the tested dose, and its active part is the aliphatic chain moiety containing a hydroxy group. Some studies have also found that both cinnamaldehyde and vanillin are anti-mutagens and DNA-damaging agents (King et al., 2007). The motility and invasive potential of many metastatic cancer cell lines has been inhibited by phytochemicals such as 6-gingerol, genistein, apigenin, ganoderic acid from the mushrooms *Ganoderma lucidum* and *Phellinus linteus* (Adams et al., 2010).

Blueberry decreased cell proliferation in HCC38, HCC1937 and MDA-MB-231 cells with no effect on the nontumuorigenic MCF-10A cell line. Additionally, black raspberries inhibited esophageal tumours in rats and modulated NFκB, AP-1, nuclear factor of activated T cells and the expression of a number of genes associated with cellular matrix, cell signaling and apoptosis (Li *et al.*, 2008). Oral intake of blueberries could be a key component of long-term breast cancer prevention strategies. It was suggested that the antiproliferative activity of fruit extracts against cancer cell lines is due to the production of H_2O_2 and resultant oxidative stress. A single serving of fresh blueberries could be an important part of dietary cancer prevention strategies (Adams *et al.*, 2010). An extract of *Mangifera pajang* kernel has been previously found to contain a high content of antioxidant phytochemicals (Naczk and Shahidi, 2006). Consistent with this, *M. pajang* kernel extract has been shown to inhibit the proliferation of liver carcinoma (HepG2), ovarian carcinoma (Caov3) and colon carcinoma (HT-29) cell lines *in vitro* (Abu Bakar *et al.*, 2010a); however, the extract did not inhibit the proliferation of normal human fibroblasts, suggesting a selective action of the kernel extract on tumour cells. Extract of the kernel of *M. pajang* contains high levels of phenolic compounds, including phenolic acids (gallic, p-coumaric, sinapic, caffeic, ferulic and chlorogenic) and flavonoids (naringin, hesperidin, rutin, luteolin and diosmin) (Abu Bakar *et al.*, 2010a). Many of these are known to inhibit growth of breast cancer cells, and so it is likely that at least some of the growth inhibitory effect of the extract can be ascribed to these phenolic compounds (Abu Bakar *et al.*, 2010b). On the other hand, similar extracts of a related *Mangifera* species, *M. indica* (mango), have been demonstrated to contain a xanthine glycoside, mangiferin, in addition to other phenolic compounds, which is a highly active cytotoxic agent (Masibo and He, 2008). To emphasise, luteolin potentiates the cytotoxicity of cisplatin in LNM35 cells and decreases the growth of LNM35 tumour xenografts in athymic mice after intraperitoneal injection. Thus, luteolin, in combination with standard anti-cancer drugs such as cisplatin, may be a promising phytochemical for the treatment of lung cancer (Attoub *et al.*, 2011).

The diferuloylmethane curcumin has been consumed for centuries in Asian countries as a dietary spice in amounts in excess of 100 mg/day without any side effects. In Southeast Asia, up to 4 g per adult/day appears to lower the incidence rate of colorectal cancer. This spice and food-colouring agent has been considered as nutraceutical because of its strong anti-inflammatory, antitumour, antibacterial, antiviral, antifungal, antispasmodic and hepato-protective roles (Montopoli *et al.*, 2009). Increasingly, preclinical and clinical evidence supports curcumin's chemopreventive and antitumour progression properties against human malignancies (Matés *et al.*, 2009a). Curcumin acts on cell signaling pathways, modulates transcription factor activities, induces apoptosis, modulates the cell cycle and cell adhesion, and inhibits angiogenesis and metastasis (Fong *et al.*, 2010). Curcumin inhibits cell proliferation, arrestes the cell-cycle progression and induces cell apoptosis in rat aortic smooth muscle cell line (A7r5). It produces similar effect on human leukemia HL-60, mouse leukemia WEHI-3 cells, and on the population of B cells from murine leukemia in vivo (Su *et al.*, 2008). While men residing in Asia show a lower incidence of CaP compared to Caucasian males, Asian men who move to and live in the United States and adopt a Western lifestyle have CaP rates indistinguishable from Caucasian males. These findings suggest that Asian diets contain ingredients that might protect against the development of CaP. Here, whether or not a combination of EGCG, genistein and quercetin, phytochemicals present in a traditional Asian diet, might exert synergy in controlling proliferation and gene expression of cancer cells (Hsieh and Wu, 2009). *In vitro*, curcumin causes cell-cycle arrest in many different tumour cells by affecting various molecular targets, such as up-regulating Cdk inhibitors and p53 and down-regulating cyclin D1, Cdk-1, cdc2, NFκB (Aggarwal *et al.*, 2006). *In vivo*, curcumin reduces incidence and multiplicity of both epithelial invasive and non-invasive adenocarcinomas (Sharma *et al.*, 2004). Curcumin causes suppression, retardation or inversion of carcinogenesis, i.e. decreasing the frequency of tongue carcinoma, and oral carcinogenesis. Curcumin combined with cisplatin or oxaliplatin cause best dose- and time-dependent increases in cell-cycle arrest and apoptosis (Figure 4.2). Curcumin is cytotoxic to ovarian cancer cells (Montopoli et al., 2009). In experimental models with mice using a soluble formulation of curcumin (dimethyl sulfoxide), curcumin injected at the tail vein crosses the blood-brain barrier and kills tumour cells inside the brain (Purkayastha et al., 2009). This investigation becomes the first in vivo demonstration of the anti-carcinogenic and anti-metastatic activity of curcumin in the brain. In summary, curcumin acts as antioxidant (scavenging free radicals), detoxifying agent (inducing Phase II enzymes) and like a multi drug resistance compound (inhibiting ABC transporters) (Fong et al., 2010). Courmarins, tannins, ketones weakly moderated tumour specific cytotoxicity against human oral squamous cell carcinomas, whereas anthracyclines, nocobactins and cyclic α,β-unsaturated compounds showed much higher tumour specific cytotoxicity (Shin et al., 2010).

Silibinin is the primary active constituent of crude extract (known as silymarin) from the seeds of milk thistle plant *Silybum marianum* (Matés et al., 2009a). Chemically, the active constituent of this extract is a flavolignan, silymarin, which in itself represents the mixture of four isomeric flavonoids: silibinin, isosilibinin, silydianin and silychristin. Silibinin is the major component (70–80%) found in silymarin and is thought to be the most biologically active (Ramakrishnan et al., 2009). Several studies have clearly shown the preclinical efficacy of both silibinin and its crude extract source, silymarin, against various epithelial cancers, and at least silibinin efficacy is currently being evaluated in cancer patients (Deep and Agarwal, 2007). Silibinin treatment strongly inhibits the growth of colorectal cancer LoVo cells and induces apoptotic death, which was associated with increased levels of cleaved caspases and cleaved poly(ADP-ribose) polymerase. Analyses of xenograft tissue showed that silibinin treatment inhibits proliferation and increases apoptosis (Figure 4.2). Together, these results suggest the potential use of silibinin against advanced human colorectal cancer (Kaur et al., 2009). Pharmacological studies have revealed that silymarin is nontoxic even at relatively high physiological doses, which suggests that it is safe to use for treatment of various diseases (Matés et al., 2009a). Some studies have shown that silymarin is a strong antioxidant and hypolipidaemic agent with a potent anticarcinogenic effect (Lah et al., 2007). Silymarin has been shown to have an inhibitory effect on ornithine decarboxylase activity (Katiyar et al., 2007). Silymarin suppresses proliferation of hepatocellular carcinoma (HCC) by inhibiting β-catenin accumulation and, hence, its target genes for cyclin D1 and c-Myc; and this may be the underlying mechanism of its antiproliferative effect. In conclusion, it has been found that the mechanism by which silymarin exhibits its growth inhibitory effect on human hepatocellular carcinoma cells *in vitro* is inhibiting cell proliferation and inducing apoptosis (Ramakrishnan et al., 2009).

Rosemary phytochemicals, such as carnosic acid, have inhibitory effects on anti-cancer drug efflux transporter P-glycoprotein and may become useful to enhance the efficacy of cancer chemotherapy (Nabekura et al., 2010). The leaves of rosemary (*Rosmarinus officinalis*) are commonly used as a spice in cooking. However, because of the presence of phenolic diterpenes and triterpenes with strong antioxidative activity, interest has grown in using rosemary as a natural antioxidant in foods. In addition to antioxidative activities, rosemary phytochemicals are reported to have antimicrobial, anti-inflammatory, and anticancer properties (Nabekura et al., 2010). Interestingly, inhibitory effects of several dietary chemopreventive and antioxidative phytochemicals, such as curcumin, RSV, tannic acid, quercetin, kaempferol, EGCG and other tea catechins, on the function of P-glycoprotein have been reported (Kitagawa et al., 2007). Natural antioxidative and chemopreventive rosemary phytochemicals, carnosic acid, carnosol and ursolic acid, have inhibitory effects on P-glycoprotein and the potential to cause food–drug interactions (Nabekura *et al.*, 2010).

Triptolide/PG490, an extract of the Chinese herb *Tripterygium wilfordii* Hook F, is a potent anti-inflammatory agent that also possesses anti-cancer activity. Triptolide is a potent inhibitor of colon cancer proliferation and migration *in vitro*. The down-regulation of multiple cytokine receptors, in combination with inhibition of *COX-2* and VEGF and positive cell-cycle regulators, may contribute to the antimetastatic action of this herbal extract (Johnson *et al.*, 2011). Herbal extracts may modify biologic responses to classic chemotherapy agents and influence multiple signaling pathways; their actions likely include antiproliferative, antiangiogenic, proapoptotic and/or antimetastatic effects (HemaIswarya and Doble, 2006). Very recently, antioxidative and antiproliferative activity of different horsetail (*Equisetum arvense* L.) extracts has been investigated. Extracts inhibited cell growth that was dependent on cell line, type of extract and extract concentration. Ethyl acetate extract exhibited the most prominent antiproliferative effect, without inducing any cell growth stimulation on human tumour cell lines (Cetojević-Simin *et al.*, 2010).

RSV functions as a fungicide produced by the plant itself to ward off potentially lethal organisms and counteract environmental stress. RSV exhibits antioxidant, anti-inflammatory and antiaging action and displays chemopreventive effects in a number of biological systems (Zamin *et al.*, 2009). In spite of the anti-cancer efficacy of RSV in preclinical models, its low bioavailability remains enigmatic and elusive. In order to explore whether RSV metabolites exert antitumour properties, some major human sulfated conjugates have recently been tested against human breast cancer cells (Matés *et al.*, 2009a). RSV affected the growth of a human cholangiocarcinoma cell line (Lentini *et al.*, 2010) and reduced formation of preneoplastic lesions (aberrant crypt foci) in rats, to decrease the incidence and size of tumours in the 1,2-dimethylhydrazine-induced model of colon cancer in rats and to prevent the formation of colon and small intestinal tumours in mice (Paul *et al.*, 2010). The anticarcinogenesis activity of RSV was first shown in a pioneering study by Jang and Pezzuto (1999), who reported that RSV was effective in all the three major stages (initiation, promotion and progression) of carcinogenesis. RSV suppresses the proliferation of a variety of human cancer cells *in vitro*, including glioma cells (Kundu and Surh, 2008) and HCC (Bishayee *et al.*, 2010). RSV is a potent antioxidant because of its ability to scavenge free radicals, spare and/or regenerate endogenous antioxidants, that is, GSH and α-tocopherol, and to promote the activities of a variety of antioxidant enzymes (Figure 4.2). The antioxidant property of RSV is also attributed to its ability to promote the activities of a variety of antioxidant enzymes. When intraperitoneally administered in rats, it was found to dose-dependently increase SOD, catalase and peroxidase activities in the brains of healthy rats (Mokni *et al.*, 2007). The long-term exposure of human lung fibroblasts to RSV results in a highly specific upregulation of MnSOD (Matés *et al.*, 2009a). In human lymphocytes, RSV increased GSH levels and the activity of glutathione peroxidase, GST and glutathione reductase. Depending on the concentration and cell type, RSV can also act as a prooxidant molecule, this effect being Cu(II)-dependent (De la Lastra and Villegas, 2007). Such a prooxidant effect could be an important action mechanism for its anti-cancer and proapoptotic properties. Compared with normal cells, cancer cells have been shown to contain elevated levels of copper and, hence, might be more sensitive to the prooxidant and cell-damaging effects of RSV. Therefore, DNA damage induced by RSV in the presence of Cu(II) might be an important pathway through which cancer cells can be killed while normal cells survive. In rats and mice, RSV toxicity is minimal, and even actively proliferating tissues are not adversely affected. The high systemic levels of RSV conjugate metabolites would warrant investigation of their potential cancer chemopreventive properties. Interestingly, RSV and quercetin, chronically administered, presented a strong synergism in inducing senescence-like growth arrest (Zamin et al., 2009). These results suggest that the combination of polyphenols can potentialise their antitumoural activity. Gliomas are the most malignant of primary tumours that affect the brain and nervous system and carry the worst clinical prognosis in both adults and children. RSV-induced inhibition of catalase might represent an additional tool for the current clinical armamentarium that might increase glioma cell kill, leading to improved survival time for patients. At present, RSV is undergoing various Phase I and Phase II interventional trials. Results from the most recent studies performed in rat and human glioma cell lines suggest that the use of RSV in combination with other bioactive food components, such as quercetin and sulforaphane, might be a viable approach for the treatment of human glioma (Gagliano et al., 2010).

Pterostilbene (trans-3,5-dimethoxy-4′-hydroxystilbene), a structurally similar compound to RSV found in blueberries, has been also shown to exhibit significant effects against cell proliferation, invasion and metastasis (Pan et al., 2007). Overall analyses indicated that pterostilbene reduced colon tumour multiplicity of non-invasive adenocarcinomas, lowered proliferating cell nuclear antigen and downregulated the expression of β-catenin and cyclin D1. In HT-29 cells, pterostilbene reduced the protein levels of β-catenin, cyclin D1 and c-Myc, altered the cellular localisation of β-catenin and inhibited the phosphorylation of p65 (Paul et al., 2010). Cyclin D1 is a very well-known cell-cycle protein targeted by β-catenin and is known to be overexpressed in colonic tumours. c-Myc is yet another important protein for cell proliferation regulated by β-catenin and Wnt pathway. Pterostilbene alone or in combination with other known chemopreventive agents can be of great importance for colon cancer prevention (Paul et al., 2010). In this regard, eugenol, a natural phenolic constituent of clove oil, cinnamon, basil and nutmeg, used primarily as a food flavouring agent has been documented to exhibit antiproliferative effects in diverse cancer cell lines as well as in B16 melanoma xenograft model. Recent studies have indicated that some of these phytochemicals are substrates and modulators of specific members of the superfamily of ABC transporting proteins. Such interactions may have implications on the pharmacokinetics of xenobiotics and the possible role of phytochemicals in the reversal of multi-drug resistance in cancer chemotherapy (Li et al., 2010).

The plant-derived anti-cancer agents are commonly classified into one of four major classes: vinca alkaloids (vinblastin, vincristine and vindesine), epipodophyllotoxins (etoposide and teniposides), texanes (paclitaxel and docetaxel) and camtothecins (camptothecin and irinotecan) (Siddiqui et al., 2009). Plant alkaloids are used as chemotherapeutic agents due to their capability to depolymerise the microtubules, inhibiting cell division. Vinca alkaloids are isolated from *Vinca rosea L.* and they are potent microtubule destabilising agents, first recognised for their myelosuppressive effects. Another alkaloid used as a chemotherapy drug for some types of cancer is taxol. It is mainly used to treat ovarian, breast and non-small cell lung cancer. Moreover, capsaicin, the major pungent ingredient in red peppers, has a profound antiproliferative effect on prostate cancer (Mori et al., 2006). Tylophorine downregulates cyclin A2, which plays an important role in G1 arrest in carcinoma cells. In addition, camptothecin, a pentacyclic alkaloid isolated from *Camptotheca acuminata Decne*, was reported to possess an interesting antitumour activity (Staker et al., 2002).

Finally, combination of carotenoids and myo-inositol was found to prevent HCC development in patients with chronic viral hepatitis and cirrhosis (Nishino, 2009). Supplemental β-caroteno has been shown to reduce precancerous lesions of the oral cavity and cervix, but not of the lung (Mayne, 1996). Agents that suppress tumour formation, such as the vitamin A metabolite all-trans retinoic acid and the isothiocyanate sulforaphane in an ultraviolet (UV) model, can block AP-1 signaling (Dickinson *et al.*, 2009). Another naturally occurring chemical, perillyl alcohol, also suppressed tumour formation in the UV model, and this effect correlated with AP-1 suppression. A study including combination of dietary elagic acid or calcium D-glucarate plus topical RSV, or grape seed extract and dietary grape seed extract plus topical RSV showed synergistic effects in reducing DMBA-induced hyperplasia (Clifford and DiGiovanni, 2010). A previous study showed that oral green tea polyphenols can suppress both UV-induced and chemically induced skin tumours in mice (Kowalczyk *et al.*, 2010).

4.6 Summary and conclusions

Fruit and vegetable consumption has been inversely associated with the risk of many pathological diseases, with the beneficial effects attributed to a variety of protective phytonutrients (Kawashima *et al.*, 2007). The mechanisms explaining this correlation have not been fully elucidated. Regardless, a consensus exists that a diet rich in fruit and vegetables is beneficial for health in preventing coronary heart disease and some forms of cancer. The nutrients responsible for the protective action are not known, but vitamins, antioxidants and flavonoids are among the likely candidates (Matés *et al.*, 2009a). Dietary supplements are generally used to increase plasma levels of these compounds. Similar results can be achieved also by increasing the proportion of vegetables and fruit in the diet. Disease prevention by dietary factors or foods consumed in small quantities and exotic plants is likely to represent one of the strategies to reduce the risk of development of malignancy in humans. The identification of new molecules able to reduce proliferative and metastatic potential of cancer cells is the goal of the newborn differentiation therapy. Noteworthy examples of diet-derived substances that have been shown to reduce experimental carcinogenesis are I3C from cruciferous vegetables, curcumin from the root of curcuma, EGCG from tea and RSV from red wine. Vegetables and fruits contain fibre, vitamins, minerals and a variety of bioactive compounds, such as carotenoids, flavonoids, indoles and sterols, all of which could account for this protective effect. On the other hand, spices, herbs and rare plants are gaining uses as pharmacologic and therapeutic agents. A better understanding of the actions of phytochemicals will facilitate its use in more specific clinical trials, may allow for synthetic derivatives to be engineered (equally effective but less toxic) and potentially offer insight into additional therapeutic uses for their antioxidant potential (Johnson *et al.*, 2011).

The chemopreventive effects of dietary phytochemicals on malignant tumours have been studied extensively because of a relative lack of toxicity. To achieve desirable effects, however, treatment with a single agent mostly requires high doses. Therefore, studies on effective combinations of phytochemicals at relatively low concentrations might contribute to chemopreventive strategies. The field of chemoprevention has expanded to include nanotechnology as a novel approach to deliver packaged chemopreventive agents in a manner which allows them to be delivered selectively to the target tissues. For example, Mukhtar and colleagues recently reported on the bioavailability of EGCG which had been packaged into nanoparticles. Results showed that EGCG delivered by nanoparticles maintained the efficacy of EGCG both as an antiangiogenic and proapoptotic agent (Siddiqui *et al.*, 2009).

These results indicated that nano-chemoprevention can provide a new approach to avoiding systemic toxicity and increasing bioavailability. Similarly, packaging RSV into solid lipid nanoparticles has been reported to improve intracellular delivery of RSV and to reduce RSV toxicity. The application of lipid- or polymer-based nano-particles or nano-shells for improved delivery of chemopreventive or chemotherapeutic agents has facilitated the delivery of agents to selective tissues and may serve to lessen systemic toxicity by reducing the amount of agent needed and/or limiting the exposure to the body (Teskac and Kristl, 2010).

Of particular significance, heyneanol, a tetramer of RSV, has comparable or better antitumour efficacy than RSV in a mouse lung cancer model (Jeong et al., 2011). The current interest in phytochemicals has been driven primarily by epidemiological studies. However, to establish conclusive evidence for the effectiveness of dietary phytochemicals in disease prevention, it is useful to better define the bioavailability of these bioactive compounds, so that their biological activity can be evaluated. The bioavailability appears to differ greatly among the various plant compounds, and the most abundant ones in our diet are not necessarily those that have the best bioavailability profile. The evaluation of the bioavailability of phytochemicals has recently been gaining increasing interest as the food industries are continually involved in developing new products, defined as functional food, by virtue of the presence of specific compounds. Despite the increasing amount of data available, definitive conclusions on bioavailability of most bioactive copmpounds are difficult to obtain and further studies are necessary. At least four critical lines of research should be explored to gain a clear understanding of the health beneficial effects of dietary phytochemicals (D'Archivio et al., 2010):

1. The potential biological activity of the metabolites of many dietary phytochemicals needs to be better investigated. In fact, metabolomic studies, including the identification and the quantification of metabolites currently represent an important and growing field of research.
2. Strategies to improve the bioavailability of the phytochemicals need to be developed. Moreover, it is necessary to determine whether these methods translate into increased biological activity.
3. Whereas *in vitro* studies shed light on the mechanisms of action of individual dietary phytochemicals, these findings need to be supported by in vivo experiments. The health benefits of dietary phytonutrients must be demonstrated in appropriate animal models of disease and in humans at appropriate doses.
4. Novel technologies, such as nanotechnology, along with a better understanding of stem cells, are certain to continue the advancement of the field of CVD and cancer chemoprevention in years to come (Mehta et al., 2010).

References

Abere, T.A., Okoto, P.E. and Agoreyo, F.O. (2010) Antidiarrhoea and toxicological evaluation of the leaf extract of Dissotis rotundifolia Triana (Melastomataceae). *BMC Complementary and Alternative Medicine*, 10, 71.

Abu Bakar, M.F., Mohamad, M., Rahmat, A., Burr, S.A. and Fry, J.R. (2010) Cytotoxicity, cell cycle arrest, and apoptosis in breast cancer cell lines exposed to an extract of the seed kernel of Mangifera pajang (bambangan). *Food Chemistry and Toxicology*, 48, 1688–1697.

Abu Bakar, M.F., Mohamed, M., Rahmat, A., Burr, S.A. and Fry, J.R. (2010) Cytotoxicity and polyphenol diversity in selected parts of Mangifera pajang and Artocarpus odoratissimus fruits. *Nutrition and Food Science*, 40, 29–38.

Adams, L.S., Phung, S., Yee, N., Seeram, N.P., Li, L. and Chen, S. (2010) Blueberry phytochemicals inhibit growth and metastatic potential of MDA-MB-231 breast cancer cells through modulation of the phosphatidylinositol 3-kinase pathway. *Cancer Research*, 70, 3594–35605.

Adlercreutz, H. and Mazur, W. (1997) Phyto-oestrogens and western diseases. *Annals of Internal Medicine*, 29, 95–120.

Adom, K.K. and Liu, R.H. (2002) Antioxidant activity of grains. *Journal of Agriculture and Food Chemistry*, 50, 6182–6187.

Adom, K.K., Sorrells, M.E. and Liu, R.H. (2005) Phytochemicals and antioxidant activity of milled fractions of different wheat varieties. *Journal of Agriculture and Food Chemistry*, 53, 2297–2306.

Aggarwal, B.B. and Shishodia, S. (2006) Molecular targets of dietary agents for prevention and therapy of cancer. *Biochemical Pharmacology*, 71, 1397–1421.

Aggarwal, S., Ichikawa, H., Takada, Y., Sandur, S.K., Shishodia, S. and Aggarwal, B.B. (2006) Curcumin (diferuloylmethane) down-regulates expression of cell proliferation and antiapoptotic and metastatic gene products trough suppression of IB kinase and Akt activation. *Molecular Pharmacology*, 69, 1195–1206.

Anderson, J.W. (2002) Whole-grains intake and risk for coronary heart disease. In L. Marquart, J.L. Slavin and R.G. Fulcher (eds) *Whole-Grain Foods in Health and Disease*, Eagan Press, St Paul (MN), pp. 187–200.

Andriambeloson, E., Kleschyov, A.L., Muller, B., Beretz, A., Stoclet, J.C. and Andriantsitohaina, R. (1997) Nitric oxide production and endothelium-dependent vasorelaxation induced by wine polyphenols in rat aorta. *British Journal of Pharmacology*, 120, 1053–1058.

Attoub, S., Hassan, A.H., Vanhoecke, B., Iratni, R., Takahashi, T., Gaben, A.M., Bracke, M., Awad, S., John, A., Kamalboor, H.A., Al, Sultan, M.A., Arafat, K., Gespach, C. and Petroianu, G. (2011) Inhibition of cell survival, invasion, tumor growth and histone deacetylase activity by the dietary flavonoid luteolin in human epithelioid cancer cells. *European Journal of Pharmacology*, 651, 18–25.

Awika, J.M., McDonough, C.M., Rooney, L.W. (2005) Decorticating sorghum to concentrate healthy phytochemicals. *Journal of Agriculture and Food Chemistry*, 53, 6230–6234.

Ayanwuyi, L.O., Yaro, A.H. and Abodunde, O.M. (2010) Analgesic and anti-inflammatory effects of the methanol stem bark extract of Prosopis africana. *Pharmacuetical Biolology*, 48, 296–299.

Banerjee, S., Li, Y., Wang, Z. and Sarkar, F.H. (2008) Multi-targeted therapy of cancer by genistein. *Cancer Letters*, 269, 226–242.

Baur, J.A. and Sinclair, D.A. (2006) Therapeutic potential of resveratrol: the in vivo evidence. *Nature Reviews Drug Discovery*, 5, 493–506.

Bazzano, L.A. (2008) Effects of soluble dietary fiber on low-density lipoprotein cholesterol and coronary heart disease risk. *Current Atherosclerosis Reports*, 10, 473–477.

Berliner, J.A. and Heinecke, J.W. (2006) The role of oxidized lipoproteins in atherogenesis. *Free Radical Biology and Medicine*, 20, 707–727.

Bharti, A.C., Takada, Y. and Aggarwal, B.B. (2004) Curcumin (diferuloylmethane) inhibits receptor activator of NF-kappaB ligand-induced NF-kappaB activation in osteoclast precursors and suppresses osteoclastogenesis. *Journal of Immunology*, 172, 5940–5947.

Bishayee, A., Ahmed, S., Brankov, N. and Perloff, M. (2011) Triterpenoids as potential agents for the chemoprevention and therapy of breast cancer. *Frontiers in Bioscience*, 16, 980–996.

Bishayee, A., Politis, T. and Darvesh, A.S. (2010) Resveratrol in the chemoprevention and treatment of hepatocellular carcinoma. *Cancer Treatment Reviews*, 36, 43–53.

Boots, A.W., Kubben, N., Haenen, G.R. and Bast, A. (2003) Oxidized quercetin reacts with thiols rather than with ascorbate: implication for quercetin supplementation. *Biochemical and Biophysical Research Communications*, 308, 560–565.

Calabrese, V., Cornelius, C., Mancuso, C., Lentile, R., Stella, A.M. and Butterfield, D.A. (2010) Redox homeostasis and cellular stress response in aging and neurodegeneration. *Methods in Molecular Biology*, 610, 285–308.

Carlsen, H., Myhrstad, M.C., Thoresen, M., Moskaug, J.O. and Blomhoff, R. (2003) Berry intake increases the activity of the gamma-glutamylcysteine synthetase promoter in transgenic reporter mice. *Journal of Nutrition*, 133, 2137–2140.

Cetojević-Simin, D.D., Canadanović-Brunet, J.M., Bogdanović, G.M., Djilas, S.M., Cetković, G.S., Tumbas, V.T. and Stojiljković, B.T. (2010) Antioxidative and antiproliferative activities of different horsetail (Equisetum arvense L.) extracts. *Journal of Medicinal Food*, 13, 452–459.

Clifford, J.L. and DiGiovanni, J. (2010) The promise of natural products for blocking early events in skin carcinogenesis. *Cancer Prevention Research*, 3, 132–135.

Conte, C., Floridi, A., Aisa, C., Piroddi, M. and Galli, F. (2004) Gamma-tocotrienol metabolism and antiproliferative effect in prostate cancer cells. *Annals of the New York Academy of Sciences*, 1031, 391–394.

D'Archivio, M., Filesi, C., Varì, R., Scazzocchio, B. and Masella, R. (2010) Bioavailability of the polyphenols: status and controversies. *International Journal of Molecular Science*, 11, 1321–1342.

Das, S. and Das, D.K. (2007) Resveratrol: a therapeutic promise for cardiovascular diseases. *Recent Patents on Cardiovascular Drug Discovery*, 2, 133–138.

De la Lastra, C.A and Villegas, I. (2007) Resveratrol as an antioxidant and pro-oxidant agent: mechanisms and clinical implications. *Biochemical Society Transactions*, 35, 1156–1160.

Deep, G. and Agarwal, R. (2007) Chemopreventive efficacy of silymarin in skin and prostate cancer. *Integrative Cancer Therapies*, 6, 130–145.

Dickinson, S.E., Melton, T.F., Olson, E.R., Zhang, J., Saboda, K. and Bowden, G.T. Inhibition of activator protein-1 by sulforaphane involves interaction with cysteine in the cFos DNA-binding domain: implications for chemoprevention of UVB-induced skin cancer. *Cancer Research*, 69, 7103–7110.

Diebolt, M., Bucher, B. and Andriantsitohaina, R. Wine polyphenols decrease blood pressure, improve NO vasodilatation and induce gene expression. *Hypertension*, 38, 159–165.

Duthie, S.J., Jenkinson, A.M., Crozier, A., Mullen, W., Pirie, L., Kyle, J., Yap, L.S., Christen, P. and Duthie, G.G. (2006) The effects of cranberry juice consumption on antioxidant status and biomarkers relating to heart disease and cancer in healthy human volunteers. *European Journal of Nutrition*, 45, 113–122.

Dykes, L. and Rooney, L.W. (2006) Sorghum and millet phenols and antioxidants. *Journal of Cereal Science*, 44, 236–251.

Elbling, L., Weiss, R.M., Teufelhofer, O., Uhl, M., Knasmueller, S., Schulte-Hermann, R., Berger, W. and Micksche, M. (2005) Green tea extract and (−)-epigallocatechin-3-gallate, the major tea catechin, exert oxidant but lack antioxidant activities. *FASEB Journal*, 19, 807–809.

El-Najjar, N., Chatila, M., Moukadem, H., Vuorela, H., Ocker, M., Gandesiri, M., Schneider-Stock, R. and Gali-Muhtasib, H. (2010) Reactive oxygen species mediate thymoquinone-induced apoptosis and activate ERK and JNK signaling. *Apoptosis*, 15, 183–195.

Ernst, E. and Pittler, M.H. (1999) Ginkgo biloba for dementia: A systematic review of double-blind placebo-controlled trials. *Clinical Drug Investigation*, 17, 301–308.

Fang, M.Z., Wang, Y., Ai, N., Hou, Z., Sun, Y., Lu, H., Welsh, W. and Yang, C.S. (2003) Tea polyphenol (−)-epigallocatechin-3-gallate inhibits DNA methyltransferase and reactivates methylation-silenced genes in cancer cell lines. *Cancer Research*, 63, 7563–7570.

Farrar, J.L., Hartle, D.K., Hargrove, J.L. and Greenspan, P. (2008) A novel nutraceutical property of select sorghum (Sorghum bicolor) brans: inhibition of protein glycation. *Phytotherapy Research*, 22, 1052–1056.

Fletcher, R.H. and Fairfield, K.M. (2002) Vitamins for chronic disease prevention in adults: Clinical applications. *Journal fo the Aerican Medical Association (JAMA)*, 187, 3127–3129.

Fong, D., Yeh, A., Naftalovich, R., Choi, T.H. and Chan, M.M. (2010) Curcumin inhibits the side population (SP) phenotype of the rat C6 glioma cell line: towards targeting of cancer stem cells with phytochemicals. *Cancer Letters*, 293, 65–72.

Frescas, D., Valenti, L. and Accili, D. (2005) Nuclear trapping of the forkhead transcription factor FoxO1 via Sirt-dependent deacetylation promotes expression of glucogenetic genes. *Journal of Biological Chemistry*, 280, 20589–20595.

Fuster, V., Badimon, J.J. and Chesebro, J.H. The pathogenesis of coronary artery disease and the acute coronary syndromes. *New England Journal of Medicine*, 326, 242–250.

Gagliano, N., Aldini, G., Colombo, G., Rossi, R., Colombo, R., Gioia, M., Milzani, A. and Dalle-Donne, I. (2010) The potential of resveratrol against human gliomas. *Anticancer Drugs*, 21, 140–150.

Gerhäuser, C., Klimo, K., Heiss, E., Neumann, I., Gamal-Eldeen, A., Knauft, J., Liu, G.Y., Sitthimonchai, S. and Frank, N. (2003) Mechanism-based in vitro screening of potential cancer chemopreventive agents. *Mutatation Research*, 523–524, 163–172.

Gilbert, E.R. and Liu, D. (2010) Flavonoids influence epigenetic-modifying enzyme activity: structure – function relationships and the therapeutic potential for cancer. *Current Medicinal Chemistry*, 17, 1756–1768.

Haddad, J.J. (2002) Antioxidant and prooxidant mechanisms in the regulation of redox(y)-sensitive transcription factors. *Cellular Signalling*, 14, 879–897.

Hallikainen, M.A., Sarkkinen, E.S. and Uusitupa, M.I.J. (2000) Plant stanols esters affect serum cholesterol concentrations of hypercholesterolemic men and women in a dose-dependent manner. *Journal of Nutrition*, 130, 767–776.

Hara, Y. (1992) The effect of tea polyphenols on cardiovacular disease. *Preventive Medicine*, 21, 333.

Hasler, C.M., Bloch, A.S., Thomson, C.A., Enrione, E. and Manning, C. (2004) Position of the American Dietetic Association: Functional foods. *Journal of the American Dietetic Association*, 104, 814–826.

Hassel, C.A. (1998) Animal models: new cholesterol raising and lowering nutrients. *Current Opinion in Lipidology*, 9, 7–10.

Hayek, T., Fuhrman, B., Vaya, J., Rosenblat, M., Belinky, P., Coleman, R., Elis, A. and Aviram, M. (1997) Reduced progression of atherosclerosis in apolipoprotein E-deficient mice following consumption of red wine, or its polyphenols quercetin or catechin, is associated with reduced susceptibility of LDL to oxidation and aggregation. *Arteriosclerosis, Thrombosis, and Vascular Biology*, 17, 2744–2752.

Heijnen, C.G.M., Haenen, G.R.M.M., Oostveen, R.M., Stalpers, E.M. and Bas, A. (2002) Protection of flavonoids against lipid peroxidation: structure activity relationship revisited. *Free Radicical Research*, 36, 575–581.

HemaIswarya, S. and Doble, M. (2006) Potential synergism of natural products in the treatment of cancer. *Phytotherapy Research*, 20, 239–249.

Hertog, M.G.L., Feskens, E.J.M. and Kromhout, D. (1997) Antioxidant flavonols and coronary heart disease risk. *The Lancet*, 349, 699.

Hossain, M.M. (2011) Therapeutic orchids: traditional uses and recent advances--an overview. *Fitoterapia*, 82, 102–140.

Howells, L.M., Mitra, A. and Manson, M.M. (2007) Comparison of oxaliplatin- and curcumin-mediated antiproliferative effects in colorectal cell lines. *International Journal of Cancer*, 121, 175–183.

Hsieh, T.C. and Wu, J.M. (2009) Targeting CWR22Rv1 prostate cancer cell proliferation and gene expression by combinations of the phytochemicals EGCG, genistein and quercetin. *Anticancer Research*, 29, 4025–4032.

Jang, M. and Pezzuto, JM. Cancer chemopreventive activity of resveratrol. (1999) *Drugs Under Experimental and Clinical Research*, 25, 65–77.

Jayaraman, P., Sakharkar, M.K., Lim, C.S., Tang, T.H. and Sakharkar, K.R. (2010) Activity and interactions of antibiotic and phytochemical combinations against Pseudomonas aeruginosa in vitro. *International Journal of Biolological Sciences*, 6, 556–568.

Jeong, S.J., Koh, W., Lee, E.O., Lee, H.J., Lee, H.J., Bae, H., Lü, J. and Kim, S.H. (2011) Antiangiogenic phytochemicals and medicinal herbs. *Phytotherapy Research*, 25, 1–10.

Johnson, S.M., Wang, X. and Evers, B.M. (2011) Triptolide inhibits proliferation and migration of colon cancer cells by inhibition of cell cycle regulators and cytokine receptors. *Journal of Surgical Research*, doi:10.1016/j.jss.2009.07.002.

Jones, J.M. and Anderson, J.W. (2008) Grain foods and health: a primer for clinicians. *Physician and Sportsmedicine*, 36, 18–33.

Kang, J., Badger, T.M., Ronis, M.J. and Wu, X. (2010) non-isoflavone phytochemicals in soy and their health effects. *Journal of Agricultural Food Chemistry*, 58, 8119–8133.

Katiyar, S.K., Korman, N.J., Mukhtar, H. and Agarwal, R. (1997) Protective effects of silymarin against photocarcinogenesis in a mouse skin model. *Journal of the National Cancer Institute*, 89, 556–565.

Kaur, M., Agarwal, C. and Agarwal, R. (2009) Anticancer and cancer chemopreventive potential of grape seed extract and other grape-based products. *Journal of Nutrition*, 139, 1806S–1812S.

Kawanishi, S., Oikawa, S. and Murata, M. (2005) Evaluation for safety of antioxidant chemopreventive agents. *Antioxidants and Redox Signaling*, 7, 1728–1739.

Kawashima, A., Madarame, T., Koike, H., Komatsu, Y. and Wise, J.A. (2007) Four week supplementation with mixed fruit and vegetable juice concentrates increased protective serum antioxidants and folate and decreased plasma homocysteine in Japanese subjects. *Asia Pacific Journal of Clinical Nutrition*, 16, 411–421.

Kim, E.C., Min, J.K., Kim, T.Y., Lee, S.J., Yang, H.O., Han, S., Kim, Y.M. and Kwon, Y.G. (2005) [6]-Gingerol, a pungent ingredient of ginger, inhibits angiogenesis in vitro and in vivo. *Biochemical and Biophysical Research Communications*, 335, 300–308.

King, A.A., Shaughnessy, D.T., Mure, K., Leszczynska, J., Ward, W.O., Umbach, D.M., Xu, Z., Ducharme, D., Taylor, J.A., Demarini, D.M. and Klein, C.B. (2007) Antimutagenicity of cinnamaldehyde and vanillin in human cells: global gene expression and possible role of DNA damage and repair. *Mutation Research*, 616, 60–69.

Kitagawa, S., Nabekura, T., Nakamura, Y., Takahashi, T. and Kashiwada, Y. (2007) Inhibition of P-glycoprotein function by tannic acid and pentagalloylglucose. *Journal of Pharmacy and Pharmacology*, 59, 965–969.

Kowalczyk, M.C., Kowalczyk, P., Tolstykh, O., Hanausek, M., Walaszek, Z. and Slaga, T.J. (2010) Synergistic effects of combined phytochemicals and skin cancer prevention in SENCAR mice. *Cancer Prevention Research*, 3, 170–178.

Kukongviriyapan, U., Luangaram, S., Leekhaosoong, K., Kukongviriyapan, V. and Preeprame, S. (2007) Antioxidant and vascular protective activities of Cratoxylum formosum, Syzygium gratum and Limnophila aromatica. *Biological and Pharmacuetical Bulletin*, 30, 661–666.

Kundu, J.K. and Surh, Y.J. Cancer chemopreventive and therapeutic potential of resveratrol: mechanistic perspectives. *Cancer Letters*, 269, 243–261.

Kunwar, R.M., Shrestha, K.P. and Bussmann, R.W. (2010) Traditional herbal medicine in far-west Nepal: a pharmacological appraisal. *Journal of Ethnobiolology and Ethnomedicine*, 6, 35.

Kuriyama, S. (2008) The relation between green tea consumption and cardiovascular disease as evidenced by epidemiological studies. *Journal of Nutrition*, 138, 1548S–1553S.

Kusari, S., Zühlke, S. and Spiteller, M. (2011) Chemometric evaluation of the anti–cancer pro-drug podophyllotoxin and potential therapeutic analogues in juniperus and podophyllum species. *Phytochemical Analysis*, 22, 128–143.

Kwon, Y.W., Masutani, H., Nakamura, H., Ishii, Y. and Yodoi, J. (2003) Redox regulation of cell growth and cell death. *Biological Chemistry*, 384, 991–996.

Lah, J.L., Cui, W. and Hu, K.Q. (2007) Effects and mechanisms of silibinin on human hepatoma cell lines. *World Journal of Gastroenterology*, 13, 5299–5305.

Lambert, J.D., Hong, J., Yang, G., Liao, J. and Yang, C.S. (2005) Inhibition of carcinogenesis by polyphenols: evidence from laboratory investigations. *American Journal of Clinical Nutrition*, 81, 284S–291S.

Lampe, J.W. (2007) Diet, genetic polymorphisms, detoxification, and health risks. *Alternative Therapies in Health and Medicine*, 13, S108–111.

Le Marchand, L. (2002) Cancer preventive effects of flavonoids. A review. *Biomedicine and Pharmacotherapy*, 56, 296–301.

Lee, H.S., Seo, E.Y., Kang, N.E. and Kim, W.K. (2008) [6]-Gingerol inhibits metastasis of MDA-MB-231 human breast cancer cells. *Journal of Nutritional Biochemistry*, 19, 313–319.

Lee, J., Shim, Y. and Zhu, B.T. Mechanisms for the inhibition of DNA methyltransferases by tea catechins and bioflavonoids. *Molecular Pharmacology*, 68 (4), 1018–1030.

Lentini, A., Tabolacci, C., Provenzano, B., Rossi, S. and Beninati, S. (2010) Phytochemicals and protein-polyamine conjugates by transglutaminase as chemopreventive and chemotherapeutic tools in cancer. *Plant Physiology and Biochemistry*, 48, 627–633.

Li, J., Zhang, D., Stoner, G.D. and Huang, C. (2008) Differential effects of black raspberry and strawberry extracts on BaPDE-induced activation of transcription factors and their target genes. Molecular Carcinogenesis, 47, 286–294.

Li, Y., Revalde, J.L., Reid, G. and Paxton, J.W. (2010) Interactions of dietary phytochemicals with ABC transporters: possible implications for drug disposition and multidrug resistance in cancer. *Drug Metabolism Reviews*, 42: 590–611.

Lin, Y. and Lin, J. (1997) (–)-Epigallocatechin-3-gallate blocks the induction of nitric oxide synthase by down-regulating lipopolysaccharide-induced activity of transcription factor nuclear factor-kappaB. *Molecular Pharmacology*, 52, 465–472.

Liu, R.H. (2004) Potential synergy of phytochemicals in cancer prevention: mechanism of action. *Journal of Nutrition*, 134, 3479S–3485S.

Loots, du T., van der Westhuizen, F.H. and Botes, L. (2007) Aloe ferox leaf gel phytochemical content, antioxidant capacity, and possible health benefits. *Journal of Agricultural and Food Chemistry*, 55, 6891–6896.

Ma, H., He, X., Yang, Y., Li, M., Hao, D. and Jia, Z. (2011) The genus Epimedium: An ethnopharmacological and phytochemical review. *Journal of Ethnopharmacology*, doi:10.1016/j.jep.2011.01.001.

Magee, P.J. and Rowland, I.R. (2004) Phyto-oestrogens, their mechanism of action: current evidence for a role in breast and prostate cancer. *British Journal of Nutrition*, 91, 513–531.

Manach, C., Scalbert, A., Morand, C., Rémésy, C. and Jimenez, L. (2004) Polyphenols: food sources and bioavailability. *American Journal of Clinical Nutrition*, 79, 727–747.

Manikandan, P., Murugan, R.S., Priyadarsini, R.V., Vinothini, G. and Nagini, S. (2010) Eugenol induces apoptosis and inhibits invasion and angiogenesis in a rat model of gastric carcinogenesis induced by MNNG. *Life Sciences*, 86, 936–941.

Manson, M.M., Foreman, B.E., Howells, L.M. and Moiseeva, E.P. (2007) Determining the efficacy of dietary phytochemicals in cancer prevention. *Biochemical Society Transactions*, 35, 1358–1363.

Masella, R., Di Benedetto, R., Varì, R., Filesi, C. and Giovannini, C. (2005) Novel mechanisms of natural antioxidant compounds in biological systems: involvement of glutathione and glutathione-related enzymes. *Journal of Nutrional Biochemistry*, 16, 577–586.

Masibo, M. and He, Q. Major mango polyphenols and their potential significance to human health. *Comprehensive Reviews in Food Science and Food Safety*, 7, 309–319.

Matés, J.M., Pérez-Gómez, C., Núñez de Castro, I., Asenjo, M. and Márquez, J. (2002) Glutamine and its relationship with intracellular redox status, oxidative stress and cell proliferation/death. *International Journal of Biochemistry and Cell Biology*, 34, 439–458.

Matés, J.M., Segura, J.A., Alonso, F.J. and Márquez, J. (2008) Intracellular redox status and oxidative stress: implications for cell proliferation, apoptosis, and carcinogenesis. *Archives of Toxicology*, 82, 273–299.

Matés, J.M., Segura, J.A., Alonso, F.J. and Márquez, J. (2009) Natural antioxidants, therapeutic prospects for cancer and neurological diseases. *Mini-Reviews in Medicinal Chemistry*, 9, 1202–1214.

Matés, J.M., Segura, J.A., Alonso, F.J. and Márquez, J. (2006) Pathways from glutamine to apoptosis. *Frontiers in Bioscience*, 11, 3164–3180.

Matés, J.M., Segura, J.A., Alonso, F.J. and Márquez, J. (2010) Roles of dioxins and heavy metals in cancer and neurological diseases using ROS-mediated mechanisms. *Free Radical Biology and Medicine*, 49, 1328–1341.

Matés, J.M., Segura, J.A., Campos-Sandoval, J.A., Lobo, C., Alonso, L., Alonso, F.J. and Márquez, J. (2009) Glutamine homeostasis and mitochondrial dynamics. *International Journal of Biochemistry and Cell Biology*, 41, 2051–2061.

Matés, J.M. (2000) Effects of antioxidant enzymes in the molecular control of reactive oxygen species toxicology. *Toxicology*, 153, 83–104.

Mattson, M.P. and Meffert, M.K. (2006) Roles for NF-kappaB in nerve cell survival, plasticity, and disease. *Cell Death and Differentiaion*, 13, 852–860.

Mattson, M.P. (2008) Dietary factors, hormesis and health. *Ageing Research Review*, 7, 43–48.

Mayne, S.T. (1996) Beta-carotene, carotenoids, and disease prevention in humans. *FASEB Journal*, 10, 690–701.

McWalter, G.K., Higgins, L.G., McLellanm L.I., Henderson, C.J., Song, L., Thornalley, P.J., Itoh, K., Yamamoto, M. and Hayes, J.D. (2004) Transcription factor Nrf2 is essential for induction of NAD(P) H:quinone oxidoreductase 1, glutathione S-transferases, and glutamate cysteine ligase by broccoli seeds and isothiocyanates. *Journal of Nutrition*, 134, 3499S–3506S.

Mehta, R.G., Murillo, G., Naithani, R. and Peng, X. (2010) Cancer chemoprevention by natural products: how far have we come? *Pharmaceutical Research*, 27, 950–961.

Michalska, M., Gluba, A., Mikhailidis, D.P., Nowak, P., Bielecka-Dabrowa, A., Rysz, J. and Banach, M. The role of polyphenols in cardiovascular disease. *Medical Science Monitor*, 16, RA110–119.

Middleton, E.J.R., Kandaswami, C. and Theoharides, T.C. (2000) The effect of plant flavonoids on mammalian cells: implications for inflammation, heart disease and cancer. *Pharmacology Review*, 52, 673–751.

Miura, Y., Chiba, T., Tomita, I., Koizumi, H., Miura, S., Umegaki, K., Hara, Y., Ikeda, M. and Tomita, T. (2001) Tea catechins prevent the development of atherosclerosis in apoprotein E-deficient mice. *Journal of Nutrition*, 131, 27–32.

Mizutani, K., Ikeda, K., Kawai, Y. and Yamori, Y. (1999) Extract of wine phenolics improves aortic biomechanical properties in stroke-prone spontaneously hypertensive rats (SHRSP). *Journal of Nutritional Science and Vitaminology*, 45, 95–106.

Moiseeva, E.P., Manson, M.M. (2009) Dietary chemopreventive phytochemicals: too little or too much? *Cancer Prevention Research*, 2, 611–616.

Mokni, M., Elkahoui, S., Limam, F., Amri, M. and Aouani, E. (2007) Effect of resveratrol on antioxidant enzyme activities in the brain of healthy rat. *Neurochemical Research*, 32, 981–987.

Montopoli, M., Ragazzi, E., Froldi, G. and Caparrotta, L. (2009) Cell-cycle inhibition and apoptosis induced by curcumin and cisplatin or oxaliplatin in human ovarian carcinoma cells. *Cell Proliferation*, 42, 195–206.

Mori, A., Lehmann, S., O'Kelly, J., Kumagai, T., Desmond, J.C., Pervan, M., McBride, W.H., Kizaki, M. and Koeffler, H.P. (2006) Capsaicin, a component of red peppers, inhibits the growth of androgen-independent, p53 mutant prostate cancer cells. *Cancer Research*, 66, 3222–3229.

Moskaug, J.Ø., Carlsen, H., Myhrstad, M.C. and Blomhoff, R. (2005) Polyphenols and glutathione synthesis regulation. *American Journal of Clinical Nutrition*, 81, 277S–283S.

Myhrstad, M.C., Carlsen, H., Nordstrom, O., Blomhoff, R. and Moskaug, J.Ø. Flavonoids increase the intracellular glutathione level by transactivation of the gamma-glutamylcysteine synthetase catalytical subunit promoter. *Free Radical Biology and Medicine*, 32, 386–393.

Nabekura, T., Yamaki, T., Hiroi, T., Ueno, K. and Kitagawa, S. (2010) Inhibition of anticancer drug efflux transporter P-glycoprotein by rosemary phytochemicals. *Pharmacology Research*, 61, 259–263.

Naczk, M. and Shahidi, F. (2006) Phenolics in cereals, fruits and vegetables: occurrence, extraction and analysis. *Journal of Pharmaceutical Research and Biomedical Analysis*, 41, 1523–1542.

Najeeb-Ur-Rehman, Mehmood, M.H., Alkharfy, K.M. and Gilani, A.U. (2011) Prokinetic and laxative activities of Lepidium sativum seed extract with species and tissue selective gut stimulatory actions. *Journal of Ethnopharmacology*, doi:10.1016/j.jep.2011.01.047.

Nakamura, Y., Yogosawa, S., Izutani, Y., Watanabe, H., Otsuji, E. and Sakai, T. (2009) *Molecular Cancer*, 8, 100.

Ning, X-H, Ding, X., Childs, K.F., Bolling, S.F. and Gallagher, K.P. (1993) Flavone improves functional recovery after ischemia in isolated reperfused rabbit hearts. *Journal of Thoracic and Cardiovascular Surgery*, 105, 541–549.

Nishino, H. (1990) Phytochemicals in hepatocellular cancer prevention. *Nutrition and Cancer*, 61, 789–791.

Oakenfull, D. and Sidhu, G.S. (1990) Could saponins be a useful treatment for hypercholesterolaemia? *European Journal of Clinical Nutrition*, 44, 79–88.

O'Keefe, J.H. Jr and Cordain, L. (2004) Cardiovascular disease resulting from a diet and lifestyle at odds with our Paleolithic genome: how to become a 21st-century hunter-gatherer. *Mayo Clinic Proceedings*, 79, 101–108.

Ortiz, D. and Shea, T.B. (2004) Apple juice prevents oxidative stress induced by amyloid-beta in culture. *Journal of Alzheimer's Disease*, 6, 27–30.

Pace-Asciak, C.R., Hahn, S., Diamandis, E.P., Soleas, G. and Goldberg, D.M. (1995) The red wine phenolics trans-resveratrol and quercetin block human platelet aggregation and eicosanoid synthesis: implications for protection against coronary heart disease. *Clinica Chimica Acta*, 31, 207–219.

Pan, M.H., Chang, Y.H., Badmaev, V., Nagabhushanam, K. and Ho, C.T. (2007) Pterostilbene induces apoptosis and cell cycle arrest in human gastric carcinoma cells. *Journal of Agricultural and Food Chemistry*, 55, 7777–7785.

Park, E.J. and Pezzuto, J.M. (2002) Botanicals in cancer chemoprevention. *Cancer and Metastasis Reviews*, 21, 231–255.

Patel, K.R., Brown, V.A., Jones, D.J., Britton, R.G., Hemingway, D., Miller, A.S., West, K.P., Booth, T.D., Perloff, M., Crowell, J.A., Brenner, D.E., Steward, W.P., Gescher, A.J. and Brown, K. (2010) Clinical pharmacology of resveratrol and its metabolites in colorectal cancer patients. *Cancer Research*, 70, 7392–7399.

Paul, S., DeCastro, A.J., Lee, H.J., Smolarek, A.K., So, J.Y., Simi, B., Wang, C.X., Zhou, R., Rimando, A.M. and Suh, N. (2010) *Carcinogenesis*, 31, 1272–1278.

Pellegrini, N., Serafini, M., Salvatore, S., Del Rio, D., Bianchi, M. and Brighenti, F. (2006)Total antioxidant capacity of spices, dried fruits, nuts, pulses, cereals and sweets consumed in Italy assessed by three different in vitro assays. *Molecular Nutrition and Food Research*, 50, 1030–1038.

Purkayastha, S., Berliner, A., Fernando, S.S., Ranasinghe, B., Ray, I., Tarig, H. and Banarjee, P. (2009) Curcumin blocks brain tumor formation. *Brain Research*, 1266, 130–8.

Ramakrishnan, G., Lo Muzio, L., Elinos-Báez, C.M., Jagan, S., Augustine, T.A., Kamaraj, S., Anandakumar, P. and Devaki, T. (2009) Silymarin inhibited proliferation and induced apoptosis in hepatic cancer cells. *Cell Proliferation*, 42, 229–240.

Rangkadilok, N., Worasuttayangkurn, L., Bennett, R.N. and Satayavivad, J. (2005) Identification and quantification of polyphenolic compounds in Longan (Euphoria longana Lam.) fruit. *Journal of Agricultural and Food Chemistry*, 53, 1387–92.

Rao, B.N. (2003) Bioactive phytochemicals in Indian foods and their potential in health promotion and disease prevention. *Asia Pacific Journal of Clinical Nutrition*, 12, 9–22.

Reiter, E., Azzi, A. and Zingg, J.M. (2007) Enhanced anti-proliferative effects of combinatorial treatment of delta-tocopherol and resveratrol in human HMC-1 cells. *Biofactors*, 30, 67–77.

Riccioni, G., Mancini, B., Di Ilio, E., Bucciarelli, T. and D'Orazio, N. (2008) Protective effect of lycopene in cardiovascular disease. *European Review for Medical and Pharmacolical Sciences*, 12,183–90.

Rice-Evans, (2001) C. Flavonoid antioxidants. *Current Medicinal Chemistry*, 8, 797–807.

Ridker, P.M., Hennekens, C.H., Roitman-Johnson, B., Stampfer, M.J. and Allen, J. (1998) Plasma concentration of soluble intercellular adhesion molecule 1 and risks of future myocardial infarction in apparently healthy men. *Lancet*, 351, 88–92.

Rosen, C.A. and Bryson, P.C. (2004) Indole-3-carbinol for recurrent respiratory papillomatosis: long-term results. *Journal of Voice*, 18, 248–253.

Ross, J.A., Potter, J.D. and Robison, L.L. (1994) Infant leukemia, topoisomerase II inhibitors, and the MLL gene. *Journal of the National Cancer Institute*, 86, 1678–1680.

Salas-Salvadó, J., Bulló, M., Pérez-Heras, A. and Ros, E. (2006) Dietary fibre, nuts and cardiovascular diseases. *British Journal of Nutrition*, 96, S46–S51.

Sarkar, F.H. and Li, Y. Using chemopreventive agents to enhance the efficacy of cancer therapy. *Cancer Research*, 66, 3347–3350.

Sato, M., Maulik, G., Ray, P.S., Bagchi, D. and Das, D.K. (2000) Cardioprotective effect of grape seed proanthocyanidin against ischemic reperfusion injury. *Journal of Mollecular and Cellular Cardiology*, 31, 1289–1297.

Scalbert, A. and Williamson, G. (2000) Dietary intake and bioavailability of polyphenols. *J Nutr* 130:2073S–85S.

Seeram NP. Berry fruits: compositional elements, biochemical activities, and the impact of their intake on human health, performance, and disease. *J Agric Food Chem* 56:627–9 (2008).

Seeram NP. Recent trends and advances in berry health benefits research. *Journal of Agricultural and Food Chemistry*, 58:3869–70 (2010).

Seifried HE, Anderson DE, Fisher EI, Milner JA. A review of the interaction among dietary antioxidants and reactive oxygen species. *Journal of Nutritional Biochemistry*, 18:567–79 (2007).

Seifried, H.E., Anderson, D.E., Milner, J.A. and Greenwald, P. (2006) Reactive oxygen species and dietary antioxidants: double-edged swords?. In: H. Panglossi (ed.) *New Developments in Antioxidant Research*, Hauppauge (NY), Nova Science Publishers Inc, pp. 1–25.

Sharma, R.A., Euden, S.A. and Platton, S. (2004) Phase I clinical trial of oral curcumin: biomarkers of systemic activity and compliance. *Clinical Cancer Research*, 10, 6847–5684.

Shin, H.K., Kim, J., Lee, E.J. and Kim, S.H. Inhibitory effect of curcumin on motility of human oral squamous carcinoma YD-10B cells via suppression of ERK and NF-kappaB activations. *Phytotherapy Research*, 24:577–82 (2010).

Siddiqui, I.A., Adhami, V.M., Bharali, D.J., Hafeez, B.B., Asim, M., Khwaja, S.I., Ahmad, N., Cui, H., Mousa, S.A. and Mukhtar, H. (2009) Introducing nanochemoprevention as a novel approach for cancer control: proof of principle with green tea polyphenol epigallocatechin-3-gallate. *Cancer Research*, 69, 1712–1716.

Singh, D., Singh, B. and Goel, R.K. (2011) Traditional uses, phytochemistry and pharmacology of Ficus religiosa: A review. *Journal of Ethnopharmacology*, doi:10.1016/j.jep.2011.01.046.

Skibola, C.F. and Smith, M.T. (2000) Potential health impacts of excessive flavonoid intake. *Free Radicical Biology and Medicine*, 29, 375–383.

Slavin, J. (2004) Whole grains and human health. *Nutritional Research Review*, 17, 99–110.

Slavin, J. (2003) Why whole grains are protective: biological mechanisms. *Proceedings of the Nutrition Society*, 62, 129–134.

Sloan, A.E. (2002) The top 10 functional food trends: The next generation. *Food Technology*, 56, 32–57.

Somasundaram, S., Edmund, N.A., Moore, D.T., Small, G.W., Shi, Y.Y. and Orlowski, R.Z. (2002) Dietary curcumin inhibits chemotherapy-induced apoptosis in models of human breast cancer. *Cancer Research*, 62, 3868–3875.

Squires, M.S., Hudson, E.A., Howells, L., Sale, S., Houghton, C.E., Jones, J.L., Fox, L.H., Dickens, M., Prigent, S.A. and Manson, M.M. Relevance of mitogen activated protein kinase (MAPK) and phosphotidylinositol-3-kinase/protein kinase B (PI3K/PKB) pathways to induction of apoptosis by curcumin in breast cells. *Biochemical Pharmacology*, 65, 361–376.

Staker, B.L., Hjerrild, K., Feese, M.D., Behnke, C.A., Burgin, Jr, A.B. and Stewart, L. (2002) The mechanism of topoisomerase I poisoning by a camptothecin analog. *Proceedings of the National Academy of Science USA*, 99, 5387–5392.

Stoclet, J.C., Andriambeloson, E. and Andriantsitohaina, R. (2000) Endothelial NO release caused by red wine polyphenols with specific structure. *Experimental and Clinical Cardiology*, 5, 24–27.

Stoner, G.D., Kresty, L.A., Carlton, P.S., Siglin, J.C. and Morse, M.A. (1999) Isothiocyanates and freeze-dried strawberries as inhibitors of esophageal cancer. *Toxicological Sciences*, 52, 95–100.

Strick, R., Strissel, P.L., Borgers, S., Smith, S.L. and Rowley, J.D. (2000) Dietary bioflavonoids induce cleavage in the MLL gene and may contribute to infant leukemia. *Proceedings of the National Academy of Science USA*, 97, 4790–4795.

Su, C.C., Yang, J.S., Lin, S.Y., Lu, H.F., Lin, S.S., Chang, Y.H., Huang, W.W., Li, Y.C., Chang, S.J. and Chung, J.G. (2008) Curcumin inhibits WEHI-3 leukemia cells in BALB/c mice in vivo. *In Vivo* 22, 63–68.

Sumiyoshi, H. (1997) New pharmacological activities of garlic and its constituents. *Nippon Yakurigaku Zasshi*, 110, 93P–97P.

Teskac, K. and Kristl, J. The evidence for solid lipid nanoparticles mediated cell uptake of resveratrol. *International Journal of Pharmacology*, 390, 61–69.

Thomasset, S.C., Berry, D.B., Garcea, G., Marczylo, T., Steward, W.P. and Gescher, A.J. Dietary polyphenolic phytochemicals--promising cancer chemopreventive agents in humans? A review of their clinical properties. *International Journal of Cancer*, 120, 451–458.

Thompson, L.U. (1994) Antioxidant and hormone-mediated health benefits of whole grains. *Critical Reviews in Food Science and Nutrition*, 34, 473–497.

Tian, L., Shi, M.M. and Forman, H.J. (1997) Increased transcription of the regulatory subunit of gamma-glutamylcysteine synthetase in rat lung epithelial L2 cells exposed to oxidative stress or glutathione depletion. *Archives of Biochemistry and Biophysics*, 342:126–133.

Tirapelli, C.R., Ambrosio, S.R., da Costa, F.B. and de Oliveira, A.M. (2008) Diterpenes: a therapeutic promise for cardiovascular diseases. *Recent Patents on Cardiovascular Drug Discovery*, 3, 1–8.

Tsao, A.S., Kim, E.S. and Hong, W.K. (2004) Chemoprevention of cancer. *Cancer Journal for Clinicians*, 54, 150–180.

Vitrac, X., Bornet, A., Vanderlinde, R., Valls, J., Richard, T., Delaunay, J.C., Mérillon, J.M. and Teissédre, P.L. (2005) Determination of stilbenes (delta-viniferin, trans-astringin, trans-piceid, cis- and trans-resveratrol, epsilon-viniferin) in Brazilian wines. *Journal of Agricultural and Food Chemistry*, 53, 5664–5669.

von Metzler, I., Krebbel, H., Kuckelkorn, U., Heider, U., Jakob, C., Kaiser, M., Fleissner, C., Terpos, E. and Sezer, O. (2009) Curcumin diminishes human osteoclastogenesis by inhibition of the signalosome-associated I kappaB kinase. *Journal of Cancer Research and Clinical Oncology*, 135, 173–179.

Vuddanda, P.R., Chakraborty, S. and Singh, S. (2010) Berberine: a potential phytochemical with multispectrum therapeutic activities. *Expert Opinion on Investigational Drugs*, 19, 1297–1307.

Wang, Y.P., Wu, J.X., Zhang, F.L. and Wang, X.R. (1993) Effect of soyasaponin and ginsenoside (stem-leave saponin) on SOD and LPO in diabetic rats. *Journal of Baiqiuen Medical University*, 19, 122–123.

Watanabe, S., Zhuo, X.G. and Kimira, M. (2004) Food safety and epidemiology: new database of functional food factors. *Biofactors*, 22, 213–219.

Wollny, T., Aiello, L., Di Tomasso, D. et al. (1999) Modulation of haemostatic function and prevention of experimental thrombosis by red wine in rats: a role for increased nitric oxide production. *British Journal of Pharmacology*, 127, 747–755.

Xiao, J.X., Peng, G.H. and Zhang, S.H. (2005) Prevention effects of soyasaponins on hyperlipidemia mice and its molecular mechanism. *Acta Nutrimenta Sinica*, 27, 147–150.

Xu, Y., Khaoustov, V.I., Wang, H., Yu, J., Tabassam, F. and Yoffe, B. (2009) Freeze-dried grape powder attenuates mitochondria- and oxidative stress-mediated apoptosis in liver cells. *Journal of Agricultural and Food Chemistry*, 57, 9324–9331.

Yang, C.S., Wang, X., Lu, G. and Picinich, S.C. Cancer prevention by tea: animal studies, molecular mechanisms and human relevance. *Nature Reviews Cancer*, 9, 429–439.

Yang, G., Zhong, L., Jiang, L., Geng, C., Cao, J., Sun, X. and Ma, Y. (2010) 6-Gingerol caused DNA strand breaks in HepG2 cells. These genotoxic effects are probably associated with oxidative stress. Genotoxic effect of 6-gingerol on human hepatoma G2 cells. *Chemico-biological Interactions*, 185, 12–17.

Yang, J. and Liu, R.H. (2009) Synergistic effect of apple extracts and quercetin 3-beta-d-glucoside combination on antiproliferative activity in MCF-7 human breast cancer cells in vitro. *Journal of Agricultural and Food Chemistry*, 57, 8581–8586.

Yankah, V.V. and Jones, P.J.H. (2001) Phytosterols and health implications-efficacy and nutritional aspects. *Inform*, 12, 899–903.

Zamin, L.L., Filippi-Chiela, E.C., Dillenburg-Pilla, P., Horn, F., Salbego, C. and Lenz, G. Resveratrol and quercetin cooperate to induce senescence-like growth arrest in C6 rat glioma cells. *Cancer Science*, 100, 1655–1662.

Zenebe, W., Pechánová, O. and Bernátová, I. (2001)_Protective effects of red wine polyphenolic compounds on the cardiovascular system. *Experimental and Clinical Cardiology*, 6, 153–158.

Zenebe, W. and Pechanova, O. (2002) Effects of red wine polyphenolic compounds on the cardiovascular system. *Bratislavske Lekarske Listy*, 103, 159–165.

Zenk, M.H. and Juenger, M. Evolution and current status of the phytochemistry of nitrogenous compounds. *Phytochemistry*, 68, 2757–2772.

Zern, T.L., Wood, R.J., Greene, C., West, K.L., Liu, Y., Aggarwal, D., Shachter, N.S. and Fernandez, M.L. (2005) Grape polyphenols exert a cardioprotective effect in pre- and postmenopausal women by lowering plasma lipids and reducing oxidative stress. *Journal of Nutrition*, 135, 1911–1917.

Zhu, B.H., Zhan, W.H., Li, Z.R., Wang, Z., He, Y.L., Peng, J.S., Cai, S.R., Ma, J.P. and Zhang, C.H. (2007) Epigallocatechin-3-gallate inhibits growth of gastric cancer by reducing VEGF production and angiogenesis. *World Journal of Gastroenterology*, 13, 1162–1169.

Part II
Sources of Phytochemicals

5 Fruit and vegetables

Uma Tiwari and Enda Cummins

UCD School of Biosystems Engineering, Agriculture and Food Science Centre, Belfield, Dublin, Ireland

5.1 Introduction

Fruit and vegetables are rich sources of phytochemicals with many reported human health promoting benefits beyond basic nutrition. There is an emerging interest among food researchers/manufacturers for developing novel food products by incorporating phytochemicals (either in raw or extracted form) in order to meet increasing consumer demands for functional foods. Fruits and vegetables contain a range of antioxidant compounds showing synergistic effects, which may contribute to protection against oxidative damage (Lako et al., 2007). A number of epidemiological studies have identified an inverse association between the consumption of fruit and vegetables with reduced risk for several chronic diseases. It is hypothesised (with still much debate) that phytochemicals play a central role in this positive effect. A range of phytochemicals have been reported in fruit and vegetables and are typically grouped based on function, chemical structure and also based on source. Classification of phytochemicals has been discussed in Chapter 2. This chapter focuses on various sources of phytochemicals present in fruit and vegetables, including an appraisal of typical concentrations and influencing factors.

5.2 Polyphenols

Polyphenols are a major group of phytochemicals and are sub-classified into two main groups: phenolics and flavonoids. A wide range of phenolic compounds and flavonoids are reported in fruit and vegetables. Both phenol and flavonoid content in fruit and vegetables can be influenced by variety, environmental and growing conditions, maturity stages and harvesting factors (Marín et al., 2004; Nazk and Shahidi, 2006; Hogan et al., 2009; Song et al., 2010). Table 5.1 details the range for total phenolic, flavonoid and anthocyanin content of different fruit and vegetables found in different scientific studies.

Handbook of Plant Food Phytochemicals: Sources, Stability and Extraction, First Edition.
Edited by B.K. Tiwari, Nigel P. Brunton and Charles S. Brennan.
© 2013 John Wiley & Sons, Ltd. Published 2013 by John Wiley & Sons, Ltd.

Table 5.1 Total phenolic, flavonoid and anthocyanin content of fruit and vegetables

Fruit	Total Phenolics (mg GAE/100g FW)	Total Flavonoids (mg quercetin or mg CTE /100g FW)	Total Anthocyanins (mg of Cya-3-glu/ 100g FW)	Reference
Strawberries	159 to 385	50 to 70.5	41.4[a]	Sun et al. (2002); Meyers et al. (2003); Chun et al. (2005); Aaby et al. (2007)
Blackberries	389 to 2809	41 to 54	52 to 126	Sellappan et al. (2002); Acosta-Montoya et al. (2010); Sariburun et al. (2010)
Blueberries	42 to 773		44 to 301	Sellappan et al. (2002); Cevallos-Casals and Cisneris-Zevallos (2003)
Raspberries	359 to 2066	24 to 103.4	0.17 to 57.60	Liu et al. (2002); Pantelidis et al. (2007); Sariburun et al. (2010)
Cranberries	506 to 549	8.3 to 12		Häkkinen et al. (1999); Sun et al. (2002)
Cherries	210 to 306		148 to 228	Pantelidis et al. (2007); Yilmaz et al. (2009)
Plums	42 to 684	55 to 366	121 to 129	Chun et al. (2003); Cevallos-Casals et al. (2006); Slimestad et al. (2009)
Peach	409 to 489		31 to 35	Cevallos-Casals et al. (2006)
Grapes	63 to 428	25 to 308	7 to 264	Chun et al. (2005); Hongan et al. (2009)
Apple	120 to 303	55 to 69		Sun et al. (2002); Chun et al. (2005)
Bananas	87 to 217	33 to 35		Sun et al. (2002); Chun et al. (2005); Isabelle et al. (2010)
Oranges	82 to 138	9 to 15		Sun et al. (2002); Chun et al. (2005); Isabelle et al. (2010)

Vegetables

Vegetable	Total Phenolics (mg GAE/100g FW)	Total Flavonoids (mg quercetin or mg CTE /100g FW)	Total Anthocyanins (mg of Cya-3-glu/ 100g FW)	Reference
Lettuce-red	322 to 571[b]	138[c]	26 to 46[d]	Ferreres et al. (1997); Llorach et al. (2008)
Lettuce-green	18 to 126[b]	24.4c		Ferreres et al. (1997); Llorach et al. (2008)
Lettuce-white	21.3[b]	4.3[c]		Ferreres et al. (1997)
Cabbage-red	131 to 679		38 to 80	Amin and Lee (2005); Podsędek et al. (2006)
Broccoli	44.5 to 82.9	2 to 5.41		Chun et al. (2005); Singh et al. (2007); Koh et al. (2009)
Onion	73.3	4.7		Chun et al. (2005); Pande and Akoh (2009)
Tomatoes	23 to 24	1.7 to1.9		Chun et al. (2005)
Potato	6.9 to 9.9			Rumbaoa et al. (2009)
Sweet potato	0.13 to 42.23	5.6 to 6.1	1.70 to 53.10	Chun et al. (2005); Teow et al. (2007)
Bell peppers	51 to 54	1		Chun et al. (2005)
Carrots	9	1		Chun et al. (2005)

GAE: gallic acid equivalent; CTE: catechin equivalent; Cya-3-glu: cyanidin 3-glucoside; FW: fresh weight; [a]Pelargonidin-3-glucoside; [b]Caffeic acid derivatives; [c]quercetin-3-glucosides; [d]cyanidin-3-rutinoside.

In general, phenolic compounds accumulate in the peel, as opposed to the pulp of the fruit and vegetables (George *et al.*, 2004; Pande and Akoh, 2009). Phenolic compounds can contribute to the astringency and bitter taste of the food product. Berries are known for their high content of phenolic compounds, however this will vary within cultivars. From a survey of 11 selected fruit, Sun *et al.* (2002) reported that total phenolics (including the soluble free and bound phenolics) was highest in cranberries, at about 527 mg/100 g FW, followed by other fruit such as apple, red grape, strawberry, peach, lemon, pear, banana, orange, grapefruit and pineapple. Kalt *et al.* (1999) studied the phenolic content of cultivars of fresh strawberries, raspberries and blueberries. They observed that the total phenolic content of blueberries was about four-fold higher when compared to strawberries and raspberries. Among blueberries, wild cultivars are known to posses higher phenolic content (600 mg Gallic Acid Equivalent (GAE)/100 g) compared to the cultivated blueberries (250–310 mg GAE/100 g) (Giovannelli and Buratti, 2009). The phenolic content of blackberries ranged an average of 833 mg of GAE/100 g in whole fruit compared to 1607 mg of GAE/100 g in seed (Iriwoharn and Wrolstad, 2004).

Among berries, the phenolic content is reported to result in development of colour of the skin. For example, Liu *et al.* (2002) observed the variations in phenolic content of dark red 'Heritage' and yellow 'Anne' raspberries. Dark red raspberries showed a higher phenolic content of about 513 mg/100 g compared to 359 mg/100 g for yellow raspberries, indicating the relationship between phenol concentration and colour formation. Cevallos-Casals and Cisneros-Zevallos (2003) showed extremely high phenolic content among red flesh sweet-potato cultivars of 945 mg Chlorogenic Acid Equivalent (CAE)/100 g FW, whereas approximately 95 mg of CAE/100 g FW was found in purple-fleshed sweet potatoes, further indicating the relationship between phenolics and colour development. Total phenols are concentrated more in the skin of fruit compared to the flesh. For example, Cevallos-Casals *et al.* (2006) showed a three- to four-fold higher phenolic concentration in the skin than in the flesh among plum cultivars, ranging from 292 to 672 mg cholorgenic acid/100 g FW. It is also reported that the seed of various fruit contains higher phenolic compounds compared to skin and flesh, for example the total phenolic content among muscadine grapes is about five times more in the seed compared to the grape skin and about 80 times more than in the pulp (Pastrana-Bonilla *et al.*, 2003). In a study, Aaby *et al.* (2005) reported that the total phenolic content of fresh strawberries ranges from 230 to 340 mg GAE/100 g FW among 'Totem' and 'Puget Reliance' cultivars, while the strawberry achenes (the seeds of a strawberries) contained a high amount of total phenolics (~3600 mg GAE/100 g FW). Similarly, apple peels are reported to contain 1.5- to 2.5-fold higher phenolic compounds compared to the whole apple fruit (Valavanidis *et al.*, 2009). Amongst the *Brassica* family, red cabbage is shown to contain a high amount of total phenolics with about 679 mg GAE/100 g FW compared to green cabbage, which was 224 mg GAE/100 g FW (Amin and Lee, 2005). In addition, the purple variety of cauliflower 'Gragitti' contained 146 mg GAE/100 g FW, which is approximately twice that of the white and green cauliflower cultivars (Volden *et al.*, 2009). A higher value of total phenolic acid was obtained for purple coloured carrots (75 mg/100 g) compared to orange, white or yellow cultivars (Alasalvar *et al.*, 2001). In a study, Ferreres *et al.* (1997) reported the variations within the white, green and red tissue of 'Lollo Roso' lettuce with phenolic content of 21.3 mg/100 g FW, 57.0 mg/100 g FW and 169.6 mg/100 g FW (Ferreres *et al.*, 1997), respectively.

The phenolic content of fruit also varies depending on the maturity stage. For example, a decrease in the ellagic acid content of strawberries has been reported with green fruit (142 mg/100 g DW), is intermediate in mid-ripe fruit (72 mg/100 g DW) and lowest in full-ripe fruit (37 mg/100 g DW) (Williner et al., 2003). A similar decrease of about 33% from red fruit to fully ripe blackberry fruit is also reported by Acosta-Montoya et al. (2010). Çelik et al. (2008) reported a decrease in total phenolic concentration from 799 to 475 mg GAE/100 g FW during the green to dark red stage of cranberry maturity. Likewise, Marín et al. (2004) indicated that immature green peppers contain four- to five-fold higher phenolic compounds compared to green, immature red and red ripe peppers.

Flavonoids and their derivatives (flavonols, flavones, flavanols, flavanones, anthocyanidins and isoflavones) are the largest group of phenolic compounds or polyphenols. The majority of flavonols and flavones mostly occur in bound form and are present in fresh fruit and vegetables (Hollman and Arts, 2000). Among flavonoids, quercetins constitute the most abundant group of phenolic phytochemicals and are widespread in fruit and vegetables. In a survey, Häkkinen et al. (1999) surveyed the flavonoid contents of 25 edible berries and reported that the quercetin level was high among all berries studied. Häkkinen et al. (1999) showed that wild berries contain higher levels of quercetin concentration (1.7–14.6 mg/100 g FW) followed by cranberry (8.3–12 mg/100 g FW) with raspberry and strawberry containing the lowest (0.6–0.8 mg/100 g FW, respectively). Sellappan et al. (2002) demonstrated that catechin was the major flavonoid among blueberries with concentrations of up to 388 mg/100 g FW. Within fruit, achenes of strawberry are reported to contain about four-fold higher levels of catechin and flavonols compared to the flesh of strawberries (Aaby et al., 2005). As with phenolic compounds there is a correlation with flavonol content and colour formation, for example the dark red raspberry variety was shown to contain high flavonoids (103 mg/100 g FW) compared to pink-red and yellow varieties (Liu et al., 2002). The total flavonoids in raspberry and blackberry cultivars is reported to vary from 15.41 to 41.08 mg Catechin Equivalent (CTE)/100 g FW in raspberries and from 29.07 to 82.21 mg CTE/100 g FW for blackberries (Sariburun et al., 2010). Catechin is reported to be the predominant flavonoid present in pomegranate cultivars and is mostly concentrated in the peel of the fruit (Pande and Akoh, 2009). The quercetin glycosides are the major flavonoids present in apple ranging from 22 to 35 mg/100 g FW (Tsao et al., 2003), whereas the total flavonoid content of fresh apple peel is reported to be 306 mg CTE/100 g (Wolfe and Liu, 2003). Conversely, in grapes quercetin content ranges from 6.3 to 22 mg/100 g FW, with about 13 times greater concentration in the whole fruit and six times greater in the skin (Pastrana-Bonilla et al., 2003).

Among vegetables, onions are well-known sources of flavonoid antioxidants, in particular quercetin and its glycosides. Marotti and Piccaglia (2002) reported that among 12 onion cultivars, including yellow, red and white types, total flavonoid content ranged from 0.12 mg/100 g (white onion) to 98.0 mg/100 g (yellow onions). The rutin (quercetin-3-O-rutinoside) and kaempferol-3-O-rutinoside are the main compounds of flavonoid present in potatoes with concentrations ranging from approximately 19.1 to 22.7 mg/100 g of DW (Andre et al., 2007). Stewart et al. (2000) reported that the majority of the rutin forms of flavonols in tomatoes are present in the skin with about 20.4 mg rutin equivalent/100 g FW. The same authors also stated that tomato flesh and seed contain very low amounts of flavonoid with about 0.012 and 0.15 mg rutin equivalent/100 g FW, respectively (Stewart et al., 2000). Leafy vegetables such as cabbage also contain a significant amount of flavanols. Podsędek et al. (2006) reported significant variations in the range of 0.03–1.38 mg/100 g of FW for white and savoy cabbages. Table 5.2 shows the reported levels of flavonols (quercetin, kaempferol and myricetin) present in some common fruit and vegetables. Flavanol composition also varies from vegetable to vegetable, for example kaempferol and myricetin derivatives were reported in *Brassica* vegetables whereas myricetin was not reported in broccoli, white cabbage, purple cabbage or cauliflower (USDA, 2007; Koh et al., 2009). Lee

Table 5.2 Flavonol content in different fruit and vegetables

Fruit or vegetable	Flavonols (mg/100g FW)			Reference
	Quercetin	Kaempferol	Myricetin	
Black currant	4.4		7.1	Häkkinen et al. (1999)
Green currant	3.2			Häkkinen et al. (1999)
Red currant	0.9			Häkkinen et al. (1999)
Berries	10.7	0.4	5.7	USDA (2007)
Cranberry	2.9 to 3.0		4.3 to 14.2	Häkkinen et al. (1999); USDA (2003)
Cranberry-seed	10.6 to 11.2			Pande and Akoh (2009)
Cranberry-peel	92.1 to 99.2			Pande and Akoh (2009)
Cranberry-pulp	66.7 to 77.1			Pande and Akoh (2009)
Blueberry	5.82 to 14.6	2.51 to 3.72	6.68 to 8.62	Sellappan et al. (2002)
Strawberry	0.7	0.8		Häkkinen et al. (1999)
Raspberry-red	0.8			Häkkinen et al. (1999)
Grapes-red (whole fruit)	3.1 to 5.5	0.8 to 1.4	1.3 to 2.6	Montealegre et al. (2006)
Grapes-bronze (whole fruit)	0.4 to 1.8	0.1 to 1.4	1.8 to 6.3	Pastrana-Bonilla et al. (2003)
Grapes-bronze (skins)	0.9 to 3.8	0.2 to 3.0	4.1 to 19.6	Pastrana-Bonilla et al. (2003)
Grapes-purple (whole fruit)	0.2 to 1.4	0.1 to 0.2	0.7 to 2.8	Pastrana-Bonilla et al. (2003)
Grapes-purple (skins)	0.5 to 3.0	0.2 to 0.4	1.8 to 6.4	Pastrana-Bonilla et al. (2003)
Spinach		7.6		USDA (2007)
Lettuce leaf	7.1			USDA (2007)
Broccoli	0.03 to 10.85	0.24 to 13.20		Koh et al. (2009)
Cabbage	2.61	1.30 to 7.03		Kim et al. (2004)
Pepper	0.089 to 4.99			Lee et al. (2005)
Tomato	2.76	0.10	0.03	USDA (2007)
Onion-white	8.93 to 10.1			Perez-Gregorio et al. (2010)
Onion-red	21.17 to 28.02			Perez-Gregorio et al. (2010)
Onion-yellow	18.97			Grzelak et al. (2009)

et al. (2005) showed that quercetin and luteolin concentrations varied with geographical location among pepper cultivars. Many studies showed a significant change in flavonoid content with an increase during the maturity of a fruit or vegetable. Acosta-Montoya *et al.* (2010) showed that following maturity flavonols reduced from 5.1 mg quercetin equivalent/100 g FW (light red fruit) to 2.0 mg quercetin equivalent/100 g FW for dark bluish-purple blackberries. Similarly, Slimestad and Verheul (2009) reported that the naringenin level increased from 1.2 mg/100 g (green fruit) to 4.0 mg/100 g FW in ripe fruit and finally declined to 2.3 mg/100 g FW in overripe tomato fruit.

Anthocyanins are naturally occurring plant metabolites belonging to the flavonoid group and comprise of intense coloured pigments which results in the orange, red, purple and blue colours of many fruit, vegetables and their plant tissues. These colour pigments of anthocyanin serve as important alternatives for synthetic food colorants in the food industry (Clifford, 2000). The colour of anthocyanins changes depending on the pH and level in fruit and vegetables. Anthocyanin lacks stability and degrades during thermal and non thermal processing and storage (Tiwari *et al.*, 2009). There are various forms of anthocyanin that are reported among fruit and vegetables, for example in strawberries the pelargonidin-3-glucoside contributes nearly 76% of the total anthocyanin content (Aaby et al., 2007). Approximately 48–81% of cyanidin-3-rutinoside is reported to be present in fig skin and pulp (Dueñas et al., 2008) whereas more than 60% is reported to be found in plum cultivars (Slimestad et al., 2009). Cyanidin-3-rutinoside is also a predominant anthocyanin found in red and blue-purple plums (Chun et al., 2003; Kim et al., 2003). For vegetables, Lewis et al. (1999) found that red tubers contained mostly pelagonidin-3-(p-coumaroyl-rutinoside)-5-glucoside whereas cyanindin 3-malonylglucoside has been isolated and identified as the main anthocyanin pigment in red lettuce tissues (Ferreres et al., 1997). Acylated anthocyanins are shown to have better stability and impart natural food colourant to food materials (Giusti and Worstad, 2003). For example, pelargonidin-3-rutinoside-5-glucoside is commonly found in red radishes (Giusti et al., 1998) and red potatoes (Rodriguez-Saona et al., 1998). Netzel et al. (2007) observed that pepper contains a high amount of anthocyanin with nearly 73% being cyanidin-3-rutinoside. Investigations showed that environmental factors also influence the level of anthocyanins. Awad et al. (2001) found that apple exposed to sun contains higher levels of anthocyanins and quercetin glycosides compared to the shaded fruit. Likewise, Schwartz et al. (2009) demonstrated that high temperatures within a geographical location can reduce the anthocyanin content in both the peels and arils of pomegranate fruit. The authors suggest that light intensity would also have influenced the levels of anthocyanin in the pomegranate peels. In addition, cultivar variations also influence the anthocyanin level. Meyers et al. (2003) showed a huge anthocyanin variance from 21.9 to 41.4 mg Cya-3-gly/100 g FW among strawberries indicating a strong difference among eight cultivars. Similarly, Giusti et al. (1998) demonstrated cultivar variations for anthocyanin content among radish cultivars grown in spring or winter. The study by Giusti et al. (1998) noted that, on average, radish contains 39.3–185 mg anthocyanin/100 g in the skin among spring cultivars whereas the red-fleshed winter cultivars contained 12.2–53 mg anthocyanin/100 g roots.

Generally anthocyanins are more concentrated in the skin than the flesh: this is clear from the fact that anthocyanins are responsible for the colour of fruit and vegetables. Cevallos-Casals et al. (2006) showed that plum skin contains a three- to nine-fold higher anthocyanin content compared to plum flesh. Total anthocyanin content of strawberries is reported to be about 53 mg/100 g FW out of which 40 mg is pelargonidin-3-glucoside/100 g of FW (Aaby et al., 2007). In the case of blackberries, the total anthocyanin content ranged from 190 mg Cya-3-glu/100 g FW in whole fruit to 13 mg Cya-3-glu/100 g FW in seeds (Iriwoharn and Wrolstad, 2004). The difference in the colour of raspberries is influenced by the anthocyanin content, for example yellow raspberries contain significantly low anthocyanin (1.3–7.8 mg Cya-3-glu/100 g FW) compared to red raspberries (35–49 mg/100 g FW) (Pantelidis et al., 2007). Likewise, the anthocyanin content of grapes varies with cultivar, skin colour and with different anatomical parts of the fruit. Pastrana-Bonilla et al. (2003) reported that the purple skinned muscadine grapes showed a higher level of anthocyanin content ranging from 65.5 to 177.0 Cya-3-glu/100 g FW. The study by Pastrana-Bonilla et al. (2003) also showed that the skin of purple grapes contains about 65 times more anthocyanins than that of bronze grapes. The total anthocyanin content in the seeds of purple grapes was about 1.3 times higher than in bronze grape seeds, and the pulps of purple grapes had, on average, 2 mg Cya-3-glu/100 g of FW (Pastrana-Bonilla et al., 2003). Duan et al. (2007) showed that the total anthocyanin content of litchi fruit ranged from 20 to 60 mg/100 g FW with nearly 85% in pelargonidin-3-glucoside, while about 94% of total anthocyanins found in litchi pericarp are in a form of cyanidin-3-rutinside. The occurrence of anthocyanin among sweet potato varieties also influences the flesh colour, which can be red or purple. The total anthocyanin content of sweet potato is reported to range from 2 to 53 mg/100 g FW (Teow *et al.*, 2007). Lewis *et al.* (1998) also suggested that the anthocyanin content of red tubers is mostly concentrated in the skin ranging from 0 to 700 mg/100 g FW, whereas in the flesh it ranges from 0 to 200 mg/100 g FW.

Several authors point out the importance of anthocyanin accumulation during ripening or maturity of fruit. Guisti *et al.* (1998) reported that no difference was noted when radish cultivars were harvested at different maturity stages, i.e. four or seven weeks after seeding. Çelik *et al.* (2008) investigated the effect of different maturity stages of cranberry fruit (*Vaccinium macrocarpon* Ait. *cv.* Pilgrim) and observed an increase in monomeric anthocynain content from 0.08 to 11.1 mg Cya-3-gal/100 g FW while progressing from a green to dark red colour. In another study, Usenik *et al.* (2009) noted a significant increase from 4.1 to 23.4 mg cyanidin 3-rutinoside/100 g FW during 25–33 day interval of plum ripening. Similarly, Acosta-Montoya *et al.* (2010) examined the influence of three ripening stages on the anthocyanin content of blackberries and they observed and increase from 9 mg Cy-3-glc equivalents/100 g for light red to 77 mg Cy-3-glc equivalents/100 g for dark bluish-purple blackberries.

Proanthocyanidins are oligomers and polymeric end products of flavonoids and are determinants of flavour and astringency in teas, wines and fruit juices. Rentzsch *et al.* (2007) compared proanthocyanins and anthocyanins and concluded that pyranoanthocyanins differ from anthocyanins mainly in analytical aspects, especially colour. The same authors also reported that proanthocyanins have higher stability at varying pH values due to the presence of a pyran ring which acts as nucleophilic attack of water and hinders the formation of carbinol base (Rentzsch *et al.*, 2007). Several authors reported that proanthocyanidin increases during storage and develops a change in chain length involving increases in the degree of polymerisation, in the proportion of (−)-epigallocatechin extension units, and in polymer-associated anthocyanins (Schwarz *et al.*, 2004). The proanthocyanidins of grapes are in the form of condensed tannins which influences the organoleptic properties in wine during the ageing process. Sanchez *et al.* (2003) demonstrated that red wine contains ~20-fold higher proanthocyanidins compared to white wine. This may be due to the absence of grape skin in the production of white wine, compared to the red wine process where the grape skin is also included during the fermentation process and which contains proanthocyanidins, ~175 mg/L. The authors conclude that the presence of proanthocyanidin are important for producing the flavour and astringency of red wine (Sanchez *et al.*, 2003). Grape seed extract contains about 42.8–87.7% proanthocyanidins (Nakamura *et al.*, 2003). In another study, Aaby *et al.* (2005) identified proanthocyanidins in strawberry flesh at levels of about 16 mg/100 g FW.

5.3 Carotenoids

Almost all fruit and vegetables contain a certain amount of carotenoid compounds, however, it is predominantly found in red, yellow, orange, purple, dark green and leafy vegetables. Carotenoids such as α-cartoene, β-cartoene, lycopene, lutein and zeaxanthine from fruit and vegetables are well recognised for their potential health benefits as discussed in Chapter 3. Table 5.3 details the different carotenoid content present in fruit and vegetables. In general, spinach, carrots and tomatoes contain a relatively high carotenoid content compared to other vegetable sources, for example the total carotenoid content in spinach ranges from 17.6 to 22.63 mg/100 g FW (Kidmose *et al.*, 2001) and in carrots and tomatoes it may range up to

Table 5.3 Carotenoid content in fruit and vegetables

Fruit and vegetables	α-carotene	β-carotene	β-cryptoxanthin	lutein	zeaxanthin	Reference
Fruit						
Apple		0.02		0.05		Kim et al. (2007)
Grape		0.02		0.02		Kim et al. (2007)
Plum		0.04	0.02	0.01		Kim et al. (2007)
Tangerine	0.01	0.04	0.27	0.01		Kim et al. (2007)
Bread fruit	0.02	0.09		0.34	0.04	Englberger et al. (2003)
Watermelon		0.50 to 0.60	0.01	0.01	0.00	Barba et al. (2006); Kim et al. (2007)
Persimmon		0.10		0.83		Holden et al. (1999); Barba et al. (2006)
Avocado	~0.10	0.01 to 0.11	0.02 to 0.05	0.26 to 0.87	~0.02	Lu et al. (2009)
Banana	0.05 to 0.18	0.04 to 0.15		0.08 to 0.20		Wall (2006)
Micronesian bananas	0.49	0.41		0.11	0.01	Englberger et al. (2003)
Papaya		0.07 to 0.45	0.26 to 1.08	0.07 to 0.33	1.1 to 4.0	Wall (2006)
Saskatoon berries		0.19 to 36.30		2.6 to 173.3	0.4 to 8.9	Mazza and Cottrel (2008)
Strawberry	0	0.01	0	0.02		Marinova and Ribarova (2007)
Blueberry	0	0.05	0.01	0.23		Marinova and Ribarova (2007)
Blackberry	0.01	0.1	0.03	0.27		Marinova and Ribarova (2007)
Raspberry	0.02	0.01	0.01	0.32		Marinova and Ribarova (2007)
Red currant	0	0.01	0	0.03		Marinova and Ribarova (2007)
Black currant	0	0.06	0	0.21		Marinova and Ribarova (2007)
Vegetables						
Pepper-red		0.75				Barba et al. (2006)
Pepper-green	0.02	0.23 to 0.31		0.61 to 0.93	0.46	Lee et al. (2005); Nizu et al. (2005); Kim et al. (2007)
Lettuce		1.20		1.34		Kim et al. (2007)
Curly lettuce		1.13 to 1.97		1.19 to 1.67		Nizu et al. (2005)
Cabbage	0.002	0.01 to 0.12		0.02 to 0.26		Kurilich et al. (1999); Singh et al. (2007)

Carotenoids (mg/100g FW)

Vegetable	Value 1	Value 2	Reference
Cauliflower		0.05	Singh et al. (2007)
Brussels sprout	0.01	0.07 to 0.08	Kurilich et al. (1999);
		0.14 to 0.90	Singh et al. (2007)
Broccoli		0.48 to 2.42	Kurilich et al. (1999);
		0.41 to 1.02	Singh et al. (2007)
Kale	0.06	4.86	Kurilich et al. (1999)
Carrot	3 to 4	5.25 to 7.95	Nizu et al. (2005);
		0.41 to 0.61	Barba et al. (2006)
Carrot red	0.11	2.51 to 4.29	0.06 to 0.58	Surles et al. (2004)
Carrot orange	1.4 to 3	9.5 to 16.1	0.18 to 0.34	Surles et al. (2004)
Carrot yellow	0.05	0.01 to 0.35	0.24 to 0.78	Surles et al. (2004)
Carrot purple	2.9 to 5.3	7.2 to 17.4	0.37 to 0.58	Surles et al. (2004)
Amaranth/drumstick leaves			23 to 27	Liu et al. (2007)
			1.1 to 1.5	
Spinach		2.95 to 3.35	5.13 to 13	Liu et al. (2007);
			2.3 to 3.1	Bunea et al. (2008)
Sesame leaf		2.58	4.45	Kim et al. (2007)
Tomato		0.26 to 0.85	0.08 to 0.12	Nizu et al. (2005); Barba et al. (2006)
Pumkin	1.44	3.77	10.62	Murkovic et al. (2002); Aruna et al. (2009)
Aubergine or Eggplant (long, green)			1.80	Aruna et al. (2009)
Green chilli			0.02	Aruna et al. (2009)
Sweet potato		0.17 to 0.23	1.90	Aruna et al. (2009)
			0.06	Teow et al. (2007)
Yam			0.01	Aruna et al. (2009)
Taro	0.43	0.80	0.16	Englberger et al. (2003);
			0.19 / 0.02	Aruna et al. (2009)

30 mg/100 g FW (Surles *et al*., 2004; Kandlakunta *et al*., 2008). The β-carotene (precursor of vitamin A) is one of the major carotenoid compounds found in various green vegetables (collard, turnip, spinach and lettuce), mangos, cantaloupe melons, peppers, pumpkin, carrots and sweet potatoes (Holden *et al*., 1999). Alasalvar *et al*. (2001) investigated the level of α- and β-carotene in 19 carrot cultivars and reported a significant variation of between 4 to 9 mg/100 g and 7 to 16 mg/100 g FW for α- and β-carotenes, respectively. This study also noted that the purple carrot cultivar contained 2.2 and 2.3 times more α- and β-carotene compared to the orange carrot cultivar (Alasalvar *et al*., 2001). Similarly, Kurilich *et al*. (1999) reported a significant variation for the β-carotene content of broccoli (0.37–2.42 mg/100 g FW). Like other phytochemicals, carotenoids are also shown to be influenced by cultivar, environmental and agronomic factors, including the maturity stage of the fruit or vegetable (Lee *et al*., 2005; Lu *et al*., 2009). Lu *et al*. (2009) observed that growing locations as well as harvesting times have an influence on the total carotenoid content of California Hass avocado (*Persea americana*). The same authors also showed that the α- and β-carotene of avocado increased 26-fold and more than 10-fold from the January to September growing period (i.e. the total carotenoids varied from 5.9 to 42.2 μg/g among San Luis Obispo avocado samples). A similar observation in variation in carotenoids was also reported in peppers by Lee *et al*. (2005) and saskatoon berries by Mazza and Cottrell (2008).

Lycopene is generally responsible for the red pigmentation of fruit and vegetables, especially tomatoes, red pepper, red carrots, watermelons, red guavas and papayas (Thompson *et al*., 2000). Lycopene lacks the β-ionone ring structure and is therefore devoid of provitamin A activity, however, it possess strong antioxidant activity. The all-*trans* isomer of lycopene is considered as thermodynamically stable, however, different food processing, especially with organic solvents, significantly degrades the lycopene (Hackett *et al*., 2004). The level of lycopene in different fruit and vegetables will also vary with cultivar, environmental conditions (including solar radiations), growing seasons and maturity stages (Brandt *et al*., 2006). Tomato has been recognised as the most important source of lycopene, a red-coloured carotenoid associated with several health benefits (Slimestad and Verheul, 2009). The lycopene content of tomato ranges from 1.0 to 6.7 mg/100 g FW (Thompson *et al*., 2000; Martínez-Valverde *et al*., 2002; Hernández *et al*., 2007), of which tomato skin (peel) ranges from 5 to 14 mg/100 g FW and the pulp ranges from 3 to 7 mg/100 g FW. Red colour tomato cultivars generally contain higher levels of lycopene (2.7–6.6 mg/100 g) compared to yellow tomato cultivars (0.8–1.24 mg/100 g) (Walia *et al*., 2010). Table 5.4 shows the lycopene content of some common fruit and vegetables. Like tomatoes, watermelons are also a known source of lycopenes (Perkins-Veazie *et al*. 2001). The lycopene content of watermelon ranges from 2.5 to 10 mg/100 g FW, of which seedless cultivars are reported to contain a higher lycopene content (5.0 mg/100 g FW) compared to seeded cultivars (Perkins-Veazie *et al*., 2001).

Lutein and zeaxanthin are oxygenated carotenoids present mainly in leafy green vegetables (spinach, collard greens, kale) and green vegetables (broccoli, brussel sprouts) (Holden *et al*., 1999). Leafy greens such as amaranth, including green-leafy amaranth and red- and green-leafy amaranth, contain relatively high amounts of lutein ranging from 14.3 to 14.7 mg/100 g (Liu *et al*., 2007) compared to leafy vegetables (1.43–5.61 μg/100 g) (Niizu and Rodriguez-Amaya, 2005). Apart from cultivar variation, the leutin content of vegetables also depends on its maturity index. Lee *et al*. (2005) observed an increase in the concentration of lutein and zeaxanthin with an increase in maturity of peppers (*Capsicum spp.*) irrespective of growing location and cultivars. Among carrots, purple colour carrots are reported to contain higher levels of lutein (9.9 and 29.9 mg/100 g) on DW (Metzger and Barnes, 2009). Lutein accounts for nearly 70% of total carotenoid content of avocado (Lu *et al*.,

Table 5.4 Lycopene content in fruit and vegetables

Fruit/vegetables	Lycopene (mg/100g FW)	Reference
Watermelon	2.3 to 8.3	Barba et al. (2006); Kim et al. (2007)
Watermelon-Seedless types	3.86 to 6.6	Perkins-Veazie et al. (2001); Barba et al. (2006)
Persimmon	0.2 to 0.4	Barba et al. (2006)
Red guava	0.9 (DW)	Ben-Amotz and Fisher (1998)
Papaya	1.1 to 4.0	Wall (2006)
Tomatoes	1.0 to 6.7	Thompson et al. (2000); Martínez-Valverde et al. (2002); Hernández et al. (2007)
Tomato seeds	2	Toor et al. (2005)
Tomato skin/peel	5 to 14	George et al. (2004); Toor et al. (2005)
Tomato pulp	2.0 to 7.0	George et al. (2004); Toor et al. (2005)
Carrot – High-βC orange	0.87 to 2.53	Surles et al. (2004)
Carrot – Red	5.5 to 6.7	Surles et al. (2004)
Pepper-red	0.55(DW)	Ben-Amotz and Fisher (1998)
Aubergine or Eggplant	2.74 (DW)	Ben-Amotz and Fisher (1998)

DW: dry weight.

2005) whereas Marinova and Ribarova (2007) reported that lutein was the predominant carotenoid content found in raspberries (*Rubus ideaus* L.) and blackberries (*Rubus fruticosus* L.), with approximately 32 mg/100 g and 27 mg/100 g, respectively. The level of lutein was reported to vary with maturation stage of the fruit, for example Mazza and Cottrell (2008) observed a low lutein value for ripe saskatoon berries (300 mg/100 g FW) compared to green or unripe berries (1000 mg/100 g FW).

5.4 Glucosinolates

Glucosinolates are a group of sulphur-containing phytochemicals present abundantly in *Cruciferous* or the *Brassicaceae* family (Cieślik et al., 2007; Barbieri et al., 2008). Glucosinolates can be classified based on the chemical structure such as aliphatic, indole and aromatic, which are derived from one of several amino acids (Griffiths et al., 2001) (for more details, see Chapter 2). These glycosinolates are generally unstable in nature and are hydrolysed to produce several secondary metabolites which are reported to reduce the risk of cancer in humans and animals (Qazi et al., 2010). The enzyme myrosinase catalysis glucosinolates by a thioglucohydrolase to produce glucose and an unstable aglycone, which are further broken down to isothiocyanates, organic thiocyanates and nitriles. Figure 5.1 shows the hydrolysis of glycosinolates by myrosinase. The processing of the cruciferous vegetables breaks down the activation of myrosinase, thus leading to hydrolysis of glucosinolate to produce break down products (Cieślik et al., 2007).

The variation in the level and types of glucosinolates in cruciferous vegetables has been attributed to cultivar, environmental or climatic conditions and agronomic factors (Kushad et al., 1999; Vallejo et al., 2003; Padilla et al., 2007). For example, the glucosinolate content of broccoli cultivars have a variable concentration of glucoraphanin and glucobrassicin depending on the growing conditions and agronomic practices (Barbieri et al., 2008). The variation within a cultivar is also based on the plant age which may be a major determinant factor for both qualitative and quantitative glucosinolate composition (Fahey et al., 2001).

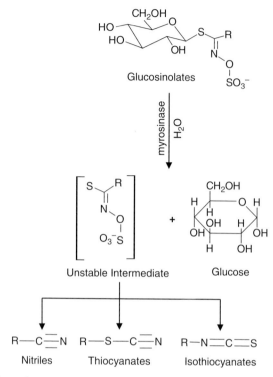

Figure 5.1 Hydrolysis of glycosinolate.

Variations in glucosinolate content of vegetables are also reported to be influenced by temperature, plant growth stages (i.e. from seedling to early flowering stages) and cultivar factors (Pereira et al., 2002; Velasco et al., 2007). Rosa and Rodriques (2001) showed the influence of temperature on the glucosinolate level with increasing myrosinase activity which would degrade glucosinolates among broccoli cultivars. Rosa and Rodriques (2001) also observed the variation in the total and individual glucosinolate levels depending on inflorescence stages of broccoli. For instance, the total glucosinolate content of the broccoli cultivar 'Shogun' contained 35.2 mmol/kg DW in primary inflorescences of the early crop whereas it increased to 47.9 mmol/kg DW in secondary inflorescences of the late crop (Rosa and Rodriques, 2001). Schonhof et al. (2004) reported that with decreasing mean daily temperature, green broccoli and cauliflower showed a strong increase in the content of glucoraphanin and glucoiberin, whereas for white cauliflower an increase of indole content was noted.

Omirou et al. (2009) showed that aliphatic glucosinolate are predominant in the leaves and florets of broccoli, in contrast indolyl glucosinolates were predominant in roots (Omirou et al., 2009). Fertiliser application (e.g. nitrogen and sulphates) has been reported to influence glucosinolate content (Li et al., 2007). For example, Gerendás et al. (2008) demonstrated the interaction of nitrate and sulphate supplies on Kohlrabi (*Brassica oleracea* L. Var. *gongylodes*). Gerendás et al. (2008) also showed that mineral compositions exert a strong interactive impact on plant growth and on the concentration of isothiocyanate in kohlrabi. There is a similar interaction between nitrate and sulphate with glucoinolate content of Indian mustard (*Brassica juncea* L.), probably due to an increase in the myrosinase activity under low nitrogen and sulphur applications (Gerendás et al., 2009).

Table 5.5 Total glucosinolate content in cruciferous vegetables

Cultivars	Total glucosinolate (μmol/g DM)	Reference
Brussels sprouts	25.10	Kushad et al. (1999)
Cauliflower	15.10	Kushad et al. (1999)
Kale	11 to 53	Cartea et al. (2008)
Kale sprouts	58.38 to 412.4	
Kale rosette leaves	4.3 to 17.33	Sun et al. (2011)
Kale blotting stems	3.45 to 30.55	
Cabbage	10.90 to 26.95	Kushad et al. (1999); Cartea et al. (2008)
Korean Chinese cabbage-seed	196 to 276	Hong et al. (2011)
Broccoli	3 to 72	Kushad et al. (1999); Barbieri et al. (2008)
Turnip	4.48 to 74.00	Padilla et al. (2007); Barbieri et al. (2008); Francisco et al. (2009); Kim et al. (2010)

It is evident that glucosinolates (total glucosinolate or as individual glucosinolates) of cruciferous vegetables are influenced by cultivar, environmental and other agronomic factors. However, the hydrolysis of glucosinolates results in break down products which can reportedly act as a precursor to reduce certain cancers.

5.4.1 Variations in glucosinolates

Table 5.5 shows the glucosinolate content of common cruciferous vegetables. Cartea et al. (2008) studied the total glucosinolate content of 153 kales and 29 cabbage varieties grown in north-western Spain. Among kale varieties, the variations in total glucosinolate content ranged from 11 to 53 μmol/g DW, whereas for cabbage varieties it ranged from 10.9 to 27 μmol/g DW. In a recent study, Sun et al. (2011) demonstrated the glucosinolate content of 27 Chinese kale (*Brassica alboglabra* Bailey) varieties, kale sprout, kale rosette leaf and kale bolting stem. They observed that the total glucosinolate contents in kale sprouts was the highest, ranging from 58.38 to 412.4 μmol/g DW compared to bolting stems (3.45–30.55 μmol/g DW) and rosette leaves (4.3–17.33 μmol/g DW). Sun et al. (2011) also reported that the gluconapin was the most abundant glucosinolate among all the edible parts of the 27 varieties, except for one of the sprout varieties. Among 113 varieties of turnip grown in north-western Spain (Padilla et al., 2007) significant variations in the total glucosinolate content (*Brassica rapa* L.) was noted from 7.5 to 56.9 μmol/g DW. Gluconapin was the major compound detected in most of the turnip varieties (Padilla et al., 2007). Similarly, Francisco et al. (2009) illustrated that gluconapin was the predominant glucosinolate, representing ~84% of the total glucosinolate found in *Brassica rapa*. Korean Chinese cabbage (*Brassica rapa* L. ssp. *pekinensis*) is also reported to contain mainly aliphatic glucosinolates (glucobrassicanapin and gluconapin) (Kim et al., 2010) ranging from 4.48 to 31.58 μmol/g DW. Among the *Brassica rapa* cultivars the bitterness is attributed to the presence of gluconapin and glucobrassicanapin (Padilla et al., 2007). In addition, Ciska et al. (2000) also identified glucobrassicin, glucoiberin and sinigrin as predominant glucosinolates among *Brassica oleracea* L (white cabbage, red cabbage, Savoy cabbage, brussel sprouts, cauliflower, kale, kohlrabi). Besides these three compounds in the *Brassica oleracea* L family, the progoitrin also dominated in red cabbage and brussel sprouts, and glucoraphanin dominated in kohlrabi and red cabbage.

5.5 Glycoalkaloids

Glycoalkaloids are secondary plant metabolites generally found in plants of *solanaceous* family such as potato, tomato and egg-plant (also known as aubergine). The glycoalkaloids consist of nonpolar lipophilic steroid glycosides along with nitrogen-containing plant steroids with a carbohydrate side chain attached to the 3-OH position (Friedman and Levin, 1995, 1998; Friedman *et al.*, 2003). The main glycoalkaloids in potato are α-chaconine and α-solanine, contributing 95% of the total glycoalkaloids whereas the major glycoalkaloids found in tomato are dehydrotomatine and α-tomatine. Similarly, the major glycoalkaloids found in aubergine or egg-plant are solamargine and solasonine, which are derivatives of solasodine and occur mainly in leaves and unripe fruit (Paczkowski *et al.*, 2001; Friedman *et al.*, 2003). Potato glycoalkaloids have been widely studied by many authors (Pęksa *et al.*, 2002; Friedman *et al.*, 2003; Kodamatani *et al.*, 2005; Finotti *et al.*, 2006) followed by tomato glycoalkaloids (Friedman and Levin, 1995, 1998). In general, the glycoalkaloids are formed when potatoes are exposed to light in the field, at harvest or during storage (Jadhav and Salunkhe, 1975). Amongst tomatoes, tomatines are naturally occurring antimicrobial compounds and are said to increase due to exposure of the plant to stress including 'phytoanticipin'; however, the role of phytoanticipin in increasing the level of tomatine is still not clear (VanEtten *et al.*, 1994). Glycoalkaloids from potatoes are reported to exhibit a cytotoxicity effect on human cancer cells (Yang *et al.*, 2006). Tomato glycoalkaloid are also reported to decrease plasma LDL cholesterol in hamsters by 41% with 0.2 g of tomatine/100 g compared to a control diet. A 59 and 44% decrease in plasma LDL cholesterol was reported when hamsters were fed with green (high-tomatine) or red (low-tomatine) at the same concentration (Friedman *et al.*, 2000). Excess accumulation of glycoalkaloids in potato can result in a bitter taste.

Glycoalkaloids concentration in potato, tomato and egg-plant may be influenced by several factors including cultivar, environmental and other agronomical factors such as harvest time and storage, where interactions between light and temperature can have an effect (Griffiths *et al.*, 1998). The glycoalkaloids content of potato and tomato also varies depending upon the location within the plant (Friedman and Levin, 1995; Friedman *et al.*, 2003). For example, Friedman and Levin (1995) observed ~700-fold variation in the α-tomatine levels from medium-small red fruit containing 0.19 mg of α-tomatine/100 g FW compared to flowers containing 130 mg/100 g FW. Similarly, Friedman *et al.* (2003) noted a high concentration of glycoalkaloids in potato sprouts and skins compared to the flesh of potato. The total glycoalkaloids concentration is reported to be higher in potato peel compared to the flesh (Friedman *et al.*, 2003). Additionally, Jansen and Flamme (2006) also reported high total glycoalkaloids of potato peel (17.19 mg/100 g FW) compared with the potato flesh (2.32 mg/100 g FW) and whole potato (4.44 mg/100 g FW). Furthermore, Friedman *et al.* (2003) reported that the ratio of α-chaconine to α-solanine glycoalkoloids is slightly higher in potato peels (~2) compared to the flesh (~1.5) of eight potato cultivars. The majority of the glycoalkaloids present in potato are concentrated in the skin and upon peeling the α-chaconine and α-solanine content decreases significantly, thus decreasing the total glycoalkaloids (Pęksa *et al.*, 2002). Green and unripe tomatos have a high tomatine content (8.65 mg/100 g FW) compared to 0.22 mg/100 g FW for ripe or red tomatoes (Friedman and Levin, 1995). Table 5.6 details the glycoalkaloids level in potato and tomato.

In another study, Griffiths *et al.* (1998) demonstrated the effect harvest time (i.e. early lifting and main lifting) had on total glycoalkaloid content of six potato cultivars ('Brodick, Eden, Pentland Crown, Pentland Dell, Record and Torridon'). The total glycoalkaloids in the main lift varied from 3.8 mg/100 g FW for Eden to the highest value of 12.6 mg/100 g FW in Pentland Dell. In addition, the authors noted that the storage temperature also affected the rate of glycoalkaloid synthesis in response to light exposure, for instance when cultivars were stored at a lower storage temperature of 4 °C a rapid accumulation of glycoalkaloids was noted in two of the cultivars (Brodick and Pentland) (Griffiths *et al.*, 1998). In another study, Kodamatani *et al.* (2005) illustrated a significant difference in α-chaconine and α-solanine content when potatoes were stored in the dark and the other potatoes were exposed to light for three weeks. Maturity stage also has an effect on the concentration of glycoalkaloids in potato tubers. Griffiths *et al.* (1994) found that total glycoalkaloids of the immature tubers were higher (65 mg/100 g) compared to the mature tubers (49 mg/100 g). A comparable effect of maturity was also reported for tomato by Friedman and Levin (1995, 1998). They observed an average of 98% α-tomatine in green unripe tomatoes compared to red ripe tomatoes (Table 5.6).

5.6 Polyacetylenes

Polyacetylenes are phytochemicals predominantly found in vegetables of the *Apiaceae* family including carrots, celery, fennel, parsley and parsnips, all of which contain a group of bioactive aliphatic C17-polyacetylenes (Zidorn *et al.*, 2005; Christensen and Brandt, 2006). The bioactivity of polyacetylenes (falcarinol-type) is reported to result in a decrease in the (pre)neoplatic lesions in the colon of rats (Kidmose *et al.*, 2004; Kobæk-Larsen *et al.*, 2005). The polyacteylenes have also been studied for their antiallergenic, anti-inflammatory, antifungal, antibacterial and reduced platelet aggregation properties (Christensen and Brandt, 2006). Zidorn *et al.* (2005) investigated the polyactylenes of some vegetables of Apiaceae. The falcarindiol was the main polyacetylene in celery (*Apium graveolens* I, *Apium graveolens* II), carrot (*Daucus carota*), fennel (*Foeniculum vulgare*), parsley (*Pastinaca satica*) and parsnip (*Petroselinum crispum*), whereas the falcarinol was not detected in parsley. Among the vegetables, the total polyacetylenes content is highest in parsnip (more than 7500 μg/g); whereas in celery and parsley it is about 2500 μg/g of dried plant material. In contrast, Zidorn *et al.* (2005) observed only trace amounts (less than 300 μg/g) of falcarinol and falcarindiol in carrot and fennel. Table 5.7 summarises the polyacetylene content in vegetables found in different studies.

The major polyacetylenes found in carrots are falcarinol, falcarindiol and falcarindiol-3-acetate, among these both falcarinol and falcarindiol have received much attention (Czepa and Hofmann, 2003; Kidmose *et al.*, 2004). Consumption of falcarinol from carrots is well documented to inhibit cancerous lesions in both animals and humans (Kobæk-Larsen *et al.*, 2005; Young *et al.*, 2007). Moreover, polyacetylenes (in particular the falcarindiol compound) contribute to bitterness in carrots (Czepa and Hofmann, 2003; Kreutzmann *et al.*, 2008).

The falcarinol content of carrots ranges from 7 to 40.6 μg/g FW (Pferschy-Wenzig *et al.*, 2009); however, this is influenced by cultivar and several environmental factors. Kidmose *et al.* (2004) indicated that the falcarindiol and falcarindiol-3 acetate contents were higher in small carrot roots (50–100/g root size) compared to large carrot root sizes (more than 250/g), whereas the authors observed no change in falcarinol with the root size. Generally, the falcarinol content is located in the phloem tissue whereas the falcarindiol and falcarindiol-3-acetate are concentrated in the periderm of the carrot. Since the periderm roots (root-hair zone proximal from growing root tip) of carrots usually take water from the soil this may dilute the falcarindiol and falcarindiol-3 aceteate with increasing root size when

Table 5.6 Glycoalkaloid level in potato and tomato

Potato, descriptions	α-Chaconine (mg/100 g FW)	α-Solanine (mg/100 g FW)	Total glycoalkaloids (mg/100 g FW)	Reference
Potato-whole	5.58	2.21	7.79	Pęksa et al. (2002)
Potato-whole	3.22	1.93	5.15	Friedman et al. (2003)
Potato-whole	2.51	0.54	3.06	Finotti et al. (2006)
Potato-peeled-whole	2.57	1.07	3.64	Pęksa et al. (2002)
Potato-flesh	1.82	1.17	2.98	Friedman et al. (2003)
Potato-skin/peel	17.31	8.93	26.24	Friedman et al. (2003)
Potato-skin/peel	134.2	137.9	272.1	Kodamatani et al. (2005)
Potato-pith (heart of potato)	2.71	1.76	4.47	Kodamatani et al. (2005)

Tomato, descriptions	α-Tomatine (mg/100 g FW)	α-dehydrotomatine (mg/100 g FW)	Total glycoalkaloids (mg/100 g FW)	Reference
Tomato-green unripe	8.65 and 19.3	1.25	20.55	Friedman and Levin (1995, 1998)
Tomato-red ripe	0.22 and 0.13	0.01	0.14	Friedman and Levin (1995, 1998)

Table 5.7 Polyacetylene content in vegetables (μg/g DW)

Vegetables	Falcarinol	Falcarindiol	Falcarindiol 3-acetate	Reference
Carrots (FW)	7.9 to 9.7	26.8 to 41.1	11.3 to 13.7	Kidmose et al. (2004)
Carrots	270 to 310	210 to 270		Zidorns et al. (2005)
Carrots	82 to 583	261 to 970	245 to 1553	Metzger and Barnes (2009)
Commercial baby carrots	136 to 148	272 to 344	233 to 239	Metzger and Barnes (2009)
Commercial market carrots	358 to 378	1073 to 1107	600 to 604	Metzger and Barnes (2009)
Celery	230 to 1680	2040 to 4670	140 to 200	Zidorns et al. (2005)
Fennel	40	240		Zidorns et al. (2005)
Parsnip	1570 to 1630	5700 to 5800		Zidorns et al. (2005)
Parsley	0	2300 to 2340		Zidorns et al. (2005)

compared to falcarinol content (Garrod and Lewis, 1979; Olsson and Svensson, 1996). Søltoft et al. (2010) conducted a comparative analysis of organically and conventionally grown carrots to study the effect of the cultivation system on the concentration of polyacetylenes in carrot roots in different years. The authors found that the concentrations of falcarindiol, falcarindiol-3-acetate and falcarinol significantly varied with year (i.e. year one were 222, 30 and 94 μg of falcarindiol equivalent/g of DW, respectively, and is 3–15% lower in year two) Søltoft et al. (2010) demonstrated the effect of several extrinsic factors on the level of polyacetylenes in carrots. Scientific evidence suggests polyacetylenes in carrots are concentrated in the carrot peel and add to bitterness, conversely processing (such as peeling) may remove significant amounts of these bioactive compounds (Metzger and Barnes, 2009).

5.7 Sesquiterpene lactones

Sesquiterpene lactones constitute an important group of secondary metabolites found in over 500 different members of the *Compositae (Asteraceae)* family. The sesquiterpene lactones occur naturally in the leaves of lettuce and chicory. Lettuce (*Lactuca sativa*. L) is important and widely used in the human diet as a healthy, low calorie salad component of meals (Tamaki et al., 1995). The sesquiterpene lactones are C_{15} terpenoids known for their antibiotic, cytotoxic and allergenic properties (Baruah et al., 1994). In a study, Han et al. (2010) reported three new sesquiterpenes such as compound 1–3 and compound 4–11 from *Lactuca sativa* L. var *anagustata*. The authors also noted a pronounced effect on radical-scavenging activities, cytotoxicity of cancer cells, human epithelial carcinoma (HeLa) and human colon carcinoma (HCT-116) cell lines (Han et al., 2010). The intense bitterness in lettuce leaves is mainly caused by the presence of sesquiterpene lactones which includes lactucin, 8-deoxylactucin and lactucopicrin (Bennett et al., 2002). Similar bitterness compounds such as lactucin and lactucopicrin are partly responsible for the bitterness in chicory (*Cichorium intybus* L.) leaves and roots (Peter et al., 1996). Tamaki et al. (1995) identified the presence of three major sesquiterpene lactone compounds such as lactucin, 8-deoxylactucin and lactucopicrin among two wild lettuces (*Lactuca saligna* and *Lactuca virosa*). The above study also reported that the sesquiterpene lactone content from the wild lettuces mainly occurred in free and glycoside bound forms. The lactucin, 8-deoxylactucin and lactucopicrin compounds are reported to be 103, 372 and 79 μg/g for *Lactuca saligna* whereas

Table 5.8 Sesquiterpene lactone content of lettuce

Cultivars	Sesquiterpene lactones (µg/g) DW			Reference
	lactucin	8-deoxylatucin	lactucopicrin	
Wild lettuce				Tamaki et al. (1995)
Wild lettuce (*Lactuca saligna*)	103.1	372.1	79	Tamaki et al. (1995)
Wild lettuce (*Lactuca virosa*)	256.9	173.5	1732.7	Tamaki et al. (1995)
Commercial lettuce cultivars/clones	11.1		13.9	Tamaki et al. (1995)
Lettuce-green	5.2	7.5	18.3	Seo et al. (2009)
Lettuce-red	7.5	10.8	23.4	Seo et al. (2009)
Lettuce-basal leaves	10.1	2.9	20.8	Seo et al. (2009)
Lettuce-mid stalk leaves	13.2	17.6	30.9	Seo et al. (2009)
Lettuce-flower stalk leaves	28.9	35.9	113.2	Seo et al. (2009)

a value of 257, 173 and 1733 µg/g was noted in *Lactuca virosa*. In addition, a 2- to 20-fold greater concentration of lactucin and lactucopicrin was determined in the wild lettuce species in comparison to the commercial varieties 'Montello' and 'Saladcrisp', which were low in lactucin and lactucopicrin and devoid of 8-deoxylactucin. The presence of the high levels of sesquiterpenoid lactones lactucin in wild lettuce may be the main cause of bitterness (Tamaki *et al.*, 1995). The cultivar variance, including genotype and the leaf location on the plant, contribute to the wide range in bitterness of sesquiterpene lactones on different parts of the lettuce (Seo *et al.*, 2009). The bitter sesquiterpene lactones varied significantly among ten cultivars including six red- and four green-pigmented lettuces (*Lactuca sativa* L. var. *crispa* L.) with the total concentration ranging from 14.6 to 67.7 µg/g DW. This study also noted that lactucopicrin was the major contributor to bitterness of the lettuce cultivars. The red and curled leaf cultivars contain high sesquiterpene lactones ranging from 2.9 to 17.2 µg/g DW for lactucin, 2.8 to 17.1 µg/g DW for 8-deoxylactucin and 8.8 to 36.1 µg/g DW for lactucopicrin. Moreover, the authors also reported a significantly high concentration of nearly 2.9, 12.4 and 5.4 times of lactucin, 8-deoxylactucin, and lactucopicrin in flower stalk leaves than the basal leaves, respectively indicating that the concentration increases acropetally in lettuce cultivars (Seo *et al.*, 2009). In addition to cultivar effects, this study shows that the proportion of individual sesquiterpene lactones can also change depending on the stage of plant development. Table 5.8 shows the amount of sesquiterpene lactones present in lettuce and in different components of lettuce.

5.8 Coumarins

Coumarin glycosides belong to the benzopyrone family of compounds and its derivatives occur abundantly in nature and are classified into simple, furanocoumarins, pyranocoumarins and pyrone-substituted coumarins. Most of the coumarins occur in fruit and vegetables belonging to the *Rutaceae* and *Umbelliferae* families and include celery, carrots and parsnips (Ostertag *et al.*, 2002) and citrus species. Nigg *et al.* (1993) indicated that the level of coumarins in citrus peel is about 13–182 times higher compared to its pulp. The citrus peel contains simple coumarins such as auraptene having mevalonate-derived side chains with various oxidation levels. These auraptene are reported to be found in large amounts in the juice sac of fruit of the trifoliate orange (about 7 mg/g) with lower concentration in the peel (about 1 mg/g) (Ogawa et al., 2000). Barreca et al. (2010) demonstrated the presence of furocoumarin (bergapten and epoxybergamottin) in sour orange or chinotto juice with about 0.91 and 0.67 mg/L for bergapten and epoxybergamottin, respectively. The same study also showed high furocoumarins in green chinotto fruit (approximately 57.4 mg/L) which significantly reduced (to 18.3 mg/L) on ripening (Barreca et al., 2010). Celery (*Apium graveolens*) contains a variable amount of furanocoumarin with the highest amount in outer celery leaves (4.5 mg/100 g) compared to the inner leaves (containing 0.9 mg/100 g) while the petioles and roots contain about 0.09–0.15 mg/100 g (Diawara et al., 1995). The bitter compound found in carrots is methyl-6-methoxy-8-hydroxy-3, 4-dihydroisocoumarin (also known as 6-methoxymellein) (Talcott and Howard, 1999). Talcott and Howard (1999) showed that carrots contain an appreciable amount of 6-methoxymellein ranging from 10.4 to 40.3 mg/100 g DW, which is associated with the extreme bitter flavour. The authors suggest that the high amount 6-methoxymellein in carrots may be due to environmental stress in the field, which induces the production of wound ethylene and subsequent phytoalexin formation (Talcott and Howard, 1999). Kidmose et al. (2004) showed a significant variation in 6-methocymellein content with effect of locations ranging from 0.03 to 0.21 mg/100 g FW. Kidmose et al. (2004) indicated 6-methoxymellein was found in small carrot roots and the content decreases significantly with an increase in root size. The reason is primarily due to the accumulation of 6-methocymellein in the periderm, thus when the root size increases, the weight of this layer, relative to total decreases resulting in a lower 6-methocymellein content per unit weight (Mercier et al., 1994).

5.9 Terpenoids

The terpenoids or isoprenoids form the part of the group of isoprene substances that are characterised by their biosynthetic origin from isopentenyl and dimethylallyl pyrophosphastes and their broadly lipophilic properties (Harbone et al., 1993). Terpenoids are divided on the basis of their C-skeleton such as monoterpene, sesquiterpene, diterpene, tritepene, tetraterpene and polyterpene (Graβman, 2005). Terpenoids are found in many constituents of essential oils, herbs, spices and some fruit and vegetables (Graβman, 2005). Simon (1982) showed a wide variation in volatile terpenoid of carrots ranging from 574 to 1852 µg/mg. The same author also reported a wide variation in raw carrot roots ranged from 497 to 2824 µg/mg indicating a five-fold variation among individual terpenoids. Alasalvar et al. (2001) noted that among white carrot varieties the terpenoid mostly comprise of about 45% of the total volatile, whereas about 24% was noted in yellow carrot varieties. In some vegetables terpenoid compounds accumulate as a result of stress. Jadhav et al. (1991) reported accumulation of rishitin, spirovetivanes derivatives and phytuberin type terpenoid compounds as stress compounds in potato.

5.10 Betalains

Betalains are water-soluble nitrogen-containing pigments, which comprise the red-violet betacyanins and the yellow betaxanthins (Stintzing et al., 2007; Azeredo, 2009; Moussa-Ayoub et al., 2011). These compounds accumulate in the flowers, fruits and vegetative tissues of plants belonging to the Caryophyllales family. The main sources of betalains are red beet, prickly pear or cactus and tuber (Strack et al., 2003; Nemzer et al., 2011). The betalains structure consists of betalamic acid, betanidin and indicaxanthin, mainly classified in two distinct groups, namely betacyanins and betaxanthins. Betacyanin contain cyclo-3, 4-dihydroxyphenylalanine (cyclo-Dopa) residue, whereas betaxanthins contain various amino acids and amine along with various betanidin conjugates (glycosides and acylglucosides) (Strack et al., 2003). Like other phytochemicals, the betalains are also reported for their anti-radical properties and exhibit strong antioxidant activity (Butera et al., 2002; Cai et al., 2003). Red beet contains high amounts of betanin (betanidin 5-O-β-glucoside) ranging from 30 to 60 mg/100 g FW, indicating lower concentrations of isobetanin, betanidin and betaxanthins in beet roots (Kanner et al., 2001). In a study, Butera et al. (2002) reported that yellow prickly pear cultivars contain about 89% of betalains and indicaxanthin. They also reported that the red prickly pear cultivar accounts for nearly 66% betanin, whereas indicaxanthin is predominant in the white prickly pear cultivar. The level of total betacyanin among red-skinned ulluco tubers (*Ullucus tuberosus*), is reported to vary from 41.2 to 70.4 μg/g FW with a significant variation in betaxanthins content (Sevenson et al., 2008). The authors conclude that the variation in betaxanthins is probably dependent on the level and proportion of amino acids available to react with betalamic acid in the tuber. In another study, Kugler et al. (2004) reported that the Swiss chard contains about 51.1 μg/g FW of betacyanins and 49.7 μg/g FW of betaxanthins. The Amaranthaceae family also contains betalain pigments such as red-violet gomphrenin type betacyanins and yellow betaxanthins, which are cited for both their natural antioxidants and natural colorants (Cai et al., 2003).

5.11 Vitamin E or tocols content in fruit and vegetables

Tocols (tocopherol and tocotrienols) are natural antioxidants present mainly in vegetable oils, nuts and grains with relatively low levels in fruit and vegetables. Generally leaves and other green parts of plants are rich in tocopherol while the tocotrienols are mostly found in the bran and germ fraction of cereals (Tiwari and Cummins, 2009). Among tocopherols, α-tocopherols are a predominant compound in fruit and vegetables compared to other forms of tocopherol (Piironen et al., 1986). Table 5.9 shows the different forms of tocopherols and tocotrienols present in fruit and vegetables. Piironen et al. (1986) noted the highest α-tocopherol values in dark green leafy vegetables and sweet pepper which were more than 1 mg/100 g FW. The above study also noted that α-tocopherol for different fruit surveyed ranged from 0.06 to 0.96 mg/100 g FW, of which the berries showed up to 4.14 mg/100 g FW (Piironen et al., 1986). Likewise, Ching and Mohamed (2001) investigated the α-tocopherol content of 62 edible plant sources and reported, among vegetables, the red *Capsicum annum* contained the highest level of α-tocopherol (15.54 mg/100 g FW) followed by celery (*Apium graveolens*) (13.64 mg/100 g FW). A comparison study on the level of α-tocopherol among bell peppers showed that red and yellow peppers contained high levels of α-tocopherol (0.06 mg/100 g DW) compared to green peppers with 0.01 mg/100 g DW (Burns et al., 2003). Similarly, Singh et al. (2007) observed significant variations in vitamin E content within various cultivars of cabbage (0.03 to 0.20 mg/100 g) and broccoli (0.22 to 0.68 mg/100 g). These variations in vitamin E content may be influenced by several factors, including cultivar, environment, harvesting stage and different methods of extraction (Kim et al., 2007; Singh et al., 2007).

Some studies reported that the α-form of tocopherol is mainly present in fruit and vegetables (Ching and Mohamed, 2001), whereas a number of studies also showed the existence

Table 5.9 Tocopherol and tocotrienol content in fruit and vegetables

Fruit/vegetables	Tocopherols (mg/100g FW)				Tocotrienols (mg/100g FW)			Reference
	α	β	γ	δ	α	β	γ	
Fruits								
Apple-green	0.40		0.02	0.01				Isabelle et al. (2010)
Apple-red	0.38		0.04	0.01				Chun et al. (2006)
Avocado	1.33 to 2.66	0.03 to 0.08	0.13 to 0.69	0.02				Chun et al. (2006); Lu et al. (2009)
Banana	0.13 to 0.21				0.02			Piironen et al. (1986); Chun et al. (2006)
Papaya	0.32		0.02					Isabelle et al. (2010)
Mango	1.10		0.02					Isabelle et al. (2010)
Durian	3.77		1.01	0.01				Isabelle et al. (2010)
Peach	0.21 to 0.96	0.01	0.05 to 0.08			0.02		Piironen et al. (1986); Chun et al. (2006)
Plum	0.28 to 0.85	0.01	0.05 to 0.08			0.02	0.22	Piironen et al. (1986); Chun et al. (2006)
Prune	0.37	0.01	0.03					Chun et al. (2006)
Pear	0.21	0.01	0.08			0.12		Chun et al. (2006)
Kiwifruit	1.31 to 1.69		0.05	0.04			0.11 to 0.18	Chun et al. (2006); Isabelle et al. (2010)
Grape-red	0.42		0.11	0.01				Isabelle et al. (2010)
Grape-green	0.05		0.53	0.02				Isabelle et al. (2010)
Blackcurrant	2.23		0.83					Piironen et al. (1986)
Redcurrant	0.82	0.16	0.32	0.15				Piironen et al. (1986)
Blackberry	1.43	0.04	1.42	0.85				Chun et al. (2006)
Blueberry	0.58 to 1.85		0.21 to 0.38	0.02			0.66	Piironen et al. (1986); Chun et al. (2006)
Cranberry	0.94 to 1.23	0.02	0.25		0.05		0.35	Piironen et al. (1986); Chun et al. (2006)
Raspberry	0.88	0.15	1.47		1.15			Piironen et al. (1986); Chun et al. (2006)
Strawberry	0.56 to 1.05		0.15	1.19	0.04	0.01		Piironen et al. (1986); Chun et al. (2006); Isabelle et al. (2010)
Gooseberry	0.73		0.11					Piironen et al. (1986)
Pomegranate-fruit	1 to 5		0.1 to 0.2	0.1 to 0.8				Pande and Akoh (2009)
Pomegranate-seed	160 to 175		79 to 93	20 to 25				Pande and Akoh (2009)
Grapefruit	0.16 to 0.32							Piironen et al. (1986); Chun et al. (2006)
Orange	0.25 to 0.36		0.14				0.06	Piironen et al. (1986); Chun et al. (2006)

(Continued)

Table 5.9 (Continued)

Fruit/vegetables	Tocopherols (mg/100g FW)				Tocotrienols (mg/100g FW)			Reference
	α	β	γ	δ	α	β	γ	
Vegetables								
Beans	0.13		0.32	0.03				Piironen et al. (1986)
Broccoli	0.46 to 2.57		0.13 to 0.31					Piironen et al. (1986); Kaur et al. (2007)
Brussel sprout	0.15 to 0.83		0.04					Piironen et al. (1986); Kurilich et al. (1999); Singh et al. (2007)
Cabbage	0.17							Kurilich et al.(1999)
Cabbage-red	0.07 to 0.86		0.005		0.05			Podsędek et al. (2006); Chun et al.(2006); Singh et al. (2007)
Cabbage-white	0.21				0.04			Chun et al. (2006)
Cauliflower	0.08 to 0.17		0.20 to 0.26		0.06			Kurilich et al. (1999); Chun et al. (2006); Singh et al. (2007)
Lettuce	0.22 to 0.63		0.13 to 0.34	0.08				Piironen et al. (1986); Kin et al. (2007)
Celery	0.26 to 0.5	0.02	0.07		0.08			Piironen et al. (1986); Chun et al. (2006)
Kale	1.92		0.23					Kurilich et al. (1999)
Sesame leaf	0.55			0.18				Kim et al. (2007)
Parsley	3.58		1.23					Piironen et al. (1986)
Spinach	1.22 to 1.96		0.21					Piironen et al. (1986); Chun et al. (2006)
Cucumber	0.04	0.01	0.04	0.01	0.08			Piironen et al. (1986); Chun et al. (2006)
Mushroom-button	0.01	0.01	0.02	0.02	0.07			Chun et al. (2006)
Onion-red	0.04	0.01	0.01					Piironen et al. (1986)
Onion-white	0.04							Chun et al. (2006)
Onion-yellow	0.04	0.01	0.01					Piironen et al. (1986)
Pea	0.03		1.6	0.04				Piironen et al. (1986)
Pepper-green	0.31		0.22	0.01				Kim et al. (2007)
Sweet pepper	2.16	0.11	0.02					Piironen et al. (1986)
Carrot	0.36 to 0.86	0.01			0.04			Piironen et al. (1986); Chun et al. (2006)
Parsnip	0.82	0.03	0.02					Piironen et al. (1986)
Potato	0.05 to 0.07					0.01		Piironen et al. (1986); Chun et al. (2006)
Sweet potato	0.18 to 0.25		0.01			0.1		Chun et al. (2006); Kim et al. (2007)
Tomato	0.53 to 0.66		0.14 to 0.20					Piironen et al. (1986); Chun et al. (2006)

of other forms of tocopherol and tocotrienols in the fruit and vegetables (Chun *et al.*, 2006). Chun *et al.* (2006) recognised higher levels of γ-tocopherol than α-tocopherol in some berries, cruciferous vegetables, mushrooms and green peas. The study demonstrated high levels of α-tocotrienols in coconut with 0.79 mg/100 g, while cranberry, cabbage, kiwi and plum were relatively high in γ-tocotrienol levels with 0.33, 0.32, 0.11 and 0.22 mg/100 g, respectively (Chun *et al.*, 2006).

Vitamin E is also influenced by increasing fruit maturation. Horvath *et al.* (2006) examined the accumulation of tocopherols and tocotrienols during seed development of grapes (*V. vinifera* L. Alphonse Lavallée). This study indicates that during the sigmodial growth period (days after flowering) of grapes, the tocopherols gradually decrease, whereas tocotrienols gradually increases during the development stage to a maximum of 61 µg/g DW. This may be due to the fact that the tocotrienols are only found in the endosperm of the grape seeds while the tocopherols are condensed in all tissues of the seed, thus during seed development the level reduced steadily (Horvath *et al.*, 2006).

5.12 Conclusions

Phytochemicals are important bioactive naturally occurring compounds present in almost all fruit and vegetables and regularly cited for their potential beneficial health effects for both humans and animals. Phytochemicals, namely polyphenols, carotenoids, glucosinolates, alkaloids or glycoalkaloids, betalains and vitamin E, are reported in many fruit and vegetables. A significant amount of these are found in the skin or peel of the fruit and vegetables. These phytochemicals may vary within different fruit and vegetables, with varying efficacy in protecting against chronic diseases (for example, cancer, CVD). Various literature sources reveal that the phytochemical levels present in fruit and vegetables are influenced by various factors such as cultivar, environmental, growing locations, agronomic and storage factors. Among these factors maturity stage of the fruit and vegetables are reported to increase the level of poplyphenols but decrease the level of carotenoids and lycopene. Thus, to maximise the consumption of phytochemicals in processed fruit and vegetable it is imperative to quantify the changes in photochemicals based on a farm to fork approach. This approach can assist in optimising the various farm level factors influencing the level of phytochemicals in harvested fruit and vegetables.

References

Aaby, K., Skrede, G. and Wrolstad, R.E. (2005) Phenolic composition and antioxidant activities in flesh and achenes of strawberries (Fragaria ananassa). *Journal of Agricultural and Food Chemistry*, 53, 4032–4040.

Aaby, K., Wrolstad, R.E., Ekeberg, D. and Skrede, G. (2007) Polyphenol composition and antioxidant activity in strawberry purees; impact of achene level and storage. *Journal of Agricultural and Food Chemistry*, 55, 5156–5166.

Acosta-Montoya, Ó., Vaillant, F., Cozzano, S., Mertz, C., Pérez, A.M. and Castro, M.V. (2010) Phenolic content and antioxidant capacity of tropical highland blackberry (*Rubus adenotrichus* Schltdl.) during three edible maturity stages. *Food Chemistry*, 119, 1497–1501.

Alasalvar, C., Grigor, J.M., Zhang, D., Quantick, P.C. and Shahidi, F. (2001) Comparison of volatiles, phenolics, sugars, antioxidant, vitamins, and sensory quality of different colored carrot varieties. *Journal of Agricultural and Food Chemistry*, 49, 1410–1416.

Amin, I. and Lee, W.Y. (2005) Effect of different blanching times on antioxidant properties in selected cruciferous vegetables. *Journal of the Science of Food and Agriculture*, 85, 2314–2320.

Andre, C.M., Oufir, M., Guignard, C., Hoffmann, L., Hausman, J.F., Evers, D. and Larondelle, Y. (2007) Antioxidant profiling of native Andean potato tubers (*Solanum tuberosum* L.) reveals cultivars with high levels of β-carotene, α-tocopherol, chlorogenic acid, and petanin. *Journal of Agricultural and Food Chemistry*, 55, 10839–10849.

Aruna, G., Mamatha, B.S. and Baskaran, V. (2009) Lutein content of selected Indian vegetables and vegetable oils determined by HPLC. *Journal of Food Composition and Analysis*, 22, 632–636.

Awad, M.A., Wagenmakers, P.S. and de Jager, A. (2001) Effects of light on flavonoid and chlorogenic acid levels in the skin of 'Jonagold' apples. *Scientia Horticulturae*, 88, 289–298.

Azeredo, H.M.C. (2009) Betalains: properties, sources, applications, and stability–a review. *International Journal of Food Science and Technology*, 44, 2365–2376.

Barba, A.I.O., Hurtado, M.C., Mata, M.C.S., Ruiz. V. F. and de Tejada M.L.S. (2006) Application of a UV–vis detection-HPLC method for a rapid determination of lycopene and β-carotene in vegetables. *Food Chemistry*, 95, 328–336.

Barbieri, G., Pernice, R., Maggio, A., De Pascalr, S. and Fogliano, V. (2008) Glucosinolates profile of *Brassica rapa* L. Subsp. *Sylvestris* L. Janch. *var.esculenta* Hort. *Food Chemistry*, 107, 1687–1691.

Barreca, D., Bellocco, E., Caristi, C., Leuzzi, U. and Gattuso, G. (2010) Flavonoid composition and antioxidant activity of juices from Chinotto (*Citus × myrtifolia* Raf.) fruits at different ripening stages. *Journal of Agricultural and Food Chemistry*, 58, 3031–3036.

Baruah, N.C., Sarma, J.C., Barua, N.C., Sarma, S. and Sharma, R.P. (1994) Germination and growth inhibitory sesquiterpene lactones and a flavone from *Tithonia diversifolia*. *Phytochemistry*, 36, 29–36.

Ben-Amotz, A. and Fisher, R., (1998) Analysis of carotenoids with emphasis on 9-cis-β-carotene in vegetables and fruits commonly consumed in Israel. *Food Chemistry*, 62, 515–520.

Bennett, M.H., Mansfield, J.W., Lewis, M.J. and Beale, M.H. (2002) Cloning and expression of sesquiterpene synthase genes from lettuce (*Lactuca sativa* L.). *Phytochemistry*, 60, 255–261.

Brandt, S., Pék, Z., Barna, É., Lugasi, A. and Helyes, L. (2006) Lycopene content and color of ripening tomatoes as affected by environmental conditions. *Journal of the Science of Food and Agriculture*, 86, 568–572.

Bunea, A., Andjelkovic, M., Carmen Socaciu, C., Bobis, O., Neacsu, M., Verhé, R. and Camp, J.V. (2008) Total and individual carotenoids and phenolic acids content in fresh, refrigerated and processed spinach (*Spinacia oleracea* L.). *Food Chemistry*, 108, 649–656.

Burns, J., Fraser, P.D. and Bramley, P.M. (2003) Identification and quantification of carotenoids, tocopherols and chlorophylls in commonly consumed fruits and vegetables. *Phytochemistry*, 62, 939–947.

Butera, D., Tesoriere, L., Di Gaudio, F., Bongiorno, A., Allegra, M., Pintaudi, A.M., Kohen, R. and Livrea, M.A. (2002) Antioxidant activities of Sicilian prickly pear (*Opuntia ficus indica*) fruit extracts and reducing properties of its betalains: betanin and indicaxanthin. *Journal of Agricultural and Food Chemistry*, 50, 6895–6901.

Cai, Y., Sun, M. and Corke, H. (2003) Antioxidant activity of betalains from plants of the Amaranthaceae. *Journal of Agricultural and Food Chemistry*, 51, 2288–2294.

Cartea, M.E., Velasco, P., Obregón, S., Padilla, G. and De Haro, A. (2008) Seasonal variation in glucosinolate content in *Brassica oleracea* crops grown in northwestern Spain. *Phytochemistry*, 69, 403–410.

Çelika, H., Özgen, M., Serçec, S. and Kayad, C. (2008) Phytochemical accumulation and antioxidant capacity at four maturity stages of cranberry fruit. *Scientia Horticulturae*, 117, 345–348.

Cevallos-Casals, B.A., Byrne, D., Okie, W.R. and Cisneros-Cevallos, L. 2006. Selecting new peach and plum genotypes rich in phenolic compounds and enhanced functional properties. *Food Chemistry*, 96, 273–280.

Cevallos-Casals, B.A. and Cisneros-Zevallos, L.A. (2003) Stoichiometric and kinetic studies of phenolic antioxidants from Andean purple corn and red-fleshed sweetpotato. *Journal of Agricultural and Food Chemistry*, 51, 3313–3319.

Ching, L.S. and Mohamed, S. (2001) α-Tocopherol content in 62 edible tropical plants. *Journal of Agricultural and Food Chemistry*, 49, 3101–3105.

Christensen, L.P. and Brandt, K. (2006) Bioactive polyacetylenes in food plants of the Apiaceae family: Occurrence, bioactivity and analysis. *Journal of Pharmacuetical and Biomedical Analysis*, 41, 683–693.

Chun, J., Lee, J., Ye, L., Exler, J. and Eitenmiller, R.R. (2006) Tocopherol and tocotrienol contents of raw and processed fruits and vegetables in the United States diet. *Journal of Food Composition and Analysis*, 19, 196–204.

Chun, O.K., Kim, D.O., Moon, H.Y., Kang, H.G. and Lee, C.Y. (2003) Contribution of individual polyphenolics to total antioxidant capacity of plums. *Journal of Agricultural and Food Chemistry*, 51, 7240–7245.

Cieślik, E., Leszczyńska, T., Filipiak-Florkiewiez, A., Sikora, E. and Pisulewski, P.M. (2007) Effects of some technological processes on glucosinolate content in cruciferous vegetables. *Food Chemistry*, 105, 976–981.

Ciska, E., Martyniak-Przybyszewska, B. and Kozlowska, H. (2000) Content of glucosinolates in cruciferous vegetables grown at the same site for two years under different climatic conditions. *Journal of Agricultural and Food Chemistry*, 48, 2862–2867.

Clifford, M.N. (2000) Anthocyanins—nature, occurrence and dietary burden. *Journal of the Science of Food and Agriculture*, 80, 1063–1072.

Czepa, A. and Hofmann, T. (2003) Structural and sensory characterization of compounds contributing to the bitter off-taste of carrots (*Daucus carota* L.) and carrot puree. *Journal of Agricultural and Food Chemistry*, 51, 3865–3873.

Diawara, M.M., Trumble, J.T., Quiros, C.F. and Hansen, R. (1995) Implications of distribution of linear furanocoumarins within celery. *Journal of Agricultural and Food Chemistry*, 43, 723–727.

Duan, X., Jian, Y., Su, X., Zhang, Z. and Shi, J. (2007) Antioxidant properties of anthocyanins extracted from litchi (*Litchi chinenesis* Sonn.) fruit pericarp tissues in relation to their role in the pericarp browning. *Food Chemistry*, 101, 1365–1371.

Englberger, L., Schierle, J., Marks, G.C. and Fitzgerald, M.H. (2003) Micronesian banana, taro, and other foods: newly recognized sources of provitamin A and other carotenoids. *Journal of Food Composition and Analysis*, 16, 3–19.

Fahey, J.W., Zalcmann, A.T. and Talalay, P. (2001) The chemical diversity and distribution of glucosinolates and isothiocyanates among plants. *Phytochemistry*, 56, 5–51.

Ferreres, F., Gil, M.I., Castañer, M. and Tómas-Barberán, F.A, (1997) Phenolic metabolites in red pigmented lettuce (*Lactuca sativa*). Changes with minimal processing and cold storage. *Journal of Agricultural and Food Chemistry*, 45, 4249–4254.

Finotti, E., Bertone, A. and Vivanti, V. (2006) Balance between nutrients and anti-nutrients in nine Italian potato cultivars. *Food Chemistry*, 99, 698–701.

Francisco, M., Moreno, D.A., Cartea, M.E., Ferreres, F., García-Viguera, C. and Velasco, P. (2009) Simultaneous identification of glucosinolates and phenolic compounds in a representative collection of vegetable *Brassica rapa*. *Journal of Chromatography A*, 1216, 6611–6619.

Friedman, M., Fitch, T.E. and Yokoyama, W.H. (2000) Lowering of plasma LDL cholesterol in hamsters by the tomato glycoalkaloid tomatine. *Food and Chemical Toxicology*, 38, 548–553.

Friedman, M. and Levin, C.E. (1995) α-Tomatine content in tomato and tomato products determined by HPLC with pulsed amperometric detection. *Journal of Agricultural and Food Chemistry*, 43, 1507–1511.

Friedman, M. and Levin, C.E. (1998) Dehydrotomatine content in tomatoes. *Journal of Agricultural and Food Chemistry*, 46, 4571–4576.

Friedman, M., Roitman, J.N. and Kozukue, N. (2003) Glycoalkaloid and calystegine contents of eight potato cultivars. *Journal of Agricultural and Food Chemistry*, 51, 2964–2973.

Garrod, B. and Lewis, B.G. (1979) Location of the antifungal compound falcarindiol in carrot root tissue. *Transactions of the British Mycological Society*, 72, 515–517.

Gerendás, J., Podestát, J., Stahl, T., Kübler, K., Brückner, H., Mersch-Sundermann, V. and Mühling, K.H. (2009) Interactive effects of sulfur and nitrogen supply on the concentration of sinigrin and allyl isothiocyanate in Indian mustard (*Brassica juncea* L.). *Journal of Agricultural and Food Chemistry*, 57, 3837–3844.

Gerendás, J., Sailer, M., Fendrich, M.-L., Stahl, T., Mersch-Sundermann, V. and Mühling, K.H. (2008) Influence of sulfur and nitrogen supply on growth, nutrient status and concentration of benzyl-isothiocyanate in cress (*Lepidium sativum* L.). *Journal of the Science of Food and Agriculture*, 88, 2576–2580.

Giovanelli, G. and Buratti, S. (2009) Comparison of polyphenolic composition and antioxidant activity of wild Italian blueberries and some cultivated varieties. *Food Chemistry*, 112, 903–908.

Giusti, M.M. and Wrolstad, R.E. (2003) Acylated anthocyanins from edible sources and their applications in food systems. *Biochemical Engineering Journal*, 14, 217–225.

Giusti, M.M., Rodríguez-Saona, L.E., Baggett, J.R., Reed, G.L., Durst, R.W. and Wrolstad, R.E. (1998) Anthocyanin pigment composition of red radish cultivars as potential food colorants. *Journal of Food Science*, 63, 219–224.

Graβman, J. (2005) Trepenoids as plant antioxidants. Vitamin and Hormones: In Gerald Litwack (ed.) *Plant Hormones* Volume 72, Academic Press, London, pp. 505–535.

Grelak, K., Milala, J., Król, B., Adamicki, F. and Badełek, E. (2009) Content of quercetin glycosides and fructooligosaccharides in onion stored in a cold room. *European Food Research Technology*, 28, 1001–1007.

Griffiths, D.W., Bain, H. and Dale, M.F.B. (1998) Effect of storage temperature on potato (*Solanum tuberosum* L.) tuber glycoalkaloid content and the subsequent accumulation of glycoalkaloids and chlorophyll in response to light exposure. *Journal of Agricultural and Food Chemistry*, 46, 5262–5268.

Griffiths, D.W., Dale, M.F.B. and Bain, H. (1994) The effect of cultivar, maturity and storage on photoinduced changes in the total glycoalkaloid and chlorophyll content of potatoes (*Solanum tuberosum*). *Plant Science*, 98, 103–109.

Griffiths, W.D., Deighton, N., Birch, A.N.E., Patrian, B., Baur, R. and Städler, E. (2001) Identification of glucosinolates on the leaf surface of plants from the Cruciferae and other closely related species. *Phytochemistry*, 57, 693–700.

Hackett, M.M., Lee, J.H., Francis, D. and Schwartz, S.J. (2004) Thermal stability and isomerization of lycopene in tomato oleoresins from different varieties. *Journal of Food Sciences*, 69, 536–541.

Han, Y-F., Cao, G-X., Gao, X-J. and Xia, M. (2010) Isolation and characterisation of the sesquiterpene lactones from *Lactuca sativa* L var. *anagustata*. *Food Chemistry*, 120, 1083–1088.

Harborne, J.B. (ed.) (1993) *The Flavonoids; Advances in Research Since 1986*, Chapman and Hall, London.

Hernández, M., Rodríguez, E. and Díaz, C. (2007) Free hydroxycinnamic acids, lycopene, and color parameters in tomato cultivars. *Journal of Agricultural and Food Chemistry*, 55, 8604–8615.

Hogan, S., Zhang, L., Li, J., Zoecklein, B. and Zhou, K. (2009) Antioxidant properties and bioactive components of Norton (*Vitis aestivalis*) and Cabernet Franc (*Vitis vinifera*) wine grapes. *LWT – Food Science and Technology*, 42, 1269–1274.

Holden, J.M., Eldridge, A.L., Beecher, G.R., Buzzard, I.M., Bhagwat, S., Davis, C.S., Douglass, L.W., Gebhardt, S., Haytowitz, D. and Schake, S.I. (1999) Carotenoid content of U.S. foods: An update of the database. *Journal of Food Composition and Analysis*, 12, 1696–1196.

Hollman, P.C.H. and Arts, I.C.W. (2000) Flavonols, flavones and flavanols—nature, occurrence and dietary burden. *Journal of the Science of Food and Agriculture*, 80, 1081–1093.

Hong, E., Kim, S-J. and Kim, G-H. (2011) Identification and quantitative determination of glucosinolates in seeds and edible parts of Korean Chinese cabbage. *Food Chemistry*, 128, 1115–1120.

Horvath, G., Wessjohann, L., Bigirimana, J., Monica, H., Jansen, M., Guisez, Y., Caubergs, R. and Horemans, N. (2006) Accumulation of tocopherols and tocotrienols during seed development of grape (*Vitis vinifera* L. cv. Albert Lavallée). *Plant Physiology and Biochemistry*, 44, 724–731.

Häkkinen, S. H., Kärenlampi, S.O., Heinonen, I. M., Mykkänen, H. M. and Törrönen, A.R. (1999) Content of the flavonols quercetin, myricetin, and kaempferol in 25 edible berries. *Journal of Agricultural and Food Chemistry*, 47, 2274–2279.

Iriwoharn, T. and Wrolstad, R.E. (2004) Polyphenolic composition of marion and evergreen blackberries. *Journal of Food Science*, 69, 233–240.

Isabelle, M., Lee, B.L., Lim, M.T., Koh, W-P., Huang, D. and Ong, C.N. (2010) Antioxidant activity and profiles of common fruits in Singapore. *Food Chemistry*, 123, 77–84.

Jadhav, S.J. (1991) Trepenoid phytoalexins in potatoes: A review. *Food Chemistry*, 41, 195–217.

Jansen, G. and Flamme, W. (2006) Coloured potatoes (*Solanum tuberosum* L.) – anthocyanin content and tuber quality. *Genetic Resources and Crop Evolution*, 53, 1321–1331.

Kalt, W., Forney, C.F., Martin, A. and Prior, R.L. (1999) Antioxidant capacity, vitamin C, phenolics, and anthocyanins after fresh storage of small fruits. *Journal of Agricultural and Food Chemistry*, 47, 4638–4644.

Kandlakunta, B., Rajendran, A. and Thingnganing, L. (2008) Carotene content of some common (cereals, pulses, vegetables, spices and condiments) and unconventional sources of plant origin. *Food Chemistry*, 106, 85–89.

Kaur, C., Kumar, K., Anil, D. and Kapoor, H.C. (2007) Variations in antioxidant activity in broccoli (*Brassica oleracea* L.) cultivars. *Journal of Food Biochemistry*, 31, 621–638.

Kidmose, U., Hansen, S.L., Christensen, L.P., Edelenbos, M., Larsen, E. and Norbaek, R. (2004) Effects of genotype, root size, storage, and processing on bioactive compounds in organically grown carrots (*Daucus carota* L.). *Journal of Food Science*, 69, S388–S394.

Kidmose, U., Knuthsen, P., Edelenbos, M., Justesen, U. and Hegelund, E. (2001) Carotenoids and flavonoids in organically grown spinach (*Spinacia oleracea* L.) genotypes after deep frozen storage. *Journal of the Science of Food and Agriculture*, 81, 918–923.

Kim, J.K., Chu, S.M., Kim, S.J., Lee, D.J., Lee, S.Y., Lim, S.H. Ha, S-H., Kweon, S.J. and Cho, H.S. (2010) Variation of glucosinolates in vegetable crops of *Brassica rapa* L. ssp. Pekinensis. *Food Chemistry*, 119, 423–428.

Kim, Y., Giraud, D.W. and Driskell, J.A. (2007) Tocopherol and carotenoid contents of selected Korean fruits and vegetables. *Journal of Food Composition and Analysis*, 20, 458–465.

Kim, Y-K., Ishii, G. and Kim, G-H. (2004) Isolation and identification of cabbages glucosinolates in edible (*Brassica campestris* L. ssp. Peckinensis). *Food Science and Technology Research*, 10, 469–473.

Kobæk-Larsen M., Christensen, L.P., Vach, W., Ritskes-Hoitinga, J. and Brandt, K. (2005) Inhibitory effects of feeding with carrots or (–)- falcarinol on development of azoxymethane-induced preneoplastic lesions in the rat colon. *Journal of Agricultural and Food Chemistry*, 53, 1823–1827.

Kodamatani, H., Saito, K., Niina, N., Yamazaki, S. and Tanaka, Y. (2005) Simple and sensitive method for determination of glycoalkaloids in potato tubers by high-performance liquid chromatography with chemiluminescence detection. *Journal of Chromatography A*, 1100, 26–31.

Koh, E., Wimalasiri, K.M.S., Chassy, A.W. and Mitchell, A.E. (2009) Content of ascorbic acid, quercetin, kaempferol and total phenolics in commercial broccoli. *Journal of Food Composition and Analysis*, 22, 637–643.

Kreutzmann, S., Christensen, L.P. and Edelenbos, M. (2008) Investigation of bitterness in carrots (*Daucus carota* L.) based on quantitative chemical and sensory analyses. *LWT – Food Science and Technology*, 41, 193–205.

Kugler, F., Stintzing, F.C. and Carle, R. (2004) Identification of betalains from petioles of differently coloured Swiss chard (*Beta vulgaris* L. ssp. cicla [L.] Alef. cv. Bright Lights) by high performance liquid chromatography–electrospray ionization mass spectrometry. *Journal of Agricultural and Food Chemistry*, 52, 2975– 2981.

Kurilich, A.C., Tsau, G.J., Brown, A., Howard, L., Klein, B.P., Jeffery, E.H., Kushad, M., Wallig, M.A. and Juvik, J.A. (1999) Carotene, tocopherol, and ascorbate contents in subspecies of *Brassica oleracea*. *Journal of Agricultural and Food Chemistry*, 47, 1576–1581.

Kushad, M.M., Brown, A.F., Kurilich, A.C., Juvik, J.A., Klein, B.P., Wallig, M.A. and Jeffery, E.H. (1999) Variation of Glucosinolates in Vegetable Subspecies of Brassica oleracea. *Journal of Agricultural and Food Chemistry*, 47, 1541–1548.

Kushad, M.M., Brown, A.F., Kurilich, A.C., Juvik, J.A., Klein, B.K., Wallig, M.A. and Jeffery, E.H. (1999) Variation of glucosinolates in vegetable subspecies of *Brassica oleracea*. *Journal of Agricultural and Food Chemistry*, 47, 1541–1548.

Lako, J., Trenerry, V.C., Wahlqvist, M., Wattanapenpaiboon, N., Sotheeswaran, S. and Premier, R. (2007) Phytochemical flavonols, carotenoids and the antioxidant properties of a wide selection of Fijian fruit, vegetables and other readily available foods. *Food Chemistry*, 101, 1727–1741.

Lako, J., Trenerry, V.C., Wahlqvist, M., Wattanapenpaiboon, N., Sotheeswaran, S. and Premier, R. (2007) Phytochemical flavonols, carotenoids and the antioxidant properties of a wide selection of Fijian fruit, vegetables and other readily available foods. *Food Chemsitry*, 101, 1727–1741.

Lee, J.J., Crosby, K.M., Pike, L.M., Yoo, K.S. and Leskovar, D.I. (2005) Impact of genetic and environmental variation on development of flavonoids and carotenoids in pepper (*Capsicum* spp.). *Scientia Horticulturae*, 106, 341–352.

Lewis, C.E., Walkel, J.R.L., Lancaster, J.E. and Sutton. K.H. (1998) Determination of anthocyanins, flavonoids and phenolic acids in potatoes. I: Coloured cultivars of *Solanum tuberosum* L. *Journal of the Science of Food and Agriculture*, 77, 45–57.

Li, S., Schonhof, I., Krumbein, A., Li, L., Stützel, H. and Schreiner, M. (2007) Glucosinolate concentration in turnip (*Brassica rapa ssp. rapifera* L.) roots as affected by nitrogen and sulfur supply. *Journal of Agricultural and Food Chemistry*, 55, 8452–8457.

Liu, M., Li, X.Q., Weber, C., Lee, C.Y., Brown, J. and Liu, R.H. (2002) Antioxidant and antiproliferative activities of raspberries. *Journal of Agricultural and Food Chemistry*, 50, 2926–2930.

Liu, Y., Perera, C.O. and Suresh, V. (2007) Comparison of three chosen vegetables with others from south East Asia for their lutein and zeaxanthin content. *Food Chemistry*, 101, 1533–1539.

Llorach, R., Martínez-Sánchez, A., Tomás-Barberán, F.A., Gill, M. I. and Ferreres, F. (2008) Characterisation of polyphenols and antioxidant properties of five lettuce varieties and escarole. *Food Chemistry*, 108, 1028–1038.

Lu, Q-Y., Zhang, Y., Wang, Y., Wang, D., Lee, R-P., Gao, K., Byrns, R. and Heber, D. (2009) California Hass avocado: profiling of carotenoids, tocopherol, fatty acid, and fat content during maturation and from different growing areas. *Journal of Agricultural and Food Chemistry*, 57, 10408–10413.

Marín, A., Ferreres, F., Tomas-Barberan, F.A. and Gil, M.I., (2004) Characterization and quantification of antioxidant constituents of sweet pepper (*Capsicum annuum* L.). *Journal of Agricultural and Food Chemistry*, 52, 3861–3869.

Marinova, D. and Ribarova, F. (2007) HPLC determination of carotenoids in Bulgarian berries. *Journal of Food Composition and Analysis*, 20, 370–374.

Marotti, M. and Piccaglia, R. (2002) Characterization of flavonoids in different cultivars of onions (*Allium cepa* L.). *Journal of Food Sciences*, 67, 1229–1232.

Martínez-Valverde, I., Periago, M.J., Provan, G. and Chesson, A. (2002) Phenolic compounds, lycopene and antioxidant activity in commercial varieties of tomato (*Lycopersicon esculentum*). *Journal of the Science of Food and Agriculture*, 82, 323–330.

Mazza, G. and Cottrell, T. (2008) Carotenoids and cyanogenic glucosides in saskatoon berries (*Amelanchier alnifolia* Nutt.). *Journal of Food Composition and Analysis*, 21, 249–254.

Mercier, J., Aurl, J. and Julien, C. (1994) Effects of food preparation on the isocoumarin, 6-methoxymellein, content of UV-treated carrots. *Food Research International*, 27, 401–404.

Metzger, B.T. and Barnes, D.M. (2009) Polyacetylene diversity and bioactivity in orange market and locally grown colored carrots (*Daucus carota* L.). *Journal of Agricultural and Food Chemistry*, 57, 11134–11139.

Meyers, K.J., Watkins, C.B., Pritts, M.P. and Liu, R.H. 2003. Antioxidant and antiproliferative activities of strawberries. *Journal of Agricultural and Food Chemistry*, 2003, 51, 6887–6892.

Montealegre, R.R., Peces, R.R., Vozmediano, J.L.C., Gascuena, C.V.J.M. and Romero, E.G. (2006) Phenolic compounds in skins and seeds of ten grape *Vitis vinifera* varieties grown in a warm climate. *Journal of Food Composition and Anallysis*, 19, 687–693.

Moussa-Ayoub, T.E., El-Samahy, S.K., Rohn, S. and Kroh, L.W. (2011) Flavonols, betacyanins content and antioxidant activity of cactus *Opuntia macrorhiza* fruits. *Food Research International*, in press, corrected proof.

Murkovic, M., Mulleder, U. and Neunteuff, H. (2002) Carotenoid content in different kinds of pumpkins. *Journal of Food Composition and Analysis*, 15, 633–638.

Naczk, M. and Shahidi, F. (2006) Phenolics in cereals, fruits and vegetables: Occurrence, extraction and analysis. *Journal of Pharmaceutical and Biomedical Analysis*, 41, 1523–1542.

Nakamura, Y., Tsuji, S. and Tonogai, Y. (2003) Analysis of proanthocyanidins in grape seed extracts, health foods and grape seed oils. *Journal of Health Science*, 49, 45–54.

Nemzer, B., Pietrzkowski, Z., Spórna, A., Stalica, P., Thresher, W., Michalowski, T. and Wybraniec, S. (2011) Betalainic and nutritional profiles of pigment-enriched red beet root (*Beta vulgaris* L.) dried extracts. *Food Chemistry*, 127(1), 42–53.

Netzel, M., Netzel, G., Tian, Q., Schwartz, S. and Konczak, I. (2007) Native Australian fruits — a novel source of antioxidants for food. *Innovative Food Science and Emerging Technologies*, 8, 339–346.

Nigg, H.N., Nordby, H.E., Beir, R.C., Dillman, A., Macias, C. and Hansen, R.C. (1993) Phytotoxic coumarins in limes. *Food Chemical and Toxicology*, 31, 331–335.

Niizu, P.Y., Rodriguez-Amaya, D.B. (2005) New data on the carotenoid composition of raw salad vegetables. *Journal of Food Composition and Analysis*, 18, 739–749.

Ogawa, K., Kawasaki, A., Yoshida, T., Nesumi, H., Nakano, M., Ikoma, Y. and Yano, M. (2000) Evaluation of auraptene content in citrus fruits and their products. *Journal of Agricultural and Food Chemistry*, 48, 1763–1769.

Olsson, K. and Svensson, R. (1996) The influence of polyacetylenes on the susceptibility of carrots to storage diseases. *Journal of Phytopathoogy*, 144, 441–447.

Omirou, M., Papadopoulou, K.K., Papastylianou, I., Constantinou, M., Karpouzas, D.G., Asimakopoulos, I. and Ehaliotis, C. (2009) Impact of nitrogen and sulfur fertilization on the composition of glucosinolates in relation to sulfur assimilation in different plant organs of broccoli. *Journal of Agricultural and Food Chemistry*, 57, 9408–9417.

Ostertag, E., Becker, T., Ammon, J., Bauer-Aymanns, H. and Schrenk, D. (2002) Effects of storage conditions on furocoumarin levels in intact, chopped, or homogenized parsnips. *Journal of Agricultural and Food Chemistry*, 50, 2565–2570.

Padilla, G., Cartea, M.E., Velasco, P., de Haro, A. and Ordás, A. (2007) Variation of glucosinolates in vegetable crops of *Brassica rapa*. *Phytochemistry*, 68, 536–545.

Pande, G. and Akoh, C.C. (2009) Antioxidant capacity and lipid characterization of six Georgia-grown pomegranate cultivars. *Journal of Agricultural and Food Chemistry*, 57, 9427–9436.

Pantelidis, G.E., Vasilakakis, M., Manganaris, G.A. and Diamantidis, G. (2007) Antioxidant capacity, phenol, anthocyanin and ascorbic acid contents in raspberries, blackberries, red currants, gooseberries and cornelian cherries. *Food Chemistry*, 102, 777–783.

Pastrana-Bonilla, E., Akoh, C.C., Sellappan, S. and Krewer, G. (2009) Phenolic content and antioxidant capacity of muscadine grapes. *Journal of Agricultural and Food Chemistry*, 51, 5497–5503.

Pęksa, A., Gołubowska, G., Rytel, E., Lisin´ska, G., and Aniołowski, K. (2002) Influence of harvest date on glycoalkaloid content of three potato varieties. *Food Chemistry*, 78, 313–317.

Pereira, F.M.V., Rosa. E., Fahey. J.W., Stephenson, K.K., Carvalho, R. and Aires, A. 2002. Influence of temperature and ontogeny on the levels of glucosinolates in broccoli (*Brassica oleracea* var. *italica*) sprouts and their effect on the induction of mammalian phase 2 enzymes. *Journal of Agricultural and Food Chemistry*, 50, 6239–6244.

Pérez-Gregorioa, R.M., García-Falcóna, M.S., Jesús Simal-Gándaraa Sofia Rodrigues, A.S. and Almeida, D.P.F. (2010) Identification and quantification of flavonoids in traditional cultivars of red and white onions at harvest. *Journal of Food Composition and Analysis*, 23, 592–598.

Perkins-Veazie, P., Collins, J.K., Pair, S.D. and Roberts, W. (2001) Lycopene content differs among red-fleshed watermelon cultivars. *Journal of the Science of Food and Agriculture*, 81, 983–987.

Perkins-Veazie, P., Collins, J.K., Pair, S.W. and Roberts, W. (2006) Lycopene content differs among red-fleshed watermelon cultivars. *Journal of the Science of Food and Agriculture*, 81, 983–987.

Peters, A. M., Haagsma, N., Gensch, K.-H. and van Amerongen, A. (1996) Production and characterization of polyclonal antibodies against the bitter sesquiterpene lactones of chicory (*Cichorium intybus* L.). *Journal of Agricultural and Food Chemistry*, 44, 3611–3615.

Pferschy-Wenzig, E-M., Getzinger, V., Kunert, O., Woelkart, K., Zahrl, J. and Bauer, R. (2009) Determination of falcarinol in carrot (*Daucus carota* L.) genotypes using liquid chromatography/mass spectrometry. *Food Chemistry*, 1083–1090.

Piironen, V., Syväoja, E.L., Varo, P., Salminen, K. and Koivistoinen, P. (1986) Tocopherols and tocotrienols in Finnish foods: vegetables, fruits and berries. *Journal of Agricultural and Food Chemistry*, 34, 742–746.

Podsędek, A., Sosnowska, D., Redzynia, M. and Anders, B. (2006) Antioxidant capacity and content of *Brassica oleracea* dietary antioxidants. *International Journal of Food Science and Technology*, 41 (Supplement 1), 49–58.

Rentzsch, M., Schwarz, M. and Winterhalter, P. (2007) Pyranoanthocyanins–an overview on structures, occurrence, and pathways of formation. *Trends in Food Science and Technology*, 18, 526–534.

Rodriguez-Saona L.E., Giusti M.M. and Wrolstad R.E. (1998) Anthocyanin pigment composition of red-fleshed potatoes. *Journal of Food Science*, 63, 458–465.

Rosa, E. and Rodriques, A.S. (2001) Total and individual glucosinolate content in 11 broccoli cultivars grown in early and late seasons. *HortScience*, 36, 56–59.

Rumbaoa, R.G.O., Cornage, D.F. and Geronimo, I.M. (2009) Phenolic content and antioxidant capacity of Philippine potato (*Solanum tuberosum*) tubers. *Journal of Food Composition and Analysis*, 22, 546–550.

Sanchez, M.C., Cao, G., Ou, B. and Prior, R.L. (2003) Anthocyanin and proanthocyanidin content in selected white and red wines. Oxygen radical absorbance capacity comparison with nontraditional wines obtained from highbush blueberry. *Journal of Agricultural and Food Chemistry*, 51, 4889–4896.

Sariburun, E., Şahin, S., Demir, C., Türkben, C. and Uylaşer (2010) Phenolic content and antioxidant activity of raspberry and blackberry cultivars. *Journal of Food Science*, 75, C328– C335.

Schonhof, I., Krumbein, A. and Brückner, B. (2004) Genotypic effects on glucosinolates and sensory properties of broccoli and cauliflower. *Nahrung*, 48, 25–33.

Schwartz, E., Tzulker, R., Glazer, I., Bar-Ya'akov, I., Wiesman, Z., Tripler, E., Bar-Ilan, I., Fromm, H., Borochov-Neori, H., Holland, D. and Amir, R. (2009) Environmental conditions affect the color, taste, and antioxidant capacity of 11 pomegranate accessions' fruits. *Journal of the Science of Food and Agriculture*, 57, 9197–9209.

Schwarz, M., Wray, V. and Winterhalter, P. (2004) Isolation and identification of novel pyranoanthocyanins from black carrot (*Daucus carota* l.) juice. *Journal of Agricultural and Food Chemistry*, 52, 5095–5101.

Sellappan, S., Akoh, C.C. and Krewer, G. (2002) Phenolic compounds and antioxidant capacity of Georgia-grown blueberries and blackberries. *Journal of Agricultural and Food Chemistry*, 50, 2432–2438.

Seo, M.W. and Yang, D.S. (2009) Sesquiterpene lactones and bitterness in Korean leaf cultivars. *HortScience*, 44, 246–249.

Sevenson, J., Smallfield, B.M., Joyce, N.I., Sansom, C.E. and Perry, N.B. (2008) Betalains in red and yellow varieties of the Andean tuber crop ulluco (*Ullucus tuberosus*). *Journal of Agricultural and Food Chemistry*, 56, 7730–7737.

Simon, P.W. (1982) Genetic variation for volatile terpenoids in roots of carrot, Daucus carota, backcrosses and F2 generations. *Phytochemistry.* 21, 875–879.

Singh, J., Upadhyay, A.K., Prasad, K., Bahadur, A. and Rai, M. (2007) Variability of carotenes, vitamin C, E and phenolics in *Brassica vegetables*. *Journal of Food Composition and Analysis*, 20, 106–112.

Slimestad, R., Vangdal, E. and Brede, C. (2009) Analysis of phenolic compounds in six Norwegian plum cultivars (*Prunus domestica* L.). *Journal of Agricultural and Food Chemistry*, 57, 11370–11375.

Slimestad, R. and Verheul, M. (2009) Review of flavonoids and other phenolics from fruits of different tomato (*Lycopersicon esculentum* Mill.) cultivars. *Journal of the Science of Food and Agriculture*, 89, 1255–1270.

Stewart, A.J., Bozonett, S., Mullen, W., Jenkins, G.I., Lean, M.E.J. and Crozier, A. (2000) Occurrence of flavonols in tomatoes and tomato-based products. *Journal of Agricultural and Food Chemistry*, 48, 2663–2669.

Stintzing, F.C. and Carle, R. (2007) Betalains – emerging prospects for food scientists. *Trends in Food Science and Technology*, 18(10), 514–525.

Strack, D., Vogt, T. and Schliemann, W. (2003) Recent advances in betalain research. *Phytochemistry*, 62, 247–269.

Sun, B., Liu, N., Zhao, Y., Yan, H. and Wang, Q. (2011) Variation of glucosinolates in three edible parts of Chinese kale (*Brassica alboglabra* Bailey) varieties. *Food Chemistry*, 124, 941–947.

Sun, J., Chu, Y.F., Wu, X. and Liu, R.H. (2002) Antioxidant and antiproliferative activities of common fruits. *Journal of Agricultural and Food Chemistry*, 50, 7449–7454.

Surles, R.L., Weng, N., Simon, P.W. and Tanumihardjo, S.A. (2004) Carotenoid profiles and consumer sensory evaluation of specialty carrots (*Daucus carota*, L.) of various colors. *Journal of Agricultural and Food Chemistry*, 52, 3417–3421.

Søltoft, M., Eriksen, M.R., Träger, A.W.B., Nielsen, J., Laursen, K.H., Husted, S., Halekoh, U. and Knuthsen, P. (2010) Comparison of polyacetylene content in organically and conventionally grown carrots using a fast ultrasonic liquid extraction method. *Journal of Agricultural and Food Chemistry*, 58, 7673–7679.

Talcott, S.T. and Howard, R. (1999) Chemical and sensory quality of processed carrot puree as influenced by stress-induced phenolic compounds. *Journal of Agricultural and Food Chemistry*, 47, 1362–1366.

Teow, C.C., Truong, V.D., McFeeters, R.F., Thompson, R.L., Pecota, K.V. and Yencho, G.C. (2007) Antioxidant activities, phenolic and β-carotene contents of sweet potato genotypes with varying flesh colours. *Food Chemistry*, 103, 829–838.

Thompson, K.A., Marshall, M.R., Sims, C.A., Wei, C.I., Sargent, S.A. and Scott, J.W. (2000) Cultivar, maturity and heat treatment on lycopene content in tomatoes. *Journal of Food Science*, 65, 791–795.

Tiwari, B.K., O'Donnell, C.P. and Cullen, P.J. (2009) Effect of non-thermal technologies on anthocyanin content of fruit juices. *Trends in Food Science and Technology*. 20, 137–145.

Tiwari, U. and Cummins, E. (2009) Nutritional importance and effect of processing on tocols in cereals. *Trends in Food Science and Technology*, 20, 511–520.

Toor, R.K, Lister, C.E. and Savage, G.P. (2005) Antioxidant activities of New Zealand-grown tomatoes. *International Journal of Food Science and Nutrition*, 56, 597–605.

Tsao, R., Yang, R., Young, C. and Zhu, H. (2003) Polyphenolic profiles in eight apple cultivars using High-performance liquid chromatography (HPLC). *Journal of Agricultural and Food Chemistry*, 51, 6347–6353.

USDA. USDA Database for the Flavonoid Content of Selected Foods, 2003. [Online]. Available: http://www.nal.usda.gov/fnic/foodcomp/Data/Flav/flav.pdf [12 January 2011]

USDA, Agricultural Research Service Database for the Flavonoid Content of selected foods, release 2.1. 2007. [Online]. Available: http://www.ars.usda.gov/Services/docs.htm?docid=6231 [12 January 2011].

Usenik, V., Stampar, F. and Veberic, R. (2009) Anthocyanins and fruit colour in plums (*Prunus domestica* L.) during ripening. *Food Chemistry*, 14, 529–534.

Valavanidis, A., Vlachogianni, T., Psomas, A., Zovoili., A. and Siatis, V. (2009) Polyphenolic profile and antioxidant activity of five apple cultivars grown under organic and conventional agricultural practices. *International Journal of Food Science and Technology*, 44, 1167–1175.

Vallejo, F., Tomas-Barberan, F. and Garcia-Viguera, C. (2003) Health promoting compounds in broccoli as influenced by refrigerated transport and retail sale period. *Journal of Agricultural and Food Chemistry*, 51, 3029–3034.

VanEtten, H.D., Mansfield, J.W., Bailey, J.A. and Farmer, E.E. (1994) Two classes of plant antibiotics: phytoalexins versus phytoanticipinins. *Plant Cell, Tissue Organ Culture*, 9, 1191–1192.

Velasco, P., Cartea, M.E., González, C., Vilar, M. and Ordás, A. (2007) Factors affecting the glucosinolate content of kale (Brassica oleracea acephala group). *Journal of Agricultural and Food Chemistry*, 55, 955–962.

Volden, J., Borge, G.I.A., Hansen, M., Wicklund, T. and Bengtsson, G.B. (2009) Processing (blanching, boiling, steaming) effects on the content of glucosinolates and antioxidant-related parameters in cauliflower (*Brassica oleracea* L. ssp. botrytis). *LWT – Food Science and Technology*, 42, 63–73.

Walia, S., Singh, M., Kaur, C., Kumar, R. and Joshi, S. (2010) Antioxidant Composition of red and orange cultivars of tomatoes (*Solanum lycopersicon* L): A Comparative Evaluation. *Journal of Plant Biochemistry and Biotechnology*, 19, 95–97.

Wall, M.M. (2006) Ascorbic acid, vitamin A, and mineral composition of banana (*Musa* sp.) and papaya (*Carica papaya*) cultivars grown in Hawaii. *Journal of Food Composition and Analysis*, 19, 434–445.

Williner, M.R., Pirovani, M.E. and Güemes, D.R. (2003) Ellagic acid content in strawberries of different cultivars and ripening stages. *Journal of the Science of Food and Agriculture*, 83, 842–845.

Wolfe, K.L. and Liu, R.H. (2003) Apple peels as a value-added food ingredient. *Journal of Agricultural and Food Chemistry*, 51, 1676–1683.

Yang, S-Ah., Paek, S-H. and Kozukue, N. (2006) α-Chaconine, a potato glycoalkaloid, induces apoptosis of HT-29 human colon cancer cells through caspase-3 activation and inhibition of ERK 1/2 phosphorylation. *Food and Chemical Toxicology*, 44, 839–846.

Yilmaz, K.U., Ercisli, S., Zengin, Y., Sengul, M. and Kafkas, E.Y. (2009) Preliminary characterisation of cornelian cherry (*Cornus mas* L.) genotypes for their physico-chemical properties. *Food Chemistry*, 114, 408–412.

Young, J.F., Duthie, S.J., Milne, L., Christensen, L.P., Duthie, G.G. and Bestwick, C.S. (2007) Biphasic effect of falcarinol on CaCo-2 cell proliferation, DNA damage, and apoptosis. *Journal of Agricultural and Food Chemistry*, 55, 618–623.

Zidorn, C., Johrer, K., Ganzera, M., Schubert, B., Sigmund, E.M., Mader, J., Greil, R., Ellmerer, E.P. and Stuppner, H. (2005) Polyacetylenes from the Apiaceae vegetables carrot, celery, fennel, parsley, and parsnip and their cytotoxic activities. *Journal of Agricultural and Food Chemistry*, 53, 2518–2523.

6 Food grains

Sanaa Ragaee,[1] Tamer Gamel,[2] Koushik Seethraman,[1] and El-Sayed M. Abdel-Aal[2]

[1] Department of Food Science, University of Guelph, Guelph, Ontario, Canada
[2] Guelph Food Research Centre, Agriculture and Agri-Food Canada, Guelph, Ontario, Canada

6.1 Introduction

Cereals and pulses are staple foods for a majority of the world's population. They supply a large portion of nutrients in the human diet. Data have shown that cereals provide approximately 120 kg/capita/year, 880 kcal/capita/day and 25.8 g protein/capita/day, while pulses supply smaller amounts at roughly 7.8 kg/capita/year 74 kcal/capita/day and 5.1 g protein/capita/day (FAO, 2010). In addition, cereal and legume grains also have been recommended for healthy eating due to their content of health-promoting constituents such as dietary fiber and antioxidants. The USDA's Dietary Guidelines recommend about three to eight servings or ounce equivalents of grains per day subject to age, sex, and level of physical activity (USDA, 2010). At least half of the recommended grain servings should come from wholegrain. This recommendation is based on a body of evidence that has shown the positive relationship between consumption of wholegrain foods and health promotion such as reduced risk of cancer (Nicodemus *et al.*, 2001; Kasum *et al.*, 2002), type II diabetes (Meyer *et al.*, 2000; Fung *et al.*, 2002), and cardiovascular disease (Jacobs *et al.*, 1998; Anderson *et al.*, 2000). Canadian and European dietary guidelines also recommend consumption of similar amounts of grains to the USDA.

Grains are rich sources of many health-enhancing and/or disease-preventing components known as bioactive compounds. These components are mainly concentrated in the outer layers of the grain, which make wholegrain products healthier than their corresponding refined ones. Bioactive compounds include a wide array of plant constituents with diverse structures and functionalities such as dietary fiber, β-glucan, phenolics, anthocyanins, carotenoids, isoflavones, lignans, sterols, etc. Many of the bioactive compounds are phytochemicals produced by plants primarily for protection against predators and diseases. Phytochemicals also have been found to protect humans against certain chronic diseases. In general, phytochemicals are natural and non-nutritive bioactive compounds produced by plants that act as protective agents against external stress and pathogenic attack (Chew *et al.*, 2009). They are secondary metabolite that are crucial for plant defence and enable plants

Handbook of Plant Food Phytochemicals: Sources, Stability and Extraction, First Edition.
Edited by B.K. Tiwari, Nigel P. Brunton and Charles S. Brennan.
© 2013 John Wiley & Sons, Ltd. Published 2013 by John Wiley & Sons, Ltd.

to overcome temporary or continuous threats integral to their environment. Phytochemicals could exhibit bioactivities such as antimutagenic, anticarcinogenic, antioxidant, antimicrobial, and anti-inflammatory properties (Okarter and Liu, 2010). Only a small number of phytochemicals in grains has been investigated closely in terms of health benefits and stability during processing. The current chapter aims to discuss phytochemicals found in cereal and legume grains in terms of their occurrence, compositional properties, and stability during processing. Emphasis is put on dietary fiber, phenolics, carotenoids, anthocyanins, isoflavones, saponins, and lignans due to their potential role in human health. Examples of food applications that have demonstrated positive health effects in humans are also provided.

6.2 Phytochemicals in cereal grains

Cereal grains are a type of fruit called caryopsis that are composed of endosperm, germ and bran. The grains are a staple food that provides the main food energy supply. They also are rich in a variety of phytochemicals including dietary fibers (β-glucan, inulin, arabinxylan, resistant starch), phenolics (phenolic acids, alkylresorcinols and flavonoids), carotenoids (lutein, zanthein), anthocyanins and deoxyanthocyanins, tocols (tocopherols and tocotrienols), lignans, γ-oryzanols, sterols, and phytate. Antioxidant properties of cereal grains are mainly attributed to phenolic compounds and other phytochemicals (Ragaee et al., 2011, 2012a). Phytochemicals found in cereals are unique and complement those in fruits and vegetables when consumed together. For example ferulic acid and diferulates are predominantly found in grains but are not present in significant quantities in fruits and vegetables (Abdel-Aal et al., 2001; Bunzel et al., 2001). The majority of phytochemicals are present in the bran/germ fraction in bound form (76% in wheat, 85% in corn, and 75% in oat) (Liu, 2007). In wheat, the bran/germ fraction contribute to 83% of total phenolic content, 79% of total flavonoid content, 78% of total zeaxanthin, 51% of total lutein, and 42% of total β-cryptoxanthin (Liu, 2007). In addition, the type and concentration of phytochemicals vary among grains and genotypes (Adom et al., 2003). The main phytochemicals in cereal grains are summarized in section 6.2.1.

6.2.1 Dietary fiber

Dietary fiber is one of the major health-enhancing components in cereals, located mostly in the outer layers (pericarp, testa, and aleurone) (Selvendran, 1984). The pericarp contains insoluble fiber along with some other antioxidants bound to the cell walls. The aleurone layer has soluble and insoluble fiber, antioxidants, vitamins, and minerals, and the testa layers are composed of soluble and insoluble fiber, phenolic compounds, and other phytochemicals (Raninen et al., 2010). The main dietary fiber components in cereals are cellulose, arabinoxylans, and β-glucan (Brennan and Cleary, 2005). Barley and oat are especially rich in β-glucan (Brennan, 2005; Wood, 2007, 2010), while the major dietary fiber constituent in wheat and rye is arabinoxylan (Ragaee et al., 2001; Kamal-Eldin et al., 2009). Concentrations and type of each class of dietary fiber depend on type of cereal and/or variety (Ragaee et al., 2001; Gebruers et al., 2008; Ragaee et al., 2012b).

6.2.1.1 β-glucan

β-glucan is an important dietary fibre fraction commonly found in cell walls of many cereal grains such as oat and barley. The health-enhancing effects of β-glucan have been extensively discussed in a review article by Wood (2010). The lowering effect of β-glucan in oat and barley products on serum cholesterol is well documented (Queenan et al., 2007; Smith et al., 2008), and a health claim in this regard has been allowed in the USA, Canada and Europe. Most of wheat and rye β-glucan is insoluble and ranges 0.5–1.4% and 2.1–3.1%, respectively (Genc et al., 2001; Ragaee et al., 2001; Li et al., 2006), while most of oat and barley β-glucan is soluble ranging 3–8% (Colleoni-Sirghie et al., 2003; Yao et al., 2007).

6.2.1.2 Arabinoxylan

Arabinoxylan (AX) is a hemicellulose found in both the primary and secondary cell walls of cereal grains and constitutes the second most abundant biopolymer in plant biomass after cellulose (Gatenholm and Tenkanen, 2004). AX consists of copolymers of two pentose sugars, arabinose and xylose. Enzymatic hydrolysis of AX (during bread or beer production or in the colon upon ingestion of AX) yields arabinoxylan-oligosaccharides, consisting of arabinoxylooligosaccharides (AXOS) and xylooligosaccharides (XOS). There is evidence that AXOS and XOS exert prebiotic effects in the colon of humans and animals through selective stimulation of beneficial intestinal microbiota (Broekaert et al., 2011). AX is the main dietary fiber fraction in rye accounting for 9.1% (Åman et al., 1997; Ragaee et al., 2001), while wheat contains 6.7% AX (Lineback and Rasper, 1988).

6.2.1.3 Inulin

Inulin and oligofructose are fructans with a degree of polymerization of 2–60 and 2–20, respectively. They both resist hydrolysis by human alimentary enzymes because of the structural conformation of their glucosidic bridge ($\beta\ 2 \rightarrow 1$). Both inulin and oligofructose are fermented exclusively in the colon by colonic bifidobacteria and bacteroides (Flickinger et al., 2003). This fermentation process results in increased fecal bacterial biomass, decreased ceco-colonic pH, and the production of a large amount of fermentation products including short chain fatty acids which exert systemic effects on lipid metabolism. Wheat flour contains 1–4% fructan on a dry weight basis which provides 78% of the North American intake of oligosaccharides (Van Loo et al., 1999). Young barley kernels contain about 22% fructan (Van Loo et al., 1999) while rye grains contain a small amount.

6.2.1.4 Resistant starch

Resistant starch is a member of dietary fiber fractions found in cereal grains. There are five types of resistant starch (Englyst et al., 1992). These include the following:

- Physically trapped starch in which the granules are trapped within grain food matrices and its concentration and distribution is affected by food processing treatments.
- Resistant starch granules found in high-amylose cereal grains such as maize and wheat. The granules have crystalline regions that are less susceptible to digestion by acid or amylase enzymes. Food processes that are able to gelatinize such starches can aid in their digestion.
- Retrograded starch formed on processing and storage due to starch retrogradation. High-amylose starch retrogrades faster than normal or high amylopectin starch.
- Physically, chemically, or enzymatically modified starch. The structure is altered to enhance gelling and thickening properties of starches, and thus becomes resistant to digestion.

- Amylose-lipid complex starch that is a complex of fatty acids or monoacylglycerols with starch. In general, modified starches function similarly to dietary fiber in the human body and escape digestion and absorption in the small intestine and become a substrate for the colonic microflora in the large bowel.

6.2.2 Phenolic compounds

Phenolics include a variety of compounds bearing one or more hydroxyl groups such as phenolic acids and analogs, flavonoids, tannins, stilbenes, curcuminoids, coumarins, lignans, quinones, etc. They are ubiquitous in all plant organs and are therefore an integral part of the human diet (Kroon and Williamson, 2005; Balasundram et al., 2006; Dai and Mumper, 2010). They have been considered powerful antioxidants *in vitro* and *in vivo*. It has been proposed that the antioxidant properties of phenolic compounds can be mediated by the following mechanisms: (1) scavenging radical ROS (reactive oxygen species); (2) suppressing free radicals formation by inhibiting some enzymes or chelating trace metals involved in their production; (3) up-regulating or protecting antioxidant defence (Dai and Mumper, 2010). They also exhibit a wide range of physiological properties such as anti-allergenic, anti-artherogenic, anti-inflammatory, anti-microbial, and anti-thrombotic, and the relationship between plant phenolics intake and the risk of oxidative stress associated diseases such as cardiovascular disease, cancer, or osteoporosis has been evident (Rice-Evans et al., 1996; Manach et al., 2004; Lee et al., 2005; Scalbert et al., 2005). Phenolics are the main source of antioxidants in cereal grains concentrated mainly in the bran/germ fraction of the wholegrain wheat flour (83%) (Adom et al., 2005). The common phenolic compounds found in cereals include phenolic acids (mainly ferulic acid), flavonoids stilbenes, coumarins, tannins, proanthocyanidins, and anthocaynins. The content of phenolic compounds in cereal grains broadly vary and is dependent on grain type, genotype, part of the grain sampled, grain handling, and processing (Adom and Liu, 2002; Adom et al., 2003, 2005; Ragaee et al., 2012a). Most of the phenolic acids are found in the insoluble bound fraction (Moore et al., 2005). Ferulic acid (trans-4-hydroxy-3-methoxycinnamic acid) is one of the major phenolic acids found in wholegrain (Abdel-Aal et al., 2001) concentrated mainly in the aleurone, pericarp, and embryo cell wall. Among selected cereal grains and cultivars corn has been found to contain the highest concentration of ferulic acid followed by wheat, oat, and rice (Adom and Liu, 2002). Vanillic acid is the second abundant phenolic acid in wheat bran followed by syringic acid and *p*-coumaric acid (Kim et al., 2006). In wheat, the bran/germ fraction contributes about 79% of the total flavonoid content (Adom et al., 2005).

6.2.2.1 Anthocyanins

Anthocyanins are natural pigments located in the outer layers of specialty cereal grains such as blue and purple wheat, blue, purple and red corn, black and red rice, and blue barley (Abdel-Aal et al., 2006), while the natural pigments found in black sorghum are deoxyanthocyanins (Awika et al., 2004). The highest concentration of anthocyanin pigments in corn was found in the pericarp, whereas the aleurone layer contained small concentrations (Moreno et al., 2005). Anthocyanins have been recognized as health-enhancing substances due to their antioxidant capacity (Nam et al., 2006), anti-inflammatory (Tsuda et al., 2002), anti-cancer (Hyun and Chung, 2004), and hypoglycemic effects (Tsuda et al., 2003).

Anthocyanin pigments in cereal grains vary from a simple (few pigments) to complex (many pigments) profile (Abdel-Aal et al., 2006). Blue or purple wheat has an intermediate anthocyanin profile with four or five major anthocyanin pigments. Black and red rice grains exhibit a simple anthocyanin profile, while blue, pink, purple, and red corns show a complex profile having more than 20 anthocyanin pigments. The predominant anthocyanin compounds are cyanidin 3-glucoside in black and red rice, purple wheat and blue, purple and red corn, pelargonidin 3-glucoside in pink corn and delphinidin 3-glucoside in blue wheat (Abdel-Aal et al., 2006). The concentration of total anthocyanins vary among different cereal grains being approximately 3276 µg/g in black rice, 94 µg/g in red rice, and 27 µg/g in wild rice (Abel-Aal et al., 2006). In the same study, eight corn grains exhibiting blue, pink, purple, and red colors were found to contain a wide range of total anthocyanins as low as 51 µg/g and as high as 1277 µg/g, in which purple corn had the highest concentration followed by sweet scarlet red corn and shaman blue corn. The concentration of anthocyanins in a large population of blue wheat lines was found to range from 35 to 507 µg/g with a mean of 183 µg/g (Abdel-Aal and Hucl, 1999). Additionally, anthocyanin concentrations were significantly influenced by growing conditions and environment in blue and purple wheat grains and the environmental effect was much stronger in the purple wheat due to the pigment location in the outer pericarp or fruit coat (Abdel-Aal and Hucl, 2003).

6.2.2.2 Carotenoids

Carotenoids constitute the yellow pigments in cereal grains. They are potent antioxidants because of the long series of alternating double and single bonds (Okarter and Liu, 2010). Their concentrations in cereal grains vary from very low in white and red wheat to relatively high in einkorn and durum wheat (Abdel-Aal et al., 2002, 2007). The common carotenoids detected in cereals include lutein, zeaxanthin, β-cryptoxanthin, β-carotene, and α-carotene. In wheat lutein is the major carotenoid present in relatively high concentration ranging from 26.4 to 143.5 µg/100 g grain, followed by zeaxanthin ranging from 8.7 to 27.1 µg/100 g grain, and then β-cryptoxanthin ranging from 1.1 to 13.3 µg/100 g grain (Adom et al., 2003). Similar carotenoids profile was found in various wheat species but their concentration was significantly higher (Abdel-Aal et al., 2002, 2007). Among all wheat species einkorn (*Triticum monococcum*) exhibited the highest level of all-trans-lutein averaging 7.41 µg/g with small amounts of all-trans-zeaxanthin, cis-lutein isomers, and α-carotene. Durum, Kamut, and Khorasan (*Triticum turgidum*) had intermediate levels of lutein (5.41–5.77 µg/g), while common bread or pastry wheat (*Triticum aestivum*) had the lowest content (2.01–2.11 µg/g) (Abdel-Aal et al., 2002). Other cereals such as corn flours contain reasonable concentrations of the different carotenoids (β-cryptoxanthin 3.7 mg/kg, lutein content was 11.5 mg/kg, and zeaxanthin content was 17.5 mg/kg) (Brenna and Berardo, 2004).

6.2.2.3 Tocols

Tocols include two groups of related compounds called tocopherols (α, β, γ, δ- tocopherols) and tocotrienols (α, β, γ, δ- tocotrienols). They are fat-soluble antioxidants and also have vitamin E activity. Tocols are mostly present in the germ fraction (Liu, 2007). The concentration and distribution of tocols vary among cereal grains (oat, corn, barley, spelt wheat, durum wheat, soft wheat, and triticale) (Panfili et al., 2003). Barley and soft wheat have relatively higher concentration of tocols while spelt wheat has lower concentration. Barley has all eight tocol isomers, while spelt, durum wheat, soft wheat, and triticale have only five isomers. The α-tocopherol is present in all grains ranging from 4 mg/kg dry matter basis in corn to 16 mg/kg dry matter basis in soft wheat. β-tocotrienol is the predominant tocol in soft wheat, triticale, and spelt, followed by α-tocopherol, β-tocopherol and α-tocotrienol. α-tocotrienol was the predominant tocol in oat, followed by α-tocopherol, β-tocotrienol, β-tocopherol, and γ-tocopherol. γ-tocopherol is only present in oat, corn, and barley, and it is the predominant tocol in corn, followed by γ-tocotrienol, α-tocopherol, and β-tocotrienol, while γ-tocotrienol is only present in corn and barley. The main tocols in different wheat species and cultivars are β-T3 ranging from 9.6 to 23.2 µg/g, followed by α-T (5.5–11.9 µg/g), α-T3 (2.5–7.4 µg/g), and β -T (2.0–6.6 µg/g) (Abdel-Aal and Rabalski, 2008). Wheat species and groups showed significant differences in their contents of the four tocols due to the differences in genotype and origin. The contents of tocols in barley (Cavallero *et al.*, 2004) and rice (Sookwong *et al.*, 2007) were also found to be influenced by genotype and growing environment. Unlike wheat, γ-T3 and β-T3 were the predominant and smallest tocols in rice, respectively (Sookwong *et al.*, 2007).

6.2.2.4 Lignans

Lignans are a group of dietary phytoestrogen compounds found in the outer layers of cereal grains (Thompson *et al.*, 1991; Tham *et al.*, 1998). Total lignan content varies among cereal grains as well as within the same cereal species depending on genetic differences and environmental conditions (Smeds *et al.*, 2009). For example, lignan content in rye wholegrain ranges from 2500 to 6700 µg/100 g, while the range is 340–2270 µg/100 g in wheat wholegrain, and 820–2550 µg/100 g in oat wholegrain (Smeds *et al.*, 2009). There are seven dietary lignans: secoisolariciresinol, matairesinol, lariciresinol, pinoresinol, syringaresinol, 7-ydroxymatairesinol, and medioresinol. When consumed, plant lignans such as secoisolariciresinol and matairesinol are converted to the mammalian lignans, enterodiol, and enterolactone, by intestinal microflora in humans which have strong antioxidant activity and weak estrogenic activity that may account for their biological effects and health benefits (Thompson *et al.*, 1991; Wang and Murphy, 1994).

6.2.3 Other phytochemicals

Alkylresorcinols and alkenylresorcinols are mainly concentrated in the bran fraction of the grain (Ross *et al.*, 2003). Rye has the most total alkylresorcinol (734 µg/g dry weight), followed by wheat (583 µg/g) and barley (45 µg/g), while alkylresorcinols are not detected in any oat products, wholegrain buckwheat grits, millet grits, long grain parboiled rice, and corn grits (Mattila *et al.*, 2005). Rye is the only grain to have detectable amounts of the 15-carbon alkylresorcinol homologue. The 19- and 21-carbon homologues are prominent in wheat. The 25- carbon homologue is prominent in barley (Ross *et al.*, 2003). About 60% of the alkylresorcinol is absorbed from the small intestine by humans. Therefore, its presence in the serum can be used as a biomarker of wholegrain cereal intake (Ross *et al.*, 2003, 2004). These compounds also have antibacterial and antifungal protection and antioxidant activity *in vitro*.

Phytosterols are mainly found in oilseeds, wholegrain cereals, nuts, and legumes, and include stanols (sitostanol, campestanol, and stigmastanol) and sterols (sitosterol, campesterol, and stigmasterol). γ-oryzanols are compounds that consist of a phenolic acid esterified to a sterol. Common γ-oryzanol compounds include cycloartenyl ferulate, 24-methylenecycloartanylferulate, and campesteryl ferulate. γ-oryzanol is found in rice, particularly in the bran fraction (3000 mg/kg of rice) (Xu and Godber, 1999) and in wheat bran (300–390 mg/kg) (Hakala *et al.*, 2002).

Phytic acid is concentrated in the bran fraction of wheat and other cereal grains. It has the ability to suppress iron-catalyzed oxidative reactions (Slavin, 2004). Although phytic acid has generally been considered an anti-nutritional factor, several studies have demonstrated its effect on the prevention of kidney stone formation, and protection against atheriosclerosis, coronary heart disease, and a number of cancers (Graf and Eaton, 1993; Jenab and Thompson, 1998). The average concentration of phytic acid in wholegrain corn and rice was reported to be 0.9% (De Boland *et al*., 1975). Jood *et al*. (1995) reported 482, 635, and 829 mg/100 g dry wholegrain of phytic acid in wheat, maize, and sorghum, respectively.

6.3 Phytochemicals in legume grains

Legumes are crops of the family *Leguminosae*, which is also called *Fabacae*. They are mainly grown for their edible seeds, and thus are named grain legumes. The expression food legumes usually means the immature pods and seeds as well as mature dry seeds used as food by humans. Based on Food and Agricultural Organization (FAO) practice, the term legume is used for all leguminous plants. Legumes such as French bean, lima bean, mung bean, chickpea, cowpea, lentil, or others, which contain a small amount of fat, are termed pulses, and legumes that contain a higher amount of fat, such as soybean and peanuts, are termed leguminous oilseeds (Riahi and Ramaswamy, 2003). The term "pulse" is limited to crops harvested solely for dry grain, thereby excluding crops harvested green for food (green peas, green beans, etc.), which are classified as vegetable crops. Also excluded are those crops used mainly for oil extraction (e.g. soybean and groundnuts) and leguminous crops (e.g. seeds of clover and alfalfa) that are used exclusively for sowing purposes (FAO, 1994). Pulses are present in almost every diet throughout the world because they are good sources of starch, dietary fiber, protein, lipid, and minerals and they are second only to the grasses (cereals) in providing food crops for world agriculture. In addition to their nutritive value, legumes contain significant quantities of health-promoting components (phytochemicals) such as phenolic compounds and phytoestrogens. Legume grains are gaining interest because they are excellent sources of bioactive compounds and can be important sources of ingredients for use in functional foods and other applications. Based on their biosynthetic origin, phytochemicals in pulses can be divided into several categories that include phenolics, alkaloids, steroids, terpenoids, etc. The common phytochemicals in pulses are discussed in this chapter.

6.3.1 Dietary fiber

Legume grains are good sources of dietary fibre (21–47 g/100 g sample) that are fermentable in the colon, and produce short chain fatty acids (SCFA) such as acetate, propionate, and butyrate (Mallillin *et al*., 2008). Dietary fiber content of dry bean, chickpea, lentil and pea are relatively high ranging 23–32, 18–22, 18–20 and 14–26%, respectively (Tosh and Yada, 2010). Soybean, jack bean, and cowpea contain even higher content of dietary fiber at levels of 54.7, 33.2 and 31.2%, respectively (Martín-Cabrejas *et al*., 2006). The main constituent groups of dietary fiber in legumes are cellulose and hemicelluloses, lignin, and pectic substances (Selvendran *et al*., 1987). The insoluble dietary fiber (IDF) components were found to be predominant in legumes ranging from 10 to 15% for lentil, chickpea, and dry pea (Berrios *et al*., 2010). Su and Chang (1995) reported a higher level of IDF fraction (72–90% of the total) in raw dry beans compared to soluble fiber (SDF). The SDF of eight whole legumes, namely Bengal gram, broad bean, cowpea, field bean, green gram, horse gram, lentil, and French bean have been found to range from 0.61 to 2.37% of total dietary fiber, with the highest being in French bean and the lowest in lentil (Khatoon and Prakash, 2004). Similarly, Berrios et al. (2010) observed that the concentration of SDF is significantly lower in lentil, chickpea, and dry pea, ranging from 0.27 to 0.75%. About 92–100% and 0–8% of the total dietary fiber found in different legume samples (black bean, red kidney bean, lentil, navy bean, black-eyed pea, split pea, and northern bean) were ISD and SDF, respectively (Bednar et al., 2001).

Milling and fractionation of pulse seeds have been used to isolate dietary fiber components for incorporation into commercial food products to enrich their fiber content and/or serve as functional ingredients (Tosh and Yada, 2010). Legume hulls contribute a significant portion of the insoluble fiber in whole pulses. Pulse hulls are rich in dietary fiber, ranging from dry weight contents of 75% (chickpea) to 87% (lentil), and 89% (pea) (Dalgetty and Baik, 2003). Field pea hulls contained 82.3% of the total dietary fiber with 8.2% hemicellulose and 62.3% cellulose (Sosulski and Wu, 1988). Reichert (1981) found that pea cotyledon cell walls are mainly composed of pectic substances (26%) and hemicelluloses (22%), whereas the hulls are primarily made of cellulose (69%).

6.3.2 Phenolic acids

The major phenolic compounds in pulses comprise mainly phenolic acids, flavonoids, and tannins. Pulses with the highest phenolic content have dark color and highly pigmented grains, such as red kidney bean (*Phaseolus vulgaris*), black gram (*Vigna mungo*), and black soybean (*Glycine max*). The dark-coat seeds with high amounts of phenolic compounds would contribute to high antioxidant capacity (Lin and Lai, 2006). The legumes, mung bean, field pea, faba bean, lentil, and pigeon pea, were found to contain 18–31 mg total phenolic acids per kg of seeds, while Navy bean, lupine, lima bean, chickpea, and cowpea, possess 55–163 mg/kg (Sosulski and Dabrowski, 1984). Lentil seeds contained the highest phenolic content (21.9 mg/g) compared to red kidney bean, soybean, and mung bean which contain 18.8, 18.7, and 17.0 mg/g, respectively (Djordjevic et al., 2010). The total phenols content (TPC) of 29 genotype of common bean (*Phaseolus vulgaris*) with diverse origin and seed coat color varied from 5.8 to 14.1 mg/g (Akond et al., 2011). In addition, soybean showed wide variations of TPC, which varied from 6.4 to 81.7 mg/g (Prakash et al., 2007).

Phenolic acids are a major class of phenolic compounds widely occurring in the plant kingdom. Phenolic acids in legume grains are mainly concentrated in the seed coat. Sosulski and Dabrowski (1984) reported that defatted flours of ten legumes (mung bean, field pea, faba bean, lentil, navy bean, lupine, lima bean, chickpea, cowpea, and pigeon pea) contain only soluble esters of trans-ferulic, trans-*p*-coumaric and syringic acids. The total phenolic acids content of common bean (*P. vulgaris* L.) has been found to be 30 mg/100 g with ferulic acid as the prevalent compound, followed by *p*-coumaric acid (Luthria and Pastor-Corrales, 2006). Garcia et al. (1998) reported the presence of caffeic, *p*-coumaric, sinapic, and ferulic acids in de-hulled soft and hard-to-cook beans (*P. vulgaris*). The de-hulled soft beans contained 45 times more methanol soluble esters of phenolic acids than hard-to-cook beans.

Generally, the abundant phenolic acids in raw leguminous seeds are ferulic acid, *p*-coumaric acid, *o*-coumaric acid, sinapic acid, caffeic acid, protocatechuic acid, vanilllic acid, and *p*-hydroxybenzoic acid (Amarowicz and Pegg, 2008; Kalogeropoulos et al., 2010). Several phenolic acids have been identified in soybeans especially the black seed coated type. Four benzoic derivatives (gallic acid, 2,3,4-trihydroxybenzoic acid, vanillic acid, and protocatechualdehyde) and 3 cinnamic-type (chlorogenic, sinapic, and trans-cinnamic acid) phenolic acids are detected in free phenolic extract of both raw and processed yellow soybean. In addition, free phenolic extract of the black soybean have additional one benzoic-type (protocatechuic acid) and one cinnamic-type phenolic acid (p-coumaric acid). The predominant phenolic acids in both yellow and black soybean have been reported to be chlorogenic and trans-cinnamic acids. In addition, nine benzoic derivatives (gallic, protocatechuic, 2,3,4-trihydroxybenzoic, p-hydroxybenzoic, gentistic, syringic, and vanillic acid, protocatechualdehyde, and vanillin) and six cinnamic analogs (caffeic, p-coumaric, m-coumaric, o-coumaric, sinapic, and trans-cinnamic acid) were found in the bound phenolics extract of both yellow and black soybean (raw and cooked) with more concentration in the black seed coat varieties (Xu and Chang, 2008).

6.3.3 Isoflavones

Isoflavones are a subclass of the more ubiquitous flavonoids. The primary isoflavones in soybeans are genistein (4',5,7-trihydroxyisoflavone) and daidzein (4',7-dihydroxyisoflavone) and their respective β-glycosides, genistein and daidzein (Akhtar and Abdel-Aal, 2006; Setchell, 1998). It has been hypothesized that isoflavones reduce the risk of cancer, heart disease, and osteoporosis, and also help relieve menopausal symptoms (Messina, 1999; McCue and Shetty, 2004; Isanga and Zhang, 2008). The dietary sources of isoflavones are almost exclusively soy foods made from whole soy beans or isolated soy proteins. The concentrations of isoflavones in soy products vary considerably ranging in most soy foods between 0.1 and 3.0 mg/g (Setchell, 1998). The isoflavones content of 48 cultivars of 16 food legume species (edible seeds) based on an isotope dilution gas chromatography-mass spectrometry technique was found to range from 37.3 to 140.3 mg/100 g in soybean (highest total concentration) followed by chickpea at range of 1.15 to 3.6 mg/100 g (Mazur et al., 1998). Reinli and Block (1996) compiled reference data on the levels of isoflavones found in a variety of food items. The content of genistein and daidzein in several soy products are 73 and 55 mg/100 g in green soybean, 32 and 19 in tempeh, 17 and 16 in soybean paste, 16.6 and 7.6 in tofu, 2.6 and 1.8 in soy milk, and 0.8 and 0.5 in soy sauce. These variations can be attributed to the various processing steps. Daidzein was not detected in 17 different types of dry bean, while genistein was found in only four samples with the highest being 1.3 mg/100 g.

6.3.4 Saponins

Saponins are a diverse group of compounds commonly found in legumes (Oakenfull and Sidhu, 1990). Saponins derive their name from the Latin word *sapo* or soap, thus relating to their common surface-active detergent properties. Saponins are categorized into two distinctive groups including steroid and triterpenoid glycosides. Steroid saponins are further divided into two groups: furostanol glycosides including protoneodioscin, protodioscin, protoneogracillin, and protogracillin; and spirostanol glycosides, which include dioscin, prosapogenin A of dioscin, and gracillin. Saponins in foods have traditionally been considered as "antinutritional factors" (Thompson, 1993) and in some cases have limited soybean utilization due to the formation of a soap-like foaming characteristic (Sarnthein-Graf and La Mesa, 2004). However, food and non-food sources of saponins have come into renewed focus in recent years due to increasing evidence of their health benefits such as cholesterol-lowering and anti-cancer properties (Milgate and Roberts, 1995; Gurfinkel and Rao, 2003).

The contribution of saponins in soybean foods to the health benefits has also been emphasized by Oakenfull (2001) and Kerwin (2004).

Saponins in soy are often referred to as soyasaponins; and they varied from 0.22 to 0.5% with more than 20 saponin compounds (Anderson *et al.*, 1995; Güçlü-Üstündağ and Mazza, 2007). Soyasapogenols A, B, C, D, and E and their corresponding glycosides, which vary in the structure of the sapogenin aglycone and their attached glycosides, have been identified in the soy extract (Haralampidis *et al.*, 2002; Isanga and Zhang, 2008). Kang and others (2010) have identified 16, 10, 4, and 6 compounds of soyasaponins under groups A, B, C, and D respectively.

Saponin is also present in other legumes and pulses but in smaller concentrations compared with soy. Ojasapogenol B has been identified as the predominant sapogenol in lima beans and jack beans (Oboh *et al.*, 1998). Peas have been found to contain saponin with the amount ranging from 1.1 g/kg in yellow peas to 2.5 g/kg in green peas, whereas the levels in lentils are 3.7–4.6 g/kg (Savage and Deo, 1989). The amount of saponin in assorted types of common bean was reported to be 0.1–3.7 g/kg dry mater in broad bean, 0.03–3.5 in field bean, 2.3 and 2.16 in haricot and kidney bean, 3.4 in moth and mung bean, and 2–16 g/kg in navy beans (Price *et al.*, 1987; Oomah *et al.*, 2011). Chickpeas contain a wide range of saponin level (2.3–60 g/kg dry mater). Fenugreek (*Trigonella foenum-graecum* L) is another member of the family *Leguminosae* that was found to be rich in saponins. Three steroidal saponins namely, diosgenin, gitogenin, and tigogenin, have been found in fenugreek seeds (Dawidar *et al.*, 1973). The Asian fenugreek seeds also contain steroidal saponins mainly in the form of diosgenin, which comprises approximately 5–6% of the seed (Petit *et al.*, 1995). Fenugreek saponin "diogenin" is able to bind bile acids and thereby limit bile salt re-absorption in the gut, consequently accelerating cholesterol degradation and decreasing plasma cholesterol concentration (Sidhu and Oakenfull, 1986). Diosgenin also inhibits cell growth and induces apoptosis in the HT-29 human colon cancer cell line *in vitro* with a dose-dependent manner (Raju *et al.*, 2004). The cholesterol-lowering effect of saponins has been demonstrated in animal and human trials (Oakenfull and Sidhu, 1990; Milgate and Roberts, 1995). The effect is attributed to inhibition of cholesterol absorption from the small intestine or to the re-absorption of bile acids (Oakenfull and Sidhu, 1990). Soybean saponins were reported to suppress the growth of colon tumor cells in vitro (Sung *et al.*, 1995). Anti-tumor-promotion and growth inhibition of tumors or tumor cell lines by soy saponins have also been reported (Koratkar and Rao, 1997).

6.3.5 Anthocyanins

Anthocyanins are natural pigments belonging to the flavonoid family. They are responsible for the blue, purple, and red color of many fruits, vegetables, and grains. Several beneficial effects have been attributed to anthocyanins largely focusing on antioxidant properties, and ocular and anti-diabetic effects of an anthocyanin rich diet (Shipp and Abdel-Aal, 2010). Several *in vitro* studies, animal models, and human trials have shown that anthocyanins possess anti-inflammatory and anticarcinogenic activity, cardiovascular disease prevention, obesity control, and diabetes alleviation properties, all of which are more or less associated with their potent antioxidant property (Pascual-Teresa and Sanchez-Ballesta, 2008; He and Giusti, 2010). Black bean and soybean in general and their seed coat in particular have been reported to contain adequate amount of anthocyanin among pulses. The major anthocyanin pigment in black bean is delphinidin 3-glucoside with the presence of small amounts of cyanidin 3-glucoside, cyanidin 3,5-diglucoside, pelargonidin 3-glucoside, and pelargonidin

3,5-diglucoside (Stanton and Francis, 1966; Tsuda et al., 1994). Another study found delphinidin 3-glucoside (56% of total anthocyanins) along with petunidin 3-glucoside (26%) and malvidin 3-glucoside (18%) in black bean (Takeoka et al., 1997). Delphinidin 3-glucoside is also the principal anthocyanin in kidney bean (*Phaseolus vulgaris* L.) along with other four anthocyanins, cyanidin 3,5-diglucoside, cyanidin 3-glucoside, petunidin 3-glucoside, and pelargonidin 3-glucoside (Choung et al., 2003). The study also reported that total anthocyanins content in six red, two black, and three brown kidney beans vary from 0.27–0.74, 2.14–2.78 and 0.07–0.10 mg/g, respectively.

A number of studies have confirmed the presence of anthocyanins in the seed coat of black soybean. The total anthocyanins in the seed coat of ten black soybeans (*Glycine max* L.) was found to range from 1.58 to 20.18 mg/g, of which three anthocyanins are identified. These anthocyanins include delphinidin-3-glucoside, cyanidin-3-glucoside, and petunidin-3-glucoside and their contents ranging 0–3.7, 0.9–16.0, and 0–1.4, respectively. Recently, anthocyanins and anthocyanidins in black soybean seed coats have been identified primarily as cyanidin 3-glucoside with the relative order of anthocyanidin as cyanidin>delphinidin>petunidin>pelargonidin, while the yellow soybean seed coat has very little anthocyanins content (Park et al., 2011).

6.3.6 Lignans

The lignans content of different food sources reported by Tham et al. (1998) confirmed that flaxseed meals and flours are the highest plant lignans source, having 675 and 526 µg/g dry matter, respectively. Among legumes lentil, soybean, kidney bean, and navy bean possess the highest lignans content with average amounts of 18.0, 8.6, 5.6, and 4.6 µg/g dry matter, respectively. The enterolactone structure has been found to form the major portion of lignan in lentil while enterodiol structure comprises the great part of soybean and dry bean lignan. It has been reported that flax seeds have extremely high contents of secoisolariciresinol and matairesinol, the most common lignans in food, being 3699 and 10.7 µg/g dry matter, respectively, while soybean and kidney bean contain 0.13–2.73 and 0.56–1.53 µg/g dry matter, respectively (Webb and McCullough, 2005).

6.3.7 Other phytochemicals

Coumestans are one of the phytoestrogens which are less common in the human diet than isoflavones. They are found in legumes, particularly food plants such as sprouts of alfalfa and mung bean (Lookhart, 1980; Mazur et al., 1998). Soy sprouts also show good level (71.1 µg/g wet weight) of coumestrol, the main coumestans compounds (Ibarreta et al., 2001).

Catechin and epicatechin are found to be predominated phenolic compounds in boiled legumes followed by chrysin, genistein, and quercetin. These flavonoids have been reported in raw leguminous seeds and their extracts (Amarowicz and Pegg, 2008). The sum of flavonoids has been found to range from 20.1 to 2109.6 mg/100 g in selected pulses, and the highest flavonoids content was observed in lentil, followed by chickpea, pinto bean, and lupin. These components provide protective benefits due to their free radical scavenging ability and inhibition of eicosanoid synthesis and platelet aggregation (Dillard and German, 2000).

Phytic acid is the main storage form of phosphorus in soybean. The phytic acid content of soybean generally ranges from 1 to 2.3% (Liener, 1994). In general legume grains have high content of phytic acid around 1.75 g/100 g and in particular lupin, pea, common bean, and cowpea having 1.38, 1.02, 0.55, and 0.42 g/100 g phytate (Hídvegi and Lásztity, 2002).

Tannins are polyphenolic substances commonly divided into two groups, condensed and hydrolysable tannins (Liener, 1994). Dietary tannins may have negative or positive effects to humans as they may depress digestibility of protein and carbohydrate and absorption of minerals or they could act as anticarcinogenic and antimutagenic agents. Soybean contains about 45 mg/100 g of tannins that are mainly located in the hull of the seeds (Liener, 1994). Mung bean contains about 3.3 mg/g tannins (Mubarak, 2005), and faba bean possesses around 1.82 mg/100 g tannins (Fernández et al., 1996).

6.4 Stability of phytochemicals during processing

Since the majority of phytochemicals are present in the outer layers of cereal grains, milling of grains into white flours will result in the removal of high portions of these components. Thus more attention should be paid to minimize the loss of phytochemicals during the milling process, in particular those exhibiting beneficial health effects. In addition, more research is required on the development of new milling technologies and new varieties to produce wholegrain foods exhibiting health-enhancing properties.

Processing of grains could have various effects on dietary fiber. Several studies have shown conflicting results. Some data indicate no significant effects on soluble and insoluble dietary fiber (Varo et al., 1983), others claim reductions (Fornal et al., 1987) or increase in dietary fibers (Theander and Westerlund, 1987; Penner and Kim, 1991). Germination of pea seeds resulted in increased contents of both insoluble and soluble dietary fiber in conjunction with a decrease in the IDF/SDF ratio (Martín-Cabrejas et al. 2003). Pérez-Hidalgo et al. (1997) observed a total dietary fiber increase of 49.5% (from 16.8 to 25.1%) after cooking and decrease of 21.4% (from 16.8 to 13.2%) after frying of chickpeas. In addition, they also observed an increase in the insoluble fiber fraction after cooking by 108% with no significant change in the level of insoluble dietary fiber after frying. Mahadevamma and Tharnathan (2004) found that various cooking processes including deep fat frying, autoclaving, popping, extrusion cooking, and roller drying of Bengal gram and green gram affect dietary fiber causing either reduction or increase depending upon process-type and fiber fraction.

Processing may open up the food matrix, thereby allowing the release of tightly bound phytochemicals from the grain structure (Fulcher and Rooney Duke, 2002). Research on cereal products showed that thermal processing might assist in releasing bound phenolic acids by breakdown cellular constituents and cell walls (Dewanto et al., 2002). In addition, browning during thermal processing may cause increase of total phenolic content and free radical scavenging capacity. This increase could be due to the dissociation of conjugated phenolic during thermal processing followed by some polymerization and/or oxidation reactions and the formation of phenolics other than those endogenous in the grains. Other reactions such as Maillard reaction (non-enzymatic browning) (Bressa et al., 1996), caramelization, and chemical oxidation of phenols could also contribute to the increase in total phenols content.

Processing may also change the ratio between various phenolic compounds due to thermal degradation. Vanillin and vanillic acid can be produced through thermal decomposition of ferulic acid (Pisarnitskii et al., 1979; Peleg et al., 1992), while p-hydroxybenzaldehyde can be formed from p-coumaric acid (Pisarnitskii et al., 1979). Some phenolic acids are heat-sensitive such as caffeic acid, which could be reduced during heat processes, while others like ferulic and *p*-coumaric acids are susceptible to thermal breakdown (Pisarnitskii *et al.*, 1979; Huang and Zayas, 1991). Degradation of conjugated polyphenolic compounds such as tannins as a result of heat stress (100 °C) could increase some phenolics such as ferulic, syringic, vanillic, and *p*-coumaric acids in wheat flour (Cheng *et al.*, 2006). Some phenolics are also known to accumulate in the cellular vacuoles (Chism and Haard, 1996), and thermal processing may release such unavailable phenolics. The processing operating conditions could also affect changes in phenolic compounds. For instance, moisture content, time, and temperature during extrusion processing would significantly determine the release of phenolic compounds (Dimberg *et al.*, 1996). Black soybean shows over three-fold higher phenolic content (6.96 mg GAE/g) than the yellow one (2.15 mg GAE/g) and thermal processes (boiling and steaming) dropped their levels by 43–63% and 10–27%, respectively (Xu and Chang, 2008).

Significant reduction in both antioxidant capacity (60–68%) and total phenolics (46–60%) in barley extrudates compared with that of the unprocessed barley flour has been reported (Altan *et al.*, 2009). Roasting can differently affect total phenolics and antioxidant capacity. For example, roasting resulted in a marked reduction in phenolic content (13.2 and 18.3%), and antioxidant capacity (27.2 and 13.5%) in yellow and white sorghum, respectively (Oboh *et al.*, 2010). A significant decrease in total phenols content (8.5–49.6%) and antioxidant capacity (16.8–108.2%) was observed after sand roasting of eight barley varieties (Sharma and Gujral, 2011). Significant increase in both antioxidant capacity and total phenols content of barley grains was obtained after roasting two layers of grains or 61.5 g in a microwave oven at 600 W for 8.5 min (Gallegos-Infante *et al.*, 2010a; Omwamba and Hu, 2010).

Significant increase in the content of free phenolics and total antioxidant capacity were found following heating canned corn in a retort at 115 °C for 10, 25, or 50 min (Dewanto *et al.*, 2002). In addition pressure cooking of corn (autoclaved for 40 min at 15 *psi*) caused substantial increase in the amount of free ferulic acid, *p*-coumaric acid, and vanillin (Steinke and Paulson, 1964). Heat treatment at high temperature (150 °C) of corn germ or other corn oil containing fractions resulted in significant reductions of γ-tocopherol, γ-tocotrienol, and δ-tocotrienol and the production of triacylglycerol oxidation products. Boiling red sorghum and finger millet at atmospheric pressure resulted in significant reduction in total extractable phenolics, while barley showed increase in total phenolic content and antioxidant capacity (Gallegos-Infante *et al.*, 2010b). Processing durum wheat into spaghetti resulted in reduction of free phenolic acids content, primarily caused by *p*-hydroxybenzoic acid decrease, and increase in bound phenolics (Hirawan *et al.*, 2010).

Baking of flat bread resulted in significant reduction in all-*trans* lutein being about 37–41% for the unfortified breads (no lutein added) and 29–33% for the lutein-fortified breads (Abdel-Aal *et al.*, 2010). The extent of reduction for natural or added lutein was considerably high and varied slightly among wheat species, einkorn, Khorasan, and durum. The degradation of carotenoids is mostly related to their well-known susceptibility to heat (Mercadante, 2007). Storage of flat bread at room temperature for up to eight weeks had a slight impact on all-*trans* lutein in the case of unfortified products, whereas the lutein-fortified products showed a linear degradation following first-order kinetics for the fortified flat breads. Canning of corn in sugar/salt brine solution at 126.7 °C for 12 min did not significantly change the contents of lutein and zeaxanthin in white and golden corn, but α-carotene significantly decreased by about 62% (Scott and Eldridge, 2005). However, the study did not measure *cis*-isomers of lutein and zeaxanthin, which were found to increase in canned vegetable (Updike and Schwartz, 2003). Lutein in wholegrain pan bread dropped to a little extent compared with flat breads (Abdel-Aal et al., 2010). The small reduction in lutein in pan bread could possibly be because of the lower concentration of lutein in the baking formulas where no lutein was added. Hidalgo et al. (2010) showed carotenoids losses of 21 and 47% for bread crumb and crust, respectively. Bread leavening had almost negligible effects on carotenoids losses, while baking resulted in a marked decrease in carotenoids. In pasta, the longer kneading step had significant effects on carotenoids losses, while the drying step did not induce significant changes (Hidalgo et al., 2010). Lipoxygenase was found to play considerable role on stability of lutein/carotenoids during dough-making where a positive correlation was found between carotenoid losses and lipoxygenase activity (Leenhardt et al., 2006). The degradation rate of lutein loss in pan bread was much higher in the high-lutein pan bread compared with the control bread which indicates that lutein degradation kinetics is concentration dependent (Abdel-Aal et al., 2010). Storage of pan bread at room temperature for up to five days resulted in an additional decrease in lutein to some extent depending on the base composite flour. Pan bread made from wheat einkorn/corn blend had a slightly higher degradation rate as compared to wheat/einkorn/corn blend. Storage of einkorn flour and bread at various temperatures (−20, 5, 20, 30, and 38 °C) for up to 239 days had major effects on carotenoids degradation, and was influenced by temperature and time following first-order kinetics (Hidalgo and Brandolini, 2008).

Einkorn alone or in blend with corn flour either unfortified or fortified with lutein was processed into cookies (Abdel-Aal et al., 2010). Stability of lutein in cookies was found to decline considerably in fortified einkorn and control cookies, whereas a moderate drop was observed for the unfortified einkorn cookies. The percentage of lutein reduction, however, was consistent at 62, 65, and 63% for unfortified einkorn, fortified einkorn, and fortified control cookie, respectively. The degradation rate is dependent on concentration of lutein as well as the baking recipe. The high decline in lutein in cookies compared with bread could be due to the high fat content in the baking recipe that may make lutein and other carotenoids more soluble and exposable to oxidation and isomerization. Cookies made from einkorn and corn composite flours, and fortified with lutein, also exhibited a sharp decline in lutein during baking process, whereas the corresponding unfortified ones had lutein reduction at a lower rate. Zeaxanthin level also reduced on baking but at a much lower rate compared with lutein, perhaps due to its lower concentration in the baking formula. Water biscuit made without adding fat and non-fat dry milk to avoid interferences with the lipophilic oxidation mechanism had lower carotenoid degradation at 31% (Hidalgo et al., 2010). Storage of cookies for up to eight weeks at ambient temperature produced almost no effect on lutein or zeaxanthin. Lutein-fortified muffins also showed a noticeable decrease in lutein similar to fortified cookies (Abdel-Aal et al., 2010). The muffin recipe also contains a high percent of fat, which may make lutein more soluble and accessible to processing conditions causing more degradation by oxidation and isomerization. The reduction percentages for lutein were 64 and 55% in unfortified and fortified muffin, and for zeaxanthin were 57 and 56%, respectively. This indicates that the extent of reduction or degradation is independent from carotenoid concentration but the degradation rate is concentration dependent. Storage of muffins for up to three days at ambient temperature had no effects on lutein or zeaxanthin content.

Blue wheat anthocyanins were found to be thermally most stable at pH 1 (Abdel-Aal and Hucl, 2003). Their degradation was slightly lower at pH 3 as compared to pH 5. Degradation of blue wheat anthocyanins would increase upon increasing temperature from 65 to 95 °C. Addition of SO_2 (500–1000 ppm for whole meals and 1000–3000 ppm for isolated anthocyanins) during heating of blue wheat had a stabilizing effect on anthocyanin pigments.

Traditional processing of legume grains such as dehulling, soaking, germination, boiling, autoclaving, and microwave cooking were found to reduce the content of tannin in mung bean seeds (*Phaseolus aureus*) (Mubarak, 2005). The tannins in uncooked raw dry seeds (3.3 mg/g) dropped by 66.7, 51.5, 45.5, and 62% in germinated, autoclaved, boiled, and microwave-cooked seeds, respectively. Fernández et al. (1996) found that tannins in faba bean (1.82 mg/100 g) became more accessible following cooking and the tannin/catechin ratio (an indicator of tannin polymerization) decreased. Soaking and cooking of five legumes (white kidney bean, red kidney bean, lentil, chickpea, and white gram) resulted in significant reduction in phytic acid and tannin contents. Maximum reduction of phytic acid (78%) and tannin (66%) was obtained with sodium bicarbonate soaking followed by cooking (Huma et al., 2008).

Changes in concentration of isoflavones and saponins in 13 pulse varieties including field pea, chickpea, and lentil was studied in whole seed, hydrated seed, and cooked seed (Rochfort et al., 2011). It was found that the concentration of isoflavones studied (genistein, daidzein, formononetin, and biochanin A) was highest in chickpea, in which soaking altered the amount of isoflavones while cooking eliminated these isoflavones.

6.5 Food applications and impact on health

Wheat, rice, corn, bean, and pea are major ingredients in the human diet. Other grain ingredients in the human diet include rye, oat, barley, sorghum, millet, buckwheat, amaranth, and triticale. These grains are good sources of phytonutrients, antioxidants, and dietary fiber exhibiting known health effects and they are present largely in the bran and hulls, and as a result wholegrain products are considered healthier foods. In addition, grain phytonutrients would complement those present in fruits and vegetables in the human diet. Indeed this makes wholegrains products promising healthy foods. Wholegrain foods, however, may exhibit poor color, taste, and textural properties, and perhaps require special processing treatments to enhance their sensory properties.

6.6 Cereal-based functional foods

Cereal grains particularly wheat, rye, oat, and barley offer great opportunities for the development of functional foods such as bread, pasta, breakfast cereals, snack bars, and others. Functional foods from selected cereal grains and their content of phytochemicals have been reported by Sidhu et al. (2007). Wholegrain bread and pasta products are commercially produced as healthier foods due to their higher content of bioactive compounds compared with those made from their corresponding refined grain flours. Such foods are in increasing demand, in particular those with improved nutritional and sensory qualities. Still more research is required to develop a wide variety of improved wholegrain food products to meet the growing demand and also to enhance product quality and satisfy consumers' needs. Many studies have shown the beneficial health effects of wholegrain foods (Anderson et al., 2000; Meyer et al., 2000; Nicodemus et al., 2001; Fung et al., 2002; Kasum et al., 2002). Wholegrain breakfast cereals have been found to be important dietary sources of antioxidants along with fruits and vegetables (Miller et al., 2000). Breads made with oat offer high satiety value and lower blood cholesterol level in human subjects (Frank et al., 2004). A number of diverse mechanisms are responsible for the protective effects of wholegrain products against chronic diseases (Slavin, 2003). They contain high levels of dietary fiber including oligosaccharides and resistant starch that escape digestion in the small intestine and are fermented in the gut producing short chain fatty acids. The short chain fatty acids serve as an energy source for the colonocytes and may alter blood lipids. Wholegrain products are rich in antioxidants that have been linked to disease and oxidative damage prevention. In addition, wholegrain products mediate insulin and glucose responses, and exhibit improvements in biomarkers such as blood lipid.

High-lutein wholegrain bakery products including bread, cookie, and muffin have been developed as staple foods to enhance lutein daily intake (Abdel-Aal et al., 2010). Lutein is the main carotenoid in wheat and accounts for 77–83% of the total carotenoids in relatively high-lutein wheat species such as einkorn, durum, Kamut, and Khorasan (Abdel-Aal et al., 2007). Specialty grains also have been employed for the production of functional and health foods. Blue, purple, or red corn is currently used for ornamentation due to its colourful appearance with only a small amount being utilized in the production of naturally coloured blue and pink tortillas as healthy additive-free foods (Abdel-Aal et al., 2006). Anthocyanin-pigmented corn especially purple corn with relatively high amounts of anthocyanins (965 µg/g) hold a promise for the development of functional foods and/or natural colorants. Purple wheat is crushed into large pieces, which are spread over the exterior of multigrain bread as a specialty food product (Bezar, 1982). Red rice has been used as a functional food in China, and is also commonly used as a food colorant in bread, ice cream and liquor (Yoshinaga, 1986). Black sorghum has also been shown to contain significant levels of anthocyanins and other phenols concentrated in the bran fraction being approximately 4.0–9.8 mg/g of anthocyanins mainly 3-deoxyanthocyanidins such as luteolinidin and apigeninidin (Awika et al., 2004). This amount is relatively high compared to pigmented fruits and vegetables (0.2–10 mg/g) on a fresh weight basis making black sorghum a good candidate as a functional food product.

6.7 Legume-based functional foods

Bean and pea are traditional foods in several parts of the world. In Latin America pulse consumption ranges from 1 kg/capita/year (Argentina) to 25 kg/capita/year (Nicaragua) with common beans accounting for 87% of the total consumption (Leterme and Muñoz, 2002). The pulse consumption in Europe is lower than other regions of the world with Spain, France, and the UK accounting for 60% of the total consumption (Schneider, 2002). In the USA only 7.9% of the population consumed beans, peas, or lentils on any given day based on dietary intake data from the 1999–2002 National Health and Nutrition Examination Survey for adults aged 19 years and over (Mitchell et al., 2009). The main sources are pinto bean, refried bean (usually made from pinto bean), baked bean, chilli, and other Mexican or Hispanic mixed dishes. The US dietary guidelines recommend about 3.5 cups per week or 0.5 cup per day (USDA, 2010).

Beans and peas provide a diverse array of nutrients and phytochemicals that have demonstrated beneficial health effects. For instance, consuming about half a cup of dry beans or peas could increase intakes of fiber, protein, folate, zinc, iron, and magnesium, and lower intakes of saturated and total fat in the diet of Americans (Mitchell et al., 2009). According to a study by Sichieri (2002), a traditional diet that relies largely on beans and rice was associated with lower risk of being overweight and obese in logistic models in Brazil. Eating beans is also inversely correlated ($r=-0.68$) with colon cancer mortality based on epidemiological studies (Correa, 1981). In addition, consumption of beans may reduce the risk of cardiovascular disease via hypocholesterolemic effects and lowering of blood pressure, body weight, and oxidative status (Winham et al., 2007).

Baked bean is a common food form, and is traditionally made in a ceramic or cast-iron bean pot. Today, bean recipes are stewed, such as canned beans, as convenience foods. Consumption of baked bean has been linked to reductions in serum cholesterol in hypercholesterolemic adults (Winham and Hutchins, 2007). Baked beans also considerably reduced total plasma cholesterol in normo-cholesterolemic adults fed one 450 g can of baked beans in tomato sauce daily for 14 days as part of their normal diet (Shutler et al., 1989).

In general, cereal and legume grains are rich sources of phytochemicals and basic nutrients that would promote beneficial health effects and constitute the foundation for healthy diet. The protective functions of phytochemicals in human health and nutrition when consumed at the required daily amount are well recognized (Anderson et al., 2007; Chan et al., 2007; Cheng et al., 2009; De Moura, 2008; Alminger and Eklund-Jonsson, 2008; Binns, 2010). These compounds possess a number of relevant biological properties that depend in part on their antioxidant capacity. They may actively contribute to the control of oxidative reactions and provide protection *in vivo* via their capacity as free radical scavengers, reducing agents, potential ability to complex with pro-oxidant metals, and as quenchers of reactive oxygen species in addition to other physiological functions.

References

Abdel-Aal, E.-S.M. and Hucl, P. (1999) A rapid method for quantifying total anthocyanins in blue aleurone and purple pericarp wheats. *Cereal Chemistry*, 76, 350–354.

Abdel-Aal, E.S.M., Hucl, P., Sosulski, F.W Graf, R., Gillot, C., and Pietrzak, L. (2001) Screening spring wheat for midge resistance in relation to ferulic acid content. *Journal of Agricultural and Food Chemistry*, 49, 3556–3559.

Abdel-Aal, E.S.M., Young, J.C., Wood, P.J., Rabalski, I., Hucl, P., and Fre´geau-Reid, J. (2002) Einkorn: a potential candidate for developing high lutein wheat. *Cereal Chemistry*, 79, 455–457.

Abdel-Aal, E.S.M. and Hucl, P. (2003) Composition and stability of anthocyanins in blue-grained wheat. *Journal of Agricultural and Food Chemistry*, 51, 2174–2180.

Abdel-Aal, E.S.M., Young, J.C., and Rabalski, I. (2006) Anthocyanin composition in black, blue, pink, purple, and red cereal grains. *Journal of Agricultural and Food Chemistry*, 54, 4696–4704.

Abdel-Aal, E.-S.M., Young, J.C., Rabalski, I., Hucl, P., and Fre´geau-Reid, J. (2007) Identification and quantification of seed carotenoids in selected wheat species. *Journal of Agricultural and Food Chemistry*, 55, 787–794.

Abdel-Aal, E.S.M., and Rabalski, I. (2008) Bioactive compounds and their antioxidant capacity in selected primitive and modern wheat species. *The Open Agriculture Journal*, 2, 7–14.

Abdel-Aal, E.S.M., Young, J.C., Akhtar, H., and Rabalski, I. (2010) Stability of lutein in wholegrain bakery products naturally high in lutein or fortified with free lutein. *Journal of Agricultural and Food Chemistry*, 58, 10109–10117.

Adom, K. and Liu, R. (2002) Antioxidant activity of grains. *Journal of Agricultural and Food Chemistry*, 50, 6182–6187.

Adom, K., Sorrells, M., and Liu, R. (2003) Phytochemical profiles and antioxidant activity of wheat varieties. *Journal of Agricultural and Food Chemistry*, 51, 7825–7834.

Adom, K., Sorrells, M., and Liu, R. (2005) Phytochemicals andantioxidant activity of milledfractions of different wheat varieties. *Journal of Agricultural and Food Chemistry*, 53, 2297–2306.

Akhtar, M.H., and Abdel-Aal, E-S.M. (2006) Recent Advances in the Analyses of Phytoestrogens and Their Role in Human Health. *Current Pharmacuetical Analysis*, 2, 183–193.

Akond, A.S.M., Khandaker, L., Berthold, J., Gates, L., Peters, K., Delong, H., and Hossain, K. (2011) Anthocyanin, total polyphenols and antioxidant activity of common bean. *American Journal of Food Technology*, 6, 385–394.

Alminger, M. and Eklund-Jonsson, C. (2008) Whole-grain cereal products based on a high-fiber barley or oat genotype lower post-prandial glucose and insulin responses in healthy humans. *European Journal of Nutrition*, 47, 294–300.

Altan, A., McCarthy, K. L., and Maskan, M. (2009) Effect of extrusion process on antioxidant activity, total phenolics and β-glucan content of extrudates developed from barley-fruit and vegetable by-products. *International Journal of Food Sciences Technology*, 1263–1271.

Åman, P., Nilsson, M., and Andersson, R. (1997) Positive health effect of rye. *Cereal Foods World*, 42, 684–688.

Amarowicz, R. and Pegg, R.B. (2008) Legumes as a source of natural antioxidants. *European Journal of Lipid Science and Technology*, 110, 865–878.

Anderson, J.W., Johnstone, B.M., and Cook-Newell, M.A. (1995) Meta-analysis of the effects of soy protein intake on serum lipids. *New England Journal of Medicine*, 333, 276–282.

Anderson, J.W., Hanna, T.J., Peng, X., and Kryscio, R.J. (2000) Whole grain foods and heart disease risk. *Journal of the American College of Nutrition*, 19, 291s–299s.

Anderson, A., Tengblad, S., Karlstrom, B., Kamal-Eldin, A., Landberg, R., Basu, S., Aman, P., and Vessby, B. (2007) Whole-grain foods do not affect insulin sensitivity or markers of lipid peroxidation and inflammation in healthy, moderately overweight subjects. *Journal of Nutrition*, 137, 1401–1407.

Awika, J.M., Rooney, L.W., and Waniska, R.D. (2004) Anthocyanins from black sorghum and their antioxidant properties. *Food Chemistry*, 90, 293–301.

Balasundram, N., Sundram, K., and Samman, S. (2006) Phenolic compounds in plants and agri-industrial by-products: Antioxidant activity, occurrence, and potential uses. *Food Chemistry*, 99, 191–203.

Bednar, G.E., Patil, A.R., Murray, S.M., Grieshop, C.M., Merchen, N.R., and Fahey, G.C. Jr. (2001) Starch and fiber fractions in selected food and feed ingredients affect their small intestinal digestibility and fermentability and their large bowel fermentability in vitro in a canine model. *Journal of Nutrition*, 131, 276–286.

Berrios, J. De J., Morales, P., Cámara, M., and Sánchez-Mata, M.C. (2010) Carbohydrate composition of raw and extruded pulse flours. *Food Research International*, 43, 531–536.

Bezar, H.J. (1982) Konini, specialty bread wheat. *New Zealand Wheat Review*, 15, 62–63.

Brenna, O.V. and Berardo, N. (2004) Application of near-infrared reflectance spectroscopy (NIRS) to the evaluation of carotenoids content in maize. *Journal of Agricultural and Food Chemistry*, 52, 5577–5582.

Brennan, C.S. (2005) Dietary fibre, glycaemic response, and diabetes. *Molecular Nutrition and Food Research*, 49, 560–570.

Brennan, C.S. and Cleary, L.J. (2005) The potential use of cereal $(1 \rightarrow 3, 1 \rightarrow 4)$-b-d-glucans as functional food ingredients. *Journal of Cereal Science*, 42, 1–13.

Binns, N. (2010) Regulatory aspects for whole grain and whole grain foods: An EU perspective. *Cereal Chemistry*, 87, 162–166.

Bressa F., Tesson, N., Rosa, M.D., Sensidoni, A., and Tubaro, F. (1996) Antioxidant effect of maillard reaction products: application to a butter cookie of a competition kinetics analysis. *Journal of Agricultural and Food Chemistry*, 44, 692–695.

Broekaert, W.F. Courtin, C.M., Verbeke, K., De Wiele, T.V., Verstraete, W., and Delcour, J.A. (2011) Prebiotic and Other Health-Related Effects of Cereal-Derived Arabinoxylans, Arabinoxylan-Oligosaccharides, and Xylooligosaccharides. *Critical Reviews in Food Science and Nutrition*, 51, 178–194.

Bunzel, M., Ralph, J., Martia, J.M., Hatfield, R.D., and Steinhart, H. (2001) Diferulates as structural components in soluble and insoluble cereal dietary fibre. *Journal of the Science of Food and Agriculture*, 81, 653–660.

Cavallero, A., Gianinetti, A., Finocchiaro, F., Delogu, G., and Stanca, A.M. (2004) Tocols in hull-less and hulled barley genotypes grown in contrasting environments. *Journal of Cereal Science*, 39, 175–180.

Chan, J. M., Wang, F., and Holly, E. A. (2007) Whole grains and risk of pancreatic cancer in a large population-based case-control study in the San Francisco Bay Area, California. *American Journal of Epidemiology*, 166, 1174–1185.

Cheng, Z., Su, L., Moore, J., Zhou, K., Luther, M., Yin J., and Yu, L. (2006) Effects of post harvest treatment and heat stress on availability of wheat antioxidants, *Journal of Agricultural and Food Chemistry*, 54, 5623–5629.

Cheng, G., Karaolis-Danckert, N., Libuda, L., Bolzenius, K., Remer, T., and Buyken, A.E. (2009) Relation of dietary glycemic index, glycemic load, and fiber and whole-grain intakes during puberty to the concurrent development of percent body fat and body mass index. *American Journal of Epidemiology*, 169, 667–677.

Chew, Y.L., Goh, J.K., and Lim, Y.Y. (2009) Assessment of in vitro antioxidant capacity and polyphenolic composition of selected medicinal herbs from *Leguminosae* family in Peninsular alaysia. *Food Chemistry*, 119, 373–378.

Chism, G.W. and Haard, N.F. (1996) Characteristics of edible plant tissues. In O.R. Fennema (9ed.) *Food Chemistry*, third edition, Dekker, New York, pp. 943–1011.

Choung, M.G., Choi, B.R., An, Y.N., Chu, Y.H., and Cho, Y.S. (2003) Anthocyaninin profile of Korean cultivated kidney bean (*Phaseolus vulgaris* L.). *Journal of Agricultural and Food Chemistry*, 51, 7040–7043.

Colleoni-Sirghie, M., Kovalenko, I.V., Briggs, J.L., Fulton, B., and White, B.J. (2003) Rheological and molecular properties of water soluble (1,3)(1,4)-β-D-glucan from high β-glucan and traditional oat lines. *Carbohydrate Polymers*, 52, 439–447.

Correa, P. (1981) Epidemiological correlation between diet and cancer frequency. *Cancer Research*, 41, 3685–3689.

Dai, J. and Mumper, R.J. (2010) Plant phenolics: extraction, analysis and their antioxidant and anticancer properties. *Molecules*, 15, 7313–7352.

Dalgetty, D.D. and Baik, B.K. 2003. Isolation and characterization of cotyledon fibres from peas, lentils and chickpeas. *Cereal Chem.* 80, 310–315.

Dawidar, A.M., Saleh, A.A., and Elmotei, S.L. (1973) Steroid sapogenin constituents of fenugreek seeds. *Planta Medica*, 24, 367–370.

De Boland, A.R., Garner, G.B., and O'Dell, B.L. (1975) Identification and properties of "Phytate"in cereal grains and oilseed products. *Journal of Agricultural and Food Chemistry*, 23, 1186–1189.

De Moura, F. (2008) *Whole Grain Intake and Cardiovascular Disease and Whole Grain Intake and Diabetes: A Review*, Life Sciences Research Organization, Bethesda, MD.

Dewanto, V., Wu, X., and Liu, R.H. (2002) Processed sweet corn has higher antioxidant activity. *Journal of Agricultural and Food Chemistry* 50, 4959–4964.

Dillard, C.J., and German, J.B. (2000) Review: Phytochemicals, nutraceuticals and human Health. *Journal of the Science of Food and Agriculture*, 80, 1744–1756.

Dimberg, L.H., Molteberg, E.L., Solheim, R., and Frölich, W. (1996) Variation in groats due to variety, storage and heat treatment. I: phenolic compounds. *Journal of Cereal Science*, 263–272.

Djordjevic, T.M., Šiler-Marinkovic, S.S., and Dimitrijevic-Brankovic, S.I. (2010) Antioxidant activity and total phenolic content in some cereals and legumes. *International Journal of Food Properties*, 14, 175–184.

Englyst, H.N., Kingman, S., and Cummings, J. (1992). Classification and measurement of nutritionally important starch fractions. *Eur. J. Clin. Nutr.* 46, S33–S50.

Fernández, M., Pez-Jurado, M.L., Aranda, P., and Urbano, G. (1996). Nutritional assessment of raw and processed Faba Bean (*Vicia faba* L.) Cultivar major in growing rats. *Journal of Agricultural and Food Chemistry* 44, 2766–2772.

Flickinger, E., Van Loo, J. and Fahey, G. (2003). Nutritional responses to the presence of inulin and oligofructose in the diets of domesticated animals: a review. *Crit Rev. Food Sci. Nutr.* 43, 19–60.

Food and Agriculture Organization of the United Nations (FAO). (1994). Pulses and derived products. In: definition and classification of commodities. Rome, Italy.

Food and Agriculture Organization of the United Nations (FAO). (2010). Production and Trade Yearbooks. Rome, Italy. Annual issue.

Fornal, L., Soral-Smietana, M., Smietana, Z., and Szpendowski, J. (1987). Chemical characteristics and physicochemical properties of the extruded mixtures of cereal starches. *Starch/Staerke*. 39, 75–78.

Frank, J., Sundberg, B., Kamal-Eldin, A., Vessby, B., Aman, P. (2004). Yeast-leavened oat breads with high or low molecular weight beta-glucans do not differ in their effect on blood concentrations of lipids, insulin, or glucose in humans. *J. Nutr.* 134, 1384–1388.

Fulcher, R.G. and Rooney Duke, T.K (2002). Whole grain structure and organization: implications for nutritionists and processors. In: "*Whole Grain Foods in Health and Disease*". L. Marquart, R.G. Fulcher and J.L. Slavin (Eds). American Association of Cereal Chemists, St. Paul, MN, USA. p. 9–45.

Fung, T.T., Hu, F.B., and Pereira, M.A. (2002). Whole-grain intake and the risk of type 2 diabetes: a prospective study in men. *Am. J. Clin. Nutr.* 76, 535–40.

Gallegos-Infante, J.A., Rocha-Guzman, N.E., Gonzalez-Laredo , R.F., and Pulido-Alonso, J. (2010a). Effect of processing on the antioxidant properties of extracts from Mexican barley (*Hordeum vulgare*) cultivar. *Food Chem.* 119, 903–906.

Gallegos-Infante, J.A., Rocha-Guzman, N.E., Gonzalez-Laredo, R.F., Ochoa-Martínez, L.A., Corzo, N., Bello-Perez, L.A., Medina-Torres, L., and Peralta-Alvarez, L.E. (2010b). Quality of spaghetti pasta containing Mexican common bean flour (*Phaseolus vulgaris* L.). *Food Chem.* 119, 1544–1549.

Garcia, E., Filisetti, T.M.C.C., Udaeta, J.E.M., and Lajolo, F.M. (1998). Hard-to-cook beans (Phaseolus vulgaris): involvement of phenolic compounds and pectates. *Journal of Agricultural and Food Chemistry* 46, 2110–2116.

Gatenholm, P., & Tenkanen, M. (2004). Preface. In P. Gatenholm, & M. Tenkanen (Eds.), *Hemicelluloses: Science and technology*. Oxford University Press, 15–16.

Gebruers, K., Dornez, E., Boros, D., Fra, A., Dynkowska, W., Bed, Z., Rakszegi, M., Delcour, J.A., and Courtin, C.M. (2008). Variation in the content of dietary fiber and components thereof in heats in the HEALTHGRAIN diversity screen. *Journal of Agricultural and Food Chemistry* 56, 9740–9749.

Genc, H., Ozdemir, M., and Demirbas, A. (2001). Analysis of mixed-linked $(1 \rightarrow 3), (1 \rightarrow 4)$-β-D-glucans in cereal grains from Turkey. *Food Chem.* 73, 221–224.

Graf, E., and Eaton, J.W. (1993). Suppression of colonic cancer by dietary phytic acid. *Nutr. Cancer.* 19, 11–19.

Güçlü-Üstündağ, Ö., and Mazza, G. (2007). Saponins: Properties, Applications and Processing. *Crit. Rev. Food Sci. Nutr.* 47, 231–258.

Gurfinkel, D.M., and Rao, A.V. (2003). Soyasaponins: The relationship between chemical structure and colon anticarcinogenic activity. *Nutr. Cancer.* 47, 24–33.

Hakala, P., Lampi, A.M., Ollilainen, V., Werner, U., Murkovic, M., Wahala, K., Karkola, S., and Piironen, V. (2002). Steryl phenolic acid esters in cereals and their milling fractions. *Journal of Agricultural and Food Chemistry* 50, 5300–5307.

Haralampidis, K., Trojanowska, M., and Osbourn, A. (2002). Biosynthesis of triterpenoid saponins in plants. *Adv. Biochem. Engin Biotech.* 75, 31–50.

He, J., and Giusti, M.M. (2010). Anthocyanins: natural colorants with health-promoting properties. *Annu. Rev. Food Sci. Technol.* 1, 163–87.

Hidalgo, A., and Brandolini, A. (2008). Kinetics of carotenoids degradation during the storage of einkorn (Triticum monococcum ssp. monococcum). *Journal of Agricultural and Food Chemistry* 56, 11300–11305.

Hidalgo, A., Brandolini, A., and Pompei, C. (2010). Carotenoids evolution during pasta, bread and water biscuit preparation from wheat flours. *Food Chem.* 121, 746–751.

Hídvegi, M, and Lásztity, R. (2002). Phytic acid content of cereals and legumes and interaction with proteins. *Period. Polytech. Chem.* 46, 59–64.

Hirawan, R., Yuin Ser, W., Arntfield, S.D., and Beta, T. (2010). Antioxidant properties of commercial, regular- and whole-wheat spaghetti. *Food Chem.* 119, 258–264.

Huang, C.J., and Zayas, J.F. (1991). Phenolic acid contributions to taste characteristics of corn germ protein flour products. *J. Food Sci.* 56, 1308–1315.

Huma, N., Anjum, M., Sehar, S., Khan, M.I., and Hussain, S. (2008). Effect of soaking and cooking on nutritional quality and safety of legumes. *Nutr. Food Sci.* 38, 570–577.

Hyun, J.W., and Chung, H.S. (2004). Cyanidin and malvidin from *Oryza sativa* cv. Heugjinjubyeo mediate cytotoxicity against human monocytic leukemia cells by arrest of G2/M phase and induction of apoptosis. *Journal of Agricultural and Food Chemistry* 52, 2213–2217.

Ibarreta, D., Daxenberger, A., and Meyer, H.H.D. (2001). Possible health impact of phytoestrogens and xenoestrogens in food. *APMIS (Acta Pathologica, Microbiologica et Immunologica Scandinavica).* 109, 161–184.

Isanga, J., and Zhang, G-N. (2008). Soybean bioactive components and their implications to health—A review. *Food Rev. Inter.* 24, 252–276.

Jacobs, D.R., Meyer, K.A., Kushi, L.K., and Folsom, A.R. (1998). Whole-grain intake may reduce the risk of ischemic heart disease death in postmenopausal women: the Iowa women's health. *Am. J. Clin. Nutr.* 68, 248–257.

Jenab, M., and Thompson, L.U. (1998). The Influence of Phytic Acid in Wheat Bran on Early Biomarkers of Colon Carcinogenesis. *Carcinogenesis.* 19, 1087–1092.

Jood, S., Kapoor, A.C. and Singh, R. (1995). Polyphenol and Phytic Acid Contents of Cereal Grains As Affected by Insect Infestation *Journal of Agricultural and Food Chemistry* 43, 435–438.

Kalogeropoulos, N., Chiou, A., Ioannou, M., Karathanos, V.K., Hassapidou, M., and Andrikopoulos, N.K. (2010). Nutritional evaluation and bioactive microconstituents (phytosterols,tocopherols, polyphenols, triterpenic acids) in cooked dry legumes usually consumed in the Mediterranean countries. *Food Chem.* 121, 682–690.

Kamal-Eldin, A., Laerke, H.N., Knudsen, K.E., Lampi, A.M, Piironen, V., Adlercreutz, H., Katina, K., Poutanen, K., and Åman, P. (2009). Physical, microscopic and chemical characterisation of industrial rye and wheat brans from the Nordic countries. *Food Nutr. Res.* 53, 1–11.

Kang, J., Badger, T.M., Ronis, M.J.J., and Wu, X. (2010). Non-isoflavone phytochemicals in soy and their health effects. *Journal of Agricultural and Food Chemistry* 58, 8119–8133.

Kasum, C.M., Jacobs, D.R.J., Nicodemus, K., and Folsom, A.R. (2002). Dietary risk factors for upper aerodigestivetract cancers. *Int. J. of Cancer.* 99, 267–272.

Kerwin, S.M. (2004). Soy saponins and the anticancer effects of soybean and soy-based foods. *Curr. Med. Chem. – Anticancer Agents.* 4, 263–272.

Khatoon, N., and Prakash, J. (2004). Nutritional quality of microwave-cooked and pressure-cooked legumes. *Int. J. Food Sci. Nutr.* 55, 441–448.

Kim, K-H., Tsao, R, Yang,R., Cui, S.W.(2006). Phenolic acid profiles and antioxidant activities of wheat bran extracts and the effect of hydrolysis conditions. *Food Chem.* 95, 466–473.

Koratkar, R., and Rao, A.V. (1997). Effect of soya bean saponins on azoxymethane-induced preneoplastic lesions in the colon of mice. *Nutr. Cancer.* 27, 206–207.

Kroon, P., and Williamson, G. (2005). Polyphenols: Dietary components with established benefits to health. *J. Sci. Food Agric.* 85, 1239–1240.

Lee, K.W., Hur, H.J., Lee, H.J., and Lee, C.Y. (2005). Antiproliferative effects of dietary phenolic substances and hydrogen peroxide. *Journal of Agricultural and Food Chemistry* 53, 1990–1995.

Leenhardt, F., Lyan, B., Rock, E., Boussard, A., Potus, J., Chanliaud, E., Remesy, C. (2006). Genetic variability of carotenoid concentration and lipoxygenase and peroxidase activities among cultivated wheat species and bread wheat varieties. *Eur. J. Agron.* 25, 170–176.

Leterme, P., and MunÜoz, L.C. (2002). Factors influencing pulse consumption in Latin Am. *Brit. J. Nutr.* 88, S251–S254.

Li, W., Cui, S.W., and Kakuda, Y. (2006). Extraction, fractionation, structural and physical characterization of wheat β-D-glucans. *Carbohydr. Polym.* 63, 408–416.

Liener, I.E. (1994). Implications of antinutritional components in soybean foods. *Crit. Rev. Food. Sci. Nutr.* 34, 31–67.

Lin, P-Y., and Lai, H-M. (2006). Bioactive compounds in legumes and their germinated products. *Journal of Agricultural and Food Chemistry* 54, 3807–3814.

Lineback, D.R., and Rasper, V.F. (1988). Wheat carbohydrates. In Y. Pomeranz (Ed.), *Wheat chemistry and technology* (pp. 277e372). St. Paul, MN: American Association of Cereal Chemists.

Liu, R.H., (2007). Whole grain phytochemicals and health. *J. Cereal Sci.* 46, 207–219.

Lookhart, G.L. (1980). Analysis of coumestrol, a plant estrogen, in animal feeds by high-performance liquid chromatography. *Journal of Agricultural and Food Chemistry* 28, 666–667.

Luthria, D.L., and Pastor-Corrales, M.A. (2006). Phenolic acid content of fifteen dry edible beans (phaseolus vulgaris L.) varieties. *J Food composition Anal.* 19, 205–211.

Mahadevamma, S., and Tharanathan, R.N. (2004). Processing of legumes: resistant starch and dietary fiber contents. *J. Food Qual.* 27, 289–303.

Mallillin, A.C., Trinidad, T.P., Raterta, R., Dagbay, K., and Loyola, A.S. (2008). Dietary fiber and fermentability characteristics of root crops and legumes. *Br. J. Nutr.* 100, 485–488.

Manach, C., Scalbert, A., Morand, C., Rémésy, C., and Jiménez, L. (2004). Polyphenols: food sources and bioavailability. *Amer. J. Clin. Nutr.* 79, 727–747.

Martín-Cabrejas, M.A., Ariza, N., Esteban, R., Mollá, E., Waldron, K., and López-Andréu, F.J. (2003). Effect of germination on the carbohydrate composition of the dietary fiber of peas (*Pisum sativum* L.). *J. Agric Food Chem.* 51, 1254–1259.

Martín-Cabrejas, M.A., Aguilera, Y., Benítez, V., Mollá, E., López-Andréu, F.J., and Esteban, R.M. (2006). Effect of industrial dehydration on the soluble carbohydrates and dietary fiber fractions in legumes. *Journal of Agricultural and Food Chemistry* 54, 7652–7657.

Mattila, P., Pihlava, J.M., and Hellstrom, J. (2005). Contents of phenolic acids, alkyl- and alkenylresorcinols and avenanthramides in commercial grain products. *Journal of Agricultural and Food Chemistry* 53, 8290–8295.

Mazur, W.M., Duke, J.A., Wähälä, K., Rasku, S., and Adlercreutz, H. (1998). Isoflavonoids and lignans in legumes: Nutritional and health aspects in humans. *Nutr. Biochem.* 9, 193–200.

McCue, P., and Shetty, K. (2004). Health benefits of soy isoflavonoids and strategies for enhancement: A review. *Crit. Rev. Food Sci. Nutr.* 44, 361–367.

Mercadante, A.Z. (2007). Carotenoids in foods: sources and stability during processing and storage. In *Food Colorants Chemical and Functional Properties*; Socaciu, C., Ed.; CRC Press: Boca Raton, FL, PP 213–240.

Messina, M.J. (1999). Legumes and soybeans: overview of their nutritional profiles and health effects. *Am. J. Clin. Nutr.* 70, suppl 439S–450S.

Meyer, K.A., Kushi, L.H., Jacob, D.R.J., Slavin, J., Sellers, T.A., and Folsom, A.R. (2000). Carbohydrates, dietary fiber, and incident type 2 diabetes in older women. *Am. J. Clin. Nutr.* 71, 921–930.

Milgate, J., and Roberts, D.C.K. (1995). The nutritional and biological significance of saponins. *Nutr. Res.* 15, 1223–1249.

Miller, H.E., Rigelhof, F., Marquart, L., Parkash, A., and Kanter, M. (2000). Antioxidant content of whole grain breakfast cereals, fruits and vegetables. *J. Am. Coll. Nutr.* 19, 312S–319S.

Mitchell, D.C., Lawrence, F.R., Hartman, T.J., and Curran, J.M. (2009). Consumption of dry beans, peas and lentils could improve diet quality in the US population. *J. AM. Dietetic Assoc.* 109, 909–913.

Moore, J., Hao, Z., Zhou, K., Luther, M., Costa, J., and Yu, L. (2005). Carotenoid, tocopherol, phenolic acid, and antioxidant properties of Maryland-grown soft wheat. *Journal of Agricultural and Food Chemistry* J. Agric. Food Chem. 53, 6649–6657.

Moreno, Y.S., Sánchez, G.S., Hernández, D.R., Lobato, N.R. (2005). Characterization of anthocyanin extracts from maize kernels. *J. Chromatogr. Sci.* 43, 483–487.

Mubarak, A.E. (2005). Nutritional composition and antinutritional factors of mung bean seeds (*Phaseolus aureus*) as affected by some home traditional processes. *Food Chem.* 89, 489–495.

Nam, S.H., Choi, S.P., Kang, M.Y., Koh, H.J., Kozukue, N., Friedman, M. (2006). Antioxidative activities of bran from twenty one pigmented rice cultivars. *Food Chem.* 94, 613–620.

Nicodemus, K.K., Jacobs, D.R.J., and Folsom, A.R. (2001). Whole and refined grain intake and risk of incident postmenopausal breast cancer. *Cancer Causes and Control.* 12, 917–925.

Oakenfull, D.G., and Sidhu, G.S. (1990). Could saponins be a useful treatment for hypercholesterolaemia?. *Eur. J. Clin. Nutr.* 44, 79–88.

Oakenfull, D. (2001). Soy protein, saponins and plasma cholesterol. *J. Nutr.* 131, 2971.

Oboh, H.A., Muzquiz, M., Burbano, C., Cuadrado, C., Pedrosa, M.M., Ayet, G., and Osagie, A.U. (1998). Anti-nutritional constituents of six underutilized legumes grown in Nigeria. *J. Chromatogr.* 9, 307–312.

Oboh, G., Ademiluyi, A.O., and Akindahunsi, A.A. (2010). The effect of roasting on the nutritional and antioxidant properties of yellow and white maize varieties. *Int. J. Food Sci. and Technol.* 45, 1236–1242.

Okarter, N., and Liu, R.H. (2010). Health Benefits of Whole Grain Phytochemicals. *Crit. Rev. Food Sci. Nutr.* 50, 193–208.

Oomah, B.D, Patras, A., Rawson, A., Singh, N., and Compos-Vega, R. (2011). Chemistry of pulses. In: Tiwari, B., Gowen, A., McKenna, B. (Eds.), *Pulse foods: processing, quality and nutraceutical applications.* Elsevier Inc. pp 9–55.

Omwamba, M., and Hu, Q. (2010). Antioxidant Activity in Barley (HordeumVulgare L.) Grains Roasted in a Microwave Oven under conditions Optimized Using Response Surface Methodology. *J. Food Sci.* 75, 66–73.

Panfili, G., Fratianni, A., and Irano, M. (2003). Normal phase highperformance liquid chromatography method for the determination of tocopherols and tocotrienols in cereals. *Journal of Agricultural and Food Chemistry* 51, 3940–3944.

Park, S.M., Kim, J., Dung, T.H., Do, L.T., Thu, D.T.A., Sung, M.K., Kim, J.S., and Yoo, H. (2011). Identification of anthocyanin from the extract of soybean seed coat. *Inter. J. Oral Biol.* 36, 59–64.

Pascual-Teresa, S.D., and Sanchez-Ballesta, M.T. (2008). Anthocyanins: from plant to health. *Phytochem. Rev.* 7, 281–299.

Peleg, H., Naim, M., Zehavi, U., Rouseff, R.L., and Nagy, S. (1992). Pathway of 4-vinylguaiacol formation from ferulic acid in model solutions of orange juice. *Journal of Agricultural and Food Chemistry* 40, 764–767.

Penner, M.H., and Kim, S. (1991).Nonstarchy polysaccharide fractions of raw, processed, and cooked Carrots. *J. Food Sci.* 56, 1593.

Pérez-Hidalgo, M.A., Guerra-Hernandez, E., and Garcia-Villanova, B. (1997). Dietary fiber in three raw legumes and processing effect on chick peas by an enzymatic–gravimetric method. *J. Food Compos. Anal.* 10, 66–72.

Petit, P.R., Sauvaire, Y.D., Hillaire-Buys, D.M., Lectone, O.M., Baissac, Y.G., and Ponsin, G.R. (1995). Steroids saponins from fenugreek seeds: extraction, purification, and pharmacological investigation on feeding behavior and plasma cholesterol. *Steroids.* 60, 674–80.

Pisarnitskii, A.F., Egorov, I.A., and Egofarova, R. Kh. (1979). Formation of volatile phenols in cognac alcohols. *Appl. Biochem. Microbiol.* 15: 103–109.

Prakash, D., Upadhyay, G., Singh, B.N., Singh, H.B. (2007). Antioxidant and free radical-scavenging activities of seeds and agri-wastes of some varieties of soybean (*Glycine max*). *Food Chem.*104, 783–790.

Price., K.R., Johnson, I.T., and Fenwick, G.R. (1987). The chemistry and biological significance of saponins in foods and feeding stuffs. *Crit. Rev. Food Sci. Nutr.* 26, 27–135.

Queenan, K.M., Stewart, M.L., Smith, K.N., Thomas, W., Fulcher, R.G., Slavin, J.L. (2007). Concentrated oat beta-glucan, a fermentable fiber, lowers serum cholesterol in hypercholesterolemic adults in a randomized controlled trial. *Nutr. J.* 6, 6–14.

Ragaee, S.M., Campbell, G.L., Scoles, G.J., McLeod, J.G., & Tyler, R.T. (2001). Studies on rye (*Secale cereale* L.) lines exhibiting a range of extract viscosities. I. Composition, molecular weight distribution of water extracts, and biochemical haracteristics of purified water extractable arabinoxylan. *J. Agric. Food Chem.* 49, 2437–2445.

Ragaee, S., Guzar, I., Dhull, N., and Seetharaman, K. (2011). Effects of fiber addition on antioxidant capacity and nutritional quality of wheat bread. *LWT- Food Sci. Technol.* 44, 2147–2153.

Ragaee, S., Abdel-Aal, E-S.M., and Seetharaman, K. (2012a). Impact of milling and thermal processing on phenolic compounds in cereal grains. *Crit. Rev. Food Sci. Nutr.* (in Press).

Ragaee, S., Guzar, I., Abdel-Aal, E-S.M., and Seetharaman, K. (2012b). Bioactive Components and Antioxidant Capacity of Ontario Hard and Soft Wheat varieties. Canadian *J. Plant Sci.* 92, 19–30.

Raju, J., Patlolla, J.M.R., Swamy, M.V. (2004). Diosgenin, a steroid saponin of *Trigonella foenum graecum* (Fenugreek), inhibits Azoxymethane-induced aberrant crypt foci formation in F344 rats and induces apoptosis in HT-29 human colon cancer cells. *Cancer Epidemiol Biomarkers Prev.* 13, 1392–1398.

Raninen, K., Lappi, J., Mykkänen, H., and Poutanen, K. (2010). Dietary fiber type reflects physiological functionality: comparison of grain fiber, inulin, and polydextrose. *Nutr. Rev.* 69, 9–21.

Reichert, R.D. (1981). Quantitative isolation and estimation of cell wall material from dehulled pea (*Pisum sativum*) flours and concentrates. *Cereal Chem.* 58, 266–270.

Reinli, K., and Block, G. (1996). Phytoestrogen content of foods—a compendium of literature values. *Nutr. Cancer.* 26, 123–148.

Riahi, E., and Ramaswamy, H.S. (2003). Structure and composition of cereal grains and legumes. In: Chakraverty, A., Mujumdar, A.S., Raghavan, G.S.V. and Ramaswamy, H.S. (eds). *Handbook of postharvest technology*. Marcel Dekker, NY, USA, pp 1–16.

Rice-Evans, C.A., Miller, N.J., and Paganga, G. (1996). Structure-antioxidant activity relationships of flavonoids and phenolic acids. *Free Radic. Biol. Med.* 20, 933–956.

Rochfort, S., Ezernieks, V., Neumann, N., and Panozzo, J. (2011). Pulses for Human Health: Changes in Isoflavone and Saponin Content with Preparation and Cooking. *Aust. J. Chem.* 64, 790–797.

Ross, A.B., Shepherd, M.J., Schupphaus, M., Sinclair, V., Alfaro, B., Kamal-Eldin, A., and Åman, P. (2003). Alkylresorcinols in cereals and cereal products. *Journal of Agricultural and Food Chemistry* 51, 4111–4118.

Ross, A.B., Kamal-Eldin, A., and Åman, P. (2004). Dietary Alkylresorcinols: absorption, bioactivities, and possible use as biomarkers of whole-grain wheat– and rye–rich foods. *Nutr. Rev.* 62, 81–95.

Sarnthein-Graf, C., and La Mesa, C. (2004). Association of saponins in water and water-gelatine mixtures. *Thermochim. Acta.* 418, 79–84.

Savage, G.P., and Deo, S. (1989). The nutritional value of peas (Pisum sativum). *A literature review. Nutr. Abstr.Rev. (Series A).* 59, 66–83.

Scalbert, A., Manach, C., Morand, C., Remesy, C., and Jimenez, L. (2005). Dietary polyphenols and the prevention of diseases. *Crit. Rev. Food Sci. Nutr.* 45, 287–306.

Schneider, A.V. (2002). Overview of the markety and consumption of pulses in Europe. *Brit. J. Nutr.* 88, S243–S250.

Scott, C.E.; Eldridge, A.L. (2005). Comparison of carotenoid content in fresh, frozen and canned corn. *J. Food Compos. Anal.* 18, 551–559.

Selvendran R. (1984). The plant cell wall as a source of dietary fiber: chemistry and structure. *Am. J. Clin. Nutr.* 39, 320–337.

Selvendran, R., R., Stevens, B.J.H., and Dupnt, S. (1987). Dietary fiber: Chemistry, analysis, and properties. *Adv. Food Res.* 1, 117–209.

Setchell, K.D.R. (1998). Phytoestrogens: the biochemistry, physiology, and implications for human health of soy isoflavones. *Am. J. Clin. Nutr.* 68, 1333S–1346S.

Sharma, P., and Gujral, H.S. (2011). Effect of sand roasting and microwave cooking on antioxidant activity of barley. *Food Res. Int.* 44, 235–240.

Shipp, J., and Abdel-Aal E.M. (2010). Food application and physiological effects of anthocyanins as functional food ingredients. *The open Food Sci. J.* 4, 7–22.

Shutler, S.M., Bircher, G.M., Tredger, J.A., Morgan, L.M., Walker, A.F., and Low, A.G. (1989). The effect of daily baked bean (*Phaseolus vulgaris*) consumotion on the plasma lipid levels of young, normo-cholestolaemic men. *Br. J. Nutr.* 61, 257–265.

Sichieri, R. (2002). Dietary patterns and their associations with obesity in the Brazilian city of Rio de Janeiro. *Obesity Res.* 10, 42–48.

Sidhu, G.S., and Oakenfull, D.G. (1986). A mechanism for the hypocholesterolemic activity of saponins. *Br. J. Nutri.* 55, 643–649.

Sidhu, J.S., Kabir, Y., and Huffman, F.G. (2007). Functional foods from cereal grains. *Int. J. Food Prop.* 10, 231–244.

Slavin, J.L. (2003). Why whole grains are protective: biological mechanisms. *P. Nutr. Soc.* 62, 129–134.

Slavin, J.L. (2004). Whole grains and human health. *Nutr. Res. Rev.* 17, 99–110.

Sookwong, P., Nakagawa, K., Murata, K., Kojima, Y., Miyazawa, T. (2007). Quantification of tocotrienol and tocopherol in various rice bran. *J. Agric Food Chem.* 55, 461–466.

Sosulski, F.W., and Dabrowski, K.J. (1984). Composition of free and hydrolyzable phenolic acids in the flours and hulls of ten legume species. *Journal of Agricultural and Food Chemistry* 32, 131–133.

Sosulski, F.W., and Wu, K.K. (1988). High-fiber breads containing field pea hulls, wheat, corn and wild oat brans. *Cereal Chem.* 65, 186–191.

Smeds, A.I., Jauhiainen, L., Tuomola, E., and Peltonen-Sainio, P. (2009). Characterization of variation in the lignan content and composition of winter rye, spring wheat, and spring oat. *Journal of Agricultural and Food Chemistry* 57, 5837–5842.

Smith K., Queenan, K.M., Thomas, W., Fulcher, R.G. and Slavin, J.L. (2008). Physiological effects of concentrated barley beta-glucan in mildly hypercholesterolemic adults. *J. Am. Coll. Nutr.* 27, 434–440.

Stanton, W.R., and Francis, B.J. (1966). Ecological significance of anthocyanins in the seed coats of the Phaseoleae. *Nature.* 211, 970–971.

Steinke, R.D., and Paulson, M.C. (1964). The production of phenolic acids which can influence flavor properties of steam-volatile phenols during the cooking and alcoholic fermentation of grain. *Journal of Agricultural and Food Chemistry* 12, 381–387.

Su, H.L., and Chang, K.C. (1995). Physicochemical and sensory properties of dehydrated bean paste products as related to bean varieties. *J. Food Sci.* 60, 764–794.

Sung, M.K., Kendall, C.W.C., Koo, M.M., and Rao, A.V. (1995). Effect of soybean saponins and gypsophilla saponin on growth and viability of colon carcinoma cells in culture. *Nutr. Cancer.* 23, 259–270.

Takeoka, G.R., Dao, L.T., Full, G.H., Wong, R.Y., Harden, L.Y., Edwards, R.H., and Berrios, J.D.J. (1997). Characterization of Black Bean (*Phaseolus vulgaris* L.) Anthocyanins. *Journal of Agricultural and Food Chemistry* 45, 3395–3400.

Tham, D.M., Gardnera, C.D., and Haskell, W.L. (1998). Potential health benefits of dietary phytoestrogens: A review of the clinical, epidemiological, and mechanistic evidence. *J. Clin. Endocr. Metab.* 83, 2223–2235.

Theander, O. and Westerlund, E., (1987). Studies on chemical modification in heat-processed starch and wheat flour. *Starch/Staerke.* 39, 88–91.

Thompson, L.U. (1993). Potential health benefits and problems associated with antinutrients in foods. *Food Res. Inter.* 26, 131–149.

Thompson, L.U., Robb, P., Serraino, M., and Cheung, F. (1991). Mammalian lignin production from various foods. *Nutr. Cancer.* 16, 43–52.

Tosh, S.M., and Yada, S. (2010). Dietary fibres in pulse seeds and fractions. characterization, functional attributes, and applications. *Food Res. Int.* 43, 450–460.

Tsuda, T., Ohshima, K., Kawakishi, S., and Osawa, T. (1994). Antioxidative pigments isolated from the seeds of Phaseolus vulgaris L. *Journal of Agricultural and Food Chemistry* 42, 248–251.

Tsuda, T., Horio, F., Osawa, T. (2002). Cyanidin 3-*O*-â-glucoside suppresses nitric oxide production during a zymosan treatment in rats. *J. Nutr. Sci. Vitaminol.* 48, 305–310.

Tsuda, T., Horio, F., Uchida, K., Aoki, H., Osawa, T. (2003). Dietary cyanidin 3-*O*-â-D-glucoside-rich purple corn color prevents obesity and ameliorates hyperglycemia. *J. Nutr.* 133, 2125–2130.

Updike, A.A., and Schwartz, S.J. (2003). Thermal processing of vegetables increase cis isomers of lutein and zeaxanthin. *Journal of Agricultural and Food Chemistry* 51, 6184–6190.

USDA, U.S. Department of Agriculture and U.S. Department of Health and Human Services. (2010). *Dietary Guidelines for Americans.* 7th Edition, Washington, DC, USA.

Van Loo, J., Cummings, J., and Delzenne, N., (1999). Functional food properties of non-digestible oligosaccharides: a consensus report from the ENDO project (DGXII AIRII-CT94-1095). *Br. J. Nutr.* 81,121–132.

Varo, P., Laine, R., and Koivistoinen, P. (1983). Effect of heat treatment on dietary fiber. Inter-laboratory study. *J. Assoc. Off. Anal. Chem.* 66, 933–938.

Wang, H-J., and Murphy, P.A. (1994). Isoflavone composition of American and Japanese soybeans in Iowa: effects of variety, crop year, and location. *Journal of Agricultural and Food Chemistry* 42, 1674–7.

Webb, M.L., and McCullough, M.L. (2005). *Dietary Lignans: Potential role in cancer prevention, nutrition and cancer*. 51, 117–131.

Winham, D.M., and Hutchins, A.M. (2007). Baked bean consumption reduces serum cholesterol in hypercholestrolemic adults. *Nutr. Res.* 27, 380–386.

Winham, D.M., Hutchins, A.M., and Johnston, C.S. (2007). Pinto Bean Consumption Reduces Biomarkers for Heart Disease Risk. *J. Am. Coll. Nutr.* 26, 243–249.

Wood, P.J. (2007). Cereal β-glucans in diet and health. *J. Cereal Sci.* 46, 230–238.

Wood, P.J. (2010). Oat and rye β-glucan: properties and function. *Cereal Chem.* 87, 315–330.

Xu, Z., and Godber, J.S. (1999). Purification and identification of components of gamma-oryzanol in rice bran Oil. *Journal of Agricultural and Food Chemistry* 47, 2724–2728.

Xu, B., and Chang, S.K.C. (2008). Total phenolics, phenolic acids, isoflavones, and anthocyanins and antioxidant properties of yellow and black soybeans as affected by thermal processing. *Journal of Agricultural and Food Chemistry* 56, 7165–7175.

Yao, N., Jannink, J.-L., and White, P.J. (2007). Molecular weight distribution of $(1 \rightarrow 3)(1 \rightarrow 4)$-β-glucan affects pasting properties of flour from oat lines with high and typical amounts of β-glucan. *Cereal Chem.* 84, 471–479.

Yoshinaga, K. (1986). Liquor with pigments of red rice. *J. Brew. Soc. Jpn.* 81, 337–342.

7 Plantation crops and tree nuts: composition, phytochemicals and health benefits

Narpinder Singh and Amritpal Kaur

Department of Food Science and Technology, Guru Nanak Dev University, Amritsar, India

Abbreviations: CHD: coronary heart disease, DPPH: 1,1-diphenyl-2-picrylhydrazyl, FRAP: ferric reducing antioxidant power, FW: fresh weight, HDL: high density lipoproteins, GAE: gallic acid equivalents, LDL: low density lipoprotein, MT: metric tones, Se: selenium, TE: trolox equivalents.

7.1 Introduction

Almonds (*Prunus amygdalus*), Brazil nuts (*Bertholletia excelsa*), cashews (*Anacardium occidentale* L.), chestnuts (*Castanea sativa*), hazelnuts (*Corylus avellana*), macadamia nuts (*Macadamia integrifolia*), pecans (*Carya illinoinensis*), pine nuts (*Pinus pinea*), pistachios (*Pistacia vera*) and walnuts (*Juglans regia* L.) are important tree nuts consumed all over the world. Amongst these tree nuts, almonds, hazelnuts, walnuts and pistachios are the most common. Almonds and chestnuts belong to the family of *Rosaceae* and *Fagaceae*, respectively, while cashew nuts and pistachios belong to the *Anacardiaceae* family. Hazel nuts belong to the *Betulaceae* or Birch family whereas walnuts and pecans belong to the family of *Juglandaceae*. Almonds are one of the most popular tree nuts in terms of world production followed by hazelnuts, cashews, walnuts and pistachios. The global production of different tree nuts during 2008–2009 and 2009–2010 is shown in Table 7.1 (USDA, 2009, 2010; INC, 2009).

Global production of almond, hazelnut and cashew nut kernels during 2008–2009 was between 872 250–884 697 MT, 571 962 MT and 538 400 MT, respectively. The production of these nuts showed a slight decline during 2009–2010. The production of pine nuts, macadamia nuts and pecan kernels was around 17 330 MT, 27 302 MT and 60 642 MT, respectively. The production of these nuts showed a slight increase during 2009–2010. Walnut in-shell production during 2008–2009 was around 1 000 995–1 117 730 MT. USA, Spain, Syria, Italy, Iran and Morocco are the major almond producing countries in the

Handbook of Plant Food Phytochemicals: Sources, Stability and Extraction, First Edition.
Edited by B.K. Tiwari, Nigel P. Brunton and Charles S. Brennan.
© 2013 John Wiley & Sons, Ltd. Published 2013 by John Wiley & Sons, Ltd.

Table 7.1 World tree nuts production 2008–2010 (MT)

Nuts	Basis	2008–2009	2009–August 2010
Almonds	Kernel	872 250[a]	760 000[a]
	Kernel	884 697[d]	853 728[d]
Brazil nuts	Kernel	22 800[d]	19 380[d]
	In-shell	12 900[d]	10 965[d]
Cashews	Kernel	538 400[d]	490 400[d]
Hazelnuts	In-shell	978 030[c]	663 500[c]
	Kernel	571 962[d]	375 000[d]
	In-shell	117 280[d]	772 000[d]
Macadamia nuts	Kernel	27 302[d]	31 351[d]
	In-shell	105 290[d]	120 000[d]
Pecans	Kernel	60 642[d]	106 536[d]
	In-shell	134 078[d]	235 930[d]
Pine nuts	Kernel	17 330[d]	18 830[d]
Pistachios	In-shell	360 300[b]	395 000[b]
	In-shell	350 300[d]	475 000[d]
Walnuts	In-shell	1 187 730[c]	1 240 780[c]
	In-shell	1 000 995[d]	990 000[d]

Sources: [a] data from USDA Foreign Agricultural Service/USDA Office of Global Analysis, August 2009; [b] data from USDA Foreign Agricultural Service/USDA Office of Global Analysis, February 2010; [c] data from USDA Foreign Agricultural Service/USDA Office of Global Analysis, October 2009; [d] data from International Nut and Dried Fruit Council Foundation (INC), XXVIII World Nut and Dried Fruit Congress Newsletter, Monaco, 29–31 May 2009.

world (FAO, 2009). Turkey and Italy are the principal producers of hazelnuts (FAO, 2009). Walnuts are widely distributed all over the world, and China, USA, Iran, Turkey and Ukraine are the main walnut producing countries (FAO, 2009). Iran, the USA, Turkey, Syria and China are the main pistachio producing countries (FAO, 2009). Brazil nuts are the largest of the commonly consumed nuts from the giant Brazil nut tree, which is a native of South America. Bolivia, Brazil, Peru, Colombia and Venezuela are the main producer of Brazil nut. Vietnam, India, Nigeria, Côte d'Ivoire and Brazil are the main cashew nut producing countries (FAO, 2009). Spain, Italy, China, Portugal and Turkey are the principal producing countries of pine kernels. Coconut is also a tree nut, however, whether it should be considered as such is a matter of some controversy. Tree nuts are rich sources of various nutrients and phytochemicals with many potential health benefits. The nutrients and phytochemicals content of different nuts vary with varieties and environment. These phytochemicals possess many functions and reduce the risk of certain types of cancer, coronary heart disease (CHD), atherosclerosis, osteoporosis, type-2 diabetes and some neurodegenerative diseases associated with oxidative stress (Surh *et al.*, 2003; Kelly *et al.*, 2006; Tapsell *et al.*, 2004; Jiang *et al.*, 2003). Tree nuts are consumed mainly as snacks and are also used as ingredients in a variety of food products. Tree nut oils are also used in many skin moisturizers and cosmetic products (Madhaven, 2001). Chestnuts are tree nuts but are rich in starch and have different nutrients profile in comparison to other common nuts. Consumers generally consider peanuts (*Arachis hypogea*) also as nuts but they are actually botanically legumes. In this chapter, the information about composition, phytochemicals and health benefits of all the common tree nuts except chestnuts and coconut is presented.

7.2 Composition

The composition of different tree nuts is compared in Table 7.2. Protein content varies between 7.5 and 21.2%. Almonds, pistachios and cashew nuts contain higher protein content and lower lipids content as compared to walnuts, hazelnuts, Brazil nuts and pine nuts. Tree nuts are rich sources of lipids and some of these have lipids as high as 75%. Tsantili *et al.* (2010) reported protein and fat content varied from 18.99 to 21.87% and 49.79 to 56.75%, respectively, in different pistachio nut varieties. Ash content that represents the inorganic matter varies from 1.14% to 4.26%; pecan and macadamia nuts have lower ash content as compared to other tree nuts. Carbohydrates content vary from 12.27 to 30.19%, the lowest in Brazil nuts and the highest in cashew nuts. Almonds, Brazil nuts, hazelnuts, macadamias, pecans, pine nuts and pistachios have low starch content (0.25–1.67 g/100 g) while cashews have higher starch content of 23.49 g/100 g (USDA, 2010). Cashew and pine nuts have lower dietary fibre content (3.3–3.7%) as compared to other three nuts. Almonds, pistachios, walnuts, hazelnuts and Brazil nuts have dietary fibre of 6.7–12.2 g/100 g (Table 7.2). Cashews, almonds and pistachios provide lower energy in comparison to other nuts due to their lower lipids content. The difference in chemical constituents of a particular tree nut has been reported to vary with variety and environmental conditions. The majority of tree nut oils contain about 70% unsaturated fatty acids which make them susceptible to oxidative rancidity. Tree nut lipids vary in fatty acid composition, however, the majority are rich in monounsaturated fatty acids (oleic acid, 18:1) and have much lower amounts of polyunsaturated fatty acids (i.e. linoleic acid, 18:2). Cashew nuts (21.12 g/100 g), macadamia nuts (18.18 g/100 g), Brazil nuts (25.35 g/100 g) and pine nuts (24.1 g/100 g) were reported to have more total saturated fatty acids than almonds (9.09 g/100 g), hazelnuts (9.11 g/100 g), pecan (8.35 g/100 g), pistachios (14.24 g/100 g) and walnuts (11.76 g/100 g) by Venkatachalam and Sathe (2006). Polyunsaturated fatty acids were observed to be the main group of fatty acids in walnut oil, ranging from 70.7 to 74.8%; monounsaturated fatty acids ranged from 15.8 to 19.6% and saturated fatty acids ranged from 8.9 to 10.1% (Amaral *et al.*, 2003). Venkatachalam and Sathe (2006) reported monounsaturated fatty acid content of nut seed oils and found the highest for hazelnuts (83.1 g/100 g), followed by macadamia nuts (77.43 g/100 g), pecan nuts (66.73 g/100 g), cashew nuts (61.68 g/100 g), almonds (61.6 g/100 g), pistachio nuts (51.47 g/100 g), Brazil nuts (29.04 g/100 g) and walnuts (15.28 g/100 g). Tsantili *et al.* (2010) reported that oleic acid content ranged from 51.6 to 67.86%, linoleic acid (18:2) content from 11.56 to 27.03%, palmitic acid (16:0) from 8.54 to 10.24% and linolenic acid (18:3) from 0.34 to 0.5% in oil from different pistachio varieties. Arranz *et al.* (2009) reported linoleic acid content of 63.19, 21.39, 7.13 and 33.04%, respectively, in oil from walnuts, almonds, hazelnuts and pistachios. Ryan *et al.* (2006) reported linoleic acid content of 42.8, 50.31, 45.41, 30.27 and 20.8%, respectively, for oil extracted from Brazil nuts, pecan nuts, pine nuts, pistachio nuts and cashew nuts. These authors reported higher stearic acid (18:0) content in oil extracted from Brazil nuts (11.77%) as compared to oil from pecan nuts (1.8%), pine nuts (4.48%), pistachio nuts (0.86%) and cashew nuts (8.70%). Oleic acid (18:1) was observed to be higher for oil from pistachio and cashew nuts as compared to oil from Brazil nuts (29.09%), pecan nuts (40.63%) and pine nuts (39.55%). Walnuts are a good source of omega-3 fatty acid and have the highest amount of this fatty acid group amongst the different tree nuts. Walnut oil is the only tree nut oil that contains an appreciable amount of α-linolenic acid. Tree nuts are cultivated for use as oil crops in many parts of the world. In the Middle East and Asia, tree nuts are important

Table 7.2 Proximate composition of different nuts (edible portion)

Nuts	Moisture content (g/100g)	Protein content (g/100g)	Lipids content (g/100g)	Ash content (g/100g)	Sugars content (g/100g)	Carbohydrates content (g/100g)	Dietary fiber (g/100g)	Energy (kcal/100g)
Almonds	9.51±0.08[a]	19.48±0.51[a]	43.36±0.62[a]	2.48±0.05[a]	2.11±0.11[a]	21.67±0.00[d]	12.2±0.194[d]	575±0.00[d]
	4.70±0.046[d]	21.22±0.044[d]	49.42±0.188[d]	2.99±0.015[d]	3.89±0.00[d]			
Brazil nuts	3.07±0.37[a]	13.93±0.40[a]	66.71±1.71[a]	3.28±0.01[a]	0.69±0.04[a]	12.27±0.00[b]	7.5±0.232[d]	656±0.00[d]
	3.48±0.170[d]	14.32±0.146[d]	66.43±0.237[d]	3.51±0.033[d]	2.33±0.078[d]			
Cashews	4.39±0.04[a]	18.81±0.06[a]	43.71±1.13[a]	2.66±0.21[a]	3.96±0.08[a]	30.19±0.00[d]	3.3±0.00[d]	553±0.00[d]
	5.20±0.00[d]	18.22±0.00[d]	43.85±0.00[d]	2.54±0.00[d]	5.91±0.00[d]			
Hazelnuts	4.19±0.04[a]	14.08±0.34[a]	61.46±0.57[a]	2.03±0.14[a]	1.41±0.05[a]	16.70±0.00[d]	9.7±0.374[d]	628±0.00[d]
	3.90±0.20[c]	15.35±0.42[c]	61.21±0.99[c]	2.24±0.03[c]				
	5.31±0.196[d]	14.95±0.156[d]	60.75±0.386[d]	2.29±0.017	4.34±0.071[d]			
Macadamia nuts	2.10±0.12[a]	8.40±0.71[a]	66.16±0.92[a]	1.16±0.04[a]	1.36±0.05[a]	13.82±0.00[d]	8.6±0.911[d]	718±0.00[d]
	1.36±0.068[d]	7.91±0.351[d]	75.77±1.147[d]	1.14±0.034	4.57±0.180			
Pecans	7.40±0.80[a]	7.50±0.24[a]	66.18±0.53[a]	1.88±0.07[a]	1.55±0.04[a]	13.86±0.00[d]	9.6±0.406[d]	691±0.00[d]
	3.52±0.114[d]	9.17±0.088[d]	71.97±0.120[d]	1.49±0.055	3.97±0.153[d]			
Pine nuts	1.47±0.29[a]	13.08±0.75[a]	61.73±0.55[a]	2.50±0.15[a]	1.82±0.07[a]	13.08±0.00[d]	3.7±0.052[d]	673±0.00[d]
	2.28±0.094[d]	13.69±0.156[d]	68.37±0.249[d]	2.59±0.032[d]	2.59±0.032[d]			
Pistachios	5.74±0.03[a]	19.80±0.49[a]	45.09±0.27[a]	3.21±0.03[a]	1.52±0.07[a]	27.51±0.00[d]	10.3±0.204[d]	562±0.00[d]
	3.91±0.169[d]	20.27±0.358[d]	45.39±1.355[d]	2.91±0.112[d]	7.66±0.178[d]			
Walnuts	2.70±0.20[a]	13.46±0.47[a]	64.50±0.45[a]	1.82±0.02[a]	2.06±0.23[a]	13.71±0.00[d]	6.7±0.549[d]	654±0.00[d]
	3.8±0.05 to	14.38±0.27 to	68.83±2.00 to	3.31±0.31to	2.61±0.094[d]			
	4.50±0.22[b]	18.03±0.29[b]	72.14±0.27[b]	4.26±0.02[b]				
	4.07±0.155[d]	15.23±0.238[d]	65.21±0.494[d]					

Sources: [a] data from Venkatachalam and Sathe (2006); [b] data from Pereira et al. (2008); [c] data from Alasalvar et al. (2003); [d] data from USDA National Nutrient Database for Standard Reference, Release 23 (2010) (assessed on 20 June, 2011).

sources of energy, essential dietary nutrients and phytochemicals (Bonvehi *et al.*, 2000). Brazil nuts are a good source of nutrients, including protein, fibre, selenium (Se), magnesium, phosphorus and thiamine. Pine nuts (8.8 mg/100 g), hazelnuts (6.17 mg/100 g), pecan nuts (4.5 g/100 g) and macadamia nuts (4.13 mg/100 g) are rich in bone-building manganese (USDA, 2010). These nuts also contain niacin, vitamin E, vitamin B6, calcium, iron, potassium, zinc and copper. Alasalvar *et al.* (2009) reported that the manganese content varied from 2.17 to 19.0 mg/100 g in hazelnut varieties.

The composition of nuts with and without seed coat also differed significantly. Endosperm of the majority of tree nuts contains about 70% unsaturated fats, which make them susceptible to rancidity. Brazil nuts have protein, which is a rich source of methionine (Antunes and Markakis, 1977). Nuts are a rich source of arginine that was observed to range from 9.15 g/100 g of protein in pistachios to 15.41 g/100 g of protein in pine nuts (Venkatachalam and Sathe, 2006). Walnuts are good sources of both antioxidants and n-3 fatty acids, with particularly high amounts of α-linolenic acid (6.3 g/100 g), whereas other nuts such as almonds, pecans, and pistachios possess much smaller amounts (0.4–0.7 g/100 g). Almonds are especially high in vitamin E and magnesium. Smeds *et al.* (2007) reported that certain tree nuts (almond, cashew and walnut) contain the polyphenolics, lignans, in amounts comparable to certain rice types (346–486 μg of lignans/100 g edible portion), but lower than the cereals like rye (10377 μg of lignan/100 g) and wheat (7548 μg of lignan/100 g). Almonds, cashews and walnuts contain between 344 and 912 μg of lignan/100 g edible portion, cashew nut was reported to be the most abundant tree nut source of lignan (912 μg/100 g) (Smeds *et al.*, 2007).

The oxalate content of nuts was reported to vary widely and it was suggested by Ritter and Savage (2007) that people who have a tendency to form kidney stones consume certain nuts in moderate levels. These authors extracted gastric soluble and intestinal soluble oxalates from the nuts using an *in vitro* assay, which involved incubations of the food samples for 2 h at 37 °C in gastric and intestinal juice. Pistachio nuts (roasted) contained relatively low levels of gastric soluble oxalate (67 mg/100 g FW). Almonds and Brazil nuts were observed to contain high levels of gastric soluble oxalate (538.5 and 492.0 mg/100 g FW, respectively). The intestinal soluble oxalate is the fraction that absorbed in the small intestine. Pecan nuts and pistachios (roasted) contained relatively low levels of intestinal soluble oxalate (155 and 76 mg/100 g FW, respectively) as compared to almonds, Brazil nuts, cashew nuts and pine nuts (222, 304, 216 and 581 mg/100 g FW, respectively). Pinenuts contained the highest levels of intestinal soluble oxalate (581 mg/100 g FW), while roasted pistachio nuts were observed to contain low level (77 mg /100 g FW).

7.3 Phytochemicals content

Nuts are a good source of phytochemicals, including phenolics, flavonoids, isoflavones, terpenes, organosulfuric compounds and vitamin E (Bravo, 1998; Kris-Etherton *et al.*, 2002). The majority of nuts have low concentrations of carotenoids, and are not an excellent source of dietary carotenoids. The β-carotene and lutein content were found to be 0.21 and 2.32 mg/100 g (dry weight), respectively, in pistachios (Kornsteiner *et al.*, 2006). Tocopherol content, lutein, zeaxanthin and Se content of different tree nuts is shown in Table 7.3. Almonds and hazelnuts are excellent sources of α-tocopherol (vitamin E). Cashews, Brazil nuts, macadamias, pecans, walnuts and pistachios are poor sources of vitamin E. Higher amount of lutein plus zeaxanthin (1405 mcg) for pistachio nuts as compared to other nuts

Table 7.3 Tocopherol, selenium and lutein+zeaxantin content of different tree nuts

Nuts	α-Tocopherol	β- and γ-Tocopherol	δ-Tocopherol	Se	Lutein+ zeaxanthin
Almonds	24.2[a] 26.22[b]	3.1[a] 0.94[b]	0.05[b]	2.5±0.361[b]	1.0[b]
Brazil nuts	1.0[a]	13.2[a]		1917±231.79[b]	–
Cashews	0.9[b]	5.1[a] 5.34[b]	0.3[a] 0.36[b]	19.9±0.00b	22[b]
Hazelnuts	31.4[a] 15.03[b]	6.9[a] 0.33[b]	0.1[a]	2.4±0.561[b]	92[b]
Macadamia nuts	0.54[b]	–	–	3.6±0.00	–
Pecans	1.4[b]	14.8[a] 24.83[b]	0.2[a] 0.47	3.8±0.114	17[b]
Pine nuts	4.1[a] 9.33[b]	8.1[a] 11.15[b]	0.3[a]	0.7±0.069[b]	9[b]
Pistachios	2.30[b]	29.3[a] 22.6[b]	0.5[a] 0.8[b]	7.0±0.00[b]	1405[b]
Walnuts	0.7[b]	21.9[a] 20.98[b]	3.8[a] 1.89[b]	4.9±0.417	9[b]

Sources: [a]Data from Kornsteiner et al. (2006), Data expressed as mg/g oil; [b]data from USDA National Nutrient Database for Standard Reference, Release 23 (2010) (assessed on 20 June, 2011), data expressed as (mcg).

was reported (Table 7.3). Proanthocyanidins were reported to be present in the majority but not in all nuts, with concentrations of 501 mg/100 g in hazelnuts, 494 mg/100 g in pecans, 237 mg/100 g in pistachios, 184 mg/100 g in almonds, 67 mg/100 g in walnuts, 16 mg/100 g in peanuts, and 9 mg/100 g in cashews (Gu et al., 2004). Brazil nuts are rich food sources of Se (1917 mcg/100 g). Cashews, almonds, hazelnuts, macadamias, pecans, pine nuts, pistachios and walnuts were reported to have Se content 0.7–19.9 mcg (Table 7.3).

Tocopherol, squalene and phytosterol content of oil from different tree nuts is shown in Table 7.4. Almond oil has the highest α-tocopherol content followed by that of hazelnut, pine nut and macademia nut (Yang et al., 2009). Squalene content was reported to be the highest for Brazil nuts (1377 μg/g oil), followed by hazelnuts (186 μg/g oil) and macadamia nuts (185 μg/g oil) by Ryan et al. (2006) and Maguire et al. (2004). Ryan et al. (2006) reported that Brazil nut has higher squalene content (1377.8 mg/g) as compared to pine (39.5 mg/g), cashew (89.4 mg/g), pistachio (91.4 mg/g), and pecan (151.7 mg/g). Tree nuts are a good source of phytosterols and amongst the various phytosterols determined in tree nuts, β-sitosterol was reported to be present in the highest amount (Phillips et al., 2005).

Pistachios nuts contain the higher total phytosterols (β-sitosterol, Campesterol, Stigma sterol, Δ^5-avenasterol, Sitostanol, Campestanol and other sterols) content of 279 mg/100 g. While almonds, macadamia, pine nuts, hazelnuts, pecans, walnuts (English) and Brazil nuts had total phytosterol content of 199, 187, 236, 121, 157, 113 and 95 mg/100 g, respectively (Table 7.5). Campestrol was observed to be higher in pine nuts, pistachios, macadamias and cashews as compared to almonds, Brazil nuts, pecans, hazelnuts and walnuts. Pine nuts and pistachio nuts have higher Δ^5-avenasterol content in comparison to other nuts. Thompson et al. (2006) analysed phytoestrogen content of 121 food samples including seven major tree nuts (almond, cashew, chestnut, hazelnut, pecan, pistachio and walnut). It was reported that tree nuts have four each of isoflavones (formononetin, daidzein, genistein

Table 7.4 Tocopherol, squalene and phytosterol content of oil extracted from different tree nuts

Nuts	Tocopherol (µg/g oil)			Squalene (µg/g oil)	Phytosterol (µg/g oil)		
	α-Tocopherol	β-Tocopherol	γ-Tocopherol		β-Sitosterol	Campesterol	Stigmasterol
Almonds	439.5±4.8[a]		12.5±2.1[a]	95.0±8.5[a]	2071.7±25.9[c]	55.0±10.8[c]	51.7±3.6[c]
Brazil nuts	82.9±9.5[b]	116.2±5.1[b]		1377.8±8.4[b]	1325.4±68.1[b]	26.9±4.4[b]	577.5±34.3[b]
Cashews	3.6±1.4[b]	57.2±6.2[b]		89.4±9.7[b]	1768.0±210.6[b]	105.3±16.0[b]	116.7±12.6[b]
Hazelnuts	310.1±31.1[c]		61.2±29.8[c]	186.4±11.6[c]	991.2±73.2[c]	66.7±6.7[c]	38.1±4.0[c]
Macadamia nuts	122.3±24.5[c]		Trace[c]	185.0±27.2[c]	1506.7±140.5	73.3±8.9[c]	38.3±2.7[c]
Pecans	12.2±3.2[b]	168.5±15.9[b]		151.7±10.8[b]	1572.4±41.0[b]	52.2±7.1[b]	340.4±29.5[b]
Pine nuts	124.3±9.4[b]	105.2±7.2[b]		39.5±7.7[b]	1841.7±125.2[b]	214.9±13.7[b]	680.5±45.7[b]
Pistachios	15.6±1.2[b]	275.4±19.8[b]		91.4±18.9[b]	4685.9±154.1[b]	236.8±24.8[b]	663.3±61.0[b]
Walnuts	20.6±8.2[c]		300.5±31.0[c]	9.4±1.8[c]	1129.5±124.6[c]	51.0±2.9[c]	55.5±11.0[c]

Sources: [a] data from Yang et al. (2009); [b] data from Ryan et al. (2006); [c] data from Maguire et al. (2004).

Table 7.5 Phytosterol composition of different tree nuts (mg/100g)

Nuts	β-sitosterol	Campesterol	Stigma sterol	Δ⁵-avenasterol	Sitostanol	Campestanol	Other sterols
Almonds	143.4[a]	4.9[a]	5.0[a]	19.7[a]	3.2[a]	3.3[a]	19.6[a]
Brazil nuts	65.5[a]	2.0[a]	6.2[a]	13.6[a]	4.1[a]	2.0[a]	3.4[a]
		5.0[b]	11.33[b]				
Cashews	112.6[a]	8.9[a]	<1.2[a]	13.7[a]	<1.2[a]	2.0[a]	13.3[a]
Hazelnuts	102.[a]	6.6[a]	<2.5[a]	2.6[a]	4.0[a]	3.0[a]	2.5[a]
Macadamia nuts	143.7[a]	9.6[a]	Nd	13.3[a]	Nd	2.9[a]	17.0[a]
Pecans	116.5[a]	5.9[a]	2.6[a]	14.6[a]	<1.7[a]	2.8[a]	14.1[a]
Pine nuts	132.0[a]	19.8[a]	<1.7[a]	40.3[a]	5.9[a]	3.8[a]	34.2[a]
Pistachios	209.8[a]	10.1[a]	2.3[a]	26.2[a]	1.3[a]	5.0[a]	24.6[a]
Walnuts, English	88.9[a]	4.9[a]	Nd	7.3[a]	<1.7[a]	2.4[a]	9.1[a]

Sources: [a] data from Phillips et al. (2005); [b] data from da Costa et al. (2010).

and glycitein) and lignans (matairesinol, lariciresinol, pinoresinol and secoisolariciresinol); and one coumestan (coumestrol). Amongst the tree nuts studied, pistachio was the richest source of total isoflavones (176.9 µg/100 g on an as is basis), total lignans (198.9 µg/100 g), and total phytoestrogens (382.5 µg/100 g). Hazelnut contained higher total isoflavones (30.2 µg/100 g), primarily genistein, as compared to pistachio and walnut and had the sixth highest total lignans (77.1 µg/100 g), primarily secoisolariciresinol, and total phyoestrogens (107.5 µg/100 g).

Nuts are a good source of phenolics (tannins, ellagic acid and curcumin) and flavonoids such as luteolin, quercetin, myricetin, kaempferol and resveratrol (Bravo, 1998; Kris-Etherton *et al.*, 2002). Almonds contain an abundance of flavonoids, including catechins, flavonols and flavonones in their aglycone and glycoside forms (Sang *et al.*, 2002). Pistachio nuts also have several flavonoids and were reported to be rich in resveratrol (Lou *et al.*, 2001), while cashew nuts contain an abundance of alkylphenols (Trevisan *et al.*, 2006). Resveratrol content of 115 µg/100 g for pistachio nuts was reported by Tokusoglu *et al.* (2005). Walnuts contain a wide variety of phenolics, tocopherols and nonflavonoids such as ellagitannins (Anderson *et al.*, 2001). Hazelnuts contained different phenolic acids such as gallic acid, caffeic acid, p-coumaric acid, ferulic acid and sinapic acid in both free and esterified forms (Shahidi *et al.*, 2007).

Tree nuts are externally covered with a thin layer of skin known as testa (seed coat). The testa contributes a bitter/astringent taste to nuts and reduces the consumer acceptability. Therefore, testa is removed from the majority of nuts before marketing or using in different food products. Testa constitutes about 1–3% of total weight of cashews. Testa is a rich source of hydrolysable tannins with polymeric proanthocyanidins as major polyphenols (Mathew and Parpia, 1970). Extracts of whole almond seed, brown skin, shell and green shell cover (hull) possess potent free radical-scavenging capacities (Amarowicz, Troszynska and Shahidi, 2005; Jahanban *et al.*, 2009; Moure, Pazos, Medina, Dominguez and Parajo, 2007; Pinelo, Rubilar, Sineiro and Nunez, 2004; Siriwardhana, Amarowicz and Shahidi, 2006; Siriwardhana and Shahidi, 2002; Wijeratne *et al.*, 2006). These activities may be related to the presence of flavonoids and other phenolic compounds in nuts. Almond hulls are a rich source of three triterpenoids (about 1% of the hulls), betulinic, urosolic and oleanolic acids (Takeoka *et al.*, 2000), as well as flavonol glycosides and phenolic acids (Sang *et al.*, 2002). Sang *et al.* (2002) isolated catechin, protocatechuic acid, vanillic acid, p-hydroxybenzoic acid and naringenin glucoside, as well as galactoside, glucoside and rhamnoglucoside of 3b-O-methylquercetin and rhamnoglucoside of kaempferol from almond hulls. As a result Almond hulls, which are mainly used in livestock feed, have been suggested as a potential source of antioxidants (Siriwardhana *et al.*, 2006; Shahidi, Zhong, Wijeratne and Ho, 2009). Pecans and walnuts were reported to have higher total antioxidant activity of 179.4 and 135.4 µmol of TE/g, respectively as compared to hazelnuts (96.45 µmol of TE/g), pistachio nuts (79.83 µmol of TE/g), almonds (44.54 µmol of TE/g), cashews (19.97 µmol of TE/g), macadamias (16.95 µmol of TE/g), brazil nuts (14.19 µmol of TE/g) and pine nuts (7.19 µmol of TE/g) by Wu *et al.* (2004). Antioxidant activity of hazelnuts, walnuts and pistachios with and without seed coat (testa) was compared by Arcan and Yemenicioğlu (2009). Yang *et al.* (2009) evaluated tree nuts for total phenolic and flavonoid contents, antioxidant and antiproliferative activities. Walnuts had the higher total phenolic and flavonoid contents (1580.5 ± 58.0 mg/100 g and 744.8 ± 93.3 mg/100 g, respectively), followed by pecan nuts (1463.9 ± 32.3 mg/100 g and 704.7 ± 29.5 mg/100 g, respectively), pistachios (571.8 ± 12.5 mg/100 g and 143.3 ± 18.7 mg/100 g, respectively) and macadamia nuts (497.8±52.6 mg/100 g and 137.9±9.9 mg/100 g, respectively). Almonds, Brazil nuts, cashews, hazelnuts and pine nuts showed total phenolics and flavonoids content between 152.9±14.1–316.4±7.0 and 45.0±5.4–107.8±6.0, respectively. Walnuts also had the highest total antioxidant activity (458.1 µmol of vitamin C equiv/g). Both soluble phenolic and flavonoid contents were observed to be positively correlated with total antioxidant activity.

It was reported that the removal of seed coat considerably reduced the total antioxidant activity of hazelnuts, walnuts and pistachios. The removal of seed coat was observed to reduce the total antioxidant activity of hazelnuts, walnuts and pistachios to the extent of 36, 90 and 55%, respectively (Arcan and Yemenicioglu, 2009). These authors reported the antioxidant activity in a one-serving portion (one-serving portion=42 g) of fresh or dry walnuts equivalent to that of a two-serving portion of black tea (one-serving portion=200 ml) and 1.2–1.7-serving portions of green and Earl Grey tea (one-serving portion=200 ml). Ethanolic extract of cashew nut testa was reported to exhibit a significant level of antioxidant activity, which was attributed to its phenolic composition (Kamath and Rajini, 2007). Cashew nut testa has been found to have higher levels of (+)-catechin and (−)-epicatechin as compared to those reported for green tea and chocolate (Trox et al., 2011). Cashew nuts with testa possess significantly higher amounts of carotenoids and tocopherols when compared to testa-free kernels. The presence of such potentially bioactive compounds in the testa-containing cashew nut kernels was suggested as an interesting economical source of natural antioxidants for use in food and nutraceutical industries (Trox et al., 2011). Tomaino et al. (2010) reported higher antioxidant activity of pistachio skin as compared to seed and this has been attributed to gallic acid, catechin, cyanidin-3-O-galactoside, eriodictyol-7-O-glucoside and epicatechin together with other unidentified compounds. They observed gallic acid, catechin, cyanidin-3-O-galactoside, eriodictyol-7-O-glucoside and epicatechin content of 1453.31, 377.45, 5865.12, 365.68 and 104.8 mg/g (fresh weight), respectively, in pistachio skin. Whereas pistachio seed was observed to have gallic acid, catechin and eriodictyol-7-O-glucoside content of 12.66, 2.41 and 31.91 mg/g (fresh weight), respectively. They reported antioxidant activity of 1.65 and 116.32 measured as mg of GAE/g (fresh weight), respectively in pistachio seeds and skins.

Free and bound phenolics and flavonoids distribution vary amongst different nuts (Table 7.6). Yang et al. (2009) reported that walnuts contain the highest soluble-free phenolic content (1325 mg/100 g), followed by pecans (1227 mg/100 g), pistachios (339 mg/100 g), cashews (86.7 mg/100 g), almonds (83 mg/100 g), Brazil nuts (46 mg/100 g), pine nuts (39 mg/100 g), and macadamia nuts (36 mg/100 g). Hazelnuts had the lowest free phenolic content of 22.5 mg/100 g. Macadamia nuts had the highest bound phenolics (462 mg/100 g) followed by peanuts (237 mg/100 g), hazelnuts (292 mg/100 g), walnuts (255 mg/100 g), pecans (293 mg/100 g), pistachios (232 mg/100 g), cashews (230 mg/100 g), almonds (130 mg/100 g), Brazil nuts (123 mg/100 g) and pine nuts (114 mg/100 g). The contribution of bound fraction was insignificant compared to the soluble phenolic fraction of cashew nuts and testa. High temperature (130 °C for 33 min) treated cashew nuts and testa showed a higher phenolic content and antioxidant activity than low temperature (70 °C for 6 h) treated samples (Chandrasekara and Shahidi, 2011). DPPH radical scavenging activity of soluble phenolics extracts of raw cashew nut kernels and testa was 3.17 and 179.3 (mg of GAE/g of defatted meal), respectively; while bound phenolics extract showed 0.13 and 81.16 (mg of GAE/g of defatted meal), respectively for kernel and testa. The DPPH radical scavenging activity of soluble phenolic extracts of kernel and testa significantly increased with increasing roasting temperature, whereas bound extracts generally showed a decrease.

Table 7.6 Total phenolics contents, flavonoids contents and antioxidant activity of different tree nuts

Nuts	Phenolics (mg/100g)			Flavonoids (mg/100g)			Total antioxidant activity (μmol of Vit. C equiv/g)	Total antioxidant activity (μmol of TE/g)
	Free form	Bound form	Total	Free form	Bound form	Total		
Almonds	83.0±1.3[a]	129.9±13[a]	212.9±12.3[a]	39.8±2.0[a]	53.7±11.9[a]	93.5±10.8[a]	25.4±2.0[a]	44.54[b]
Brazil nuts	46.2±5.7[a]	123.1±18.4[a]	169.2±14.6[a]	29.2±7.2[a]	78.6±9.2[a]	107.8±6.0[a]	16.0±1.2[a]	14.19[b]
Cashews	86.7±8.1[a]	229.7±15.1[a]	316.8±7.0[a]	42.1±3.8[a]	21.6±5.2[a]	63.7±2.1[a]	29.5±2.7[a]	19.97[b]
Hazelnuts	22.5±1.1[a]	292.2±48.4[a]	314.8±47.3[a]	13.9±2.3[a]	99.8±28.5[a]	113.7±30.2[a]	7.1±0.9[a]	96.45[b]
Macadamia nuts	36.2±2.6[a]	461.7±51.2[a]	497.8±52.6[a]	9.4±0.7[a]	128.5±9.3[a]	13.9±9.9[a]	13.4±0.4[a]	16.95[b]
Pecans	1227.3±8.4[a]	236.6±28.1[a]	1463.9±32.3[a]	639.3±17.0[a]	65.4±12.7[a]	704.7±29.5[a]	427.0±21.6[a]	179.4[b]
Pine nuts	39.1±0.6[a]	113.8±14.3[a]	152. ±14.1[a]	13.0±1.5[a]	32.0±6.8[a]	45.0±5.4[a]	14.6±1.1[a]	7.19[b]
Pistachios	339.6±15.1[a]	232.1±13.3[a]	571.8±12.5[a]	87.4±14.0[a]	55.9±13.6[a]	143.3±18.7[a]	75.9±1.2[a]	79.83[b]
Walnuts	1325.1±37.4[a]	255.4±25.0[a]	1580.5±58.0[a]	535.4±71.5[a]	209.4±22.1[a]	744.8±93.3[a]	458.1±14.0[a]	135.41[b]

Sources: [a]data from Yang et al. (2009); [b]data from Wu et al. (2004).

The soluble extracts of testa treated at high temperature had a higher DPPH radical scavenging activity than that of low temperature treated testa. Mathew and Parpia (1970) reported the presence of catechin and epicatechin as predominant polyphenolics in cashew nut testa. High temperature treated testa had a higher flavonoid content to that in the raw testa. This increase has been attributed to the liberation and isomerization of such compounds during heat treatment of cashew nuts and testa. Locatelli et al. (2010) reported that roasting at 180 °C for 20 min brought about higher total phenol content of the soluble extract than roasting at the same temperature for 10 min of hazelnut skin. Amaral et al. (2006) studied the effects of roasting of hazel nuts at different temperatures (125–200 °C for 5–30 min)) on phytosterols and observed a modest decrease in the total levels of the beneficial phytosterols (maximum of 14.4%) and vitamin E (maximum of 10.0%) compounds during roasting. A negligible increase of the potentially harmful *trans* fatty acids was also observed. Bolling et al. (2010) reported that processing and storage change the polyphenol and antioxidant activity of almond skin. They reported that dry roasted (135 °C for 14 min) almonds had 26% less total phenols and 34% less ferric reducing antioxidant power (FRAP) than raw. Storage of almonds at 4 °C and 23 °C for 15 months resulted in gradual increase in flavonoids and phenolic acids, up to 177 and 200%, respectively. However, FRAP and total phenols were found to increase to 200 and 190% of initial values after 15 months. Thus, roasting decreased total phenols and FRAP of almond skin but not flavonoids and phenolic acids, whilst storage for up to 15 months doubled flavonoids and phenolic acids.

Bleaching of pistachio shells is done to improve the appearance by increasing the whiteness. This practice is not permitted and actually illegal in many countries. The effects of bleaching (0.1–50% hydrogen peroxide) on phenolic levels and antioxidative capacities in raw and roasted nuts were reported by Seeram et al. (2006). Bleaching decreased total anthocyanin levels and antioxidative capacity of raw and roasted nuts. Raw nuts preserved phenolic levels and antioxidant capacity better than roasted nuts, suggesting contributing effects of other substances and/or matrix effects that are destroyed by the roasting process.

7.4 Health benefits

Many epidemiologic and clinical studies have associated frequent consumption of nuts with reduced risk of CHD (Kelly and Sabaté, 2006; Fraser et al., 1992; Hu et al., 1998, 1999; de Lorgeril et al., 2001; Sabaté et al., 2001) and various types of cancer (Jenab et al., 2004; González and Salas-Salvadó, 2006). Nuts are a good source of dietary fibre, which was reported to be higher than legumes, whole grain bread, fruits and vegetables (Salas-Salvado et al., 2006). A lower risk of type-2 diabetes with higher intakes of dietary fibre and lower glycemic loads has already been reported (Chandalia et al., 2000; Luscombe et al., 1999). Brazil nut has higher levels of phytonutrients and its consumption has been associated with many health benefits, mainly including cholesterol-lowering effects, antioxidant activity and antiproliferative effects. Brazil nuts are considered to be the best source of Se from plant-based foods, which is needed for proper thyroid and immune function. Brazil nut has good antioxidant activity and this has been attributed to its high Se content. It is an essential cofactor for glutathione peroxidase, which prevents lipid peroxidation and cell damage (Patrick, 2004). The role of selenium is a chemopreventive agent for a variety of cancers (Patrick, 2004). Selenium and vitamin E work synergistically. Selenium prevents free radical production by reducing peroxide concentrations in the cell whereas vitamin E neutralizes the free radicals when produced (Patrick, 2004).

Tree nuts and their oils are known to contain several bioactive and health-promoting substances. Epidemiological evidence has indicated that the consumption of tree nuts may exert several cardioprotective effects, which were speculated to arise from their lipid component that includes unsaturated fatty acids, phytosterols and tocols (Hu and Stampfer, 1999). Studies have also shown that dietary consumption of tree nut oils may exert even more beneficial effects than consumption of whole tree nuts, possibly due to the replacement of dietary carbohydrates with unsaturated lipids and/or other components present in the oil extracts (Hu and Stampfer, 1999). However, traditionally tree nuts were not considered as very healthy because of their high lipid content. Walnuts are receiving increasing interest as a healthy foodstuff because their regular consumption has been reported to decrease the risk of CHD (Blomhoff *et al.*, 2006; Davis *et al.*, 2007). The health benefits of walnuts are usually attributed to their chemical composition, being good sources of essential fatty acids and tocopherols (Amaral *et al.*, 2003, 2005). Linoleic acid is the major fatty acid in walnuts, followed by oleic, linolenic, palmitic and stearic (Amaral *et al.*, 2003; Ruggeri *et al.*, 1998; Savage *et al.*, 1999); its high content of poly unsaturated fatty acids, it has been suggested, can reduce the risk of heart disease by decreasing total and LDL-cholesterol and increasing HDL-cholesterol (Davis *et al.*, 2007; Tapsell *et al.*, 2004). In addition, walnuts have other components that may be beneficial for health including plant protein, dietary fibre, melatonin (Reiter *et al.*, 2005), plant sterols (Amaral *et al.*, 2003), folate, tannins and polyphenols (Anderson *et al.*, 2001; Li *et al.*, 2006). The chemical constituents, particularly the oil content and the fatty acid and tocopherols have been found to vary significantly among different walnut cultivars and environmental conditions (Amaral *et al.*, 2005). Nut consumption lowered the risk of CHD, which was partly explained by the cholesterol-lowering effect. The favourable fatty acid composition and lipid lowering effect of nuts have been demonstrated in experimental studies with almonds (Hyson *et al.*, 2002), macadamia nuts (Curb *et al.*, 2000), pecans (Morgan and Clayshulte, 2000), pistachios (Edwards *et al.*, 1999) and walnuts (Ros, 2000). Walnuts are good sources of both antioxidants and n-3 fatty acids, in particular high amounts of α-linolenic acid (6.3 g/100 g), whereas other nuts such as almonds, pecans and pistachios possess much smaller amounts (0.4–0.7 g/100 g). Brazil nuts are particularly rich in the antioxidant compound Se, while pecans are rich in bone-building manganese. Ryan *et al.* (2006) reported that Brazil nuts are a good source of squalene (1377.8 mg/100 g), which is a straight-chain terpenoid hydrocarbon and is a precursor of steroids and also plays an important role in the synthesis of cholesterol and vitamin D in the human body. It has been reported that squalene significantly decreases total cholesterol, LDL cholesterol and triacylglycerols levels in hypercholesterolemic patients (Miettinen *et al.*, 1994; Chan *et al.*, 1996). Tree nuts are a good source of phytosterols, which interferes with cholesterol absorption and results in reduction of serum LDL cholesterol levels (Thompson *et al.*, 2005). Epidemiologic and experimental studies have suggested that the phytosterols may offer protection from colon, breast and prostate cancers (Award and Fink, 2000, 2001).

References

Alasalvar, C., Amaral. J. S. and Satir, G., and Shahidi, F. (2009) Lipid characteristics and essential minerals of native Turkish hazelnut varieties (*Corylus avellana* L.). *Food Chemistry*, 113, 919–925.

Alasalvar, C., Karamac, M., Amarowicz, R. and Shahidi, F. (2006) Antioxidant and antiradical activities in extracts of hazelnut kernel (*Corylus avellana* L.) and hazelnut green leafy cover. *Journal of Agricultural and Food Chemistry*, 54, 4826–4832.

Alasalvar, C., Shahidi, F., Liyanapathirana, C.M. and Ohshima, T. (2003) Turkish Tombul Hazelnut (*Corylus avellana* L.). 1. Compositional Characteristics. *Journal of Agricultural and Food Chemistry*, 51, 3790–3796.

Amaral, J.S., Alves, MR., Seabra, R.M. and Oliveira, B.P.P. (2005) Vitamin E composition of walnuts (*Juglans regia* L.): A 3-year comparative study of different cultivars. *Journal of Agricultural and Food Chemistry*, 53, 5467–5472.

Amaral, J.S., Casal, S., Pereira, J.A., Seabra, R.M. and Oliveira, B.P.P. (2003) Determination of sterol and fatty acid compositions, oxidative stability, and nutritional value of six walnut (*Juglans regia* L.) cultivars grown in Portugal. *Journal of Agricultural and Food Chemistry*, 51, 7698–7702.

Amaral, J.S., Casal, S., Seabra, S.M. and Oliveira, B.P. (2006). Effects of roasting on hazelnut lipids. *Journal of Agricultural and Food Chemistry*, 54, 1315–1321.

Amarowicz, R., Troszyñska, A. and Shahidi, F. (2005) Antioxidant activity of almond seed extract and its fractions. *Journal of Food Lipids*, 12, 344–358.

Anderson, K.J., Teuber, S.S., Gobeille, A., Cremin, P., Waterhouse, A.L. and Steinberg, F.M. (2001) Walnut polyphenolics inhibit in vitro human plasma and LDL oxidation, *Journal of Nutrition*, 131, 2837–2842.

Antunes, A.J. and Markakis, P. (1977) Protein Supplementation of Navy Beans with Brazil Nuts. *Journal of Agricultural and Food Chemistry*, 25, 1096–1098.

Arcan, I. and Yemenicioğlu, A. (2009) Antioxidant activity and phenolic content of fresh and dry nuts with or without the seed coat. *Journal of Food Composition and Analysis*, 22, 184–188.

Arranz, S., Cert, R., Pérez-Jiménez, J., Cert, A. and Saura-Calixto, F. (2008) Comparison between free radical scavenging capacity and oxidative stability of nut oils. *Food Chemistry*, 110, 985–990.

Awad, A.B. and Fink, C.S. (2000) Phytosterols as anticancer dietary components: Evidence and mechanism of action, *Journal of Nutrition*, 130, 2127–2130.

Awad, A.B., Downie, A., Fink, C.S. and Kim, U. (2000) Dietary phytosterols inhibits the growth and metastasis of MDA-MB-231 human breast cancer cells grown in SCID mice. *Anticancer Research*, 20, 821–824.

Awad, A.B., Williams, H. and Fink, C.S. (2001) Phytosterols reduce in vitro metastic ability of MDA-MB-231 human breast cancer cells, *Nutrition and Cancer*, 40, 157–164.

Blomhoff, R., Carlsen, M.H., Anderson, L.F. and Jacobs, D.R. Jr. (2006) Health benefits of nuts: Potential role of antioxidants, *British Journal of Nutrition*, 96 (Suppl. 2), 52S–60S.

Bolling, B.W., Blumberg, J.B. and Oliver Chen, C.Y. (2010) The influence of roasting, pasteurisation, and storage on the polyphenol content and antioxidant capacity of California almond skins. *Food Chemistry*, 123, 1040–1047.

Bonvehi, J.S., Coll, F.V. and Rius, I.A. (2000) Liquid chromatographic determination of tocopherols and tocotrienols in vegetable oils, formulated preparations, and biscuits. *The Journal of AOAC International*, 83, 627–634.

Bravo, L. (1998) Polyphenols: chemistry, dietary sources, metabolism, and nutritional significance. *Nutrition Reviews*, 56, 317–33.

Chan, P., Tomlinson, B., Lee, C. B. and Lee, Y. S. (1996) Effectiveness and safety of low-dose pravastatin and squalene, alone and in combination, in elderly patients with hypercholesterolemia. *The Journal of Clinical Pharmacology*, 36, 422–427.

Chandalia, M., Garg, A., Lutjohann, D., von Bergmann, K., Grundy, S.M. and Brinkley, L.J. (2000) Beneficial effects of high dietary fibre intake in patients with type 2 diabetes mellitus. *The New England Journal of Medicine*, 342, 1392–1398.

Chandrasekara, N. and Shahidi, F. (2011) Effect of roasting on phenolic content and antioxidant activities of whole cashew nuts, kernels, and testa. *Journal of Agricultural and Food Chemistry*, 59, 5006–5014.

Curb, J.D., Wergowske, G., Dobbs, J.C., Abbott, R.D. and Huang, B. (2000) Serum lipid effects of a highmonounsaturated fat diet based on macadamia nuts. *Archives of Internal Medicine*, 160, 1154–1158.

da Costa, P.A., Ballus, C.A., Teixeira-Filho, J. and Godoy, H.T. (2010) Phytosterols and tocopherols content of pulps and nuts of Brazilian fruits. *Food Research International*, 43, 1603–1606.

Davis, L., Stonehouse, W., Loots, D.T., Mukuddem-Petersen, J., van der Westhuizen, F.H., Hanekom, S.M. and Jerling, J.C. (2007) The effects of high walnut and cashew nut diets on the antioxidant status of subjects with metabolic syndrome. *European Journal of Nutrition*, 46, 155–164.

De Lorgeril, M., Salen, P., Laporte, F., Boucher, F. and De Leiris, J. (2001) Potential use of nuts for the prevention and treatment of coronary heart disease: From natural to functional foods. *Nutrition, Metabolism and Cardiovascular Diseases*, 11, 362–371.

Edwards, K., Kwaw, I., Matud, J. and Kurtz, I. (1999) Effect of pistachio nuts on serum lipid levels in patients with moderate hypercholesterolemia, *Journal of the American College of Nutrition*, 18, 229–232.

FAO (2009). http://faostat.fao.org.

Fraser, G.E., Sabaté, J., Beeson, W.L. and Strahan, T.M. (1992) A possible protective effect of nut consumption on risk of coronary heart disease. The adventist health study. *Archives of Internal Medicine*, 152, 1416–1424.

Gola, U., Nohr, D. and Biesalski, H.K. (2011) Catechin and epicatechin in testa and their association with bioactive compounds in kernels of cashew nut (*Anacardium occidentale* L.). *Food Chemistry*, 128, 1094–1099.

Goli, A.H., Barzegar, M. and Sahari, M A. (2005) Antioxidant activity and total phenolic compounds of pistachio (*Pistachia vera*) hull extracts. *Food Chemistry*, 92, 521–525.

González, C.A. and Salas-Salvadó, J. (2006) The potential of nuts in the prevention of cancer. *British Journal of Nutrition*, 96, 87–94.

Gu, L., Kelm, M.A., Hammerstone, J.F., Beecher, G., Holden, J., Haytowitz, D., Gebhardt, S. and Prior, R.L. (2004) Concentrations of proanthocyanidins in common foods and estimations of normal consumption. *Journal of Nutrition*, 134, 613–617.

Hu, F.B. and Stampfer, M.J. (1999) Nut consumption and risk of coronary heart disease: A review of epidemiologic evidence. *Current Atherosclerosis Reports*, 3, 204–209.

Hu, F.B., Stampfer, M.J., Manson, J.E., Rimm, E.B., Colditz, G.A., Rosner, B.A., Speizer, F.E., Hennekens, C.H. and Willett, W.C. (1998) Frequent nut consumption and risk of coronary heart disease in women: Prospective cohort study. *British Medical Journal*, 317, 1341–1345.

Hyson, D.A., Schneeman, B.O. and Davis, P.A. (2002) Almonds and almond oil have similar effects on plasma lipids and LDL oxidation in healthy men and women. *Journal of Nutrition*, 132, 703–707.

International Nut and Dried Fruit Council Foundation, XXVIII World Nut and Dried Fruit Congress Newsletter, Monaco, 29–31 May, 2009.

Jarvi, A.E., Karlstrom, B.E., Granfeldt, Y.E., Bjorck, I.E., Asp, N.G. and Vessby, B.O. (1999) Improved glycemic control and lipid profile and normalized fibrinolytic activity on a low-glycemic index diet in type 2 diabetic patients. *Diabetes Care*, 22, 10–18.

Jenab, M., Ferrari, P., Slimani, N., et al. (2004) Association of nut and seed intake with colorectal cancer risk in the European prospective investigation into cancer and nutrition, *Cancer Epidemiol. Biomarkers Prevention*, 13, 1595–1603.

Jiang, Q. and Ames, B.N. (2003) γ-Tocopherol, but not α-tocopherol, decreases proinflammatory eicosanoids and inflammation damage in rats. *FASEB Journal*, 17, 816–822.

Jiang, R., Manson, J.E., Stampfer, M.J., Liu, S., Willett, W.C. and Hu, F.B. (2002) Nut and peanut butter consumption and risk of type 2 diabetes in women. *Journal of the American Medical Association*, 288, 2554–2560.

Kamath, V. and Rajini, P.S. (2007) The efficiency of cashew-nut (*Anacardium occidentale* L.) skin extract as a free radical scavenger. *Food Chemistry*, 103, 428–433.

Kelly Jr, J.H. and Sabaté, J. (2006) Nuts and coronary heart disease: An epidemiological perspective. *British Journal of Nutrition*, 96, 61S–67S.

Kendall, C.W., Marchie, A., Parker, T.L., Augustin, L.S., Ellis, P.R., Lapsley, K.G., Ternus, M. and Jenkins, D.J. (2003) Effect of nut consumption on postprandial starch digestion-a dose response study. *Annals of Nutrition and Metabolism*, 47, 636.

Kornsteiner, M., Wagner, K.H. and Elmadfa, I. (2006) Tocopherols and total phenolics in 10 different nut types. *Food Chemistry*, 98, 381–387.

Kris-Etherton, P.M., Lefevre, M., Beecher, G.R., Gross, M.D., Keen, C.L. and Etherton, T.D. (2004) Bioactive compounds in nutrition and health-research methodologies for establishing biological function: The antioxidant and antiinflammatory effects of flavonoids on atherosclerosis. *Annual Reviews on Nutrition*, 24, 511–538.

Li, L., Tsao, R., Yang, R., Liu, C.M., Zhu, H.H. and Young, J.C. (2006) Polyphenolic profiles and antioxidant activities of heartnut (Juglans ailanthifolia var. cordiformis) and Persian walnut (*Juglans regia* L.). *Journal of Agricultural and Food Chemistry*, 54, 8033–8040.

Locatelli, M., Travaglia, F., Coisson, J.D., Martelli, A., Stevigny, C. and Arlorio, M. (2010) Total antioxidant activity of hazelnut skin (Nocciola piemonte PGI): impact of different roasting conditions. *Food Chemistry*, 119, 1647–1655.

Luscombe, N.D., Noakes, M. and Clifton, P.M. (1999). Diets high and low in glycemic index versus high monounsaturated fat diets: Effects on glucose and lipid metabolism in NIDDM. *European Journal of Clinical Nutrition*, 53, 473–478.

Madhaven, N. (2001) Final report on the safety assessment of *Corylus avellana* (Hazel) seed oil, *Corylus americana* (Hazel) seed oil, *Corylus avellana* (Hazel) seed extract, *Corylus americana* (Hazel) seed extract, *Corylus avellana* (Hazel) leaf extract, *Corylus americana* (Hazel) leaf extract, and *Corylus rostrata* (Hazel) leaf extract. *International Journal of Toxicology*, 20, 15–20.

Maguire, L.S., O'Sullivan, S.M., Galvin, K., O'Connor, T.P. and O'Brien, N.M. (2004) Fatty acid profile, tocopherol, squalene and phytosterol content of walnuts, almonds, peanuts, hazelnuts and the macadamia nut. *International Journal of Food Science and Nutrition*, 55, 171–178.

Mathew, A.G. and Parpia, H.A.B. (1970) Polyphenols of cashew nut kernel testa. *Journal of Food Science*, 35, 140–143.

Mattes, R.D. (2008) The energetics of nut consumption, *Asia Pacific Journal of Nutrition*, 17, 337–339.

Miettinen, T.A., and Vanhanen, H. (1994) Serum concentration and metabolism of cholesterol during rapeseed oil and squalene feeding. *American Journal of Clinical Nutrition*, 59, 356–363.

Morgan, W.A. and Clayshulte, B.J. (2000) Pecans lower low-density lipoprotein cholesterol in people with normal lipid levels. *Journal of the American Dietetic Association*, 100, 312–318.

Moure, A., Pazos, M., Medina, I., Dominguez, H. and Parajo, J.C. (2007) Antioxidant activity of extracts produced by solvent extraction of almond shells acid hydrolysates, *Food Chemistry*, 101, 193–201.

Patrick, L. (2004) Selenium biochemistry and cancer: A review of the literature. *Alternative Medicine Review*, 9, 239–258.

Pereira, J.A., Oliveira, I., Sousa, A., Ferreira, I., Bento, A. and Estevinho, L. (2008) Bioactive properties and chemical composition of six walnut (*Juglans regia* L.) cultivars. *Food and Chemical Toxicology*, 46, 2103–2111.

Phillips, K.M., Ruggio, D.M. and Ashraf-Khorassani, M. (2005) Phytosterol composition of nuts and seeds commonly consumed in the United States. *Journal of Agricultural and Food Chemistry*, 53, 9436–9445.

Reiter, R.J., Manchester, L.C. and Tan, D.X. (2005) Melatonin in walnuts: Influence on levels of melatonin and total antioxidant capacity of blood. *Journal of Nutrition*, 21, 920–924.

Ritter, M.M.C. and Savage, G.P. (2007) Soluble and insoluble oxalate content of nuts. *Journal of Food Composition and Analysis*, 20, 169–174.

Ros, E. (2000) Substituting walnuts for monounsaturated fat improves the serum lipid profile of hypercholesterolemic men and women. A randomized crossover trial. *Annals of Internal Medicine*, 132, 538–546.

Ruggeri, S., Cappelloni, M., Gambelli, L. and Carnovale, E. (1998) Chemical composition and nutritive value of nuts grown in Italy. *Italian Journal of Food Science*, 10, 243–252.

Ryan, E., Galvin, K., O'Connor, T.P., Maguire, A.R. and O'Brien, N.M. (2006) Fatty acid profile, tocopherol, squalene and phytosterol content of Brazil, pecan, pine, pistachio and cashew nuts. *International Journal of Food Sciences and Nutrition*, 57, 219–228.

Sabaté, J., Radak, T. and Brown Jr, J. (2001) The role of nuts in cardiovascular disease prevention. In Wildman, R.E.C. (ed.) *Handbook of Nutraceuticals and Functional Foods*, CRC Press, Boca Raton, FL, pp. 477–495.

Salas-Salvado, J., Bullo, M., Perez-Heras, A. and Ros, E. (2006) Dietary fibre, nuts and cardiovascular diseases. *British Journal of Nutrition*, 96 (Suppl. 2), 45S–51S.

Salmeron, J., Ascherio, A., Rimm, E.B., Colditz, G.A., Spiegelman, D., Jenkins, D.J., Stampfer, M.J., Wing, A.L. and Willett, W.C. (1997) Dietary fiber, glycemic load, and risk of NIDDM in men. *Diabetes Care*, 20, 545–550.

Salmeron, J., Manson, J.E., Stampfer, M.J., Colditz, G.A., Wing, A.L. and Willett, W.C. (1997) Dietary fiber, glycemic load, and risk of non-insulin-dependent diabetes mellitus in women. *Journal of the American Medical Association*, 277, 472–477.

Sang, S., Lapsley, K., Jeong, W.S., Lachance, P.A., Ho, C.T. and Rosen, R.T. (2002) Antioxidative phenolic compounds isolated from almond skins (*Prunus amygdalus* Batsch). *Journal of Agricultural and Food Chemistry*, 50, 2459–2463.

Savage, G.P., Dutta, P.C. and Mcneil, D.L. (1999) Fatty acid and tocopherol contents and oxidative stability of walnut oils. *Journal of the American Oil Chemists' Society*, 76, 1059–1063.

Seeram, N.P., Zhang, Y., Henning, S.M., Lee, R., Niu, Y., Lin, G. and Heber, D. (2006) Pistachio skin phenolics are destroyed by bleaching resulting in reduced antioxidative capacities. *Journal of Agricultural and Food Chemistry*, 54, 7036–7040.

Shahidi, F., Alasalvar, F. and Liyana-Pathirana, C.M. (2007) Antioxidant phytochemicals in hazelnut kernel (*Corylus avellana* L.) and hazelnut byproducts. *Journal of Agricultural and Food Chemistry*, 55, 1212–1220.

Shahidi, F., Zhong, Y., Wijeratne, S.S.K. and Ho, C.T. (2009) Almond and almond products: Nutraceutical components and health effects. In C. Alasalvar and F. Shahidi (eds) *Tree Nuts: Nutraceuticals, Phytochemicals, and Health Effects*, CRC Press, Boca Raton, FL, pp. 127–141.

Siriwardhana, S.S.K.W. and Shahidi, F. (2002) Antiradical activity of extracts of almond and its by-products. *Journal of the American Oil Chemists Society*, 79, 903–908.

Smeds, A.I., Eklund, P.C., Sjoholm, R E., Willfor, S.M., Nishibe, S., Deyama, T. and Holmbom, B.R. (2007) Quantification of a broad spectrum of lignans in cereals, oilseeds, and nuts. *Journal of Agricultural and Food Chemistry*, 55, 1337–1346.

Surh, Y.J. (2003) Cancer chemoprevention with dietary phytochemicals. *Nature Review Cancer*, 3, 768–780.

Takeoka, G.R. and Dao, L.T. (2003) Antioxidant constituents of almond [*Prunus dulcis* (Mill.) D.A. Webb] hulls. *Journal of Agricultural and Food Chemistry*, 51, 496–501.

Tapsell, L.C., Gillen, L.J., Patch, C.S., Batterham, M., Owen, A., Bare, M. and Kennedy, M. (2004) Including walnuts in a low-fat/modified-fat diet improves HDL cholesterol-to-total cholesterol ratios in patients with type 2 diabetes. *Diabetes Care*, 27, 2777–2783.

Thompson, G.R. and Grundy, S.M. (2005) History and development of plant sterol and sterol esters for cholesterol-lowering purposes. *The American Journal of Cardiology*, 96, 3S–9S.

Thompson, L.U., Boucher, B.A., Liu, Z., Cotterchio, M. and Kreiger, N. (2006) Phytoestrogen content of foods consumption in Canada, including isoflavones, lignans, and coumestan. *Nutrition and Cancer*, 54, 184–201.

Tokusoglu, O., Unal, M.K. and Yemis, F. (2005) Determination of the phytoalexin resveratrol (3,5,4'-trihydroxystilbene) in peanuts and pistachios by high-performance liquid chromatographic diode array (HPLC-DAD) and gas chromatography-mass spectrometry (GC-MS). *Journal of Agricultural and Food Chemistry*, 53, 5003–5009.

Tomaino, A., Martorana, M., Arcoraci, T., Monteleone, D., Giovinazzo, C. and Saija, A. (2010) Antioxidant activity and phenolic profile of pistachio (*Pistacia vera* L., variety Bronte) seeds and skins. *Biochimie*, 92, 1115–1122.

Trevisan, M.T.S., Pfundstein, B., Haubner, R., Würtele, G., Spiegelhalder, B., Bartsch, H. and Owen, R.W. (2006) Characterization of alkyl phenols in cashew (*Anacardium occidentale*) products and assay of their antioxidant capacity. *Food and Chemical Toxicology*, 44, 188–197.

Trox, J., Vadivel, V., Vetter, W., Stuetz, W., Kammerer, D.R., Carle, R., Scherbaum, V., Tsantili, E., Takidelli, C., Christopoulosa, M.V., Lambrineab, E., Rouskasc, D. and Roussosa, P.A. (2010) Physical, compositional and sensory differences in nuts among pistachio (*Pistachia vera* L.) varieties. *Scientia Horticulturae*, 125, 562–568.

USDA (2009) Foreign Agricultural Service/USDA Office of Global Analysis, August, October.

USDA (2010) Foreign Agricultural Service/USDA Office of Global Analysis, February.

USDA (2010) National Nutrient Database for Standard Reference, Agricultural Research Service, U.S. Department of Agriculture, Washington, DC, Release 23, (assessed on 20 June, 2011)

Venkatachalam, M. and Sathe, S.K. (2006) Chemical composition of selected edible nut seeds. *Journal of Agricultural and Food Chemistry*, 54, 4705–4714.

Wijeratne, S.S.K., Abou-Zaid, M.M. and Shahidi, F. (2006) Antioxidant polyphenols in almond and its co products. *Journal of Agricultural and Food Chemistry*, 54, 312–318.

Wijeratne, S.S.K., Amarowicz, R. and Shahidi, F. (2006) Antioxidant activity of almonds and their by-products in food model systems. *Journal of the American Oil Chemists' Society*, 83, 223–230.

Wu, X., Beecher, G.R., Holden, J. M., Haytowitz, D.B., Gebhardt, S.E. and Prior, R.L. (2004) Lipophilic and hydrophilic antioxidant capacities of common foods in the United States. *Journal of Agricultural and Food Chemistry*, 52, 4026–4037.

Yang, J., Liu, R.H. and Halim, L. (2009) Antioxidant and antiproliferative activities of common edible nut seeds. *LWT - Food Science and Technology*, 42, 1–8.

8 Food processing by-products

Anil Kumar Anal

Food Engineering and Bioprocess Technology, Asian Institute of Technology, Klongluang, Pathumthani, Thailand

8.1 Introduction

The food industry produces large volumes of waste, both solid and liquid, resulting from the production, preparation, and consumption of food. These wastes pose increasing disposal and potential severe pollution problems and represent a loss of valuable biomass and nutrients. Besides their pollution and hazard aspects, in many cases, food processing wastes might have a potential for conversion into value-added products. Food processing wastes are those end products of various food processing industries that have not been recycled or used for other purposes. They are the non-product flows of raw materials whose economic values are less than the cost of collection and recovery for reuse; and are therefore discarded as wastes. These wastes could be considered valuable by-products if there were processed by appropriate technical means and if the value of the subsequent products were to exceed the cost of reprocessing (Schieber et al., 2001). The composition of wastes emerging from the food processing factories is extremely varied and depends on both the nature of the product and the production technique employed. For instance, waste from the meat industry contains high amounts of fat and proteins while waste from the canning industry contains high concentrations of sugars and starches. Fruits from temperate zones are usually characterized by a large edible portion and moderate amounts of waste material such as peels, seeds, and stones. In contrast, considerably higher ratios of by-products arise from tropical and subtropical fruit processing. Due to increasing production, disposal represents a growing problem since the plant material is usually prone to microbial spoilage, thus limiting further exploitation. One the other hand, cost of drying, storage, and shipment of by-products are economically limiting factors (Lowe and Buckmaster, 1995). Therefore, agro-industrial waste is often utilized as animal feed or fertilizer. However, demand for feed may be varying and dependent on agricultural yields. The problem of disposing by-products is further aggravated by legal restrictions. Thus, efficient, inexpensive, and environmentally sound utilization of these materials is becoming more important due to profitability.

Handbook of Plant Food Phytochemicals: Sources, Stability and Extraction, First Edition.
Edited by B.K. Tiwari, Nigel P. Brunton and Charles S. Brennan.
© 2013 John Wiley & Sons, Ltd. Published 2013 by John Wiley & Sons, Ltd.

8.2 Phytochemicals from food by-products

Due to the high consumption of the edible parts of fruits such as orange, apple, peach, olive, etc. which are mainly commercialized in processed form, fruit wastes (consisting of peels, pomace, and seeds) are produced in large quantities in markets. These materials could be a restrictive factor in the commercialization of these products if it they are not usefully recovered, because they represent significant losses with respect to the raw materials, which considerably increases the price of the processed products and also causes a severe problem in the community as they gradually ferment and give off odors. This might be due to their lack of commercial application, however, nowadays these by-products can be converted to different high-added value compounds particularly the fiber fraction. For example, orange and lemon sub-products, which are abundant and cheap, also constitute an important source of fiber since they are very rich in pectins. Among many other bioactive compounds, significant amounts of pectins and polyphenols can be recovered from apple by-products; and different types of fibers are isolated from grapes, after the extraction of their juice, as well as from guava skin and pulp (Clifford, 2001). Since these fibers are associated with antioxidant compound acid derivatives, they constitute a multiple and complete dietary supplement. Other fibers of interest are those rich in highly branched pectins that can be isolated from the mango skin. Other waste products are those coming from the kiwi that contain about 25% fiber as a percentage of dry matter and from the pineapple shell that has a high percentage of insoluble fiber (70% total fiber), which is mainly composed of neutral sugars, such as xylose and glucose, and presents a great antioxidant capacity. Olives that are largely destined for the production of olive oil also leave a by-product that is rich in different bioactive components, including dietary fiber.

8.2.1 Biowaste from tropical fruit and vegetables

Large amounts of fruit and vegetable processing wastes are produced from packing plants, canneries, etc., which may be disposed of in several ways including immediate use for landfill or drying to a stable condition (about 10% moisture) in order to use as animal feed out of season, or which, alternatively, may be processed biotechnologically in order to produce single cell protein (SCP). Industry continues to make progress in solving waste problems through recovery of by-products and waste materials such as peel, pulp, or molasses by the employment of the fermentation process. The protein content of fruit and vegetable processing wastes with an adequate level of fermentable carbohydrates can be increased to 20–30% by using solid substrate fermentation. The composition of some fruits and vegetables indicates that many have a significant proportion of fermentable sugars. Of these, oranges, carrots, apple, and peas have been successfully utilized as a substrate in the fermentation. Vinegar, citric acid, and acetic acid are produced from the by-products of the fruit and vegetable industry.

8.2.2 Citrus peels and seeds

Due to the large amounts being processed into juice, a considerable by-product industry has evolved to utilize the residual peels, membranes, seeds, and other wastes. Residues of citrus juice production are a source of dried pulp and molasses, fiber, pectin, cold-pressed oils, essences, D-limonene, juice pulps, and pulp wash, ethanol, seed oil, limnoids, and flavonoids (Askar, 1998; Braddock, 1995; Ozaki et al., 2000). Citrus juice processing is one of the important food industries of the world, yielding an enormous quantity of processing residues. Juice recovery from citrus fruit is about 40–55%, with the processing residue consisting of peel and rag, pulp wash, seeds, and citrus molasses. Most of the citrus fruits peels contain fibers and pectin, which can easily be recovered. The main flavonoids found in citrus species are hesperidin, narirutin, naringin, and eriocitrin (Mouly et al., 1994). Peels and other solid residues of citrus waste mainly contain heperidin and eriocitrin, while the naringin and eriocitrin are predominantly found in liquid residues (Coll et al., 1998). Citrus seeds and peels have been reported as having high concentrations of flavonoids and have been tested for their antioxidative properties (Benavente-Garcia, 1997; Manthey et al., 2001). Citrus by-products and wastes also contain large amount of coloring materials in addition to their complex polysaccharide contents. Hence, they are a potential source of natural clouding agents for many beverages. Sreenath et al. (1995) reported that citrus by-products could be utilized as natural sources for the production of beverage clouding agents using fermentation techniques, pectinolytic treatments, and alcohol extraction. They also evaluated the strength and stability of prepared clouds in model test beverage systems to determine their similarities to commercially available beverage cloud types.

8.2.3 Mango peels and kernels

Mango (*Mangifera indica* L., Anacardiaceae) is one of the most important tropical fruits. Major wastes from mango processing are peels and stones, amounting to 35–60% of the total fruit weight (Larrauri et al., 1996). Mango is one of the world's most popular tropical fruits with total production worldwide being around 25 million metric tons a year followed by banana, pineapple, papaya, and avocado. Because it is a seasonal fruit, approximately 20% of fruits are processed into various other forms such as puree, juices, nectar, pickles, canned slices, and dried fruits. Mango consists of 33–85% edible pulp, with 9–40% inedible kernel and 7–24% inedible peel. So, during industrial processing of mango, peel is a major by-product that is discarded as waste without any commercial purpose and is becoming a source of pollution. However, in recent studies, few scientific investigations have examined the importance of mango peels as a dietary fiber and natural antioxidant source.

Dietary fiber content ranged from 45 to 78% of mango peel and was found at a higher level in ripe peels. Dietary fiber in mango peel has recently been shown as a favorable source of high quality polysaccharides, because it not only has high starch, cellulose, hemicellulose, lignin, and pectin content but also has low lipid content. In addition, *in vitro* starch studies predicted low glycaemic responses from mango peel fiber (Ajila et al., 2007; Vergara-Valencia, 2007).

Mango peel extract offers a rich and inexpensive source of valuable compounds such as antioxidant compounds and dietary fiber, thus it shows potential as a functional food or value added ingredient. Therefore, mango peel if conveniently processed, could furnish useful products that may balance out waste treatment costs and also decrease the cost of main products. This new source will potentially be a functional food or value added ingredient in the future in our dietary system. There is scope for the isolation of these active ingredients and also use of mango peel as an ingredient in processed food products such as bakery products, breakfast cereals, pasta products, bars, and beverages. For example, incorporation of mango peel powder in macaroni not only increased the polyphenol, carotenoid, and dietary fiber contents but also exhibited improved antioxidant activity. The studies on cooking quality, textural, and sensory evaluations showed that these macaroni had good acceptability (Ajila *et al.*, 2010). Beside this, the kernel has also been found to be a potentially good source of nutrients for human and animal feed with 44.4% moisture content, 6.0% protein, 12.8% fat, 32.8% carbohydrate, 2.0% crude fiber, 2.0% ash, and 0.39% tannin (Elegbede *et al.*, 1995).

Mango seed kernel fat is a source of edible oil and has attracted attention due to having higher amounts of unsaturated fatty acids. Mango seed kernels may also be used as antioxidants. The major phenolics in mango seed kernels are gallic, egallic acids, gallates, and gallotanins (Puravankara *et al.*, 2000; Arogba, 2000). Ethanolic extracts of mango seed kernels displayed a broad antimicrobial spectrum and were more effective against Gram-positive than against Gram-negative bacteria (Kabuki *et al.*, 2000). Mango peels were also reported to be a good source of dietary fiber containing high amounts of extractable polyphenols (Larrauri *et al.*, 1996; Larrauri *et al.*, 1997; Larrauri, 1999). Mango latex which is deposited in fruit ducts and removed with the fruit at harvest has been shown to be a source of monoterpenes (John *et al.*, 2003).

8.2.4 Passion fruit seed and rind

Passion fruit, which botanically belongs to the family of Passifloraceae, of the genus Passiflora with the scientific name *Passiflora edulis*, is native to subtropical wild regions of South America probably originating in Paraguay. Over 500 cultivars exist; however, two main types, purple and yellow passion fruits, are widely cultivated. The ripening fruit is oval-shaped (average weight 35–50 g) with thick rind, smooth waxy surface and fine white specks. However, fruits with wrinkled surfaces actually have more flavor and are rich in sugar. Inside, it consists of membranous sacs containing light orange-colored, pulpy juice with numerous tiny, hard, dark brown or black, pitted seeds. The waste resulting from passion fruit processing consists of more than 75% of the raw material. The rind constitutes 90% of the waste and is a source of pectin (20% of the dry weight). Passion fruit seed oil is rich in linoleic acid (65%) (Askar, 1998).

Beside the pleasant taste of sweet and tart, passion fruit is rich in health benefiting plant nutrients, low in sodium and very low in saturated fat and cholesterol. It is also a good source of potassium, vitamin A, and vitamin C and a very good source of dietary fiber (10.4 g or 27% is contained in 100 g of fruit pulp).

Chi-Fai and Huang (2004) reported on the evaluation and compared the composition, physicochemical properties, and *in vitro* hypoglycemic effect of different fiber-rich fractions prepared from the seeds of passion fruits indigenous to Taiwan (hybrid, Tai-Nong-1). In this study, the contents of seed and pulp in the fresh passion fruit were about 11.1 ± 0.35 and 88.9 ± 0.35 g/100 g, respectively.

Table 8.1 illustrates that the edible passion fruit seed was rich in insoluble fiber-rich fractions (insoluble dietary fiber, alcohol-insoluble solids, and water-insoluble solids) which are mainly composed of cellulose, pectic substances, and hemicellulose. The result of this study also revealed that these fiber-rich fractions had water- and oil-holding capacities (2.07–3.72 g/g relatively greater than 0.9–1.3 g/g of some orange by-product fibers) comparable with those of cellulose, while their bulk densities and cation-exchange capacities were significantly higher than those of cellulose. Moreover, the *in vitro* study indicated that all insoluble fiber-rich fractions showed significant effects in absorbing glucose and retarding amylase activity, it is speculated that these fiber-rich fractions might have potential benefit

Table 8.1 Chemical composition of the raw and defatted passion fruit seed

Composition	g/100g raw seed (dry weight)	g/100g defatted seed (dry weight)
Moisture	6.60±0.28	–
Crude protein	8.25±0.58	10.8±0.75
Crude lipid	24.5±1.58	–
Total dietary fiber (TDF)	64.8±0.05	85.9±0.07
Insoluble dietary fiber (IDF)	64.1±0.02	84.9±0.03
Soluble dietary fiber (SDF)	0.73±0.07	0.97±0.09
Ash	1.34±0.08	1.77±0.11
Carbohydrate	1.11	1.53

for controlling postprandial serum glucose, and potential applications as low calorie bulk ingredients for fiber enrichment and dietetic snacks.

Beside this, the yellow passion fruit rind is the by-product from the juice industry available in large quantities (Yapo and Koffi, 2008). By using AOAC enzymatic–gravimetric method, the total dietary fiber in alcohol-insoluble material from yellow passion fruit rind was more than 73% dry matter of which insoluble dietary fiber accounted for more than 60% (w/w). The determination of dietary fiber using the hydrolysis method revealed that non-starchy polysaccharides were the predominant components (about 70%, w/w), of which cellulose appeared to be the main fraction. The water holding and oil holding capacities of the fiber-rich material were more than 3 g of water/g of fiber and over 4 g of oil/g of fiber, respectively. So, dietary fiber from yellow passion fruit rind, prepared as alcohol-insoluble material, may be suitable to protect against diverticular diseases.

8.2.5 Pomegranate peels, rinds and seeds

The pomegranate fruit can be consumed directly as fresh seed and fresh juice. Pomegranate contains highly colored grains which give a delicious juice. The presence of anthocyanins is responsible for the red color of its juice and other products of pomegranate fruit. Polyphenols are the major class phytochemicals in pomegranate fruit, including flavonoids (anthocyanins), condensed tannins (pro-anthocyanin), and hydrolysable tannins (ellagitannins and gallotannins) (Guo *et al.*, 2003; Guo *et al.*, 2006; Barzegar *et al.*, 2007; Al-Zoreky, 2009).

8.2.6 Mangosteen rind and seeds

The mangosteen is one of the most praised tropical fruits: known as mangosteen (English), mangostan (Spanish), mangostanier (French), manggis (Malaysian), manggustan (Philippine), mongkhut (Cambodian), and mangkhut (Thai). *Garcinia mangostana* L. has been known in name as mangosteen, in the family *Guttiferae*. Mangosteen fruit is approximately 3.5–7 cm across and weighs about 60–150 g. The woody skin of its pericarp (rind) varies from thin to thick, about 6–10 mm. Peels are pale green when immature and dark purple when ripe. Juice from mangosteens are produced from whole fruit or mixed polyphenols extracted from the inedible rind. This formulation improves phytochemical value in beverages. The juice has a purple color and astringency due to pigments from the pericarp including xanthonoids. It is produced and sold in dietary supplement forms, such as juice or capsule. The juice of mangosteen is often combined with other juices, such as grape, blueberry, raspberry, apple,

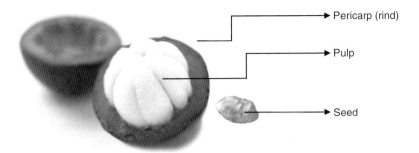

Figure 8.1 Composition of mangosteen.

Table 8.2 Proximate composition of mangosteen seed

Component	Amount (%)
Moisture	13.08
Carbohydrate	43.50
Crude protein	6.47
Crude fat	21.18
Crude fiber	13.70
Ash	1.99

cherry, and strawberry to improve the taste. The marketing promotions for these products present the advantages of compounds from mangosteen as (1) providing antioxidants against free radicals in the human body, (2) reducing inflammation, (3) reducing allergies, (4) maintaining immune system health, and (5) preventing cancer.

The pericarp of mangosteen varies in thickness. Its color is a yellowish-white to reddish-purple color depending on level of maturity (Figure 8.1), being reddish-purple when ripe, and it comprises bitter substances, mostly tannins and xanthones. The pericarp of mangosteens contains more xanthones than other fruits with medicinal properties and is traditionally used to treat diarrhoea and skin infections.

One to three larger segments of pulp contain recalcitrant seeds. Mangosteen seeds are not true seeds because they develop from the inner carpel wall, sometimes polyembryonic as an underdeveloped embryo. The seed of mangosteen is apomictic seed that is different from other common fruits. The apomictic seeds are viable for a short period of about three days if dried. Therefore, seeds must be kept moist to remain viable until planting and germination. The best way to keep them is in moist material or within the fruit, which can keep viability.

Ajayi and coworker (2006) reported seeds of mangosteen had high amount of carbohydrate, about 43.5%, and lipid, 21.18%, while protein content was low at only about 6.57% (Table 8.2). The seed powder is a good source of minerals because it contains high levels of potassium, magnesium, and calcium.

8.2.6.1 Bioactive compounds and edible color from pericarp and seeds

Generally, compounds from mangosteen have similar properties to other fruits, and been used as a health supplements and to support treatment of diseases. Mangosteen pericarp consists of an array of polyphenols including mostly xanthones and tannins which create astringency. Bioactive compounds in mangosteen pericarp are mostly in groups of polyphenolic compounds including xanthones, anthocyanin, proanthocyanidins, and catechin. Chaovanalikit and Mingmuang (2008) reported that the internal mangosteen pericarp has the highest of all phenolic compounds: 3404 mg GAE/100 g, 2930.49 mg GAE/100 g on external mangosteen pericarp, and 133.29 mg GAE/100 g in pulp.

Xanthones are one of the biologically active compounds and are unique among the group of polyphenolic compounds. The mangosteen pericarp contains more xanthones than other fruit sources. In nature, xanthones are found in very restricted families of plants, the majority being in the Gentianaceae and Guttiferae families. Xanthones can be briefly categorized into five groups including: (1) simple oxygenated xanthones, (2) prenylated, (3) xanthone glycosides, (4) xanthonolignoids, and (5) miscellaneous xanthones (Sultanbawa, 1980; Jiang et al., 2004). About 50 types of xanthones are found in the mangosteen and all of them have the same structural backbones. High-performance liquid chromatography (HPLC) is used to detect and classify the type of xanthones in mangosteen pericarp. The important xanthones in mangosteen, are α-mangostin, β-mangostin, 3-isomangostin, 9-hydroxycalabaxanthone, gartanin, and 8-desoxygartanin (Ji et al., 2007; Walker, 2007). There are numerous potential medicinal properties of xanthones, such as antiallergic, antituberculotic, anti-inflammatory, antiplatelet, and anticonvulsant properties (Marona et al., 2001). Xanthones have the molecular formula $C_{13}H_8O_2$ and six-carbon conjugated ring structure characterized by multiple double carbon bonds that confer stability on the compounds. The various xanthones found are unique because the side chains can be attached to the carbon molecules due to their versatility. Xanthone content increases in amount and type of compounds depending on the ripening levels of the fruit. Chaovanalikit and Mingmuang (2008) reported that external mangosteen pericarp has the highest anthocyanin content of 179.49 mg Cyn-3-Glu/100 g, with 19.71 mg Cyn-3-Glu/100 g in internal mangosteen pericarp but none in the pulp. Moreover, the anthocyanins in the external pericarp of mangosteen are composed of six compounds: cyaniding-sophoroside, cyaniding-glucoside-pentoside, cyaniding-glucoside, cyaniding-glucoside-X, cyaniding-X_2 and cyaniding X, where X is an unidentified residue (Palapol et al., 2008). Cyanidin-3-sophoroside and cyaniding-3-glucoside are the main compounds, which increase the color of the as it ripens.

Proanthocyanidin, a specific type of polyphenol, called flavonoids (flavan-3-ols), is one of the interesting components in grape seeds. There are many sources of proanthocyanidins, especially grapes, cranberries, and others. This substance is most abundantly found in grape seeds and mangosteen pericarps. It occurs naturally as a plant metabolite in fruits, vegetables, nuts, seeds, and flowers (Bagchi et al., 1997). Normally levels of proanthocyanidins are high in the outer shells of seeds and the bark of trees, helping to prevent degradation of some elements in plants due to oxygen and light.

The seeds of mangosteen are reported to contain about 21.18% oil (Ajayi et al., 2006). Oil that is extracted from mangosteen seeds is liquid at room temperature and golden-orange in color. The seeds contain both essential and non-essential fatty acids that, as can be seen from preliminary toxicological evaluation, are not harmful to the heart and liver of rats; hence the seed oil can be useful as edible oil. Fatty acids that were found in seeds of mangosteen are shown in Table 8.3.

The most abundant fatty acid is palmitic acid, that is, saturated fatty acid. Moreover, other unsaturated fatty acids are found in seeds including stearic acid, oleic acid, linoleic acid, gadoleic acid, and eicosadienoic acid. The most widespread unsaturated fatty acid is oleic acid which is about 34% of total oil from mangosteen seeds. In conclusion, the total unsaturated fatty acid is about 34%, while the total saturated fatty acid is 49.5% and unknown fatty acids are 5.14%.

Table 8.3 Fatty acids composition of mangosteen seed

Common name	Lipid name	Amount (%)
Palmitic acid	C16: 0	49.5
Stearic acid	C18: 1	1.33
Palmitoleic acid	C16: 1	ND*
Oleic acid	C18: 1	34.2
Linoleic acid	C18: 2	1.03
Linolenic acid	C18: 3	ND*
Arachidic acid	C20: 0	8.77
Gadoleic acid	C20: 1	0.10
Eicosadienoic acid	C20: 2	0.11
Behenic acid	C22: 0	ND*
Lingnoceric acid	C24: 0	ND*
Unknown	–	5.14
Total saturated fatty acid		59.6
Total unsaturated fatty acid		35.3

*ND: not detected.

8.3 By-products from fruit and vegetables

8.3.1 Apple pomace

Apple pomace has been used in production of pectin. In comparison to citrus pectin, apple pectin is characterized by superior gelling properties. However, the slight hue of apple pectin caused by enzymatic browning may lead to limitations with respect to use in very light colored-foods. Attempts at bleaching apple pomace by alkaline peroxide resulted in the loss of the polyphenols and in pectin degradation (Renard et al., 1997). Apple pomace has been shown to be a good source of polyphenols, which are predominantly localized in the peels and are extracted into the juice to a minor extent. Major compounds isolated and identified include catechins, hydroxycinnamtes, phloretin glycosides, and quercetin. Some phenolic compounds from apple pomace have been found to exhibit stronger antioxidant activity *in vitro* (Lu and Foo, 1997; Lu and Fu, 1998; Lu and Foo, 2000). Anthocyanins are found in the vacuoles of epidermal and subepidermal cells of the skin of red apple varieties (Lancaster, 1992; Soji et al., 1999; Alonso-Salcer et al., 2001). Enhanced release of phenolics by enzymatic liquification with pectinase and cellulases represents an alternative approach to utilizing apple pomace (Will et al., 2000).

8.3.2 By-products from grapes

Apart from oranges, grapes (*Vitis* sp., *Vitaceae*) are the world's largest fruit crop with more than 60 million tons produced annually. About 80% of the total crop is used in wine making and pomace represents approximately 20% of the weight of the grapes processed. A great range of products such as ethanol, tartrates, citric acid, grape seed oil, hydrocolloids, and dietary fiber are recovered from grape pomace (Bravo and Saura-Calixto, 1998; Nurgel ad Canbas, 1998). Anthocyanins, catechins, flavonol glycosides, phenolic acids, alcohols, and stillbenes are the principal phenolic constituents of grape pomace. Catechin, epicatechin, epicatechin gallate, and epigallocatechin are the major constitutive units of grape skin tannins (Souquet et al., 1996). Aminoethylthio-flavan-3-ol conjugates have been obtained from grape pomace by thiolysis of polymeric proanthocyanidins in the presence of cysteamine (Torres and Bobet, 2001). Grape seeds and skins are excellent sources of proanthocyanidins, flavonols, and falvan-3-ols (Souquet et al., 1996; Souquet et al., 2000). Procyanidins are the predominant proanthocyanidins in grape seeds, while procyanidins and predelphinidins are dominant in grape skins and stems. A number of stillbenes, namely trans-and cis-reservatrols (3,5,4'-trihydroxystilbene), trans- and cis-piceids (3-O-β-D-glucosides of resveratrol), trans-and cis-astringins (3-O-β-D-glucosides of 3'-hydroxyresveratrol). cis-resveratrolosides (4'-O-β-D-glucosides of resveratrol), and pterostilbene (a dimethylated derivative of stilbene) have been detected in both grape leaves and berries (Souquet et al., 2000; Cheynier and Rigaud, 1986).

8.3.3 Banana peels

Banana represents one of the most important fruit crops, with a global annual production of more than 50 million tons. Worldwide production of cooking bananas amounts to nearly 30 million per year. Peels constitute up to 30% of the ripe fruit. Attempts at utilization of banana waste include the biotechnological production of protein (Chung and Meyers, 1979), ethanol (Tewari et al., 1986), α-amylases (Krishna and Chandrasekaran, 1996), and cellulases (Krishna 1999). Banana peel contains a lot of phytochemical compounds, mainly antioxidants. The total amount of phenolic compounds in banana (*Musa acuminata* Colla AAA) peel ranges from 0.90 to 3.0 g/100 g DW (Someya et al., 2002). Ripened banana peel also contains other compounds, such as the anthocyanins delphinidin, cyanidin, and catecholamines (Kanazawa and Sakakibara, 2000). Furthermore, carotenoids, such as β-carotene, α-carotene, and different xanthophylls have been identified in banana peel in the range of 300–400 μg lutein equivalents/100 g (Subagio et al., 1996), as well as sterols and triterpenes, such as β-sitosterol, stigmasterol, campesterol, cycloeucalenol, cycloartenol, and 24-methylene cycloartanol (Knapp and Nicholas, 1969).

8.3.4 Tomato

During processing of tomato juice, about 3–7% of the raw material is lost as waste. Tomato pomace consists of the dried and crushed skins and seeds of the fruits. The seeds account for approximately 10% of the fruit and 60% of the total waste, respectively, and are a source of protein (35%) and fat (25%). Due to an abundance of unsaturated fatty acids, tomato seed oil is getting unique interest (Askar, 1998). Lycopene is the principal carotenoid causing the characteristic red hue of tomatoes. Most of the lycopene is associated with the water-insoluble fraction and the skin (Sharma and Maguer, 1996). Supercritical CO_2 extraction of lycopene and β-carotene from tomato paste waste resulted in recoveries up to 50% when ethanol was added as a solvent (Baysal et al., 2000).

8.3.5 Carrot

Despite considerable improvements in processing techniques, including the use of depolymerizing enzymes, mash heating, and decanter technology, a major part of valuable compounds such as carotenes, uronic acids, and neutral sugars is still retained in the pomace, which is usually disposed of as feed or fertilizer. Juice yield is reported to be only 60–70%, and up to 80% of carotene may be lost with the pomace (Sims et al., 1993). Stoll et al. (2001) found 2 g/kg dry matter of total carotene content of pomace, depending on the processing conditions. Attempt has been made to incorporate carrot pomace into various food such as bread, cake, dressing, and pickles; and in the production of functional drinks.

8.3.6 Mulberry leaves

There are three different kinds of mulberry: (1) red mulberry (*Morus rubra*), with dark purple edible fruit, (2) black mulberry (*Morus* nigra), with dark foliage and fruit, and (3) white mulberry (*Morus alba*), which is thin, glossy, and light green in color, with quite a variable leaf shape even on the same tree. White mulberry is primarily used for raising silkworms, which utilize the leaves as their main food source. These leaves are highly nutritious and the fruits boast high medicinal value in their amino acids, vitamin C, and antioxidants; the leaves can also be effective in regulating fat and boosting metabolism (Bae and Suh, 2007). There are many uses for the mulberry leaf and fruit. In China, Japan, and European countries this plant has a huge market for its medicinal and cosmetic value. Consuming mulberry leaf tea can be relaxing for the body and mind and is generally recommended for treating diabetes and hypertension (Herald, 2005).

8.4 Tuber crops and cereals

8.4.1 Cassava

Cassava or tapioca is an important economic root crop grown in Southeast Asia as well as in tropical Africa and Central America. Cassava leaves are a by-product of cassava roots' harvest (depending on the varieties), which is rich in proteins, minerals, vitamins B_1, B_2, C, and carotenes (Eggum, 1970; Ravindran and Blair, 1992; Adewusi and Bradbury, 1993; Aletor and Adeogum, 1995). The protein content of cassava leaves is high for a non-leguminous plant. Although cassava leaves are rich in protein, other minerals such as crude fiber may limit their nutritive value for monogastric animals. Rogers and Milner (1963) reported a range of 4.0–15.2%. Immature cassava leaves were evidently used in the above analyses, since values as high 29% have been reported for mature leaves (Ravindran et al., 1982). Roger and Milner (1963) were the first to conduct detailed analyses of amino acid content of cassava leaves. They analyzed the leaves of 20 Jamaican and Brazilian cultivars obtained from ten month old healthy cassava and found that protein from the leaf was deficient in methionine, possibly marginal in tryptophan, but rich in lysine. In addition, there are natural compounds such as toxicant or antinutritional compound, cyanogenic glycosides, and tannins (Ravindran, 1993). The toxic properties associated with fresh cassava leaves are due to the hydrocyanic acid that is liberated when their cyanogenic glycosides, namely linamarin and lotaustralin, are hydrolyzed by endogenous enzymes. These strong complexes reduce the digestibility of protein and may have inhibited the activities of proteolytic enzymes like pepsin and trypsin (Chavan et al., 1979). Tannins are reduced by soaking in alkali solutions, for example sodium hydroxide, potassium hydroxide, and sodium carbonate, and with heating.

8.4.2 Defatted rice bran

After rice is harvested and dried, the first stage of processing is de-husking, followed by milling where the bran layer is removed to produce white rice. The bran layer commonly termed as rice bran consists of the aleurone layer, part of the embryo, germ, and endosperm. Rice bran is a good source of nutrients and it is well-known that a major fraction contains

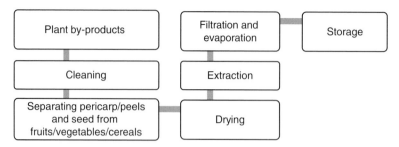

Figure 8.2 Basic steps for extraction of bioactive compounds from food plant by-products.

approximately 12–15% protein, 15–20% fat, and 7–12% fiber. Defatted rice bran is the by-product of rice bran oil extraction. It generally contains 12–19% protein, 0.5–7% fat, and 5.5–14.5% fiber. Defatted rice bran is free flowing, light in weight, and has tendency to form dust. In the proper process, it is light in color but will be dark red after being desolved under too drastic conditions. Defatted rice bran is normally used as animal feed with a low economic value. Besides, it is utilized for food supplements, such as binder in sausage and raw material for hydrolyzed vegetable protein. Moreover, it is widely used in bakery products such as doughnuts, pancakes, muffins, breads, and cookies because it can improve quantity of dough, and increase amino acid, vitamin, and mineral content. In breakfast cereals and wafers defatted rice bran is used to improve absorption capacity, appearance, and flavor.

8.5 Extraction of bioactive compounds from plant food by-products

Steps for extraction of bioactive compounds or other components are different for every matrix depending on the selection and suitability for various materials. Figure 8.2 shows the basic steps for extracting compounds among which important steps are the drying and extraction processes. There are many methods of drying and extraction that can be used in a laboratory and in industry. Basic principles for extraction of bioactive compounds are identical for whole foods and for byproducts. Detailed extraction techniques are discussed in Chapters 17 and 18. Table 8.4 outlines some of the extraction methods and conditions for extracting bioactive compounds from by-products of the food industry.

8.6 Future trends

Agro-waste cannot be regarded as waste but will likely become an additional resource to augment existing natural materials. Recycling, reprocessing, and eventual utilization of plant food processing residues offer potential of returning these by-products to beneficial uses rather than merely discharging into the environment causing detrimental environmental effects. The reprocessing of those wastes could involve rendering the recovered by-products suitable for beneficial use; promotion in suitable markets ensuring the profitability; suitable and economic reprocessing technology; and creation of an overall enterprise that is acceptable and economically feasible. The exploitation of by-products of plant food processing as a source of functional compounds and their application in food is a promising field, requiring cross-cutting technologies.

Table 8.4 Effects of extraction methods and extraction conditions

Extraction methods	Samples	Results	References

Microwave-assisted	Plants	• when compared to conventional method (soxhlet):
		  ○ extraction time was reduced
		  ○ less solvent was used
		  ○ amount of extracted phenolic compounds was increased
			Proestos and Komaitis (2008)

Microwave-assisted	Citrus Mandarin peels	• high extraction efficiency • high antioxidant activity • short extraction time	Zhanga et al. (2009)

Microwave-assisted	Green tea leaves	• more effective than the conventional extraction in
		  ○ extraction time
		  ○ extraction efficiency
		  ○ the percentages of tea polyphenols
			Liu et al. (2003)

Ultrasound-assisted	Orange peel	• high total phenolic content • the sonication power was the most influential factor in the UAE process followed by temperature and ethanol:water ratio.	Chemat et al. (2010)
Ultrasound-assisted	Citrus peel	• yields of phenolic compounds increased with both ultrasonic time and temperature increased • temperature is the most sensitive on stability of phenolic compounds • the optimal ultrasound condition was different one compound from another	Ye et al. (2009)
Ultrasound-assisted	Coconut shell powder	• high amounts of phenolics can be extracted from coconut shell • extraction time was the most significant parameter for the process	Rodrigues and Pinto (2007)
Temperature-controlled water bath shaker	Mengkudu	• extraction time (20–120 min) had significant effect on total phenolic content • excess extraction time indeed reduced the yield of phenolic compounds. • heat has been found to enhance the recovery of phenolic compounds	Tan et al. (2010)

(Continued)

Table 8.4 (Continued)

Extraction methods	Samples	Results	References
Thermostatic rotary shaker	Grape marc	• both time and temperature highly influenced antioxidants yields, with higher yields at 60°C • yield increased with length of maceration but at 60°C there seemed to be a reduction beyond 20 h due to thermal degradation	Spigno et al. (2007)
Solvent extraction using an orbital shaker		• total phenolics increased significantly when the extraction time was increased • After 60 min, increasing the extraction time did not significantly improve the recoveries	Campos et al. (2007)

References

Adewusi, S.R.A. and Bradbury, J.H. (1993) Carotenoid in cassava; comparison of open column and HPLC method of analysis. *Journal of the Science of Food Agriculture*, 62, 375–383.

Ajayi, I.A., Oderinde, R.A., Ogunkoya, B.O., Egunyomi, A., and Taiwo, V.O. (2006) Chamical analysis and preliminary toxicological evaluation of *Garcinia mangostana* seeds and seed oil. *Journal Food Chemistry*, 101, 999–1004.

Ajila, C.M., Naidu, K.A., Bhat, S.G., and Prasada Rao, U.J.S (2007) Bioactive compounds of mango peel extract. *Food Chemistry*, 105, 982–988.

Ajila, C.M., Aalawi, M., Leelavathi, K., and Prasad Rao, U.J.S. (2010) Mango peel powder: A potential source of antioxidative and dietary fiber in macaroni preparations. *Innovative Food Science and Emerging Technologies*, 11, 219–224.

Aletor, O., Oshodi, A.A., and Ipinmoroti, K. (2002) Chemical composition of common eafy vegetables and functional properties of their leaf protein concentrates. *Food Chemistry*, 78, 63–68.

Alonso-Salcer, R.M., Korta, E., Barranco, A., Berrueta, L.A., Gallo, B., and Vicente, J. (2001) *Journal of Agriculture and Food Chemistry*, 49, 3761–3767.

Alapol, Y., Ketsa, S., Stavenson, D., Cooney, D.M., Allan, A.C., and Ferguson, I.B. (2008) Colour development and quality of mangosteen *(Garinia mangostana* L.) fruit during ripening and after harvest. *Journal Postharwest Biology and Technology*, 51, 349–353.

Al-Zoreky, N.S. (2009) Antimicrobial activity of pomegranate (Punica granatum L.) fruitpeels. *International Journal of Food Microbiology*, 134, 244–248.

Andres, A., Bilbao, C., and Fito, P. (2004) Drying kinetics of apple cylinders under combined hot air-microwave dehydration. *Journal of Food Engineering*, 63, 71–78.

Argoba, S.S. (2000) Mango (Mangifera indica) kernel: chromatographic analysis of the tannin, and stability study of the associated polyphenol oxidase activity. *Journal of Food Composition and Analysis*, 13, 149–156.

Arsdel, W.B.V., Copley, M.J., and Morganm, A.I., Jr. (1973) Food Dehydration. In W.B. Van Arsdel, M.J. Copley, and A.I. Morgan, Jr., *Drying Methods and Phenomena*, second edition, volumes 1 and 2, The AVI Publishing Company, Inc. Westport, Connecticut.

Askar, A. (1998) Importance and characteristics of tropical fruits. *Fruit Processing*, 8, 273–276.

Bae, S.H. and Suh H.J. (2007) Antioxidant activities of five different mulberry cultivars in Korea. *LWT – Food science and Technology*, 40, 955–962.

Bagchi, D., Krohn R.L., and Bagchi M. (1997) Oxygen free radical scavenging abilities of vitamins C and E, and a grape seed proanthocyanidin extract in vitro. *Molecular Pathology and Pharmacology*, 95, 179–189.

Bagchi, D., Bagchi, M., Stohs, S.J., Ray, S.D., Sen, C.K., and Pruess, H.G. (2002) Cellular protection with proanthocyanidins derived from grape seeds. *Annals of New York Academy of Science*, 957, 260–270.

Barbieri, S., Elustondo, M., and Urbicain, M. (2004) Retention of aroma compounds in basil dried with low pressure superheated steam. *Journal of Food Engineering*, 65, 109–115.

Baysal, T., Ersus, S., and Starmans, D.A.J. (2000) Supercritical CO_2 extraction of β-carotene and lycopene from tomato paste waste. *Journal of Agriculture and Food Chemistry*, 48, 5507–5511.

Balasundram, N., Sundram, K., and Samman, S. (2006) Phenolic compounds in plants and agri-industrial by-prodcuts: Antioxidant activity, Occurrence and potential uses. *Food Chemistry*, 99, 191–203.

Barzegar, M., Yasoubi, P., Sahari, M.A., and Azizi, M.H. (2007) Total Phenolic Contents and Antioxidant Activity of Pomegranate (*Punica granatum* L.) Peel Extracts. *Journal of Agricultural Science and Technology*, 9, 35–42.

Benavente-Garcia, O., Castillo, J., Marin, F. R., Ortuno, A., and Del Rio, A. (1997) Uses and properties of citrus flavonoids. *Journal of Agricultural and Food Chemistry*, 45, 4505–4515.

Boudhrioua, N., Bahloul, N., Slimen, I.B., and Kechaou, N. (2009) Comparison on the total phenol contents and the color of fresh and infrared dried olive leaves. *Industrial Crops and Products*, 29, 412–419.

Broillard, R. (1982) Chemical structure of anthocyanins. In *Anthocyanins as Food Colors*, Markakis, P. (ed.), Academic Press, New York, pp. 1–40.

Bravo, L. (1998) Polyphenols: chemistry, dietary sources, metabolism, and nutritional significance. *Nutrition Reviews*, 56, 317–333.

Braddock, R.J. (1995) By-products of citrus fruit. *Food Technology*, 49, 74–77.

Bravo, L. and Saura-Claxito, F. (1998) Characterization of dietary fiber and the in vitro indigestible fraction of grape pomace. *American Journal of Enology and Viticulture*, 49, 135–141.

Cai, J., Liu, X., Li, Z., and An, C. (2003) Study on extraction technology of strawberry pigments and its physicochemical properties. *Food and Fermentation Industries*, 29, 69–73.

Chi-Fai, C. and Huang Y. (2004) Characterization of passion fruit seed fibers—A potential fiber source. *Food Chemistry*, 85, 189–194.

Chaisawadi, S., Thongbutr, D., Kulamai, S., Methawiriyasilp, W., and Juntawong, P. (2005) Clean production of freeze-dried Kaffir lime powder. 31st Congress on Science and Technology of Thailand at Suranaree University of Technology, 18–20 October, 2005.

Chaovanalikit, A. and Mingmuang, A. (2008) Anthocyanin and total phenolic contents of mangosteen and its juices. *SWU Scientific Journal*, 23, 68–78.

Chavan, J.K., Kadam, S.S, Ghonsikar, C.P., and Salunkhe, D.K. (1979) Removal of tannins and improvement of in vitro protein digestibility of sorghum seeds by soaking in alkali. *Journal of Food Science*, 44, 1319–1321.

Chemat, F., Khan, M.K., Abert-Vian, M., Fabiano-Tixier, A-S., and Dangles, O. (2010) Ultrasound-assisted extraction of polyphenols (flavanone glycosides) from orange (*Citrus sinensis* L.) peel. *Food Chemistry*, 119, 851–858.

Cheynier, V. and Rigaud, J. (1986) *American Journal of Enology and Viticulture*, 37, 248–252.

Chung, S.L. and Meyers, S.P. (1979) Bioprotein from banana waste. *Developments in Industrial Microbiology*, 20, 723–731.

Clifford, M.N. (2000) Nature, occurrence and dietary burden. *Journal of the Science of Food and Agriculture*, 80, 1063–1072.

Coll, M.D., Coll, L., Laencine, J., and Tomas-Barberan, F.A. (1998) Recovery of flavanons from wastes of industrially processed lemons. *Zeitschrift für Lebensmittel- Unterschung und – Forschung*, 206, 404–407.

Cui, Z.W., Sun, L.J., Chen, W., and Sun, D.W. (2008) Preparation of dry honey by microwave-vacuum drying. *Journal of Food Engineering*, 84, 582–590.

Decareau, R.V. (1985) *Microwaves in the Food Processing Industry*. Food Science and Technology, A Series of Monographs, Academic Press, Inc., New York.

Delgado-Vargas, F., Jiménez, A.R. and Paredes-López, O. (2000) Natural pigments: carotenoids, anthocyanins, and betalains – characteristics, biosynthesis, processing and stability. *Critical Reviews in Food Science and Nutrition*, 40, 173–289.

Delgado-Vargas F. and Paredes-López O. (2003) Anthocyanins and betalains. In F. Delgado-Vargas and O. Paredes-Lopez (eds) *Natural Colorants for Food and Nutraceutical Uses*, CRC Press, Boca Raton, pp. 167–219.

Earl, R.L. (1969) *Unit Operations in Food Engineering*, Pegamon Press, Ltd., Great Britain.
Eder A. (2000) Pigments. In Nollet M.L.L. (ed.) *Food Analysis by HPLC*. Marcel Dekker, New York, pp. 845–880.
Eggum, B.O. (1970) The protein quality of cassava leaves. *British Journal Nutrition*, 24(3), 761–768.
Elegbede, J.A., Achoba, I.I., and Richard, H. (1995) Nutrient composition of mango (*Mangfifera indica* L.) seed kernel from Nigeria. *Journal of Food Biochemistry*, 19, 391–398.
Femenia, A., Garau, M.C., Simal, S., and Rossello, C. (2007) Effect of air-drying temperature on physico-chemical properties of dietary fibre and antioxidant capacity of orange (*Citrus aurantium* v. Canoneta) by-products. *Food Chemistry*, 104, 1014–1024.
Giri, S.K. and Prasad, S. (2007) Drying kinetics and rehydration characteristics of microwave-vacuum and convective hot-air dried mushrooms. *Journal of Food Engineering*, 78, 512–521.
Guo, C., Li, Y., Yang, J., Wei, J., Xu, J., and Cheng, S. (2006) Evaluation of antioxidant properties of pomegranate peel extract in comparison with pomegranate pulp extract. *Food Chemistry*, 96, 254–260.
Guo, C.J., Yang, J.J., Wei, J.Y., Li, Y.F., Xu, J., and Jiang, Y.G. (2003) Antioxidant activities of peel, pulp and seed fractions of common fruits as determined by FRAP assay. *Nutrition Research*, 23, 1719–1726.
Harborne, J.B. and Turner, B.L. (1984) *Plant Chemosystematics*, Academic Press, London, UK.
Harborne, J.B. (1998) *Phenolic Compounds in Phytochemical Methods – A Guide to Modern Techniques of plant Analysis*, third edition, Chapman & Hall, New York, pp. 66–74.
Hemmerle, H. *et al.* (1997) Chlorogenic acid and synthetic chlorogenic acid derivatives: Novel inhibitors of hepatic glucose-6-phopshate translocase. *Journal of Medical Chemistry*, 40, 137.
Heredia, F.J., Francia-Aricha E.M., Rivas-Gonzalo J.C., Vicario I.M., and Santos-Buelga C. (1998) Chromatic characterization of anthocyanins from red grapes - I. pH effect. *Food Chemistry*, 63, 491–498.
Horbowicz, M., Kosson, R., Grzesiuk, A., and Dębski, H., (2008) Anthocyanins of fruits and vegetables-their occurrence, analysis and role in human nutrition. *Vegetable Crops Research Bulletin*, 68, 5–22.
Hu, Qing-guo., Zhang, Min., Mujumdar, Arun S., Xiao, Gong-nian, and Sun, Jin-cai. (2006) Drying of edamames by hot air and vacuum microwave combination. *Journal of Food Engineering*, 77, 977–982.
Jaiswal, V., DerMarderosian, A., and Poter, J.R. (2009) Anthocyanins and polyphenol oxidase from dried arils of pomegranate (*Punica granatum* L.). *Food Chemistry*, 118 (1), 11–16.
Ji, X., Avula, B., and Khan, I. A. (2007) Quantitative and qualitative determination of six xanthones in *Garcinia mangostana* L. by LC–PDA and LC–ESI-MS. *Journal of Pharmaceutical and Biomedical Analysis*, 43, 1270–1276.
Jiang, D.J., Dai, Z., and Li, Y.J. (2004) Pharmacological effects of xanthones as cardiovascular protective agents. *Cardiovascular Drug Reviews*, 22, 91–102.
John, K.S., Bhat, S.G., and Prasad Rao, U.J.S. (2003) Biochemical characterization of sap (latex) of few Indian mango varieties. *Phytochemistry*, 62, 13–19.
Juare, L. (1995). Estudio agronomico sobre la utilizacion de la yucca como forraje. Lima, *Estacion Experimental Agricola La Molina*. Bol. 58.
Kabuki, T., Nakajima, H., Arai, M., Ueda, S., Kuwabara, Y., and Dosako, S. (2000) Characterization of novel antimicrobial compounds from Mango (*Maginifera indica* L.) kernel seeds. *Food Chemistry*, 47, 71, 61–66.
Kathirvel, K., Naik, K.R., Gariepy, Y., Orsat, V., and Raghavan, G.S.V. (2006) Microwave Drying – A promising alternative for the herb processing industry. Written for presentation at the CSBE/SCGAB 2006 Annual Conference Edmonton Alberta, 16–19 July, 2006. The Canadian Society for Bioengineering.
Katsube, T., Tsurunaga, Y., Sugiyama, M., Furuno, T., and Yamasaki, Y. (2009) Effect of air-drying temperature on antioxidant capacity and stability of polyphenolic compounds in mulberry (*Morus alba* L.) leaves. *Food Chemistry*, 113, 964–969.
Knapp, F.F. and Nicholas, H.J. (1969) Sterols and triterpenes of banana peel. *Phytochemistry*, 8(1), 207–214.
Krishna, C. (1999) Production of bacterial celluloses by solid state bioprocessing of banana wastes. *Bioresource Technology*, 69, 231–239.
Krishna, C. and Chandrasekaran, M. (1996) Banana waste as substrate for alpha-amylase production by Bacillus subtilis (CBTK106) under solid-state fermentation. *Applied Microbiology and Biotechnology*, 46, 106–111.
Kumar, D.G.P, Hebbar, H.U., Sukumar, D., and Ramesh, M.N. (2005) Infrared and hot-air drying of onions. *Journal of Food Engineering*, 29, 132–150.
Kroon, P. *et al.* (1997) Release of covalently bound freulic acid from fiber in the human colon. *Journal of Agricultural and Food Chemistry*, 45, 661.
Lancaster, J.E. (1992) Regulation of skin color in apples. *CRC Critical Review Plant Science*, 10, 487–502.

Lancaster P.A. and Brook, J.E. (1983) Cassava leaves as human food. *Economic Botany*, 37, 331–348.

Larrauri, J.A. (1999) New approaches in the preparation of high dietary fiber powders from fruit by-products. *Trends in Food Science and Technology*, 10, 3–8.

Larrauri, J.A., Ruperez, P., and Saura-Calixto, F. (1997) Mango peels with high antioxidant activity. *Zeitschrift für Lebensmittel- Unterschung und – Forschung*, 205, 39–42.

Larrauri, J.A., Rupereze, P., Borroto, B., and Saura-Calixto, F. (1996) Mango peels as a new topical fibre: Preparation and characterization. *Lebensmittel-Wiseenchaft und- Technologie*, 29, 729–733.

Lempereur, I., Rouau, X., and Abecassis, J. (1997) Genetic and agronomic variation in arabionoxylan and ferulic acid contents of durum wheat (Triticum durum L.) grain and its milling fractions, *Journal of Cereal Science*, 25, 103.

Lowe, E.D. and Buckmaster, D.R. (1995) Dewatering makes big difference in compost strategies. *Biocycle*, 36, 78–82.

Lu, Y. and Foo, L.Y. (1997) Identification and quantification of major polyphenols in apple pomace. *Food Chemistry*, 59, 187–194.

Lu, Y. and Foo, l.Y. (1998). Constitution of some chemical components of apple seed. *Food Chemistry*, 61, 29–33.

Lu, Y. and Foo, L.Y. (1999) The polyphenol constitutents of grape pomace. *Food Chemistry*, 65, 1–8.

Lu, Y. and Foo, L.Y. (2000) Antioxidant and radical scavenging activities of polyphenols from apple pomace, *Food Chemistry*, 68, 81–85.

Marona H., Pekala E., Filipek B., Maciag D., and Sznelar E. (2001) Pharmacologicalproperties of some aminoalkanolic deverivatives of xanthone. *Pharmaxie*, 56, 567–572.

Maskan, Medini (2001) Drying, shrinkage and rehydration characteristics of kiwifruits during hot air and microwave drying. *Journal of Food Engineering*, 48, 177–182.

Mazza, G. and Miniati, E. (1993) *Anthocyanins in Fruits, Vegetables, and Grains*, CRC Press, Boca Raton, FL.

Mouly, P.P., Arzouyan, C.R., Gaydou, E.M., and Estienne, J.M. (1994) Differentiation of citrus juices by factorial discriminant analysis using liquid chromatography of flavones glycosides. *Journal of Agriculture and Food Chemistry*, 42, 70–79.

Madhavi, D.L., Deshpande, S.S., and Salunkhe, D.K. (1996) *Food Antioxidants: Technological, Toxicological, and Health Perspectives*, Marcel Dekker, Inc., New York.

Manach, C., Mazur, A., and Scalbert, A. (2005a) Polyphenols and prevention of cardiovascular diseases. *Current Opinions in Lipidology*, 16, 77–84.

Manach, C., Williamson, G., Morand, C., Scalbert, A., and Remesy, C. (2005b) Bioavailability and bioefficacy of polyphenols in humans. I. Review of 97 bioavailability studies. *American Journal of Clinical Nutrition*, 81(suppl), 230S–242S.

Murakami, Shinichi (2003) Mulberry leaves powder manufacturing method. Freepatentsonline. http://www.freepatentsonline.com/6536689.html.

Naczk, M., and Shahidi, F. (2006) Phenolics in cereals, fruits and vegetables: Occurrence, extraction and analysis. *Journal of Pharmaceutical and Biomedical Analysis*, 41, 1523–1542.

Niamnuy, C. and Devahastin, S. (2005) Drying kinetics and quality of coconut dried in a fluidized bed dryer. *Journal of Food Engineering*, 66, 267–271.

Nurgel, C. and Canbas, A. (1998) Production of tartaric acid from pomace of some Anatolian grape cultivars. *American Journal of Enology and Viticulture*, 49, 95–99.

Ozaki, Y., Miyake, M., Inaba, N., Ayano, S., Ifuku, Y., and Hasegawa, S. (2000) Limnoid glucosides of sastuma mandarin (*Citrus unshiu Marcov.*) and its processing products. In M.A. Berhow, S. Hasegawa, and G.D. Manners (eds) *Citrus Limnoids: Functional Chemicals in Agriculture and Food*, ACS Symposium Series 758, Washington DC, pp. 107–119.

O'Riordan, D., Harbourne, N., Marete, E., and Jacquier, J.C. (2009) Effect of drying methods on the phenolic constituents of meadowsweet (*Filipendula ulmaria*) and willow (*Salix alba*). *LWT – Food Science and Technology*, 42, 1468–1473.

Pan, X., Niu, G., and Liu, H. (2003) Microwave-assisted extraction of tea polyphenols and tea caffeine from green tea leaves. *Chemical Engineering and Processing*, 42, 129–133.

Paul.W. and Riechel, T.L. (1998) High molecular weight plant polyphenolics (tannins) as antioxidants. *Journal of Agricultural and Food Chemistry*, 46, 1887–1892.

Palakajornsak, Y. (2004) *Extraction and stability if anthocyanins from mangosteenpeel*, Thesis, Silpakorn University, Thailand, p. 103.

Proestos, C. and Komaitis, M. (2008) Application of microwave-assisted extraction to the fast extraction of plant phenolic compounds. *LWT – Food Science and Technology*, 41, 652.

Prabhanjan, D.G., Ramaswamy, H.S., and Raghava, G.S. (1995) Microwave-assisted convective air drying of thin layer carrots. *Journal of Food Engineering*, 25, 283–293.

Puravankara, D., Boghra, V., and Sharma, R.S. (2000) Effect of anti-oxidant principles isolated from mango (Magifera indica L.) seed kernels on oxidative stability of buffalo ghee (butter-fat). *Journal of the Science of Food and Agriculture*, 74, 331–339.

Ravindran, V. (1993) Cassava leaves as Animal Feed: Potential and Limitations. *Journal of the Science of Food and Agriculture*, 61, 141–150.

Ravindran V., Kornegay, E.T., Webb, K.E., and Rajaguru, A.S.B. (1982) Nutrient characterization of some feed stuffs of Sri Lanka. *Journal of the National Agriculture Society of Ceylon*, 19, 19–32.

Renard, C.M.C.G., Rohou, Y., Hubert, C., della Valle, G., Thibault, F., and Savina, J.P. (1997) Bleaching of apple pomace by hydrogen peroxide in alkaline conditions: Optimization and characterization of the products. *Lebensmittel-Wissenschaft und-Technologie*, 30, 398–405.

Robichaud J. L. and Noble A. C. (1990) Astringency and bitterness of selected phenolics in wine. *Journal of the Science of Food and Agriculture*, 53(3), 343–352.

Roger D.J. (1959) Cassava leaves protein. *Economic Botany*, 13, 261–263.

Rogers D.J. and Milner, M. (1963) Amino acid profile of manioc leaf protein, in relation to nutritive value. *Economic Botany*, 17, 211–216.

Sato M., Maulik G., Ray P.S., Bagchi D., and Das D.K. (1999) Cardioprotective effects of grape seed proanthocyanidin againstischemic reperfusion injury. *Journal of Molecular and Cellular Cardiology*, 32, 1289–1297.

Sharma, G.P. and Prasad, S. (2001) Drying of garlic (*Allium sativum*) cloves by microwave-hot air combination. *Journal of Food Engineering*, 50, 99–105.

Souquet, J.M., Cheynier, V., Brossaud, F., and Moutounet, M. (1996) Polymeric proanthocyanidins from grape skins. *Phytochemistry*, 43, 509–512.

Souquet, J.M., Labarbae, B., Le Guernevé, Cheynier, V., and Moutounet, M. (2000) *Journal of Agriculture and Food Chemistry*, 48, 1076–1080.

Soysal, Y. (2004) Microwave drying characteristics of parsley. *Biosystems Engineering*, 89(2), 167–173.

Sreenath, H.K., Carndall, P.G., and Baker, R.A. (1995) Utilization of citrus by-products and wastes as beverage clouding agents. *Journal of Fermentation and Bioengineering*, 80, 190–194.

Samman, S., Lyons Wall, P.M., and Cook, N.C. (1998) Flavonoids and coronary heart disease: Dietary perspectives. In C.A. Rice Evans and L. Packer (eds) *Flavonoids in Health and Diseases*, Marcel & Dekker, USA, pp. 469–482.

Schieber, A., Stintzing, F.C., and Carle, R. (2001) By-products of plant food processing as a source of functional compounds-recent developments. *Trends in Food Science and Technology*, 12, 401–413.

Sharma, S.K. and Maguer, M.L. (1996) Lycopene in tomatoes and tomato pulp fractions. *Italian Journal of Food Science*, 2, 107–113.

Samman, S., Sandstrom, B., Toft, B., Bukhave, K., Jensen, M., Srensen, S., *et al.* (2001) Green tea or rosemary extract added to foods reduces nonheme-iron absorption. *American Journal of Clinical Nutrition*, 73, 607–612.

Soji, T., Yanagida, A., and Kanda, T. (1999) *Journal of Agriculture and Food Chemistry*, 47, 2885–2890.

Someya S., Yoshiki Y., and Okubo, K. (2002) Antioxidant compounds from bananas (Musa Cavendish). *Food Chemistry*, 79(3), 351–354.

Sultanbawa, M.U.S. (1980) Xanthonoids of tropical plants. *Tetrahedron*, 36, 1465–1506.

Subagio, A., Morita, N., and Sawada, S. (1996) Carotenoids and their fatty-acid esters in banana peel. *Journal of Nutritional Science and Vitaminology*, 42(6), 553–566.

Tasirin, S.M., Kamarudin, S.K., Jaafar, K., and Lee, K.F. (2006) The drying kinetics of bird's chilies in a fluidized bed dryer. *Journal of Food Engineering*, 79, 2, 695–705.

Tewari, H.K., Marwaha, S.S., and Rupal, K. (1986) Ethanol from banana peels. *Agriculture Wastes*, 16, 135–146.

Torres, J.L. and Bobet, R. (2001) New flavanol derivatives from grape byproducts. Antioxidant aminiethylthio-flavan-3-ol-conjugats from a polymeric waste fraction used as a source of flavanols. *Journal of Agriculture and Food Chemistry*, 49, 4627–4634.

Vega-Gálvez, A., Di Scala, K., Rodriguez, K., Lemus-Mondaca, R., Miranda, M., and Lopez, J. (2009) Effect of air-drying temperature on physico-chemical properties, antioxidant capacity, colour and total phenolic content of red pepper (*Capsicum annuum*, L. var. Hungarian). *Food Chemistry*, 47, 647–653.

Vergara-Valencia, N., Granados-Perez, E., Agama-Acevedo, E., Tovar, T., Ruales, J., and Bello-Perez, L.A. (2007) Fibre concentrate from mango fruit: Characteriziation, associated antioxidant capacity and application as a bakery product ingredient. *LWT – Food Science and Technology*, 40, 722–729.

Venir, E., Torre, M.D., Stecchini, M.L., Maltini, E., and Nardo, P.D. (2007) Preparation of freeze-dried yoghurt as a space food. *Journal of Food Engineering*, 80, 402–407.

Walker, E.B. (2007) HPLC analysis of selected xanthones in mangosteen fruit. *Journal of Separation Science*, 30, 1229–1234.

Yapo, B.M. and Koffi, K.L. (2008) Dietary fiber components in yellow passion fruit rind: A potential fiber source. *Journal of Agriculture and Food Chemistry*, 56, 5880–5883.

Yeoh, H.H. and M.Y. Chew (1976) Protein content and amino acid composition of cassava leaf. *Phytochemistry*, 15, 1597–1599.

Yongsawatdigul, J. and Gunasekaran, S. (1996) Microwave-vacuum drying of cranberries: Part I. Energy use and efficiency. *Journal of Food Processing and Preservation*, 20, 121–143.

Zhanga, X. *et al.* (2009) Optimized microwave-assisted extraction of phenolic acids from citrus mandarin peels and evaluation of antioxidant activity in vitro. *Separation and Purification Technology*, 70, 63–70.

Part III
Impact of Processing on Phytochemicals

9 On farm and fresh produce management

Kim Reilly
Horticulture Development Unit, Teagasc, Kinsealy Research Centre, Dublin, Ireland

9.1 Introduction

There is overwhelming evidence that the level of bioactive compounds in a plant food crop is not fixed and can vary substantially depending on how the crop is grown. Agricultural practices which can affect phytochemical content include (1) choice of crop and cultivar; (2) tissue type and developmental stage at harvest; (3) fertilizer supply; (4) seasonal and environmental effects; (5) biotic and abiotic stresses; and (6) mode of production (organic and conventional practices). Some of these factors such as temperature, solar irradiation or stress treatments would be difficult or uneconomic to use as practical strategies to increase desired plant phytochemicals in field grown fruits, vegetables or grains. However factors such as cultivar selection, fertilizer regime and post-harvest treatment could readily be incorporated into existing production practices to produce crops which are optimized in phytochemical content. A number of other treatments such as water stress, light, salinity and temperature although not easily applicable to field crops may find future applications in greenhouse grown crops such as lettuce or salad leaves, tomatoes and herbs; or in production of edible sprouted seeds, where inputs are more easily controlled.

The effect of on-farm and fresh produce management practices on bioactive content will be discussed using glucosinolates, polyacetylenes and phenolic compounds as examples; and vegetable crops especially rich in these compounds, namely broccoli (*Brassica oleracea* var. *italica*), carrot (*Daucus carota*) and onion (*Allium cepa*) as model crops. Numerous peer reviewed publications support a protective role in human health for plant phenolic compounds (especially flavonoids); glucosinolates from *Brassica* species; polyacetylenes from carrot and flavonols and cysteine sulfoxides from onion (reviewed in Christensen and Brandt, 2006; Desjardins, 2008; Lee and Lee, 2006; Zhang and Tang, 2007). Onion is believed to be the major source of human flavonol intake. It is one of the most important vegetable crops globally with an estimated annual production of over 72 million tonnes. Estimated world production of *Brassica oleraceae* crops in 2009 was almost 71 million tonnes, whilst carrot and turnip production was almost 28 million tonnes (FAO, 2007).

Handbook of Plant Food Phytochemicals: Sources, Stability and Extraction, First Edition.
Edited by B.K. Tiwari, Nigel P. Brunton and Charles S. Brennan.
© 2013 John Wiley & Sons, Ltd. Published 2013 by John Wiley & Sons, Ltd.

9.2 Pre-harvest factors affecting phytochemical content

The phytochemical profile of a plant is strongly dependant on genetic components. The range, type and level of individual bioactive compounds vary between different crops, between different species of the same genus, between different groups of the same species or sub-species, and between different accessions or cultivars. Fruits, vegetables and herbs are a rich source of terpenoids, carotenoids, polyphenols and sulphur containing compounds. Cereals and pulses are a good source of plant sterols; oil seed crops such as olive, linseed, sunflower and rape (canola) are rich in fatty acids and sterols (Table 9.1). Phytoestrogens (isoflavones and lignans) are predominantly found in legume crops such as soya, other pulses and some seed crops (reviewed in Buttriss, 2005).

The *Brassicaceae* (*Cruciferae*) include a number of important cultivated crops including various members of the widely consumed *Brassica oleracea* group (Table 9.2). Other cultivated *Brassicaceae* include turnip (*B. rapa* subsp. *rapa*), oriental cabbage (*B. rapa* subsp. *pekinensis*), oil seed rape/canola (*B. napus* subsp. *oleifera*), swede (*B. napus* subsp. *napo-Brassica*), radish (*Raphanus sativus*), watercress (*Nasturtium officinale*), mustards (*B. juncea* and *Sinapsis alba*), horseradish (*Amoracia rusticana*) and rocket (*Eruca sativa*). A characteristic feature of the *Brassica*ceae is the production of glucosinolates, although glucosinolates are also produced in some dicot plant families (Fahey, Zalcmann and Talalay, 2002). In the intact plant cell glucosinolates and the enzyme myrosinase (thioglucosidase, EC 3.2.1.147, previously EC 3.2.3.1) are physically separated, with myrosinase sequestered in the vacuoles of specialized myrosin cells (Andreasson, Jorgensen, Hoglund, Rask and Meijer, 2001;

Table 9.1 Plant food sources of bioactive compounds

Phytochemical class	Dietary sources
Terpenoids	
α- and β-carotene	Carrots, squashes, oranges, tomato, mangoes, green leafy vegetables
Lycopene	Tomato, watermelon
Lutein	Yellow peppers
Sterols	Seeds and grains, nuts, avocados
Other terpenoids	Herbs and spices
Phenolic compounds	
Flavonols	Onion, kale, broccoli, berries, tea
Flavan-3-ols	Cocoa seeds, green tea, apples
Flavones	Parsley, thyme, celery
Flavanones	Citrus fruit
Anthocyanidins	Berries
Phenolic acids	Tea, grapes, coffee beans
Stilbenes	Grapes
Isoflavones	Legumes
Lignans	Grains, linseed
Sulphur containing compounds	
Glucosinolates	Broccoli and other *Brassicas*
S-alkyl cysteine sulphoxides	Onion, garlic
Polyacetylenes	Carrots, parsley
Fatty acids	Seed and nut oils, soya, nuts

Table 9.2 Predominant glucosinolates in commonly consumed members of the *Brassicaceae*

Plant name	Binomial name	Haploid chromosome number	Predominant glucosinolates reported	Reference
Broccoli	B. oleracea var. italica	n = 9	Glucoraphanin Glucobrassicin	(Carlson et al., 1987b; J. Valverde; Kushad et al., 1999; Mithen, Lewis, Heaney and Fenwick, 1987; Vallejo et al., 2002; Ververk, 2010)
Cauliflower	B. oleracea var. botrytis	n = 9	Glucobrassicin Sinigrin Glucoiberin	(Carlson et al., 1987b; Kushad et al., 1999; Mithen et al., 1987; Ververk, 2010)
Brussel's sprouts	B. oleracea var. gemmifera	n = 9	Glucobrassicin Sinigrin	(Carlson et al., 1987b; Kushad et al., 1999; Song, Lori and Thornalley, 2006; Ververk, 2010)
Cabbage	B. oleracea var. capitata	n = 9	Sinigrin Glucobrassicin Glucoiberin	(Cartea et al., 2008b; K. Sones, 1984; Kushad et al., 1999; Mithen et al., 1987; Ververk, 2010)
Savoy cabbage	B. oleracea var. sabauda	n = 9	Glucoiberin	(K. Sones, 1984)
Kale	B. oleracea var. acephala	n = 9	Sinigrin Glucobrassicin Glucoiberin	(Carlson et al., 1987b; Cartea et al., 2008b; Kushad et al., 1999; Velasco et al., 2007; Ververk, 2010)
Kohlrabi	B. oleracea var. gongylodes	n = 9	Glucoerucin	(Carlson et al., 1987b)
Collard	B. oleracea var. viridis	n = 9	Glucobrassicin	(Carlson et al., 1987b)
Chinese kale	B. oleracea var albogflabra	n = 9	Gluconapin	(Mithen et al., 1987; Sun et al.)
Turnip	B. rapa subsp. rapa	n = 10	Progoitrin Gluconasturtiin	(Carlson et al., 1987a; K. Sones, 1984; Li et al., 2007)
Pak-choi	B. rapa subsp. chinensis	n = 10	Glucobrassicin	(Mucha-Pelzer, Mewis and Ulrichs, 2010)
Oriental cabbage	B. rapa subsp. pekinensis	n = 10	Progoitrin	(Mithen et al., 1987)
Oil seed rape	B. napus subsp. oleifera	n = 19	Progoitrin	(Song et al., 2006)
Leaf rape	B. napus subsp. pabularia	n = 19	Glucobrassicanapin	(Cartea et al., 2008a)
Swede	B. napus subsp. napobrassica	n = 19	Progoitrin	(K. Sones, 1984)
Mustard	B. juncea	n = 18	Sinigrin	(Carlson et al., 1987b)
Radish	Raphanus sativus	n = 9	Glucoerucin	(Carlson, Daxenbichler, Vanetten, Hill and Williams, 1985)
Watercress	Nasturtium officinale	n = 16	Gluconasturtiin	(Fenwick and Heaney, 1983)
Rocket	Eruca sativa	n = 11	Glucoerucin	(Cataldi, Rubino, Lelario and Bufo, 2007)
Horseradish	Armoracia rusticana	n = 16	Sinigrin	(Li et al., 2004)
White mustard	Sinapis alba	n = 12	Sinigrin	(Song et al., 2006)

Mithen, 2001). Following cellular disruption the glucosinolates are released and hydrolysed by endogenous myrosinase into a range of breakdown products. The breakdown product formed is dependent on the initial glucosinolate, pH, availability of ferrous ions and the activity of epithiospecifier protein (ESP) (Halkier and Du, 1997) – a heat sensitive co-factor of myrosinase – which directs glucosinolate hydrolysis towards nitrile, rather than isothiocyanate, formation (Matusheski and Jeffery, 2001; Matusheski, Swarup, Juvik, Mithen, Bennett and Jeffery, 2006). A number of *Brassicas* including radish (*Raphanus sativus*), white mustard (*Sinapis alba*), horseradish (*Armoracia rusticana*) and daikon (*Raphanus sativus* var. *niger*) produce isothiocyanates only, whilst other *Brassicas* produce both isothiocyanates and nitriles. This trait is related to differences in the presence and expression of ESP or an ESP like protein (Matusheski *et al.*, 2006; Mithen, Faulkner, Magrath, Rose, Williamson and Marquez, 2003). Phenolic compounds found in broccoli include flavonols such as quercetin and kaempferol glycosides, and phenolic acids such as hydroxycinnamic and chlorogenic acid. Anthocyanins are also found and accumulate to higher levels in purple broccoli (Moreno, Perez-Balibrea, Ferreres, Gil-Izquierdo and Garcia-Viguera).

A number of detailed studies have been carried out to evaluate the content of bioactive compounds in *Brassica oleracea* groups including broccoli, brussel sprouts, cabbage, cauliflower and kale grown individually or under uniform cultural conditions (Carlson, Daxenbichler, Vanetten, Kwolek and Williams, 1987b; Cartea, Rodriguez, de Haro, Velasco and Ordas, 2008a; Kushad *et al.*, 1999; Padilla, Cartea, Velasco, de Haro and Ordas, 2007; Schonhof, Krumbein and Brückner, 2004; Schonhof, Krumbein, Schreiner and B, 1999; Schreiner, 2005; Singh, Upadhyay, Prasad, Bahadur and Rai, 2007; Vallejo, Garcia-Viguera and Tomas-Barberan, 2003; Vallejo, Tomas-Barberan and Garcia-Viguera, 2002; Ververk, 2010). Predominant glucosinolates reported in different *Brassica* crops are shown in Table 9.2. In broccoli (*Brassica oleracea* var. *italica*) the predominant glucosinolates detected are glucoraphanin and glucobrassicin. In contrast in turnip (*Brassica rapa* subsp. *rapa*) gluconasturtiin and progoitrin have been reported as the major glucosinolates in the root, with gluconapin predominant in the leaf (consumed as turnip greens) (Carlson, Daxenbichler, Tookey, Kwolek, Hill and Williams, 1987a; K. Sones, 1984; Li, Schonhof, Krumbein, Li, Stutzel and Schreiner, 2007; Padilla *et al.*, 2007). In brussel sprouts, cabbage, cauliflower and kale the predominant glucosinolates reported are sinigrin, glucoiberin and glucobrassicin (Kushad *et al.*, 1999; Ververk, 2010). In a recent study examining a range of *Brassica* vegetables grown in a single location the level of total glucosinolates found varied from 14 to 625 µmol/100 g FW and the overall level of total glucosinolates was highest in brussel spouts. Levels of glucoraphanin, the precursor of sulforaphane, ranged from 0 to 141 µmol/100 g FW and were higher in broccoli (27–141 µmol/100 g FW). Cauliflower and kohlrabi contained relatively low levels of glucosinolates (Ververk, 2010). Levels of myrosinase activity have been reported to be higher in broccoli, brussel sprouts and cauliflower than in other groups such as kale and cabbage (Charron and Sams, 2004).

Considerable variation between broccoli (*B. oleracea* var. *italica*) varieties has also been reported. Most studies report glucoraphanin or less commonly glucobrassicin as the predominant glucosinolate in broccoli (Carlson *et al.*, 1987b; Jones, Imsic, Franz, Hale and Tomkins, 2007; Krumbein, Schonhof, Rühlmann and S, 2001b; Kushad *et al.*, 1999; Pereira, Rosa, Fahey, Stephenson, Carvalho and Aires, 2002; Rodrigues and Rosa, 1999; Schonhof *et al.*, 2004; Verkerk, Dekker and Jongen, 2001). In a study by Schonhof *et al.* (2004) glucobrassicin was the predominant glucosinolate in the purple variety Viola (Schonhof *et al.*, 2004), whilst neoglucobrassicin was the predominant glucosinolate detected in three cultivars including cv. Marathon in the study by Vallejo *et al.* (2003). In a detailed study which evaluated the glucosinolate profile of 50 broccoli accessions grown under uniform cultural

Figure 9.1 Diversity of broccoli (B.oleracea L. var. italica).
Anti-clockwise from top: Green heading type (cv. Parthenon), Purple heading type (cv. Sicilian Purple Early Autumn), White sprouting types (cv. Sprouting Early White and cv. Early Broccoli), Purple sprouting types (cv. TZ 7033 and cv. Cardinal), Traditional landrace (HRI accession 8721).

conditions using HPLC, Kushad *et al.* (1999) reported glucoraphanin as the predominant glucosinolate, with levels ranging from 0.8 μmol/g DW in the cultivar EV6-1 to 21.7 μmol/g DW in the cultivar Brigadier – a 27-fold difference (Kushad *et al.*, 1999). Cultivated forms of *B. oleracea* are believed to have originated in the Middle East, from where they were introduced to Italy, which is regarded as the centre of diversity for both *botrytis* (cauliflower) and *italica* (broccoli) groups (Gray, 1982; Massie, Astley and King, 1996). Broccoli (*B.oleracea* var. *italica*) and cauliflower (*B.oleracea* var. *botrytis*) are fully cross compatible and can be readily cross pollinated to produce intermediate forms (Malatesta, 1996). Cultivated broccoli varieties within the *B. oleracea* var. *italica* group show considerable morphological diversity (Figure 9.1). Floret colour may be green, white or purple. In heading or crown types the floret occurs as a single large primary floret, whilst in sprouting types smaller primary and numerous secondary florets are produced. In a study in which a green heading type (cv. Ironman), a white sprouting type (cv. TZ4039) and three purple sprouting types (cv.s Red Admiral, TZ5052 and TZ6002) were compared, levels of total glucosinolates ranged from 62 μmol/g DW in the white sprouting variety to 109 μmol/g DW in the purple sprouting variety TZ5052. Glucobrassicin was the predominant glucosinolate, except in the white sprouting variety where sinigrin was predominant. This profile is more similar to that of cauliflower and suggests the variety may be of *botrytis* × *italica* parentage (J. Valverde, pers. com.). Green crown type broccoli has been reported to have higher total glucosinolate content and higher levels of glucoraphanin than other broccoli types such as purple broccoli varieties, which contain higher levels of glucoiberin (Ververk, 2010) or indolyl glucosinolates (Schonhof *et al.*, 1999). Levels of phenolic compounds in broccoli are likewise affected by variety although fewer studies have been carried out with a smaller number of cultivars examined under uniform cultural conditions (Gliszczynska-Swiglo, Kaluzewicz, Lemanska, Knaflewski and Tyrakowska, 2007; Robbins, Keck, Banuelos and Finley, 2005; Vallejo *et al.*, 2003; Vallejo *et al.*, 2002). Kaempferol and quercetin are commonly reported as the major phenolic compounds in broccoli with lower levels of phenolic acids. In a study on 12 broccoli varieties grown in Spain levels of flavonoids ranged from 12.3 mg/kg to 65.4 mg/kg and were highest in cultivars Marathon, Lord and I-9809 (Vallejo *et al.*, 2002). In a smaller study by the same authors examining three commercial varieties (Marathon, Monterrey and Vencedor), levels of phenolics were similar across these three varieties (Vallejo *et al.*, 2003). However in a Polish study which evaluated three varieties (Marathon, Lord, and Fiesta) over three years quercetin content was higher in florets of cv. Lord, whilst kaempferol was highest in florets of cv. Fiesta. Levels of total flavonols (quercetin + kaempferol) ranged from 57 to 273 mg/kg FW depending on cultivar and year (Gliszczynska-Swiglo *et al.*, 2007). In a US field trial where two varieties were cultivated, flavonoids were higher in cv. Majestic than in cv. Legacy (Robbins *et al.*, 2005).

Field studies on 113 varieties of turnip greens (*B. rapa*) (Padilla *et al.*, 2007), 36 varieties of nabicol (*B. napus* var. *pabularia*) (Cartea *et al.*, 2008a; Padilla *et al.*, 2007), 27 varieties of Chinese kale (*B. oleracea* var. *alboglabra*) (Sun, Liu, Zhao, Yan and Wang), 28 varieties of cabbage (*B. oleracea* var. *capitata*) (K. Sones, 1984), 60 oilseed rape varieties (*B. napus* var. *oleifera*) (Davik and Heneen, 1993) and 27 horseradish varieties (*Amoracia rusticana*) (Li and Kushad, 2004) showed a wide range of glucosinolate levels between different accessions of these crops. Similar variation was found in an evaluation of antioxidant compounds, vitamin C, β-carotene, lutein, α-tocopherol and total phenolics in a range of cabbage, cauliflower, Brussel's sprout, Chinese cabbage and broccoli cultivars grown under uniform cultural conditions (Singh *et al.*, 2007). In this study higher levels of vitamin C, β-carotene, lutein, α-tocopherol and total phenolics were found in broccoli than in the other *Brassicas*.

Carrot (*Daucus carota*) is a member of the *Apiaceae* (or *Umbelliferae*) family which also includes celery (*Apium graveolens* var. *dulce*), fennel (*Foeniculum vulgare*), parsnip (*Pastinaca sativa*), dill (*Anethum graveolens*), caraway (*Carum carvi*), cumin (*Cuminum cyminum*) and parsley (*Petroselinum crispum*). Non-cultivated or non-edible species include wild carrot/Queen Anne's lace (*Daucus carota*), hogweed (*Heracleum sphondylium*), cow parsley/wild chervil (*Anthriscus silvestris*) and poison hemlock (*Conium maculatum*). In addition to the familiar orange carrot, white, yellow, purple, red and black varieties also occur. Important phytochemicals found in carrot include the polyacetylenes falcarinol, falcarindiol and falcarindiol-3-acetate; and the isocoumarin 6-methoxymellein (6-MM). Falcarinol (synonym: panaxynol, (9Z)-heptadeca-1,9-dien-4,6-diyn-3-ol) is the most potent and best studied of the carrot polyacetylenes (Kidmose, Hansen, Christensen, Edelenbos, Larsen and Norbaek, 2004; Zidorn et al., 2005). Falcarindiol has recently been identified as the main compound responsible for the bitter flavour in fresh and stored carrots. Black and purple carrot varieties also accumulate anthocyanins and five major cyanidin based anthocyanins have been described in carrot (Netzel et al., 2007).

Variation in carotenoid profile has been reported with some carrot cultivars primarily accumulating lutein whilst others primarily accumulated α and β carotene (Schreiner, 2005). Levels of individual polyacetylenes differ between different carrot varieties (Alasalvar, Grigor, Zhang, Quantick and Shahidi, 2001; Christensen and Kreutzmann, 2007; Czepa and Hofmann, 2004; Hansen, Purup and Christensen, 2003; Kidmose et al., 2004; Mercier, Ponnampalam, Berard and Arul, 1993; Metzger and Barnes, 2009), and between different members of the *Apiaceae* family (Degen, Buser and Stadler, 1999; Zidorn et al., 2005). In an analysis of five members of the *Apiaceae*, falcarinol was found in all investigated taxa except parsley. Falcarindiol was found in all taxa and was the main polyacetylene except in carrot, where falcarinol was the predominant polyacelylene. Levels of total polyacetylenes and of falcarinol were higher in celery (*Apium graveolens*) and parsnip (*Pastinaca sativa*) than in carrot. However as carrot is consumed more frequently it is likely to be the major source of polyacetylenes in the Western diet (Zidorn et al., 2005). In an earlier analysis of 12 members of the *Apiaceae* the highest levels of falcarindiol were found in caraway (*Carum carvi*) and hogweed (*Heracleum sphondylium*). Levels of falcarinol were extremely high in poison hemlock (*Conium maculatum*). Amongst the cultivated species levels of falcarinol were higher in chervil (*Anthriscus cerefolium*), dill (*Anethum graveolens*) and parsnip (*Pastinaca sativa*) (Degen et al., 1999). Levels of carotenoids, phenolics and antioxidant capacity are reported as higher in purple carrot varieties than in other varieties (Alasalvar et al., 2001; Sun, Simon and Tanumihardjo, 2009). In a field study of 27 carrot varieties grown under uniform cultural conditions levels of falcarinol ranged from 0.70 to 4.06 mg/100 g FW (Pferschy-Wenzig, Getzinger, Kunert, Woelkart, Zahrl and Bauer, 2009).

Onions (*Allium cepa*) belong to the species *Allium* which also contains vegetables such as shallots (*Allium cepa* var. *aggregatum*), scallion/Welsh onion (*Allium fistulosum*), garlic (*Allium sativum*), wild garlic (*Allium ursinum*), leek (*Allium porrum*) and chives (*Allium schoenoprasum*). Two classes of phytochemical found in onion – the sulphur containing alk(en)yl sulfoxides and the flavonols – are believed to show health promoting activity. The main flavonols found in onion are quercetin, quercetin 4'-glucoside, quercetin 3,4'diglucoside, kaempferol and kaempferol glucosides. Isorhamnetin and rutin have also been identified in some cultivars (Marotti and Piccaglia, 2002; Slimestad, Fossen and Vagen, 2007). Red onions are commonly reported to contain higher levels of total flavonols than yellow or white varieties. Quercetin and its derivatives, quercetin-3,4'-O-diglucoside (QDG) and quercetin-4'-O-monoglucoside (QMG), are thought to make up over 90% of the flavonoid content in onion. In red onions anthocyanins are also present. These are primarily cyanadin glucosides acylated with malonic acid or non-acylated (Desjardins, 2008; Pérez-Gregorio, García-Falcón, Simal-Gándara, Rodrigues and Almeida, 2010; Slimestad et al., 2007). Levels of flavonols vary considerably between onion cultivars with reported levels ranging from 2549 quercetin equivalents per kg fresh weight (FW) in the red onion cultivar Karmen to less than 1 quercetin equivalents in the white onion cultivar Contessa (Slimestad et al., 2007). A study by Lombard et al. (2005) examined the total flavonol content in five onion varieties and found significantly higher concentrations in red skinned onion varieties compared to yellow varieties (Davis, Epp and Riordan, 2004). In a detailed study of 75 onion cultivars levels of quercetin were higher in red, pink and yellow onions (in the range

54–286 mg/kg FW) whilst white onions contained only trace amounts of quercetin (Patil, Pike and Yoo, 1995c). Similar results are found in other studies (Marotti et al., 2002; Pérez-Gregorio et al.). It has also been suggested that long day onion cultivars of Rijnsburger type from Northern Europe have higher levels of quercetin glucosides than short day onions of North American and Japanese origin (Slimestad et al., 2007). In a study on 16 European onion varieties levels of quercetin glycosides in the edible parts were highest in the red onion cv. Red Baron and yellow skinned variety cv. Ailsa Craig (Beesk, Perner, Schwarz, George, Kroh and Rohn, 2010). Variation in the level of bioactivity of different onion or *Allium* varieties has also been demonstrated. In an analysis of ten onion and shallot varieties levels of total flavonoids and total phenolic content varied significantly between varieties and was strongly correlated with antioxidant activity and with inhibition of proliferation of HepG(2) and Caco-2 cells (Yang, Meyers, Van der Heide and Liu, 2004). A 50-fold variability in onion induced anti-platelet activity among cultivated and wild accessions in the genus *Allium* has been demonstrated (Goldman, Kader and Heintz, 1999).

Similar differences in the levels of measured bioactive components between varieties are reported across cultivated food crops. Examples include β-glucans, phenolics and terpenoids in cereals (Tiwari and Cummins, 2008; Ward et al., 2008); isoflavones in soya (*Glycine max*) (Lee et al., 2010; Prakash, Upadhyay, Singh and Singh, 2007); lignans in oilseeds such as flax (*Linum ussitatum*) (Schmidt et al., 2010); carotenoids and lycopene in fruits such as watermelon (*Citrullus lanatus*) and tomato (*Solanum lycopersicum*) (Perkins-Veazie, Collins, Davis and Roberts, 2006; Taber, Perkins-Veazie, Li, White, Roderniel and Xu, 2008); and phenolic compounds in apple (*Malus domestica*) (Carbone, Giannini, Picchi, Lo Scalzo and Cecchini, 2011) and grapefruit *(Citrus paradisi)* (Girennavar, Jayaprakasha, Jifon and Patil, 2008).

9.2.1 Tissue type and developmental stage

A wide variety of different plant tissues at different stages of development are consumed as food. For example, root (carrots, turnips, swedes, parsnips, potato); leaf (cabbage, lettuce, spinach, tea); stem (leeks, celery); bulb (onion, garlic); immature flower (broccoli, cauliflower); fruit (apples, pears, plums, berries, citrus fruit, tomatoes, peppers, cucumbers, aubergine); and seeds (cereals, nuts, legumes, linseed, sunflower seed). As might be expected the tissue type consumed and the level of maturity can influence the level and types of bioactive in the food. For example the carotenoid lycopene gives the red colour to tomatoes and watermelon and levels increase as the fruit matures. Seeds including cereals, legumes and nuts have evolved to provide food resources to the developing embryo of angiosperm plants and contain higher levels of sterols and fatty acids than other plant organs.

In broccoli levels of total glucosinolates and the profile of individual glucosinolates varies depending on tissue type and developmental stages. Levels of glucosinolates are frequently reported to be higher in the earlier stages of plant growth: levels are higher in un-germinated broccoli seed than in seedlings, and levels in seedlings are higher than in florets. Reported levels of total glucosinolates in seed are in the region of 500 mg/100 g (cv. Marathon) (Perez-Balibrea, Moreno and Garcia-Viguera, 2008).

In sprouted broccoli seedlings (cv. Marathon) total glucosinolate levels were in the range 29.2 ± 2.7 to $81.7\pm3.3\,\mu mol\,g^{-1}$ DW ($9.7\pm0.5\,\mu mol\,g^{-1}$ FW and $4.6\pm0.4\,\mu mol\,g^{-1}$ FW) depending on seedling age and growth temperature. Levels of glucosinolates were higher in un-germinated seeds and levels progressively declined as the sprouts grew and developed. Levels of glucoraphanin were in the range 17.4 ± 1.5 to $49.5\pm1.9\,\mu mol\,g^{-1}$ DW depending on seedling age and growth temperature (Pereira *et al.*, 2002). A similar decline in glucosinolate content with sprout age was noted by Perez-Balibrea *et al.* (2008) and levels of total glucosinolates, total phenolics and vitamin C were observed to be higher in cotyledons than in roots or stems of the seedlings (Perez-Balibrea *et al.*, 2008).

In studies on cultivated broccoli plants highest levels of glucosinolates have been reported in the floret with lower levels reported in the leaves and roots. Levels of total glucosinolates in the floret declined during development, mainly due to a decrease in the indole glucosinolates glucobrassicin and neoglucobrassicin. However levels of glucoraphanin were unchanged during head development (Schonhof *et al.*, 1999). In a similar study, levels of total glucosinolates in florets were higher in the first two developmental stages (corresponding to 42 and 49 days after transplanting) and declined as the florets matured. Recorded levels of total glucosinolates at commercial maturity were in the range 19.6–56.4 µmol g^{-1} DW depending on cultivar and fertilization regime. Levels of glucoraphanin at commercial maturity were in the range 0.9–1.9 µmol g^{-1} DW again depending on cultivar and fertilization regime (Vallejo *et al.*, 2003). Levels of total glucosinolates and levels of 9 out of 11 individual glucosinolates measured were lower in post maturation florets in cv. Tokyodome, although slight increases in levels of hydroxyglucobrassicin and neoglucobrassicin were found (Rodrigues *et al.*, 1999). In this study levels of glucosinolates were found to differ between primary and secondary florets, with primary florets containing higher levels of glucoraphanin, glucoiberin, progoitrin, glucoalyssin, gluconapin and gluconasturtiin than secondary florets. Levels of the carotenoids β–carotene and lutein, and of chlorophyll, are also reported to increase during development of the floret (Schonhof *et al.*, 1999). Levels of phenolic compounds in broccoli have been reported to be up to ten times higher in the leaves than the stalks (Dominguez-Perles, Martinez-Ballesta, Carvajal, Garcia-Viguera and Moreno).

In carrot levels of phenolic compounds are reported to be higher in the outer layers of the carrot (Olsson and Svensson, 1996). Raman spectroscopy has been used to localise the tissue distribution of polyacetylenes (Baranska and Schulz, 2005). Highest levels of total polyacetylenes were detected in the outer part of the root in the pericyclic parenchyma and in the phloem adjacent to the secondary cambium. These data are in agreement with the observation that peeled carrots contain up to 50% less falcarindiol, a polyacelylene compound with strong anti-fungal activity associated with bitter flavour in carrots (Czepa *et al.*, 2004). In a study which examined 16 carrot accessions, high levels of falcarindiol (31.9–91.5 µg/g FW) were detected in the peel, with levels of 6.0–19.2 µg/g FW in the phloem. In contrast falcarinol levels were lower and were concentrated in the phloem. Levels of falcarinol ranged 1.3–5.3 µg/g FW in the peel and 2.8–12.2 µg/g FW in the phloem (Olsson *et al.*, 1996). Similar results are reported elsewhere (Christensen *et al.*, 2007) with carrot peel containing up to ten times more falcarindiol than the corresponding peeled roots in six varieties examined. In this study falcarinol was more evenly distributed across the root. Thus peeled carrots should retain the bulk of the health promoting compound falcarinol, whilst falcarindiol which has been associated with bitter taste would be largely removed. The polyacetylene falcarinol had lower anti-fungal activity than falcarindiol (Olsson *et al.*, 1996) but has been more widely reported as beneficial in human health (Brandt *et al.*, 2004). Levels of both falcarindiol and falcarindiol-3-acetate were found to be significantly higher in small/immature (50–100 g) than in large (>250 g) carrot roots in an analysis of six Nantes type carrot varieties, however levels of falcarinol were unaffected (Kidmose *et al.*, 2004). Similarly, in a three year field study on two carrot varieties (Bolero and Kampe) harvested at different maturity stages (103–104 days, 117–118 days, 131–133 days and 146–147 days), maturity had no effect on levels of falcarinol in fresh carrots (Kjellenberg, Johansson, Gustavsson and Olsson).

In onion, highest levels of quercetin and kaempferol glycosides are commonly found in the outer dry skins with lower levels detected in the inner edible rings (Beesk, Perner, Schwarz, George, Kroh and Rohn; Chu, Chang and Hsu, 2000; Patil and Pike, 1995a; Pérez-Gregorio *et al.*, 2010), and levels are reported to decrease from the apex to the base (root part) of the bulb. In contrast anthocyanin distribution is relatively uniform (Pérez-Gregorio *et al.*). Some studies report higher levels of flavonols and anthocyanins in smaller onion bulbs, however in other studies bulb size had no significant effect on quercetin glycoside content (Mogren, Caspersen, Olsson and Gertsson, 2008; Patil *et al.*, 1995a). In a study to evaluate the antioxidant potential of wild *Allium* species (*A. neapolitanum, A. roseum, A. subhirsutum* and *A. sativum*) growing in Italy, Nencini and colleagues report significantly higher levels of antioxidant activity as measured by FRAP test and a DPPH assay in the flowers or leaves, with lowest antioxidant capacity consistently reported for the bulbs (Nencini, Cavallo, Capasso, Franchi, Giorgio and Micheli, 2007).

9.2.2 Fertilizer application – nitrogen, phosphorus, potassium, sulphur and selenium

Nitrogen (N), phosphorus (P), potassium (K) and sulphur (S) are generally considered as plant macro-nutrients; and their application as fertilizer generally increases crop yield and nutritional quality. However excess N fertilizer in particular can cause undesirable effects such as increased nitrate levels in leafy vegetables, reduced quality, reduced vitamin C content and reduced shelf life in some crops.

In terms of bioactive content N, P and K fertilizer application has shown different and sometimes contradictory results for different phytochemical classes and different crops. In *Brassica oleraceae* crops a number of studies have shown that decreased N application results in higher accumulation of phenolic compounds and of some glucosinolates; whilst higher levels of N fertilization promote formation of carotenoids and chlorophylls (reviewed in Schreiner, 2005). However field studies with carrot have indicated that levels of total phenolics were increased in response to increasing N fertilization (Smolen and Sady, 2009). Levels of quercetin in onion were shown to be unaffected by either the type or amount of nitrogen fertilization (Mogren *et al.*, 2008). In barley (*Hordeum vulgare*) increased N application is reported to increase levels of β-glucans (Tiwari *et al.*, 2008), whilst in flax (*Linum usitatissimum*) neither N, P, K or S had a significant effect on lignan levels (Westcroft, N.D., 2002). In soya (*Glycine max*) P, K, S and B (boron) fertilizer had no effect on isoflavone content (Seguin and Zheng, 2006) although a response to K fertilizer on K deficient soils has been reported (Vyn, Yin, Bruulsema, Jackson, Rajcan and Brouder, 2002). Phosphurus and/or potassium fertilizer has been reported to increase levels of lycopene in fruits such as tomato, watermelon and grapefruit in some studies (Paliyath, 2002; Perkins-Veazie P., 2002) but not others (Oke, Ahn, Schofield and Paliyath, 2005).

In a recent study the effect of applied N and S on phytochemical accumulation in florets of broccoli cv. Marathon was investigated (Jones *et al.*, 2007). Nitrogen was applied at 0, 15, 30 or 60 kg/ha and S at 50 or 100 kg/ha. In this study highest levels of flavonoids, and of the sulforaphane precursor glucoraphanin, were obtained at low N application rates. Nitrogen application at levels above 30 kg/ha caused an increase in glucobrassicin content of up to 44%, whilst levels of glucoraphanin declined by 18–34% and levels of the flavonols quercetin and kaempferol declined by 20–38%. However crop yields declined significantly (up to 40%) at N levels below 60 kg/ha. Similar effects of N fertilization are noted in other field studies on broccoli (Fortier, 2010; Krumbein et al., 2001b; Li et al., 2007; Omirou et al., 2009) and other Brassicas (Li, 2010; Li et al., 2007). This suggests that there could be considerable potential to produce mini broccoli heads with enhanced levels of phenolic compounds and glucoraphanin at low N application rates, although levels of carotenoids in such broccoli may be reduced.

Sulphur supplementation has been demonstrated to increase glucoraphanin content in a range of Brassica species including broccoli and to increase alliin content in onion and garlic (Krumbein et al., 2001b; Schonhof, Klaring, Krumbein, Claussen and Schreiner, 2007; Schonhof et al., 1999). Sulphur applied at levels up to 600 mg per plant to broccoli grown in soil free media resulted in a significant increase in glucoraphanin content (Krumbein et al., 2001b). However field trial based studies have been disappointing and suggest that applying S to S-sufficient soils has only a minimal impact on glucosinolate accumulation. Vallejo et al. (2003) examined the effect of S application at levels of 15 and 150 kg/ha on three broccoli cultivars. Whilst significant differences were observed in glucosinolate contents of immature broccoli florets in response to S fertilization, no significant differences were observed in mature florets (Vallejo et al., 2003). In a similar study S applied as gypsum at levels of 23 kg/ha resulted in a significant increase in glucoraphanin content in cv. Marathon but had no significant effect on two other cultivars (Rangkadilok, Nicolas, Bennett, Eagling, Premier and Taylor, 2004). In the study of Jones et al. (2007) S application at levels of 50 and 100 kg/ha had no significant effect of glucosinolate or flavonol accumulation (Jones et al., 2007).

Given the high content of glucoraphanin found in broccoli sprouted seed some authors have investigated the effect of N and S application during growth of broccoli and other Brassica sprouts (Aires, Rosa and Carvalho, 2006; Kestwal, 2010). In the study by Aires et al. (2006) broccoli cv. Marathon seeds were grown in Petri dishes on rockwool disks and watered with pure water supplemented with different combinations of potassium nitrate (KNO_3) and potassium sulphate (K_2SO_4) from 6 days after sowing. Sprouts were harvested and analysed 11 days after sowing. However this was found to have a significant detrimental effect on accumulation of aliphatic glucosinolates including glucoraphanin. This may have been due to salt stress at the concentrations used (Aires et al., 2006). In the study by Kestwal et al. (2010) broccoli, radish and cabbage seeds were sprouted in soil supplemented with S as sodium thiosulphate ($Na_2S_2O_3$) at S concentrations equivalent to 20 to 60 kg/ha. This range was selected following an initial experiment to determine the optimal treatment range where sprout growth was not significantly adversely affected. Sprouts were harvested for analysis at 12 days after sowing. In this study levels of total glucosinolates including glucoraphanin were increased in S supplemented radish, broccoli and cabbage sprouts. Levels of total phenolics were higher in S supplemented radish, but not broccoli or cabbage sprouts, and antioxidant activity was higher in S supplemented radish and broccoli but not cabbage (Kestwal, 2010).

The mineral selenium (Se) has been implicated in reduced risk of cardiovascular disease and several cancers. In most plant species selenium (Se) can be toxic to the plant, however Brassica and Allium species are able to utilize Se and are referred to as seleniferous plants or 'selenium accumulators' (Irion, 1999). In most soils worldwide Se is deficient and selenium enriched Brassica and Allium crops can be grown by supplementing the soil with Se. Given the potential health benefits of a 'super broccoli' containing higher levels of both sulforaphane and Se, attempts have been made to increase levels of both sulforaphane and

Se in broccoli. However these efforts have been frustrated since there appears to be an inverse relationship between Se and glucoraphanin accumulation. Broccoli and other crucifers typically contain relatively low amounts of Se (0.1–0.3 µg/g DW) (Robbins et al., 2005). In supplementation experiments where sodium selenate solution was added to broccoli plants from one week prior to floret development onwards, accumulation of Se to as much as 950 µg/g DW was achieved. Little effect was observed on total glucosinolate levels but a significant decrease in levels of sulforaphane and some phenolic, particularly cinnamic, acids was observed (Finley, Sigrid-Keck, Robbins and Hintze, 2005; Robbins et al., 2005). In studies where both Se and S were applied to hydroponically grown *Brassica oleracea* plants an interaction between S and Se metabolism was observed. Plants exposed to increased levels of S (as sulphate, 37 ppm) showed increased accumulation of glucosinolates with levels of glucoiberin and glucoraphanin 11 and 16% higher than controls. Plants exposed to Se (as selenate, at 0.5, 0.75, 1.0 and 1.5 ppm) showed reduced accumulation of glucoiberin, glucoraphanin and other glucosinolates with increasing Se. At 1.5 ppm Se levels of glucoiberin and glucoraphanin were reduced by 58 and 68% respectively compared to controls. In combined Se/S treatments, levels of Se in leaf tissue were 178 µg Se g^{-1} and levels of glucoraphanin were only moderately reduced compared to controls. Thus the authors conclude that it may be feasible to produce selenium enriched *Brassica* crops that maintain adequate levels of glucoraphanin by selenate fertilization (Toler, Charron, Kopsell, Sams and Randle, 2007).

There have been fewer field studies on the impact of fertilizer application on bioactive content in onions and carrots. In onion S application can increase yield and bulb size, and, as might be expected, led to increased levels of the S containing alk(en)yl sulfoxides and increased pungency (measured as pyruvate content in macerated tissue) (Forney, 2010). Selenium enriched garlic has been produced and has been shown to have higher bioactivity when grown in Se rich soil. Increased activation of phase II enzymes and enhanced production of Se-methyl-selenocysteine (an inhibitor of tumourigenisis) in the Se enriched plants have been demonstrated (Arnault and Auger, 2006; Irion, 1999).

9.2.3 Seasonal and environmental effects – light and temperature

The induction of several enzymes of the phenylpropanoid pathway by light is well known (Martin, Hailing and Schwinn, 2000) and fruits and vegetables grown in full sun have been reported to contain higher levels of flavonoids. For example, shading has been reported to reduce anthocyanin content in lettuce (Kleinhenz, French, Gazula and Scheerens, 2003), and in fruits including grapes, kiwi, apple and pears (Solomakhin and Blanke, 2010; Steyn, Wand, Jacobs, Rosecrance and Roberts, 2009). Exposure to sunlight is known to enhance production of flavonols in onion bulbs (Patil, Pike and Hamilton, 1995b; Rodrigues, Perez-Gregorio, Garcia-Falcon, Simal-Gandara and Almeida; Schreiner, 2005). In a five year study which examined the effect of climatic conditions on flavonoid content in two Portuguese landrace onion varieties, total and individual flavonoid levels varied significantly between years, with highest levels observed in hot, dry years (Rodrigues et al.). In a three year field study which examined levels of the flavonols kaempferol and quercetin in three broccoli varieties (cv.s Marathon, Lord and Fiesta) the level of total solar radiation over the growing period had a significant effect on both flavonols with higher levels under increased radiation (Gliszczynska-Swiglo et al., 2007). In some instances the interplay between climatic factors may result in differential regulation of different phytochemicals.

For example high light levels can increase total flavonoid synthesis. Anthocyanin synthesis requires light and is stimulated at lower temperatures and inhibited at higher temperatures, however other flavonoid compounds are less responsive to temperature (Mori, Goto-Yamamoto, Kitayama and Hashizume, 2007a; Steyn et al., 2009). Seasonal effects on plant bioactives are likely due to this type of interaction with high light combined with lower temperature predominant in spring season and high light combined with higher temperature conditions predominant in summer season.

A number of studies carried out on broccoli and cauliflower cultivars (Charron et al., 2004; Krumbein and Schonhof, 2001a; Schonhof et al., 2007; Schonhof et al., 2004; Schonhof et al., 1999) have indicated that increasing irradiation combined with lower daily temperatures, led to increased levels of glucoraphanin and glucoiberin. In purple broccoli varieties the glucosinolate content was unaffected. In addition, low daily mean temperatures promoted synthesis of lutein, β-carotene and ascorbic acid in broccoli (Schonhof et al., 1999; Schreiner, 2005). In a study on greenhouse grown broccoli (cv. Marathon), levels of alkenyl glucosinolates such as gluconapoeiferin and progoitrin were unaffected by temperature or irradiation. In contrast alkyl glucosinolates such as glucoiberin and glucoraphanin showed increased accumulation at lower temperature (<12 °C). Of the indole glucosinolates glucobrassicin was increased by high temperature (>18 °C) and low radiation (Krumbein et al., 2001a; Schonhof et al., 2007). The authors found a significant correlation between alkyl glucosinolates in broccoli florets and levels of the stress indicator proline in leaves and postulate that stress responses may play a role in glucosinolate accumulation. In greenhouse grown broccoli higher levels of glucosinolates in broccoli leaves from plants grown at 12 °C and 32 °C as compared to those grown at 22 °C were found suggesting that temperature stress may be responsible for increased glucosinolate content (Charron et al., 2004). In other *Brassica oleracea* crops including cabbage and kale, field based trials have indicated that there is a significantly higher total glucosinolate content in spring sown crops and variations in the level of individual phytochemicals (Cartea, Velasco, Obregon, Padilla and de Haro, 2008b). Levels of myrosinase activity (measured as activity/FW and specific activity) in a range of *Brassicas* showed a response to temperature and photosynthetic photon flux (PPF) (Charron, Saxton and Sams, 2005). Activity FW was generally higher where daily mean temperatures and PPF in the two weeks prior to harvest were lower. The authors suggest that light may affect myrosinase activity indirectly via modulation of ascorbic acid – since myrosinase is inhibited by high concentrations of ascorbic acid and the accumulation of ascorbate is itself is increased by light (Yabuta et al., 2007).

Pronounced seasonal effects with large year on year variation in bioactive content are commonly reported in the literature for many crops, and the factors causing this variation are often poorly understood. In carrot levels of polyacetylenes were significantly different in different harvest years indicating a seasonal effect on falcarinol and falcarindiol (Kjellenberg et al.); however the underlying mechanism is unclear. Similarly levels of soybean isoflavones vary significantly between years with a large genotype x environment interaction (Murphy et al., 2009; Seguin et al., 2006).

Levels of bioactives in sprouted seeds are also temperature and/or light responsive. Increases in response to light treatments have been reported for isoflavones in sprouted soya seedlings (Phommalth, Jeong, Kim, Dhakal and Hwang, 2008); bioactive componants and antioxidant capacity in wheatgrass, and phenolic content and antioxidant capacity in alfalfa, broccoli and radish (Oh and Rajashekar, 2009). For commercial production of sprouted broccoli seeds, the seeds are commonly grown at 20–28 °C. Sprouted broccoli (cv. Marathon) seedlings grown under a 30/15 °C (day/night) temperature regime showed significantly higher total glucosinolate levels, specific increases in glucoraphanin content and corresponding increased induction of phase II enzymes than sprouted seed grown at 22/15 and 18/12 °C temperature regimes (Pereira *et al.*, 2002). Mean recorded glucoraphanin levels in experimental sprouts on the sixth day after sowing were 49.5 µmol g^{-1} DW and glucoraphanin made up 61.3% of total glucosinolate content. When sprouted seed was grown at either supra- or sub-optimal constant temperatures of either 33.1 °C or 11.3 °C glucoraphanin and total glucosinolate contents were also increased, although sprout growth was negatively affected by non-optimal temperature. In addition the authors raised the concern that seeds sprouted at higher temperature or longer time would be more susceptible to microbial contamination and thus such practices may not be suitable for commercial production. In an additional study it was reported that phytochemical content of sprouted broccoli seeds (cv. Marathon) was light responsive, with sprouted seed grown under a 16h light/8h dark photoperiod showing more enhanced levels of glucosinolates, phenolic compounds and vitamin C than dark grown sprouts (Perez-Balibrea *et al.*, 2008). Levels of total glucosinolates, total phenolics and vitamin C were 33, 61 and 83% higher respectively.

9.2.4 Biotic and abiotic stress

Abiotic stresses include water stress, salinity and temperature stress. Biotic stresses include wounding, pathogenesis, insect or animal herbivory and treatment with elicitors which mimic these responses, as well as competition with neighbouring plants. Phenylalanine ammonia lyase (PAL) the key entry point enzyme for synthesis of phenolic compounds is well-known to be up-regulated in response to biotic and abiotic stresses including UV light, low temperature, nutrient deficiency, wounding and pest or pathogen attack (Naoumkina, Zhao, Gallego-Giraldo, Dai, Zhao and Dixon). The polyacetylenes and the glucosinolates are also considered as defensive compounds within the plant and numerous studies show their regulation in response to abiotic and biotic stresses (Mercier *et al.*, 1993; Naoumkina *et al.*, 2010; Olsson *et al.*, 1996; Rosa, 1997; Rosa and Rodrigues, 1999).

Drought/water stress and salt stress have been reported to increase a number of phenolic compounds, terpenes, alkaloids, glucosinolates and other compounds in a range of fruits, vegetables, herbs and pulses (reviewed in Selmar, 2008). Application of water deficit irrigation treatments have been studied in crops including lettuce (Oh, Carey and Rajashekar, 2010), citrus fruit (Navarro, Perez-Perez, Romero and Botia, 2010) and peaches (Tavarini *et al.*, 2011). Drought treatments have also been reported to increase isoflavone content in soybean (Seguin *et al.*, 2006) and β-glucans in cereals (Guler, 2003). A doubling of glucosinolate content in broccoli with reduced water supply has been reported (Schonhof *et al.*, 2007). In carrot changes in polyacetylene profile and the content of individual polyacetylenes have also shown a response to water stress, although results are contradictory. In a greenhouse pot trial three novel polyacetylene compounds were found only in stressed carrots subjected to drought or waterlogged conditions. Levels of eight other polyacetylenes including falcarinol, falcarindiol and falcarindiol-3-acetate were lower in control samples, although an earlier field study by the same authors showed higher levels of polyacetylenes in field grown drought stressed carrots (Lund and White, 1990). Both drought and salt stress cause production of ROS within the plant and result in increased levels of secondary metabolites, including phytochemicals. Some of these plant secondary compounds can function as free radical scavengers and osmo-protectants (reviewed in Selmar, 2008). Studies on tomato (*Solanum lycopersicum*) have indicated that moderate salt stress can increase levels of bioactive compounds such as lycopene by up to 85% depending on cultivar (Dorais, 2007;

Kubota and Thomson, 2006). Commonly however the biomass of drought or salt stressed plants is considerably reduced. Four recent studies have shown that salt stress can increase levels of glucosinolates and phenolic compounds in *Brassicas*. Total glucosinolate content and total phenolic content were significantly increased and myrosinase activity was inhibited in radish sprouts germinated under a 100 mM NaCl treatment (Yuan, Wang, Guo and Wang). In hydroponically grown Pak-choi levels of total glucosinolates were increased significantly by 50 mM NaCl, however under 100 mM NaCl the content of indole glucosinolates increased whilst aromatic glucosinolates decreased (Hu and Zhu). In a greenhouse study on the effect of salt stress (80 mM NaCl) on three broccoli varieties (cv.s Marathon, Nubia and Viola) the salt stress treatment significantly increased levels of glucosinolates in leaf and stalk tissue of the purple variety Viola but not the green broccoli varieties Marathon or Nubia (Dominguez-Perles *et al.*). In this study salt treatment significantly affected levels of phenolic compounds in some tissues but not others and in some varieties but not others, indicating a significant variety x salt stress and tissue type x salt stress interaction on phenolic accumulation. Salt stress (40 mM and 80 mM NaCl) caused a significant increase in levels of total glucosinolates in floret tissue of cv. Marathon (Lopez-Berenguer, Martinez-Ballesta, Moreno, Carvajal and Garcia-Viguera, 2009). Floret vitamin C content was unaffected. Phenolic compounds in the floret showed a complex response with some such as sinapic acid derivatives increased at 40 mM but not 80 mM NaCl and flavonoids decreased at 80 mM NaCl.

Temperature stress has been explored as a method to increase bioactive content particularly in sprouted seeds as previously discussed. Chilling shock treatment of alfalfa, broccoli and radish sprouts was found to significantly increase phenolic content (Oh *et al.*, 2009).

Biotic stress results when an organism interacts with other living things in its environment. The carrot polyacetylenes were originally of research interest due to their role in defence and pathogenesis responses. Both falcarinol and falcarindol are implicated in variability of resistance to carrot root fly (*Psila rosae*) amongst different carrot cultivars. They act together with other compounds such as the phenolic compound methyl-isoeugenol as oviposition stimulants (Degen *et al.*, 1999). In addition the carrot polyacetylenes, falcarindiol in particular, are implicated in resistance to storage pathogens (Mercier *et al.*, 1993; Olsson *et al.*, 1996). Increased accumulation of carrot phenolic compounds and increased expression of PAL in response to mechanical wounding, ethylene and methyl jasmonates treatment and elicitor treatment have been reported (Heredia and Cisneros-Zevallos, 2009; Jayaraj, Rahman, Wan and Punja, 2009; Seljasen, Bengtsson, Hoftun and Vogt, 2001).

The glucosinolates in *Brassica* species can be induced in response to pathogen attack, herbivory and in response to elicitors or plant hormones involved in defence responses including salicylic acid, jasmonic acid and methyl jasmonate (Abdel-Farid *et al.*; Krumbein, 2010; Rosa, 1997). Schonhof *et al.* (1999) report that the synthesis of glucosinolates in broccoli could be induced by wounding or mechanical stress such as leaf damage (Schonhof *et al.*, 1999), however attempts to induce glucosinolates in other *Brassica* crops were unsuccessful (Mithen, 2001). There is a complex relationship between glucosinolates and pests, and it is currently understood that whilst glucosinolate breakdown products may repel generalist pests, some glucosinolates, in particular aliphatic glucosinolates, may act as attractants towards specialized pests (Mithen, 2001; Velasco, Cartea, Gonzalez, Vilar and Ordas, 2007). Spacing effects with other plants are also apparent. Schonhof *et al.* (1999) found that high plant density in broccoli cultivation (97 000 plants per ha) (equivalent to 9.7 plants per m^2) could increase glucoraphanin content by up to 37% in comparison with lower density planting; indole glucosinolates were not affected (Schonhof *et al.*, 1999).

In onion and other *Allium* plants a high level of arbuscular mycorrhizal colonization is common and this association can result in increases in yield especially in low nutrient soils. Quercetin mono- and di-glucoside concentrations in onion bulb can be significantly increased by application of arbuscular mycorrhizal fungal innocula due to induction of plant defence responses (Perner *et al.*, 2008).

9.2.5 Means of production – organic and conventional agriculture

European Union Council Regulation No. 2092/91 (EU, 1991) defines a number of parameters for a plant product to be considered organic including: a ban on synthetic pesticides, herbicides and mineral fertilizers; a ban on genetically modified cultivars; and lower nitrogen levels than conventional agriculture (a maximum limit for manure application of 170 kg N ha^{-1} year^{-1} (Rembialkowska, 2007). Within the EU the directive is interpreted by national certification bodies such as the Soil Association in the UK, or the Irish Organic Trust and IOFGA (Irish Organic Farmers and Growers Association) in Ireland. Certification standards of these bodies can be more stringent than regulation 2092/91, but may not be less stringent. Three main types of studies – market purchase studies, paired farm surveys and field trials – have been used to compare nutritional or, less frequently, phytochemical content, between organic and conventionally grown fruits and vegetables. Market purchase studies require multiple sampling over extended time to compensate for variation due to seasonal, annual, handling and variety effects. Paired farm surveys can give information on varieties and treatments used in crop production but are reliant on a sufficient number of paired matched farm systems, whilst field trial studies can be difficult to design in such a way that they give statistically reliable data. A number of long-term field studies of organic agriculture have been set up (for review see Raupp, 2006).

A limited number of research studies have compared nutritional content in organic and conventionally grown vegetables rather than fruit, with very few examining phytochemical content (reviewed in Dangour, Dodhia, Hayter, Allen, Lock and Uauy, 2009; Zhao, Carey, Wang and Rajashekar, 2006). In general the evidence suggests little difference in the nutritional content of organically cultivated crops with the exception that levels of nitrates are lower and levels of vitamin C and dry matter content may higher than in conventionally grown crops (Brandt and Molgaard, 2001; Rembialkowska, 2007; Williams, 2002; Woese, Lange, Boess and Bogl, 1997). Some reports suggest increased levels of phytochemicals in organically grown crops (Young, Zhao, Carey, Welti, Yang and Wang, 2005; Zhao *et al.*, 2006) and some authors have suggested that phytochemicals which can be considered as defence related secondary metabolites could be considerably higher in organic vegetables (Brandt *et al.*, 2001). Several studies have evaluated antioxidant levels rather than measuring individual phytochemicals (reviewed in Benbrook, 2005). It is often unclear to what extent reported differences may be due to factors such as low nitrogen, use of disease resistant cultivar types or increased pest damage in organic systems. In addition crops cultivated using organic production methods typically have significantly lower yield than conventional counterparts, with average yield reductions of up to 20% (Rembialkowska, 2007). In an investigation of polyphenolic content, antioxidant activity and anti-mutagenic activity of five green vegetables – Chinese cabbage (*Brassica rapa* subsp. *pekinensis*), spinach (*Spinacia oleracea*), Welsh onion (*A. fistulosum*), green pepper (*Capiscum annuum* var. *annuum*) and the Japanese vegetable 'qing-gen-cai' the antioxidant activity, anti-mutagenic activity and composition of flavonoids including quercetin were higher in the organically cultivated vegetables (Ren, Endo and Hayashi, 2001). In a study by Young *et al.* (2005) leaf lettuce (*Lactuca sativa*), collard greens (*Brassica oleracea* var. *viridis* cv. Top Bunch) and Pak-choi (*Brassica rapa* var. *chinensis* cv. Mei Qing) were cultivated on adjacent plots, and levels of individual and total phenolics were quantified. In this study levels of kaempferol-3-O-glucoside were significantly higher in organically cultivated collard greens, but levels of other phenolics and total phenolic content were not significantly different in collards or leaf lettuce. In the case of Pak-choi, levels of total phenolics were significantly higher in organically cultivated plants; however the authors attribute this to a greater damage to the organic plants by flea beetle (Young *et al.*, 2005). A market purchase study by Meyer and Adam (2008) found significant differences in glucosinolate content between organic and conventional broccoli and red cabbage, with higher levels of glucobrassicin and neoglucobrassicin in organic samples. No significant differences in glucoraphanin content were found, however gluconapin was present at lower levels in organic red cabbage (Meyer and Adam, 2008). In an analysis of polyacetylenes in carrot (cv. Bolero) grown under one conventional and two organic treatments as part of the Danish VegQure rotation experiment no difference in levels of falcarinol were found over a two year field trial. In this study levels of applied nutrients were 120, 18 and 58 kg/ha of N, P and K for the conventional treatment and either green manure or 54, 4 and 20 kg/ha of N, P and K for the organic treatments (Soltoft *et al.*). The recent meta-analysis of organic foods by Dangour *et al.* (2009) found that organically produced crops had a significantly higher content of phosphorus and higher titratable acidity, whilst conventionally cultivated crops had a significantly higher content of nitrogen. No differences were found in levels of vitamin C, soluble solids, magnesium, potassium, zinc, copper, calcium or in levels of phenolic compounds. Five rejection criteria were used in this meta-analysis: provision of a definition of organic production methods used including the name of the certification body; specification of the crop variety or livestock breed; a statement of the nutritionally relevant substance analysed; description of analytical methods used; and statement of methods used for statistical analysis. Fifty-five studies were included in their analysis – 24 field trials, 27 paired farm surveys and four market purchase studies. The analysis includes studies on phytochemical content in only two vegetable crops – the study of Young *et al.* (2005) and Meyer *et al.* (2008) described previously.

9.2.6 Other factors

An influence of soil type on phytochemical accumulation including glucosinolates and phenolic compounds is commonly mentioned anecdotally in the literature (Cartea *et al.*, 2008b; Gliszczynska-Swiglo *et al.*, 2007; Jones *et al.*, 2007; Kjellenberg *et al.*; Patil *et al.*, 1995b; Schonhof *et al.*, 1999) and in a study by Jones *et al.* (2007) higher levels of glucosinolates were found in broccoli florets of the cultivar Marathon grown in light clay soils as compared to those grown in sandy loam type soils (Jones *et al.*, 2007). However such observations are complicated to interpret as crops grown in different areas will also experience different climatic and other agronomic conditions.

Application of amino acid precursors of glucosinolates have been studied in *Brassicas*. Foliar fertilisation or leafstalk infusion of methionine (the precursor of alkenyl glucosinolates) resulted in an increase in total and individual glucosinolate content (Krumbein, 2010; Scheuner, Schmidt, Krumbein, Schonhof and Schreiner, 2005). Foliar application of elicitors in soybean was found to increase levels of isoflavones, however the response varied depending on cultivar and year (Al-Tawaha, Seguin, Smith and Beaulieu, 2006).

9.3 Harvest and post-harvest management practices

The effect of harvesting and on-farm post-harvest management practices on phytochemical content will depend on the crop and the impact of factors such as the degree of mechanical injury caused during harvest and transport, water loss and oxygen stress at wound sites, temperature at harvest and during storage; and on how these factors affect the synthesis, retention or breakdown of individual phytochemicals. Mechanical injury results in cellular disruption and can allow enzymes such as myrosinase (EC 3.2.1.147), peroxidases (EC 1.11.1.7) and polyphenol oxidase (EC 1.10.3.1) to come into contact with their substrates. Water loss and oxygen entry can trigger stress and defence responses including modulation of the phenylpropanoid pathway leading to altered expression of phenolic compounds. Lower temperatures would be expected to reduce enzyme activity as well as inhibit the growth of spoilage organisms. Harvest and post-harvest treatments of fruits and vegetables commonly rely on reducing injury, water loss and temperature, and have been largely designed to maintain visual appearance – for example, preventing yellowing of green produce due to chlorophyll breakdown, preventing browning due to oxidation of phenolic compounds by polyphenol oxidases and preventing loss of turgor (Jones, Faragher and Winkler, 2006; Yamauchi and Watada, 1993). The impact of harvest and storage techniques on phytochemicals has only recently begun to be explored. In general phenolic compounds are considered to be relatively stable at cool temperature storage. Anthocyanins especially in fruit can increase at temperatures above 1 °C but may be lost at high temperature, which may be associated with water loss (Mori, Goto-Yamamoto, Kitayama and Hashizume, 2007b). Low temperature can reduce loss of organo-sulphur compounds such as glucosinolates in *Brassicas* and cysteine sulfoxides in onion (Jones *et al.*, 2006). A number of crops including tea, coffee, cocoa, roots such as cassava, nuts and grains undergo specific post-harvest treatments including drying and fermentation which will affect phytochemical content. For example tea is produced from the leaves and buds of the tea plant (*Camelis sinesis*). Black, green and oolong tea are produced by different post-harvest treatment of the crop. Green tea is produced from leaves dried shortly after harvest and is high in catechins. For black tea production the leaves are allowed to wilt, sometimes crushed and bruised and allowed to fully oxidize, a process referred to as 'fermentation'. As a result the catechins are oxidized to theaflavins and a number of other compounds. Oolong tea is intermediate between green and black tea.

A detailed discussion of the effect of post-harvest treatments across different cultivated crop plants is beyond the scope of this chapter, which will focus specifically on post-harvest management of onion, broccoli and carrot.

9.3.1 Harvest and post-harvest management of onion

Bulb formation in long day onions grown in Northern Europe is initiated as the day length begins to shorten in mid-summer and mature bulbs are mechanically harvested at 50–100% leaf fall-down. In the UK and Ireland commercially grown onion bulbs are generally mechanically harvested in late August to mid-September from a March sowing and are cured by forced air drying at 25–28 °C and 65–75% RH for ten days to six weeks. Commonly grown varieties include the yellow (brown) variety Hyskin and the red variety Red Baron. Curing seals the neck of the onion and forms a dry outer skin, which reduces moisture loss and disease. Forced air curing reduces the incidence of neck rot caused by *Botrytis allii* and bacterial soft rots caused by *Erwinia* and *Pseudemonas* species. In addition the dried outer skins can be easily removed by mechanical cleaning after curing, resulting in a cleaner and darker skin finish demanded by consumers (Cho, Bae and Lee, 2010; Downes, Chope and Terry, 2009). Subsequently, onions may be kept in cold storage at around 1–4 °C in the dark to induce dormancy and prevent sprouting, however sprouting commonly initiates within one to three weeks after removal from cold storage (Sorensen and Grevsen, 2001). Maleic hydrazide (Fazor) can be used to prevent sprouting in bulb onions by prolonging natural dormancy and is applied to field onions a week before harvest. Traditionally and in hot dry climates, onions can be left to cure in the field in windrows or mesh bags and this has been reported to increase quercetin content (Mogren, Olsson and Gertsson, 2006; Olsson, Gustavsson and Vagen, 2010; Patil *et al.*, 1995b). The effect of curing temperature on flavonols and anthocyanin content in brown (yellow) and red onion skin has been investigated (Downes *et al.*, 2009). Two brown (yellow) (cv.s Wellington and Sherpa) and a red onion (cv. Red Baron) were cured at 20 °C, 24 °C or 28 °C for six weeks followed by cold storage at 1 °C for seven months. Samples were analysed immediately after curing and at seven months after storage. In this study quercetin levels in the skin were not affected by curing temperature but levels of quercetin 4-glucoside, quercetin 3,4-diglucoside and anthocyanins were significantly higher in cv. Red Baron cured at 20 °C. In a study of different storage methods on onion, bulbs stored at 5 °C, 24 °C and 30 °C for up to five months showed an initial rise in total quercetin levels followed by a decline. Changes were most pronounced under the 24 °C treatment. Onion bulbs stored under controlled atmosphere did not show significant changes in quercetin content over the five month storage period (Patil *et al.*, 1995c). In a study on two onion varieties (cv.s Red Baron and Crossbow) cured at 28 °C for ten days and stored for six months at ≤4 °C an initial drop in the level of quercetin monoglucosides (which the authors attribute to removal of the outer dry skin) occurred and thereafter there was little change in levels of quercetin monoglucoside or quercetin 3,4'-O-diglucoside (Price, Bacon and Rhodes, 1997). Total anthocyanins are reported to decrease in red onion (cv. Tropea) during storage, with higher levels of loss of anthocyanins at higher temperature (Gennaro *et al.*, 2002). During onion storage the enzyme alliinase (S-alk(en)yl-L-cysteine sulfoxide lyase, E.C.4.4.1.4) catalyses the breakdown of cysteine sulfoxides into the flavour compounds pyruvate, ammonia and volatile suphur compounds. Levels of pyruvate, allinase activity and cysteine sulfoxides in onion bulbs (cv. Hysam) increased during storage at 0.5 °C over nine weeks under normal atmosphere conditions (Uddin and MacTavish, 2003). A slight decline in phenolic content in onion at the end of cold storage has been noted in some studies (Benkeblia, 2000; Price *et al.*, 1997) and there is an inverse relationship between total phenolic content and sprout development (Benkeblia, 2000). The effect of a post curing heat treatment (36 °C for 24 or 96 h) on onion flavonols has recently been investigated as a method of increasing shelf life (Olsson *et al.*, 2010). Three onion varieties, Recorra, Hyred and Red Baron, were cured at RT or in the field for two weeks and then heat treated prior to cold storage at 2 °C for up to eight months. Neither storage nor heat treatment had a significant effect on total flavonoid content, however levels of quercetin 3,4-diglucoside increased in the 24 hour heat treated cv.s Red Baron and Hyred. A lower content of total flavonols was found in all varieties after eight months of cold storage following the 96 hour heat treatment and the authors suggest this may be due to negative effects of heat treatment on onion metabolism. UV Irradiation is currently used as a post-harvest treatment in several products for sterilisation, and to inhibit sprouting and delay maturity. A number of studies have indicated that post-harvest UV irradiation can increase the levels of α-tocopherol and flavonoids in several fruits and vegetables including onion

(Higashio, Hirokane, Sato, Tokuda and Uragami, 2007; Patil, 2004; Rodov, Tietel, Vinokur, Horev and Eshel). In onion, short wave UV irradiation was shown to significantly increase levels of both free and total quercetin, and could reduce incidence of spoilage moulds including *Penicillium allii* and survival of human pathogens such as *Escherichiae coli* (Higashio, Hirokane, Sato, Tokuda and Uragami, 2005; Patil, 2004; Rodov *et al.*, 2010).

9.3.2 Harvest and post-harvest management of broccoli

In Ireland and the UK commercially grown green broccoli is normally produced using modular transplants which can be sown from mid-February to June and transplanted in the field from mid-April to late July. Florets are harvested by hand when the florets are 250–600 g with tight unopened flowers. To increase shelf life the crop is cooled to below 6 °C within 12 hours and kept at holding temperatures of 3–5 °C and high humidity. Commonly grown varieties in Ireland are cv.s. Ironman, Steel, Parthenon, Manaco, and Monterey. The variety Marathon was widely grown in several countries including Ireland but has largely been superseded by newer varieties.

Broccoli is normally harvested in the early morning to allow time for processing and packing on the same day, however a recent study has indicated that evening harvest could better maintain quality and may affect phytochemical content. In this study broccoli florets (cv. Iron) were harvested at 8 am, 1 pm and 6 pm and quality parameters as well as levels of total phenolics and antioxidant capacity were measured over five days at 20 °C storage. Chlorophyll loss was significantly accelerated in florets harvested at 8 am. Levels of total phenolics and antioxidant capacity were significantly lower in 8 am harvested florets on day three of storage, but differences were not significant on other days (Hasperue, Chaves and Martinez). Several studies (Howard, Jeffery, Wallig and Klein, 1997; Jones *et al.*, 2006; Leja, Mareczek, Starzynska and Rozek, 2001; Rangkadilok *et al.*, 2002; Rodrigues *et al.*, 1999; Winkler, Faragher, Franz, Imsic and Jones, 2007) have examined the effect of post-harvest handling and storage conditions on glucosinolate and/or phenolic compounds in broccoli. Levels of glucoraphanin, quercetin and kaempferol in broccoli cv. Marathon were not significantly affected by post-harvest storage treatments designed to simulate commercial storage and marketing. Florets were stored at 1–4 °C at 99% relative humidity (RH) for 2–28 days to simulate initial storage and transport conditions, and were then kept at 8–20 °C and 70–99% RH in order to simulate marketing conditions (Winkler *et al.*, 2007). Storage of both primary and secondary broccoli florets (cv. Tokyodome) at either room temperature (~20 °C) or at 4 °C for five days showed that higher temperature storage led to a significant reduction in total and individual glucosinolates although levels of hydroxyglucobrassicin and gluconasturtiin increased (Rodrigues *et al.*, 1999). Under refrigerated (4 °C) storage the decrease in total glucosinolates was considerably lower at 16 and 4% for primary and secondary inflorescences respectively. Levels of glucoraphanin declined by 82 and 89% in primary and secondary inflorescences under the RT storage treatment, and by 31 and 10% under the refrigeration treatment (Rodrigues *et al.*, 1999). In a similar study broccoli florets (cv. Marathon) were stored at either 4 °C or 20 °C in open boxes or in plastic bags. Storage at 20 °C in both systems caused a significant decrease in glucoraphanin by day seven, although the decline was more rapid in the open box system. In broccoli stored in open boxes a 55% loss of glucoraphanin was observed by day three, in broccoli stored in bags a 56% loss was observed by day seven. At 4 °C little decrease in glucoraphanin content was observed for either system (Rangkadilok *et al.*, 2002). Levels of sulforaphane in florets of broccoli (cv. Arcadia) have been determined over a 21 day storage period at 4 °C (Howard *et al.*, 1997). Levels of sulforaphane measured in fresh broccoli samples were 36.7–49.4 mg/100 g depending on year of harvest. After five days of storage at 4 °C levels of sulforaphane had declined by 33% and by 21 days of storage levels had declined by 49–55%. Levels of total phenolic compounds, flavonoids and antioxidant capacity have been reported to increase during storage of broccoli florets under both 5 °C and 20 °C storage, with changes in ROS scavenging enzymes also reported (Leja et al., 2001). Research to date indicates that storage factors currently used to maintain visual appearance and nutritional quality in broccoli, that is, low temperature and/or high RH, can maintain reasonable levels of glucosinolates and other bioactive compounds such as phenolic compounds. In a review of post-harvest treatments on glucosinolate content in broccoli the authors suggest that if broccoli is stored at 4 °C there is little benefit in maintaining high RH, however where broccoli is stored at room temperature high RH should be maintained by use of packaging in order to maintain both glucosinolates levels and visual appearance (Jones et al., 2006).

9.3.3 Harvest and post-harvest management of carrot

Commercial harvesting practices for carrot commonly include mechanical lifting, topping to remove the leaves, and grading followed by brushing, tumble-washing and hydro-cooling. Storability is improved at low temperature and high RH. Main-crop carrots in the UK and Ireland are generally harvested in October and November when the carrot roots are fully mature. The variety cv. Nairobi is widely grown in the UK and Ireland. Mechanical harvesters may be either 'top lifters', which lift the crop by the foliage, or 'share lifters', which run in the soil lifting the crop that is then separated from the soil by mechanical shaking and sieving. The mechanical force of harvesting and transport operations and severing of the foliage would be expected to induce plant wound and stress responses and consequent increases in phenolic compounds.

A detailed study on the effect of hand or machine harvesting and simulated transport on five carrot varieties (cv.s. Bolero, Panter, Yukon, Napa and Newburg) showed increased mechanical stress led to increased respiration and increased ethylene synthesis. Levels of the phytoalexin phenolic compound 6-methoxymellein (6MM) were increased in response to ethylene, as were other phenolic compounds such as chlorogenic and isochlorogenic acid, whilst sugars decreased (Heredia et al., 2009; and references therein). In this study machine harvesting did not induce significant changes compared to hand harvesting, however the severity of post-harvest handling had a significant effect (Heredia et al., 2009). Increased respiration and increased levels of phenolic compound accumulation in carrots related to the severity of post-harvest handling, storage and processing treatments have been widely reported (Barry-Ryan and O'Beirne, 2000; Barry-Ryan, Pacussi and O'Beirne, 2000; Kenny and O'Beirne, 2010; Ruiz-Cruz, Islas-Osuna, Sotelo-Mundo, Vazquez-Ortiz and Gonzalez-Aguilar, 2007), however such responses can be slowed by low temperature storage (Hager and Howard, 2006). Increases in phenolic compounds are associated with increased antioxidant potential, however oxidation of phenolic compounds can result in undesirable browning during storage in carrot and other crops; and accumulation of certain phenolic compounds such as isocoumarins can result in bitter flavour (Hager et al., 2006; Heredia et al., 2009; Ruiz-Cruz et al., 2007). In a recent study the effect of wounding intensity, methyl jasmonate and ethylene treatment on accumulation of total and individual phenolic compounds, antioxidant capacity and PAL enzyme activity in carrot (cv. Choctaw) were examined (Heredia et al., 2009). The relative proportions of chlorogenic acid, dicaffeoyl-quinic acid, ferulic acid, isocoumarins and antioxidant capacity differed under different stress combinations and the authors suggest that environmental modification could be used to enhance the phenolic profile in stored and processed carrots.

Few studies have been carried out to evaluate the effect of storage on carrot polyacetylenes (Hansen *et al.*, 2003; Kidmose *et al.*, 2004; Kjellenberg *et al.*). In a study in which raw carrots (cvs. Bolero, Rodelika and Fancy) were stored at 1 °C and 98% RH falcarinol content was initially in the range 22.3–24.8 mg/kg. Levels were largely unchanged during the first month but subsequently declined by nearly 35% after 120 days storage, which the authors attribute to a change in the balance between biosynthesis and degradation (Hansen *et al.*, 2003). However in a study in which polyacetylene content in carrot roots of six Nantes cultivars (cv.s. Bolero, Fancy, Duke, Express, Line 1 and Cortez) stored at 1 °C for four months was evaluated levels of falcarinol, falcarindiol and falcarindiol-3-acetate increased significantly during storage. (Kidmose *et al.*, 2004). In the most recent study (Kjellenberg *et al.*) levels of falcarinol, falcarindiol and falcarindiol-3-acetate in roots of two carrot varieties (cv.s Kampe and Bolero) were reported to stabilize during storage with an increase noted in samples that were initially low and a decrease in samples initially high in polyacetylenes (Kjellenberg *et al.*).

9.4 Future prospects

Fruits and vegetables are already 'functional foods', however there is considerable potential to increase their health promoting effects by nudging plant metabolism towards increased synthesis and retention of particular phytochemicals during cultivation and storage; by promoting consumption of plant groups and plant tissues known to be rich in important phytochemicals; and in particular by identifying and/or breeding varieties high in beneficial phytochemicals. Novel uses of crop plants will include use of the crop itself or of crop wastes as functional products or ingredients.

9.4.1 Growing bio-fortified crops – optimized agronomic and post-harvest practices

As discussed previously a number of studies have examined methods to modify the growing environment to produce optimum levels of phytochemicals. This approach uses crop and bioactive specific research and knowlege to identify the key points in production at which levels of the phytochemical of interest can be enhanced or retained. Key issues are feasibility of extrapolating lab based studies to a production scale, and economic feasibility – how much would it cost the producer, and how much would consumers be willing to pay. In Australia and New Zealand a 'Vital Vegetable' project has developed optimized varieties, cultivation and storage procedures for enhanced phytochemical content in vegetable crops and the first product, a high sulforophane broccoli called Booster™ is now available commercially in Australia and New Zealand. In Ireland producers Keogh and sons have developed a brand of selenium enriched potatoes marketed as 'Selena potatoes' (PotatoPro, 2009). Greenhouse grown crops such as tomatoes, strawberries and herbs are high value crops where inputs are more easily controlled and offer considerable potential in this regard.

9.4.2 Edible sprouts

Edible sprouts represent an excellent opportunity to develop phytochemically enriched foods either as a 'ready to eat' food or as a source of functional food ingredients. There is huge potential to optimize the variety used and to manipulate the sprouting conditions (light, temperature, stress treatments) in order to produce sprouted seeds which are high in bioactives of interest. As discussed earlier, considerable research to date has focused on optimizing bioactive content in sprouted seeds of *Brassica* species and soybean and these perhaps represent the most likely candidates for development of optimized commercial products. In the USA a number of products based on glucoraphanin and/or sulforaphane enhanced broccoli sprouts have been developed and patented including BroccoSprouts[R], Brassic[R] tea with SGS™ ('sulforaphane glucosinolate') and a supplement Xymogen Oncoplex SGS (BPP).

9.4.3 Variety screening and plant breeding for bio-fortified crops

It is clear that genetic factors play a major role in controlling phytochemical content. Given the wide variation in phytochemical profiles between different varieties in most crop plants studied to date there is considerable potential to increase levels of key bioactive compounds by a) identifying existing varieties which contain higher levels of phytochemicals of interest and b) using plant breeding approaches. Most modern crop varieties have been extensively bred for specific traits such as yield or quality and contain only a small percentage of the genetic diversity available in the wider genepool. Older 'heritage' varieties and seed bank accessions could be a valuable resource for breeding crops with enhanced phytochemical levels. Breeding approaches have been used to increase levels of flavonoids in onion and carotenes and anthocyanins in carrot (Crosby, Jifon, Pike and Yoo, 2007; Kim, Binzel, Yoo, Park and Pike, 2004; Murthy, Jayaprakasha, Pike and Patil, 2007), and there is considerable potential to breed phytochemical enhanced *Brassica* crops.

Ancestral cross pollination and hybridization between the six major *Brassica* groups is described by the 'Triangle of U' (U, 1935) as represented in Figure 9.2. Combination of the genomes of the three diploid species *B.rapa* (A genome, n=10), *B.nigra* (B genome, n=8) and *B.oleracea* (C genome, n=9) gave rise to the allotetraploid species *B.juncea* (AB genome, n=18), *B.napus* (AC genome, n=19) and *B.carinata* (BC genome, n=17). Within each species cross pollination and the formation of fertile offspring is relatively common (SUTTON, 1908). For example broccoli (*B.oleracea* var. *italica*) and cauliflower (*B.oleracea* var. *botrytis*) are fully cross compatible and can be readily cross pollinated to produce intermediate forms (Malatesta, 1996), including Romanesco and Tenderstem types. Hybridisation between *Brassica* species is less common in nature but techniques such as embryo rescue and somatic hybridization can be used to assist traditional breeding approaches and enable transfer of novel alleles into elite lines (Allender and King, 2010).

Although precise comparisons between wild and cultivated broccoli species is complicated by differences in floret morphology some authors have estimated that florets of cultivated broccoli lines contain 3–10 µmol g^{-1} DW of glucosinolates whilst wild species contain 50–100 µmol g^{-1} DW glucosinolates (Mithen *et al.*, 2003). The development of hybrid broccoli varieties with higher levels of glucoraphanin, the precursor of the sulforaphane, by introgression of a wild ancestor *Brassica villosa* has been described (Faulkner, Mithen and Williamson, 1998; Mithen *et al.*, 2003; Sarikamis, Marquez, MacCormack, Bennett, Roberts and Mithen, 2006). Following mild cooking the high glucosinolate broccoli lines produced three-fold higher levels of sulforaphane than conventional varieties (Gasper *et al.*, 2007; Mithen *et al.*, 2003). Hybrid broccoli lines have been licensed to a commercial company for development (R. Mithen, personal communication) and are expected to be marketed in the USA and UK as 'Beneforte' broccoli (Dixon, 2011).

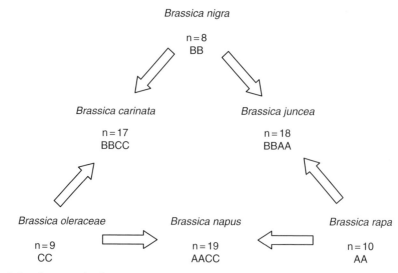

Figure 9.2 The 'Triangle of U'.
Relationship between diploid and allotetraploid Brassica species. Hybridization between ancestral members with A, B or C genomes gave rise to tetraploid species with four genomes, two from each parent. 'n' is the haploid chromosome number, i.e. the number of chromosomes present in pollen or ovule.

As noted earlier the epithiospecifier protein (ESP), together with ferrous iron, plays an important role in directing hydrolysis of glucosinolates towards nitrile rather than isothiocyanate formation. In a study in *Arabidiopsi thaliana*, enhanced nitrile production was found in transgenic plants over-expressing ESP compared to wild type plants (Zabala et al., 2005). The ESP gene from broccoli has been cloned and the recombinant protein expressed in *E.coli* (Matusheski et al., 2006). A polyclonal antibody to the recombinant protein was used to examine ESP expression in members of the *Brassicaceae*, and reactive bands (indicating ESP activity) were found in broccoli and cabbage, but not in daikon or horseradish. Since daikon and horseradish produce isothiocyanates only and do not produce nitriles as breakdown products of glucosinolates the clear implication is that in these crops ESP is either not present or inactive. The authors also examined ESP activity in floret tissue of 20 commercial broccoli varieties using an assay based on hydrolysis of epiprogoitrin. There was a considerable variation across varieties with levels of ESP activity ranging from 17.1 to 46 (expressed as mole percentage of epithionitrile formed from epiprogoitrin) (Matusheski et al., 2006). The wide variability in ESP activity levels in broccoli varieties suggests that it may be possible to develop broccoli lines with reduced ESP activity and thus enhanced potential for sulforaphane production by traditional breeding approaches.

9.4.4 Novel uses for crops and crop wastes

Plant tissues that are currently discarded during harvest or processing represent a significant source of bioactive compounds. For many *Brassica* crops including broccoli, cauliflower and Brussel's sprouts less than 50% of the biomass is used for human consumption with the remainder discarded, re-incorporated into the soil or used for fodder (Rosa et al., 1999). Given the health promoting and antioxidant properties of many crop wastes, numerous studies have investigated the potential for production of novel products (e.g. juices) or of functional ingredients from these waste streams (Desjardins, 2008; Dominguez-Perles *et al.*; Makris, Boskou and Andrikopoulos, 2007; Morra and Borek; Roldan, Sanchez-Moreno, de Ancos and Cano, 2008; Wijngaard, Rossle and Brunton, 2009). In some instances a clear beneficial effect of a plant bioactive is demonstrated but levels generally consumed in the diet are too low to have an effect. For example plant sterols and stanols are present in nuts, seeds, grains and avocados, but large volumes would need to be consumed to exert a positive effect on blood cholesterol levels. This has led to the development of a number of novel functional food products including spreads, drinks and yoghurts which contain complete servings of plant sterols derived from crops including sunflower, soya and pine oil. (Lazzeri, Leoni and Manici, 2004). *Brassica* products have also been investigated for their herbicidal, antimicrobial and insecticidal activity (Morra *et al.*, 2010). One potential use of *Brassica* wastes based on the antimicrobial properties of glucosinolates and their breakdown products has been as a biofumigation agent for control of soil borne pathogens as an alternative to methyl bromide soil fumigation. Approaches using both dried and fresh material, and the use of *Brassica* species as green manures and break crops have been investigated in a number of studies (reviewed in Conaway *et al.*, 2005). Commercial products have been developed, for example a commercial green manure BQ Mulch™ consisting of a mixture of *Brassica* species has been developed and marketed in New Zealand for control of soil nematodes and soil pathogens such as P*hytophthora* and *Pythium* (Marsh and Du L.C., 2007; Stirling and Stirling, 2003) and there are approaches to develop *Brassica* derived biocidal dried plant pellets for biofumigation

References

Abdel-Farid, I.B., Jahangir, M., Mustafa, N.R., van Dam, N.M., van den Hondel, C., Kim, H.K., Choi, Y.H., and Verpoorte, R. (2010) Glucosinolate profiling of Brassica rapa cultivars after infection by Leptosphaeria maculans and Fusarium oxysporum. *Biochemical Systematics and Ecology*, 38(4), 612–620.

Aires, A., Rosa, E., and Carvalho, R. (2006) Effect of nitrogen and sulfur fertilization on glucosinolates in the leaves and roots of broccoli sprouts (Brassica oleracea var. italica). *Journal of the Science of Food and Agriculture*, 86(10), 1512–1516.

Al-Tawaha, A.M., Seguin, P., Smith, D.L., and Beaulieu, C. (2006) Foliar application of elicitors alters isoflavone concentrations and other seed characteristics of field-grown soybean. *Canadian Journal of Plant Science*, 86(3), 677–684.

Alasalvar, C., Grigor, J.M., Zhang, D.L., Quantick, P.C., and Shahidi, F. (2001) Comparison of volatiles, phenolics, sugars, antioxidant vitamins, and sensory quality of different colored carrot varieties. *Journal of Agricultural and Food Chemistry*, 49(3), 1410–1416.

Allender, C.J., and King, G.J. (2010) Origins of the amphiploid species Brassica napus L. investigated by chloroplast and nuclear molecular markers. *BMC Plant Biology*, 10, 54.

Andreasson, E., Jorgensen, L.B., Hoglund, A.-S., Rask, L., and Meijer, J. (2001) Different Myrosinase and Idioblast Distribution in Arabidopsis and Brassica napus. *Plant Physiology*, 127(4), 1750–1763.

Arnault, I., and Auger, J. (2006) Seleno-compounds in garlic and onion. *Journal of Chromatography A*, 1112(1–2), 23–30.

Baranska, M., and Schulz, H. (2005) Spatial tissue distribution of polyacetylenes in carrot root. *Analyst*, 130(6), 855–859.

Barry-Ryan, C., and O'Beirne, D. (2000) Effects of peeling methods on the quality of ready-to-use carrot slices. *International Journal of Food Science and Technology*, 35(2), 243–254.

Barry-Ryan, C., Pacussi, J. M., and O'Beirne, D. (2000) Quality of shredded carrots as affected by packaging film and storage temperature. *Journal of Food Science*, 65(4), 726–730.

Beesk, N., Perner, H., Schwarz, D., George, E., Kroh, L.W., and Rohn, S. (2010) Distribution of quercetin-3,4'-O-diglucoside, quercetin-4 '-O-monoglucoside, and quercetin in different parts of the onion bulb (Allium cepa L.) influenced by genotype. *Food Chemistry*, 122(3), 566–571.

Benbrook, C. (2005) Elevating antioxidant levels in food through organic farming and food proccessing: an organic centre state of science review. http://www.organic-center.org/reportfiles/Antioxidant_SSR.pdf.

Benkeblia, N. (2000) Phenylalanine ammonia-lyase, peroxidase, pyruvic acid and total phenolics variations in onion bulbs during long-term storage. *Lebensmittel-Wissenschaft Und-Technologie-Food Science and Technology*, 33(2), 112–116.

BPP, Brassica Protection Products LLC. http://www.brassica.com/.

Brandt, K., Christensen, L.P., Hansen-Moller, J., Hansen, S.L., Haraldsdottir, J., Jespersen, L., Purup, S., Kharazmi, A., Barkholt, V., Frokiaer, H., and Kobaek-Larsen, M. (2004) Health promoting compounds in vegetables and fruits: A systematic approach for identifying plant components with impact on human health. *Trends in Food Science and Technology*, 15(7–8), 384–393.

Brandt, K., and Molgaard, J.P. (2001) Organic agriculture: does it enhance or reduce the nutritional value of plant foods? *Journal of the Science of Food and Agriculture*, 81(9), 924–931.

Buttriss, A.D.a.J. (2005) *Synthesis Report No 4: Plant Foods and Health: Focus on Plant Bioactives*.

Carbone, K., Giannini, B., Picchi, V., Lo Scalzo, R., and Cecchini, F. (2011) Phenolic composition and free radical scavenging activity of different apple varieties in relation to the cultivar, tissue type and storage. *Food Chemistry*, 127(2), 493–500.

Carlson, D.G., Daxenbichler, M.E., Tookey, H.L., Kwolek, W.F., Hill, C.B., and Williams, P.H. (1987a) Glucosinolates in turnip tops and roots – cultivars grown for greens and or roots. *Journal of the American Society for Horticultural Science*, 112(1), 179–183.

Carlson, D.G., Daxenbichler, M.E., Vanetten, C.H., Hill, C.B., and Williams, P.H. (1985) Glucosinolates in radish cultivars. *Journal of the American Society for Horticultural Science*, 110(5), 634–638.

Carlson, D.G., Daxenbichler, M.E., Vanetten, C.H., Kwolek, W.F., and Williams, P.H. (1987b) glucosinolates in crucifer vegetables – broccoli, brussels-sprouts, cauliflower, collards, kale, mustard greens, and kohlrabi. *Journal of the American Society for Horticultural Science*, 112(1), 173–178.

Cartea, M.E., Rodriguez, V.M., de Haro, A., Velasco, P., and Ordas (2008a) Variation of glucosinolates and nutritional value in nabicol (Brassica napus pabularia group). *Euphytica*, 159(1–2), 111–122.

Cartea, M.E., Velasco, P., Obregon, S., Padilla, G., and de Haro, A. (2008b) Seasonal variation in glucosinolate content in Brassica oleracea crops grown in northwestern Spain. *Phytochemistry*, 69, 403–410.

Cataldi, T.R.I., Rubino, A., Lelario, F., and Bufo, S.A. (2007) Naturally occuring glucosinolates in plant extracts of rocket salad (Eruca sativa L.) identified by liquid chromatography coupled with negative ion electrospray ionization and quadrupole ion-trap mass spectrometry. *Rapid Communications in Mass Spectrometry*, 21(14), 2374–2388.

Charron, C.S., and Sams, C.E. (2004) Glucosinolate content and myrosinase activity in rapid-cycling Brassica oleracea grown in a controlled environment. *Journal of the American Society for Horticultural Science*, 129(3), 321–330.

Charron, C.S., Saxton, A.M., and Sams, C.E. (2005) Relationship of climate and genotype to seasonal variation in the glucosinolate-myrosinase system. I. Glucosinolate content in ten cultivars of Brassica oleracea grown in fall and spring seasons. *Journal of the Science of Food and Agriculture*, 85(4), 671–681.

Cho, J., Bae, R.N., and Lee, S.K. (2010) Current Research Status of Postharvest Technology of Onion (Allium cepa L.). *Korean Journal of Horticultural Science and Technology*, 28(3), 522–527.

Christensen, L.P., and Brandt, K. (2006) Bioactive polyacetylenes in food plants of the Apiaceae family: Occurrence, bioactivity and analysis. *Journal of Pharmaceutical and Biomedical Analysis*, 41(3), 683–693.

Christensen, L.P., and Kreutzmann, S. (2007) Determination of polyacetylenes in carrot roots (Daucus carota L.) by high-performance liquid chromatography coupled with diode array detection. *Journal of Separation Science*, 30(4), 483–490.

Chu, Y.H., Chang, C.L., and Hsu, H.F. (2000) Flavonoid content of several vegetables and their antioxidant activity. *Journal of the Science of Food and Agriculture*, 80(5), 561–566.

Conaway, C.C., Wang, C.X., Pittman, B., Yang, Y.M., Schwartz, J.E., Tian, D.F., McIntee, E.J., Hecht, S.S., and Chung, F.L. (2005) Phenethyl isothiocyanate and sulforaphane and their N-acetylcysteine conjugates inhibit malignant progression of lung adenomas induced by tobacco carcinogens in A/J mice. *Cancer Research*, 65(18), 8548–8557.

Crosby, K., Jifon, J., Pike, L., and Yoo, K.S. (2007) Breeding vegetables for optimum levels of phytochemicals. *Proceedings of the 1st International Symposium on Human Health Effects of Fruits and Vegetables*, 219–224.

Czepa, A., and Hofmann, T. (2004) Quantitative studies and sensory analyses on the influence of cultivar, spatial tissue distribution, and industrial processing on the bitter off-taste of carrots (Daucus carota L.) and carrot products. *Journal of Agricultural and Food Chemistry*, 52(14), 4508–4514.

Dangour, A.D., Dodhia, S.K., Hayter, A., Allen, E., Lock, K., and Uauy, R. (2009) Nutritional quality of organic foods: a systematic review. *American Journal of Clinical Nutrition*, 90(3), 680–685.

Davik, J., and Heneen, W.K. (1993) Identification of oilseed turnip (brassica-rapa l var oleifera) cultivar groups by their fatty-acid and glucosinolate profiles. *Journal of the Science of Food and Agriculture*, 63(4), 385–390.

Davis, D.R., Epp, M.D., and Riordan, H.D. (2004) Changes in USDA food composition data for 43 garden crops, 1950 to 1999. *Journal of the American College of Nutrition*, 23(6), 669–682.

Degen, T., Buser, H.R., and Stadler, E. (1999) Patterns of oviposition stimulants for carrot fly in leaves of various host plants. *Journal of Chemical Ecology*, 25(1), 67–87.

Desjardins, Y. (2008) Onion as a Nutraceutical and Functional Food. *Chronica Horticulturae*, 48(2), 8–14.

Dixon, G. (2011) Given the health benefits – why is the vegetable market not booming. *The Vegetable Farmer*, May 2011, 30–31.

Dominguez-Perles, R., Martinez-Ballesta, M.C., Carvajal, M., Garcia-Viguera, C., and Moreno, D.A. (2010) Broccoli-Derived By-Products-A Promising Source of Bioactive Ingredients. *Journal of Food Science*, 75(4), C383–C392.

Dorais, M. (2007) Effect of cultural management on tomato fruit health qualities. *Proceedings of the 1st International Symposium on Human Health Effects of Fruits and Vegetables*, 279–293.

Downes, K., Chope, G.A., and Terry, L.A. (2009) Effect of curing at different temperatures on biochemical composition of onion (Allium cepa L.) skin from three freshly cured and cold stored UK-grown onion cultivars. *Postharvest Biology and Technology*, 54(2), 80–86.

EU (1991) Council Regulation No. 2092/91 of 24th June on organic production of agricultural producrs and indications referring thereto on agricultural products and foodstuffs. *OJL* 198 22.7.p.1.

Fahey, J.W., Zalcmann, A.T., and Talalay, P. (2002) The chemical diversity and distribution of glucosinolates and isothiocyanates among plants (vol 56, pg 5, 2001). *Phytochemistry*, 59(2), 237–237.

FAO (2007) FAOstat database http://faostat.fao.org/.

Faulkner, K., Mithen, R., and Williamson, G. (1998) Selective increase of the potential anticarcinogen 4-methylsulphinylbutyl glucosinolate in broccoli. *Carcinogenesis*, 19(4), 605–609.

Fenwick, G.R., and Heaney, R.K. (1983) glucosinolates and their breakdown products in cruciferous crops, foods and feedingstuffs. *Food Chemistry*, 11(4), 249–271.

Finley, J.W., Sigrid-Keck, A., Robbins, R.J., and Hintze, K.J. (2005) Selenium enrichment of broccoli: Interactions between selenium and secondary plant compounds. *Journal of Nutrition*, 135(5), 1236–1238.

Forney, C.F., Jordan, M.A., and Best., K., (2010) Sulphur fertilization affects onion quality and flavour chemistry during storage. *Acta Horticulturae*, 877, 163–168.

Fortier, E., Desjardins, Y., Tremblay, N., Belec, C., and Cote, M. (2010) Influence of Irrigation and Nitrogen Fertilization on Broccoli Phenolics Concentration. *Acta Horticulturae*, 856, 55–61.

Gasper, A., Traka, M., Bacon, J.R., Smith, J.A., Taylor, M.A., Hawkey, C.J., Barrett, D.A., and Mithen, R.F. (2007) Consuming broccoli does not induce genes associated with xenobiotic metabolism and cell cycle control in human gastric mucosa. *Journal of Nutrition*, 137(7), 1718–1724.

Gennaro, L., Leonardi, C., Esposito, F., Salucci, M., Maiani, G., Quaglia, G., and Fogliano, V. (2002) Flavonoid and carbohydrate contents in Tropea red onions: Effects of homelike peeling and storage. *Journal of Agricultural and Food Chemistry*, 50(7), 1904–1910.

Girennavar, B., Jayaprakasha, G.K., Jifon, J.L., and Patil, B.S. (2008) Variation of bioactive furocoumarins and flavonoids in different varieties of grapefruits and pummelo. *European Food Research and Technology*, 226(6), 1269–1275.

Gliszczynska-Swiglo, A., Kaluzewicz, A., Lemanska, K., Knaflewski, M., and Tyrakowska, B. (2007) The effect of solar radiation on the flavonol content in broccoli inflorescence. *Food Chemistry*, 100(1), 241–245.

Goldman, I.L., Kader, A.A., and Heintz, C. (1999) Influence of production, handling, and storage on phytonutrient content of foods. *Nutrition Reviews*, 57(9), S46–S52.

Gray, A.R. (1982) taxonomy and evolution of broccoli (brassica-oleracea var italica). *Economic Botany*, 36(4), 397–410.

Guler, M. (2003) Barley grain beta-glucan content as affected by nitrogen and irrigation. *Field Crops Research*, 84(3), 335–340.

Hager, T.J., and Howard, L.R. (2006) Processing effects on carrot phytonutrients. *Hortscience*, 41(1), 74–79.

Halkier, B.A., and Du, L.C. (1997) The biosynthesis of glucosinolates. *Trends in Plant Science*, 2(11), 425–431.

Hansen, S.L., Purup, S., and Christensen, L.P. (2003) Bioactivity of falcarinol and the influence of processing and storage on its content in carrots (Daucus carota L). *Journal of the Science of Food and Agriculture*, 83(10), 1010–1017.

Hasperue, J.H., Chaves, A.R., and Martinez, G.A. (2011) End of day harvest delays postharvest senescence of broccoli florets. *Postharvest Biology and Technology*, 59(1), 64–70.

Heredia, J.B., and Cisneros-Zevallos, L. (2009) The effect of exogenous ethylene and methyl jasmonate on pal activity, phenolic profiles and antioxidant capacity of carrots (Daucus carota) under different wounding intensities. *Postharvest Biology and Technology*, 51(2), 242–249.

Higashio, H., Hirokane, H., Sato, F., Tokuda, S., and Uragami, A. (2005) Effect of UV irradiation after the harvest on the content of flavonoid in vegetables. *Proceedings of the 5th International Postharvest Symposium*, Vols 1–3(682), 1007–1012.

Higashio, H., Hirokane, H., Sato, F., Tokuda, S., and Uragami, A. (2007) Enhancement of functional compounds in Allium vegetahles with UV radiation. *Proceedings of the 1st International Symposium on Human Health Effects of Fruits and Vegetables*, 357–361.

Howard, L.A., Jeffery, E.H., Wallig, M.A., and Klein, B.P. (1997) Retention of phytochemicals in fresh and processed broccoli. *Journal of Food Science*, 62(6), 1098.

Hu, K.L., and Zhu, Z.J. (2010) Effects of different concentrations of sodium chloride on plant growth and glucosinolate content and composition in pakchoi. *African Journal of Biotechnology*, 9(28), 4428–4433.

Irion, C.W. (1999) Growing alliums and brassicas in selenium-enriched soils increases their anticarcinogenic potentials. *Medical Hypotheses*, 53(3), 232–235.

Jayaraj, J., Rahman, M., Wan, A., and Punja, Z.K. (2009) Enhanced resistance to foliar fungal pathogens in carrot by application of elicitors. *Annals of Applied Biology*, 155(1), 71–80.

Jones, R.B., Faragher, J.D., and Winkler, S. (2006) A review of the influence of postharvest treatments on quality and glucosinolate content in broccoli (Brassica oleracea var. italica) heads. *Postharvest Biology and Technology*, 41(1), 1–8.

Jones, R.B., Imsic, M., Franz, P., Hale, G., and Tomkins, R.B. (2007) High nitrogen during growth reduced glucoraphanin and flavonol content in broccoli (Brassica oleracea var. italica) heads. *Australian Journal of Experimental Agriculture*, 47, 1498–1505.

Kenny, O. and O'Beirne, D. (2010) Antioxidant phytochemicals in fresh-cut carrot disks as affected by peeling method. *Postharvest Biology and Technology*, 58(3), 247–253.

Kestwal, R.M., Lin, J.C., Bagal-Kestwal, D., Chiang, B.H. (2010) Glucosinolate fortification of cruciferous sprouts by sulfur supplementation during cultivation to enhance anti-cancer activity. *Food Chemistry*, doi 10.1016/j.foodchem.2010.11.152. Accepted manuscript.

Kidmose, U., Hansen, S.L., Christensen, L.P., Edelenbos, M., Larsen, E., and Norbaek, R. (2004) Effects of genotype, root size, storage, and processing on bioactive compounds in organically grown carrots (Daucus carota L.). *Journal of Food Science*, 69(9), S388–S394.

Kim, S., Binzel, M.L., Yoo, K.S., Park, S., and Pike, L.M. (2004) Pink (P), a new locus responsible for a pink trait in onions (Allium cepa) resulting from natural mutations of anthocyanidin synthase. *Molecular Genetics and Genomics*, 272(1), 18–27.

Kjellenberg, L., Johansson, E., Gustavsson, K.E., and Olsson, M.E. (2010) Effects of Harvesting Date and Storage on the Amounts of Polyacetylenes in Carrots, Daucus carota. *Journal of Agricultural and Food Chemistry*, 58(22), 11703–11708.

Kleinhenz, M.D., French, D.G., Gazula, A., and Scheerens, J.C. (2003) Variety, shading, and growth stage effects on pigment concentrations in lettuce grown under contrasting temperature regimens. *Horttechnology*, 13(4), 677–683.

Krumbein, A. and Schonhof, I. (2001a) Influence of temperature and irradiation on glucosinolates in broccoli heads. *Biologically-Active Phytochemicals in Food* (269), 477–479.

Krumbein, A., Schonhof, I., Rühlmann, J., and S.S.W. (2001b) Influence of sulphur and nitrogen supply on flavour and health-affecting compounds in Brassicaceae. In: W. Horst, *Plant Nutrition – Food security and Sustainability of AgroEecosystems*, Netherlands: Kluwer Academic Publishers, pp. 294–295.

Krumbein, A., Schonhof, I., Smetanska, I., Scheuner, E., Ruhlmann, J., and Schreiner, M. (2010) Improving levels of bioactive compounds in *Brassica* vegetables by crop management strategies. *Acta Horticulturae*, 856, 37–47.

Kubota, C. and Thomson, C.A. (2006) Controlled environments for production of value-added food crops with high phytochemical concentrations: Lycopene in tomato as an example. *Hortscience*, 41(3), 522–525.

Kushad, M.M., Brown, A.F., Kurilich, A.C., Juvik, J.A., Klein, B.P., Wallig, M.A., and Jeffery, E.H. (1999) Variation of glucosinolates in vegetable crops of Brassica oleracea. *Journal of Agricultural and Food Chemistry*, 47(4), 1541–1548.

Lazzeri, L., Leoni, O., and Manici, L.M. (2004) Biocidal plant dried pellets for biofumigation. *Industrial Crops and Products*, 20(1), 59–65.

Lee, K.W. and Lee, H.J. (2006) The roles of polyphenols in cancer chemoprevention. *Biofactors*, 26(2), 105–121.

Lee, S.J., Seguin, P., Kim, J.J., Moon, H.I., Ro, H.M., Kim, E.H., Seo, S.H., Kang, E.Y., Ahn, J.K., and Chung, I.M. (2010) Isoflavones in Korean soybeans differing in seed coat and cotyledon color. *Journal of Food Composition and Analysis*, 23(2), 160–165.

Leja, M., Mareczek, A., Starzynska, A., and Rozek, S. (2001) Antioxidant ability of broccoli flower buds during short-term storage. *Food Chemistry*, 72(2), 219–222.

Li, J., Zhu, Z., and Guo, S. (2010) Effect of Nitrogen and Sulphur Application on Antioxidant substances in Leaf Mustard. *Acta Horticulturae*, 856, 83–89.

Li, S.M., Schonhof, I., Krumbein, A., Li, L., Stutzel, H., and Schreiner, M. (2007) Glucosinolate concentration in turnip (Brassica rapa ssp rapifera L.) roots as affected by nitrogen and sulfur supply. *Journal of Agricultural and Food Chemistry*, 55(21), 8452–8457.

Li, X. and Kushad, M.M. (2004) Correlation of glucosinolate content to myrosinase activity in horseradish (Armoracia rusticana). *Journal of Agricultural and Food Chemistry*, 52(23), 6950–6955.

Lopez-Berenguer, C., Martinez-Ballesta, M.D., Moreno, D.A., Carvajal, M., and Garcia-Viguera, C. (2009) Growing Hardier Crops for Better Health: Salinity Tolerance and the Nutritional Value of Broccoli. *Journal of Agricultural and Food Chemistry*, 57(2), 572–578.

Lund, E.D. and White, J.M. (1990) Polyacetylenes in normal and water-stressed orlando gold carrots (daucus-carota). *Journal of the Science of Food and Agriculture*, 51(4), 507–516.

Makris, D.P., Boskou, G., and Andrikopoulos, N.K. (2007) Polyphenolic content and in vitro antioxidant characteristics of wine industry and other agri-food solid waste extracts. *Journal of Food Composition and Analysis*, 20(2), 125–132.

Malatesta, M. and Davey, J.C. (1996) Cultivar identification within broccoli, brassica oleracea l. var. italica plenck and cauliflower, brassica oleracea var. BOTRYTIS L. *Acta Horticulturae*, 407, 109–114.

Marotti, M. and Piccaglia, R. (2002) Characterization of flavonoids in different cultivars of onion (Allium cepa L.). *Journal of Food Science*, 67(3), 1229–1232.

Marsh, A. and Du, L.,C. (2007) Crop rotation affects populations of soil borne pathogens in potato soil. *New Zealand Plant Protection Society Journal*, 60, 310.

Martin, C., Hailing, J., and Schwinn, K. (2000) Mechanisms ans applications of transcriptional control of phenylpropanoid metabolism. In: J.A. Saunders, J.T. Romeo, and B.F. Matthews, *Regulation of Phytochemicals by Molecular Techniques*, Oxford: Pergamon Press, pp. 155–169.

Massie, I.H., Astley, D., and King, G.J. (1996) Patterns of genetic diversity and relationships between regional groups and populations of Italian landrace cauliflower and broccoli (Brassica oleracea L var botrytis L and var italica Plenck). In: J.S. Dias, I. Crute, and A.A. Monteiro, *International Symposium on Brassicas – Ninth Crucifer Genetics Workshop*, pp. 45–53.

Matusheski, N.V. and Jeffery, E.H. (2001) Comparison of the bioactivity of two glucoraphanin hydrolysis products found in broccoli, sulforaphane and sulforaphane nitrile. *Journal of Agricultural and Food Chemistry*, 49(12), 5743–5749.

Matusheski, N.V., Swarup, R., Juvik, J.A., Mithen, R., Bennett, M., and Jeffery, E.H. (2006) Epithiospecifier protein from broccoli (Brassica oleracea L. ssp italica) inhibits formation of the anticancer agent sulforaphane. *Journal of Agricultural and Food Chemistry*, 54(6), 2069–2076.

Mercier, J., Ponnampalam, R., Berard, L.S., and Arul, J. (1993) polyacetylene content and uv-induced 6-methoxymellein accumulation in carrot cultivars. *Journal of the Science of Food and Agriculture*, 63(3), 313–317.

Metzger, B.T. and Barnes, D.M. (2009) Polyacetylene Diversity and Bioactivity in Orange Market and Locally Grown Colored Carrots (Daucus carota L.). *Journal of Agricultural and Food Chemistry*, 57(23), 11134–11139.

Meyer, M. and Adam, S.T. (2008) Comparison of glucosinolate levels in commercial broccoli and red cabbage from conventional and ecological farming. *European Food Research and Technology*, 226(6), 1429–1437.

Mithen, R., Faulkner, K., Magrath, R., Rose, P., Williamson, G., and Marquez, J. (2003) Development of isothiocyanate-enriched broccoli, and its enhanced ability to induce phase 2 detoxification enzymes in mammalian cells. *Theoretical and Applied Genetics*, 106(4), 727–734.

Mithen, R.F. (2001) Glucosinolates and their degradation products. *Advances in Botanical Research, Vol 35*, 35, 213–262.

Mithen, R.F., Lewis, B.G., Heaney, R.K., and Fenwick, G.R. (1987) glucosinolates of wild and cultivated brassica species. *Phytochemistry*, 26(7), 1969–1973.

Mogren, L.M., Caspersen, S., Olsson, M.E., and Gertsson, U. (2008) Organically fertilized onions (Allium cepa L.): Effects of the fertilizer placement method on quercetin content and soil nitrogen dynamics. *Journal of Agricultural and Food Chemistry*, 56, 361–367.

Mogren, L.M., Olsson, M.E., and Gertsson, U.E. (2006) Quercetin content in field-cured onions (Allium cepa L.): Effects of cultivar, lifting time, and nitrogen fertilizer level. *Journal of Agricultural and Food Chemistry*, 54(17), 6185–6191.

Moreno, D.A., Perez-Balibrea, S., Ferreres, F., Gil-Izquierdo, A., and Garcia-Viguera, C. (2010) Acylated anthocyanins in broccoli sprouts. *Food Chemistry*, 123(2), 358–363.

Mori, K., Goto-Yamamoto, N., Kitayama, M., and Hashizume, K. (2007a) Effect of high temperature on anthocyanin composition and transcription of flavonoid hydroxylase genes in 'Pinot noir' grapes (Vitis vinifera). *Journal of Horticultural Science and Biotechnology*, 82(2), 199–206.

Mori, K., Goto-Yamamoto, N., Kitayama, M., and Hashizume, K. (2007b) Loss of anthocyanins in red-wine grape under high temperature. *Journal of Experimental Botany*, 58(8), 1935–1945.

Morra, M. J. and Borek, V. (2010) Glucosinolate preservation in stored Brassicaceae seed meals. *Journal of Stored Products Research*, 46(3), 206–207.

Mucha-Pelzer, T., Mewis, I., and Ulrichs, C. (2010) Response of Glucosinolate and Flavonoid Contents and Composition of Brassica rapa ssp chinensis (L.) Hanelt to Silica Formulations Used as Insecticides. *Journal of Agricultural and Food Chemistry*, 58(23), 12473–12480.

Murphy, S.E., Lee, E.A., Woodrow, L., Seguin, P., Kumar, J., Rajcan, I., and Ablett, G.R. (2009) Genotype x Environment Interaction and Stability for Isoflavone Content in Soybean. *Crop Science*, 49(4), 1313–1321.

Murthy, K.N.C., Jayaprakasha, G.K., Pike, L.M., and Patil, B.S. (2007) Potential of 'BetaSweet' carrot in cancer prevention. *Hortscience*, 42(4), 887–887.

Naoumkina, M.A., Zhao, Q., Gallego-Giraldo, L., Dai, X.B., Zhao, P.X., and Dixon, R.A. (2010) Genome-wide analysis of phenylpropanoid defence pathways. *Molecular Plant Pathology*, 11(6), 829–846.

Navarro, J.M., Perez-Perez, J.G., Romero, P., and Botia, P. (2010) Analysis of the changes in quality in mandarin fruit, produced by deficit irrigation treatments. *Food Chemistry*, 119(4), 1591–1596.

Nencini, C., Cavallo, F., Capasso, A., Franchi, G.G., Giorgio, G., and Micheli, L. (2007) Evalution of antitoxidative properties of Allium species growing wild in Italy. *Phytotherapy Research*, 21(9), 874–878.

Netzel, M., Netzel, G., Kammerer, D.R., Schieber, A., Carle, R., Simons, L., Bitsch, I., Bitsch, R., and Konczak, L. (2007) Cancer cell antiproliferation activity and metabolism of black carrot anthocyanins. *Innovative Food Science and Emerging Technologies*, 8(3), 365–372.

Oh, M.M., Carey, E.E., and Rajashekar, C.B. (2010) Regulated Water Deficits Improve Phytochemical Concentration in Lettuce. *Journal of the American Society for Horticultural Science*, 135(3), 223–229.

Oh, M.M. and Rajashekar, C.B. (2009) Antioxidant content of edible sprouts: effects of environmental shocks. *Journal of the Science of Food and Agriculture*, 89(13), 2221–2227.

Oke, M., Ahn, T., Schofield, A., and Paliyath, G. (2005) Effects of phosphorus fertilizer supplementation on processing quality and functional food ingredients in tomato. *Journal of Agricultural and Food Chemistry*, 53(5), 1531–1538.

Olsson, K. and Svensson, R. (1996) The influence of polyacetylenes on the susceptibility of carrots to storage diseases. *Journal of Phytopathology-Phytopathologische Zeitschrift*, 144(9–10), 441–447.

Olsson, M.E., Gustavsson, K.E., and Vagen, I.M. (2010) Quercetin and Isorhamnetin in Sweet and Red Cultivars of Onion (Allium cepa L.) at Harvest, after Field Curing, Heat Treatment, and Storage. *Journal of Agricultural and Food Chemistry*, 58(4), 2323–2330.

Omirou, M.D., Papadopoulou, K.K., Papastylianou, I., Constantinou, M., Karpouzas, D.G., Asimakopoulos, I., and Ehaliotis, C. (2009) Impact of Nitrogen and Sulfur Fertilization on the Composition of Glucosinolates in Relation to Sulfur Assimilation in Different Plant Organs of Broccoli. *Journal of Agricultural and Food Chemistry*, 57(20), 9408–9417.

Padilla, G., Cartea, M.E., Velasco, P., de Haro, A., and Ordas, A. (2007) Variation of glucosinolates in vegetable crops of Brassica rapa. *Phytochemistry*, 68(4), 536–545.

Paliyath, G., Schfield, A., Oke, M., and Taehyun, A. (2002) Phosphorus fertilization and biosynthesis of functional food ingredients. *Proceedings of the Symposium on Fertilizing Crops for Functional Food*.

Patil, B.S. (2004) Irradiation applications to improve functional components of fruits and vegetables. *Irradiation of Food and Packaging: Recent Developments*, 875, 117–137.

Patil, B.S. and Pike, L.M. (1995a) Distribution of quercetin content in different rings of various colored onion (allium-cepa l) cultivars. *Journal of Horticultural Science*, 70(4), 643–650.

Patil, B.S., Pike, L.M., and Hamilton, B.K. (1995b) Changes in quercetin concentration in onion (allium-cepa l) owing to location, growth stage and soil type. *New Phytologist*, 130(3), 349–355.

Patil, B.S., Pike, L.M., and Yoo, K.S. (1995c) Variation in the quercetin content in different colored onions (allium-cepa l). *Journal of the American Society for Horticultural Science*, 120(6), 909–913.

Pereira, F.M.V., Rosa, E., Fahey, J.W., Stephenson, K.K., Carvalho, R., and Aires, A. (2002) Influence of temperature and ontogeny on the levels of glucosinolates in broccoli (Brassica oleracea var. italica) sprouts and their effect on the induction of mammalian phase 2 enzymes. *Journal of Agricultural and Food Chemistry*, 50(21), 6239–6244.

Perez-Balibrea, S., Moreno, D.A., and Garcia-Viguera, C. (2008) Influence of light on health-promoting phytochemicals of broccoli sprouts. *Journal of the Science of Food and Agriculture*, 88(5), 904–910.

Pérez-Gregorio, R.M., García-Falcón, M.S., Simal-Gándara, J., Rodrigues, A.S., and Almeida, D.P.F. (2010) Identification and quantification of flavonoids in traditional cultivars of red and white onions at harvest. *Journal of Food Composition and Analysis*, 23(6), 592–598.

Perkins-Veazie, P., Collins, J.K., Davis, A.R., and Roberts, W. (2006) Carotenoid content of 50 watermelon cultivars. *Journal of Agricultural and Food Chemistry*, 54(7), 2593–2597.

Perkins-Veazie P.,a.R.W. (2002) Can potassium application affect the mineral and antioxidant content of horticultural crops?, *Proceedings of the Symposium on Fertilizing Crops for Functional Food*.

Perner, H., Rohn, S., Driemel, G., Batt, N., Schwarz, D., Kroh, L.W., and George, E. (2008) Effect of nitrogen species supply and mycorrhizal colonization on organosulfur and phenolic compounds in onions. *Journal of Agricultural and Food Chemistry*, 56(10), 3538–3545.

Pferschy-Wenzig, E.M., Getzinger, V., Kunert, O., Woelkart, K., Zahrl, J., and Bauer, R. (2009) Determination of falcarinol in carrot (Daucus carota L.) genotypes using liquid chromatography/mass spectrometry. *Food Chemistry*, 114(3), 1083–1090.

Phommalth, S., Jeong, Y.S., Kim, Y.H., Dhakal, K.H., and Hwang, Y.H. (2008) Effects of Light Treatment on Isoflavone Content of Germinated Soybean Seeds. *Journal of Agricultural and Food Chemistry*, 56(21), 10123–10128.

PotatoPro (2009) Peter Keogh and Sons Launch 'Selena' Potatoes Naturally Enriched With Selenium

Prakash, D., Upadhyay, G., Singh, B.N., and Singh, H.B. (2007) Antioxidant and free radical-scavenging activities of seeds and agri-wastes of some varieties of soybean (Glycine max). *Food Chemistry*, 104(2), 783–790.

Price, K.R., Bacon, J.R., and Rhodes, M.J.C. (1997) Effect of storage and domestic processing on the content and composition of flavonol glucosides in onion (Allium cepa). *Journal of Agricultural and Food Chemistry*, 45(3), 938–942.

Rangkadilok, N., Nicolas, M.E., Bennett, R.N., Eagling, D.R., Premier, R.R., and Taylor, P.W.J. (2004) The effect of sulfur fertilizer on glucoraphanin levels in broccoli (B-oleracea L. var. italica) at different growth stages. *Journal of Agricultural and Food Chemistry*, 52(9), 2632–2639.

Rangkadilok, N., Tomkins, B., Nicolas, M.E., Premier, R.R., Bennett, R.N., Eagling, D.R., and Taylor, P.W.J. (2002) The effect of post-harvest and packaging treatments on glucoraphanin concentration in broccoli (Brassica oleracea var. italica). *Journal of Agricultural and Food Chemistry*, 50(25), 7386–7391.

Raupp, J., Pekrun, C., Oltmans, M., and Kopke, U., (2006) *Long-term Term Field Experiments in Organic Agricultural Research*. Verlag Dr. Köster, Berlin, Germany.

Rembialkowska, E. (2007) Quality of plant products from organic agriculture. *Journal of the Science of Food and Agriculture*, 87, 2757–2762.

Ren, H.F., Endo, H., and Hayashi, T. (2001) Antioxidative and antimutagenic activities and polyphenol content of pesticide-free and organically cultivated green vegetables using water-soluble chitosan as a soil modifier and leaf surface spray. *Journal of the Science of Food and Agriculture*, 81(15), 1426–1432.

Robbins, R.J., Keck, A.S., Banuelos, G., and Finley, J.W. (2005) Cultivation conditions and selenium fertilization alter the phenolic profile, glucosinolate, and sulforaphane content of broccoli. *Journal of Medicinal Food*, 8(2), 204–214.

Rodov, V., Tietel, Z., Vinokur, Y., Horev, B., and Eshel, D. (2010) Ultraviolet Light Stimulates Flavonol Accumulation in Peeled Onions and Controls Microorganisms on Their Surface. *Journal of Agricultural and Food Chemistry*, 58(16), 9071–9076.

Rodrigues, A.S., Perez-Gregorio, M.R., Garcia-Falcon, M.S., Simal-Gandara, J., and Almeida, D.P.F. (2011) Effect of meteorological conditions on antioxidant flavonoids in Portuguese cultivars of white and red onions. *Food Chemistry*, 124(1), 303–308.

Rodrigues, A.S. and Rosa, E.A.S. (1999) Effect of post-harvest treatments on the level of glucosinolates in broccoli. *Journal of the Science of Food and Agriculture*, 79(7), 1028–1032.

Roldan, E., Sanchez-Moreno, C., de Ancos, B., and Cano, M.P. (2008) Characterisation of onion (Allium cepa L.) by-products as food ingredients with antioxidant and antibrowning properties. *Food Chemistry*, 108, 907–916.

Rosa (1997) Glucosinolates in Crop Plants In: R. Heaney, G. Fenwick, and C. Portas, *Horticultural Reviews*, vol. 19, Chichester: John Wiley & Sons, Ltd., pp. 99–215.

Rosa, E.A.S. and Rodrigues, P.M.F. (1999) Towards a more sustainable agriculture system: The effect of glucosinolates on the control of soil-borne diseases. *Journal of Horticultural Science and Biotechnology*, 74(6), 667–674.

Ruiz-Cruz, S., Islas-Osuna, M.A., Sotelo-Mundo, R.R., Vazquez-Ortiz, F., and Gonzalez-Aguilar, G.A. (2007) Sanitation procedure affects biochemical and nutritional changes of shredded carrots. *Journal of Food Science*, 72(2), S146–S152.

Sarikamis, G., Marquez, J., MacCormack, R., Bennett, R.N., Roberts, J., and Mithen, R. (2006) High glucosinolate broccoli: a delivery system for sulforaphane. *Molecular Breeding*, 18(3), 219–228.

Scheuner, E.T., Schmidt, S., Krumbein, A., Schonhof, I., and Schreiner, M. (2005) Effect of methionine foliar fertilization on glucosinolate concentration in broccoli and radish. *Journal of Plant Nutrition and Soil Science-Zeitschrift Fur Pflanzenernahrung Und Bodenkunde*, 168(2), 275–277.

Schmidt, T.J., Hemmati, S., Klaes, M., Konuklugil, B., Mohagheghzadeh, A., Ionkova, I., Fuss, E., and Alfermann, A.W. (2010) Lignans in flowering aerial parts of Linum species – Chemodiversity in the light of systematics and phylogeny. *Phytochemistry*, 71(14–15), 1714–1728.

Schonhof, I., Klaring, H.P., Krumbein, A., Claussen, W., and Schreiner, M. (2007) Effect of temperature increase under low radiation conditions on phytochemicals and ascorbic acid in greenhouse grown broccoli. *Agriculture Ecosystems and Environment*, 119(1–2), 103–111.

Schonhof, I., Krumbein, A., and Brückner, B. (2004) Genotypic effects on glucosinolates and sensory properties of broccoli and cauliflower. *Food*, 48, 25–33.

Schonhof, I., Krumbein, A., Schreiner, M., and B. B.G. (1999) *Bioactive Substances in Cruciferous Products*. The Royal Society of Chemistry Cambridge UK. Special publication 229, 222–226.

Schreiner, M. (2005) Vegetable crop management strategies to increase the quantity of phytochemicals. *European Journal of Nutrition*, 44(2), 85–94.

Seguin, P. and Zheng, W. (2006) Potassium, phosphorus, sulfur, and boron fertilization effects on soybean isoflavone content and other seed characteristics. *Journal of Plant Nutrition*, 29(4), 681–698.

Seljasen, R., Bengtsson, G.B., Hoftun, H., and Vogt, G. (2001) Sensory and chemical changes in five varieties of carrot (Daucus carota L) in response to mechanical stress at harvest and post-harvest. *Journal of the Science of Food and Agriculture*, 81(4), 436–447.

Selmar, D. (2008) Potential of salt and drought stress to increase pharmaceutical significant secondary compounds in plants. *Landbauforschung Volkenrode*, 58(1–2), 139–144.

Singh, J., Upadhyay, A.K., Prasad, K., Bahadur, A., and Rai, M. (2007) Variability of carotenes, vitamin C, E and phenolics in Brassica vegetables. *Journal of Food Composition and Analysis*, 20(2), 106–112.

Slimestad, R., Fossen, T., and Vagen, I.M. (2007) Onions: A source of unique dietary flavonoids. *Journal of Agricultural and Food Chemistry*, 55, 10067–10080.

Smolen, S. and Sady, W. (2009) The effect of various nitrogen fertilization and foliar nutrition regimes on the concentrations of sugars, carotenoids and phenolic compounds in carrot (Daucus carota L.). *Scientia Horticulturae*, 120(3), 315–324.

Solomakhin, A. and Blanke, M.M. (2010) Can coloured hailnets improve taste (sugar, sugar: acid ratio), consumer appeal (colouration) and nutritional value (anthocyanin, vitamin C) of apple fruit? *Lwt-Food Science and Technology*, 43(8), 1277–1284.

Soltoft, M., Eriksen, M.R., Trager, A.W.B., Nielsen, J., Laursen, K.H., Husted, S., Halekoh, U., and Knuthsen, P. (2010) Comparison of Polyacetylene Content in Organically and Conventionally Grown Carrots Using a Fast Ultrasonic Liquid Extraction Method. *Journal of Agricultural and Food Chemistry*, 58(13), 7673–7679.

Sones, K., Heaney, R.K. and Fenwick, G.R. (1984) The glucosinolate content of uk vegetables – cabbage (Braccicae oleracea), swede (B.napus) and turnip (B.campestris). *Food Additives and Contaminants*, 1(3), 289–296.

Song, L.J., Lori, R., and Thornalley, P.J. (2006) Purification of major glucosinolates from Brassicaceae seeds and preparation of isothiocyanate and amine metabolites. *Journal of the Science of Food and Agriculture*, 86(8), 1271–1280.

Sorensen, J.N. and Grevsen, K. (2001) Sprouting in bulb onions (Allium cepa L.) as influenced by nitrogen and water stress. *Journal of Horticultural Science and Biotechnology*, 76(4), 501–506.

Steyn, W.J., Wand, S.J.E., Jacobs, G., Rosecrance, R.C., and Roberts, S.C. (2009) Evidence for a photoprotective function of low-temperature-induced anthocyanin accumulation in apple and pear peel. *Physiologia Plantarum*, 136(4), 461–472.

Stirling, G.R. and Stirling, A.M. (2003) The potential of Brassica green manure crops for controlling root-knot nematode (Meloidogyne javanica) on horticultural crops in a subtropical environment. *Australian Journal of Experimental Agriculture*, 43(6), 623–630.

Sun, B., Liu, N., Zhao, Y., Yan, H., and Wang, Q. (2011) Variation of glucosinolates in three edible parts of Chinese kale (Brassica alboglabra Bailey) varieties. *Food Chemistry*, 124(3), 941–947.

Sun, T., Simon, P.W., and Tanumihardjo, S.A. (2009) Antioxidant Phytochemicals and Antioxidant Capacity of Biofortified Carrots (Daucus carota L.) of Various Colors. *Journal of Agricultural and Food Chemistry*, 57(10), 4142–4147.

Sutton, A.W. (1908) Brassica Crosses. *Journal of the Linnean Society of London, Botany*, 38, 337–349.

Taber, H., Perkins-Veazie, P., Li, S., White, W., Roderniel, S., and Xu, Y. (2008) Enhancement of tomato fruit lycopene by potassium is cultivar dependent. *Hortscience*, 43(1), 159–165.

Tavarini, S., Gil, M.I., Tomas-Barberan, F.A., Buendia, B., Remorini, D., Massai, R., Degl'Innocenti, E., and Guidi, L. (2011) Effects of water stress and rootstocks on fruit phenolic composition and physical/chemical quality in Suncrest peach. *Annals of Applied Biology*, 158(2), 226–233.

Tiwari, U. and Cummins, E. (2008) A predictive model of the effects of genotypic, pre- and postharvest stages on barley beta-glucan levels. *Journal of the Science of Food and Agriculture*, 88(13), 2277–2287.

Toler, H.D., Charron, C.S., Kopsell, D.A., Sams, C.E., and Randle, W.M. (2007) Selenium and sulfur increase sulfur uptake and regulate glucosinolate metabolism in Brassica oleracea. *Proceedings of the 1st International Symposium on Human Health Effects of Fruits and Vegetables*, 311–315.

U N (1935). Genome analysis in Brassica with special reference to the experimental formation of B.napus and peculiar mode of fertilization. *Japanese Journal of Botany*, 7, 389–452.

Uddin, M. and MaecTavish, H.S. (2003) Controlled atmosphere and regular storage-induced changes in S-alk(en)yl-L-cysteine sulfoxides and alliinase activity in onion bulbs (Allium cepa L. cv. Hysam). *Postharvest Biology and Technology*, 28(2), 239–245.

Vallejo, F., Garcia-Viguera, C., and Tomas-Barberan, F.A. (2003) Changes in broccoli (Brassica oleracea L. var. italica) health-promoting compounds with inflorescence development. *Journal of Agricultural and Food Chemistry*, 51(13), 3776–3782.

Vallejo, F., Tomas-Barberan, F.A., and Garcia-Viguera, C. (2002) Potential bioactive compounds in health promotion from broccoli cultivars grown in Spain. *Journal of the Science of Food and Agriculture*, 82(11), 1293–1297.

Velasco, P., Cartea, M.E., Gonzalez, C., Vilar, M., and Ordas, A. (2007) Factors affecting the glucosinolate content of kale (Brassica oleracea acephala group). *Journal of Agricultural and Food Chemistry*, 55(3), 955–962.

Verkerk, R., Dekker, M., and Jongen, W.M.F. (2001) Post-harvest increase of indolyl glucosinolates in response to chopping and storage of Brassica vegetables. *Journal of the Science of Food and Agriculture*, 81(9), 953–958.

Ververk, S., Tebbenhoff, S., Dekker, M., (2010) Variation and distribution of glucosinolates in 42 cultivars of Brassica oleracea vegetable crops. *Acta Horticulturae*, 856, 63–66.

Vital Vegetables http://vitalvegetables.com.au/consumer/index.html.

Vyn, T.J., Yin, X.H., Bruulsema, T.W., Jackson, C.J.C., Rajcan, I., and Brouder, S.M. (2002) Potassium fertilization effects on isoflavone concentrations in soybean [Glycine max (L.) Merr.]. *Journal of Agricultural and Food Chemistry*, 50(12), 3501–3506.

Ward, J.L., Poutanen, K., Gebruers, K., Piironen, V., Lampi, A.M., Nystrom, L., Andersson, A.A.M., Aman, P., Boros, D., Rakszegi, M., Bedo, Z., and Shewry, P.R. (2008) The HEALTHGRAIN cereal diversity screen: concept, results, and prospects. *Journal of Agricultural and Food Chemistry*, 56(21), 9699–9709.

Westcott, N.D., Muir, A.D. Lafond, G., McAndrew, D.W., May, W., Irvine, B., Grant, C., Shirtliffe, S. and Bruulsema, T.W. (2002). Factors affecting the concentration of nutraceutical lignan in flaxseed. *Proceedings of the Symposium on Fertilizing Crops for Functional Food*.

Wijngaard, H.H., Rossle, C., and Brunton, N. (2009) A survey of Irish fruit and vegetable waste and by-products as a source of polyphenolic antioxidants. *Food Chemistry*, 116(1), 202–207.

Williams, C.M. (2002) Nutritional quality of organic food: shades of grey or shades of green? *Proceedings of the Nutrition Society*, 61(1), 19–24.

Winkler, S., Faragher, J., Franz, P., Imsic, M., and Jones, R. (2007) Glucoraphanin and flavonoid levels remain stable during simulated transport and marketing of broccoli (Brassica oleracea var. italica) heads. *Postharvest Biology and Technology*, 43(1), 89–94.

Woese, K., Lange, D., Boess, C., and Bogl, K.W. (1997) A comparison of organically and conventionally grown foods – Results of a review of the relevant literature. *Journal of the Science of Food and Agriculture*, 74(3), 281–293.

Yabuta, Y., Mieda, T., Rapolu, M., Nakamura, A., Motoki, T., Maruta, T., Yoshimura, K., Ishikawa, T., and Shigeoka, S. (2007) Light regulation of ascorbate biosynthesis is dependent on the photosynthetic electron transport chain but independent of sugars in Arabidopsis. *Journal of Experimental Botany*, 58(10), 2661–2671.

Valverde, L., Reilly, K., Brunton, N. Gaffney, M., Sorensen, J., and Sorensen, H. Analysis of the major glucosinolates present in 5 varieties of Broccoli (Brassica oleracea L. italica) cultivated in controlled field trials. *Submitted*.

Yamauchi, N. and Watada, A.E. (1993) Pigment changes in parsley leaves during storage in controlled or ethylene containing atmosphere. *Journal of Food Science*, 58(3), 616.

Yang, J., Meyers, K.J., Van der Heide, J., and Liu, R.H. (2004) Varietal differences in phenolic content and antioxidant and anti proliferative activities of onions. *Journal of Agricultural and Food Chemistry*, 52(22), 6787–6793.

Young, J.E., Zhao, X., Carey, E.E., Welti, R., Yang, S.S., and Wang, W.Q. (2005) Phytochemical phenolics in organically grown vegetables. *Molecular Nutrition and Food Research*, 49(12), 1136–1142.

Yuan, G.F., Wang, X.P., Guo, R.F., and Wang, Q.M. (2010) Effect of salt stress on phenolic compounds, glucosinolates, myrosinase and antioxidant activity in radish sprouts. *Food Chemistry*, 121(4), 1014–1019.

Zabala, M.D., Grant, M., Bones, A.M., Bennett, R., Lim, Y.S., Kissen, R., and Rossiter, J.T. (2005) Characterisation of recombinant epithiospecifier protein and its over-expression in Arabidopsis thaliana. *Phytochemistry*, 66(8), 859–867.

Zhang, Y. and Tang, L. (2007) Discovery and development of sulforaphane as a cancer chemopreventive phytochemical. *Acta Pharmacologica Sinica*, 28(9), 1343–1354.

Zhao, X., Carey, E.E., Wang, W.Q., and Rajashekar, C.B. (2006) Does organic production enhance phytochemical content of fruit and vegetables? Current knowledge and prospects for research. *Horttechnology*, 16(3), 449–456.

Zidorn, C., Johrer, K., Ganzera, M., Schubert, B., Sigmund, E.M., Mader, J., Greil, R., Ellmerer, E.P., and Stuppner, H. (2005) Polyacetylenes from the Apiaceae vegetables carrot, celery, fennel, parsley, and parsnip and their cytotoxic activities. *Journal of Agricultural and Food Chemistry*, 53(7), 2518–2523.

10 Minimal processing of leafy vegetables

Rod Jones and Bruce Tomkins*

Future Farming Systems Research Department of Primary Industries, Knoxfield, Victoria, Australia

10.1 Introduction

Minimal processing can be defined as fresh produce that has been sliced, shredded, diced or peeled before packaging, and differs from other processing methods as heating is not involved. Therefore plant tissues remain viable, albeit in many cases in a wounded state (Barry-Ryan and O'Beirne, 1999). Market research has indicated that consumers widely perceive fresh produce as more nutritious than processed (Richards, 2003), and minimally processed fruits and vegetables are therefore considered more desirable. In many instances this perception is true as several nutrient phytochemicals in plants are destroyed or reduced during processing (Jones *et al.*, 2006). However, it is also recognised that nutritional value and phytochemical content of plant tissue is usually reduced during normal aging and senescence after harvest (Kays and Paull, 2004). Less well understood is the impact of size reduction on the fate of phytochemicals in plant tissue. The aim of this chapter is to review the impact of minimal processing on the phytochemical content of the most commonly consumed fresh-cut product: leafy vegetables in salad mixes. For the purposes of this chapter, phytochemicals for human health can be defined as 'non-nutrient chemicals found in plants that have biological activity against chronic diseases' (Kushad *et al.*, 2003). In addition, we will focus on ascorbic acid, as this compound makes a major contribution, along with phenolics, to the antioxidant capacity (as measured in vitro by ORAC) in leafy vegetables.

In plants, phytochemicals serve a wide range of functions including pigmentation (anthocyanins, lycopene), pest and disease defence (glucosinolates, cysteine sulfoxides), and

---

*This chapter is a publication funded by Vital Vegetables, a Trans Tasman research project jointly supported by Horticulture Australia Ltd, New Zealand Institute for Crop and Food Research Ltd, the New Zealand Foundation for Research Science and Technology, the Australian Vegetable and Potato Growers Federation Inc, New Zealand Vegetable and Potato Growers Federation Inc and the Victorian Department of Primary Industries.

Handbook of Plant Food Phytochemicals: Sources, Stability and Extraction, First Edition.
Edited by B.K. Tiwari, Nigel P. Brunton and Charles S. Brennan.
© 2013 John Wiley & Sons, Ltd. Published 2013 by John Wiley & Sons, Ltd.

prevention of UV light-induced oxidative stress (flavonols) (Kays and Paull, 2004). Phytochemicals have been linked to many positive health effects in humans including some cancers, coronary heart disease, diabetes, high blood pressure, inflammation, infection, psychotic diseases, ulcers and macular degeneration (Le Marchand, 2002; Nijveldt et al., 2001; Steinmetz and Potter, 1996). Many phytochemical types, such as polyphenolics, carotenoids and organosulphur compounds, are thought to be involved in this protection and they may act synergistically, or have different modes of action (Chen et al., 2007). Over the past 20 years there has been an increased interest in phytochemicals for human health and the potential economic advantages of creating novel food products based on elevated levels of health-promoting phytochemicals. There has also been an increased stimulus to re-examine the post-harvest practices for fruits and vegetables, in order to ascertain whether conditions that result in the preservation of visual and organoleptic parameters also impact on the preservation of phytochemicals. A range of minimally processed products are available in the market. Hence this chapter will focus on the effects of minimal processing procedures on the post-harvest fate of phytochemicals and ascorbic acid contained in common ingredients of salad mixes.

10.2 Minimally processed products

It has been estimated that up to 80% of the minimally processed market in the US is made up of salad mixes (Cook, 2004). A wide variety of leafy vegetables are used in salad mixes (Table 10.1), with a similar wide variation in phytochemical content, ascorbic acid and antioxidant capacity. Commonly consumed lettuce varieties, such as oakleaf and coral, are high in phenolic compounds and are also a source of carotenoids such as lutein and zeaxanthin

Table 10.1 Common salad mix ingredients used in USA and Australian markets

Common name/s	Botanical name
Lettuce – oakleaf	Lactuca sativa
Lettuce – coral	Lactuca sativa
Lettuce – frisee	see endive
Lettuce – Batavia	Lactuca sativa
Lettuce – Cos	Lactuca sativa
Lettuce – Iceberg	Lactuca sativa
Lettuce – Butter	Lactuca sativa
Spinach	Spinacia oleracea
Chard/Beet	Beta vulgaris
Endive/Frisee (chicory)	Chicorium endiva
Radicchio	Chicorium intybus
Mustard	Brassica juncea or Sinapis alba
Pea tendrils	Pisium sativum
Mizuna	Brassica rapa v. nipponsicia and japonica
Mibuna	Brassica rapa v. nipponsicia and japonica
Tatsoi	Brassica rapac v. group Taatsai
Pak choi	Brassica rapac v. group Pak choi
Rocket – wild	Diplotaxis tenuifolia
Rocket – *Arugula*	Eruca vesicaria v. sativa
Deltona	Lactuca sativa
Kale	Brassica oleracea v.sabellica

(Wilson et al., 2004). Red lettuce, for example, can be up to five times higher in antioxidant capacity than similar green varieties (Wilson et al., 2004), due mainly to higher levels of anthocyanins and other phenolics. Spinach, another common ingredient, is relatively high in ascorbic acid (Yadav and Sehgal, 1995). Other ingredients, being from the *Brassicaceae*, contain glucosinolates. Water cress contains high levels of carotenoids and the glucosinolate nasturtiin (Cruz et al., 2009), rocket contains high quercetin, kaempferol and isorhamnetin content, and the glucosinolate glucoerucin (Jin et al., 2009), while mizuna and mibuna are high in phenolics (Martinez-Sanchez et al., 2008). With such a wide range of different ingredients at hand it is theoretically possible to produce a number of salad mixes with specific health benefits. Apart from salad mixes, ready to use products such as carrot batons, diced onions, soup mixes and fruit salads are now readily available. The demand for these products derives from their convenience as consumers are spending less and less time preparing fruits and vegetables prior to consumption.

10.3 Cutting and shredding

Antioxidant capacity in lettuce leaves is derived primarily from phenolic compounds and ascorbic acid (Reyes et al., 2007), so induction of the phenylpropanoid pathway that synthesises phenolics by cutting or shredding should increase antioxidant content. This was found to be the case in 'Iceberg' leaves, where cutting caused an increase in Phenylalanine Ammonia Lyase (PAL) activity of approximately ten-fold, and a concomitant increase in phenolic content and antioxidant activity (Reyes et al., 2007). Ascorbic acid content declined, however, and the authors hypothesised that the inherent low ascorbic acid content in Iceberg is used up quickly after wounding and is unavailable to deal with the increase in Reactive Oxygen Species (ROS). Phenolics are therefore rapidly synthesised to partially control this wound-induced increase in ROS (Reyes et al., 2007).

This situation is not seen in all cases and cutting lettuce tissues can have a variety of effects. Phenolic compounds increased significantly in the mid-rib tissues of Iceberg lettuce, contributing to browning symptoms (Ke and Saltveit, 1989). Similarly, excised lettuce leaf discs showed an increase in phenolics if they were from the mid-rib region (Tomas-Barberan et al., 1997a). However there was little effect of cutting found on caffeic acid derivatives or flavonols in red or green lettuce tissues, while anthocyanins declined significantly. Similarly, cyanidin glycosides declined after shredding and 48 h storage at 22 °C in the red 'Lollo Rosso' and 'Red Oak' lettuce leaves (DuPont et al., 2000). However anthocyanins can also increase during minimal processing as the tissue is still alive and able to continue to synthesise these compounds. Cutting red lettuce leaves induced cyanidin glycoside production in the midrib during the first seven days of storage (Ferreres et al., 1997).

Gil et al. (1998) found soluble phenolic compounds doubled in the mid-rib tissues of the red lettuce Lollo Rosso after wounding and storage in air at 5°C which contributed to enhanced browning. Cutting Lamb's lettuce leaves, however, caused a decrease in both phenolic content and ascorbic acid (Ferrante et al., 2009). Shredding also resulted in significant losses of flavonoids in a range of lettuces, with losses varying from 6% for Lollo Rosso to 94% for green oak after 48 h at 22 °C (DuPont et al., 2000). There was also a significant decline in flavonoid compounds in endives after shredding (DuPont et al., 2000). The content of many phenolic compounds in fruit and vegetables can decline significantly during processing (Tomas-Barberan and Espin, 2001). This is most likely due to phenolic leaching during washing caused by the significant tissue damage shredding entails, compared with minimal cutting.

Some common salad mix ingredients (e.g. rocket, mibuna, mizuna; Table 10.1) are members of the *Brassicaceae* family and as such, contain glucosinolates. Any processing step that involved cutting, chopping or disruption of cellular integrity caused a loss of total glucosinolates, as this resulted in the mixture of glucosinolates with the enzyme myrosinase (Jones *et al.*, 2006). There is little published information on the effects of cutting on glucosinolate content in leafy *Brassicas*, but studies in broccoli offer some clues. After chopping and storage of both broccoli and cabbage at room temperature (approximately 20 °C) there were significant reductions in aliphatic glucosinolates (e.g. glucoraphanin) but an increase in some indole glucosinolates, such as glucobrassicin (Verkerk *et al.*, 2001). As leafy *Brassicas* are washed thoroughly after cutting it is reasonable to assume some reduction in glucosinolate content will occur due to leaching, but the extent will depend on degree of tissue damage. Proper temperature management after cutting should minimise glucosinolate reduction thereafter (Jones *et al.*, 2006).

Ascorbic acid generally declined rapidly after harvest and during processing (Lee and Kader, 2000). Cutting method can also significantly impact on rate of ascorbic acid loss. Manual tearing of Iceberg lettuce leaves resulted in better ascorbic acid retention than machine cutting, while blunt blades used in cutters resulted in greater loss of ascorbic acid than if sharp stainless steel blades were used (Barry-Ryan and O'Beirne, 1999). This is likely due to less cellular damage caused by sharp blades resulting in lower ascorbic acid leakage and enzymatic degradation due to loss of cellular compartmentation.

There are no known reports on the effect of cutting on carotenoids in leafy vegetables. Carotenoids, such as lutein and zeaxanthin, are inherently more stable than phenolics in the post-harvest environment and are not subject to leaching as they are hydrophobic (Jones *et al.*, 2006). It is therefore reasonable to assume content would not be significantly affected by cutting, but more work is required in this area.

10.4 Wounding physiology

Cutting fresh leafy produce induces a wound response in tissues that has a wide range of effects, including increased respiration (Martinez-Sanchez *et al.*, 2008) and ethylene synthesis, and activation of the phenylpropanoid pathway (Saltveit, 2000b) that can result in increased phenolic synthesis, and resultant antioxidant capacity. This wound response is an integral part of healing in plants as it results in elevated production of compounds that are involved in wound repair and defence against pathogens, specifically lignin and suberin (Hawkins and Boudet, 1996). The production of these, and other, compounds results in lignification, which is ubiquitous in all plants (Dyer *et al.*, 1989). Of particular interest to this review is that lignin is synthesised via the phenylpropanoid pathway via initiation of Phenylalanine Ammonia-Lyase (PAL; EC 4.3.1.5; Dyer *et al.*, 1989), which also results in increased phenolic synthesis and antioxidant capacity. Figure 10.1 represents a simplified schematic of the phenylpropanoid pathway, showing how initiation of PAL can result in lignin production (from 4-Coumarate), and phenolic accumulation. The phenolic compounds quercetin, kaempferol, isorhamnetin and anthocyanins are all commonly found in leafy vegetables and contribute significantly to antioxidant capacity (Rochfort *et al.*, 2006). Wounding is also thought to increase phenylalanine synthesis by stimulation of the shikimate pathway, so it would appear the response involves both initiation of PAL and increased production of the amino acid this enzyme acts upon (Figure 10.1; Dyer *et al.*, 1989).

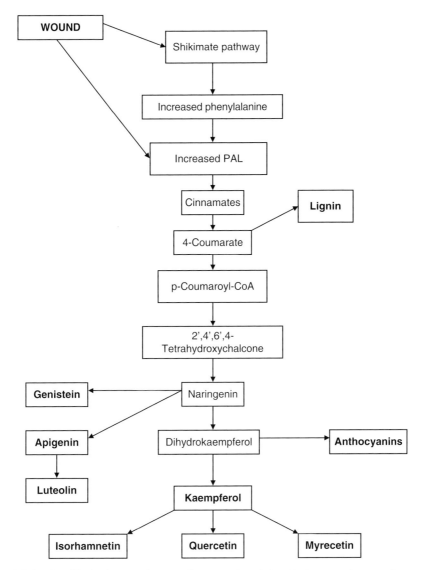

Figure 10.1 Simplified schematic diagram illustrating the relationship between the wound response in plant tissues and the shikimate and phenylpropanoid pathways.
Wounding initiates phenylalanine induction via the shikimate pathway, and enhances PAL activity, which, in turn, results in increased lignin production (from 4-Coumarate) and production of a range of phenolic compounds with antioxidant activity (in bold).

Activation of the phenylpropanoid pathway by wounding in lettuce leaves is primarily via induction of PAL (Dixon and Paiva, 1995; Tomas-Barberan *et al.*, 1997b), leading to an increase in soluble phenolic compounds. Wounded Iceberg lettuce tissue showed a 6–12-fold increase in PAL within 24 h of wounding, while phenolic content rose within 48 h (Saltveit, 2000a). In Iceberg, Romaine and Butterleaf lettuces caffeic acid derivatives were the major phenolics induced by wounding (Tomas-Barberan *et al.*, 1997a). These phenolic compounds are then thought to be readily oxidised by Polyphenol Oxidase (PPO), which leads to visual browning (Ke and Saltveit, 1989), but cutting had no effect on either PPO or peroxidase (POD) activities in a range of lettuce types (Degl'Innocenti et al., 2007), indicating that endogenous activity was sufficient to produce browning once increased PAL activity resulted in enhanced substrate content. In addition wounding caused cellular decompartmentalisation that allowed mixing of phenolic compounds at the cut surfaces with PPO (Tomas-Barberan et al., 1997b).

The wounding mechanism appears to be similar between lettuce types, but differences in susceptibility to browning could be caused by changes in enzymatic activity that lead to increases in phenolics, i.e. PAL activity (Saltveit, 2000a). Although wounding leaf tissue also leads to a transient increase in ethylene, it is not thought this is responsible for the increase in phenolics (Ke and Saltveit, 1989; Tomas-Barberan et al., 1997a).

Reactive oxygen species (ROS) are produced during senescence and wounding and may act as messengers during episodes of stress (Desikan et al., 2001). Hydrogen peroxide, for example, acted as a secondary messenger in tomato leaves after wounding by activating defence genes (Orozco-Cardenas et al., 2001). However, hydrogen peroxide is also known to oxidise phenols, producing browning in lettuce leaves, and this oxidation is catalysed by PPO and POD (Degl'Innocenti et al., 2007). Hence wounding can cause an increased incidence of browning in a range of salad mix ingredients.

There are two key steps to controlling the wound response in leaf tissues: inhibiting the induction of the wound signal; and/or direct inhibition of PAL activity (Saltveit, 2000a). The nature of the wound signalling compound(s) is not known, but ethylene, jasmonic acid, salicylic acid and ascorbic acid are known not to be involved (Saltveit, 2000a). PAL inhibitors, however, are well known and are dealt with in the section on browning inhibition.

As the wound response in plant tissue results in both the production of lignin and phenolic compounds via the phenylpropanoid pathway, fresh cut salad mixes would be expected to be higher in antioxidant capacity than intact leaves. However, this is not certain, as there are no known studies comparing antioxidant capacity in fresh cut and intact lettuce leaves, and the wound response is transitory. Not all lettuce mixes, contain tissues that are wounded other than during harvest. The popular 'Baby leaf' mixes are an example of this where all ingredients used are small and immature and therefore do not need size reduction through cutting after harvest. These ingredients behave differently to cut or shredded leaves as a) the phenylproponoid pathway may not be as highly induced, and b) there are few cut surfaces for antioxidant compounds (phenolics, ascorbic acid) to leach from during processing. More work is required to determine the antioxidant capacity of minimally processed mixes compared with intact leaves.

The process of peeling minimally processed ready to eat products also constitutes a unit operation which can induce a wound response. For example, industrially produced ready to use carrot disks are peeled using mechanical abrasion, which induces a wound response and thus phytochemical content. Initially, machine peeling resulted in a greater accumulation of phenolic compounds and increased total antioxidant activity compared to that of hand peeled carrot disks (Kenny and O'Beirne, 2010). However these higher levels were not maintained during storage.

10.5 Browning in lettuce leaves

Not all salad mix ingredients exhibit browning symptoms after cutting. While radicchio (also called chicory or escarole) and lettuce showed extensive browning during storage, symptoms were not seen until the end of the storage period in rocket leaves (Degl'Innocenti et al., 2007).

Red lettuce varieties also tend to be more resistant to browning symptoms (Degl'Innocenti et al., 2005). This resistance to browning is thought to be related to endogenous ascorbic acid content in the tissue, and is therefore also related to the 'health status' of lettuces. A green lettuce susceptible to browning showed a rapid loss in ascorbic acid during 72 h storage at 4 °C, while ascorbic acid increased in a resistant red variety stored under the same conditions (Degl'Innocenti et al., 2005). The authors hypothesised that endogenous ascorbic acid had a protective effect against browning in lettuce leaves, and the rapid loss of ascorbic acid in the green variety removed that protection, while in the resistant red variety ascorbic acid content increased, conferring resistance (Degl'Innocenti et al., 2005). Exogenous ascorbic acid is known to inhibit PPO and browning in plants (Alscher et al., 1997), and protected rocket leaves from browning by inhibiting PPO activity by reducing cytostolic pH (Degl'Innocenti et al., 2007). It is possible that browning resistant leaves in general may be higher in ascorbic acid and, therefore, have a higher inherent antioxidant capacity.

Inhibition of browning in leaf tissue can be achieved by slowing PAL enzyme activity with cold temperatures (<4°C), low O_2/high CO_2 atmospheres, or the application of exogenous inhibitors (Saltveit, 2000a). Cycloheximide is particularly effective at inhibiting PAL action, but cannot be used commercially (Saltveit, 2000a). Other, less toxic, post-harvest PAL inhibitors have been identified. For example, browning in lettuce tissue was inhibited by $CaCl_2$, acetic acid or 2,4-D, with all three treatments significantly depressing PAL activity (Tomas-Barberan et al., 1997b). Heat shock is an intriguing post-harvest browning inhibition treatment that can be very effective in lettuce tissues (Saltveit, 2000a). A short heat treatment (e.g. 90 second at 45 °C) effectively inhibited browning in Iceberg lettuce as the tissue preferentially synthesised heat shock proteins over PAL (Saltveit, 2000a). The wound signal appeared to dissipate before cells recovered from the heat shock, so PAL activity did not increase. All treatments that result in reduced PAL activity and/or inhibit the subsequent rise in phenolic content will also reduce antioxidant activity in lettuce tissues, leading to an inherent disparity between high antioxidant content and poor visual quality.

10.6 Refrigerated storage

Phenolic compounds and antioxidant capacity are generally stable during cool storage (<4 °C) of most leafy vegetables provided proper cool temperature control is adhered to (Jones et al., 2006), but performance during storage is variable depending on degree of cutting, variety and species. Storage of whole lettuce heads for 16 days at 4 °C induced an increase in total phenolics (Zhao et al., 2007), but lettuce flavonol glycosides declined 7–46% during seven days at 1 °C, and the rate of decline was cultivar dependent (DuPont et al., 2000). Total flavonoids in intact spinach leaves did not change during storage at 4 °C for seven days, but increased after three days in cut leaves (Bottino et al., 2009).

Results are also variable when cut leaves are cool stored. For example, 4 °C storage for 72 h maintained antioxidant capacity and phenolic content in fresh-cut radicchio, but both increased transiently in lettuce and rocket leaves (Degl'Innocenti et al., 2008). Similarly, total flavonoid content did not change in cut spinach leaves after three or seven days storage in air at 10 °C (Gil et al., 1999), but antioxidant capacity declined, due to a decrease in ascorbic acid. Total phenolics and antioxidant capacity both increased in cut Iceberg and Romaine lettuce leaves after 48 h storage at 10 °C (Kang and Saltveit, 2002), and a linear correlation was seen between phenolic content and antioxidant capacity as measured by FRAP and DPPH. Cut Lamb's lettuce stored at 4 °C for eight days also showed an increase in total phenolics, including anthocyanins, while carotenoids declined (Ferrante et al., 2009). In the lettuce variety Lollo Rosso, results were dependent on tissue colour (Ferreres et al., 1997). Storage at 5 °C for 7–14 days after cutting caused an increase in phenolics and anthocyanins in mid-ribs, but no changes were observed in phenolics in green or red tissues, while anthocyanins declined (Ferreres et al., 1997).

Little is known of the behaviour of carotenoids or glucosinolates contained in leafy vegetables during cool storage. Bunea et al. (2008) analysed phenolic and carotenoid compounds after cool storage or freezing of cut spinach and reported that while phenolics declined by approximately 20% during storage at 4 °C or −18 °C, only one day at 4 °C was sufficient to cause a 48% decline in violaxanthin content and a 40% loss of beta-carotene. Storage of shredded kale at 7–9 °C for five days caused a significant decrease in both beta-carotene and lutein (de Azevedo and Rodriguez-Amaya, 2005). It is therefore reasonable to assume that carotenoids in leafy vegetables decline significantly during cool storage. The behaviour of glucosinolates in leafy *Brassica*s during cool storage is also not clear. Total glucosinolates in cut rocket leaves increased after three days storage at either 4 °C or 15 °C (Kim and Ishii, 2007), but it is not known what effect cooling has on individual glucosinolates. In broccoli florets glucoraphanin content declined by 82% after five days at 20 °C, but by only 31% at 4 °C (Rodrigues and Rosa, 1999). Similarly, Rangkadilok et al. (2002) reported a 50% decrease in glucoraphanin in 'Marathon' heads after seven days at 20 °C, but no decrease after seven days at 4 °C. Indole glucosinolates, however, increased in concentration during nine days storage at 10 °C in 'Marathon' florets (Hansen et al., 1995), and total glucosinolates did not change significantly, indicating that the rise in indole glucosinolates may have masked any decline in alkenyl forms such as glucoraphanin. It is likely, therefore, that glucosinolates in leafy *Brassica*s decline during cool storage but the rate of loss is inhibited by temperatures ≤4 °C.

There is a marked tendency for ascorbic acid to decline during post-harvest storage, with lower temperatures acting to alleviate degradation (Lee and Kader, 2000). Ascorbic acid degraded between 2.7 and 2.9 times faster in lettuce leaves held at 8 °C or 15 °C, compared with 0 °C (Moreira et al., 2006). Levels were also better retained in cut rocket leaves stored at 4 °C compared with 15 °C (Kim and Ishii, 2007). Ascorbic acid in spinach leaves declined significantly when stored for 72 h at 4 °C, while total flavonoids did not change (Bottino et al., 2009). Despite the null effect of storage on total flavonoids, antioxidant capacity, as measured by FRAP, declined, reflecting the decline in ascorbic acid.

10.7 Modified atmosphere storage

Once minimally processed salad lines are cut and washed, they are most commonly packed in bags capable of modifying gas atmospheres. Modified Atmosphere Packaging (MAP) occurs when actively respiring produce is placed in sealed plastic bags with differential permeability that results in relatively low O_2 (<2%) and high CO_2 (>10%) atmospheres (Kays and Paull, 2004). These conditions are thought to generally result in greater antioxidant capacity retention in fruits and vegetables (Kalt, 2005), but this does not appear to be the case in salad mix ingredients. An atmosphere of 2–3% O_2 and 12–14% CO_2 inhibited the increase in phenolics in lettuce Lollo Rosso mid-rib tissue that was seen when leaves were stored in air, and also inhibited subsequent browning (Gil et al., 1998). A significant decline in phenolics was recorded in MAP-stored green and red lettuce tissues, particularly in red tissue which exhibited increased losses of anthocyanins and total phenolics compared with air storage, indicating that MAP was effective in preventing browning but may have adverse effects on phytochemical retention (Gil *et al.*, 1998). Total phenolic content in spinach leaves did not change when stored in either air or MAP for seven days at 10 °C (Gil *et al.*, 1999). Antioxidant capacity declined during MAP, however, partly due to a marked decrease in ascorbic acid content (Gil *et al.*, 1999). However, storage under low O_2/high CO_2 atmospheres for eight days caused marked declines in both flavonoids and glucosinolates contained in rocket leaves (Martinez-Sanchez *et al.*, 2006). In contrast, ascorbic acid was retained in rocket leaves stored for eight days in either air or low O_2/high CO_2 atmospheres (Martinez-Sanchez *et al.*, 2006). Flushing with N_2 resulted in better ascorbic acid retention in Iceberg lettuce than low O_2/high CO_2 atmospheres (Barry-Ryan and O'Beirne, 1999).

Little is known of the effect of MAP on glucosinolate content in leafy *Brassica*s, but reports on broccoli florets offer some direction. When broccoli heads were stored at 4 °C there was no difference in the glucoraphanin levels between air and MAP after ten days storage (Rangkadilok *et al.*, 2002). At 20 °C, however, broccoli stored in air lost 50% of its glucoraphanin in seven days, while under MAP there was no significant decrease in glucoraphanin over ten days (Rangkadilok *et al.*, 2002). In comparison with the glucosinolate content of freshly harvested broccoli, glucoraphanin content of Marathon broccoli heads stored for seven days at 1 °C under MAP decreased by approximately 48% (Vallejo *et al.*, 2003). A further 17% was lost after three days at 15 °C. If temperatures rise above 4 °C, as they commonly do in the retail environment, then both atmospheres and RH are important factors in maintaining glucosinolate levels in *Brassica*s. At higher temperatures, CA studies show that O_2 levels below 1.5% and CO_2 above 6% maintained or improved glucosinolate levels (Hansen *et al.*, 1995; Rangkadilok *et al.*, 2002). We can conclude, therefore, that MAP may be useful in maintaining glucosinolate content after harvest of leafy *Brassica*s, providing that the atmospheres reached and/or RH achieved were sufficient to have prevented membrane degradation and subsequent mixing of glucosinolates with myrosinase. Based on these studies, we conclude that MAP alone is insufficient to adequately maintain phytochemical and ascorbic acid content in minimally processed salad ingredients, and proper temperature management is of primary importance.

10.8 Conclusions

Minimal processing can have a marked effect on both phytochemical and ascorbic acid content in salad mixes. Cutting and shredding commonly induce a wound response in leafy tissues that results in induction of the phenylpropanoid pathway via PAL, and a concomitant increase in phenolic compounds that leads to higher *in vitro* antioxidant capacity. However, cutting and shredding also lead to a loss of visual quality by enhanced browning and antioxidant loss via leaching during the washing process. Refrigerated storage maintains antioxidant capacity in salad ingredients, providing temperatures are kept at 4 °C or lower. Modified atmosphere packaging reduced antioxidant capacity compared with leaves stored in air. In most commercial salad mixes available today, the use of baby leaves which undergo minimal cutting, refrigeration during marketing and MA packs adequately maintain ascorbic acid content, but may result in lower antioxidant content compared with whole lettuce heads of the same variety and age. More research is required to clarify this.

References

Alscher, R.G., Donahue, J.L. and Cramer, C.L. (1997) Reactive oxygen species and antioxidants: relationships in green cells. *Physiologia Plantarum*, 100, 224–233.

Barry-Ryan, C. and O'Beirne, D. (1999) Ascorbic acid retention in shredded Iceberg lettuce as affected by minimal processing. *Journal of Food Science*, 64, 498–500.

Bottino, A., degl'Innocenti, E., Guidi, L., Graziani, G. and Fogliano, V. (2009) Bioactive compounds during storage of fresh-cut spinach: the role of endogenous ascorbic acid in the improvement of product quality. *Journal of Agricultural and Food Chemistry*, 57, 2925–2931.

Bunea, A., Andjelkovic, M., Socaciu, C., Bobis, O., Neascu, M., Verhe, R. and Van Camp, J. (2008) Total and individual carotenoids and phenolic acids content in fresh, refrigerated and processed spinach (*Spinacia oleracea* L.). *Food Chemistry*, 108, 649–656.

Chen, L., Vigneault, C., Raghavan, G.S.V. and Kubow, S. (2007) Importance of the phytochemical content of fruits and vegetables to human health. *Stewart Postharvest Review*, 3, 1–5.

Cook, R. (2004) Trends in the marketing of fresh produce and fresh-cut products. In: Fresh-cut Products: Maintaining Quality and Safety. University of California Davis, CA. http://postharvest.ucdavis.edu/datastorefiles/234-680.pdf. Accessed 23/09/10.

Cruz, R.M.S., Vieira, M.C. and Silva, C.L.M. (2009) Effect of cold chain temperature abuses on the quality of frozen watercress (*Nasturtium officinale* R.Br.). *Journal of Food Engineering*, 94, 90–97.

de Azevedo, C.H. and Rodriguez-Amaya, D.B. (2005) Carotenoid composition of kale as influenced by maturity, season and minimal processing. *Journal of the Science of Food and Agriculture*, 85, 591–597.

Degl'Innocenti, E., Guidi, L., Pardossi, A. and Tognini, F. (2005) Biochemical study of leaf browning in minimally processed leaves of lettuce (*Lactuca sativa* L. Var. *Acephela*). *Journal of Agricultural and Food Chemistry*, 53, 9980–9984.

Degl'Innocenti, E., Pardossi, A., Tognini, F. and Guidi, L. (2007) Physiological basis of sensitivity to enzymatic browning in lettuce, escarole and rocket salad when stored as fresh-cut products. *Food Chemistry*, 104, 2209–2215.

Degl'Innocenti, E., Pardossi, A., Tattini, M. and Guidi, L. (2008) Phenolic compounds and antioxidant power in minimally processed salad. *Journal of Food Biochemistry*, 32, 642–653.

Desikan, R., Mackerness, S.A.H., Hancock, J.T. and Neill, S.J. (2001) Regulation of the Arabidopsis transcriptome by oxidative stress. *Plant Physiology*. 127, 159–172.

Dixon, R.A. and Paiva, N.L. (1995) Stress-induced phenylpropanoid metabolism. *The Plant Cell*, 7, 1085–1097.

DuPont, M.S., Mondin, Z., Williamson, Z. and Price, K.R. (2000) Effect of variety, processing and storage on the flavonoid glycoside content and composition of lettuce and endive. *Journal of Agriculture and Food Chemistry*, 48, 3957–3964.

Dyer, W.E., Henstrand, J.M., Handa, A.K. and Herrmann, K. (1989) Wounding induces the first enzyme of the shikimate pathway in Solonaceae. *Proceedings of the National Academy of Science USA*, 86, 7370–7373.

Ferrante, A., Martinetti, L. and Maggiore, T. (2009) Biochemical changes in cut vs. intact lamb's lettuce (*Valerianella olitoria*) leaves during storage. *International Journal of Food Science and Technology*, 44, 1050–1056.

Ferreres, F., Gil, M., Castaner, M. and Tomas-Barberan, F.A. (1997) Phenolic metabolites in red pigmented lettuce (*Lactuca sativa*) changes with minimal processing and cold storage. *Journal of Agriculture and Food Chemistry*, 45, 4249–4254.

Gil, M.I., Ferreres, F. and Tomas-Barberan, F.A. (1999) Effect of postharvest storage and processing on the antioxidant constituents (flavonoids and vitamin C) of fresh-cut spinach. *Journal of Agriculture and Food Chemistry*, 47, 2213–2217.

Gil, M.I., Castaner, M., Ferreres, F., Artes, F. and Tomas-Barberan, F.A. (1998) Modified-atmosphere packaging of minimally processed 'Lollo Rosso' (*Lactuca sativa*). Phenolic metabolites and quality changes. *Zeitschrift fur Lebensmittel Untersuchung und Forschung*, 206, 350–335.

Hansen, M., Moller, P., Sorensen, H. and Cantwell de Trejo, M. (1995) Glucosinolates in broccoli stored under controlled atmosphere. *Journal of the American Society of Horticultural Science*, 120, 1069–1074.

Hawkins, S. and Boudet, A. (1996) Wound-induced lignin and suberin deposition in a woody angiosperm (*Eucalyptus gunnii* Hook.): histochemistry of early changes in young plants. *Protoplasma*, 191, 96–104.

Jin, J., Koroleva, O.A., Gibson, T., Swanston, J., Magan, J., Zhang, Y., et al (2009) Analysis of phytochemical composition and chemoprotective capacity of rocket (*Eruca sativa* and *Diplotaxis tenuifolia*) leafy salad following cultivation in different environments. *Journal of Agricultural and Food Chemistry*, 57, 5227–5234.

Jones, R.B., Premier, R. and Tomkins, R.B. (2006) The effects of post-harvest handling conditions on phytochemicals important for human health contained in fruits and vegetables. In Nouredine, B., and Norio, S. (eds.) *Advances in Postharvest Technologies for Horticultural Crop*, Kerala: Research Signpost, pp. 1–20.

Kalt, W. (2005) Effects of production and processing factors on major fruit and vegetable antioxidants. *Journal of Food Science*, 70, R11–R19.

Kang, H.M. and Saltveit, M.E. (2002) Antioxidant capacity of lettuce leaf tissue increases after wounding. *Journal of Agricultural and Food Chemistry*, 50, 7536–7541.

Kays, S.J. and Paull, R.E. (2004) Postharvest Biology. In Kays, S.J., and Paull, R.E. (eds.) *Postharvest Biology*, Atlanta Georgia: Exon Press.

Ke, D. and Saltveit, M.E. (1989) Wound-induced ethylene production, phenolic metabolism and susceptibility to russett spotting in iceberg lettuce. *Physiologia Plantarum*, 76, 412–418.

Kenny, O. O'Beirne, D. (2010). Antioxidant phytochemicals in fresh-cut carrot disks as affected by peeling method. *Postharvest Biology and Technology*, 58, 247–253

Kim, S.J. and Ishii, G. (2007) Effect of storage temperature and duration on glucosinolate, total vitamin C and nitrate contents in rocket salad (*Eruca sativa* Mill.). *Journal of the Science of Food and Agriculture*, 87, 966–973.

Kushad, M.M., Masiunas, J., Kalt, W., Eastman, K. and Smith, M.A.L. (2003) Health promoting phytochemicals in vegetables. *Horticultural Reviews*, 28, 125–185.

Le Marchand, L. (2002) Cancer preventive effects of flavonoids - a review. *Biomedicine and Pharmacotherapy*, 56, 296–301.

Lee, S.K. and Kader, A.A. (2000) Preharvest and postharvest factors influencing vitamin C content of horticultural crops. *Postharvest Biology and Technology*, 20, 207–220.

Martinez-Sanchez, A., Gil-Izquierdo, A., Gil, M.I. and Ferreres, F. (2008) A comparative study of flavonoid compounds, vitamin C, and antioxidant properties of baby leaf Brassicaceae species. *Journal of Agricultural and Food Chemistry*, 56, 2330–2340.

Martinez-Sanchez, A., Allende, A., Bennett, R.N., Ferreres, F. and Gil, M. (2006) Microbial, nutritional and sensory quality of rocket leaves as affected by different sanitizers. *Postharvest Biology and Technology*, 42, 86–97.

Moreira, M.d.R., Ponce, A.G., del Valle, C.E. and Roura, S.I. (2006) Ascorbic acid retention, microbial growth, and sensory acceptability of lettuce leaves subjected to mild heat shocks. *Journal of Food Science*, 71, S188–S192.

Nijveldt, R.J., van Noord, E., van Hoorn, D.E.C., Boelens, P.G., van Norren, K. and van Leeuwen, P.A.M. (2001) Flavonoids: a review of probable mechanisms of action and potential applications. *American Journal of Clinical Nutrition*, 74, 418–425.

Orozco-Cardenas, M.L., Narvaez-Vasquez, J. and Ryan, C.A. (2001) Hydrogen peroxide acts as a second messenger for the induction of defense genes in tomato plants in response to wounding, systemin and methyl jasmonate. *The Plant Cell*, 13, 179–191.

Rangkadilok, N., Tomkins, R.B., Nicolas, M.E., Premier, R.R., Bennett, R.N., Eagling, D.R. and Taylor, P.W.J. (2002) The effect of post-harvest and packaging treatments on glucoraphanin concentration in broccoli (*Brassica oleracea var. italica*). *Journal of Agricultural and Food Chemistry*, 50, 7386–7391.

Reyes, L.F., Villareal, J.E. and Cisneros-Zevallos, L. (2007) The increase in antioxidant capacity after wounding depends on the type of fruit or vegetable tissue. *Food Chemistry*, 101, 1254–1262.

Richards, D. (2003) Horticultural products as functional foods – a consumers' perspective. P. 65. Proceedings of the Australasian Postharvest Conference, 1–3 October 2003, Brisbane, Australia.

Rico, D., Martin-Diana, A.B., Barat, J.M. and Barry-Ryan, C. (2007) Extending and measuring the quality of fresh-cut fruit and vegetables: a review. *Trends in Food Science and Technology*, 18(7), 373–386.

Rochfort, S.J., Imsic, M., Jones, R.B., Tomkins, R.B. and Trenerry, C.V. (2006) Characterisation of Flavonol Conjugates in Immature Leaves of Pak Choi (Brassica rapa L. ssp. chinensis L. Hanelt.) by HPLC-DAD and LC-MS/MS. *Journal of Agricultural and Food Chemistry*, 54, 4855–4860.

Rosa, E.A.S., Heany, R.K., Fenwick, G.R. and Portas, C.A.M. (1997) Glucosinolates in Crop Plants. In Janick, J. (ed.) *Horticultural Reviews*, Chichester, John Wiley & Sons, Ltd., pp. 99–215.

Saltveit, M.E. (2000a) Wound induced changes in phenolic metabolism and tissue browning are altered by heat shock. *Postharvest Biology and Technology*, 21, 61–69.

Saltveit, M.E. (2000b) Wound induced changes in phenolic metabolism and tissue browning are altered by heat shock. *Postharvest Biology and Technology*, 21, 61–69.

Steinmetz, K.A. and Potter, J.D. (1996) Vegetables, fruit and cancer prevention: a review. *Journal of the American Dietetic Association*, 96, 1027–1039.

Tomas-Barberan, F.A. and Espin, J.C. (2001) Phenolic compounds and related enzymes as determinants of quality in fruits and vegetables. *Journal of the Science of Food and Agriculture*, 81, 853–876.

Tomas-Barberan, F.A., Loaiza-Velarde, J.G., Bonfanti, A. and Saltveit, M.E. (1997a) Early wound- and ethylene-induces changes in phenylpropanoid metabolism in harvested lettuce. *Journal of the American Society for Horticultural Science*, 122, 399–404.

Tomas-Barberan, F.A., Gil, M., Castaner, M., Artes, F. and Saltveit, M.E. (1997b) Effect of selected browning inhibitors on phenolic metabolism in stem tissue of harvested lettuce. *Journal of Agricultural and Food Chemistry*, 45, 583–589.Vallejo, F., Tomas-Barberan, F.A. and Garcia-Viguera, C. (2003) Health-promoting compounds in broccoli as influenced by refrigerated transport and retail sale period. *Journal of Agricultural and Food Chemistry*, 51, 3029–3034.

Verkerk, R., Dekker, M., Jonden, W.M.F. and Lindsay, D.G. (2001) Post-harvest increase of indolyl glucosinolates in response to chopping and storage of Brassica vegetables. *Journal of the Science of Food and Agriculture*, 81, 953–958.

Wilson, P.E., Morrison, S.C., Hedges, L.J., Kerkhofs, N.S. and Lister, C.E. (2004) Phenolics contribute significantly to higher antioxidant activity of red lettuce compared to green lettuce. In *XXII International Conference on Polyphenols* Helsinki, Finland, pp. 273–274.

Yadav, S.K. and Sehgal, S. (1995) Effect of home processing on ascorbic acid and B-carotene content of spinach (*Spinacia oleracia*) and amaranth (*Amaranthus tricolor*) leaves. *Plant Foods for Human Nutrition*, 47, 125–131.

Zhao, X., Carey, E.E., Young, J.E., Wang, W. and Iwamoto, T. (2007) Influences of organic fertilization, high tunnel environment, and postharvest storage on phenolic compounds in lettuce. *HortScience*, 42, 71–76.

11 Thermal processing

Nigel P. Brunton
School of Agriculture and Food Science, University College Dublin, Dublin, Ireland

11.1 Introduction

Whilst a wealth of new technologies (and these are reviewed elsewhere in this book) are available that can be used to render a food safe or improve organoleptic properties, the application of heat is still the most common form of processing applied to all foods. From an industrial perspective most manufacturers of plant foods employ thermal processing in some form before their foods appear on supermarket shelves. As outlined in Chapters 3 and 4 the importance of phytochemicals from plant foods has long been recognised and therefore a wealth of information exists as to how thermal processing can affect these important components. However, to keep a pace with consumer and industrial trends thermal processing techniques are continuously evolving. Therefore there is a need to keep abreast with how recent advances affect the phytochemical content of plant foods. A principle objective therefore of the present chapter is to review and critically evaluate contemporary work in this area with view to providing plant food manufacturers and researchers with a state of the art view of the area. As alluded to in many of the chapters of the 'the handbook' many thousands of phytochemicals have been identified and to give an overview of how thermal processing affects them all is beyond the remit of a single book chapter. Therefore I have adopted the approach of selecting five phytochemical groups as case studies these are (1) Polyphenols and anthocyanins, (2) Carotenoids, (3) Glucosinolates/ Isothiocyanates, (4) Polyacetylenes and (5) Ascorbic acid. The rationale for the selection of these groups in based on diversity of chemical and physical properties, emerging significance and depth and volume of knowledge presently available. The proceeding chapter is divided into five sections based on the nature of the thermal strategy adopted. The order of the sections is based on severity of thermal challenge starting with the least severe (blanching) and finishing with most severe (frying).

Handbook of Plant Food Phytochemicals: Sources, Stability and Extraction, First Edition.
Edited by B.K. Tiwari, Nigel P. Brunton and Charles S. Brennan.
© 2013 John Wiley & Sons, Ltd. Published 2013 by John Wiley & Sons, Ltd.

11.2 Blanching

Blanching, especially of vegetables, is an essential step for maintaining the quality of the final products as it inactivates enzymes that would otherwise lead to visual and/or organoleptic deterioration of the final product. Blanching is typically carried out prior to subsequent thermal processing by immersing the vegetable in water at 90–100°C for a relatively short period of time. In general it is used in industrial process where there is a lag between processing steps that could lead to losses in quality due to enzymatic activity. Many authors have investigated the effect of blanching on the phytochemical content of plant foods and Table 11.1 lists some recent examples of studies in this area. Whilst blanching is generally considered to be a mild thermal treatment, in many cases investigators have reported that it leads to significant losses in levels of phytochemical groups. This is especially true for ascorbic acid which is generally considered to be the most thermally labile of phytonutrients present in plant foods. For example, losses of this compounds from 55% in a selection of cruciferous vegetables (Cieślik, Leszczyńska, Filipiak-Florkiewicz, Sikora and Pisulewski, 2007) to 34% in kale has been reported (Korus and Lisiewska, 2011). As outlined, losses of this nature are not surprising given the thermally labile nature of the compound and the fact that it is reasonably hydrophilic, therefore losses via leaching into the surrounding water would be expected. The severity of the blanching conditions can also affect the stability of ascorbic acid. Castro *et al.* (2008) reported that ascorbic acid content decreased progressively to about 45 and 30% of the initial value as the severity of blanching conditions increased from 70–98°C and 1–2.5 min. Most authors have not investigated thermal degradation products of ascorbic acid, however it is likely that products such as furfural, 2-furoic acid, 3-hydroxy-2-pyrone (Yuan and Chen, 1998) that have been reported for model solutions are formed.

Whilst glucosinolates are not themselves considered to impart health promoting properties, in most cases investigators have measured the levels of these compounds in blanched plant foods rather than their biologically active degradation products isothiocyanates. The assumption therefore is that there is a direct relationship between glucosinolate content and isothiocyante levels, which may not be the case as thermal inactivation of myrosinase, the enzyme responsible for conversion of glucosinolates to isthiocyanates, may also have occurred. Nevertheless numerous reports have indicated that glucosinolates are susceptible to losses when subjected to blanching. In two related studies Volden *et al.* (2008, 2009) reported losses of glucosinolates of up to 37% following blanching. Significantly in both studies the authors reported that blanching had a greater influence on glucosinolate content than the other thermal treatments examined, which included boiling and steaming. Few authors have studied the thermal degradation of glucosinolates in plant foods, however Oerlemans *et al.* (2006) investigated the relative thermal stability of indole and aliphatic glucosinolates in red cabbage in which myrosinase had been inactivated to eliminates its role in the degradation process. The authors concluded that following blanching at 90°C for 3 min both indole and aliphatic glucosinolates had similar predicted degradation rates. Polyphenols are also susceptible to loss following blanching either via thermal degradation or leaching. In fact in some cases higher losses of polyphenols than ascorbic acid have been reported. For example, Mayer-Miebach *et al.* (2003) reported that polyphenol losses of up to 50% occurred when endive was blanched at low temperatures (50–55°C, 5–10 min). As is the case for ascorbic acid more severe blanching conditions result in greater losses of polyphenols with Jaiswal *et al.* (2012) reporting that total phenolic

Table 11.1 Effect of blanching on levels of phytochemicals in plant foods

Food	Bioactive compounds	Salient results	Reference
Bell peppers	Ascorbic acid	Not affected by thermal blanching, with the exception of the more severe condition (98 C/150 s)	Castro, Saraiva, Domingues, and Delgadillo (2011)
Sweet green and red bell pepper fruits (*Capsicum annuum L.*)	Ascorbic acid	Ascorbic acid content decreased progressively as blanching conditions were more severe to about 45% and 30% of the initial value	Castro et al. (2008)
Selection of cruciferous vegetables	Total glucosinolates	Blanching led to losses of total glucosinolates from 2.7 to 30.0%	Cieślik, Leszczyńska, Filipiak-Florkiewicz, Sikora and Pisulewski (2007)
Irish York cabbage	Total Phenols and flavonoid content	Total phenolic and flavonoid content retained ranged from 19.6–24.5% to 22.0–25.7%, respectively	Jaiswal, Gupta, and Abu-Ghannam (2012)
Kale (*Brassica oleracea L.* var. *acephala*) leaves	Vitamin C, total phenols and antioxidative activity	34% decrease in vitamin C, 51% decrease in polyphenols and a 33% decrease in antioxidative activity	Korus and Lisiewska (2011)
Endive	Polyphenol content and antioxidant capacity	Reduced polyphenol content and antioxidant capacity to 50% compared to the salad washed with cold water	Mayer-Miebach, Gärtner, Großmann, Wolf and Spieß (2003)
Kintoki carrots	Lycopene and β-carotene	β-carotene slightly decreased but at 90°C but lycopene stable from 50–90°C	Mayer-Miebach and Spieß (2003)
Selection of vegetables	Phenolics, Sterols and carotenoids	Significant losses (20–30%) of antioxidant activity and total phenolics. Carotenoids and sterols were not affected by blanching	Puupponen-Pimiä, Häkkinen, Aarni, Suortti, Lampi, Eurola, et al. (2003)
Organically grown carrots	Polyacetylenes, carotenoids and isocoumarins	Level of falcarinol increased and that of falcarindiol and falcarindiol-3-acetate decreased. No changes were observed in the content of carotenoids and 6-methoxymellein	Kidmose, Hansen, Christensen, Edelenbos, Larsen and Nørbæk (2004)
Cauliflower (*Brassica oleracea L.* ssp. *botrytis*)	Indole glucosinolates (GLS) L-ascorbic acid (L-AA), total phenols (TP), anthocyanins, FRAP and ORAC	Reduced total aliphatic and indole glucosinolates (GLS) by 31% and 37%, respectively. L-ascorbic acid (L-AA), total phenols (TP), anthocyanins, FRAP and ORAC by 19, 15, 38, 16 and 28%, respectively	Volden, Bengtsson and Wicklund (2009)
Brassica oleracea L. ssp. *botrytis*	Glucosinolates (GLS), total phenols (TP), total monomeric anthocyanins (TMA), L-ascorbic acid (L-AA)	Blanching, reduced total aliphatic and indole glucosinolates (GLS) by 31% and 37%, respectively. L-ascorbic acid (L-AA), total phenols (TP), anthocyanins, FRAP and ORAC were reduced by 19, 15, 38, 16 and 28%, respectively	Volden, Borge, Hansen, Wicklund and Bengtsson (2009)
Red cabbage (*Brassica oleracea L.* ssp. *capitata f. rubra*)	Glucosinolates (GLS), total phenols (TP), total monomeric anthocyanins (TMA), L-ascorbic acid (L-AA)	Levels were reduced: TP, 43%; TMA 59%, FRAP 42%, ORAC 51%, L-AA 48%. Total GLS reduced by 64%	Volden, Borge, Bengtsson, Hansen, Thygesen and Wicklund (2008)
White cabbage (*Brassica oleracea var. capitata*)	Glucosinolates	Reduction of 74 and 50% in Predikant Heckla species respectively	Wennberg, Ekvall, Olsson and Nyman (2006)

and flavonoid content retention ranged from 19.6–24.5% to 22.0–25.7% in Irish York cabbage. In this case the authors did not speculate as to the thermal fate of the lost phenolics, however they did suggest that leaching was the major route to loss. Polyphenols in most cases are mildly polar and therefore will be solubilised when immersed in hot water. Interestingly however Jaiswal et al. (2012) reported that at high temperatures an increase of 7–12% in the levels of polyphenols was observed when they are expressed on a dry weight basis. This phenomenon has been reported for other phytochemicals and most authors have attributed it to a loss of soluble solids into the leaching water without a corresponding of loss of the phytochemical into the water resulting in a net increase in levels of the compound when expressed on a dry weight basis. However in most cases this has been shown to occur for hydrophobic molecules such as carotenoids and not a largely polar entity such as a polyphenol. In fact Kidmose et al. (2004) reported that levels of the hydrophobic polyacetylene falcarinol increased in blanched organically grown carrots compared to their fresh counterparts even when expressed on a fresh weight basis. However other polyacetylenes measured (Falcarindiol and falcarindiol-3-actetate) were reported to decrease significantly in blanched samples. Similar to polyacetylenes, carotenoids are hydrophobic molecules and this combined with a degree of heat stability means that most studies have concluded that carotenoids are either unaffected by blanching (Kidmose, Hansen, Christensen, Edelenbos, Larsen and Nørbæk, 2004) or decrease slightly (Mayer-Miebach and Spieß, 2003).

In recognition of the deleterious effect blanching can have on phytochemical content some authors have investigated the potential of other enzyme inactivation routes with a view to increasing retention of these compounds during this crucial step. Rawson et al. (2011) reported that replacing water immersion based blanching with ultrasound pre-treatment could significantly improve the retention of polyacetylenes in freeze and hot air dried carrot disks. Other alternatives to water immersion based blanching such as superheated steam and hot water spray (Sotome, Takenaka, Koseki, Ogasawara, Nadachi, Okadome, et al., 2009) are also available and could help minimise leaching based losses of phytochemicals in plant foods.

11.3 Sous vide processing

Sous vide processing, which involves thermal treatment of foods in vacuumised packs at temperatures of 90°C, is often considered a minimal processing strategy, however it has the ability to impart an appreciable shelf of up to 25 days at 4°C to a plant food. In many cases sous vide processing can deliver these shelve lives whilst having less of an effect on quality attributes such as colour, nutritional quality and flavour. In theory sous vide should have many advantages with respect to retention of phytochemicals over other methods because (1) the influence of leaching into the surrounding water is eliminated as foodstuffs are not directly in contact with water and (2) oxidatively labile compounds should be protected as foods are heated and stored under a vacuum . However, examination of Table 11.2 reveals that to date sous vide cooking has not delivered on its considerable potential for retention of phytochemicals in plant foods. Table 11.2 lists seven recently conducted studies on the effect of sous vide processing on a number of phytochemical groups and in only one case the technique was reported to have no effect on phytochemical content (Rawson, Koidis, Rai, Tuohy and Brunton, 2010). Even in the study where no effect was observed sous vide gave no additional degradation of the phytochemical studied (polyacetylenes) following blanching. In all other cases substantial reduction in levels of phytochemicals were observed.

Table 11.2 Effect of sous vide processing on levels of phytochemicals in plant foods

Food	Bioactive compounds	Salient results	Reference
Parsnips	Polyacetylenes	SV processing did not result in additional significant losses in polyacetylenes compared to blanched samples	Rawson, Koidis, Rai, Tuohy and Brunton (2010)
Swede	L-Ascorbic acid, Total phenolics, DPPH, FRAP	Percentage recoveries after sous-vide were, 62.6, 81.3, 65.1 and 69.7 for L-Ascorbic acid, Total phenolics, DPPH and FRAP respectively	Baardseth, Bjerke, Martinsen and Skrede (2010)
Green beans	L-Ascorbic acid, Total phenolics, DPPH, FRAP	Percentage recoveries after sous-vide were, 66.8, 89.2, 75.1 and 74.2 for L-Ascorbic acid, Total phenolics, DPPH and FRAP respectively.	Baardseth, Bjerke, Martinsen and Skrede (2010)
Apple purees	Total Phenolic Index and levels of individual phenols	Total phenolic index decreased by 36%, levels of chlorogenic acid decreased by 47% in sous-vide processed purees.	Keenan, Brunton, Butler, Wouters and Gormley (2011)
Carrot disks	Anti-radical power (DPPH method) and Total Phenols	ARP decreased significantly by 20% in sous-vide processed samples as compared to uncooked. TP decreased by 29.2% in sous vide processed samples.	Patras, Brunton and Butler (2010)
Carrot disks	Polyacetylenes	Following SV processing there was a significant decrease ($p < 0.05$) in the levels of all the three polyacetylenes.	Rawson, Brunton and Tuohy (2012)

In the most severe case sous vide processing resulted in a 47% reduction in chlorogenic acid in sous vide processed apple purees (Keenan, Brunton, Butler, Wouters and Gormley, 2011). Sous vide processing has also been shown to result in significant reduction in antioxidant capacity, total phenolic content and levels of ascorbic acid in green beans and swede (Baardseth, Bjerke, Martinsen and Skrede, 2010) and carrot disks (Patras, Brunton and Butler, 2010). The reason why the use of sous vide processing has not increased retention of phytochemicals in plant foods remains unclear and in common with other thermal strategies most studies to date have concentrated on merely quantifying the effect of the method on phytochemical content rather than developing and understanding the underlying causes of the effects observed. More detailed investigation are therefore necessary which concentrate on degradation routes, hence providing recommendations for preserving phytochemicals in plant foods subjected to sous vide processing.

11.4 Pasteurisation

Although in theory pasteurisation can apply to a food in any form, it usually refers to the application of heat to reduce viable pathogens in liquids and for the purposes of this chapter only pasteurisation of liquids will be considered. Pasteurisation is usually carried out at temperatures below boiling (although this depends on the food to which it is applied) and the purpose is to increase shelf life without having an adverse effect on the eating quality of the food. Table 11.3 lists a sample of recent studies concerned with the effect pasteurisation has on the phytochemical properties of mostly plant based beverages. A wide variety of responses

Table 11.3 Effect of pasteurisaton on levels of phytochemicals in plant foods

Food	Bioactive compounds	Salient result	Reference
Yellow banana Peppers (*Capsicum annuum*)	Ascorbic acid, quercetin, luteolin	Processing reduced ascorbic acid content by 63%. Quercetin and luteolin contents declined 45%.	Lee and Howard (1999)
Yellow passion fruit (*Passiflora edulis*)	Total Phenolic, total carotenoids and ascorbic acid	Total phenolic and carotenoids unchanged but 25% decrease in ascorbic acid	Talcott, Percival, Pittet-Moore and Celoria (2003)
Pomegranate juice	Total phenolics	Total phenol reduced by 7.1% with no clarification	Alper, Bahçeci and Acar (2005)
Iranian pomegranate (*Punica granatum L.*)	Total anthocyanins and levels of individual anthocyanins	Total anthocyanins reduced by 14% pasteurisation but diglucoside ACs increased slightly	Alighourchi, Barzegar and Abbasi (2008)
Fuit Smoothie	Total antioxidant capacity and Total phenols and ascorbic acid	Reduced Total antioxidant capacity and Total phenols and ascorbic acid	Keenan, Rößle, Gormley, Butler and Brunton (2012)
Highland blackberry (*Rubus adenotrichus*)	Total phenolics and levels of individual anthocyanins	Slight increase of total phenolic compounds (11%), No significant change in total or individual anthocyanins,	Gancel, Feneuil, Acosta, Pérez and Vaillant (2011)
Cruciferae sprouts	Aliphatic, indolic and aryl glucosinolates	Pasteurisation of rapeseed sprouts caused a decrease in the content of all aliphatic, indolic and aryl GLS by 49%, 59% and 100% respectively	Piskuła and Kozłowska (2005)
Selection of *Brassicas*	Isothiocyanates	Isothiocyanates reduced in pasteurised samples with reduction ranging from 49.2% in broccoli to 17% in brussel sprouts	Tříska, Vrchotová, Houška and Strohalm (2007)

to pasteurisation have been reported ranging from no change to significant decreases. It should be noted also that processing of plant based beverages usually involves other unit processes apart from pasteurisation, which can also affect phytochemical content.

In common with other thermal strategies pasteurisation usually results in a decrease in ascorbic acid content. Table 11.3 lists a number of studies where decreases ranging from 25% (Lee and Howard, 1999) to 63% (Talcott, Brenes, Pires and Del Pozo-Insfran, 2003) have been reported. In the case of pasteurisation of beverages reductions of this nature are purely a reflection of the heat and oxidative lability of the compound as the beverage is not in direct contact with the heating medium and therefore no leaching can occur. Fruit juices are the most popular item in the plant based beverage category and some of these products are well recognised sources of ascorbic acid. Therefore losses of this compound as a result of pasteurisation could undermine the marketability of the product. Despite this no effective strategy seems to be available to limit losses as a response to pasteurisation.

As is often the case with polyphenols and anthocyanins in plant foods such a variety of responses to pasteurisation have been reported that it is difficult to come to a definite conclusion on the subject. For example, Alighourchi *et al*. (2008) reported that pasteurisation reduced total anthocyanins by 14% in an Iranian pomegranate juice. Alper *et al*. (2005) also reported that phenolic content was reduced by 7.1% in a pomegranate juice. Keenan *et al*. (2012) reported that pasteurisation of a fruit smoothie reduced total antioxidant capacity and total phenols. In contrast, Gancel *et al*. (2011) reported that there was a slight increase in total phenolic compounds (11%) and no significant change in total or individual anthocyanins. The major route to enzymatic degradation of polyphenols is of course via the action of polyphenol oxidases (PPO) and maceration of whole fruits and vegetables will place cell content in contact with this extracellular enzyme. Therefore some of the degradation reported may be due to degradation of polyphenols by PPO prior to pasteurisation. In fact Keenan *et al*. (2012) reported that PPO activity in a fresh fruit smoothie increased significantly in the first 10 h after preparation.

The effect of pasteurisation on levels of glucosinolates appears to be more straight forward, with most authors reporting that pasteurisation reduced levels of this phytochemical group (Piskuła and Kozłowska, 2005). Similar to polyphenol oxidase maceration of foods during the preparation of plant food beverages places the enzyme responsible for breakdown of glucosinolates in contact with its substrate, thus resulting in a reduction in glucosinolate content. However it is this breakdown product itself that is active against Phase I and Phase II enzymes. Taking glucoraphin as an example, the active breakdown product is sluphoraphane but depending on the action of epithiospecifier protein (ESP) either the active isothiocyanate sulurophane is formed or the less active sulurophane nitrile. When conditions are favourable for ESP activity more of the nitrile is formed. The principle is important here when discussing pasteurisation as the temperatures required to deliver this heat treatment are close to those required to inactivate ESP. Therefore some authors have shown that heat treatments at temperatures 60–70°C for 5–10 min favour formation of sulurophane but not the nitrile (Matusheski, Juvik and Jeffery, 2004). At temperatures above 100°C (for 5–15 min) no isothiocyantes are formed as myrosinase itself is inactivated. The question therefore arises as to the heat stability of isothiocyantes themselves as many are volatile and thus susceptible to loss by evaporation. It would appear that isothiocyanate stability is mostly a function of the matrix in which it is found. For example Rose *et al*. (2000) reported that methylthioalkyl isothiocyanates from watercress were not present in aqueous extracts due to their volatility. However Ji *et al*. (2005) found that the methylthioalkyl isothiocyanate phenylethyl isothiocyanate (PEITC) was stable in aqueous buffers at pH 7.4. Thus it is possible that juices made from cruciferous vegetables could contain significant amounts of isothiocyanates providing they were made within 24 h and were refrigerated. In fact it has been reported that while isothiocyanates were reduced by levels of 17–49% in a range of cruciferous vegetables they were present in the pasteurised samples (Tříska, Vrchotová, Houška, and Strohalm, 2007).

11.5 Sterilisation

The objective of sterilisation is to render a foodstuff safe for long-term storage at ambient temperatures. Whilst consumers are demanding more fresh-like products, especially for plant foods, a significant proportion of foods are still processed to sterilisation temperatures (121°C) and then stored in brines or syrups in jars or aluminium cans. Table 11.4 lists a selection of recent studies on the effect of sterilisation on the content of a selection of phytochemicals in plant foods. Given that sterilisation could be regarded as the most severe of the heat treatments reviewed in this chapter readers may suspect that it is the most deleterious to phytochemical content. However a brief examination of Table 11.4 reveals that a range of responses have been reported ranging from severe reduction to increases to isomerisation. A particularly diverse range of responses have been reported for polyphenols and anthocyanins. Korus and Lisiewska (2011) reported that antioxidative activity was reduced in canned kale by 57, 73 and 45% respectively and that losses of polyphenol constituents were also significant, ranging from 64% for caffeic acid to 82% for p-coumaric acid. In contrast Sablani et al. (2010) reported that phenolic contents and antioxidant activity of both raspberries and blueberries generally increased by up to 50 and 53% respectively in organically grown berries. Chaovanalikit and Wrolstad (2004) also reported that there was an apparent increase in total anthocyanins in particular cyanidin and pelargonidin glucosides in canned cherries. The loss of polyphenols following a severe heat treatment such as canning is easy to rationalise given the probability for heat induced degradation and leaching, however increases in polyphenol content are less easy to understand. A number of explanations have been put forward for this phenomenon including (1) increased extraction efficiency after canning, (2) complete inactivation of PPO and (3) depolymerisation of high-molecular-weight phenolics. Explanation 2 seems the least likely explanation as it would not necessarily result in increased levels of polyphehols. However no experimental evidence has been offered for hypotheses 1 and 3 and therefore a satisfactory explanation is still not available. A variety of responses for carotenoids to sterilisation have been reported, however it is probably fair to say that in general carotenoids are reasonably resistant to sterilisation and in some cases increases have been reported (Edwards and Lee, 1986; Seybold, Fröhlich, Bitsch, Otto and Böhm, 2004). However, perhaps the most remarkable finding with regard to the effect of severe heat treatments such as sterilisation on carotenoids is that it can actually increase bio-availability. This is because heating generally favours the formation of cis-carotenoid isomers (Shi and Le Maguer, 2000; Shi, Maguer, Kakuda, Liptay and Niekamp, 1999), which are more bio-available because cis-isomers are more soluble in bile acid micelles and may be preferentially incorporated into chylomicrons (Boileau, Merchen, Wasson, Atkinson and Erdman Jr, 1999). Perhaps because glucosinolate containing vegetables are infrequently subjected to canning, few studies have examined the effect of sterilisation on glucosinolate content. However, Oerlemans et al. (2006) concluded that canning reduced glucosinolate levels in red cabbage by 73%. An earlier investigation showed a decline in available glucosinolates in canned cabbage as compared to fresh and frozen cabbage (Dekker and Verkerk, 2003).

Table 11.4 Effect of sterilisation on levels of phytochemicals in plant foods

Food	Bioactive compounds	Salient result	Reference
Raspberries and blueberries	Total anthocyanins, phenolic content and antioxidant activity	After canning, total anthocyanins decreased by up to 44%, while phenolic contents and antioxidant activity of both raspberries and blueberries generally increased by up to 50 and 53% respectively	Sablani et al. (2010)
Broccoli florets	Ascorbic acid	After canning or storage in jars ascorbic acid decreased to 43.34 and 39.01%, respectively	Faten, Sober and El-Malak (2009)
Cherries	Total anthocyanins and levels of individual polyphenols	There was an apparent increase in total anthocyanins. The proportions of cyanidin and pelargonidin rutinosides in cherries slightly decreased with canning, whereas cyanidin and pelargonidin glucosides increased	Chaovanalikit and Wrolstad (2004)
Blueberries	Anthocyanins, flavonols and ORAC	Canned samples had greater levels of anthocyanins, flavonols and ORAC than fresh berries 1 d after processing	Brownmiller, Howard and Prior, (2008)
Red cabbage	Glucosinolates	Canning degraded all measured glucosinolates by 73%	Oerlemans, Barrett, Suades, Verkerk and Dekker (2006)
Kale	Vitamin C, polyphenols and antioxidative activity	Levels of vitamin C, polyphenols and antioxidative activity were reduced by 57%, 73% and 45% respectively. Losses of polyphenol constituents were also significant, ranging from 64% for caffeic acid to 82% for p-coumaric acid	Korus and Lisiewska (2011)
Corn	Lutein, zeaxanthin, and total carotenoids	Levels of lutein, zeaxanthin, and total carotenoids were similar to their respective fresh counterparts	Scott and Eldridge (2005)
Green peas and carrots	Pro-vitamin A carotenoids	Canned carrots and green peas had a higher carotenoid content than fresh samples	Edwards and Lee (1986)
Canned tomato products	B-carotene and lycopene	On a dry weight basis, contents of lycopene increased or decreased depending on the origin of the tomatoes used, whereas the β-carotene contents decreased or were quite stable. In contrast to lycopene, β-carotene isomerised due to thermal processing	Seybold, Fröhlich, Bitsch, Otto and Böhm (2004)

11.6 Frying

Plant foods (excluding starchy foods such as potatoes) are infrequently subjected to frying and therefore the effect of this thermal practice on phytochemicals is less well studied than other thermal processing methods. Table 11.5 summarises the limited number of recent studies with regard to the effect of frying on levels of phytochemicals in plant foods. During frying, a complex series of various chemical reactions takes place, such as thermoxidation, hydrolysis, polymerisation and fission (Fritsch, 1981). Whilst the temperature of the heat

Table 11.5 Effect of frying on levels of phytochemicals in plant foods

Food	Bioactive compounds	Salient result	Reference
Black seeded bean Cultivars	Total Phenolics, anthocyanins and tannins	Non-uniform but generalised reduction in levels of all three phytochemical groups in refried beans	Almanza-Aguilera, Guzmán-Tovar, Mora-Avilés, Acosta-Gallegos and Guzmán-Maldonado (2008)
Blue potato green beans, mango chips and sweet-potato chips	Total monomeric anthocyanins, carotenoids	Anthocyanin (mg/100 g d.b.) of vacuum-fried blue potato chips was 60% higher. Final total carotenoids (mg/g d.b.) were higher by 18% for green beans, 19% for mango chips, and by 51% for sweet-potato chips	Da Silva and Moreira (2008)
Broccoli florets	Phenolics, vitamin C and glucosinolates	Phenolics and vitamin C more affected than glucosinolates. Frying with extra virgin oil preserved vitamin C best	Moreno, López-Berenguer and García-Viguera (2007)
Broccoli, brussel sprouts, cauliflower and green cabbage	Glucosinolates	Levels not affected by stir frying	Song and Thornalley (2007)
Potatoes, green peppers, zucchinis and eggplants	Polyphenols	Retention of polyphenols of 25 and 70% in vegetables and Olive oil respectively	Kalogeropoulos, Mylona, Chiou, Ioannou and Andrikopoulos (2007)

medium (oil) is much higher than in other thermal processing techniques foods are generally only subjected to this process for a relatively short period of time. It would appear however that despite this frying of plant foods causes dramatic decreases in phytochemical content. Some authors have examined the ability of modified frying techniques such as vacuum frying to reduce phytochemical loss during frying. Da Silva and Moreira (2008) compared the ability of vacuum frying to retain anthocyanins and total phenolos in a range of plant foods. Whilst the authors concluded that vacuum increased frying retention of these compounds, when compared to conventional frying retention was still very low (≥50% in most cases) in green beans, potatoes, mango and sweet potato. Other authors have also reported retention levels of this order for polyphenols in refried black beans (Almanza-Aguilera, Guzmán-Tovar, Mora-Avilés, Acosta-Gallegos and Guzmán-Maldonado, 2008) and zucchinis (courgette) and egg plants (aubergine) (Kalogeropoulos, Mylona, Chiou, Ioannou and Andrikopoulos, 2007). The influence of frying oil on polyphenol content in plant foods has also been investigated with Kalogeropoulos et al. (2007) reporting that frying in olive oil resulted in 50% greater retention of polyphenols than frying in vegetable oil. Moreno et al. (2007) also reported that frying in olive oil retained vitamin C better than frying in vegetable oil in broccoli florets. When plant food are fried, phenolic anti-oxidants are lost by steam distillation (Fritsch, 1981) and, furthermore, are consumed by reacting with lipid free radicals, originally formed by the action of oxygen on unsaturated fatty acids, to form relatively stable products which interrupt the propagation stage of oxidative chain reactions. During pan-frying the anti-oxidant loss is expected to occur to a greater extent as a result of higher surface-to-volume ratio, higher temperatures and contact with atmospheric oxygen, as the potatoes remain partly uncovered by oil. Indeed Song and Thornalley (2007) reported that glucosinolates were unaffected by stir frying in broccoli, brussel sprouts, cauliflower and green cabbage.

11.7 Conclusion

Thermal processing encompasses a suite of techniques but is currently the most commonly employed method for domestic and industrial processing of plant foods. Thermal processing techniques have been shown to elicit a range of responses on phytochemical content in plant foods. It is dangerous therefore to make generalisations and recommendations as to thermal methods suitable for retention of phytochemicals as to some extent the effect is dependent on the chemical identity of the phytochemical and the matrix in which it is contained. For example, the severity of the heat process is not always reflected in the effect it has on phytochemical content. Blanching is a relatively mild thermal process, however in most cases it results in a reduction in phytochemical content. At the other end of the scale sterilisation has been shown to increase the bioavailability of carotenoids by inducing isomerisation. As stated elsewhere in the book, ascorbic acid is the most heat labile of the phytochemicals commonly encountered and in general any thermal processing results in a reduction of this compound. For the other phytochemical groups reviewed here it is not possible to come to a uniform conclusion as all have been shown to increase, decrease or be unaffected by thermal processing. A number of factors that can either decrease or increase phytochemical content appear to determine the final phytochemical content in a plant food. These include the severity of the heat process, the thermal stability of the phytochemical, the solubility of the phytochemical in the surrounding medium, the binding of the phytochemical in the food matrix and the oxidative lability of the phytochemical. Thermal processing techniques are available that can insulate plant foods from some of these effects but not all. There is scope therefore in cases where phytochemical content is severely affected by thermal processing for the use of non-thermal processing in series or in combination with conventional thermal processing. Whilst some authors have reported on the degradation pathways resulting in reductions in phytochemical content to date, most studies have concentrated on quantifying the response of the phytochemical to processing. There is therefore a need to thoroughly investigate these pathways with view to understanding the chemical and enzymatic pathways involved.

References

Alighourchi, H., Barzegar, M., and Abbasi, S. (2008) Anthocyanins characterization of 15 Iranian pomegranate (*Punica granatum* L.) varieties and their variation after cold storage and pasteurization. *European Food Research and Technology*, 227(3), 881–887.

Almanza-Aguilera, E., Guzmán-Tovar, I., Mora-Avilés, A., Acosta-Gallegos, J., and Guzmán-Maldonado, S. (2008) Phytochemical content of black seeded bean cultivars after cooking and frying. *Annual Report-Bean Improvement Cooperative*, 51, 104.

Alper, N., Bahçeci, K. S., and Acar, J. (2005) Influence of processing and pasteurization on color values and total phenolic compounds of pomegranate juice. *Journal of Food Processing and Preservation*, 29(5-6), 357–368.

Baardseth, P., Bjerke, F., Martinsen, B. K., and Skrede, G. (2010) Vitamin C, total phenolics and antioxidative activity in tip-cut green beans (Phaseolus vulgaris) and swede rods (Brassica napus var. napobrassica) processed by methods used in catering. *Journal of the Science of Food and Agriculture*, 90(7), 1245–1255.

Boileau, A. C., Merchen, N. R., Wasson, K., Atkinson, C. A., and Erdman Jr, J. W. (1999) Cis-lycopene is more bioavailable than trans-lycopene in vitro and in vivo in lymph-cannulated ferrets. *The Journal of Nutrition*, 129(6), 1176–1181.

Brownmiller, C., Howard, L. and Prior, R. (2008) Processing and storage effects on monomeric anthocyanins, percent polymeric color, and antioxidant capacity of processed blueberry products. *Journal of Food Science*, 73(5), H72–H79.

Castro, S. M., Saraiva, J. A., Lopes-da-Silva, J. A., Delgadillo, I., Loey, A. V., Smout, C., and Hendrickx, M. (2008) Effect of thermal blanching and of high pressure treatments on sweet green and red bell pepper fruits (Capsicum annuum L.). *Food Chemistry*, 107(4), 1436–1449.

Castro, S. M., Saraiva, J. A., Domingues, F. M. J. and Delgadillo, I. (2011) Effect of mild pressure treatments and thermal blanching on yellow bell peppers (Capsicum annuum L.). *LWT-Food Science and Technology*, 44(2), 363–369.

Chaovanalikit, A., and Wrolstad, R. (2004) Anthocyanin and polyphenolic composition of fresh and processed cherries. *Journal of Food Science*, 69(1), FCT73–FCT83.

Cieślik, E., Leszczyńska, T., Filipiak-Florkiewicz, A., Sikora, E., and Pisulewski, P. M. (2007) Effects of some technological processes on glucosinolate contents in cruciferous vegetables. *Food Chemistry*, 105(3), 976–981.

Da Silva, P. F., and Moreira, R. G. (2008) Vacuum frying of high-quality fruit and vegetable-based snacks. *LWT – Food Science and Technology*, 41(10), 1758–1767.

Dekker, M., and Verkerk, R. (2003) Dealing with variability in food production chains: a tool to enhance the sensitivity of epidemiological studies on phytochemicals. *European Journal of Nutrition*, 42(1), 67–72.

Edwards, C., and Lee, C. (1986) Measurement of provitamin A carotenoids in fresh and canned carrots and green peas. *Journal of Food Science*, 51(2), 534–535.

Faten, B., Sober, S. and El-Malak, G. (2009) Effect of some preservation processes on the phytochemical compounds with antioxidant activities of broccoli. *Arab Universities Journal of Agricultural Sciences*, 17(2), 351–359.

Fritsch, C. (1981) Measurements of frying fat deterioration: A brief review. *Journal of the American Oil Chemists' Society*, 58(3), 272–274.

Gancel, A.-L., Feneuil, A., Acosta, O., Pérez, A. M., and Vaillant, F. (2011) Impact of industrial processing and storage on major polyphenols and the antioxidant capacity of tropical highland blackberry (Rubus adenotrichus). *Food Research International*, 44(7), 2243–2251.

Jaiswal, A. K., Gupta, S., and Abu-Ghannam, N. (2012) Kinetic evaluation of colour, texture, polyphenols and antioxidant capacity of Irish York cabbage after blanching treatment. *Food Chemistry*, 131(1), 63–72.

Ji, Y., Kuo, Y. and Morris, M. E. (2005) Pharmacokinetics of dietary phenethyl isothiocyanate in rats. *Pharmaceutical research*, 22(10), 1658–1666.

Kalogeropoulos, N., Mylona, A., Chiou, A., Ioannou, M. S., and Andrikopoulos, N. K. (2007) Retention and distribution of natural antioxidants (α-tocopherol, polyphenols and terpenic acids) after shallow frying of vegetables in virgin olive oil. *LWT – Food Science and Technology*, 40(6), 1008–1017.

Keenan, D. F., Brunton, N., Butler, F., Wouters, R., and Gormley, R. (2011) Evaluation of thermal and high hydrostatic pressure processed apple purees enriched with prebiotic inclusions. *Innovative Food Science andamp; Emerging Technologies*, 12(3), 261–268.

Keenan, D. F., Rößle, C., Gormley, R., Butler, F., and Brunton, N. P. (2012) Effect of high hydrostatic pressure and thermal processing on the nutritional quality and enzyme activity of fruit smoothies. *LWT – Food Science and Technology*, 45(1), 50–57.

Kidmose, U., Hansen, S. L., Christensen, L. P., Edelenbos, M., Larsen, E., and Nørbæk, R. (2004) Effects of Genotype, Root Size, Storage, and Processing on Bioactive Compounds in Organically Grown Carrots (Daucus carota L.). *Journal of Food Science*, 69(9), S388–S394.

Korus, A., and Lisiewska, Z. (2011) Effect of preliminary processing and method of preservation on the content of selected antioxidative compounds in kale (Brassica oleracea L. var. acephala) leaves. *Food Chemistry*, 129(1), 149–154.

Lee, Y., and Howard, L. (1999) Firmness and Phytochemical Losses in Pasteurized Yellow Banana Peppers (Capsicum a nnuum) As Affected by Calcium Chloride and Storage. *Journal of agricultural and food Chemistry*, 47(2), 700–703.

Matusheski, N. V., Juvik, J. A., and Jeffery, E. H. (2004) Heating decreases epithiospecifier protein activity and increases sulforaphane formation in broccoli. *Phytochemistry*, 65(9), 1273–1281.

Mayer-Miebach, E., Gärtner, U., Großmann, B., Wolf, W., and Spieß, W. E. L. (2003) Influence of low temperature blanching on the content of valuable substances and sensory properties in ready-to-use salads. *Journal of Food Engineering*, 56(2–3), 215–217.

Mayer-Miebach, E., and Spieß, W. E. L. (2003) Influence of cold storage and blanching on the carotenoid content of Kintoki carrots. *Journal of Food Engineering*, 56(2–3), 211–213.

Moreno, D. A., López-Berenguer, C., and García-Viguera, C. (2007) Effects of Stir-Fry Cooking with Different Edible Oils on the Phytochemical Composition of Broccoli. *Journal of Food Science*, 72(1), S064–S068.

Oerlemans, K., Barrett, D. M., Suades, C. B., Verkerk, R., and Dekker, M. (2006). Thermal degradation of glucosinolates in red cabbage. *Food Chemistry*, 95(1), 19–29.

Patras, A., Brunton, N. P., and Butler, F. (2010) Effect of water immersion and sous-vide processing on antioxidant activity, phenolic, carotenoid content and color of carrot disks. *Journal of Food Processing and Preservation*, 34(6), 1009–1023.

Piskuła, M., and Kozłowska, H. (2005) Biologically active compounds in Cruciferae sprouts and their changes after thermal treatment. *Polish Journal of Food and Nutrition Sciences*, 14(4), 375–380.

Puupponen, P. R., Hakkinen, S. T., Aarni, M., Suortti, T., Lampi, A. M., Eurola, M., Piironen, V., Nuutila, A. M. and Oksman, C. K. M. (2003) Blanching and long-term freezing affect various bioactive compounds of vegetables in different ways. *Journal of the Science of Food and Agriculture*, 83(14), 1389–1402.

Rawson, A., Koidis, A., Rai, D. K., Tuohy, M., and Brunton, N. (2010) Influence of Sous Vide and Water Immersion Processing on Polyacetylene Content and Instrumental Color of Parsnip (Pastinaca sativa) Disks. *Journal of Agricultural and Food Chemistry*, 14, 58(13), 7740–7747.

Rawson, A., Tiwari, B. K., Tuohy, M. G., O'Donnell, C. P. and Brunton, N. (2011) Effect of ultrasound and blanching pretreatments on polyacetylene and carotenoid content of hot air and freeze dried carrot discs. *Ultrasonics Sonochemistry*, 18(5), 1172–1179.

Rawson, A., Tuohy, M.G. and Brunton, N.P. (2012) An investigation of the effects of thermal and non-thermal processing methods on polyacetylenes from Apiaceae. Thesis submitted to NUIG Galway. PP 1–222.

Rose, P., Faulkner, K., Williamson, G. and Mithen, R. (2000) 7-Methylsulfinylheptyl and 8-methylsulfinyloctyl isothiocyanates from watercress are potent inducers of phase II enzymes. *Carcinogenesis*, 21(11), 1983–1988.

Sablani, S. S., Andrews, P. K., Davies, N. M., Walters, T., Saez, H., Syamaladevi, R. M., and Mohekar, P. R. (2010) Effect of thermal treatments on phytochemicals in conventionally and organically grown berries. *Journal of the Science of Food and Agriculture*, 90(5), 769–778.

Scott, C. E. and Eldridge, A. L. (2005) Comparison of carotenoid content in fresh, frozen and canned corn. *Journal of Food Composition and Analysis*, 18(6), 551–559.

Seybold, C., Fröhlich, K., Bitsch, R., Otto, K., and Böhm, V. (2004) Changes in Contents of Carotenoids and Vitamin E during Tomato Processing. *Journal of Agricultural and Food Chemistry*, 52(23), 7005–7010.

Shi, J., and Le Maguer, M. (2000) Lycopene in tomatoes: chemical and physical properties affected by food processing. *Critical Reviews in Food Science and Nutrition*, 40(1), 1–42.

Shi, J., Maguer, M. L., Kakuda, Y., Liptay, A., and Niekamp, F. (1999) Lycopene degradation and isomerization in tomato dehydration. *Food Research International*, 32(1), 15–21.

Song, L., and Thornalley, P. J. (2007) Effect of storage, processing and cooking on glucosinolate content of Brassica vegetables. *Food and Chemical Toxicology*, 45(2), 216–224.

Sotome, I., Takenaka, M., Koseki, S., Ogasawara, Y., Nadachi, Y., Okadome, H., and Isobe, S. (2009) Blanching of potato with superheated steam and hot water spray. *LWT – Food Science and Technology*, 42(6), 1035–1040.

Talcott, S. T., Brenes, C. H., Pires, D. M., and Del Pozo-Insfran, D. (2003) Phytochemical stability and color retention of copigmented and processed muscadine grape juice. *Journal of Agricultural and Food Chemistry*, 51(4), 957–963.

Tříska, J., Vrchotová, N., Houška, M., and Strohalm, J. (2007) Comparison of total isothiocyanates content in vegetable juices during high pressure treatment, pasteurization and freezing. *High Pressure Research*, 27(1), 147–149.

Volden, J., Bengtsson, G. B. and Wicklund, T. (2009) Glucosinolates, l-ascorbic acid, total phenols, anthocyanins, antioxidant capacities and colour in cauliflower (Brassica oleracea L. ssp. botrytis); effects of long-term freezer storage. *Food Chemistry*, 112(4), 967–976.

Volden, J., Borge, G. I. A., Hansen, M., Wicklund, T. and Bengtsson, G. B. (2009) Processing (blanching, boiling, steaming) effects on the content of glucosinolates and antioxidant-related parameters in cauliflower (Brassica oleracea L. ssp. botrytis). *LWT-Food Science and Technology*, 42(1), 63–73.

Wennberg, M., Ekvall, J., Olsson, K., and Nyman, M. (2006) Changes in carbohydrate and glucosinolate composition in white cabbage (Brassica oleracea var. capitata) during blanching and treatment with acetic acid. *Food Chemistry*, 95(2), 226–236.

Yuan, J.-P., and Chen, F. (1998) Degradation of Ascorbic Acid in Aqueous Solution. *Journal of Agricultural and Food Chemistry*, 46(12), 5078–5082.

12 Effect of novel thermal processing on phytochemicals

Bhupinder Kaur,* Fazilah Ariffin, Rajeev Bhat, and Alias A. Karim

Food Biopolymer Research Group, Food Technology Division, School of Industrial Technology, Universiti Sains Malaysia, Penang, Malaysia

12.1 Introduction

The preference of consumers towards high-quality foods with longer shelf life has brought about a revolution in food processing technologies. New alternative food processing technologies are emerging that can meet the demand for better quality food products.

Novel thermal processing technologies such as ohmic heating and dielectric heating are promising alternatives to conventional methods of heat processing. Dielectric heating includes radio frequency and microwave heating. The forms of heating used in these technologies are volumetric where the thermal energy is generated directly inside the food. This helps in overcoming excessive cooking times and improves energy and heating efficiency (Pereira and Vicente, 2010). The food industry is ready to adopt cost effective technologies that offer better quality and safe products.

In the case of fruits and vegetables, they need to be processed with minimal damage to their nutritive compounds. There is a need for food processing to ensure longevity of the foods as they are to be used in different types of foods, made into various food products, and, if possible through processing, be made available all the year round instead of just seasonally.

Thermal processing is the most widely used method for preserving and extending the useful shelf life of foods. The conventional food processing methods were carried out primarily for food safety and shelf stability. However, today more emphasis is placed on high-quality and value-added foods with convenient end use (Awuah *et al.*, 2007). Rickman *et al.* (2007) did a nutritional comparison of fresh, frozen, and canned fruits and vegetables in relation to vitamins B and C and phenolic compounds. Others who have worked on conventional thermal processing on fruits and vegetables include Lee *et al.* (1976), Elkins (1979), Lathrop and Leung (1980), Abou-Fadel and Miller (1983), Murcia *et al.* (2000), Dewanto *et al.* (2002), and Gorinstein *et al.* (2009).

*Dr. Bhupinder Kaur is the recipient of USM Post-Doctoral Fellowship in Research.

> **Box 12.1** Possible effects of food processing on the overall antioxidant potential of foods
>
> No effect
> Loss of naturally occurring antioxidants
> Improvement of antioxidant properties of naturally occurring compounds
> Formation of novel compounds having antioxidant activity (i.e. Maillard reaction products)
> Formation of novel compounds having pro-oxidant activity (i.e. Maillard reaction products)
> Interactions among different compounds (e.g. lipids and natural antioxidants, lipids and Maillard reaction products)
>
> (Reprinted from *Trends in Food Science & Technology*, 10, Nicoli, M.C., Anese, M., & Parpinel, M., Influence of processing on the antioxidant properties of fruit and vegetables, 94–100, Copyright (1999) with permission from Elsevier).

Although novel thermal technologies have some advantages with respect to overall quality of fruits and vegetables, there is concern about the bioactive phytochemicals. This chapter is written in order to generate a better understanding of the effect of novel thermal processing methods on phytochemicals found in fruits and vegetables.

12.2 An overview of different processing methods for fruits and vegetables

Most fruits and some vegetables are consumed raw, but the rest go through various types of processing to make them edible, not easily perishable, and available year round. Therefore the processing of fruits and vegetables is for economic, safety, and quality reasons.

The most widely used method for destroying microorganisms and imparting foods with a lasting shelf life is thermal processing (Ranesh, 1999). However, this unavoidably degrades the vitamin and nutrient levels to some extent. Different processing methods are used to treat fruits and vegetables before they become consumable. Some of the processing methods include vacuum drying, freeze drying, cooking, blanching, pasteurization, and sterilization. The novel thermal methods used to carry out these processes include ohmic heating, radio frequency, and microwave heating.

Food processing methods are expected to affect the content, activity, and bioavailability of bioactive compounds found in fruits and vegetables (Nicoli *et al.*, 1999). The effect of heat in food processing does not always cause a loss of quality and health properties. For example, β-carotene was found to increase with moderate heating or by way of enzymatic disruption of the vegetables' cell wall structures (Southon, 1998; Weat and Castenmiller, 1998). Some of the possible effects of food processing on the overall antioxidant potential of foods were tabulated by Nicoli *et al.* (1999) as shown in Box 12.1. Prevention of oxidative changes can be achieved by maintaining the structural integrity of foods as contact with oxygen sensitive components is reduced. Food processes that prevent exposure of antioxidants to oxygen to the greatest possible extent have become a requirement in the industry (Lindley, 1998).

12.3 Novel thermal processing methods

Novel thermal methods used to carry out food processes include ohmic heating, radio frequency, and microwave heating. Table 12.1 lists the novel thermal processing methods available and their principle advantages and disadvantages.

Table 12.1 Novel thermal processing methods: their principle advantages and disadvantages. Reprinted from *Trends in Food Science & Technology*, 10, Nicoli, M.C., Anese, M. & Parpinel, M., Influence of processing on the antioxidant properties of fruit and vegetables, 94–100, copyright (1999) with permission from Elsevier.

Novel thermal processing methods	Principles	Advantages	Disadvantages	References
Ohmic heating (also known as Joule heating, electrical resistance heating, direct electrical resistance heating, electroheating or electroconductive heating)	• It is a direct heating method where the food itself is a conductor of electricity. • Low frequency between 50–60 Hz is used. • The rate of heating is proportional to the square of the electric field strength, the electrical conductivity and the type of food being heated.	• No moving part in heat exchanger. • No need for hot heat transfer surfaces. • Rapid and uniform heat treatment with minimum heat damage. • Ideal for shear-sensitive products. • Promotes increased nutrient retention and reduce damage to particulates. • Produces high quality products. • Reduces risk of fouling. • Quiet operation, low maintenance costs and easier control. • High energy efficiency as 90% of the electrical energy is converted into heat.	• There is a possibility for the metal ions to be released into the conducting solution and eventually into foods if the electrode materials are not inert. • Non-uniform heating as the conductivity of each particulate in a food system varies.	Parrott, 1992; Ruan et al., 2002; Sastry, 2005; Vicente et al., 2006.
Microwave heating	• Heating by radiation. • Dielectric heating mechanism dominates up to moderated temperatures. • Polar molecules (dominant one being water) try to align themselves to the rapidly changing direction of the electric field. The molecule "relaxes" with the changing of direction of the field and absorbed energy is dissipated to the surroundings, ie., inside the food. • Volumetric heating, materials absorb microwave energy directly and internally and convert it into heat. • Frequencies used are 2450 or 900 MHz.	• Homogeneous very quick heat processing, leading to small quantity changes. • Requires smaller floor space.	• Distribution of the energy within food can vary due to limited penetration depth of microwaves. • Water content affects the heating performance of foods.	Ehlermann, 2002; Ramaswamy and Marcotte, 2006; Vadivambal and Jayas, 2007.

(Continued)

Table 12.1 (Continued)

Novel thermal processing methods	Principles	Advantages	Disadvantages	References
Radio-frequency heating	• Heating of food is done by transmitting electromagnetic energy through food placed between an electrode and the ground. • Frequencies between 13.56, 27.12, 40.68 MHz are used. • Transfer of energy is through air gaps and nonconducting packaging materials. • High electric field intensities are needed for rapid heating in foods.	• Rapid, volumetric, uniform heating throughout a medium. • Saving energy through heat efficiency increase. • No pollution, as there are no combustion by-products. • The product moisture profile is evened therefore reduces checking the uneven stresses. • Increases production without increasing the floor plant length. • Reduces flashing off of volatile flavouring therefore allowing minimal quantities to be used.	• Uniformity of heat distribution within foods of mixed composition is disputable.	Ramaswamy and Marcotte, 2006; Zhao, 2006; Birla et al., 2008; Marra et al., 2009

Ohmic heating involves direct electrical resistance heating of a food product under the passage of an electric current. The energy dissipation within the food product is given by $Q = I^2 R$, where R is the food product offering resistance and I is the alternating current. Therefore, the applicability of ohmic heating depends on the electrical conductivity of the food product. Foods generally are good candidates for ohmic heating as they contain a minimal amount of free water with dissolved ionic salts (Ramaswamy and Marcotte, 2006). The square of the electric field strength, the electrical conductivity, and the type of food being heated directly affect the rate of heating (Ruan et al., 2002). Industrial ohmic heating plants have been established for thermal processing of tomato sauces and pastes, diced and sliced peach and apricot, diced pears and apples, low-acid vegetable purees, strawberries, fruit preparation, plum peeled tomato and tomato dices, and vegetable sauces (Leadley, 2008).

A table of dielectric properties of fruits and vegetables was published by Sosa-Morales et al. (2010) where the temperature, moisture content, dielectric constant frequency, and loss factor frequency was given. These dielectric properties include permittivity, dielectric constant, loss factor, penetration depth, and electrical conductivity (Sosa-Morales et al., 2010). Dielectric heating includes microwave heating and radio frequency.

Microwave processing is heating by radiation and not by convection or conduction (Ehlermann, 2002). Electromagnetic energy comes in photons, which are discrete and in very small quantities. These photons must match the energy difference between several allowed atomic energy states of the electrons in the treated materials for energy to be absorbed. Therefore food which mainly contains water can be heated. Polar molecules which are dominantly water molecules, try to align themselves to the rapidly changing direction of the electric field. When the field changes direction, the molecule relaxes and energy absorbed is dissipated into the food (Ehlermann, 2002; Ramaswamy and Marcotte, 2006). Some of the changes that take place in the agricultural products after being microwave treated are discussed by Vadivambal and Jayas (2007).

Radio frequency heating involves application of high-voltage current signal to a set of parallel electrodes. The food to be heated is placed between the electrodes and the current flows through the food. Polar molecules in the food align and rotate in opposite direction to match the electrical current applied. Interaction between polar molecules and neighboring molecules results in lattice and frictional losses as they rotate, thus causing heat to occur. The higher the frequency the greater the energy imparted to the food (Zhao, 2006). Radio frequency heating applications have been discussed by Zhao et al. (2000), Piyasena et al. (2003), and Marra et al. (2009).

12.4 Effect of novel processing methods on phytochemicals

Phytochemicals are plant chemicals that have been defined as the bioactive non-nutrient plant compounds found in fruits, vegetables, grains, and other plant foods and have been directly linked to the reduction in the risk of major chronic diseases (Liu, 2003). Phytochemicals can be classified as carotenoids, phenolics, alkaloids, nitrogen-containing compounds, and organosulfur compounds, with the most studied being the phenolics and carotenoids. As there are many phytochemicals, each compound works differently and sometimes they overlap. Some of their possible actions are acting as antioxidants, regulation of hormone metabolism, stimulation of enzyme activities in detoxification, oxidation, and reduction, interference with DNA replication, and anti-bacterial and anti-viral effect

Table 12.2 Fruit and vegetables used in the study of the effect of novel thermal processing methods

Novel thermal processing methods	Fruit or vegetables treated	References
Ohmic heating	Spinach puree	Yildiz et al., 2010
	Pea puree	Icier et al., 2006
	Beet root pieces	Mizrahi, 1996
	Artichoke byproduct	Icier, 2010
	Carrot pieces	Lemmens et al., 2009
	Orange juice	Vikram et al., 2005
	Pomegranate juice	Yildiz et al., 2009
Microwave heating	Apple puree	Oszmiański et al., 2008
	Apple mash	Gerard and Roberts, 2004
	Dried cranberry	Leusink et al., 2010
	Cranberry press cake	Raghavan and Richards, 2007
	Grape seeds	Hong et al., 2001
	Strawberry fruit	Wojdyło et al., 2009
	Asparagus	Sun et al., 2007
	Broccoli	López-Berenguer et al., 2007;
	Carrot	Howard et al., 1997
	Cherry tomato	Liu et al., 1998
	Cauliflower, peas, spinach,	Heredia et al., 2010
	Swiss chard	Natella et al., 2010
	Potatoes	Barba et al., 2008; Phillippy et al., 2004
	Pigmented potatoes	Mulinacci et al., 2008
	Purple sweet potato	Lu et al., 2010; Steed et al., 2008
	Chinese purple corn cob	Yang and Zhai, 2010
	Barley grain	Omwamba and Hu, 2010
	Olive oil	Cerretani et al., 2009
	American ginseng root	Popovich et al., 2005
Radio frequency	Carrot	Zhong et al., 2004; Orsat et al., 2001
	Potato	Zhong et al., 2004
	Apple puree	Manzocco et al., 2008

(Liu, 2004). As antioxidants they help in the prevention of oxidative damage to biological macromolecules in the presence of reactive oxygen species which can lead to many human diseases (Lindley, 1998). The works of Sun et al. (2002) and Chu et al. (2002) showed that the phytochemical extracts from fruits and vegetables had potent antioxidant and antiproliferative effects. The potent antioxidant and anti-cancer activities in the human body are attributed to the additive and synergistic effects of phytochemicals in fruits and vegetables (Liu, 2004). Table 12.2 shows the fruits and vegetables that have been used to study the effect of novel thermal processing methods.

In sections 12.4.1, 12.4.2, and 12.4.3 we discuss the effects of ohmic heating, microwave heating, and radio frequency on the phytochemicals.

12.4.1 Ohmic heating

Ohmic heating possesses the benefits of conventional thermal processing besides having a potential to improve on the retention of vitamins and nutrients (Ruan et al., 2002; Jaeger et al., 2010). It is a high-temperature short-time (HTST) method which can heat an 80% solids food product from 25 °C to 129 °C in about 90 s thus decreasing the possibility of overprocessing due to high temperature (Zuber, 1997).

Ohmic heating applied to vegetable purees resulted in higher retention of color attributes and β-carotene compared to conventional heating. Ohmic heating had an enhancing effect on β-carotene biosynthesis and formation of chlorophyll derivatives. Adjustment of the voltage gradient in ohmic heating could also be used to heat a spinach puree faster than conventional heating (Yildiz et al., 2010).

Vikram et al. (2005) found that with ohmic heating there was better vitamin retention at all temperatures; however microwave heating led to lower degradation in color with visual color being used as an index of the carotenoid content in orange juice.

In a comparative study of hot water blanching and ohmic heating blanching for diced beet root, it was found that the leaching of betanine and betalamic acid was reduced by one order of magnitude in ohmic heating blanching by removing the need to dice the beet root and shortening the process time (Mizrahi, 1996). Ohmic blanching using 30 V/cm in pea puree inactivated peroxidize enzyme in less time than water blanching, however best color quality was obtained with ohmic blanching at 50 V/cm resulted in a critical inactivation time of 54 s (Icier et al., 2006). Ohmic blanching of artichoke by-product, resulted in the highest retention of vitamin C and the total phenolic content at 40 V/cm voltage gradient at 85 °C compared to water blanching at 85 °C and 100 °C (Icier, 2010).

A study by Lemmens et al. (2009) on thermal pretreatments (conventional heating, microwave heating and ohmic heating) of carrot pieces using different heating techniques showed that almost no influence of these pretreatments on the β-carotene content of the sample occured.

Ohmic heating had the same effect on total phenolic contents of pomegranate juice as conventional heating, however it resulted in less browning during heat treatment than conventional heating. Ohmic heating can be used as an alternative heating method providing rapid and uniform heating (Yildiz et al., 2009).

12.4.2 Microwave heating

Fast uniform heating throughout the food at a lower temperature thus reducing cooking time is the advantage of microwave heating, wherein the loss of heat-sensitive vitamins is minimized. The extent of chemical reactions may be reduced and retention of nutrients enhanced (Ehlermann, 2002).

Sterilization of fresh green asparagus using a pilot-scale 915 MHz microwave-circulated water combination heating system gave rise to significantly greater retention of antioxidant activity as well as a greener color compared to pressurised hot-water heating and steam heating in a retort (Sun et al., 2007). A study by Lin et al. (1998) on carrot slices found that the total loss of α- and β-carotene for air dried samples was higher than for vacuum-microwave dried samples. Cherry tomato heated with microwave energy at 3 W/g and 80 °C resulted in a greater level of isomerization of lycopene but a lower percentage of residual total lycopene. There was an increase in luminosity and the color of the tomato changed towards orange tones indicating greater presence of lycopene cis-isomers in these samples (Heredia et al., 2010). Microwave processing of potatoes at power input of 500 W was found to be the best compromise in terms of short baking time and reduced water and phenolic losses (Barba et al., 2008). Cauliflower, peas, spinach, and Swiss chard showed no decrease or a smaller decrease of their total phenolic content after microwaving than after boiling (Natella et al., 2010).

Phytate present in potato was found to be stable when it was cooked by microwave (Phillippy et al., 2004). Although phytate has a role as an antioxidant and anticarcinogen

(Jenab and Thompson, 2002) it can also decrease the bioavailability of critical nutrients such as zinc, iron, calcium, and magnesium in foods such as whole grains, nuts, and legumes (Weaver and Kannan, 2002).

Extraction of anthocyanin from purple sweet potato was found to be higher in samples prepared by microwave baking and extracted using acidified electrolyzed water where the percentage of extraction was 35.0% compared to without microwave baking which was only 7.8% (Lu et al., 2010). Due to its strong penetrating power, selectivity, and high heating efficiency, the instantaneous transmission of microwave heating caused the plant cells to be broken easily, which in return sped up the extraction rate and effectively improved the yield (Came, 2000; Pensado and Casais, 2000). Microwave assisted solvent extraction (MASE) has been known to be an effective extraction method for tea polyphenols and tea caffeine (Pan et al., 2003), extraction of phenolic compounds from grape seeds (Hong et al., 2001) and extraction of total phenols in cranberry press cake (Raghavan and Richards, 2007). Microwave-assisted extraction was found to be highly efficient and rapid in extracting anthocyanins from Chinese purple corn cob. The highest total anthocyanin content was obtained at an extraction time of 19 min, a solid to liquid ratio of 1:20 and a microwave irradiation power of 555 W (Yang and Zhai, 2010).

To produce apple purees with high phenolic contents, it has been recommended that ascorbic acid should be added and that they should be heated in a microwave oven, as microwave energy has the advantage of heating solids rapidly and uniformly, therefore minimizing phenolic oxidation by inactivating the enzymes more quickly (Oszmiański et al., 2008). Microwave heat treatment at four heat treatments (40 °C, 50 °C, 60 °C, and 70 °C) of Fuji and McIntosh apple mashes increased extraction of phenolics and flavonoids from apple mash resulting in apple juice with increased concentrations of total phenolics and flavonoids (Gerard and Roberts, 2004).

Roasting barley grains was found to increase both antioxidant activity as well as total phenolic content. In a response surface methodology study the optimum conditions for microwave roasting of barley grains was found to be 600 W microwave power, 8.5 min roasting time, and 61.5 g or two layers of grains. These three factors significantly influenced the radical scavenging activity of barley grain, independently and interactively (Omwamba and Hu, 2010).

Microwave drying caused the highest decrease in total phenolic content and antioxidant activity for the *Phyllanthus amarus* plant when compared to other drying methods such as sun drying and oven drying (Lim and Murtijaya, 2007). A study by Sultana et al. (2008) on the effect of different cooking methods on total phenolic content found that microwave treatment was most deleterious as compared to boiling and frying. Domestic processing of broccoli florets using microwave cooking mostly affected the vitamin C bioactive phytochemical. The most stable phytonutrients were the different mineral nutrients. The losses in phenolics, glucosinolates, and minerals were mainly due to leaching into the cooking water. Losses of phytochemicals can be prevented with shorter cooking times and avoiding cooking with water (López-Berenguer et al., 2007). Phenolic compounds in both extra virgin olive oil and olive oil decreased with increase in microwave heating time (Cerretani et al., 2009). Blanching, storage, and microwave cooking were found to decrease the concentrations of phytochemicals in fresh and frozen broccoli (Howard et al., 1997). Antioxidant content was found to be highest in steamed, followed by boiled, and least in microwave cooked vegetables and the content decreased with longer cooking time (Wachtel-Galor et al., 2008).

However the effect of microwave heating on the content of phenolics and antioxidant activities of tartary buckwheat flour was not as severe as pressured steam heating (Zhang et al.,

2010). Microwave heating using a 60 kW, 915 MHz continuous flow system was applied on pumpable purees from purple-flesh sweet potatoes. It was found that the total phenolics increased and total monomeric anthocyanins decreased slightly whereas the antioxidant activity did not change significantly as a result of microwave processing (Steed *et al.*, 2008). Microwave heating did not cause any changes in the phenolic contents of pigmented potatoes (*Solanum tuberosum* L.) and the anthocyanin content decreased only slightly (Mulinacci *et al.*, 2008). Jiménez-Monreal *et al.* (2009) concluded that griddling and microwave cooking produced the lowest losses in its antioxidant activity with pressure cooking and boiling lead to the greatest losses, with frying occupying an intermediate position.

Vacuum microwave drying of strawberry fruits at 240 W gave rise to higher levels of vitamin C, anthocyanins and phenolic compounds, and antioxidant activity than strawberries dried at other vacuum microwave powers such as 360 W and 480 W, and other drying methods such as freeze drying, vacuum drying, and convection drying (Wojdyło *et al.*, 2009). Vacuum microwave drying and freeze drying resulted in similar retention of anthocyanins and antioxidant activity recovered in dried cranberries compared to hot air drying (Leusink *et al.*, 2010). Extraction efficiency and actual retention of individual ginsenoside in North American ginseng root material can be improved by using the freeze-drying and vacuum microwave drying method (Popovich *et al.*, 2005). Böhm *et al.* (2006) observed that although vacuum microwave drying is a relatively new technique, it needs to be optimized to obtain nutritionally relevant compounds with least deterioration.

12.4.3 Radio frequency

Using a continuous flow radio frequency unit, Zhong *et al.* (2004) processed carrot and potato cubes using a 1% CMC solution as carrier. A small temperature gradient was observed inside the carrots and potato cubes that were heated using a short residence time. Radio frequency-treated carrot sticks maintained color, taste, and vacuum of the packages, which was not the case for the chlorinated water-treated samples or hot-water-treated carrots. However the authors of this work (Orsat *et al.*, 2001) concluded that radio frequency heating should not be the sole treatment to improve stability and food safety of minimally processed ready-to-eat carrot sticks but should be a part of an integrated approach, which would include proper packaging and adequate refrigeration. Radio frequency blanched apples, processed to purees had comparable color and sensory attributes to conventionally water blanched apple puree. Radio frequency also efficiently inactivated the polyphenoloxidase and lipoxygenase enzymes in model systems (Manzocco *et al.*, 2008).

12.5 Challenges and prospects/future outlook

There is still a lot of work and scope for novel thermal processing methods in the food industry. The data we have now is limited in the area of nutritional effect of these novel thermal processing methods. The literature available on the subject of phytochemicals and treatment of fruits and vegetables with these novel thermal processing methods is relatively new. As these methods gain importance in the food industry, more research should be done regarding their effects on the phytochemicals found in food. In recent years there has been a dedicated interest in phytochemicals as a much needed nutritional compound in foods for health and longevity.

12.6 Conclusion

The area of study in novel thermal processing methods affecting the phytochemicals in fruit and vegetables is relatively new. There is a positive effect of these methods on phytochemicals from the little research work that has been done thus far. More work is required in this field to fully utilize these methods in the food industry as there is potential for these methods to maintain or enhance the presence of phytochemicals in fruit and vegetables.

References

Abou-Fadel, O.S. and Miller, L.T. (1983) Vitamin retention, color and texture in thermally processed green beans and Royal Ann cherries packed in pouches and cans. *Journal of Food Science*, 48, 920–923.

Awuah, G.B., Ramaswamy, H.S., and Economides, A. (2007) Thermal processing and quality: Principles and overview. *Chemical Engineering and Processing*, 46, 584–602.

Barba, A.A., Calabretti, A., d'Amore, M., Piccinelli, A.L., and Rastrelli, L. (2008) Phenolic constituents levels in cv. Agria potato under microwave processing. *LWT – Food Science and Technology*, 41, 1919–1926.

Birla, S.L., Wang, S., Tang, J., and Tiwari, G. (2008) Characterization of radio frequency heating of fresh fruits influenced by dielectric properties. *Journal of Food Engineering*, 89, 390–398.

Böhm, V., Kühnert, S., Rohm, H., and Scholze, G. (2006) Improving the nutritional quality of microwave-vacuum dried strawberries: A preliminary study. *Food Science and Technology International*, 12, 67–75.

Came, V. (2000) Microwave-assisted solvent extraction of environmental samples. *Trends in Analytical Chemistry*, 19, 229–249.

Cerretani, L., Bendini, A., Rodriguez-Estrada, M.T., Vittadini, E., and Chiavaro, E. (2009) Microwave heating of different commercial categories of olive oil: Part I. Effect on chemical oxidative stability indices and phenolic compounds. *Food Chemistry*, 115, 1381–1388.

Chu, Y.-F., Sun, J., Wu, X., and Liu, R.H. (2002) Antioxidant and antiproliferative activities of vegetables. *Journal of Agricultural and Food Chemistry*, 50, 6910–6916.

Dewanto, V., Wu, X., Adom, K.K., and Liu, R.H. (2002) Thermal processing enhances the nutritional value of tomatoes by increasing total antioxidant activity. *Journal of Agricultural and Food Chemistry*, 50, 3010–3014.

Ehlermann, D.A.E. (2002) Microwave processing. In: C.J.K. Henry and C. Chapman (eds) *The Nutrition Handbook for Food Processors*, CRC Press, New York, pp. 396–406.

Elkins, E.R. (1979) Nutrient content of raw and canned green beans, peaches, and sweet potatoes. *Food Technology*, 33, 66–70.

Fratianni, A., Cinquanta, L., and Panfili, G. (2010) Degradation of carotenoids in orange juice during microwave heating. *LWT – Food Science and Technology*, 43, 867–871.

Gerard, K.A. and Roberts, J.S. (2004) Microwave heating of apple mash to improve juice yield and quality. *LWT – Food Science and Technology*, 37, 551–557.

Gorinstein, S., Jastrzebski, Z., Leontowicz, H., Leontowicz, M., Namiesnik, J., Najman, K., Park, Y.-S., Heo, B.-G., Cho, J.-Y., and Bae, J.-H. (2009) Comparative control of the bioactivity of some frequently consumed vegetables subjected to different processing conditions. *Food Control*, 20, 407–413.

Heredia, A., Peinado, I., Rosa, E., and Andrés, A. (2010) Effect of osmotic pre-treatment and microwave heating on lycopene degradation and isomerisation in cherry tomato. *Food Chemistry*, 123, 92–98.

Hong, N., Yaylayan, V.A., Raghavan, G.S., Pare, J.R., and Belanger, J.M. (2001) Microwave-assisted extraction of phenolic compounds from grape seed. *Natural Products Letters*, 15, 197–204.

Howard, L.A., Jeffery, E.H., Wallig, M.A., and Klein, B.P. (1997) Retention of phytochemicals in fresh and processed broccoli. *Journal of Food Science*, 62, 1098–1104.

Icier, F. (2010) Ohmic blanching effects on drying of vegetable byproduct. *Journal of Food Process Engineering*, 33, 661–683.

Icier, F., Yildiz, H., and Baysal, T. (2006). Peroxidase inactivation and colour changes during ohmic blanching of pea puree. *Journal of Food Engineering*, 74, 424–429.

Jaeger, H., Janositz, A., and Knorr, D. (2010) The Maillard reaction and its control during food processing. The potential of emerging technologies. *Pathologie Biologie*, 58, 207–213.

Jenab, M. and Thompson, L. U. (2002) Role of phytic acid in cancer and other diseases. In: N.R. Reddy and S.K. Sathe (eds) *Food Phytates*, CRC Press, Boca Raton, FL, pp. 225–248.

Jiménez-Monreal, A.M., García-Diz, L., Martínez-Tomé, M., Mariscal, M., and Murcia, M.A. (2009) Influence of cooking methods on antioxidant activity of vegetables. *Journal of Food Science*, 74, H97–H103.

Lathrop, P.J. and Leung, H.K. (1980) Thermal degradation and leaching of vitamin C from green peas during processing. *Journal of Food Science*, 45, 995–998.

Leadley, C. (2008) Novel commercial preservation methods. In: G. Tucker (ed.) *Food Biodeterioration and Preservation*, Blackwell Publishing Ltd., Oxford, UK, pp. 211–244.

Lee, C.Y., Downing, D.L., Iredale H.D., and Chapman, J.A. (1976) The variations of ascorbic acid content in vegetable processing. *Food Chemistry*, 1, 15–22.

Leusink, G.J., Kitts, D.D., Yaghmaee, P., and Durance, T. (2010) Retention of antioxidant capacity of vacuum microwave dried cranberry. *Journal of Food Science*, 75, C311–C316.

Lim, Y.Y. and Murtijaya, J. (2007) Antioxidant properties of Phyllanthus amarus extracts as affected by different drying methods. *LWT – Food Science and Technology*, 40, 1664–1669.

Lin, T.M., Durance, T.D., and Scaman, C.H. (1998) Characterization of vacuum microwave, air and freeze dried carrot slices. *Food Research International*, 31, 111–117.

Lindley, M.G. (1998) The impact of food processing on antioxidants in vegetable oils, fruits and vegetables. *Trends in Food Science and Technology*, 9, 336–340.

Liu, R.H. (2003) Health benefits of fruit and vegetables are from additive and synergistic combinations of phytochemicals. *American Journal of Clinical Nutrition*, 78, 517S–520S.

Liu, R.H. (2004) Potential synergy of phytochemicals in cancer prevention: Mechanism of action. *Journal of Nutrition*, 134, 3479S–3485S.

López-Berenguer, C., Carvajal, M., Moreno, D.A., and Garcia-Viguera, C. (2007) Effects of microwave cooking conditions on bioactive compounds present in broccoli inflorescences. *Journal of Agricultural and Food Chemistry*, 55, 10001–10007.

Lu, L.-Z., Zhou, Y.-Z., Zhang, Y.-Q., Ma,Y.-L., Zhou, L.-X., Li, L., Zhou, Z.-Z., and He, T.-Z. (2010) Anthocyanin extracts from purple sweet potato by means of microwave baking abd acidified electrolysed water and their antioxidation *in vitro*. International *Journal of Food Science and Technology*, 45, 1378–1385.

Manzocco, L., Anese, M., and Nicoli, M.C. (2008) Radio frequency inactivation of oxidative food enzymes in model systems and apple derivatives. *Food Research International*, 41, 1044–1049.

Marra, F., Zhang, L., and Lyng, J.G. (2009) Radio frequency treatment of foods: Review of recent advances. *Journal of Food Engineering*, 91, 497–508.

Mizrahi, S. (1996) Leaching of soluble solids during blanching of vegetables by ohmic heating. *Journal of Food Engineering*, 29, 153–166.

Mulinacci, N., Ieri, F., Giaccherini, C., Innocenti, M., Andrenelli, L., Canova, G., Saracchi, M., and Casiraghi, M.C. (2008) Effect of cooking on the anthocyanins, phenolic acids, glycoalkaloids, and resistant starch content in two pigmented cultivars of *Solanum tuberosum* L.. *Journal of Agricultural and Food Chemistry*, 56, 11830–11837.

Murcia, M.A., López-Ayerra, B., Martinez-Tomé, M., Vera, A.M., and García-Carmona, F. (2000) Evolution of ascorbic acid and peroxidise during industrial processing of broccoli. *Journal of the Science of Food and Agriculture*, 80, 1882–1886.

Natella, F., Belelli, F., Ramberti, A., and Scaccini, C. (2010) Microwave and traditional cooking method: Effect of cooking on antioxidant capacity and phenolic compounds content of seven vegetables. *Journal of Food Biochemistry*, 34, 796–810.

Nicoli, M.C., Anese, M., and Parpinel, M. (1999) Influence of processing on the antioxidant properties of fruit and vegetables. *Trends in Food Science and Technology*, 10, 94–100.

Omwamba, M. and Hu, Q. (2010) Antioxidant activity in barley (*Hordeum Vulgare* L.) grains roasted in a microwave oven under conditions optimized using response surface methodology. *Journal of Food Science*, 75, C66–C73.

Orsat, V., Gariépy, Y., Raghavan, G.S.V., and Lyew, D. (2001) Radio-frequency treatment for ready-to-eat fresh carrots. *Food Research International*, 34, 527–536.

Oszmiański, J., Wolniak, M., Wojdyło, A., and Wawer, I. (2008) Influence of apple puree preparation and storage on polyphenol contents and antioxidant activity. *Food Chemistry*, 107, 1473–1484.

Pan, X., Niu, G., and Liu, H. (2003) Microwave-assisted extraction of tea polyphenols and tea caffeine from green tea leaves. *Chemical Engineering and Processing*, 42, 129–133.

Parrott, D.L. (1992) Use of OH for aseptic processing of food particulates. *Food Technology*, 45, 68–72.

Pensado, L. and Casais, C. (2000) Optimization of the extraction of poly cyclic aromatic hydrocarbons rom wood samples by the use of microwave energy. *Journal of Chromatography A*, 86, 505–513.

Pereira, R.N. and Vicente, A.A. (2010) Environmental impact of novel thermal and non-thermal technologies in food processing. *Food Research International*, 43, 1936–1943.

Phillippy, B.Q., Lin, M., and Rasco, B. (2004) Analysis of phytate in raw and cooked potatoes. *Journal of Food Composition and Analysis*, 17, 217–226.

Piyasena, P., Dussault, C., Koutchma, T., Ramaswamy, H.S., and Awnah G.B. (2003) Radio frequency heating of foods: principles, applications and related properties. A review. *Critical Reviews in Food Science and Nutrition*, 43, 587–606.

Popovich, D.G., Hu, C., Durance, T.D., and Kitts, D.D. (2005) Retention of ginsenosides in dried ginseng root: Comparison of drying methods. *Journal of Food Science*, 70, S355–S358.

Raghavan, S. and Richards, M.P. (2007) Comparison of solvent and microwave extracts of cranberry press cake on the inhibition of lipid oxidation in mechanically separated turkey. *Food Chemistry*, 102, 818–826.

Ramaswamy, H. and Marcotte, M. (2006) *Food Processing: Principles and Applications*, CRC Taylor & Francis, New York, pp. 163–165.

Ranesh, M.N. (1999) Food preservation by heat treatment. In: R.S. Rahman, editor. Handbook of Food Preservation, Marcel Dekker, New York, pp. 95–172.

Rickman, J.C., Barrett, D.M., and Bruhn, C.M. (2007) Nutritional comparison of fresh, frozen and canned fruits and vegetables. Part I. Vitamins C and B and phenolic compounds. *Journal of the Science of Food and Agriculture*, 87, 930–944.

Ruan, R., Ye, X., Chen, P., Doona, C., and Taub, I. (2002) Ohmic heating. In: C.J.K. Henry and C. Chapman (eds) *The Nutrition Handbook for Food Processors*, CRC Press, New York, pp. 407–422.

Sastry S.K. (2005) Advances in ohmic heating and moderate electric field (MEF) processing. In: G.V. Barbosa-Cánovas, M.S. Tapia, M.P. Cano (eds) *Novel Food Processing Technologies*, CRC Press, New York, pp. 491–499.

Sosa-Morales, M.E., Valerio-Junco, L., López-Malo, A., and Garcia, H.S. (2010) Dielectric properties of foods: Reported data in the 21st century and their potential applications. *LWT – Food Science and Technology*, 43, 1169–1179.

Southon, S. (1998) Increased consumption of fruits and vegetables within the EU: Potential health benefits. In: V. Gaukel, W.E.L. Spiess (eds) *European Research towards Safer and Better Foods*, Druckerei Grasser: Karlsruhe, Germany, pp. 158–159.

Steed, L.E., Truong, V.-D., Simunovic, J., Sandeep, K.P., Kumar, P., Cartwright, G.D., and Swartzel, K.R. (2008) Continuous flow microwave-assisted processing and aseptic packaging of purple-fleshed sweetpotato purees. *Journal of Food Science*, 73, E455–E462.

Sultana, B., Anwar, F., and Iqbal, S. (2008) Effect of different cooking methods on the antioxidant activity of some vegetables from Pakistan. *International Journal of Food Science and Technology*, 43, 560–567.

Sun, J., Chu, Y.-F., Wu, X., and Liu, R.H. (2002) Antioxidant and antiproliferative activities of fruits. *Journal of Agricultural and Food Chemistry*, 50, 7449–7454.

Sun, T., Tang, J., and Powers, J.R. (2007) Antioxidant activity and quality of asparagus affected by microwave-circulated water combination and conventional sterilization. *Food Chemistry*, 100, 813–819.

Vadivambal, R. and Jayas, D.S. (2007) Changes in quality of microwave-treated agricultural products- a review. *Biosystems Engineering*, 98, 1–16.

Vicente A.A., de Castro I., and Teixeira, J.A. (2006) Ohmic heating for food processing. In: D.-W. Sun (ed.) *Thermal Food Processing: New Technologies and Quality Issues*, CRC Taylor & Francis, New York, pp. 425–468.

Vikram, V.B., Ramesh, M.N., and Prapulla, S.G. (2005) Thermal degradation kinetics of nutrients in orange juice heated by electromagnetic and conventional methods. *Journal of Food Engineering*, 69, 31–40.

Wachtel-Galor, S., Wong, K.W., and Benzie, I.F.F. (2008) The effect of cooking on *Brassica* vegetables. *Food Chemistry*, 110, 706–710.

Weaver, C.M. and Kannan, S. (2002) Phytate and mineral bioavailability . In N.R. Reddy and S.K. Sathe (eds) *Food Phytates*, CRC Press, Boca Raton, FL, pp. 211–223.

West, C. and Castenmiller, J. (1998) Food processing and bioavailability of food antioxidants. In: *Proceedings of FAIR CT95–0158 Symposium Natural Antioxidants in Processed Foods – Effects on Storage Characteristics and Nutritional Value*, Agricultural University, Frederiksberg, Denmark.

Wojdyło, A., Figiel, A., and Oszmiański, J. (2009) Effect of drying methods with the application of vacuum microwaves on the bioactive compounds, color, and antioxidant activity of strawberry fruits. *Journal of Agricultural and Food Chemistry*, 57, 1337–1343.

Yang, Z. and Zhai, W. (2010) Optimization of microwave-assisted extraction of anthocyanins from purple corn (*Zea mays* L.) cob and identification with HPLC-MS. *Innovative Food Science and Emerging Technologies*, 11, 470–476.

Yildiz, H., Bozkurt, H., and Icier, F. (2009) Ohmic and conventional heating of pomegranate juice: Effects on rheology, color, and total phenolics. *Food Science and Technology International*, 15, 503–512.

Yildiz, H., Icier, F., and Baysal, T. (2010) Changes in β-carotene, chlorophyll and color of spinach puree during ohmic heating. *Journal of Food Process Engineering*, 33, 763–779.

Zhang, M., Chen, H., Li, J., Pei, Y., and Liang, Y. (2010) Antioxidant properties of tartary buckwheat extracts as affected by different thermal processing methods. *LWT – Food Science and Technology*, 43, 181–185.

Zhao, Y. (2006) Radio frequency dielectric heating. In: D.-W. Sun (ed.) *Thermal Food Processing: New Technologies and Quality Issues*, CRC Taylor & Francis, New York, pp. 469–492.

Zhao, Y., Flugstad, B., Kolbe, E., Park, J.E., and Wells, J.H. (2000) Using capacitive (radio frequency) dielectric heating in food processing and preservation- a review. *Journal of Food Process Engineering*, 23, 25–55.

Zhong, Q., Sandeep, K.P., and Swartzel, K.R. (2004) Continuous flow radio frequency heating of particulate foods. *Innovative Food Science and Emerging Technologies*. 5, 475–483.

Zuber, F. (1997) Ohmic heating: A new technology for stabilising ready-made dishes. *Viandes et Produits Carnes*, 18, 91–95.

13 Non thermal processing

B.K. Tiwari,[1] PJ Cullen,[2] Charles S. Brennan[3] and Colm P. O'Donnell[4]

[1] Food and Consumer Technology, Manchester Metropolitan University, Manchester, UK
[2] School of Food Science and Environmental Health, Dublin Institute of Technology, Dublin, Ireland
[3] Faculty of Agriculture and Life Sciences, Lincoln University, New Zealand
[4] UCD School of Biosystems Engineering, University College Dublin, Belfield, Dublin, Ireland

13.1 Introduction

Thermal processing of food remains the most widely adopted technology for shelf life extension and preservation. However, growing consumer demand for nutritious foods, which are minimally and naturally processed, has resulted in continued interest in non-thermal technologies. Non-thermal technologies encompass all preservation treatments that are effective at ambient or sub lethal temperatures and are generally found to be more energy efficient. The temperature of foods is held below the temperature range normally used in thermal processing, thereby minimising negative effects on bioactive compounds present in food. A number of novel thermal and non-thermal preservation techniques are being developed to satisfy consumer demand with regard to the nutritional and sensory aspects of foods. Ensuring food safety and at the same time meeting such demands, has resulted in increased interest in non-thermal preservation techniques for inactivating microorganisms and enzymes in foods (P. Cullen, 2011; Vega-Mercado, Martin-Belloso, Qin, Chang, Marcela Góngora-Nieto, Barbosa-Canovas, et al., 1997). This chapter summarises potential non-thermal food preservation techniques currently under investigation. Ensuring food safety, while at the same time preserving bioactive compounds, is a challenge due to variations in intrinsic and extrinsic processing parameters of foods. Novel non-thermal preservation techniques considered in this chapter include high pressure, pulsed electric field, ultrasound, irradiation, dense phase carbon dioxide and ozone processing of solid, semi-solid and liquid foods. The effects of non-thermal techniques on the stability of phytochemical compounds are also discussed.

13.2 Irradiation

Irradiation treatment generally involves the exposure of food products (raw or processed) to ionising or non-ionising radiation for the purpose of food preservation. The ionising radiation source could be high-energy electrons, X-rays (machine generated) or gamma rays (from Cobalt-60 or cesium-137), while the non-ionising radiation is electromagnetic radiation that does not carry sufficient energy/quanta to ionise atoms or molecules, represented mainly by UV-A (315–400 nm), UV-B (280–315 nm) and UV-C (200–280 nm). Irradiation of food products typically causes minimal modification in the flavour, colour, nutrients, taste and other quality attributes of food (M. Alothman, R. Bhat and A. A. Karim, 2009). However, the levels of modification (in flavour, colour, nutrients, taste etc.) may vary depending on the product, irradiation dose and on the type of radiation source employed (gamma, X-ray, UV, electron beam) (R. Bhat and Sridhar, 2008; R. Bhat, Sridhar and Tomita-Yokotani, 2007). Depending upon the radiation dose, foods may be pasteurised to reduce or eliminate food-borne pathogens. Inactivation of microorganisms by irradiation is primarily due to DNA damage, which destroys the reproductive capabilities and other functions of the cell (DeRuiter and Dwyer, 2002). Tables 13.1 and 13.2 lists reported applications and effects of irradiation on bioactive compounds in selected food products.

13.2.1 Ionising radiation

Application of gamma radiation has been investigated for a wide range of foods and food products including fruit juices (Alighourchi, Barzegar and Abbasi, 2008; D. Kim, Song, Lim, Yun and Chung, 2007), fresh cut fruit and vegetables (M. Alothman, R. Bhat and A. A. Karim, 2009; Fan and Sokorai, 2011; Jimenez, Alarcon, Trevithick-Sutton, Gandhi and Scaiano, 2011; J.-H. Kim, Sung, Kwon, Srinivasan, Song, Choi, et al., 2009). Irradiation induces negligible or subtle losses of nutrients and sensory qualities in food compared to thermal processing as it does not substantially raise the temperature of food during processing (Wood and Bruhn, 2000). However, Alighourchi, Barzegar and Abbasi (2008) reported a significant reduction in total and individual anthocyanin content in pomegranate juice after irradiation at higher doses (3.5–10 kGy). Similarly, Jimenez et al. (2011) observed inconsistent changes in the oxygen radical absorbance capacity values and total phenolic content of irradiated fresh cut spinach, where a significant decrease in the ascorbic acid content of irradiated spinach during storage at 4 °C was found compared to untreated fresh samples. Irradiation effects on anthocyanin pigments depend upon the nature of anthocyanin, for example, diglycosides are reported to be relatively stable to irradiation compared to monoglycosides. Conversely, Ayed, Yu and Lacroix (1999) reported that the anthocyanin content in grape pomace increases with irradiation dose, with an optimum at 6 kGy. Increase in the anthocyanin content can be attributed to the release of bound pigment as a result of cell wall degradation. Some studies suggest that the decrease or increase in bioactive compounds is not dose dependent. For example, Zhu, Cai, Bao and Corke (2010) observed a decrease in phenolic compounds (*p*-coumaric acid, ferulic acid and sinapinic acid) and anthocyanins (cyanidin-3-glucoside and peonidin-3-glucoside) in black, red and white rice. They observed that at most irradiation doses a significant reduction in total phenolic acid and anthocyanin content of black rice was found. However, they also observed a significant increase in total anthocyanins and phenolic acids in black rice at doses of 6 and 8 kGy.

13.2.2 Non ionising radiation

Application of UV radiation to whole fruit and vegetables and their products such as juice (Guan, Fan and Yan, 2012; Keyser, Muller, Cilliers, Nel, and Gouws, 2008) has been reported for the inactivation of microorganisms. UV radiated food products are

Table 13.1 Effect of ionising irradiation on bioactive compounds of selected food and food materials. Reproduced from Alothman et al. (2009). Effects of radiation processing on phytochemicals and antioxidants in plant produce. Trends in Food Science and Technology, 20(5), 201–212. With permission from Elsevier.

Food	Bioactive compounds	Salient results	Reference
Grape pomace	Anthocyanin	Pomace was γ-irradiated at 0–9 kGy. Low doses of irradiation (below 2 kGy) prevented the loss of anthocyanin while higher doses decreased the content of anthocyanin	(Ayed, Yu & Lacroix, 1999)
Mushrooms (Agaricus bisporus)	Phenolic compounds	Phenolic compounds increased after 2 kGy dose of γ-irradiation when samples compared with control over 9 days storage period.	(Beaulieu, Béliveau, D'Aprano & Lacroix, 1999),
Clementines peel (Citrus clementina Hort. Ex. Tanaka)	Flavanones, polymethoxylated flavones, flavonoids, p-coumaric acid	The difference in the accumulation of the analyzed compounds was significant over 49 days storage at 3°C for the γ-irradiated irradiated fruits (mean dose of 0.3 kGy)	(Oufedjikh, Mahrouz, Amiot & Lacroix, 2000)
Strawberries	Phenolic acids and Flavonoids	γ-irradiation (1–10 kGy) led to the degradation of cinnamic, p-coumaric, gallic, and hydroxybenzoic acids. Ellagic acid derivatives and quercitin concentrations were not affected by γ-irradiation (1–6 kGy). Catechin and kaempferol components diminished noticeably due to the treatment.	(Breitfellner, Solar & Sontag, 2002a, 2002b)
Artichoke (Cynara scolymus Linné)	Flavonoids, phenolic compounds, tannins, β-carotene	Gamma irradiation doses (0, 10, 20, 30 kGy) did not induce any significant changes in flavonoids, tannins, phenolic contents while a slight decrease in β-carotene content was observed	(Koseki, Villavicencio, Brito, Nahme, Sebastião, Rela et al., 2002)
Fresh-cut vegetables (Romaine, iceberg lettuce, endive)	Phenolic compounds	Gamma irradiated (0, 0.5, 1, & 2 kGy) showed significant increase in the total phenolic content and antioxidant capacity corresponding to the increased treatment time.	Fan (2005)
Tomato	p-hydroxybenz-aldehyde, p-coumaric acid, ferulic acid, rutin, naringenin	The γ-irradiation treatment (2, 4, and 6 kGy) markedly reduced the concentration of the phenolic compounds	(Schindler, Solar & Sontag, 2005)
Brazilian mushroom (Agaricus blazei)	Phenolic compounds	Doses of γ-irradiation between 2.5 and 20 kGy increased the antioxidant activity of the extracts	(Huang & Mau, 2006)
Carrot and kale juice	Phenolic compounds	Over 3 days of cold storage (10°C), total phenolic compounds and antioxidant activity increased significantly at 10 kGy of γ-irradiation	(Song, Kim, Jo, Lee, Kim & Byun, 2006)

(Continued)

Table 13.1 (Continued)

Food	Bioactive compounds	Salient results	Reference
Almond skin	Phenolic compounds	Gamma irradiation (0–16 kGy) showed significant increase in the total phenolics and antioxidant activity	(Harrison & Were, 2007)
Mango (Mangifera indica L.)	Phenolic compounds, ascorbic acid, carotenoids	Electron beam irradiation (1–3.1 kGy) did not affect the total phenolic content of the fruits, while there was a significant increase in flavonols after 18 days storage period for the irradiated fruits (at 3.1 kGy). Ascorbate content of the fruits decreased when the dose exceeded 1.5 kGy. No major changes in the carotenoids content were recorded.	(Reyes & Cisneros-Zevallos, 2007)
Citrus unshiu pomaces	Phenolic compounds	37.9 kGy dose of e-Beam treatment increased the total phenolic compounds, DPPH radical-scavenging activity, and the reducing power of the extracts	Kim, Lee, Lee, Nam & Lee (2008)
Turmeric (Curcuma longa L.)	Curcuminoids	The antioxidant activity (expressed as TBA value) after 10 kGy gamma irradiation treatment did not affect the curcuminoids like curcumin, demethoxy curcumin, and bisdemethoxy curcumin	Chatterjee, Desai & Thomas (1999)
Sweet basil (Ocimum basilicum Linné)	Flavonoids, phenolic compounds, tannins, β-carotene	Gamma irradiation up to 30 kGy (10 intervals) did not show any significant changes in the flavonoids, tannins, and phenolic contents, while it slightly decreased the β-carotene content of the extracts	Koseki et al. (2002)
Rosemary (Rosmarinus officinalis Linné)	Flavonoids, phenolic compounds, tannins, β-carotene	Extracts of the γ-irradiated samples (0, 10, 20, and 30 kGy) had similar flavonoid content, less tannin content, less phenolic content, and slightly less β-carotene content	Koseki et al. (2002)
Basil, bird pepper, black pepper, cinnamon, nutmeg, oregano, parsley, rosemary, and sage	Vitamin C, carotenoids	Application of the commercial practicing dose (10 kGy) of γ-radiation caused significant loss in vitamin C in black pepper, cinnamon, nutmeg, oregano, and sage, while it caused a decrease in carotenoids in cinnamon, oregano, parsley, rosemary, bird pepper, and sage when all compared with control samples	Calucci et al. (2003)
Black pepper (Piper nigrum L.)	Phenolic compounds	DPPH free radical-scavenging activity decreased after doses of γ-irradiation (5–30 kGy) over a storage period up to five months	Suhaj, Rácová, Polovka & Brezová (2006)

Green tea byproducts and green tea leaf extracts	Phenolic compounds	γ-irradiation (20 kGy), revealed insignificant changes in the phenolic content with no affect on the antioxidant capacity of the samples when measured by two methods: DPPH scavenging activity, and FRAP	Lee, Jo, Sohn, Kim & Byun (2006)
Rosemary (*Rosmarinus officinalis* L.)	Phenolic compounds	Antioxidant capacity and total phenolic content of the extracts increased on γ-irradiation treatment depending on the extraction solvent	Pérez et al. (2007)
Niger seeds (*Nigella sativa* L.)	Phenolic compounds	γ-irradiation dose (2–16 kGy) enhanced the DPPH free radical-scavenging activity, total phenolics content, and the extraction yield depending on the solvent used	Khattak et al. (2008)
Rice bran	Vitamin E vitamers, oryzanol	The increased γ-irradiation dose (5, 10, and 25 kGy) had a deleterious effect on both vitamin E vitamers and oryzanol content.	Shin & Godber (1996)
Almond skin	Phenolic compounds	γ-irradiated samples (0–16 kGy) significantly increased the total phenolics in addition to the antioxidant activity	Harrison & Were (2007)
Cashew nuts	Tocopherols (vitamin E)	γ-irradiation doses (0.25–1.00 kGy) decreased the antioxidative activity which further decreased during storage period	Sajilata & Singhal (2006)
Velvet beans (*Mucuna pruriens* L. Dc.)	Phenolic compounds	Phenolics increased (0, 2.5, 5, 7.5, 10, 15 and 30 kGy) on γ-irradiation	Bhat, Sridhar, Bhushan et al. (2007)

Table 13.2 Effect of non ionising irradiation on bioactive compounds of selected food and food materials. Modified from Alothman et al. (2009). Effects of radiation processing on phytochemicals and antioxidants in plant produce. *Trends in Food Science and Technology*, 20(5), 201–212. With permission from Elsevier.

Food	Bioactive compounds	Salient results	Reference
Strawberries	Anthocyanins	UVC doses at 0.25 and 1.0 kJ/m^2 increased anthocyanins concentrations in fresh strawberries	Baka, Mercier, Corcuff, Castaigne & Arul (1999)
Table grapes cultivar Napoleon	Phenolic compounds, resveratrol	UVB, UVC irradiation showed increase in the resveratrol content of the irradiated grapes while it did not induce any significant changes in the other phenolic compounds	Cantos, García-Viguera, de Pascual-Teresa & Tomás-Barberán (2000)
Grapes (table grapes) cultivar Napoleon	Resveratrol, vitamin C	UV Irradiation (254 nm) increased resveratrol concentration 11-folds higher than that in control grapes after 3 days storage. Vitamin C concentration remained unaltered after 1-week storage time	Cantos, Espín & Tomás-Barberán (2001)
Pomegranate arils (*Punica granatum* cv. 'Mollar of Elche')	Anthocyanins	Exposure to UVC (0.56–13.62 kJ/m^2) showed insignificant changes in the anthocyanins as well as the antioxidant capacity	López-Rubira, Conesa, Allende & Artés (2005)
Peppers (*Capsicum annum* L. cv. Zafiro)	Carotenoids, phenolic compounds	Peppers exposed to 7 $kJ\ m^{-2}$ UVC light showed lower total phenolic content and higher antioxidant capacity (DPPH scavenging activity) with insignificant effect on the carotenoids content.	Vicente et al. (2005)
Broccoli florets (*Brassica oleracea* L. cv. Cicco)	Phenolic compounds, flavonoids	UVC (4–14 $kJ\ m^{-2}$) treated broccoli florets displayed lower total phenolic and total flavonoid content along with higher antioxidant capacity compared to the control samples	Costa, Vicente, Civello, Chaves & Martínez (2006)
Fresh-cut mangoes	Phenolic compounds, flavonoids, ascorbic acid, β-carotene	Fresh-cut mangoes UVC irradiated for 0, 10, 20, and 30 min, showed increase in phenolic compounds and flavonoids contents with the increase in treatment time, while both β-carotene and ascorbic acid decreased	González-Aguilar, Villegas-Ochoa et al. (2007)
Mango 'Haden'	Phenolic compounds, flavonoids	UVC exposure (2.46 and 4.93 $kJ\ m^{-2}$) increased both total phenolic and total flavonoids content	González-Aguilar, Zavaleta-Gatica et al. (2007)
Apple fruit (*Malus domestica* Borkh., cv. Aroma)	Chlorogenic acid, ascorbic acid, anthocyanins, flavonols, quercetin glycosides, phenolic compounds	UVB exposure increased the antioxidant capacity and all the other analyzed content of the peel, while no changes occurred in the flesh portion	Hagen et al. (2007)

Plant	Compounds	Effect	Reference
Broccoli (*Brassica oleracea* var. *Italica*)	Phenolic compounds, ascorbic acid	Exposure to UV-C (8 kJ m^{-2}) increased total phenolic and ascorbic acid contents, and the antioxidant capacity	Lemoine, Civello, Martínez & Chaves (2007)
Strawberries	Anthocyanins, phenolic compounds	UV-C treatment for different durations (1, 5, & 10 min) increased the antioxidant capacity and the concentrations of anthocyanins and phenolic compounds	Erkan, Wang & Wang (2008)
Blueberries (*Vaccinium corymbosum*, cvs. Collins, Bluecrop)	Anthocyanins, phenolic compounds	2 or 4 kJ/m^2 UV-C exposures did not change the total phenolic content while it increased the total anthocyanins content and FRAP values (Bluecrop cv.). For Collins cultivar, there was no significant changes when compared with the control fruits	Perkins-Veazie, Collins & Howard (2008)
Peanut hulls	Phenolic compounds, luteolin	The amount of phenolic compounds and luteolin decreased with the increasing time of UV-C irradiation (0, 3, and 6 days)	Duh & Yen (1995)
Soybean	Phenolic compounds	UV-C irradiated plants showed faster phenolics accumulation than those of the non-irradiated. There was an increase in the isorhamnetin- and quercetin-based flavonoids concentrations in the UV-C exposed plants	Winter & Rostás (2008)

reported to have lower levels of some phytochemicals during storage. For example, Guan, Fan and Yan (2012) observed that UV-C doses of 0.45–3.15 kJ m^{-2} applied to mushrooms resulted in a reduction in the antioxidant activity, total phenolics and ascorbic acid content compared to non-radiated samples during the first seven days of storage at 4 °C. UV radiation is also reported to have a negative influence on anthocyanins. Bakowaska et al. (2003) reported a strong negative influence of UV irradiation on the complex of cyanidin-3-glucoside with copigment compared to thermal treatment at 80 °C. However, the presence of certain copigments can inhibit the degradation effect of UV on anthocyanins improving the cyanidin-copigment complex (Kucharska, Oszmianski, Kopacz and Lamer-Zarawska, 1998). Literature reveals that most of the applications of irradiation are limited to solid foods and there is scarcity of information regarding treatment of fruit juices. Application of UV radiation on orange, guava and pineapple juice (Keyser, Muller, Cilliers, Nel and Gouws, 2008) has been reported for the inactivation of microorganisms. Alothman, Bhat and Karim (2009) investigated the effect of UV-C treatment on total phenol, flavonoid and vitamin C content of fresh-cut honey pineapple, banana 'pisang mas' and guava. On average, the samples received a UV radiation dose of 2.158 J/m^2. In their study, total phenol and flavonoid contents of guava and banana increased significantly with treatment time ($p<0.05$). In pineapple, the increase in total phenol content was not significant ($p>0.05$), but the flavonoid content increased significantly after 10 min of treatment. In contrast UV-C treatment decreased the vitamin C content of all three fruits. A separate study conducted by López-Rubira et al. (2005) demonstrated insignificant changes in anthocyanins and antioxidant activity of pomegranate arils after exposure to UV-C (0.56–13.62 kJ/m^2). González-Aguilar et al. (2007) observed significant increase the antioxidant capacity of UV-C irradiated fresh cut mango during storage at 5 °C for 15 days even though they observed a significant decrease in β-carotene and ascorbic acid contents.

Costa et al. (2006) observed higher retention rates for chlorophyll content of UV-C irradiated broccoli florets with improved antioxidant activity associated with total phenol and flavonoid content compared to untreated broccoli florets. Similar increases in antioxidants (total phenol and ascorbic acid) because of UV-C treatment of broccoli was observed by Lemoine, Chaves and Martínez (2010). However, they observed either an increase or no significant change in total phenol and flavonoid content during storage. Irradiation of plant tissues with UV has been shown to have positive interactions, indicating an increase in the enzymes responsible for flavonoid biosynthesis. UV irradiation of fruits is also reported to induce anthocyanin biosynthesis. For example, Kataoka and Beppu (2004) observed an increase in the anthocyanin content in peach with an increase in irradiation dose up to 7.3 W m^{-2}. Enhancement of anthocyanin synthesis as a result of UV light is reported due to an increase in phenylalanine ammonia lyase activity which is involved in phenol synthesis in apple (Faragher and Chalmers, 1977). Similar increases in anthocyanin content in cherries (Takos, Jaffé, Jacob, Bogs, Robinson and Walker, 2006) and pears (D. Zhang, Yu, Bai, Qian, Shu, Su, et al., 2011) were reported. UV treatment of grapes during post harvest treatment is reported to produce stilbene-enriched (resveratrol and piceatannol) red wine. Cantos et al. (2003) reported an increase in resveratrol and piceatannol content of wine by 2- and 1.5-fold respectively, when compared to the control wine without affecting other key quality parameters. Similarly, Jagadeesh et al. (2011) observed higher levels of ascorbic acid and total phenolic content for UV treated mature green tomato fruit exposed to UV stored at 13 °C and 95% RH. However, they observed a significant reduction in the lycopene content of the tomatoes.

13.3 High pressure processing

High pressure (HP) processing is employed as a potential non thermal preservation technique for microbial and enzyme inactivation while minimising effects on nutritional and quality parameters. High hydrostatic pressure (HHP) processing uses water as a medium to transmit pressures from 300 to 700 MPa to foods resulting in a reduction in microbial numbers (Meyer, Cooper, Knorr and Lelieveld, 2000) and enzyme activity (Weemaes, Ludikhuyze, Van den Broeck and Hendrickx, 1998) leading to an extension of product shelf life. However, processing conditions employed for achieving food safety may have negative effects on phytochemical content. Table 13.3 lists reported applications of HHP for various foods along with reported effects on bioactive constituents. HHP processing offers many advantages over conventional techniques and is particularly useful for producing homogeneous products, such as smoothies (Keenan, Roessle, Gormley, Butler and Brunton, 2012). This has been attributed to the instantaneous transmission of isostatic pressure to the product, independent of size, shape and food composition (Patterson, Quinn, Simpson and Gilmour, 1996). It has been shown that food processed in this way maintains its original freshness, flavour and taste, while colour changes are minimal (Dede, Alpas and Bayındırlı, 2007). Despite alternations to the structure of high-molecular-weight molecules such as proteins and carbohydrates, HHP does not typically affect smaller molecules associated with the sensory, nutritional and health promoting properties. These molecules include volatile compounds, pigments and vitamins. Therefore, HPP imparts fresh-like characteristics and preserves the nutritional value of food (Barba, Esteve and Frigola, 2011; De-Ancos, Gonzales and Cano, 2000; Ferrari, Maresca and Ciccarone, 2010; Frank, Koehler and Schuchmann, 2012; Keenan, Brunton, Gormley, Butler, Tiwari and Patras, 2010; Meyer, Cooper, Knorr and Lelieveld, 2000; Oey, Van der Plancken, Van Loey and Hendrickx, 2008; Patras, Brunton, Da Pieve, Butler and Downey, 2009; Patras, Brunton, Da Pieve and Butler, 2009; Plaza, Colina, de Ancos, Sanchez-Moreno and Cano, 2012).

High pressure processed juices have shown better retention of bioactive compounds during storage compared to thermally processed tomato juices (Hsu, Tan and Chi, 2008). Hsu *et al.* (2008) reported a significant increase of up to 60% for lycopene and 62% for total carotenoid content during high pressure processing (300–500 MPa/25 °C/10 min) compared to fresh and thermally processed (98 °C/15 min) tomato juice. Whereas, during storage at 25 °C for 28 days they observed no significant decrease in total carotenoid content and lycopene content in HP processed tomato juice. However, a decrease of about 18.4 and 12.5% was reported for thermally processed tomato juice.

There are many reports concerning the preservation of phytochemicals in high pressure processed food and food products. For example, Patras *et al.* (2009) observed a significant increase in total phenol content and a decrease in anthocyanin content of strawberry and blackberry purées at 600 MPa compared to unprocessed purée. In another study, the effects of pressure treatments of 350 MPa on orange juice carotenoids, β-carotene, ά-carotene, zeaxanthin, lutein and β-cryptoxanthin, associated with pro-vitamin A and radical-scavenging capacity values, resulted in significant increases of 20–43% in the carotenoid content of fresh orange juice (De-Ancos, Gonzales and Cano, 2000). Similarly, Plaza *et al.* (2012) studied the effect of high pressure processing on carotenoid content of persimmon fruit. They observed that a high pressure treatment at 200 MPa for 6 min significantly increased the extractability of carotenoids by up to 86% for astringent persimmon fruits. Anthocyanins content of raspberry (Suthanthangjai, Kajda and Zabetakis, 2005), strawberry (Zabetakis,

Table 13.3 Effect of HHP processing of bioactive compounds of some food and food materials

Food	Bioactive compounds	Processing conditions	Effect	Reference
Strawberry puree	Polyphenols, ascorbic acid and pelargonidin-3-glucoside	400, 500, 600 MPa/15 min/10–30°C	Ascorbic acid content (5.3% decrease at 600 MPa) pelargonidin-3-glucoside(1% decrease at 600 MPa) Total phenol content (9.8% increase at 600 MPa)	(Patras, Brunton, Da Pieve & Butler, 2009)
Blackberry purees	Polyphenols and cyanidin-3-glycoside	400, 500, 600 MPa/15 min/10–30°C	Pelargonidin-3-glucoside(0.95% decrease at 600 MPa) Total phenol content (4.9% increase at 600 MPa)	(Patras, Brunton, Da Pieve & Butler, 2009)
Tomato puree	Total phenolic content Ascorbic acid Total carotenoid	400, 500, 600 MPa/15 min/10–20°C	Total carotenoids (172% increase at 600 MPa) Total phenol content (3.09% increase at 600 MPa) Ascorbic acid content (6.2% decrease at 600 MPa)	(Patras, Brunton, Da Pieve, Butler & Downey, 2009)
Carrot puree	Total phenolic content Total carotenoid	400, 500, 600 MPa/15 min/10–20°C	Total carotenoids (58.5% increase at 600 MPa) Total phenol content (0.5% decrease at 600 MPa)	(Patras, Brunton, Da Pieve, Butler & Downey, 2009)
Blue berry juice	ascorbic acid, total phenolics, anthocyanin stability	200, 400 and 600 MPa and treatment times (5, 9 and 15 min	Vitamin C (6.9% decrease at 600MPa/15 min). Total phenolic content (23% increase at 400 MPa/15 min The total and monomeric anthocyanin (16% increase at 400 MPa/15 min)	(Barba, Esteve & Frigola, 2011)
Fresh carrots, green beans and broccoli	Total carotenoids	400 and 600 MPa for 2 min	No significant change in α Carotene, β Carotene and Lutein	McInerney, Seccafien, Stewart & Bird, 2007)
Persimmon fruit (Astringent)	Carotenoid content and vitamin A	200–400 MPa/25°C/ 1–6 min	Leutin (55.8% increase at 200 MPa/6 min). Zeaxanthin (32.5% increase at 200 MPa/6 min). Lycopene (16.3% increase at 200 MPa/6 min). β-Carotene (80.8% increase at 200 MPa/6 min). Vitamin A (22.4% increase at 200 MPa/6 min).	(Plaza, Colina, de Ancos, Sanchez-Moreno & Cano, 2012)

Food	Compounds	Conditions	Main findings	Reference
Fruit smoothies (Strawberries, apples, apple juice, bananas and oranges)	total phenols (TP), anthocyanins and ascorbic acid	450 MPa/20°C/5 min or 600 MPa/20°C/10 min and thermal processing (70°C for 10 min) stored for 10 h at 4°C.	Phenolic contents (15% increase at 450 MPa compared to 600 MPa). Ascorbic acid (35 and 44% lower for thermally processed compared to fresh and HHP processed samples). Anthocyanins no significant changes 48.56, 61.35 and 24.39% losses in ascorbic acid content within 0.5 and 1 h storage for fresh, thermal and 450 MPa). Superior quality smoothies at 450 MPa compared to thermal; 600 MPa was appropriate for maintaining long term storage in relation to inactivation of enzymes	(Keenan, Rößle, Gormley, Butler & Brunton, 2012)
Raspberry	Cyanidin-3-glucoside (C3G) Cyanidin-3-sophoroside (C3S)	200 to 800 MPa 18–22°C, 15 min Storage temperature: 4, 20, 30°C for 9 days	Greater stability at 800 MPa for C3G and C3S at storage temperature of 4°C	Suthanthangjai et al. (2005)
Strawberry	Pelargonidin-3-glucoside (P3G) Pelargonidin-3-rutinoside (P3R)	200 to 800 MPa 18–22°C, 15 min Storage temperature: 4, 20, 30°C for 9 days	Greater stability at 800 MPa for P3G and P3R at storage temperature of 4°C	Zabetakis et al. (2000)
Blackcurrant	Delphinidin-3-rutinoside (D3R) Cyanidin-3-rutinoside (C3R)	200 to 800 MPa 18–22°C, 15 min Storage temperature: 5, 20, 30°C for 7 days	Greater stability at 600 MPa and 800 MPa for D3R at storage temperature of 5°C	Kouniaki et al. (2004)
Muscadine grape juice	Delphinidin 3,5-diglucoside Petunidin 3,5-diglucoside Peonidin 3,5-diglucoside Malvidin 3,5-diglucoside	400 and 550 MPa for 15 min	Total anthocyanin loss of 70% at 400 MPa and 46% at 550 MPa	Del Pozo-Insfran et al. (2007)

Koulentianos, Orruno and Boyes, 2000) and blackcurrant (Kouniaki, Kajda, and Zabetakis, 2004) processed at a pressure of 800 MPa for 15 min are reported to be stable compared to unprocessed samples. Improved stability of anthocyanins at higher pressure is mainly due to inactivation of enzymes associated with the degradation of bioactive compounds. Enzymes such as polyphenoloxidase, peroxidase and β-glucosidase have been associated with the degradation of anthocyanins (Fennema and Tannenbaum, 1996). Garcia-Palazon *et al.* (2004) reported that the stability of strawberry and red raspberry anthocyanins namely pelargonidin-3-glucoside and pelargonidin-3-rutinoside at 800 MPa for 15 min at moderate temperatures 18–22 °C is mainly due to complete inactivation of polyphenoloxidase. However it must be noted that the effects of HHP processing parameters such as pressure, temperature, time and physicochemical properties of food have varying effects on enzymes responsible for the stability of bioactive compounds processed using HHP (Ogawa, Fukuhisa, Kubo and Fukumoto, 1990; Tiwari, O'Donnell and Cullen, 2009).

13.4 Pulsed electric field

The use of pulsed electric field (PEF) as a novel pasteurisation method is especially suitable for the pasteurisation of fluid foods, in which microorganisms are inactivated by applying short (in general 1–300 μs), high electric field (10–60 kV cm^{-1}) between two electrodes (Fox, Esveld and Boom, 2007). Because PEF processing is controlled at ambient temperature for very short treatment times of microseconds, it provides fresh-like foods which are safe and have extended shelf life (Qin, Pothakamury, Barbosa-Cánovas and Swanson, 1996). PEF has been demonstrated to be effective against various pathogenic and spoilage microorganisms and enzymes without appreciable loss of flavour, colour or bioactive compounds (Cserhalmi, Sass-Kiss, Tóth-Markus and Lechner, 2006; Elez-Martínez and Martín-Belloso, 2007; Elez-Martinez, Soliva-Fortuny and Martin-Belloso, 2009; Plaza, Sanchez-Moreno, De Ancos, Elez-Martinez, Martin-Belloso and Pilar Cano, 2011; Sanchez-Moreno, De Ancos, Plaza, Elez-Martinez and Pilar Cano, 2009) (Table 13.4). Recently (Y. I. Zhang, Gao, Zhang, Shi and Xu (2010) reported that PEF-treated (bipolar pulse 3 μs wide, at an intensity of 32 kV/cm) longan juice retained greater amounts of vitamin C and flavour compounds than thermally treated juice. Elez-Martínez and Martín-Belloso (2007) have reported that pulses applied in bipolar mode, as well as decreasing the field strength, treatment time, pulse frequency and width, led to higher levels of vitamin C retention ($p<0.05$) in both orange juice and 'gazpacho' soup. Shivashankara, Isobe, Al-Haq, Takenaka and Shiina, (2004) studied the ascorbic acid content in Irwin mango fruits stored at 5 °C after a high electric field pre-treatment, observing that ascorbic acid decreased after 20 days of storage. To this end, all the studies indicate that vitamin C content significantly depends on the PEF treatment time and electric field strength applied during PEF-processing of the juice, so that the lower the treatment time and electric field strength, the greater the vitamin C retention. Various studies have shown the validity of PEF technology for inactivating microorganisms in more complex foods, such as a mixed orange juice and milk beverage (Rivas, Rodrigo, Company, Sampedro and Rodrigo, 2007; Rivas, Sampedro, Rodrigo, Martínez and Rodrigo, 2006; Sampedro, Geveke, Fan and Zhang, 2009; Sampedro, Rivas, Rodrigo, Martínez and Rodrigo, 2006), fruit (orange, kiwi and pineapple) juice soymilk beverage (Morales-de la Peña, Salvia-Trujillo, Rojas-Graü and Martín-Belloso, 2010) and blends of orange and carrot juice (Rivas, Rodrigo, Martinez, Barbosa-Cánovas and Rodrigo, 2006). A pre-treatment of PEF is reported to increase anthocyanin concentration in grape juice (Knorr, 2003).

Table 13.4 Effect of high-intensity pulsed electric field treatments on some health-related compounds in food systems. Modified from Soliva-Fortuny et al. (2009). Effects of Pulsed Electric Fields on bioactive compounds in Foods: a review. *Trends in Food Science and Technology, 20*, 544–556. With permission from Elsevier.

Food material	Bioactive compound	Treatment conditions	Salient results	Reference
Orange juice	Flavonoids	35 kV/cm, 750 μs	No changes in either individual flavanones nor in total content	Sánchez-Moreno et al. (2005)
Orange juice	Carotenoids	25–40 kV/cm, 30–340 μs	No significant changes in overall content. Better stability of individual compounds compared to thermal pasteurisation	Cortés, Esteve, et al. (2006)
Orange-carrot juice blend		25–40 kV/cm, 30–340 μs	Rise in carotenoids content with increasing treatment time Increase of compounds with provitamin A effect at 25 and 30 kV/cm compared to heat treatments	Torregrosa et al. (2005)
Tomato juice		40 kV/cm, 57 μs	No changes in lycopene with respect to thermal treatment	Min, Jin & Zhang (2003)
Milk	Vitamin B1	18-3–27.1 kV/cm, up to 400 μs	Very low or negligible reductions	Bendicho, Espachs, et al. (2002)
Milk	Vitamin B2	18-3–27.1 kV/cm, up to 400 μs	Very low or negligible reductions	Bendicho, Espachs, et al. (2002)
Protein fortified orange juice-based beverage	Vitamin C	28 kV/cm, 100–300 μs	Loss increasing from 4 to 13% as treatment time increased	Sharma et al. (1998)
Orange juice		87 kV/cm, 40 instant charge reversal pulses, 50°C	Very low or negligible reductions	Hodgins et al. (2002)
		15–35 kV/cm, 100–1000 μs	Loss ranging from 1.8 to 12.5%	Elez-Martínez and Martín-Belloso (2007)
Grape juice		reversal pulses, 50°C	Very low or negligible reductions	Wu, Mittal & Griffiths (2005)
Apple juice and cider		22–35 kV/cm, 94–166 μs	Very low or negligible reductions	Evrendilek et al. (2000)
'Gazpacho' soup		15–35 kV/cm, 100–1000 μs	Loss ranging from 2.9 to 15.7%	Elez-Martínez & Martín-Belloso (2007)
Milk	Vitamin D, Vitamin E	18-3–27.1 kV/cm, up to 400 μs	Very low or negligible reductions	Bendicho, Espachs, et al. (2002)

Corrales *et al.* (2008) demonstrated that PEF treatment enhances (17%) the extraction of anthocyanins compared to conventional methods and is 10% greater than HHP. However, Plaza *et al.* (2011) observed that PEF (35 kV/cm, 750 μs) treatment of orange juice retained similar level of carotenoids and flavanones to those of untreated juice while an increase in extractability with HP treatment (400 MPa, 40 °C, 1 min) of orange juice was observed. Similarly, Morales-de la Peña *et al.* (2010) investigated the effect of PEF on vitamin C in an orange, kiwi, pineapple and soymilk based drink immediately after treatment and concluded that levels were not different from the thermally processed juice. However, the beneficial effects of the PEF treatment were noticeable over a storage period of 31 days, as an 800 μs treatment at 35 kV/cm showed significantly greater retention than both a 1400 μs treatment and a thermal treatment. In general, longer exposure PEF treatment times may induce reductions in the product retention of vitamin C due to product heating.

Comparative analysis of high pressure, pulsed electric field and thermally processed orange juice indicates that HP processed orange juice shows higher retention of carotenoid content (45.19%) and vitamin A (30.89%) compared to freshly squeezed orange juice (Figure 13.1a,b). Whereas, PEF and LPT (low pasteurisation temperature) processed orange juices show non-significant changes in carotenoid content and vitamin A compared to freshly squeezed orange juice (Figure 13.1a,b) (Lucia Plaza, Sanchez-Moreno, De Ancos, Elez-Martinez, Martin-Belloso and Pilar Cano, 2011). During refrigerated storage at 4 °C, HP processed orange juice showed higher retention rates for individual carotenoids and vitamin A compared to freshly squeezed, LPT and PEF processed juice (Figure 13.1c,d).

13.5 Ozone processing

The use of ozone as a disinfecting agent has widespread application in food processing and preservation. Ozone processing of food may provide microbial food safety with several advantages over conventional disinfectant agents such as chlorine, chlorine dioxide, calcium hypochlorite, sodium chlorite, peroxyacetic acid and sodium hypochlorite. Ozone application has been reported at various post-harvest stages of fruits and vegetable processing with objectives of pathogenic and spoilage microorganism inactivation along with destruction of pesticides and other chemical residues. Both aqueous and gaseous ozone is employed for surface decontamination of whole fruits and vegetables via washing or storage in ozone-rich atmospheres (P.J. Cullen, Tiwari, O'Donnell and Muthukumarappan, 2009; P.J. Cullen, Valdramidis, Tiwari, Patil, Bourke and O'Donnell, 2010). In 2001 ozone was approved in the US as a direct additive in food products (Rice, Graham and Lowe, 2002) which triggered the application of ozone for processing of various fruit juices (P.J. Cullen, Valdramidis, Tiwari, Patil, Bourke and O'Donnell, 2010). Microbial studies to date show reductions of spoilage and pathogenic species most commonly associated with food products including fruit and vegetable juices can be achieved. However, ozone processing is reported to have significant effects on the bioactive constituents due to its strong oxidising activity. Greater impact of ozone on bioactive compounds is observed in the case of ozone processed juices compared to whole fruit and vegetables. For example, Tiwari *et al.* (2008) observed a 50% reduction in ascorbic acid content in orange juice within 2 min, whereas Zhang *et al.* (2005) reported no significant difference between ascorbic acid contents for ozonated and non-ozonated celery samples. Moreover, increase in ascorbic acid levels in spinach (Luwe, Takahama and Heber, 1993), pumpkin leaves (Ranieri, DUrso, Nali, Lorenzini and Soldatini, 1996) and strawberries (Perez, Sanz, Rios, Olias and Olias, 1999) in response to ozone

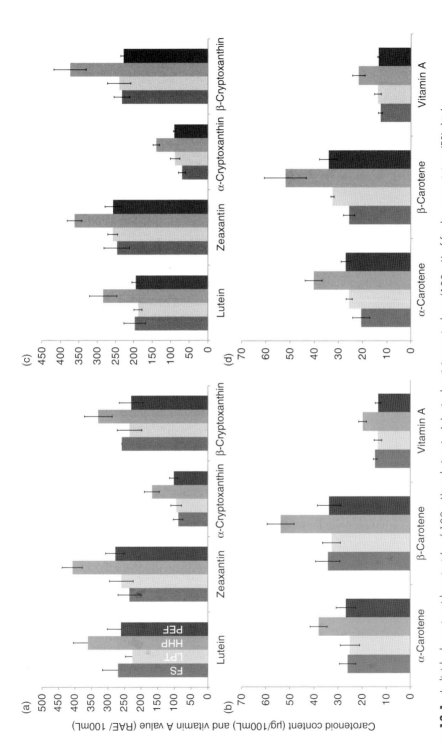

Figure 13.1 Individual carotenoid content (μg/ 100 mL) and vitamin A (retinol activity equivalents/ 100 mL) of fresh orange juices (FS), high-pressure (HP: 400MPa/40°C/1 min), pulsed electric field (PEF: 35 kV.cm^{-1}/750 μs) and low pasteurisation temperature (LPT: 70°C/30 s) processed juice immediately after processing (a,b) and after storage for 40 days at 4°C (c,d).

exposure have also been documented. Decomposition of ascorbic acid in broccoli florets was reported after ozone treatment by Lewis et al. (1996); however, only a slight decrease in vitamin C content was reported in lettuce (Beltran, Selma, Marin and Gil, 2005). Ozone treatments have been reported to have minor effects on anthocyanin contents in strawberries (Perez, Sanz, Rios, Olias and Olias, 1999) and blackberries (Barth, Zhou, Mercier and Payne, 1995). Anthocyanin content in blackberries stored in air and at 0.1 ppm ozone was found to remain stable, however it was shown to fluctuate in 0.3 ppm ozone treated samples during storage (Barth, Zhou, Mercier and Payne, 1995).

Ozonation of fruit juices rich in anthocyanins such as strawberry and blackberry juice causes a significant reduction in these pigments. A significant reduction of 98.2% in the pelargonidin-3-glucoside content of strawberry juice was reported at an ozone concentration of 7.8%w/w processed for 10 min (Tiwari, O'Donnell, Patras, Brunton and Cullen, 2009a). Reductions of >90% in the cyanidin-3-glucoside content of blackberry juice were reported under similar treatment conditions (Tiwari, O'Donnell, Patras, Brunton and Cullen, 2009a). Studies on ozonation of fresh cut honey pineapple, banana and guava indicated an increase in the total phenol and flavonoid contents of pineapple and banana, while the reverse was reported for guava (Alothman, Kaur, Fazilah, Bhat and Karim, 2010). However, significant decreases in the vitamin C content of fresh cut pineapple, banana and guava were reported. Similarly, Tzortzakis, Borland, Singleton and Barnes (2007) reported an increase in beta-carotene, lutein and lycopene contents of tomatoes stored in an ozone-enriched (1.0 μmol mol^{-1}) atmosphere at 13 °C.

These studies indicate that the effect of ozone on phytochemicals is matrix dependent. Higher degradation is attributed to greater exposure to bioactive constituents in liquid medium compared to whole fruits where penetration of ozone is limited to surfaces. Storage of fruits and vegetables in ozone rich atmosphere is reported to preserve phenolic constituents of grapes during long-term storage and simulated retail display conditions (Artés-Hernández, Aguayo, Artes and Tomás-Barberán, 2007). Artes-Hernandez et al. (2007) observed that the application of ozone during storage increased the total flavan-3-ol content and continuous 0.1 μL L^{-1} O$_3$ exposure during storage also preserved the total amount of hydroxycinnamates, while both treatments investigated the flavonol content sampled at harvest. Barboni, Cannac and Chiaramonti (2010) compared the effect of ozone rich storage and air storage over a period of seven months on the vitamin C content of kiwi fruit. Gaseous ozone concentration was 4 mg/h in the chamber at a temperature of 0 °C and a humidity of 90–95%. The authors did not observe any significant change in ascorbic acid content of kiwi fruit over a seven month storage period at an ozone concentration of 4 mg/h in the chamber (2 m^3) and a storage temperature of 0 °C.

The degradation of phytochemicals including anthocyanins, phenolic compounds and ascorbic acid during ozone treatment could be due to direct reaction with ozone or indirect reactions of secondary oxidators such as •OH, HO$^{2\bullet}$, •O$_2^-$ and •O$_3^-$. Such secondary oxidators may lead to electrophilic and nucleophilic reactions occurring with aromatic compounds that are substituted with an electron donor (e.g. OH$^-$) having high electron density on the carbon compounds in ortho and para positions. Direct reaction is described by the Criegee mechanism (Criegee, 1975) where ozone molecules undergo 1–3 dipolar cyclo addition with double bonds present, leading to the formation of ozonides (1,2,4-trioxolanes) from alkenes and ozone with aldehyde or ketone oxides as decisive intermediates, all of which have finite lifetimes (Criegee, 1975). This leads to the oxidative disintegration of ozonide and formation of carbonyl compounds, while oxidative work-up leads to carboxylic acids or ketones. Ozone attacks OH radicals, preferentially to the double bonds in organic compounds leading to the formation of unstable ozonide which subsequently disintegrates. The degradation mechanism for anthocyanin based on Criegee in strawberry juice was proposed by Tiwari et al. (2009a). It is also reported that ascorbic acid degradation in the case of whole or fresh cut fruit and vegetables may also be due to the activation of ascorbate oxidase, responsible for the degradation of ascorbic acid (Alothman, Kaur, Fazilah, Bhat and Karim, 2010).

13.6 Ultrasound processing

Ultrasound processing has emerged as an alternative non thermal food processing option to conventional thermal approaches for pasteurisation and sterilisation of food products (O'Donnell, Tiwari, Bourke and Cullen, 2010). Power ultrasound has shown promise as an alternative technology to thermal treatment for food processing (Mason, Riera, Vercet and Lopez-Bueza, 2005) and has been identified as a potential technology to meet the US Food and Drug Administration (USFDA) requirement of a five log reduction of *E. coli* in fruit juices (Tiwari and Mason, 2012). It has been reported to be effective against food-borne pathogens found in a range of juices, including orange juice (Valero, Recrosio, Saura, Munoz, Martí and Lizama, 2007) and guava juice (Cheng, Soh, Liew and Teh, 2007).

Ultrasound processing on its own or in combination with heat and/or pressure is an effective processing tool for microbial inactivation and phytochemical retention. However the literature indicates that it can negatively modify some food properties including flavour, colour or nutritional value. Ultrasound treatment of liquid foods in general has a minimal effect on the bioactive compounds during processing and results in improved stability during storage when compared to thermal treatment. Rawson et al. (2011) investigated the effect of thermosonication on the bioactive compounds of freshly squeezed watermelon juice. They observed a higher retention of ascorbic acid and lycopene at low amplitude levels and temperatures. They also observed a slight increase in lycopene at low amplitude level. Similarly, Tiwari, O'Donnell, Patras and Cullen (2008) reported a slight increase (1–2%) in the pelargonidin-3-glucoside content of sonicated strawberry juice at lower amplitude levels and treatment times, which may be due to the extraction of bound anthocyanins from the suspended pulp. Whereas at higher amplitude levels and treatment times a maximum of 5% anthocyanin degradation was reported. Cheng, Soh, Liew and Teh, (2007) reported a significant increase in the ascorbic acid content of Guava juice during sonication from 110 ± 0.5 (fresh) to 119 ± 0.8 (sonication) and to 125 ± 1.1 (combined sonication and carbonation) mg/100 mL, which could be due to cavitation effects caused by carbonation and sonication, respectively. Cheng et al. (2007) also observed that during carbonation, sample temperature decreased substantially which could have disfavoured ascorbic acid degradation. Similarly, Bhat et al. (2011) observed a significant increase in the bioactive constituents of sonicated kasturi lime (*Citrus microcarpa*) in a sonication tub at a frequency of 25 kHz. They observed an increase of about 6.7% for ascorbic acid, 27.4% for total phenolics, 42.3% for total flavonoids and 127.4% for total flavanols at 60 min. Low power sonication tends to increase the level of bioactive compounds in sonicated food materials due to enhanced extraction of bound pigments as a result of cell wall disruption. In some cases sonication treatment also enhances the antioxidant activity of treated samples. This is attributed to the addition of sonochemically generated hydroxyl radicals (OH^-) to the aromatic ring of the phenolic compounds at the *ortho*- or *para*-positions of phenolic compounds (Ashokkumar, Sunartio, Kentish, Mawson, Simons, Vilkhu, et al., 2008).

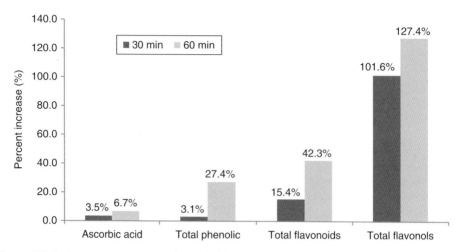

Figure 13.2 Percentage increase in the level of bioactive compound over control (untreated) due to sonication at 25 kHz.
Source: adapted from Bhat et al. (2012).

Weak ultrasonic irradiation was reported to promote an increase in the amount of phenolic compounds found in red wine (Masuzawa, Ohdaira and Ide, 2000). Literature also indicates ultrasound processing enhances extraction of phenolic and other bioactive compounds from grape must or wine (Cocito, Gaetano and Delfini, 1995). Ultrasound assisted extraction of bioactive compounds and anthocyanins were recently reviewed by Vilkhu, et al. (2008). Zhao et al. (2006) reported a degradation of (all-E)-astaxanthin into unidentified colourless molecule(s) during extraction using sonication with increased power levels and treatment times. Similarly, the degradation of the pelargonidin-3-glucoside content of strawberry juice (Tiwari, O'Donnell, Patras and Cullen, 2008) and the cyanidin-3-glucoside content of blackberry juice (Tiwari, O'Donnell, Muthukumarappan and Cullen, 2009a) was found during sonication. Figure 13.2 shows the effect of sonication on retention of blackberry and strawberry juice anthocyanins during ultrasonic processing. As can be seen from this figure, the degradation of these anthocyanins is minimal, with a retention rate of over 98%. The reported degradation of anthocyanins is mainly due to cavitation, which involves the formation, growth and rapid collapse of microscopic bubbles. The degradation of quality and nutritional parameters results from the extreme physical conditions which occur within the bubbles during cavitational collapse at micro-scale (Suslick, 1988) and several sonochemical reactions occurring simultaneously or in isolation. The chemical effects produced by cavitation generate high local temperatures (up to 5000 K), pressures (up to 500 MPa) and mechanical action between solid and liquid interfaces (Suslick, Hammerton and Cline, 1986). The anthocyanin degradation could also be due to the presence of other compounds such as ascorbic acid and can be related to oxidation reactions, promoted by the interaction of free radicals formed during sonication (Portenländer and Heusinger, 1992). Hydroxyl radicals produced by cavitation are involved in the degradation of anthocyanins by opening of rings and formation of chalcone mainly due to temperature rises occurring during sonication (Sadilova, Carle and Stintzing, 2007). The interaction of ascorbic acid with anthocyanin pigments results in mutual degradation (Markakis, Livingston and Fellers, 1957). This is also reported for strawberry juice (Tiwari, O'Donnell, Patras, Brunton and Cullen, 2009b).

Ultrasound treatment of fruit juices is reported to have a minimal effect on the ascorbic acid content during processing and results in improved stability during storage when compared to thermal treatment. This positive effect of ultrasound is assumed to be due to the effective removal of occluded oxygen from the juice (Knorr, Zenker, Heinz and Lee, 2004), a critical parameter influencing the stability of ascorbic acid (Solomon, Svanberg and Sahlström, 1995, 2009b). Tiwari et al. (2009b) reported a maximum degradation of 5% in the ascorbic acid content of orange juice when sonicated at the highest acoustic energy density (0.81 W/mL) and treatment time (10 min). During storage at 10 °C sonicated juice was found to have a higher retention of ascorbic acid compared to thermally processed and control samples. However, for sonicated strawberry juice, a higher reduction of ca. 15% was found. Ascorbic acid degradation during sonication may be due to free radical formation (Portenländer and Heusinger, 1992). Hydroxyl radical formation is found to increase with degassing. Sonication cavities can be filled with water vapour and gases such as O_2 and N_2 dissolved in the juice (Korn, Machado Primo and Santos de Sousa, 2002). The interactions between free radicals and ascorbic acid may occur at the gas–liquid interface. In summary, ascorbic acid degradation may follow one or both of the following pathways:

Ascorbic acid → thermolysis (inside bubbles) and triggering of Maillard reaction.
Ascorbic acid → reaction with OH^- → HC–OH and production of oxidative products on the bubble surface.

Thus, sonication can be related to advanced oxidative processes since both pathways are associated with the production and use of hydroxyl radicals. The cavitation bubble is mainly responsible for the degradation of volatile organic compounds due to the production of hydroxyl radicals and subsequently reacts with organic compounds in the water shell around the bubble (Petrier, Combet and Mason, 2007).

13.7 Supercritical carbon dioxide

Supercritical or dense phase carbon dioxide processing is a collective term for liquid CO_2 and supercritical CO_2 or high pressurised carbon dioxide (HPCD). It is a non-thermal alternative to heat pasteurisation for liquid foods and it is attracting much interest in the food industry (Del Pozo-Insfran, Balaban and Talcott, 2006). $SCCO_2$ extraction has been extensively applied in the fruit and vegetable industry for the extraction of different phytochemicals with desired functionalities (antioxidants, anti-depressants, antimicrobial etc.). Recent studies highlighting the presence of health promoting compounds in fruit and vegetables have stimulated the demand for process technologies capable of extracting such compounds in an environment friendly manner. Apart from extraction of bioactive compounds from fruit and vegetables, $SCCO_2$ has unique properties that make it an appealing medium for food preservation. $SCCO_2$ has strong potential as an antimicrobial agent as it is non-toxic and easily removed by simple depressurisation and out gassing. $SCCO_2$ has significant lethal effects on microorganisms in food and inactivates spoilage enzymes with a minimal effect on end product quality (Damar and Balaban, 2006; Kincal, Hill, Balaban, Portier, Sims, Wei, et al., 2006; Liu, Gao, Peng, Yang, Xu and Zhao, 2008). Plaza (2011) observed no significant change in total phenolic content of guava puree processed using dense phase carbon dioxide (30.6 MPa, 8% CO_2 and 6.8 min, 35 °C). Ferrentino et al. (2009)

investigated the effect of continuous dense phase carbon dioxide (DPCD) on red grapefruit juice. The authors used pressures of 13.8, 24.1 and 34.5 MPa and residence times of 5, 7 and 9 min as variables at constant temperature (40 °C), and CO_2 level (5.7%). A storage study was performed on the fresh juice and DPCD treated at these conditions. The treatment and the storage did not affect the total phenolic content of the juice. Slight differences were detected for the ascorbic acid content and the antioxidant capacity. The experimental results showed that the treatment can maintain the antioxidant content of grape juice. Application of $SCCO_2$ has been reported for various food products including fruit juices such as apple cider (Gasperi, Aprea, Biasioli, Carlin, Endrizzi, Pirretti et al., 2009; Gunes, Blum and Hotchkiss, 2006; Liao, Hu, Liao, Chen and Wu, 2007); orange juice (Kincal et al., 2006); grapefruit juice (Ferrentino, Plaza, Ramirez-Rodrigues, Ferrari and Balaban, 2009); and grape juice (Gunes, Blum and Hotchkiss, 2006). These studies indicated minimal changes in key quality parameters. In a study conducted by Del Pozo-Insfran et al. (2006) no significant changes in total anthocyanin content was reported for DPCD processed muscadine grape juice compared to a 16% loss observed in thermally processed juice. Enhanced anthocyanin stability was also observed in DPCD processed juice during storage for ten weeks at 4 °C. The greater stability of DPCD processed juice could be due to the prevention of oxidation by removal of dissolved oxygen. The exact mechanism for anthocyanin stability is difficult to establish. However, Del Pozo-Insfran et al. (2007) demonstrated that anthocyanin stability is also dependent on the PPO inactivation potential of DPCD treatment and governed by extrinsic control parameters of pressure and CO_2 concentration gradient. Parton et al. (2007) tested a continuous $SCCO_2$ system for liquid foods ranging from orange juice to tomato paste. They reported that the low temperatures used during $SCCO_2$ processing often resulted in improved retention of heat sensitive nutritionally important compounds.

13.8 Conclusions

Ensuring food safety and at the same time meeting the demand for nutritious foods, has resulted in increased interest in non-thermal preservation techniques. Research to date indicates that non-thermal techniques have the potential to enhance the retention of bioactive compounds without compromising food safety. In most cases storage conditions for the processed product play an important role in stability of bioactive compounds. Selection of appropriate extrinsic storage conditions for the processed product are necessary to retain optimum levels of bioactive compounds in food. A key issue for the industrial adoption of these non-thermal techniques is process optimisation. There is a need for focused studies on the stability of bioactive compounds using combined approaches. Combinations of various natural plant extracts or antimicrobials agents in response to consumer demand for 'greener' additives can be explored further to provide improved stability of phytochemicals through copigmentation while at the same time providing synergy for microbial inactivation. Combinations of thermal and non-thermal techniques such as thermosonication, manosonication, manothermosonication and pressure assisted thermal sterilisation have great potential in improving retention of bioactive compounds in food and food products. The impact of product formulation, extrinsic storage parameters and intrinsic product parameters on the efficacy of novel applications of combined non-thermal systems also requires further study. Overall, studies have shown enhanced stability of anthocyanins by novel non-thermal preservation techniques such as HHP, PEF, DPCD, irradiation and ultrasound.

References

Alighourchi, H., Barzegar, M., and Abbasi, S. (2008) Anthocyanins characterization of 15 Iranian pomegranate (Punica granatum L.) varieties and their variation after cold storage and pasteurization. *European Food Research and Technology*, 227(3), 881–887.

Alothman, M., Bhat, R., and Karim, A. (2009a) UV radiation-induced changes of antioxidant capacity of fresh-cut tropical fruits. *Innovative Food Science and Emerging Technologies*, 10(4), 512–516.

Alothman, M., Bhat, R., and Karim, A. A. (2009b) Effects of radiation processing on phytochemicals and antioxidants in plant produce. *Trends in Food Science and Technology*, 20(5), 201–212.

Alothman, M., Kaur, B., Fazilah, A., Bhat, R., and Karim, A. A. (2010) Ozone-induced changes of antioxidant capacity of fresh-cut tropical fruits. *Innovative Food Science and Emerging Technologies*, 11(4), 666–671.

Artés-Hernández, F., Aguayo, E., Artes, F., and Tomás-Barberán, F. A. (2007) Enriched ozone atmosphere enhances bioactive phenolics in seedless table grapes after prolonged shelf life. *Journal of the Science of Food and Agriculture*, 87(5), 824–831.

Ashokkumar, M., Sunartio, D., Kentish, S., Mawson, R., Simons, L., Vilkhu, K., and Versteeg, C. (2008) Modification of food ingredients by ultrasound to improve functionality: A preliminary study on a model system. *Innovative Food Science and Emerging Technologies*, 9(2), 155–160.

Ayed, N., Yu, H.L., and Lacroix, M. (1999) Improvement of anthocyanin yield and shelf-life extension of grape pomace by gamma irradiation. *Food Research International*, 32(8), 539–543.

Baka, M., Mercier, J., Corcuff, R., Castaigne, F., and Arul, J. (1999) Photochemical treatment to improve storability of fresh strawberries. *Journal of Food Science*, 64 (6), 1068–1072.

Bakowska, A., Kucharska, A.Z., and Oszmianski, J. (2003) The effects of heating, UV irradiation, and storage on stability of the anthocyanin-polyphenol copigment complex. *Food Chemistry*, 81(3), 349–355.

Barba, F.J., Esteve, M.J., and Frigola, A. (2011) Physicochemical and nutritional characteristics of blueberry juice after high pressure processing. *Food Research International*, http://dx.doi.org/10.1016/j.bbr.2011.1003.1031.

Barboni, T., Cannac, M., and Chiaramonti, N. (2010) Effect of cold storage and ozone treatment on physicochemical parameters, soluble sugars and organic acids in Actinidia deliciosa. *Food Chemistry*, 121(4), 946–951.

Barth, M.M., Zhou, C., Mercier, J., and Payne, F.A. (1995) Ozone storage effects on anthocyanin content and fungal growth in blackberries. *Journal of Food Science*, 60(6), 1286–1288.

Beltran, D., Selma, M.V., Marin, A., and Gil, M.I. (2005) Ozonated water extends the shelf life of fresh-cut lettuce. *Journal of Agricultural and Food Chemistry*, 53(14), 5654–5663.

Bhat, R., Kamaruddin, N. S. B. C., Min-Tze, L., and Karim, A. A. (2011) Sonication improves kasturi lime (Citrus microcarpa) juice quality. *Ultrasonics Sonochemistry*, 18(6), 1295–1300.

Bhat, R. and Sridhar, K. R. (2008) Nutritional quality evaluation of electron beam-irradiated lotus (Nelumbo nucifera) seeds. *Food Chemistry*, 107(1), 174–184.

Bhat, R., Sridhar, K.R., and Tomita-Yokotani, K. (2007) Effect of ionizing radiation on antinutritional features of velvet bean seeds (Mucuna pruriens). *Food Chemistry*, 103(3), 860–866.

Beaulieu, M., Béliveau, M., D'Aprano, G., and Lacroix, M. (1999) Dose rate effect of γ irradiation on phenolic compounds, polyphenol oxidase, and browning of mushrooms (Agaricus bisporus). *Journal of Agricultural and Food Chemistry*, 47 (7), 2537–2543.

Bendicho, S., Espachs, A., Arantegui, J., and Martin, O. (2002) Effect of high intensity pulsed electric fields and heat treatments on vitamins of milk. *Journal of Dairy Research*, 69 (1), 113–123.

Breitfellner, F., Solar, S., and Sontag, G. (2002a) Effect of gamma irradiation on flavonoids in strawberries. *European Food Research and Technology*, 215 (1), 28–31.

Breitfellner, F., Solar, S., and Sontag, G. (2002b) Effect of γ-irradiation on phenolic acids in strawberries. *Journal of Food Science*, 67 (2), 517–521.

Calucci, L., Pinzino, C., Zandomeneghi, M., Capocchi, A., Ghiringhelli, S., Saviozzi, F., Tozzi, S., and Galleschi, L. (2003) Effects of γ-irradiation on the free radical and antioxidant contents in nine aromatic herbs and spices. *Journal of Agricultural and Food Chemistry*, 51 (4), 927–934.

Cantos, E., Espin, J. C., and Tomas-Barberan, F. A. (2001) Postharvest induction modeling method using UV irradiation pulses for obtaining resveratrol-enriched table grapes: A new "functional" fruit? *Journal of Agricultural and Food Chemistry*, 49 (10), 5052–5058.

Cantos, E., Garcia-Viguera, C., de Pascual-Teresa, S., and Tomas-Berberan, F. A. (2000) Effect of postharvest ultraviolet irradiation on resveratrol and other phenolics of cv. *Napoleon table grapes. Journal of Agricultural and Food Chemistry*, 48 (10), 4606–4612.

Cantos, E., Espin, J.C., Fernandez, M.J., Oliva, J., and Tomas-Barberan, F.A. (2003) Postharvest UV-C-irradiated grapes as a potential source for producing stilbene-enriched red wines. *Journal of Agricultural and Food Chemistry*, 51(5), 1208–1214.

Chatterjee, S., Desai, S. R. P., and Thomas, P. (1999) Effect of γ-irradiation on the antioxidant activity of turmeric (Curcuma longa L.) extracts. *Food Research International*, 32 (7), 487–490.

Cheng, L., Soh, C., Liew, S., and Teh, F. (2007) Effects of sonication and carbonation on guava juice quality. *Food Chemistry*, 104(4), 1396–1401.

Cocito, C., Gaetano, G., and Delfini, C. (1995) Rapid extraction of aroma compounds in must and wine by means of ultrasound. *Food Chemistry*, 52(3), 311–320.

Cortes, C., Esteve, M. J., Rodrigo, D., Torregrosa, F., and Frigola, A. (2006) Changes of colour and carotenoids contents during high intensity pulsed electric field treatment in orange juices. *Food and Chemical Toxicology*, 44 (11), 1932–1939.

Costa, L., Vicente, A.R., Civello, P.M., Chaves, A.R., and Martínez, G.A. (2006) UV-C treatment delays postharvest senescence in broccoli florets. *Postharvest Biology and Technology*, 39(2), 204–210.

Criegee, R. (1975) Mechanism of ozonolyisis *Angewandte Chemie-International Edition in English*, 14(11), 745–752.

Cserhalmi, Z., Sass-Kiss, A., Tóth-Markus, M., and Lechner, N. (2006) Study of pulsed electric field treated citrus juices. *Innovative Food Science and Emerging Technologies*, 7(1), 49–54.

Cullen, P. (2011) *Novel thermal and non-thermal technologies for fluid foods*: Academic Press.

Cullen, P.J., Tiwari, B.K., O'Donnell, C.P., and Muthukumarappan, K. (2009) Modelling approaches to ozone processing of liquid foods. *Trends in Food Science and Technology*, 20(3–4), 125–136.

Cullen, P.J., Valdramidis, V.P., Tiwari, B.K., Patil, S., Bourke, P., and O'Donnell, C. P. (2010) Ozone processing for food preservation: An overview on fruit juice treatments. *Ozone: Science and Engineering*, 32(3), 166–179.

Damar, S. and Balaban, M.O. (2006) Review of dense phase CO_2 technology: Microbial and enzyme inactivation, and effects on food quality. *Journal of Food Science*, 71(1), R1–R11.

De-Ancos, B., Gonzales, E., and Cano, M.P. (2000) Effect of high pressure treatment on the carotenoid composition and the radical scavenging activity of persimmon fruit purees. *Journal of Agricultural and Food Chemistry*, 48, 3542–3548.

Dede, S., Alpas, H., and Bayındırlı, A. (2007) High hydrostatic pressure treatment and storage of carrot and tomato juices: Antioxidant activity and microbial safety. *Journal of the Science of Food and Agriculture*, 87(5), 773–782.

Del Pozo-Insfran, D., Balaban, M.O., and Talcott, S.T. (2006) Microbial stability, phytochemical retention, and organoleptic attributes of dense phase CO2 processed muscadine grape juice. *Journal of Agricultural and Food Chemistry*, 54(15), 5468–5473.

Del Pozo-Insfran, D., Del Follo-Martinez, A., Talcott, S.T., and Brenes, C.H. (2007) Stability of copigmented anthocyanins and ascorbic acid in muscadine grape juice processed by high hydrostatic pressure. *Journal of Food Science*, 72(4), S247–S253.

DeRuiter, F.E. and Dwyer, J. (2002) Consumer acceptance of irradiated foods: dawn of a new era? *Food Service Technology*, 2(2), 47–58.

Duh, P. D., and Yen, G. C. (1995) Changes in antioxidant activity and components of methanolic extracts of peanut hulls irradiated with ultraviolet-light. *Food Chemistry*, 54 (2), 127–131.

Elez-Martínez, P. and Martín-Belloso, O. (2007) Effects of high intensity pulsed electric field processing conditions on vitamin C and antioxidant capacity of orange juice and gazpacho, a cold vegetable soup. *Food Chemistry*, 102(1), 201–209.

Elez-Martinez, P., Suarez-Recio, M., and Martin-Belloso, O. (2007) Modeling the reduction of pectin methyl esterase activity in orange juice by high intensity pulsed electric fields. *Journal of Food Engineering*, 78 (1), 184–193.

Elez-Martinez, P., Soliva-Fortuny, R., and Martin-Belloso, O. (2009) Impact of High-Intensity Pulsed Electric Fields on Bioactive Compounds in Mediterranean Plant-based Foods. *Natural Product Communications*, 4(5), 741–746.

Erkan, M., Wang, S. Y., and Wang, C. Y. (2008) Effect of UV treatment on antioxidant capacity, antioxidant enzyme activity and decay in strawberry fruit. *Postharvest Biology and Technology*, 48 (2), 163–171.

Evrendilek, G. A., Yeom, H. W., Jin, Z. T., and Zhang, Q. H. (2004) Safety and quality evaluation of a yogurt-based drink processed by a pilot plant pef system. *Journal of Food Process Engineering*, 27 (3), 197–212.

Fan, X. (2005) Antioxidant capacity of fresh-cut vegetables exposed to ionizing radiation. *Journal of the Science of Food and Agriculture*, 85 (6), 995–1000.

Fan, X.T. and Sokorai, K.J.B. (2011) Changes in Quality, Liking, and Purchase Intent of Irradiated Fresh-Cut Spinach during Storage. *Journal of Food Science*, 76(6), S363–S368.

Faragher, J. and Chalmers, D. (1977) Regulation of anthocyanin synthesis in apple skin. III. Involvement of phenylalanine ammonia-lyase. *Functional Plant Biology*, 4(1), 133–141.

Fennema, O.R. and Tannenbaum, S.R. (1996) Introduction to food chemistry. In: Fennema, R.O., Karel, M., Sanderson, G.W., Tannenbaum, S.R., Walstra, P., and Witaker, J.R. (eds) *Food Chemistry*. Marcel Dekker Inc., New York, pp. 1–64.

Ferrari, G., Maresca, P., and Ciccarone, R. (2010) The application of high hydrostatic pressure for the stabilization of functional foods: Pomegranate juice. *Journal of Food Engineering*, 100(2), 245–253.

Ferrentino, G., Plaza, M., Ramirez-Rodrigues, M., Ferrari, G., and Balaban, M. (2009) Effects of dense phase carbon dioxide pasteurization on the physical and quality attributes of a red grapefruit juice. *Journal of Food Science*, 74(6), E333–E341.

Fox, M., Esveld, D., and Boom, R. (2007) Conceptual design of a mass parallelized PEF microreactor. *Trends in Food Science and Technology*, 18(9), 484–491.

Frank, K., Koehler, K., and Schuchmann, H. P. (2012) Stability of anthocyanins in high pressure homogenisation. *Food Chemistry*, 130(3), 716–719.

Garcia-Palazon, A., Suthanthangjai, W., Kajda, P., and Zabetakis, I. (2004) The effects of high hydrostatic pressure on β-glucosidase, peroxidase and polyphenoloxidase in red raspberry (*Rubus idaeus*) and strawberry (*Fragaria ananassa*) *Food Chemistry*, 88(1), 7–10.

Gasperi, F., Aprea, E., Biasioli, F., Carlin, S., Endrizzi, I., Pirretti, G., and Spilimbergo, S. (2009) Effects of supercritical CO2 and N2O pasteurisation on the quality of fresh apple juice. *Food Chemistry*, 115(1), 129–136.

González-Aguilar, G.A., Villegas-Ochoa, M.A., Martínez-Téllez, M.A., Gardea, A.A., and Ayala-Zavala, J.F. (2007) Improving Antioxidant Capacity of Fresh-Cut Mangoes Treated with UV-C. *Journal of Food Science*, 72(3), S197–S202.

Gonzalez-Aguilar, G. A., Zavaleta-Gatica, R., and Tiznado-Hernandez, M. E. (2007) Improving postharvest quality of mango 'Haden' by UV-C treatment. *Postharvest Biology and Technology*, 45 (1), 108–116.

Guan, W.Q., Fan, X.T., and Yan, R.X. (2012) Effects of UV-C treatment on inactivation of Escherichia coli O157:H7, microbial loads, and quality of button mushrooms. *Postharvest Biology and Technology*, 64(1), 119–125.

Gunes, G., Blum, L., and Hotchkiss, J. (2006) Inactivation of Escherichia coli (ATCC 4157) in diluted apple cider by dense-phase carbon dioxide. *Journal of Food Protectionand# 174;*, 69(1), 12–16.

Hagen, S. F., Borge, G. I. A., Bengtsson, G. B., Bilger, W., Berge, A., Haffner, K., and Solhaug, K. A. (2007) Phenolic contents and other health and sensory related properties of apple fruit (Malus domestica Borkh., cv. Aroma): *Effect of postharvest UV-B irradiation. Postharvest Biology and Technology*, 45 (1), 1–10.

Harrison, K., and Were, L. M. (2007) Effect of gamma irradiation on total phenolic content yield and antioxidant capacity of Almond skin extracts. *Food Chemistry*, 102 (3), 932–937.

Hodgins, A. M., Mittal, G. S., and Griffiths, M. W. (2002) Pasteurization of fresh orange juice using low-energy pulsed electrical field. *Journal of Food Science*, 67 (6), 2294–2299.

Huang, S. J., and Mau, J. L. (2006) Antioxidant properties of methanolic extracts from Agaricus blazei with various doses of γ-irradiation. *LWT – Food Science and Technology*, 39 (7), 707–716.

Hsu, K.-C., Tan, F.-J., and Chi, H.-Y. (2008) Evaluation of microbial inactivation and physicochemical properties of pressurized tomato juice during refrigerated storage. *LWT – Food Science and Technology*, 41(3), 367–375.

Jagadeesh, S.L., Charles, M.T., Gariepy, Y., Goyette, B., Raghavan, G S.V., and Vigneault, C. (2011) Influence of Postharvest UV-C Hormesis on the Bioactive Components of Tomato during Post-treatment Handling. *Food and Bioprocess Technology*, 4(8), 1463–1472.

Jimenez, L., Alarcon, E., Trevithick-Sutton, C., Gandhi, N., and Scaiano, J.C. (2011) Effect of gamma-radiation on green onion DNA integrity: Role of ascorbic acid and polyphenols against nucleic acid damage. *Food Chemistry*, 128(3), 735–741.

Kataoka, I. and Beppu, K. (2004) UV Irradiance Increases Development of Red Skin Color and Anthocyanins inHakuho'Peach. *HortScience*, 39(6), 1234–1237.

Keenan, D.F., Brunton, N.P., Gormley, T.R., Butler, F., Tiwari, B.K., and Patras, A. (2010) Effect of thermal and high hydrostatic pressure processing on antioxidant activity and colour of fruit smoothies. *Innovative Food Science and Emerging Technologies*, 11(4), 551–556.

Keenan, D.F., Roessle, C., Gormley, R., Butler, F., and Brunton, N.P. (2012) Effect of high hydrostatic pressure and thermal processing on the nutritional quality and enzyme activity of fruit smoothies. *Lwt-Food Science and Technology*, 45(1), 50–57.

Keyser, M., Muller, I.A., Cilliers, F.P., Nel, W., and Gouws, P.A. (2008) Ultraviolet radiation as a non-thermal treatment for the inactivation of microorganisms in fruit juice. *Innovative Food Science and Emerging Technologies*, 9(3), 348–354.

Khattak, K. F., Simpson, T. J., and Ihasnullah. (2008) Effect of gamma irradiation on the extraction yield, total phenolic content and free radical-scavenging activity of Nigella staiva seed. *Food Chemistry*, 110 (4), 967–972.

Kim, D., Song, H., Lim, S., Yun, H., and Chung, J. (2007) Effects of gamma irradiation on the radiation-resistant bacteria and polyphenol oxidase activity in fresh kale juice. *Radiation Physics and Chemistry*, 76(7), 1213–1217.

Kim, J. W., Lee, B. C., Lee, J. H., Nam, K. C., and Lee, S. C. (2008) Effect of electron-beam irradiation on the antioxidant activity of extracts from Citrus unshiu pomaces. *Radiation Physics and Chemistry*, 77 (1), 87–91.

Kim, J.-H., Sung, N.-Y., Kwon, S.-K., Srinivasan, P., Song, B.-S., Choi, J.-i., Yoon, Y., Kim, J. K., Byun, M.-W., Kim, M.-R., and Lee, J.-W. (2009) gamma-Irradiation Improves the Color and Antioxidant Properties of Chaga Mushroom (Inonotus obliquus) Extract. *Journal of Medicinal Food*, 12(6), 1343–1347.

Kincal, D., Hill, W., Balaban, M., Portier, K., Sims, C., Wei, C., and Marshall, M. (2006) A Continuous High-Pressure Carbon Dioxide System for Cloud and Quality Retention in Orange Juice. *Journal of Food Science*, 71(6), C338–C344.

Knorr, D., Zenker, M., Heinz, V., and Lee, D.U. (2004) Applications and ultrasonics in food potential of processing. *Trends in Food Science and Technology*, 15(5), 261–266.

Korn, M., Machado Primo, P., and Santos de Sousa, C. (2002) Influence of ultrasonic waves on phosphate determination by the molybdenum blue method. *Microchemical journal*, 73(3), 273–277.

Koseki, P. M., Villavicencio, A. L. C. H., Brito, M. S., Nahme, L. C., Sebastião, K. I., Rela, P. R., Almeida-Muradian, L. B., Mancini-Filho, J., and Freitas, P. C. D. (2002) Effects of irradiation in medicinal and eatable herbs. *Radiation Physics and Chemistry*, 63 (3–6), 681–684.

Kouniaki, S., Kajda, P., and Zabetakis, I. (2004) The effect of high hydrostatic pressure on anthocyanins and ascorbic acid in blackcurrants (Ribes nigrum). *Flavour and Fragrance Journal*, 19(4), 281–286.

Kucharska, A., Oszmianski, J., Kopacz, M., and Lamer-Zarawska, E. (1998) Application of flavonoids for anthocyanins stabilization. In).

Lee, N. Y., Jo, C., Sohn, S. H., Kim, J. K., and Byun, M. W. (2006) Effects of gamma irradiation on the biological activity of green tea byproduct extracts and a comparison with green tea leaf extracts. *Journal of Food Science*, 71 (4), C269–C274.

Lemoine, M.L., Chaves, A. R., and Martínez, G.A. (2010) Influence of combined hot air and UV-C treatment on the antioxidant system of minimally processed broccoli (Brassica oleracea L. var. Italica). *LWT-Food Science and Technology*, 43(9), 1313–1319.

Liao, H., Hu, X., Liao, X., Chen, F., and Wu, J. (2007) Inactivation of Escherichia coli inoculated into cloudy apple juice exposed to dense phase carbon dioxide. *International Journal of Food Microbiology*, 118(2), 126–131.

Liu, X., Gao, Y., Peng, X., Yang, B., Xu, H., and Zhao, J. (2008) Inactivation of peroxidase and polyphenol oxidase in red beet (Beta vulgaris L.) extract with high pressure carbon dioxide. *Innovative Food Science and Emerging Technologies*, 9(1), 24–31.

López-Rubira, V., Conesa, A., Allende, A., and Artés, F. (2005) Shelf life and overall quality of minimally processed pomegranate arils modified atmosphere packaged and treated with UV-C. *Postharvest Biology and Technology*, 37(2), 174–185.

Luwe, M.W.F., Takahama, U., and Heber, U. (1993) Role of Ascorbate in Detoxifying Ozone in the Apoplast of Spinach (Spinacia-Oleracea L) Leaves. *Plant Physiology*, 101(3), 969–976.

Markakis, P., Livingston, G., and Fellers, C. (1957) Quantitative aspects of strawberry pigment degradation. *Food Research*, 22(2), 117–130.

Mason, T., Riera, E., Vercet, A., and Lopez-Bueza, P. (2005) Application of ultrasound. *Emerging technologies for food processing*, 32, 3–351.

Masuzawa, N., Ohdaira, E., and Ide, M. (2000) Effects of ultrasonic irradiation on phenolic compounds in wine. *Japanese Journal of Applied Physics*, 39, 2978.

McInerney, J. K., Seccafien, C. A., Stewart, C. M., and Bird, A. R. (2007) Effects of high pressure processing on antioxidant activity, and total carotenoid content and availability, in vegetables. *Innovative Food Science & Emerging Technologies*, 8 (4), 543–548.

Meyer, R.S., Cooper, K.L., Knorr, D., and Lelieveld, H.L.M. (2000) High-pressure sterilization of foods. *Food Technology*, 54(11), 67–129.

Min, S., Jin, Z. T., and Zhang, Q. H. (2003) Commercial scale pulsed electric field processing of tomato juice. *Journal of Agricultural and Food Chemistry*, 51 (11), 3338–3344.

Morales-de la Peña, M., Salvia-Trujillo, L., Rojas-Graü, M., and Martín-Belloso, O. (2010) Impact of high intensity pulsed electric field on antioxidant properties and quality parameters of a fruit juice-soymilk beverage in chilled storage. *LWT–Food Science and Technology*, 43(6), 872–881.

O'Donnell, C. P., Tiwari, B. K., Bourke, P., and Cullen, P. J. (2010) Effect of ultrasonic processing on food enzymes of industrial importance. *Trends in Food Science and Technology*, 21(7), 358–367.

Oey, I., Van der Plancken, I., Van Loey, A., and Hendrickx, M. (2008) Does high pressure processing influence nutritional aspects of plant based food systems? *Trends in Food Science and Technology*, 19(6), 300–308.

Ogawa, H., Fukuhisa, K., Kubo, Y., and Fukumoto, H. (1990) Pressure Inactivation of Yeasts, Molds, and Pectinesterase in Satsuma Mandarin Juice – Effects of Juice Concentration, Ph, and Organic-Acids, and Comparison with Heat Sanitation. *Agricultural and Biological Chemistry*, 54(5), 1219–1225.

Oufedjikh, H., Mahrouz, M., Amiot, M. J., and Lacroix, M. (2000) Effect of γ-irradiation on phenolic compounds and phenylalanine ammonia-lyase activity during storage in relation to peel injury from peel of Citrus clementina Hort. ex. Tanaka. *Journal of Agricultural and Food Chemistry*, 48 (2), 559–565.

Parton, T., Elvassore, N., Bertucco, A., and Bertoloni, G. (2007) High pressure CO_2 inactivation of food: A multi-batch reactor system for inactivation kinetic determination. *The Journal of supercritical fluids*, 40(3), 490–496.

Patras, A., Brunton, N., Da Pieve, S., Butler, F., and Downey, G. (2009) Effect of thermal and high pressure processing on antioxidant activity and instrumental colour of tomato and carrot purees. *Innovative Food Science and Emerging Technologies*, 10(1), 16–22.

Patras, A., Brunton, N.P., Da Pieve, S., and Butler, F. (2009) Impact of high pressure processing on total antioxidant activity, phenolic, ascorbic acid, anthocyanin content and colour of strawberry and blackberry purees. *Innovative Food Science and Emerging Technologies*, 10(3), 308–313.

Patterson, M., Quinn, M., Simpson, R., and Gilmour, A. (1996) High pressure inactivation in foods of animal origin. *Progress in Biotechnology*, 13, 267–272.

Pérez, A.G., Sanz, C., Rios, J.J., Olias, R., and Olias, J.M. (1999) Effects of ozone treatment on postharvest strawberry quality. *Journal of Agricultural and Food Chemistry*, 47(4), 1652–1656.

Pérez, M. B., Calderón, N. L., and Croci, C. A. (2007) Radiation-induced enhancement of antioxidant activity in extracts of rosemary (Rosmarinus officinalis L.). *Food Chemistry*, 104 (2), 585–592.

Perkins-Veazie, P., Collins, J. K., and Howard, L. (2008) Blueberry fruit response to postharvest application of ultraviolet radiation. *Postharvest Biology and Technology*, 47 (3), 280–285.

Petrier, C., Combet, E., and Mason, T. (2007) Oxygen-induced concurrent ultrasonic degradation of volatile and non-volatile aromatic compounds. *Ultrasonics Sonochemistry*, 14(2), 117–121.

Plaza, L., Colina, C., de Ancos, B., Sanchez-Moreno, C., and Cano, M. P. (2012) Influence of ripening and astringency on carotenoid content of high-pressure treated persimmon fruit (Diospyros kaki L.). *Food Chemistry*, 130(3), 591–597.

Plaza, L., Sanchez-Moreno, C., De Ancos, B., Elez-Martinez, P., Martin-Belloso, O., and Pilar Cano, M. (2011) Carotenoid and flavanone content during refrigerated storage of orange juice processed by high-pressure, pulsed electric fields and low pasteurization. *Lwt-Food Science and Technology*, 44(4), 834–839.

Plaza, M.L. (2011) *Quality of guava puree by dense phase carbon dioxide treatment*. University of Florida.

Portenlänger, G., and Heusinger, H. (1992) Chemical reactions induced by ultrasound and γ-rays in aqueous solutions of L-ascorbic acid. *Carbohydrate research*, 232(2), 291–301.

Qin, B., Pothakamury, U., Barbosa-Cánovas, G., and Swanson, B. (1996) Nonthermal pasteurization of liquid foods using high-intensity pulsed electric fields. *Critical reviews in food science and nutrition*, 36(6), 603.

Ranieri, A., DUrso, G., Nali, C., Lorenzini, G., and Soldatini, G. F. (1996) Ozone stimulates apoplastic antioxidant systems in pumpkin leaves. *Physiologia Plantarum*, 97(2), 381–387.

Rawson, A., Tiwari, B.K., Patras, A., Brunton, N., Brennan, C., Cullen, P.J., and O'Donnell, C. (2011) Effect of thermosonication on bioactive compounds in watermelon juice. *Food Research International*, 44(5), 1168–1173.

Reyes, L. F., and Cisneros-Zevallos, L. (2007) Electron-beam ionizing radiation stress effects on mango fruit (Mangifera indica L.) antioxidant constituents before and during postharvest storage. *Journal of Agricultural and Food Chemistry*, 55 (15), 6132–6139.

Rice, R., Graham, D., and Lowe, M.T. (2002) Recent ozone applications in food processing and sanitation. *Food Safety Magazine*, 8(5), 10–17.

Rivas, A., Rodrigo, D., Company, B., Sampedro, F., and Rodrigo, M. (2007) Effects of pulsed electric fields on water-soluble vitamins and ACE inhibitory peptides added to a mixed orange juice and milk beverage. *Food Chemistry*, 104(4), 1550–1559.

Rivas, A., Rodrigo, D., Martinez, A., Barbosa-Cánovas, G., and Rodrigo, M. (2006a) Effect of PEF and heat pasteurization on the physical-chemical characteristics of blended orange and carrot juice. *LWT-Food Science and Technology*, 39(10), 1163–1170.

Rivas, A., Sampedro, F., Rodrigo, D., Martínez, A., and Rodrigo, M. (2006b) Nature of the inactivation of Escherichia coli suspended in an orange juice and milk beverage. *European Food Research and Technology*, 223(4), 541–545.

Sadilova, E., Carle, R., and Stintzing, F. C. (2007) Thermal degradation of anthocyanins and its impact on color and in vitro antioxidant capacity. *Molecular nutrition and food research*, 51(12), 1461–1471.

Sajilata, M. G., and Singhal, R. S. (2006) Effect of irradiation and storage on the antioxidative activity of cashew nuts. *Radiation Physics and Chemistry*, 75 (2), 297–300.

Sampedro, F., Geveke, D.J., Fan, X., and Zhang, H.Q. (2009) Effect of PEF, HHP and thermal treatment on PME inactivation and volatile compounds concentration of an orange juice-milk based beverage. *Innovative Food Science and Emerging Technologies*, 10(4), 463–469.

Sampedro, F., Rivas, A., Rodrigo, D., Martínez, A., and Rodrigo, M. (2006) Effect of temperature and substrate on PEF inactivation of Lactobacillus plantarum in an orange juice–milk beverage. *European Food Research and Technology*, 223(1), 30–34.

Sanchez-Moreno, C., De Ancos, B., Plaza, L., Elez-Martinez, P., and Pilar Cano, M. (2005) Nutritional Approaches and Health-Related Properties of Plant Foods Processed by High Pressure and Pulsed Electric Fields. *Critical Reviews in Food Science and Nutrition*, 49(6), 552–576.

Schindler, M., Solar, S., and Sontag, G. (2005) Phenolic compounds in tomatoes. Natural variations and effect of gamma-irradiation. *European Food Research and Technology*, 221 (3–4), 439–445.

Sharma, S. K., Zhang, Q. H., and Chism, G. W. (1998) Development of a protein fortified fruit beverage and its quality when processed with pulsed electric field treatment. *Journal of Food Quality*, 21 (6), 459–473.

Shin, T. S., and Godber, J. S. (1996) Changes of endogenous antioxidants and fatty acid composition in irradiated rice bran during storage. *Journal of Agricultural and Food Chemistry*, 44 (2), 567–573.

Shivashankara, K., Isobe, S., Al-Haq, M. I., Takenaka, M., and Shiina, T. (2004) Fruit antioxidant activity, ascorbic acid, total phenol, quercetin, and carotene of Irwin mango fruits stored at low temperature after high electric field pretreatment. *Journal of Agricultural and Food Chemistry*, 52(5), 1281–1286.

Soliva-Fortuny, R., Balasa, A., Knorr, D., and Martin-Belloso, O. (2009) Effects of pulsed electric fields on bioactive compounds in foods: A review. *Trends in Food Science and Technology*, 20: 544–556.

Solomon, O., Svanberg, U., and Sahlström, A. (1995) Effect of oxygen and fluorescent light on the quality of orange juice during storage at 8 C. *Food Chemistry*, 53(4), 363–368.

Song, H. P., Kim, D. H., Jo, C., Lee, C. H., Kim, K. S., and Byun, M. W. (2006) Effect of gamma irradiation on the microbiological quality and antioxidant activity of fresh vegetable juice. *Food Microbiology*, 23 (4), 372–378.

Suhaj, M., Rácová, J., Polovka, M., and Brezová, V. (2006) Effect of γ-irradiation on antioxidant activity of black pepper (Piper nigrum L.). *Food Chemistry*, 97 (4), 696–704.

Suslick, K.S. (1988) Ultrasound: its chemical, physical, and biological effects.

Suslick, K.S., Hammerton, D. A., and Cline, R. E. (1986) Sonochemical hot spot. *Journal of the American Chemical Society*, 108(18), 5641–5642.

Suthanthangjai, W., Kajda, P., and Zabetakis, I. (2005) The effect of high hydrostatic pressure on the anthocyanins of raspberry (Rubus idaeus) *Food Chemistry*, 90(1–2), 193–197.

Takos, A.M., Jaffé, F.W., Jacob, S.R., Bogs, J., Robinson, S.P., and Walker, A.R. (2006) Light-induced expression of a MYB gene regulates anthocyanin biosynthesis in red apples. *Plant Physiology*, 142(3), 1216.

Torregrosa, F., Cortes, C., Esteve, M. J., and Frigola, A. (2005) Effect of high-intensity pulsed electric fields processing and conventional heat treatment on orange-carrot juice carotenoids. *Journal of Agricultural and Food Chemistry*, 53 (24), 9519–9525.

Tiwari, B.K. and Mason, T.J. (2012) Chapter 6 – Ultrasound Processing of Fluid Foods. In P. J. Cullen, K. T. Brijesh, B. K. T. Vasilis ValdramidisA2 – P.J. Cullen and V. Vasilis (Eds.), *Novel Thermal and Non-Thermal Technologies for Fluid Foods*, (pp. 135–165). San Diego: Academic Press.

Tiwari, B.K., Muthukumarappan, K., O'Donnell, C.P., and Cullen, P.J. (2008) Kinetics of freshly squeezed orange juice quality changes during ozone processing. *Journal of Agricultural and Food Chemistry*, 56(15), 6416–6422.

Tiwari, B.K., O'Donnell, C.P., and Cullen, P.J. (2009) Effect of non thermal processing technologies on the anthocyanin content of fruit juices. *Trends in Food Science and Technology*, 20(3–4), 137–145.

Tiwari, B. K., O'Donnell, C. P., Muthukumarappan, K., and Cullen, P. J. (2009a) Anthocyanin and colour degradation in ozone treated blackberry juice. *Innovative Food Science and Emerging Technologies*, 10(1), 70–75.

Tiwari, B.K., O'Donnell, C.P., Muthukumarappan, K., and Cullen, P.J. (2009b) Effect of sonication on orange juice quality parameters during storage. *International Journal of Food Science and Technology*, 44(3), 586–595.

Tiwari, B.K., O'Donnell, C.P., Patras, A., Brunton, N., and Cullen, P.J. (2009a) Effect of ozone processing on anthocyanins and ascorbic acid degradation of strawberry juice. *Food Chemistry*, 113(4), 1119–1126.

Tiwari, B.K., O'Donnell, C.P., Patras, A., Brunton, N., and Cullen, P.J. (2009b) Stability of anthocyanins and ascorbic acid in sonicated strawberry juice during storage. *European Food Research and Technology*, 228(5), 717–724.

Tiwari, B.K., O'Donnell, C.P., Patras, A., and Cullen, P.J. (2008) Anthocyanin and ascorbic acid degradation in sonicated strawberry juice. *Journal of Agricultural and Food Chemistry*, 56(21), 10071–10077.

Tzortzakis, N., Borland, A., Singleton, I., and Barnes, J. (2007) Impact of atmospheric ozone-enrichment on quality-related attributes of tomato fruit. *Postharvest Biology and Technology*, 45(3), 317–325.

Valero, M., Recrosio, N., Saura, D., Munoz, N., Martí, N., and Lizama, V. (2007) Effects of ultrasonic treatments in orange juice processing. *Journal of Food Engineering*, 80(2), 509–516.

Vega-Mercado, H., Martin-Belloso, O., Qin, B.L., Chang, F.J., Marcela Góngora-Nieto, M., Barbosa-Canovas, G. V., and Swanson, B. G. (1997) Non-thermal food preservation: pulsed electric fields. *Trends in Food Science and Technology*, 8(5), 151–157.

Vilkhu, K., Mawson, R., Simons, L., and Bates, D. (2008) Applications and opportunities for ultrasound assisted extraction in the food industry – A review. *Innovative Food Science and Emerging Technologies*, 9(2), 161–169.

Vicente, A. R., Pineda, C., Lemoine, L., Civello, P. M., Martinez, G. A., and Chaves, A. R. (2005) UV-C treatments reduce decay, retain quality and alleviate chilling injury in pepper. *Postharvest Biology and Technology*, 35 (1), 69–78.

Weemaes, C.A., Ludikhuyze, L.R., Van den Broeck, I., and Hendrickx, M.E. (1998) Kinetics of combined pressure-temperature inactivation of avocado polyphenoloxidase. *Biotechnology and bioengineering*, 60(3), 292–300.

Winter, T. R., and Rostas, M. (2008) Ambient ultraviolet radiation induces protective responses in soybean but does not attenuate indirect defense. *Environmental Pollution*, 155 (2), 290–297.

Wood, O.B. and Bruhn, C. (2000) Food irradiation. *Journal of the American Dietetic Association*, 100(1), 246–253.

Wu, Y., Mittal, G. S., and Griffiths, M. W. (2005) Effect of pulsed electric field on the inactivation of microorganisms in grape juices with and without antimicrobials. *Biosystems Engineering*, 90 (1), 1–7.

Zabetakis, I., Koulentianos, A., Orruno, E., and Boyes, I. (2000) The effect of high hydrostatic pressure on strawberry flavour compounds. *Food Chemistry*, 71(1), 51–55.

Zhang, D., Yu, B., Bai, J., Qian, M., Shu, Q., Su, J., and Teng, Y. (2011) Effects of high temperatures on UV-B/visible irradiation induced postharvest anthocyanin accumulation in 'Yunhongli No. 1'(*Pyrus pyrifolia* Nakai) pears. *Scientia Horticulturae*.

Zhang, L.K., Lu, Z.X., Yu, Z.F., and Gao, X. (2005) Preservation of fresh-cut celery by treatment of ozonated water. *Food Control*, 16(3), 279–283.

Zhang, Y.I., Gao, B.E.I., Zhang, M., Shi, J., and Xu, Y. (2010) Pulsed Electric Field Processing Effects on Physicochemical Properties, Flavor Compounds and Microorganisms of Longan Juice. *Journal of Food Processing and Preservation*, 34(6), 1121–1138.

Zhao, L., Zhao, G., Chen, F., Wang, Z., Wu, J., and Hu, X. (2006) Different effects of microwave and ultrasound on the stability of (all-E)-astaxanthin. *Journal of Agricultural and Food Chemistry*, 54(21), 8346–8351.

Zhu, F., Cai, Y.-Z., Bao, J., and Corke, H. (2010) Effect of γ-irradiation on phenolic compounds in rice grain. *Food Chemistry*, 120(1), 74–77.

Part IV
Stability of Phytochemicals

14 Stability of phytochemicals during grain processing

Laura Alvarez-Jubete[1] and Uma Tiwari[2]

[1] Food Science Department, School of Food Science and Environmental Health, Dublin Institute of Technology, Dublin, Ireland
[2] School of Biosystems Engineering, University College Dublin, Dublin, Ireland

14.1 Introduction

Epidemiological studies have shown that whole grain consumption is associated with reduced risk of chronic diseases including cardiovascular disease and cancer (Seal, 2006; Slavin, 2004). Whole grains are a rich source of many nutrients and bioactive phytochemicals. In addition to being high in dietary fibre, resistant starch and oligosaccharides, they also contain a wide array of protective compounds such as phenolic compounds, tocopherols, tocotrienols, carotenoids, plant sterols and lignans (Slavin, 2004). Furthermore, special emphasis has been placed on the potential synergistic effect with recent evidence suggesting that the complex mixture of phytochemicals present in whole grains may be more beneficial than the addition of the individual isolated components (Liu, 2004; Liu, 2007).

Since whole grains are mostly commonly processed one way or another before consumption, it is necessary to evaluate the impact of processing on the nutritional value of grains to properly assess their importance as healthful foods (Slavin, Jacobs and Marquart, 2001). While evidence from animal models and human studies support the role of processing in enhancing the nutritive value of grains, mainly by increasing nutrient bioavailability, processing is often regarded as a negative element in nutrition, decreasing the content of important nutrients and phytochemicals (Slavin *et al.*, 2001). Thus, it is important to identify those grain processing technologies that facilitate grain consumption and improve nutrient bioavailability while providing maximum retention of nutrients and phytochemicals.

There is a higher concentration in nutrients and phytochemicals in the outer bran and germ of the grain compared to the endosperm (Liu, 2007). Thus, milling of grains to separate the bran and the germ from the starchy endosperm to produce refined white flour causes a significant reduction in the content of nutrients and phytochemicals. This process has by far the greatest impact on the phytochemical content of grains. Also, heat processing such as baking can negatively affect the content of organic compounds such as vitamin E, carotenoids and polyphenol compounds (Alvarez-Jubete, Holse, Hansen, Arendt and Gallagher, 2009; Alvarez-Jubete, Wijngaard, Arendt and Gallagher, 2010; Leenhardt

Handbook of Plant Food Phytochemicals: Sources, Stability and Extraction, First Edition.
Edited by B.K. Tiwari, Nigel P. Brunton and Charles S. Brennan.
© 2013 John Wiley & Sons, Ltd. Published 2013 by John Wiley & Sons, Ltd.

et al., 2006; Vogrincic, Timoracka, Melichacova, Vollmannova and Kreft, 2010). On the other hand, thermal processing of cereals, such as baking, can also result in the synthesis of substances with antioxidant properties, such as some Maillard reaction products in bread crust (Lindenmeier and Hofmann, 2004; Michalska, Amigo-Benavent, Zielinski and del Castillo, 2008).

This chapter presents a review of the available literature on commonly used grain processing techniques and their implications in relation with the stability and degradation of some very important grain phytochemicals. In particular, special attention will be paid to the importance of optimising process parameters to prevent or minimise losses of phytochemicals as well as adequately choosing the most adequate substrate (e.g. type of grain, grain species or variety, matrix ingredients, etc.) for each particular grain processing technique.

14.2 Germination

Germination has been used traditionally to modify the functional and nutritive properties of cereals. For instance, barley malting is a widely known controlled germination process used to produce malt for brewing purposes and food applications (Kaukovirta-Norja, Wilhelmson and Poutanen, 2004). In addition to causing a softening of the cereal kernel, germination also typically results in an increase in nutrient content and availability, and a decrease in the levels of antinutritive compounds (Kaukovirta-Norja *et al.*, 2004). Germination of a grain starts with soaking of the grain in water which in turn leads to the resumption of metabolic activity by the grain or seed. In particular, many enzymes are synthesised to degrade macromolecules, thus leading to changes in structure as well as synthesis of compounds, some of them with potential bioactivity (Kaukovirta-Norja *et al.*, 2004). Several important grain phytochemicals including phytates, sterols, phytoestrogens, phenolic compounds as well as antioxidative properties have been studied during the sprouting process of a variety of grains.

A great body of the information available to date on the effect of germination on grain phytochemicals is focused on the study of polyphenol compounds, and also total phenol content and total antioxidant capacity. In general, phenolic compounds have been shown to increase with germination in a number of studies. Alvarez-Jubete, Wijngaard, Arendt and Gallagher (2010) showed that sprouting resulted in an increase in the polyphenol content of the pseudocereal grains amaranth, quinoa and buckwheat. According to the authors, kaempferol and quercetin glycosides in quinoa sprouts reached 56.0 and 66.6 µmol/100 g dry weight basis compared with 36.7 and 43.4 µmol/100 g dry weight basis in quinoa grains. In the case of buckwheat, the main increases due to sprouting were reported in the levels of catechin, 3-coumaric acid and luteolin and apigenin glycosides. The increase in polyphenol content upon sprouting of the pseudocereal seeds may be attributed to the many metabolic changes that take place upon sprouting of seeds, mainly due to the activation of endogenous enzymes (Chavan and Kadam, 1989). It is also likely that germination may increase the extractability of polyphenol compounds, by releasing bound polyphenols therefore making them extractable in solvents such as methanol. Kim, Kim and Park (2004) also found that in buckwheat grains the content of two quercetin glycosides, rutin and quercetin, and that of two other unknown compounds, particularly increased as sprouting day progressed, whereas the content of chlorogenic acid was found to increase only moderately. In a subsequent study, S. Kim, Zaidul, Suzuki, Mukasa, Hashimoto, Takigawa, *et al.* (2008) compared the phenolic composition of common (*Fagopyrum esculentum* Moench) and tartary buckwheat

(*Fagopyrum tataricum* Gaertn.) sprouts. The main phenolic compounds determined in the sprouts included chlorogenic acid, four C-glycosylflavones (orientin, isoorientin vitexin, isovitexin), rutin and quercetin. A significant increase upon germination in the quantities of phenolic compounds was noted for both buckwheat species. The main difference reported between the two buckwheat species was in the level of the important bioactive rutin. According to the authors, rutin contents in tartary buckwheat grains and sprouts were much higher than those present in common buckwheat. When comparing common buckwheat and tartary buckwheat sprouts for their levels of anthocyanins, Kim, Maeda, Sarker, Takigawa, Matsuura-Endo, Yamauchi, *et al.* (2007) also found significant differences in the types and amounts of anthocyanins present. In the same study, the authors highly recommended the use of a new variety/line of tartary buckwheat called Hokkai T10 for the production of sprouts rich in dietary anthocyanins, with their associated health benefits including improved cardiovascular function.

Avenanthramides, a type of phenolic compounds found only in oats, have also been shown to increase significantly upon germination (Kaukovirta-Norja *et al.*, 2004). In their review, Kaukovirta-Norja, Wilhelmson and Poutanen (2004) also described how the content in avenanthramides of oat sprouts can be modified depending on the variety used. In particular, they highlighted the high content in avenanthramides in hull-less oat varieties, which they considered as indicative of the importance of these compounds in plant protection. In contrast, phytoestrogens such as lignans, an important group of phenolic compounds with reported beneficial biological effects, has not been shown to be influenced by germination of oats (Kaukovirta-Norja *et al.*, 2004). Liukkonen, Katina, Wilhelmsson, Myllymaki, Lampi, Kariluoto, *et al.* (2003), however, noted a slight increase in the amounts of lignans for rye grain following germination. In the same study, the levels of alk(en)ylresorcinols in rye were not shown to be significantly affected by germination.

Germination also increases the levels of total phenol content and antioxidant capacity of grains. This is to be expected since germination increases the levels of phenolic compounds in grains which have demonstrated antioxidant capacity *in vitro*. Alvarez-Jubete, Wijngaard, Arendt and Gallagher (2010) studied the effect of germination on the antioxidative properties of the pseudocereals amaranth, quinoa and buckwheat. The authors reported that total phenol content doubled following sprouting, and quadrupled in the case of amaranth. Buckwheat sprouts showed the highest total phenol content (670.2 mg gallic acid equivalents/100 g dry weight basis), followed by quinoa (147.2 mg gallic acid equivalents/100 g dry weight basis) and amaranth (82.2 mg gallic acid equivalents/100 g dry weight basis). Accordingly, antioxidant capacity (measured by the radical DPPH scavenging capacity assay and the ferric ion reducing antioxidant power (FRAP) assay) was also reported to increase following sprouting, although interestingly, the difference was not found to be significant. Similarly, antioxidant capacity was highest in buckwheat sprouted seeds compared with amaranth and quinoa sprouts ($p<0.01$). In addition to pseudocereals, oat sprouts and rye grain have also been investigated for their antioxidant capacity. In their review, Kaukovirta-Norja, Wilhelmson and Poutanen (2004) reported that studies in the area show an increase in antioxidant capacity and total phenol content of oats with germination. Also, a good correlation between antioxidant capacity and total phenol content was found, suggesting that a significant part of the antioxidant capacity may be due to phenolic compounds. In the case of rye grain, Liukkonen, *et al.* (2003) reported that the antioxidant capacity (DPPH radical scavenging capacity) of germinated rye grains remained practically similar to that of the native rye grains although the total phenol content of methanolic extracts of rye grain (easily extractable or free phenolics) increased notably during germination. The effect

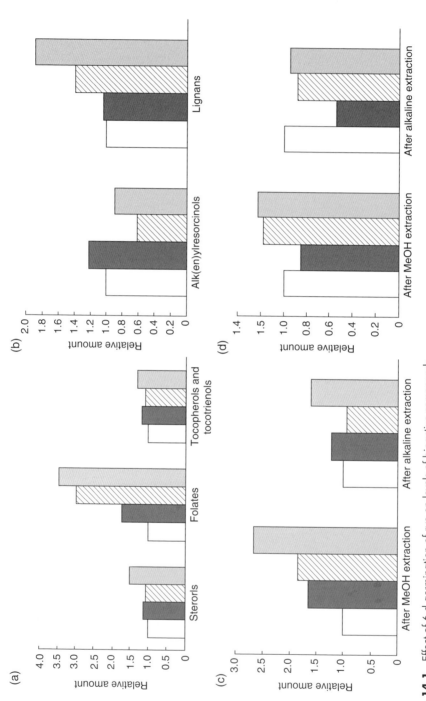

Figure 14.1 Effect of 6 d germination of rye on levels of bioactive compounds.
Source: Liukkonen et al. (2003). Process-induced changes on bioactive compounds in whole grain rye. Proceedings of the Nutrition Society, 62(01), 117–122. Reproduced with permission.

of six days germination on the bioactive compounds of rye, as previously published by Liukkonen *et al.* (2003), is presented in Figure 14.1.

The effect of germination on other phytochemicals in rye grain such as vitamin E compounds, folates and sterols has been shown to differ depending on the type of compound under study (Liukkonen *et al.*, 2003) (Figure 14.1). Similarly to the results obtained in the case of polyphenol compounds, the folate content of rye grains increased significantly upon germination. Moreover, folate levels were at least tripled when germination was conducted at 15 or 25 °C. On the other hand tocopherol and tocotrienol content, as well as sterol levels, in rye grains were not found to be modified significantly by germination of rye during six days at 5, 10 or 25 °C as can be seen in Figure 14.1. In addition, the effect of germination on grain phytochemicals seems to depend not only on the particular phytochemical compound, but also on the type of grain under study. For instance, as already discussed, no effect of germination was observed on the levels of sterols in rye grain. Yet, in the case of oats, sterols were found to increase by up to 20% following germination (Kaukovirta-Norja *et al.*, 2004). Sterols are minor lipids in plants with important biological functions and increasing their levels in foods might be of interest from a nutritional point of view. In addition, these sterols in oat sprouts were shown to be heat-stable during a drying process at different temperatures (Kaukovirta-Norja *et al.*, 2004), which suggests that they may resist subsequent food processing conditions.

Furthermore, the germination process can be optimised to minimise losses of important compounds. An interesting example is that one of β-glucan. β-glucans are types of phytochemicals known to be negatively affected by germination as they are generally broken down during germination. However, the germination process can be optimised in terms of temperature and duration time to minimise losses of high molecular weight β-glucans in grains such as oats (Wilhelmson, Oksman-Caldentey, Laitila, Suortti, Kaukovirta-Norja and Poutanen, 2001). Since the decrease in the molecular weight of β-glucan is very slow initially during germination, a short germination schedule (72 hour, 15 °C) may be employed to produce germinated oat with higher content of retained β-glucan (55–60%) (Wilhelmson *et al.*, 2001).

In summary, germination offers the possibility of modifying the texture and flavour of grains at the same time as modifying their content and/or availability in key phytochemicals such as folates and phenolic compounds. Most widely known germination processes are directed towards obtaining malt for brewing process, however, germinated grains can also be consumed directly, without the need of further processing, due to their characteristic softer structure (Kaukovirta-Norja *et al.*, 2004). In addition, sprouted grains can also be used as attractive novel ingredients in the development of new food products due to their flavour, texture and nutritive properties. For instance, germinated grains can be included in a variety of cereal-based formulations such as extrusion and baking formulations for the development of new products that are appealing to the consumer, palatable and that contain increased amounts of bioactive compounds (Kaukovirta-Norja *et al.*, 2004).

14.3 Milling

Milling of grains generally results in the removal of the bran and germ layers which are rich in fibre and phytochemical compounds, thus causing significant losses in the form of by-products such as hulls and polish waste (Tiwari and Cummins, 2009b). The milling process depends on the nature of the grain and it includes several unit operations such as cleaning, grading, tempering or conditioning, hull removal (de-hulling), pearling etc., followed by milling, to obtain various milling fractions. Research studies show that the majority of phytochemicals are concentrated in the bran and germ fractions of the grain (Adom, Sorrells and Liu, 2003; Anton, Ross, Beta, Fulcher and Arntfield, 2008; Oomah, Cardador-Martinez and Loarca-Pina, 2005; Peterson, 1995; Siebenhandl et al., 2007). Nutritional and clinical studies indicate that whole grains offer distinct advantages over refined flour due to the presence of various bioactive compounds. Similar to any other processing technique milling processing can have both desirable and undesirable effects on the bioactive compounds present in the system. Table 14.1 shows the different milling fractions of several cereals and legumes along with their phytochemical content.

Carotenoids are an important type of grain phytochemical which are located mainly in the outer layers of the grain. As a result, the bran/germ fractions of whole wheat flour contain more carotenoids, i.e. more lutein (164.1–191.7 µg/100 g), zeaxanthin (19.36–26.15 µg/100 g) and β-cryptoxanthin (8.91–10.03 µg/100 g), than the endosperm fractions (Adom, Sorrells and Liu, 2005). Also, the levels of carotenoids in flour can then be modified by the degree of milling, which is defined as the extent to which the bran layers are removed during milling. For instance, increasing the degree of milling in five different rice grain varieties resulted in decreased levels of carotenoid compounds (Lamberts and Delcour, 2008a). In particular, removal of the outer layer by approximately 5% reduces β-carotene and lutein levels by more than 50 and 20% (except for the variety Loto) respectively. Zeaxanthin levels also decreased with increasing degree of milling. Bran layer removal (DOM>9%) further decreased the levels of carotenoids and resulted in lutein, β-carotene, and zeaxanthin levels, lower than 20 ng/g (with the exception of the Loto variety).

Tocols (tocopherols and tocotrienols) are mainly concentrated in the outer aleurone, sub-aleurone and germ of cereal grains such as barley, wheat, corn and oats (Tiwari et al., 2009b). Thus, similar to carotenoids, tocols are also affected by milling, especially during pearling or de-hulled milling (Butsat and Siriamornpun, 2010; Panfili, Fratianni, Di Criscio and Marconi, 2008). In a study by Wang, Xue, Newman and Newman (1993), the authors observed that a pearling fraction consisting of 20% of the original rye kernel weight had the highest concentrations of α-tocotrienol, α-tocopherol and total tocols compared to whole hull-less barley grain. Similar results can be found upon pearling of oats and wheat (Borrelli, De Leonardis, Platani and Troccoli, 2008; Peterson, 1994). Also, the tocol level in the final product depends on the milling methodology employed. For instance, Butsat and Siriamornpun (2010) employed different methodologies to mill rice and found significant differences in the tocol composition of the bran/germ fractions obtained using the different methodologies. In addition to the milling methodology employed, the degree of milling also plays a major role in influencing the tocol content of grains. As is to be expected, increasing the milling time decreased tocol levels in rice (Chen and Bergman, 2005).

In comparison with carotenoids and tocols, the fate of phenolic compounds during the milling process has been widely reported in the literature. In a study by Glitso and Knudsen (1999), the presence of phenolic compounds in several milling fractions of rye was analysed. Phenolic acids were reported to be concentrated mainly in the pericarp/testa (743 mg/100 g dm), with levels decreasing markedly in aleurone (201 mg/100 g dm) and endosperm fractions (19 mg/100 g dm). Similarly, Heinio, Liukkonen, Myllymaki, Pihlava, Adlercreutz, Heinonen et al. (2008) noted that total phenolic acid content in different rye milling fractions was highest in the bran fraction. In another study, Liyana-Pathirana and Shahidi (2007) evaluated the effect of milling on the total phenolic content and antioxidant capacity of two wheat cultivars namely *Triticum turgidum* and *Triticum aestivum*. Several

Table 14.1 Phytochemical content of milling fractions

Milling fractions	Phenolic acids	Flavonoids	Anthocyanins
cereals/ legumes			
Rice[a]			
Bran	2.30–2.83		
Husk	1.57–1.97		
Brown rice	0.90–1.17		
Milled rice	0.43–0.70		
Rice-black			
Whole grain	12.99–18.79[a]	2.11–3.44[b]	1.84–1.03[c]
Endosperm	1.7–2.75[a]	0.15–0.33[b]	0.05–0.18[c]
Rice bran	86.23–111.65[a]	15.82–22.03[b]	6.66–10.73[c]
Wheat-purple; blue			
Bran + shorts	7.47–8.97; 7.43–7.80[d]		0.04–0.15; 0.42–0.49[c]
Middling	1.36–1.55; 6.02–638[d]		
Flour	0.67–0.95; 1.64–2.03[d]		0.01; 0.02[c]
Whole meal	1.82–2.12; 0.58–0.71[d]		0.11–0.36; 0.14–0.16[c]
Barley-six; two rowed[d]			
Bran + shorts	9.10–9.33; 7.60–8.83		
Middling	5.53–5.93; 6.17–6.87		
Flour	1.31–2.08; 3.06–3.22		
Whole meal	0.82–1.64; 0.77–1.35		
Oat[e]			
Groat	0.23–0.26		
Hulls	0.24–0.28		
Oats	0.25–0.26		
Millet[a]			
Whole meal	20–26		
Seed coat	49–79		
Seed coat after water wash	70–110		
Water extractable fraction	4.8–5.2		
After pulverising (+180 µm) fraction	115–141		
After pulverising (−180 µm) fraction	22–30		
Common Bean			
Whole bean flour	7.93[b]	0.68[f]	0.20[h]
Dehulled beans	4.40[b]	0.51[f]	0.19[h]
Hull 1	42.40[b]	1.69[f]	0.63[h]
Hull 2	60.65[b]	2.43[f]	0.60[h]
Residue 1	13.67[b]	0.71[f]	0.24[h]
Residue 2	18.83[b]	0.91[f]	0.36[h]
Powder	12.42[b]	0.91[f]	0.30[h]
Green lentils			
Whole grain	10.31[b]	0.59[g]	0.03[h]
Hull	82.95[b]	1.07[g]	0.11[h]
Residue	11.40[b]	1.10[g]	0.02[h]
Red lentils			
Whole grain	12.62[b]	0.55[g]	0.03[h]
Hull	87.16[b]	1.18[g]	0.11[h]
Residue	12.89[b]	0.93[g]	0.03[h]

(Continued)

Table 14.1 (Continued)

Milling fractions	Phenolic acids	Flavonoids	Anthocyanins
Yellow lentils			
Whole grain	3.45[b]	0.30[g]	0.02[h]
Hull	6.89[b]	0.81[g]	0.14[h]
Residue	5.69[b]	0.52[g]	0.13[h]

DW: dry weight ; FW: fresh weight
[a]GAE: Gallic acid equivalents, mg/g DW; [b]CE: Catechin equivalents, mg/g DW; [c]mg/g of total anthocyanins, DW; [d]FAE: Ferrulic acid equivalents mg/g; [e]GAE: Gallic acid equivalents, mg/g FW; [f]RE: rutin equivalent, mg/g DW; [g]CAE: Caffeic acid equivalents, mg/g DW; [g]Cya-3-glu: Cyanidin-3-glucoside equivalent, mg/g DW.
Source: adapted from Emmons and Peterson (1999); Abdel-Aal and Hucl (2003); Oomah et al. (2005); Chetan and Malleshi (2007); Siebenhandl et al. (2007); Butsat and Siriamornpun (2010); Kong and Lee (2010); Oomah et al. (2010).

fractions were produced: bran, shorts, flour and semolina. Among the milling fractions obtained, the lowest phenolic content was measured in the semolina fraction (9 ~ 140 μFAE/g) whereas the highest level (~2858 μFAE/g) was detected in the bran portion, thus indicating that bran (outer layers) of the grain contain higher level of phenolic compounds compared to refined flour (endosperm). Similarly, Adom, Sorrells and Liu (2005) reported that approximately 83% of the total phenolic content of wheat is located in the bran and germ. In buckwheat milling fractions, both free and bound phenolic levels were found to be nearly 30-fold higher in the outermost flour fraction of buckwheat compared to the innermost fraction (Hung and Morita, 2008). Similar observations have been reported for oat and barley milling fractions (Gray et al., 2000; Madhujith, Izydorczyk and Shahidi, 2006). Regarding legume grains, several authors have reported that the phenolic content is also affected by the milling process. In a study by Cardador-Martinez, Loarca-Pina and Oomah (2002), the total phenolic content of different dry bean (*Phaseolus vulgaris*) milling fractions including hull, whole and de-hulled beans was evaluated. The authors observed that the hull fraction of the bean exhibited a 37-fold greater phenolic content compared to whole bean flour or de-hulled beans. Similarly, Oomah, Cardador-Martinez and Loarca-Pina (2005) reported a higher concentration of phenolic compounds in hull compared to other milling fractions of common bean (*Phaseolus vulgaris*).

It is worth mentioning that the phenolic content of milling fractions of cereals and pulses also varies significantly depending on the milling methodology and other processing conditions (Table 14.1). For instance, during abrasive milling of grain, grains are decorticated (e.g. bran is removed by successive milling), which may lead to a reduction in the phenolic content (Awika, McDonough and Rooney, 2005; Cardador-Martinez et al., 2002; Fares, Platani, Baiano and Menga, 2010; Oomah, Caspar, Malcolmson and Bellido, 2011). During decortication of sorghum grains, Awika, McDonough and Rooney (2005) reported that the phenol concentration increased during removal of the first or second layer of bran in some cultivars. However, on repeated removal of layers, total phenolic content was reduced significantly as shown in Figure 14.2. On the other hand, de-branning of rice cultivars showed that the first bran fraction contained significantly higher level of total phenolic acids (487.6 mg ferulic acid equivalent/kg) compared to that of second, third and fourth rice bran milling fractions (327.1, 355.4 and 257.4 mg ferulic acid equivalent/kg, respectively) (Abdul-Hamid, Sulaiman, Osman and Saari, 2007).

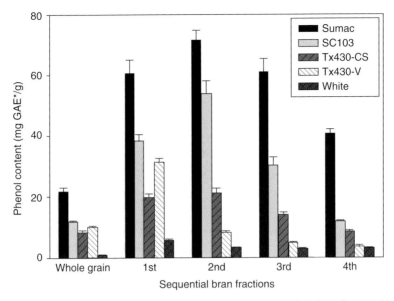

Figure 14.2 The effect of decortication on phenol concentration in sorghum bran fractions (*indicates dry-weight basis). Reproduced from Awika et al. (2005). Fermentation induced changes in the nutritional value of native or germinated rye. *GAE: galic acid equivalents, dry basis; error bars represent standard deviation.

With regards to individual phenolic compounds, ferulic acid is the most common phenolic acid present in the cell wall of cereal grains. Since this fraction is removed during the de-hulling process, the ferulic acid content of milled wheat fractions is about 50–70-fold higher in bran/germ fractions compared to endosperm fractions (Adom et al., 2005). Also in rye bran, a rich source of phenolic acids, the concentration of ferulic acid and its dehydrodimers is approximately 10–20 times higher in the bran compared to the endosperm (Andreasen, Christensen, Meyer and Hansen, 2000). In the case of barley, the predominant phenolic acid present is p-coumaric as opposed to ferulic acid as in wheat and rye (Siebenhandl et al., 2007).

Flavonoids are another group of important phenolic compounds found only in certain grains and pseudocereals. As previously described for other phenolic compounds, flavonoids are also present in the bran/germ fractions of grains (Adom et al., 2005; Hung et al., 2008; Oomah and Mazza, 1996). In the case of buckwheat, a known rich source of flavonoids, the concentration of flavonoids in buckwheat hull (~74 mg/100 g) is higher in comparison to buckwheat whole grain (18.8 mg/100 g) (Dietrych-Szostak and Oleszek, 1999). Similarly, the bran/germ fraction of wheat contributes 79% of the total flavonoid content of whole grain (Adom et al., 2005). In their study, Adom, Sorrells and Liu (2005) noted that flavonoid content of bran/germ fractions of a number of wheat varieties varied in the range 740–940 µmol of catechin equivalents/100 g and it was 10–15-fold higher compared with the flavonoid content of the respective endosperm fractions (60–80 µmol of catechin equiv/100 g of flour). Kong and Lee (2010) also noted that total flavonoid content in two black rice cultivars was lower in endosperm fractions in comparison with bran fractions.

Anthocyanins are another group of phenolic compounds which are also important pigments in cereals and pulses. These compounds are also concentrated in the outer layers of the grain. The level of anthocyanins in Blue wheat bran has been measured to be 46 mg/100 g, whereas whole wheat meal contains only about 16 mg/100 g (Abdel-Aal and Hucl, 1999).

Coloured grains such as black sorghum are known to have significantly more anthocyanin pigments in comparison to other non-coloured sorghums. Thus, the bran of black sorghum is a good source of anthocyanins (4.0–9.8 mg luteolinidin equivalents/g) (Awika et al., 2005). Similarly, Kong and Lee (2010) reported that high levels of anthocyanins are found in bran milling fractions of black rice compared to refined flour (endosperm) fractions.

In summary, these studies show that during the milling process, the outer fractions of grains rich in phytochemicals such as carotenoids, tocols, phenolic acids and flavonoids, are removed. Thus, to avail of the potential health benefits associated with these health beneficial compounds, it is recommended to consume whole grains over highly refined white flours. On the other hand, the degree of milling and/or methodology employed may also be optimised depending on the grain and the phytochemical compounds of interest for maximum retention on the resultant fractions for human consumption.

14.4 Fermentation

Cereal fermentation is one of the oldest biotechnological processes used for the production of both beer and bread (Poutanen, Flander and Katina, 2009). During cereal fermentation, grains are hydrated at room temperature and both endogenous and added enzymes as well as micro-organisms including yeast and lactic acid bacteria start to modify the grain constituents (Katina et al., 2007). In bread production, fermentation is used to produce leavened dough by the leavening agent, which converts the fermentable sugars present in the dough into ethanol and CO_2. The most commonly used leavening agent for industrial production of white bread is baker's yeast, also called *Saccharomyces cerevisiae*. However, the use of a sourdough starter as in traditional baking such as rye bread making is being increasingly recognised (Poutanen et al., 2009). In a sourdough starter a lactic acid bacteria exists in symbiotic combination with yeasts, resulting in the production of lactic and acetic acids during fermentation, which in turn results in the modification of many important quality and nutritional parameters in comparison to yeast-based breads.

In particular, sourdough fermentation has been used traditionally to improve the flavour and structure of rye and wheat breads (Katina, Arendt, Liukkonen, Autio, Flander and Poutanen, 2005). Thus, a very important characteristic of sourdough fermentation is that it facilitates consumption of whole grains, by improving the texture and palatability of whole grain products, such as rye breads, without removing the bran and germ layers which are rich in important nutrients and phytochemicals (Katina et al., 2005). In addition, sourdough fermentation has the potential to improve important nutritional properties of grains including improved mineral bioavailability and lower glycemic index. Also, sourdough fermentation can increase or decrease levels of several bioactive compounds depending on the nature of the compound and the type of sourdough process (Katina et al., 2005). The information available on the impact of sourdough fermentation on bioactive compounds is however limited.

Liukkonen et al. (2003) studied the effect of sourdough baking on the levels of several bioactive compounds. They showed that sourdough fermentation resulted in more than double the levels of folates and total phenol content of methanol extracts. It is important to note that the observed increase in the levels of total phenols of the methanolic extracts is most likely due to an increase in the levels of free phenols, as the level of total phenol in the alkaline extracts was not modified upon fermentation. Processing can result in a release of polyphenols bound to insoluble residues which can then be extracted with solvents such as methanol (Bonoli, Verardo, Marconi and Caboni, 2004; Waldron, Parr, Ng and Ralph, 1996).

The amounts of tocopherols and tocotrienols were significantly reduced, possibly due to oxidation by atmospheric oxygen. On the other hand, the levels of sterols, alk(en)ylresorcinols, lignans, phenolic acids and total phenol content of alkaline extracts changed only slightly. Sourdough fermentation also increased the antioxidant capacity (measured as DPPH radical scavenging activity) of methanol extracts, most likely due to the increased levels of easily extractable phenolic compounds following fermentation.

An increase in the amounts of folates during the fermentation phase of rye and wheat has also been reported (Kariluoto, Vahteristo, Salovaara, Katina, Liukkonen and Piironen, 2004). Interestingly, it was also found that the leavening agent was an important factor affecting the process, and that baker's yeast contributed markedly to the final folate content by synthesising folates during fermentation (Kariluoto et al., 2004). Moreover, the synthesis of folate by yeast during fermentation can increase the final folate content by up to threefold whereas the effect of sourdough bacteria may be negligible (Kariluoto, Liukkonen, Myllymaki, Vahteristo, Kaukovirta-Norja and Piironen, 2006).

The effect of combining fermentation with germination of rye to determine whether if these two processes have synergistic effects on bioactive compounds has also been evaluated (Katina et al., 2007). Katina et al. (2007) showed that both pre-processing of rye before fermentation (germination) and type of fermentation, had marked effects on the levels of potentially bioactive compounds of rye grain. The effect of the most effective sourdough fermentation (with *S. cerevisiae*) and germination on the levels of bioactive compounds in rye is presented in Figure 14.3 (Katina et al., 2007). Yeast fermentation or mixed fermentation (yeast is present as well as lactic acid bacteria) was reported to be the major factor determining the increased levels of folates, sterols, lignans, free ferulic acids and for preserving alk(en)yl-resorcinols during fermentation of both native and germinated rye. This was attributed to the considerably higher pH of yeast fermentations (pH 4.5–6.0), which may be optimum for the activity of cell wall degrading enzymes derived from the grain itself or from indigenous microbes. Also, the further increase in the levels of bioactive compounds following the use of germinated grain as raw material for fermentation was explained on the basis that germinated grain may not only provide higher amounts of fermentable sources, such as sugars and nitrogen sources, but may also provide additional cell wall degrading enzymes, synthesised during germination as well as enzyme-active microbes, which can remain active during further fermentation steps (Katina et al., 2007). Therefore, by optimising a number of bioprocessing methods, the natural bioactivity of wholemeal rye can be further improved. The resultant product can then be used as an ingredient in breads, breakfast cereals and snack foods to improve their nutritional quality (Katina et al., 2007).

Conversely, fermentation has also been shown to have an adverse effect on the content and molecular weight of β-glucans present in barley and oat flours. Degutyte-Fomins, Sontag-Strohm and Salovaara (2002) reported that fermentation of oat bran using rye sourdough starter increased the solubility and degradation of β-glucans, effects that were attributed to the activity of β-glucanase. Lambo, Oste and Nyman (2005) when studying the ability of different lactic acid bacteria to influence the content, viscosity and molecular weight of β-glucans in barley and oat concentrates found that total fibre concentrations for all samples and maximum viscosity for oat samples was decreased after fermentation. Interestingly, the authors also reported that molecular weights were not significantly affected in this study. These results may suggest that the level of acidity obtained upon sourdough fermentation, as well as the chemical composition and enzyme activity of sourdough preferment, may have a marked effect on important characteristics of β-glucans (Tiwari and Cummins, 2009a). Yeast fermentation has also been shown to reduce molecular weight of

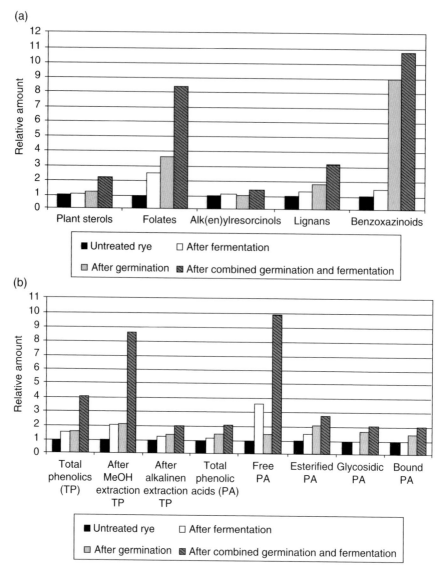

Figure 14.3 (a) Summary of effects of the most effective sourdough fermentation (with *S. cerevisiae*) and germination of rye on the level of folates, sterols, alk(en)ylresorcinols, lignans and benzoxazinoids; (b) summary of effects of the most effective sourdough fermentation (with *S. cerevisiae*) and germination of rye on the total amount of phenolic compounds and phenolic acids. Reproduced from Katina (2007). Fermentation-induced changes in the nutritional value of native or germinated rye. *Journal of Cereal Science* 46(3), 348–355. With permission from Elsevier.

β-glucans. Andersson, Armo, Grangeon, Fredriksson, Andersson and Aman (2004) studied the effect of factors such as fermentation time on the properties of $(1\rightarrow3, 1\rightarrow4)$-β-glucan in barley and/or composite wheat flour. The average molecular weight distribution was found to decrease with increasing fermentation time, thus suggesting that $(1\rightarrow3, 1\rightarrow4)$-β-glucan was degraded by endogenous β-glucanases in the barley and/or composite wheat flour. It was thus concluded that to retain high molecular weight $(1\rightarrow3, 1\rightarrow4)$-β-glucan, fermentation time should be kept as short as possible.

In summary, sourdough fermentation has been used traditionally to improve the flavour and structure of whole grain rye and wheat breads. Also, sourdough fermentation has the potential to improve important nutritional properties of grains including mineral bioavailability and glycemic index. As described in this section, sourdough fermentation can increase or decrease levels of several bioactive compounds depending on the nature of the compound and the type of sourdough process. Some of the phytochemicals most affected by sourdough fermentation include folates, phenolic compounds (improved extractability) and β-glucans. Yeast fermentation has also been shown to significantly increase the levels of folates and decrease the molecular weight of β-glucans.

14.5 Baking

Bread making is one of the most common and ancient techniques used to process grains and their respective flours into food products for human consumption. Bread, and other related bakery products, is thus a staple in many countries across the world. Bread and bakery products therefore represent a significant portion of our daily food intake. In addition to being an excellent source of energy due to its high starch content, bread can also provide a great variety of compounds with a great potential to beneficially affect human health, such as fibre, minerals, vitamins and also bioactive compounds such as tocopherols, carotenoids, polyphenols and phytosterols. The nutritional quality of the final baked bread will ultimately depend on a number of factors. In particular, the nutritional quality of the flour or flour mix used will largely determine the final nutrient profile of the baked product. For instance, the use of whole meal flours over white or more refined flours will most likely result in breads and bakery products with a higher content of fibre, minerals, vitamins and phytochemicals. In addition, several specific parameters of the baking process such as mixing time, kneading time, fermentation time, baking time and baking temperature, which can affect the degradation of many of the heat and oxygen sensitive bioactive compounds present in the system, will also have an influence on the levels of phytochemicals present in the final baked product.

In the case of folates, the effect of baking on these compounds has partly been covered in section 14.4 on fermentation. As previously commented, fermentation results in a marked increase in the levels of folates in both rye and wheat grains (Kariluoto et al., 2004; Liukkonen et al., 2003). Baking, on the other hand, has been shown to result in a decrease in the levels of folate compounds. In particular, folate losses of approximately 25% have been reported following baking. It is important to note though that these losses were nonetheless compensated by the observed synthesis during fermentation (Kariluoto et al., 2004).

The effect of baking on β-glucans has also been partly covered in section 14.4. As already stated, fermentation was found to induce an adverse effect on the content and molecular weight of β-glucans present in flours such as barley and oats. With regards to mixing, increasing mixing time has also been shown to decrease the average molecular weight distribution of $(1 \rightarrow 3, 1 \rightarrow 4)$-β-glucans (Andersson et al., 2004). Trogh, Courtin, Andersson, Aman, Sorensen and Delcour (2004) also reported that $(1 \rightarrow 3, 1 \rightarrow 4)$-β-D-glucan was degraded during proofing possibly due to the activity of endogenous β-glucanases. They also found that molecular weight decreased further during baking, but not as rapidly as during fermentation, probably due to heat induced inactivation of β-glucanases. In a study by Flander, Salmenkallio-Marttila, Suortti and Autio (2007), the proportion of very high molecular weight β-glucan in whole meal oat bread was found to decrease during the baking process whereas the proportion of lower molecular weight β-glucan was increased. As in previous studies, the degradation of β-glucan in oat bread was attributed to the β-glucanase activity of wheat flour. These data indicate that it is important to preferentially use flours with low $(1\rightarrow 3, 1\rightarrow 4)$-β-D-glucan hydrolysing activities and/or to reduce processing time to obtain soluble $(1\rightarrow 3, 1\rightarrow 4)$-β-D-glucans of high molecular weight and viscosity that have the capacity to exert beneficial physiological effects.

Carotenoids are antioxidant compounds which are in turn themselves susceptible to oxidation and degradation. Evaluating their stability during bread making is therefore vital to assess the potential role of baked cereal foods as sources of these important bioactive compounds. Certain parameters of the bread making process have been shown to significantly affect carotenoids stability. Of the three major steps in the bread making process, highest losses in carotenoid content have been reported to occur during dough making, followed by baking, while negligible losses have been recorded during fermentation (Leenhardt et al., 2006). One factor that is known to play a major role in carotenoid degradation during kneading is the presence of lipoxygenase (LOX), and strong positive correlations have been found between carotenoid losses and LOX activity in different wheat varieties (Leenhardt et al., 2006). For instance, carotenoid losses during kneading have been found to be significantly higher in bread wheat samples (26 and 23% for bread and water biscuit kneading, respectively) compared with einkorn wheat (9 and 6% for bread and water biscuit kneading, respectively). This effect may be attributed to a higher LOX activity in wheat bread compared to einkorn (Hidalgo, Brandolini and Pompei, 2010). Therefore, high-carotenoid wheat genotypes which also express very low LOX activities, such as einkorn wheat, may be employed for the production of high-carotenoid bread. As already mentioned, leavening or fermentation results in minimal reduction in carotenoid content. This result has been attributed to the protective effect of baker's yeast, the leavening agent used in bread making (Hidalgo et al., 2010). Yeast consumes oxygen during dough fermentation, thereby reducing oxygen availability within the dough system and preventing LOX-mediated carotenoid degradation (Leenhardt et al., 2006). Baking, on the other hand, may strongly reduce carotenoids in bread crust, although this effect may be minimal in the case of bread crumb. According to a study by Hidalgo, Brandolini and Pompei (2010), total carotenoids degradation from flour to final product was of 28% in water biscuits, 24% in bread crumb and 55% in bread crust for bread wheat; and 32, 20 and 43%, respectively for einkorn wheat. Figure 14.4 presents a chromatogram of einkorn wheat flour and bread crust that shows apparent reductions in the levels of carotenoids following baking. Since carotenoids are exceptionally sensitive to heat, the different carotenoid losses between bread crust and crumb can be explained on the basis of the different time–temperature processing conditions of each of the respective systems (Hidalgo et al., 2010). In summary, a careful optimisation of the most crucial stages of bread making affecting the content of carotenoids can result in the production of breads with a higher carotenoid content, which may be desirable for nutrition, and also to improve the organoleptic profile of the baked product. In particular, a reduction in kneading time and intensity may decrease carotenoids loss by limiting oxygen incorporation and thus limiting degradation of carotenoids during bread making. Also, the use of grain genotypes with high content in carotenoids which also express low lipoxygenase activity is recommended.

Similarly to carotenoid compounds, vitamin E compounds are antioxidant compounds which are in turn themselves susceptible to oxidation and degradation. Factors such as light, oxygenation and heat all normally present during bread making can thus accelerate vitamin

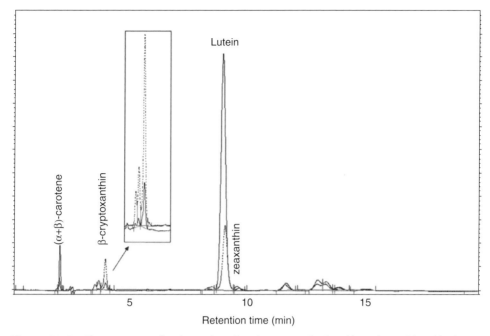

Figure 14.4 Chromatogram of einkorn wheat flour (continuous line) and bread crust (dotted line). Reprinted from Hidalgo (2010). Carotenoids evolution during pasta, bread and water biscuit preparation from wheat flours. *Food Chemistry*, 121(3), 746–751. With permission from Elsevier.

E destruction (Leenhardt *et al.*, 2006). The degradation of tocols during bread making can therefore be significantly modulated by modifying different parameters of the bread making process such as kneading time (Leenhardt *et al.*, 2006). However, contrary to carotenoids, tocols do not seem to be degraded by LOX-catalysed oxidation. Despite large varietal differences for LOX activity (whole grain flours of einkorn, durum wheat and bread wheat), no varietal differences were reported for vitamin E activity losses during kneading (Leenhardt *et al.*, 2006). Also, no significant differences in vitamin E losses after bread making have been recorded between three wheat species (einkorn, durum wheat and bread wheat), with loss values in all cases below 30%. Wennermark and Jagerstad (1992) had also previously reported vitamin E losses following wheat bread making of 20–40% depending on the bread making method. It was concluded that tocol losses during bread making could thus be attributed to direct oxygenation during dough making and heat destruction during baking (Leenhardt *et al.*, 2006).

Several other factors such as the initial total content of tocols, initial tocol profile and occurrence of antioxidant compounds other than tocols (such as flavonoids) in the initial dough systems have also been shown to significantly affect the final tocol content of the baked bread. The stability of tocols in gluten-free bread systems has also been evaluated. In a study by Alvarez-Jubete, Holse, Hansen, Arendt and Gallagher (2009), the tocopherol content of amaranth, quinoa and buckwheat as affected by baking was examined. The vitamin E losses for gluten-free control, amaranth and buckwheat breads were ≈ 30%, whereas in quinoa bread losses were 13.6%. Also prepared in this study were 100% pseudobreads (100% Q and 100% B breads) and vitamin E losses were lower compared with the respective 50% counterparts (50% Q and 50% B breads). Lowest losses amongst all of the breads studied were recorded for the 100% quinoa bread (7.5%). The authors found a significant negative correlation ($R^2 = -0.84$) between the initial vitamin E content of breads (calculated as if no loss had occurred) and the degree of loss. Thus indicating that, the higher the initial content of vitamin E, the lower the degree of loss (%) obtained. In addition, significant negative correlations were also observed between the degree of vitamin E loss (%) and α-, and γ-tocopherol initial content ($R^2 = -0.79$ and -0.84, respectively), whereas no correlations were obtained for the other tocol compounds. The authors concluded that the observed variation in vitamin E recovery for the different types of breads studied could be partly explained on the basis of differences in the initial α- and γ-tocopherol content, thus suggesting that, in addition to α-tocopherol, γ-tocopherol activity as an antioxidant is also important in these systems. According to Alvarez-Jubete, Holse, Hansen, Arendt and Gallagher (2009), another likely factor responsible for some of the variation observed in the degree of loss among the different grain samples was the presence of compounds with antioxidant capacity other than tocols, such as polyphenol compounds. The presence of flavonoid compounds in quinoa and buckwheat grains, mainly kaempferol and quercetin glycosides in quinoa, and quercetin glycosides and catechins in buckwheat have been reported previously (Alvarez-Jubete *et al.*, 2010; Dietrych-Szostak *et al.*, 1999; Dini, Tenore and Dini, 2004; Watanabe, 1998). Flavonoid compounds spare liposoluble antioxidants such as vitamin E from oxidation when present in the same sample matrix (Scalbert, Manach, Morand, Remesy and Jimenez, 2005).

The effect of rye flour extraction rate on the tocopherol and tocotrienol content of rye breads made by a classical sourdough fermentation method has also been investigated in a recent study by Michalska, Ceglinska, Amarowicz, Piskula, Szawara-Nowak and Zielinski (2007). In this study, the total tocopherol content (α-T, β-T, γ-T, and δ-T) in rye breads varied from 3.42 to 1.19 µg/g dry matter depending on the extraction rate of the flour. Similarly, total tocotrienol content varied from 3.38 to 0.38 µg/g dry matter. The main tocopherol in rye breads was α-T with α-T3 and β-T3 as the main tocotrienols. As expected, the highest levels of tocopherols and tocotrienols were found in bread with a flour extraction rate of 100%. The authors of this study also noted a considerable loss of these bioactive compounds during the baking process, with the content in flours being approximately three-fold higher compared to the respective baked breads. In addition, the total tocol content of the rye breads was also compared to commercial wheat bread roll. The authors found that only the rye bread made with a flour extraction rate of 100% had a similar content of tocopherols and tocotrienols compared to wheat bread whereas the rest of the rye breads (95, 90 and 70% extraction rate) showed lower contents of these compounds. The authors attributed this result to the distinctive bread making process involved in the preparation of the rye breads. The longer fermentation time associated with the sourdough method in which the rye flour is mixed with water and allowed to ferment is most likely the factor responsible for the greater loss of these sensitive bioactive compounds in the sourdough rye breads compared to commercial wheat bread. These results are in agreement with a previous study by Liukkonen *et al.* (2003), where the authors also reported a significant loss in vitamin E compounds during rye sourdough bread making. Similarly to Michalska, Ceglinska, Amarowicz, Piskula, Szawara-Nowak and Zielinski (2007), Liukkonen *et al.* (2003) also identified sourdough fermentation as the main step for the degradation of tocopherols and tocotrienols in rye sourdough baking.

A great amount of the research conducted on the effect of baking on grain phytochemicals has focused on the study of phenolic compounds. Grains are known to be moderate sources of polyphenol compounds such as phenolic acids and flavonoids (Manach, Scalbert, Morand, Remesy and Jimenez, 2004). Polyphenol compounds have attracted much attention over the past decade as their intake has been associated with decreased risk of diseases related to oxidative stress such as cancer and cardiovascular disease (Scalbert *et al.*, 2005). However, polyphenol compounds, and in particular flavonoid compounds, have been shown in a number of studies to be heat sensitive (Dietrych-Szostak *et al.*, 1999; Im, Huff and Hsieh, 2003; Kreft, Fabjan and Yasumoto, 2006), and may thus be negatively affected during thermal processing such as baking. The extent of polyphenol loss will be mainly determined by the type of substrate and extraction rate, and also by the processing conditions of the baking process.

For instance, rye flour extraction rate has been shown to affect the phenolic acid profile of rye breads made by sourdough fermentation (Michalska *et al.*, 2007). The main phenolic acid compounds present in rye breads were caffeic, ferulic, p-coumaric and sinapic acids, with ferulic and sinapic acids as the predominant compounds. As expected, rye bread based on lower extraction rate flours showed the lowest contents of phenolic acids (Michalska *et al.*, 2007). Also, the stability of phenolic acids and flavonoid compounds during the bread making process has been assessed in grains such as rye, and also in certain pseudocereals. The change in the content and composition of phenolic acids and ferulic acid dehydrodimers during the rye bread making process was evaluated in a study by Boskov Hansen, Andreasen, Nielsen, Larsen, Bach Knudsen, Meyer *et al.* (2002). The most predominant phenolic compounds found were ferulic acid and ferulic acid dehydrodimers. The total amount of free and ester-bound phenolic acids and ferulic acid dehydrodimers was slightly lower in the processed samples compared to that in whole meal flour (1575 µg/g and 1472 µg/g in the whole meal and bread crumb respectively). Interestingly, it was reported that only dough mixing resulted in a significant decrease in the content of ferulic acid whereas no significant changes in any of the phenolic compounds present were recorded during dough mixing, dough proofing or baking (Boskov Hansen *et al.*, 2002). Liukkonen *et al.* (2003) also studied the effect of sourdough baking on the levels of phenolic acids in rye with similar results. The fermentation phase changed very slightly the levels of lignans, alkenylresorcinols and phenolic acids. Also, during baking changes in these compounds were found to be neglible. Therefore, the phenolic acid content in the final baked bread was the same, or slightly higher, than that measured in whole meal rye flour (Liukkonen *et al.*, 2003).

The stability of phenolic acids and flavonoid compounds in the pseudocereals amaranth, quinoa and buckwheat grains during bread making has also been evaluated (Alvarez-Jubete *et al.*, 2010). Polyphenol content was generally found to be reduced in the bread samples when compared to the original grains. In particular, quercetin and kaempferol glycosides content in 100% quinoa breads was 17.1 and 19.2 µmol/100 g, compared with 43.4 and 36.7 µmol/100 g in quinoa seeds. In the case of buckwheat, quercetin glycosides content decreased significantly with bread making, resulting in an increase in quercetin content through hydrolysis. Also, the phenolic acids present in quinoa and buckwheat grains were degraded during baking and their content decreased significantly from grain to bread. However, despite the negative impact of baking on the polyphenol content of pseudocereals, the breads made using quinoa and buckwheat flour still contained flavonoids in significant quantities.

Similarly, Vogrincic, Timoracka, Melichacova, Vollmannova and Kreft (2010) studied the impact of bread making and baking procedure on rutin and quercetin content of tartary buckwheat (*Fagopyrum tataricum*) bread and breads made of mixtures of tartary buckwheat and wheat flour. The total phenol, rutin and quercetin levels in dough and bread loaves prepared using different levels of buckwheat and wheat flours are summarised in Table 14.2. As expected, rutin and quercetin concentrations increased with increasing percentage of tartary buckwheat flour as these compounds were not present in wheat

Table 14.2 Total phenol, rutin and quercetin in dough and bread loaves using different levels of tartary buckwheat and wheat flours. Reprinted with permission from *Journal of Agricultural and Food Chemistry*, 58(8). Vogrincic et al. (2010). Degradation of Rutin and Polyphenols during the preparation of Tartary Buckwheat Bread, 4883–4887. Copyright American Chemical Society.

Tf/Wf ratio (%)		dough		bread loaf	
		35 min	60 min	inside	crust
0:100 (T_0)	polyphenols	0.70 ± 0.03 a	0.82 ± 0.01 a	0.64 ± 0.02 a	0.61 ± 0.06 a
	rutin	ND a	ND a	ND a	ND a
	quercetin	ND a	ND a	ND a	ND a
30:70 (T_30)	polyphenols	4.04 ± 0.03 a	3.48 ± 0.03 b	3.40 ± 0.11 b	3.50 ± 0.11 b
	rutin	0.32 ± 0.01 a	ND b	ND b	ND b
	quercetin	1.26 ± 0.15 a	1.50 ± 0.03 b	1.53 ± 0.01 b	1.52 ± 0.01 b
50:50 (T_50)	polyphenols	8.59 ± 0.27 a	7.11 ± 0.33 b	5.11 ± 0.34 c	4.72 ± 0.24 d
	rutin	0.54 ± 0.02 a	0.31 ± 0.02 b	ND c	ND c
	quercetin	2.47 ± 0.05 a	2.65 ± 0.04 a	2.50 ± 0.05 a	2.54 ± 0.05 a
100:0 (T_100)	polyphenols	10.99 ± 0.44 a	12.66 ± 0.06 b	7.84 ± 0.37 c	7.63 ± 0.39 c
	rutin	1.01 ± 0.02 a	0.64 ± 0.01 b	0.44 ± 0.01 c	0.47 ± 0.02 c
	quercetin	5.13 ± 0.03 a	5.12 ± 0.07 a	5.00 ± 0.09 b	4.83 ± 0.06 c

flour. In agreement with the results by Alvarez-Jubete, Wijngaard, Arendt and Gallagher (2010), rutin content was reported to decrease during the bread making process, whereas quercetin content increased as a result of rutin hydrolysis. In particular, 85% of rutin was transformed into quercetin during dough mixing. The sole addition of water and yeast seem to have facilitated the degradation of rutin into quercetin, possibly by the rutin degrading enzymes naturally present in the flour. The concentration of rutin continued to decrease during proofing and its concentration in dough after 60 min of rising was significantly lower compared to that measured at 35 min of rising. Quercetin content, on the other hand, remained more stable during proofing and no significant changes were observed, except for the dough containing 30% tartary buckwheat flour which experienced a significant increase in quercetin content after 60 min in comparison to 35 min. During baking rutin continued to decrease and, as a result, rutin was only detectable in bread made of 100% tartary buckwheat flour. Similar to what happened during proofing, quercetin content remained constant during baking. Finally, no significant differences were detected in both rutin and quercetin concentrations between crust and crumb of the bread (Vogrincic et al., 2010).

As the *in vitro* antioxidant capacity of a sample is derived from its content in antioxidant compounds, baked products formulated with flours rich in antioxidant compounds such as phenolic compounds will consequently be characterised by high *in vitro* antioxidant capacities. In addition, the use of high extraction rate flours will also generally result in higher antioxidant capacity compared to those products formulated with lower extraction rate flours. Also, thermal processing techniques that can cause a degradation of antioxidant compounds, such as baking, will most likely result in final baked products with decreased antioxidant capacity. The effect of flour extraction rate on the antioxidative properties of traditional rye bread was studied by Michalska, Ceglinska, Amarowicz, Piskula, Szawara-Nowak and Zielinski (2007). As expected, those breads formulated using rye flours with extraction rates of 100–90% were found to have the highest total phenol content when compared to bread made using flour of 70% extraction rate. Moreover, the content of total phenols was about two- to three-fold higher when compared to standard wheat roll. When examined for their free radical scavenging activity against $ABTS^+$ cation radical,

80% methanol extracts of rye breads formulated on flours with extraction rates of 100–90% had the highest scavenging activity in comparison to the free radical scavenging activity of trolox. In particular, trolox equivalent antioxidant capacity (TEAC) of rye bread formulated on flour with an extraction rate of 70% was approximately 40% lower when compared to that of whole meal rye bread. Also, TEAC for methanol extracts of rye bread were almost three-fold higher when compared to TEAC of wheat roll. When the DPPH scavenging capacity of the rye breads was studied, it was found that breads formulated on flours with extraction rates ranging 100–90% showed about 25% higher radical scavenging activity when compared to bread formulated on flour with a extraction rate of 70%. The lowest radical DPPH scavenging activity was noted for wheat roll which was decreased by approximately 60% when compared to whole meal rye bread.

Liukkonen *et al.* (2003) studied the effect of sourdough baking on the total phenolic content and antioxidant activity (measured as DPPH radical scavenging capacity) of rye. The sourdough fermentation phase more than doubled the amount of total phenol content measured in methanolic extracts (easily extractable phenolic compounds), most likely due to release of bound phenolics following fermentation. However, this effect was slightly diluted with the addition of fresh 'unfermented' whole meal flour to the sourdough. Accordingly, an increase in antioxidant activity was also detected following the sourdough fermentation phase, probably due to the increase in the amount of total phenol content upon fermentation. Baking resulted in a slight increase in the total phenol content of alkaline-extractable phenolic compounds (bound phenolic compounds) whereas a slight decrease was observed in the methanolic fraction. Thus, in comparison to rye whole meal flour, rye sourdough breads had similar total phenol content, which indicates that these compounds are stable during rye sourdough baking. Also, antioxidant capacity of the rye breads was similar to that of the whole meal rye flour.

Another grain that has been extensively studied because of its antioxidative properties is the pseudocereal buckwheat. The antioxidative properties of buckwheat are derived mainly from its high content in phenolic acids and flavonoid compounds. The effect of bread making on the antioxidant properties of methanolic extracts from buckwheat, as well as amaranth and quinoa, were evaluated in a study by Alvarez-Jubete, Wijngaard, Arendt and Gallagher (2010). It was found that gluten-free breads containing 50% of pseudocereal flour had significantly higher total phenol content and antioxidant capacity compared to a gluten-free control based on rice flour and potato starch. Highest values were found in breads containing buckwheat. In addition, when breads were made of 100% quinoa or 100% buckwheat flour, total phenol content and higher antioxidant capacity increased significantly in comparison to those breads containing only 50% of pseudocereal flour. Regarding stability of these properties during processing, following a comparison of the measured total phenol content in breads with the expected values (calculated using the approximation that the pseudocereal flour is the only ingredient contributing to total phenol content in bread) the authors concluded that some degradation may have occurred. This effect was reported to be particularly pronounced in the case of buckwheat, where total phenol content reduction from buckwheat seeds to buckwheat bread was 323–64.5 mg Gallic acid equivalents/100 g dry weight basis. Degradation of antioxidant compounds during quinoa bread making appears to have occurred also, however, to a smaller extent. Despite the loss of total phenol content and antioxidant activity following bread making, all of the breads containing pseudocereals showed significantly higher antioxidant capacity when compared with the gluten-free control. Vogrincic, Timoracka, Melichacova, Vollmannova and Kreft (2010) also evaluated the impact of bread making on the antioxidant activity of tartary buckwheat (*Fagopyrum tataricum*) with similar results. Several doughs and breads were produced containing 0, 30, 50 and 100% tartary buckwheat flour respectively, with wheat flour as a composite. As expected, for both dough and breads, total phenol content was found to rise when the percentage of tartary buckwheat flour used increased. Also, total phenol content was found to be reduced by the bread making process. In general, it was found that baking was the bread making step that caused a greater effect on the total phenol content, whereas proofing had a slight effect, increasing the level in some cases and decreasing it in others. Interestingly, total phenol content in crust and crumb did not differ greatly. Antioxidant activity also increased in buckwheat dough and breads with a growing percentage of tartary buckwheat flour used. In particular, antioxidant activities of breads containing 0, 30, 50 and 100% of tartary buckwheat flour were approximately 4, 35, 55 and 85%, respectively. It was also found that antioxidant activity remained stable or decreased slightly during the bread making process in those systems containing buckwheat. In the case of the 100% wheat bread, DPPH scavenging activity was found to slightly increase during the bread making process, possibly due to the formation of maillard reaction products.

Maillard reaction products are formed during baking predominantly on bread crust as a result of a reaction between proteins and carbonyl groups of reducing sugars or other food components. Maillard reaction products have been typically considered as having a detrimental effect on health. However, a number of studies have demonstrated that some of these compounds also possess antioxidant properties and may thus affect the final antioxidative properties of baked products, especially in the crust.

Michalska, Ceglinska, Amarowicz, Piskula, Szawara-Nowak and Zielinski (2007) evaluated the effect of bread making on the formation of Maillard reaction products contributing to the overall antioxidant activity of rye bread. They showed that changes due to Maillard reaction affected bread crust principally. They also demonstrated that advanced Maillard reaction products (MRPs) resulted in good scavengers of peroxyl and ABTS radicals whereas early MRPs seemed to be correlated with antioxidant activity. They thus concluded that baking favoured the formation of some antioxidant compounds. Previously, Lindenmeier and Hofmann (2004) had also reported on the influence of baking conditions and precursor supplementation on the amounts of the antioxidant pronyl-L-lysine in bakery products. Although this antioxidant was not present in untreated flour, high amounts of pronyl-L-lysine were detected in bread crust, whereas only low amounts were present in the crumb. Interestingly, the amounts of pronyl-L lysine were found to be greatly affected by important parameters of the baking process such as baking time and temperature. In particular, increasing the baking time from 70 to 210 min or increasing the baking temperature from 220 to 260 °C led to a five- or three-fold increase, respectively, in the level of this antioxidant in the crust. Also, the type of ingredients used in the formulation was found to have a major influence on the synthesis of pronyl-L-lysine. For instance, substituting 5% of the flour with lysine-rich protein casein or with 10% of glucose increased the levels of the antioxidant by more than 200%. Finally, following quantitative analyses of commercial bread samples collected from German bakeries they also found that that the decrease of the pH value associated with sourdough fermentation resulted in the production of high amounts of pronyl-L-lysine in baking products. In summary, the authors concluded that the amounts of the antioxidant and chemopreventive compound pronyl-L-lysine in bakery products can be strongly influenced by adjusting both baking parameters and formulation.

In summary, it is possible to formulate baked products with a significant content in phytochemical compounds. However, due to the labile nature of some of these compounds, processing such as baking can lead to a reduction on the concentration of these bioactive compounds. β-glucans, carotenoids, tocols and flavonoids have been shown to be significantly reduced following baking. Thus, it is important that the initial level of phytochemical compounds in the flours is sufficiently high so that the content in the final baked product after processing remains adequate. To this end, the use of flours with a high extraction rate is recommended. In addition, the optimisation of the different baking steps, such as mixing and fermentation times, as well as opting for the most adequate grain variety, may also result in a final baked product with high levels of phytochemicals.

14.6 Roasting

Extensive heat treatment has been shown to cause degradation of heat-labile compounds such as flavonoids. As with most other heat processing techniques, the extent of degradation of phytochemical compounds will mostly depend on the intensity of the process, mainly length and temperature, as well as on the substrate under study.

Roasting has been employed in research studies to evaluate the effect of high temperatures on properties of grains such as total antioxidant capacity and total phenol content. Sensoy, Rosen, Ho and Karwe (2006) evaluated the effect of roasting on total phenolic content and antioxidant capacity of buckwheat. They reported that roasting white or dark buckwheat flour at 200°C for 10 min did not affect the total phenol content, as measured using the Folin-Ciocalteu assay. On the other hand, a slight decrease in antioxidant capacity, measured using the DPPH radical scavenging assay, was recorded. The effect of roasting temperature and length on buckwheat total phenol content, total flavonoid content and antioxidant capacity was also studied by Zhang, Chen, Li, Pei and Liang (2010). The content of total flavonoids was found to decrease significantly ($p<0.05$) with the increase of the roasting temperature from 80°C to 120°C and the roasting time from 20 min to 40 min. However, for total phenolics the only significant differences were found when the intensity of the roasting treatment was highest, 120°C for 40 min, indicating that flavonoid compounds are more instable during intense heat treatment. In particular, when the highest intensity treatment was applied, 120°C for 40 min, total flavonoids and total phenolics content were noted to decrease by 33 and 9% respectively in comparison with untreated tartary buckwheat flour. The raw tartary buckwheat extracts had high antioxidant properties with scavenging rates of 93.13% on hydroxyl radicals, 92.64% on superoxide radicals and the inhibitory rate of 34.28% on lipid peroxidation. However, antioxidant capacity decreased significantly ($p<0.05$) upon roasting. The trends observed for the decrease in antioxidant activities were in accordance with those observed for total flavonoids and total phenolics ($r=0.8401$ and 0.9909, respectively). Since a higher correlation was obtained between antioxidant activity and total phenol content, the authors concluded that phenolics were the main antioxidant compounds in tartary buckwheat. In the same study, the authors also evaluated the effect of microwave roasting (700 W for 10 min). Similar results were obtained, with total flavonoids being decreased to a higher extent than total phenolics. Also, antioxidant activity based on the scavenging of hydroxyl and superoxide anion radicals, and inhibition of liposome peroxidation, were decreased after microwave roasting of tartary buckwheat flour. The authors also suggested that the less severe reduction in total phenolics compared to total flavonoids upon thermal treatment could be attributed to the formation of Maillard reaction product. These Maillard reaction products may have reacted with Fiolin-Ciocalteau reagent masking the real decrease in total phenolics.

In another study, microwave oven roasting conditions were optimised to obtain barley grains with high antioxidant activity, measured as the ability to scavenge 1,1-diphenyl-2-picrylhydrazyl (DPPH) free radical and total phenol content according to Folin-Ciocalteu assay (Omwamba and Hu, 2010). Three processing factors were optimised in this study, temperature, time and amount of grain. All three factors under study were found to influence antioxidant activity both individually and interactively. The optimum condition for obtaining roasted barley with high antioxidant activity was found to be at 600 W microwave power, 8.5 min roasting time, and 61.5 g or two layers of grains.

Since heat treatments affect negatively certain phytochemicals, such as phenolic compounds, it is important to optimise thermal processing parameters to prevent or minimise losses of these compounds and to deliver a final food product with an adequate content in phytochemical compounds.

14.7 Extrusion cooking

Extrusion cooking is a very important food processing technology for cereal foods. It is used for the production of breakfast cereals, ready-to-eat foods, baby foods, snack foods, texturised vegetable protein, pet foods, dried soups and dry beverage mixes. In addition to improving digestibility, extrusion cooking of food grains also improves bioavailability of nutrients in comparison to conventional cooking. As discussed in Chapter 6, the presence of phytochemicals in grains is important because of their associated potential health benefits. In addition, phytochemicals in grains may help to prevent lipid oxidation, improving shelf life and consumer acceptance of extruded snacks, by acting as free radical terminators, chelators of metal catalysts, and singlet oxygen quenchers. For example, Viscidi, Dougherty, Briggs and Camire (2004) observed that the addition of ferulic acid and benzoin at levels of 1.0 g/kg or higher generally resulted in delayed onset of oxidation in oat based extrudates.

An important quantity of bioactive compounds is lost during extrusion processing as these compounds are sensitive towards a number of processing variables. Critical extrusion process variables such as temperature, screw speed and moisture content may induce desirable modifications, improving palatability and technological properties of extruded products. However, these conditions may also have positive or negative influence on the bioactive compounds of the extrudates. Several studies have shown that extrusion processing significantly reduces measurable bioactive compounds in food products. Table 14.3 summarises some of the data available on the effect of extrusion on the phytochemical content of some cereals and pseudocereals.

Vitamin E compounds (tocopherols and tocotrienols) are known to be affected by the extrusion cooking process (Tiwari et al., 2009b). It has been hypothesised that the stability of tocols may be negatively affected by high temperature during extrusion cooking (Shin, Godber, Martin and Wells, 1997). Zielinski, Kozlowska and Lewczuk (2001) reported a decrease of 30% in tocopherol and tocotrienols in cereals including oat, barley, wheat, rye and buckwheat following extrusion.

The fate of phenolic compounds during the extrusion process of legume flours has also been assessed. Anton, Fulcher and Arntfield (2009) evaluated the replacement of corn flour in corn-based extruded snacks with several common bean (*Phaseolus vulgaris*) flours at different levels to produce extruded puffed snacks. Total phenol content was found to decrease significantly (up to 10 and 70% for navy bean and small red bean, respectively) when the extrusion cooking temperature was set at 160 °C. In a study using legume flour exclusively, Abd El-Hady and Habiba (2003) evaluated the effect of extrusion process variables such as barrel temperature (140–180 °C) and feed moisture (18 and 22%) on total phenol content of whole meal of peas, chickpeas, faba and kidney beans. They noted a significant decrease in

Table 14.3 The effect of extrusion on the phytochemical content of some cereals

Extrusion	Total Phenolics (µg/gDW)		Tocols (µg/gDW)					
	Free	Ester bound	α-T	β-T	δ-T	α-T3	β-T3	γ-T3
Wheat	2.08	7.93	6.39	2.05		2.83	16.54	2.7
After extrusion 120°C	10.29	17.53	0.46	1.32		0.23	6.61	0.79
160°C	9.88	17.9	0.02	0.97		0.02	5.22	0.91
200°C	11.31	19.34	0.84	1.28		0.52	6.69	0.86
Buckwheat								
Groat			0.74–0.82	28.39–31.37		0.08–0.1		
After extrusion 120°C			0.28–0.32	9.88–11.36		0.07–0.09		
160°C			0.27–0.31	9.54–10.56		0.05–0.07		
200°C			0.32–0.38	10.22–11.76		0.06–0.08		
Barley	1.54	4.94	2.85	0.11	0.19	10.61	2.27	
After extrusion 120°C	5.07	13.39	0.12	0.03	0.08	1.75	0.77	
160°C	5.41	13.97	0.2	0.04	0.11	1.64	0.92	
200°C	5.31	15.06	0.13	0.04	0.10	1.11	0.86	
Rye	5.06	49.53	11.04	2.07	0.03	8.69	5.95	
After extrusion 120°C	10.86	24.33	0.95	0.57	0.02	2.05	2.49	
160°C	11.06	31.66	1.31	0.65	0.02	2.7	2.63	
200°C	11.77	28.51	1.95	0.76	0.02	3.33	2.66	
Oat	5.95	24.19	2.14	0.46		1.38	7.61	
After extrusion 120°C	20.29	28.61	1.13	0.45		1.55	4.19	
200°C	23.63	26.11	0.22	0.17		0.01	0.6	

DW: dry weight; FW: fresh weight.
Source: Zieliński et al. (2001); Zieliński et al. (2006)

total phenol content of extruded products, effect which was mainly attributed to the individual effects of both temperature and moisture as no interaction effect was noted for feed moisture and barrel temperature.

In cereal grains, Zielinski, Kozlowska and Lewczuk (Zielinski et al., 2001) reported a significant loss of phenolic acids in wheat, barley, rye and oat grains following extrusion cooking at temperatures of 120–200 °C. In a subsequent study, H. Zielinski, Michalska, Piskula and Kozlowska (2006) examined the effect of processing temperature on the extrusion cooking of buckwheat groats and reported a significant decrease in the total phenolic compounds from 4.08 mg/g dry matter (groats) to 1.17, 0.83 and 1.41 mg/g dm for extrudates at 120, 160 and 200 °C, respectively. However, a significant increase in some free/bound phenolic acids such as syringic, ferulic and coumaric acids during extrusion of buckwheat (*Fagopyrum esculentum*) was also noted in the same study (Table 14.3). The increase in these phenolic compounds was attributed to an increased release of these bioactive compounds from the matrix following extrusion (Zielinski et al., 2006). Yagci and Gogus (2009) also showed an increase of about three-fold in total phenolic content during extrusion cooking of rice-based snacks.

The effect of extrusion cooking on the total phenolic content of grains has also been shown to be cultivar specific according to Korus, Gumul and Czechowska (2007). In their study, Korus, Gumul and Czechowska (2007) studied the effect of extrusion processing on several polyphenol compounds such as myrecetin, quercetin, kaempferol, cyanidin, chlorogenic acid, caffeic acid, ferulic acid and *p*-coumaric acid. An increase of 14% in the amount of phenolics in dark-red bean extrudates compared to raw flour was noted, whereas in black-brown and cream coloured beans a decrease of 19 and 21%, respectively, was observed. The authors noted that the overall increase observed in dark-red beans during extrusion was mainly brought about by increases in quercetin (by 84%) and ferulic acid (by 40%) along with significant decreases in chlorogenic and caffeic acids by 33 and 9% respectively.

Extrusion cooking of grains has also been reported to have positive or negative effects on anthocyanin content depending on processing conditions and feed characteristics. Significant losses have been reported for anthocyanins during extrusion cooking, losses which have been mainly attributed to the high temperature employed in the process. White, Howard and Prior (2010) investigated the changes in the anthocyanin, flavonol and procyanidin contents of cranberry pomace/corn starch blends during extrusion cooking. They observed significant losses (46–64%) in the anthocyanin content of the extrudates. Furthermore, these losses were found to increase with barrel temperature. Anthocyanin losses of low magnitudes (10%) have also been reported for extrusion cooking of blueberry, cranberry, raspberry, grape powders and corn blends (Camire, Dougherty and Briggs, 2007). Camire, Chaovanalikit, Dougherty and Briggs (2002) observed that, despite losses during extrusion, a sufficient amount of the colorants remained after extrusion, and that a purple colour corn meal extruded product could be obtained following extrusion of corn with blueberry and grape. Notwithstanding the reported decrease in anthocyanins following extrusion, a considerable increase in biologically important monomer and dimer forms of procyanidin has also been reported (Khanal, Howard and Prior, 2009). Khanal, Howard and Prior (2009) observed a considerable increase in monomer and dimer contents, most probably due to the conversion of some higher-level oligomers and polymers into lower oligomer during extrusion. In the same study, the monomer content of grape pomace extruded at barrel temperature of 170 °C and screw speed of 200 rpm increased by approximately 120%, whereas total anthocyanin content was found to decrease significantly (18–53%). Thus, the challenge for the food processor remains in the optimisation of appropriate methodology to reduce the loss of phytochemicals such as vitamin E and phenolic compounds during extrusion cooking.

14.8 Parboiling

Rice parboiling is a hydrothermal treatment consisting of soaking, heating and drying. Parboiling changes both physicochemical and organoleptic properties of the rice grain. It reduces rice stickiness, increases hardness and darkens the colour (Lamberts, Rombouts, Brijs, Gebruers and Delcour, 2008b). According to the literature, colour changes during parboiling can be due to the migration of husk and/or bran pigments, enzymatic browning and non-enzymatic browning of the Maillard type (Lamberts *et al.*, 2008b). Research on the effect of parboiling on the phytochemical content of rice has been mainly dedicated to the study of carotenoid stability and migration. The carotenoid content and composition of raw and parboiled brown and milled rice was studied by Lamberts *et al.* (2008a). Analyses of the colour of rice flour samples with different extraction rates demonstrated that yellow and red pigments are concentrated in the bran and outer endosperm. As a result, all pigments were removed for degrees of milling of 15% or higher. The colour-determining components present in the fractions of different rice cultivars were identified as the carotenoids β-carotene, lutein and zeaxanthin, with β-carotene and lutein being the predominant compounds. Parboiling brown rice was found to reduce carotenoid levels to trace levels, thus suggesting that compounds other than carotenoids are responsible for the colour of milled parboiled rice. Thus, the decreased brightness and increased red and yellow colour intensities of parboiled rice were attributed by the authors to Maillard reactions and/or physicochemical changes of different rice components occurring during parboiling.

14.9 Conclusions

In summary, it is possible to develop cereal-based foods rich in phytochemical compounds. However, due to the labile nature of some of these compounds, processes such as baking and extrusion can result in a significant decrease in compounds such as β-glucans, carotenoids, tocols and flavonoids. On the other hand, germination and fermentation are cereal processing techniques that may result in increased levels of key phytochemicals such as folates and phenolic compounds. Milling significantly reduces the levels of phytochemicals in grains, as these compounds are mainly concentrated on the bran/germ fractions. Therefore, the use of flours with a high extraction rate is recommended. In addition, optimisation of the different processing steps, such as germination, mixing and fermentation times, as well as opting for the most adequate grain variety, may also result in a final cereal-based product with high levels of phytochemicals.

References

Abd El-Hady, E.A. and Habiba, R.A. (2003) Efffect of soaking and extrusion conditions on antinutrients and protein digestibility of legume seeds. *Lebensmittel-Wissenschaft Und-Technologie-Food Science and Technology*, 36(3), 285–293.

Abdel-Aal, E.S.M., and Hucl, P. (1999) A rapid method for quantifying total anthocyanins in blue aleurone and purple pericarp wheats. *Cereal Chemistry*, 76(3), 350–354.

Abdul-Hamid, A., Sulaiman, R.R.R., Osman, A. and Saari, N. (2007) Preliminary study of the chemical composition of rice milling fractions stabilized by microwave heating. *Journal of Food Composition and Analysis*, 20(7), 627–637.

Adom, K.K., Sorrells, M.E., and Liu, R.H. (2003) Phytochemical profiles and antioxidant activity of wheat varieties. *Journal of Agricultural and Food Chemistry*, 51(26), 7825–7834.

Adom, K.K., Sorrells, M.E., and Liu, R.H. (2005) Phytochemicals and antioxidant activity of milled fractions of different wheat varieties. *Journal of Agricultural and Food Chemistry*, 53(6), 2297–2306.

Alvarez-Jubete, L., Holse, M., Hansen, A., Arendt, E.K. and Gallagher, E. (2009) Impact of Baking on Vitamin E Content of Pseudocereals Amaranth, Quinoa, and Buckwheat. *Cereal Chemistry*, 86(5), 511–515.

Alvarez-Jubete, L., Wijngaard, H., Arendt, E.K. and Gallagher, E. (2010) Polyphenol composition and in vitro antioxidant activity of amaranth, quinoa buckwheat and wheat as affected by sprouting and baking. *Food Chemistry*, 119(2), 770–778.

Andersson, A.A.M., Armo, E., Grangeon, E., Fredriksson, H., Andersson, R. and Aman, P. (2004) Molecular weight and structure units of (1 –>3, 1 –>4)-beta-glucans in dough and bread made from hull-less barley milling fractions. *Journal of Cereal Science*, 40(3), 195–204.

Andreasen, M.F., Christensen, L.P., Meyer, A.S., and Hansen, A. (2000). Ferulic acid dehydrodimers in rye (Secale cereale L.). *Journal of Cereal Science*, 31(3), 303–307.

Anton, A.A., Fulcher, R.G., and Arntfield, S.D. (2009). Physical and nutritional impact of fortification of corn starch-based extruded snacks with common bean (Phaseolus vulgaris L.) flour: Effects of bean addition and extrusion cooking. *Food Chemistry*, 113(4), 989–996.

Anton, A.A., Ross, K.A., Beta, T., Fulcher, R.G., and Arntfield, S.D. (2008). Effect of pre-dehulling treatments on some nutritional and physical properties of navy and pinto beans (Phaseolus vulgaris L.). *Lwt-Food Science and Technology*, 41(5), 771–778.

Awika, J.M., McDonough, C.M., and Rooney, L.W. (2005). Decorticating sorghum to concentrate healthy phytochemicals. *Journal of Agricultural and Food Chemistry*, 53(16), 6230–6234.

Bonoli, M., Verardo, V., Marconi, E., and Caboni, M.F. (2004). Phenols in barley (Hordeum vulgare L.) flour: Comparative spectrophotometric study among extraction methods of free and bound phenolic compounds. *Journal of Agricultural and Food Chemistry*, 52(16), 5195–5200.

Borrelli, G.M., De Leonardis, A.M., Platani, C., and Troccoli, A. (2008). Distribution along durum wheat kernel of the components involved in semolina colour. *Journal of Cereal Science*, 48(2), 494–502.

Boskov Hansen, H., Andreasen, M.F., Nielsen, M.M., Larsen, L.M., Bach Knudsen, K.E., Meyer, A.S., Christensen, L.P. and Hansen, A. (2002) Changes in dietary fibre, phenolic acids and activity of endogenous enzymes during rye bread-making. *European Food Research and Technology*, 214, 33–42.

Butsat, S. and Siriamornpun, S. (2010) Antioxidant capacities and phenolic compounds of the husk, bran and endosperm of Thai rice. *Food Chemistry*, 119(2), 606–613.

Camire, M.E., Chaovanalikit, A., Dougherty, M.P., and Briggs, J. (2002) Blueberry and grape anthocyanins as breakfast cereal colorants. *Journal of Food Science*, 67(1), 438–441.

Camire, M.E., Dougherty, M.P. and Briggs, J.L. (2007) Functionality of fruit powders in extruded corn breakfast cereals. *Food Chemistry*, 101(2), 765–770.

Cardador-Martinez, A., Loarca-Pina, G. and Oomah, B.D. (2002) Antioxidant activity in common beans (Phaseolus vulgaris L.). *Journal of Agricultural and Food Chemistry*, 50(24), 6975–6980.

Chavan, J.K. and Kadam, S.S. (1989) Nutritional improvement of cereals by sprouting. *Critical Reviews in Food Science and Nutrition*, 28(5), 401–437.

Chen, M.H. and Bergman, C.J. (2005) Influence of kernel maturity, milling degree, and milling quality on rice bran phytochemical concentrations. *Cereal Chemistry*, 82(1), 4–8.

Degutyte-Fomins, L., Sontag-Strohm, T. and Salovaara, H. (2002) Oat bran fermentation by rye sourdough. *Cereal Chemistry*, 79(3), 345–348.

Dietrych-Szostak, D. and Oleszek, W. (1999) Effect of processing on the flavonoid content in buckwheat (Fagopyrum esculentum Moench) grain. *Journal of Agricultural and Food Chemistry*, 47(10), 4384–4387.

Dini, I., Tenore, G.C., and Dini, A. (2004). Phenolic constituents of Kancolla seeds. *Food Chemistry*, 84(2), 163–168.

Fares, C., Platani, C., Baiano, A. and Menga, V. (2010) Effect of processing and cooking on phenolic acid profile and antioxidant capacity of durum wheat pasta enriched with debranning fractions of wheat. *Food Chemistry*, 119(3), 1023–1029.

Flander, L., Salmenkallio-Marttila, M., Suortti, T. and Autio, K. (2007) Optimization of ingredients and baking process for improved wholemeal oat bread quality. *LWT-Food Science and Technology*, 40(5), 860–870.

Glitso, L.V. and Knudsen, K.E.B. (1999) Milling of whole grain rye to obtain fractions with different dietary fibre characteristics. *Journal of Cereal Science*, 29(1), 89–97.

Gray, D.A., Auerbach, R.H., Hill, S., Wang, R., Campbell, G.M., Webb, C. and South, J.B. (2000) Enrichment of oat antioxidant activity by dry milling and sieving. *Journal of Cereal Science*, 32(1), 89–98.

Heinio, R.L., Liukkonen, K.H., Myllymaki, O., Pihlava, J.M., Adlercreutz, H., Heinonen, S.M. and Poutanen, K. (2008) Quantities of phenolic compounds and their impacts on the perceived flavour attributes of rye grain. *Journal of Cereal Science*, 47(3), 566–575.

Hidalgo, A., Brandolini, A. and Pompei, C. (2010) Carotenoids evolution during pasta, bread and water biscuit preparation from wheat flours. *Food Chemistry*, 121(3), 746–751.

Hung, P.V., and Morita, N. (2008) Distribution of phenolic compounds in the graded flours milled from whole buckwheat grains and their antioxidant capacities. *Food Chemistry*, 109(2), 325–331.

Im, J., Huff, H.E. and Hsieh, F. (2003) Effects of Processing Conditions on the Physical and Chemical Properties of Bucwheat Grit Cakes. *Journal of Agricultural and Food Chemistry*, 51, 659–666.

Kariluoto, S., Liukkonen, K.H., Myllymaki, O., Vahteristo, L., Kaukovirta-Norja, A. and Piironen, V. (2006) Effect of germination and thermal treatments on folates in rye. *Journal of Agricultural and Food Chemistry*, 54(25), 9522–9528.

Kariluoto, S., Vahteristo, L., Salovaara, H., Katina, K., Liukkonen, K.H., and Piironen, V. (2004) Effect of baking method and fermentation on folate content of rye and wheat breads. *Cereal Chemistry*, 81(1), 134–139.

Katina, K., Arendt, E., Liukkonen, K.H., Autio, K., Flander, L. and Poutanen, K. (2005) Potential of sourdough for healthier cereal products. *Trends in Food Science and Technology*, 16(1–3), 104–112.

Katina, K., Liukkonen, K.H., Kaukovirta-Norja, A., Adlercreutz, H., Heinonen, S.M., Lampi, A.M., Pihlava, J.M. and Poutanen, K. (2007) Fermentation-induced changes in the nutritional value of native or germinated rye. *Journal of Cereal Science*, 46(3), 348–355.

Kaukovirta-Norja, A., Wilhelmson, A. and Poutanen, K. (2004) Germination: a means to improve the functionality of oat. *Agricultural and Food Science*, 13(1–2), 100–112.

Khanal, R.C., Howard, L.R. and Prior, R.L. (2009) Procyanidin Content of Grape Seed and Pomace, and Total Anthocyanin Content of Grape Pomace as Affected by Extrusion Processing. *Journal of Food Science*, 74(6), H174–H182.

Kim, S., Zaidul, I.S.M., Suzuki, T., Mukasa, Y., Hashimoto, N., Takigawa, S., Noda, T., Matsuura-Endo, C. and Yamauchi, H. (2008) Comparison of phenolic compositions between common and tartary buckwheat (Fagopyrum) sprouts. *Food Chemistry*, 110(4), 814–820.

Kim, S.J., Maeda, T., Sarker, M.Z.I., Takigawa, S., Matsuura-Endo, C., Yamauchi, H., Mukasa, Y., Saito, K., Hashimoto, N., Noda, T., Saito, T. and Suzuki, T. (2007) Identification of anthocyanins in the sprouts of buckwheat. *Journal of Agricultural and Food Chemistry*, 55(15), 6314–6318.

Kim, S.L., Kim, S.K. and Park, C.H. (2004) Introduction and nutritional evaluation of buckwheat sprouts as a new vegetable. *Food Research International*, 37(4), 319–327.

Kong, S. and Lee, J. (2010) Antioxidants in milling fractions of black rice cultivars. *Food Chemistry*, 120(1), 278–281.

Korus, J., Gumul, D. and Czechowska, K. (2007) Effect of extrusion on the phenolic composition and antioxidant activity of dry beans of Phaseolus vulgaris L. *Food Technology and Biotechnology*, 45(2), 139–146.

Kreft, I., Fabjan, N. and Yasumoto, K. (2006) Rutin content in buckwheat (Fagopyrum esculentum Moench) food materials and products. *Food Chemistry*, 98(3), 508–512.

Lamberts, L. and Delcour, J.A. (2008a) Carotenoids in Raw and Parboiled Brown and Milled Rice. *Journal of Agricultural and Food Chemistry*, 56(24), 11914–11919.

Lamberts, L., Rombouts, I., Brijs, K., Gebruers, K. and Delcour, J.A. (2008b) Impact of parboiling conditions on Maillard precursors and indicators in long-grain rice cultivars. *Food Chemistry*, 110(4), 916–922.

Lambo, A.M., Oste, R. and Nyman, M. (2005) Dietary fibre in fermented oat and barley beta-glucan rich concentrates. *Food Chemistry*, 89(2), 283–293.

Leenhardt, F., Lyan, B., Rock, E., Boussard, A., Potus, J., Chanliaud, E. and Remesy, C. (2006) Wheat lipoxygenase activity induces greater loss of carotenoids than vitamin E during breadmaking. *Journal of Agricultural and Food Chemistry*, 54(5), 1710–1715.

Lindenmeier, M. and Hofmann, T. (2004) Influence of baking conditions and precursor supplementation on the amounts of the antioxidant pronyl-L-lysine in bakery products. *Journal of Agricultural and Food Chemistry*, 52(2), 350–354.

Liu, R.H. (2004) Potential Synergy of Phytochemicals in Cancer Prevention: Mechanism of Action. *J. Nutr.*, 134(12), 3479S–3485.

Liu, R.H. (2007) Whole grain phytochemicals and health. *Journal of Cereal Science*, 46(3), 207–219.

Liukkonen, K.-H., Katina, K., Wilhelmsson, A., Myllymaki, O., Lampi, A.-M., Kariluoto, S., Piironen, V., Heinonen, S.-M., Nurmi, T., Adlercreutz, H., Peltoketo, A., Pihlava, J.-M., Hietaniemi, V. and Poutanen, K. (2003) Process-induced changes on bioactive compounds in whole grain rye. *Proceedings of the Nutrition Society*, 62(01), 117–122.

Liyana-Pathirana, C.M. and Shahidi, F. (2007) The antioxidant potential of milling fractions from breadwheat and durum. *Journal of Cereal Science*, 45(3), 238–247.

Madhujith, T., Izydorczyk, M. and Shahidi, F. (2006) Antioxidant properties of pearled barley fractions. *Journal of Agricultural and Food Chemistry*, 54(9), 3283–3289.

Manach, C., Scalbert, A., Morand, C., Remesy, C. and Jimenez, L. (2004) Polyphenols: food sources and bioavailability. *American Journal of Clinical Nutrition*, 79(5), 727–747.

Michalska, A., Amigo-Benavent, M., Zielinski, H. and del Castillo, M.D. (2008) Effect of bread making on formation of Maillard reaction products contributing to the overall antioxidant activity of rye bread. *Journal of Cereal Science*, 48(1), 123–132.

Michalska, A., Ceglinska, A., Amarowicz, R., Piskula, M.K., Szawara-Nowak, D. and Zielinski, H. (2007) Antioxidant contents and antioxidative properties of traditional rye breads. *Journal of Agricultural and Food Chemistry*, 55(3), 734–740.

Omwamba, M. and Hu, Q.H. (2010) Antioxidant Activity in Barley (Hordeum Vulgare L.) Grains Roasted in a Microwave Oven under Conditions Optimized Using Response Surface Methodology. *Journal of Food Science*, 75(1), C66–C73.

Oomah, B.D., Cardador-Martinez, A. and Loarca-Pina, G. (2005) Phenolics and antioxidative activities in common beans (Phaseolus vulgaris L). *Journal of the Science of Food and Agriculture*, 85(6), 935–942.

Oomah, B.D., Caspar, F., Malcolmson, L.J. and Bellido, A.S. (2011) Phenolics and antioxidant activity of lentil and pea hulls. *Food Research International*, 44(1), 436–441.

Oomah, B.D. and Mazza, G. (1996) Flavonoids and antioxidative activities in buckwheat. *Journal of Agricultural and Food Chemistry*, 44(7), 1746–1750.

Panfili, G., Fratianni, A., Di Criscio, T. and Marconi, E. (2008) Tocol and beta-glucan levels in barley varieties and in pearling by-products. *Food Chemistry*, 107(1), 84–91.

Peterson, D.M. (1994) Barley tocols – effects of milling, malting, and mashing. *Cereal Chemistry*, 71(1), 42–44.

Peterson, D.M. (1995) Oat tocols – concentration and stability in oat products and distribution within the kernel. *Cereal Chemistry*, 72(1), 21–24.

Poutanen, K., Flander, L. and Katina, K. (2009) Sourdough and cereal fermentation in a nutritional perspective. *Food Microbiology*, 26, 693–699.

Scalbert, A., Manach, C., Morand, C., Remesy, C. and Jimenez, L. (2005) Dietary Polyphenols and the Prevention of Diseases. *Critical Reviews in Food Science and Nutrition*, 45(4), 287–306.

Seal, C.J. (2006) Whole grains and CVD risk. *Proceedings of the Nutrition Society*, 65(1), 24–34.

Sensoy, I., Rosen, R.T., Ho, C.T. and Karwe, M.V. (2006) Effect of processing on buckwheat phenolics and antioxidant activity. *Food Chemistry*, 99(2), 388–393.

Shin, T.S., Godber, J.S., Martin, D.E. and Wells, J.H. (1997) Hydrolytic stability and changes in E vitamers and oryzanol of extruded rice bran during storage. *Journal of Food Science*, 62(4), 704–and.

Siebenhandl, S., Grausgruber, H., Pellegrini, N., Del Rio, D., Fogliano, V., Pernice, R. and Berghofer, E. (2007) Phytochemical profile of main antioxidants in different fractions of purple and blue wheat, and black barley. *Journal of Agricultural and Food Chemistry*, 55(21), 8541–8547.

Slavin, J. (2004) Whole grains and human health. *Nutrition Research Reviews*, 17(1), 99–110.

Slavin, J.L., Jacobs, D. and Marquart, L. (2001) Grain processing and nutrition. *Critical Reviews in Biotechnology*, 21(1), 49–66.

Tiwari, U. and Cummins, E. (2009a) Factors Influencing beta-Glucan Levels and Molecular Weight in Cereal-Based Products. *Cereal Chemistry*, 86(3), 290–301.

Tiwari, U. and Cummins, E. (2009b) Nutritional importance and effect of processing on tocols in cereals. *Trends in Food Science and Technology*, 20(11–12), 511–520.

Trogh, I., Courtin, C.M., Andersson, A.A.M., Aman, P., Sorensen, J.F. and Delcour, J.A. (2004) The combined use of hull-less barley flour and xylanase as a strategy for wheat/hull-less barley flour breads with increased arabinoxylan and (1– >3,1– >4)-beta-D-glucan levels. *Journal of Cereal Science*, 40(3), 257–267.

Viscidi, K.A., Dougherty, M.P., Briggs, J. and Camire, M.E. (2004) Complex phenolic compounds reduce lipid oxidation in extruded oat cereals. *Lebensmittel-Wissenschaft Und-Technologie-Food Science and Technology*, 37(7), 789–796.

Vogrincic, M., Timoracka, M., Melichacova, S., Vollmannova, A. and Kreft, I. (2010) Degradation of Rutin and Polyphenols during the Preparation of Tartary Buckwheat Bread. *Journal of Agricultural and Food Chemistry*, 58(8), 4883–4887.

Waldron, K.W., Parr, A.J., Ng, A. and Ralph, J. (1996) Cell wall esterified phenolic dimers: Identification and quantification by reverse phase high performance liquid chromatography and diode array detection. *Phytochemical Analysis*, 7(6), 305–312.

Wang, L., Xue, Q., Newman, R.K. and Newman, C.W. (1993) Enrichment of tocopherols, tocotrienols, and oil in barley fractions by milling and pearling. *Cereal Chemistry*, 70(5), 499–501.

Watanabe, M. (1998) Catechins as antioxidants from buckwheat (Fagopyrum esculentum Moench) groats. *Journal of Agricultural and Food Chemistry*, 46(3), 839–845.

Wennermark, B. and Jagerstad, M. (1992) Breadmaking and storage of various wheat fractions affect Vitamin E. *Journal of Food Science*, 57(5), 1205–1209.

White, B.L., Howard, L.R. and Prior, R.L. (2010) Polyphenolic Composition and Antioxidant Capacity of Extruded Cranberry Pomace. *Journal of Agricultural and Food Chemistry*, 58(7), 4037–4042.

Wilhelmson, A., Oksman-Caldentey, K.M., Laitila, A., Suortti, T., Kaukovirta-Norja, A. and Poutanen, K. (2001) Development of a germination process for producing high beta-glucan, whole grain food ingredients from oat. *Cereal Chemistry*, 78(6), 715–720.

Yagci, S. and Gogus, F. (2009) Effect of Incorporation of Various Food By-products on Some Nutritional Properties of Rice-based Extruded Foods. *Food Science and Technology International*, 15(6), 571–581.

Zhang, M., Chen, H., Li, J., Pei, Y. and Liang, Y. (2010) Antioxidant properties of tartary buckwheat extracts as affected by different thermal processing methods. *LWT – Food Science and Technology*, 43(1), 181–185.

Zielinski, H., Kozlowska, H. and Lewczuk, B. (2001) Bioactive compounds in the cereal grains before and after hydrothermal processing. *Innovative Food Science andamp; Emerging Technologies*, 2(3), 159–169.

Zielinski, H., Michalska, A., Piskula, M.K. and Kozlowska, H. (2006) Antioxidants in thermally treated buckwheat groats. *Molecular Nutrition and Food Research*, 50(9), 824–832.

15 Factors affecting phytochemical stability

Jun Yang,[1] Xiangjiu He,[2] and Dongjun Zhao[3]

[1] Frito-Lay, North America R & D, Plano, TX, USA
[2] School of Pharmaceutical Sciences, Wuhan University, Hubei, Wuhan, China
[3] Department of Food Science, Cornell University, Ithaca, NY, USA

15.1 Introduction

Food products, especially fruits, vegetables, teas, nuts, legumes, and whole grains, contain a group of naturally occurring compounds named "phytochemicals" or "phytonutrients", which are biologically active organic substances that impart colors, flavors, aromas, odors, and protection against diseases. These compounds, including phenolics, thiols, carotenoids, ascorbic acid, tocopherols, sulforaphane, indoles, isothiocyanates, and glucosinolates, may help protect human cellular systems from oxidative damage through a variety of mechanisms, and thus lower the risk of chronic diseases in human beings. Since phytochemicals not only possess unique health benefits, but also can be utilized as natural colorants, they are drawing tremendous attention. However, phytochemical stability is affected by many variables, including pH, oxygen, temperature, time of processing, light, water activity (a_w), enzymes, structure, self-association, concentration, metallic ions, atmospheric composition, copigments, the presence of antioxidants, and storage conditions, suggesting that these molecules are unstable and highly susceptible to degradation and decompositiion. Anthocyanins are a group of naturally-occurring and water-soluble flavonoids responsible for their diverse color characteristics, including red, blue, and purple colors in fruits, vegetables, and grains as well as food products derived from them. The daily intake of anthocyanins is estimated to be 12.5 mg per capita in the United States (Wu et al., 2006). Anthocyanins cannot only be utilized in the food industry as natural colorants but also possess potential health benefits in prevention of certain chronic diseases. At present, grape skin and red cabbage are the predominant concentrated sources of anthocyanin colorants (Francis, 2000). As food colorants, anthocyanins' stability is of great concern since they are usually less stable and more sensitive to condition changes such as pH in comparison with synthetic colorants. In recent years, many studies have concentrated on the health benefits of anthocyanins from different perspectives such as biological activities and bioavailability. Anthocyanins have been considered as health promoting compounds due to their antioxidant activity (Satue-Gracia et al., 1997; Yang et al., 2009), anti-inflammatory (Tsuda et al., 2002), anti-cancer (Chen et al., 2006), antiarteriosclerosis (Xia et al., 2006), inhibiting oxidation of

Handbook of Plant Food Phytochemicals: Sources, Stability and Extraction, First Edition.
Edited by B.K. Tiwari, Nigel P. Brunton and Charles S. Brennan.
© 2013 John Wiley & Sons, Ltd. Published 2013 by John Wiley & Sons, Ltd.

low-density lipoprotein and liposomes (Satue-Gracia et al., 1997), hyperlipidemia (Kwon et al., 2007), and hypoglycemic effects (Sasaki et al., 2007).

Anthocyanins are important in the human diet and their compositional characteristics, functionality, and stability need to be better understood. Generally, the color stability of anthocyanins is influenced by numerous parameters such as environmental, processing, and storage conditions, suggesting that these molecules are unstable and highly susceptible to degradation (Markakis, 1982; Skrede et al., 1992; Francis, 2000; Malien-Aubert et al., 2001). For example, they are able to chelate metal ions such as Fe, Cu, Al, and Sn present in the media or packaging, resulting in a change of color. Additionally, the major drawback in the use of anthocyanins as food colorants is their low stability and higher cost. Consequently, it is of great interest to conduct research on the stability and composition of anthocyanins in foods, although some studies on the stability of anthocyanins have been published (Bassa and Francis 1987; Inami et al., 1996; Shi et al., 1992).

Betalains are a class of red and yellow indole-derived, water-soluble nitrogenous pigments found in plants of the Caryophyllales and some higher order of fungi such as the Basidiomycetes (Strack et al., 1993). As natural colorants, betalains can be extracted from red beet (Stintzing and Carle 2004), cactus pears (Stintzing et al., 2001), and Amaranthaceae plants (Cai et al., 2005). In comparison with anthocyanins, betalains retain their appearance over the broad pH range 3–7, but degrade below pH 2 and above pH 9 (Jackman and Smith, 1996). The health effects of betalains could be contributed to anti-inflammatory activities (Gentile et al., 2004), antiradical and antioxidant activities (Stintzing et al., 2005), and inhibition of lipid oxidation and peroxidation (Kanner et al., 2001).

Betalain stability is influenced by a variety of factors, including concentration, structure such as glucosylation and acylation, pH, temperature, a_w, O_2, N_2, light, enzymes, matrix constituents, antioxidants, chelating agents, metals, and storage conditions. For instance, color retention during and after processing in foods containing betalain could be considerably improved by exclusion or removal of undesirable factors such as metal ions, light, oxygen and by addition of food additives such as antioxidants and chelating agents.

Carotenoids are one of the major phytochemicals in nature, which are widely distributed in plants as well as in the animal kingdom. In the past decade the biological functions of carotenoids have drawn considerable interest. For instance, carotenoids have been shown to exert protective effects against cardiovascular and eye diseases, as well as skin and stomach cancers (Canfield et al., 1993). In addition, several carotenoids, such as α- and β-carotene, possess vitamin A activity. Numerous studies have also shown that carotenoids may act as antioxidants through a mechanism of quenching singlet oxygen (Palozza and Krinsky, 1992) or free radicals (Jørjensen and Skibsted, 1993). Obviously, the significance of carotenoids to mankind as neutraceuticals is indisputable.

The structural characteristic of all carotenoids is its polyenoic chain, which is responsible for their physical and chemical properties, and provides this group of natural compounds with their coloring and antioxidant activities, as well as their biological functions (Rascón et al., 2011). The structures break down with attack by free radicals, such as singlet molecular oxygen and other reactive species. The common degradation pathways are isomerization, oxidation, and fragmentation of carotenoid molecules. This degradation can be induced by heat, light, oxygen, acid, transition metal, or interactions with radical species (Boon et al., 2010). Heat, light, and acids promote isomerization of the *trans*-form of carotenoids to the *cis*-form. Light, enzymes, pro-oxidant metals, and co-oxidation with unsaturated lipids, on the other hand, induce oxidation. Pyrolysis occurs under intense heat with expulsion of low molecular weight molecules. Generally, all of above can influence the color of foods as well as their nutritional value (Rao and Rao, 2007). A better understanding of the factors that influence the stability of carotenoids will help limit their degradation during processing. Also some attempts can be made to increase the stability of the carotenoids.

Catechins are a group of low molecular weight flavan-3-ols isomers including four major compounds, (−)-epicatechin (EC), (−)-epigallocatechin (EGC), (−)-epicatechin gallate (ECG), and (−)-epigallocatechin gallate (EGCG), and four minor compounds, (+)-catechin (C), (−)-catechin gallate, (−)-gallocatechin, and (−)-gallocatechin gallate, which are present in a variety of foods, such as tea, wine, fruits, and chocolate. Catechins are water soluble, colorless, and astringent. The basic structure of catechins is composed of two benzene rings (A and B rings) and a dihydropyran heterocycle (C ring) with a hydroxyl group on carbon 3. EC has an *ortho*-dihydroxyl group in the B ring at carbons 3' and 4', while EGC has a trihydroxyl group in the B ring at carbons 3', 4', and 5'. ECG differs from EC with a gallate moiety esterified at carbon 3 of the C ring, while EGCG has both trihydroxyl group in the B ring at carbons 3', 4', and 5', and a gallate moiety esterified at carbon 3 of the C ring. The concentrations of catechins in foods highly depend on the food sources and vary to a large extent. Tea has the highest level of catechins among all the food sources. Generally EGCG is the most abundant catechin in tea leaves and green tea, oolong tea, and black tea, followed by EGC, ECG, and EC, while GC and C are minor components (Shahidi *et al*., 2004). Green tea contains 30–42% catechins on a dry basis compared to black tea which contains 3–10% (Arts *et al*., 2000).

Glucosinolates are a group of plant secondary metabolites present in all families of *Brassica*, such as rapeseed, cabbage, cauliflower, brussel sprouts, turnip, calabrese/broccoli, Chinese cabbage, radishes, mustard seed, horse radish, and so on. A large body of epidemiological evidence has indicated that the protective effects of *Brassica* vegetables against cancers of the alimentary tract and lungs may be partly due to their high content of glucosinolates (Jones *et al*., 2006). Glucosinolates constitute a wide class of natural compounds from which approximately a hundred have been identified to date. They possess a common chemical structure consisting in a β-D-1-thioglucopyranose moiety bearing on the anomeric site an *O*-sulfated thiohydroximate function, and they only differ by their side chain, which can be aliphatic, aromatic, or heterocyclic (indolic) (van Eylen *et al*., 2008; Lopez-Berenguer *et al*., 2007). Glucosinolates are chemically stable until they come into contact with the enzyme myrosinase, which is stored compartmentalized from glucosinolates in plant tissue (Kelly *et al*., 1998). They become accessible to myrosinase when the plant tissue is disrupted (Rodrigues and Rosa, 1999). The hydrolysis gave rise to an unstable aglycone intermediate, thiohydroxamate-*O*-sulfonate, which is spontaneously converted to different classes of breakdown products including isothiocyanates, thiocyanates, nitriles, epithionitriles, hydroxynitriles, and oxazolidine-2-thiones (Rungapamestry *et al*., 2006). One of the principal forms of chemoprotection, however, is thought to arise from isothiocyanates formation, which may influence the process of carcinogenesis partly by inhibiting Phase I and inducing Phase II xenobiotic metabolizing enzyme activity (Song and Thornalley, 2007).

The extent of hydrolysis of glucosinolates and the nature and composition of the breakdown products formed are known to be influenced by various characteristics of the hydrolysis medium. Intrinsic factors such as coexisting myrosinase and its cofactors ascorbic acid, epithiospecifier protein (ESP), ferrous ions as well as extrinsic factors such as pH and temperature can affect the hydrolysis of glucosinolates (Fenwick and Heaney, 1983; van Poppel *et al*., 1999). Although some of the hydrolysis products are believed to have a health beneficial effect, they can also have an undesirable effect on odor and taste. For instance, bitterness can be caused by gluconapin, sinigrin and 5-vinyloxazolidine-2-thione. Allyl isothiocyanate brings about a pungent and lachrymatory response upon chewing and cutting of *Brassica*, which is important to consumer acceptance of the health promoting vegetables (Ludikhuyze et al., 2000). Therefore, studying the factors influencing the stability of glucosinolates is highly desirable.

The most interesting and researched isoflavones reported in literature include genistein, daidzein, and glycitein, mainly found in soybeans and soy products, and coumestrol, formonometin, and biochanin A, found in a variety of plants, such as alfalfa, chickpeas, clove seeds, and some pulses (Murphy et al., 1999). Soybeans and soy products are the predominant sources of isoflavones consumed in the human diet worldwide. Genistein, daidzein, and glycitein are the basic chemical structures of aglycons of soy isoflavones. Their derivatives containing a β-glucoside group are genistin, daidzin, and glycitin, respectively, also found in soy beans and soy products. The glucoside conjugates, having either an acetyl or a malonyl β-glucoside, acetyle genistin, acetyl daidzin, acetyl glycitin, malonylgenistin, malonyldaidzin, and malonylglycitin, are also of interest.

15.2 Effect of pH

Anthocyanins are sensitive to pH. In general, anthocyanins are observed to fade at pH values above 2. However, acylation with hydroxycinnamic acids does not only bring about distinct bathochromic and hyperchromic shifts, but also promotes stability at near neutral pH values, which is explained by intramolecular copigmentation due to the stacking of the hydrophobic acyl moiety and the flavylium nucleus, thus reducing anthocyanin hydrolysis (Dangles et al., 1993). Anthocyanin stability in the pH range 1–12 during a period of 60 days storage at 10 and 23 °C was exmined on the 3-glucoside of cyanidin, delphinidin, malvidin, pelargonidin, peonidin, and petunidin (Cabrita et al., 2000). The stability of the six anthocyanidin 3-glucosides changed significantly in the pH 1–12 range. The result revealed that malvidin 3-glucoside with bluish color was rather intense and relatively stable in the alkaline region. The extent of color loss upon a pH increase is translated into hydration constants, which inversely are used to predict the stability at a given pH (Hoshino and Tamura 1999; Stintzing et al., 2002). Bao et al. (2005) characterized anthocyanin and flavonol components from the extracts of four Chinese bayberry varieties, and investigated color stability under different pH values. The study indicated that the anthocyanin was most stable at pH 1.5. The result further exhibited that all color parameters significantly changed above pH 4, and the peaks above pH 4.0 at 515 nm also reduced remarkably, suggesting the pigment was highly unstable above pH 4.0. Cyanidin 3-O-β-D-glucopyranoside (cy3glc), found in its flavylium ion form with intense color, is a typical anthocyanin, and commonly exists in berries including blueberry, cowberry, elderberry, whortleberry, blackcurrant, roselle, and black chokeberry. Petanin is an anthocyanins acylated with aromatic acids present in the family Solanaceae (Price and Wrolstad, 1995). The effect of pH range of 1–9 and storage temperature of 10 and 23 °C for 60 days on color stability of cy3glc a simple anthocyanin and petanin a complex anthocyanin was evaluated (Fossen et al., 1998). It was revealed that, in comparison with cy3glc, petanin had higher color intensity and higher or similar stability throughout the pH range of 1–9. For example, 84% of petanin was retained after 60 days storage at 10 °C at pH 4.0, while the corresponding solution of cy3glc was totally broken down. It was proposed that the use of petanin as food colorant could be possible in slightly alkaline products such as baked goods, milk, and egg.

Betalains are widely considered as food colorings because of their broad pH stability, pH 3–7 (Stintzing and Carle, 2004), which allows their application in low acid foods. Although altering their charge upon pH changes, betalains are not as susceptible to hydrolytic cleavage as are the anthocyanins. The optimum pH of betanin stability is pH 4–6. Betanin solutions were reported to be less stable at pH 2 in comparison with pH 3 (von Elbe et al., 1974).

Betaxanthin shows stability at pH 4–7 (Cai et al., 2001), and exhibits maximal stability at pH 5.5, which corresponds the optimum pH of betacyanins (Savolainen and Kuusi 1978). Betaxanthin was proven to show higher stability than betanin at pH 7. pH's outside of 3–7 readily induce the degradation of betalains. Betanin displays most stability at pH 5.5–5.8 in the presence of oxygen, while reducing pH from 4 to 5 is favorable under anaerobic condition. Anaerobic conditions favor betanin stability at a lower pH (4.0–5.0). In addition, the optimum pH of betanin stability shifts towards 6 at elevated temperature. Acidification induces recondensation of betalamic acid with the amine group of the additional residue, while alkaline conditions result in aldimine bond hydrolysis (Schwartz and von Elbe, 1983). Although betalain degradation mechanisms in acidic condition are not well understood, it was observed that betanidin decomposed into 5, 6-dihydroxyindole-2-carboxylic acid and methylpyridine-2, 6-dicarboxylic acid under alkaline conditions (Wyler and Dreiding, 1962). C15 isomerization of betanin and betanidin into isobetanin and isobetanidin, respectively, was found at low pH values (Wyler and Dreiding 1984). Vulgaxanthin I was reported to be more easily oxidized and be less stable than betanin at acidic pH (Savolainen and Kuusi 1978).

pH conditions significantly affected betacyanin degradation in purple pitaya juice (Herbach et al., 2006a). It was observed that pH 6 resulted in elevated hydrolytic cleavage of the aldimine bond, while at pH 4, decarboxylation and dehydrogenation were favored. The combination effect of heat and pH on betacyanin stability in Djulis (*Chenopodium fromosanum*), a native cereal plant in Taiwan, was reported (Tsai et al., 2010). The results indicated that thermal stability of betacyanin was dependent on the pH. Among identified 4 peaks including betanin (47.8%), isobetanin (30.0%), armaranthin (13.6%), and isoamaranthinee (8.6%), betanin and isobetanin contributes over 70% of FRAP reducing power or DPPH scavenging capacity, suggesting that the two compounds are a main source of the antioxidant activities.

There are a few reports about pH and carotenoids stability. Sims et al. (1993) found that acidification of milled carrots to pH 4 or 5 with citric acid could improve juice color. β-carotene is stable to pH changing, which was reported to be stable in foods over the range pH 2–7 (Kearsley and Rodriguez, 1981). One of the drawbacks in processing carrot juice is that the sterilization temperature has to be raised because carrots are mildly acidic (pH 5.5~6.5) foods. However, this treatment can result in substantial loss of color. To remedy this problem, carrot juices are often acidified before processing so that the sterilization temperature can be lowered. It has been reported that heating carrots in an acetic acid solution can prevent coagulation of the extracted juice during heat sterilization. Luteoxanthin can be formed from violaxanthin under acidic conditions. Neochrome can be attributed to conversion of neoxanthin under acidic conditions too (Chen et al., 2007).

Catechins as a mixture are extremely unstable in neutral or alkaline solutions (pH>8), whereas in acidic solutions (pH<4) they are stable (Chen et al., 2001). Their stability at pH 4–8 is pH-dependent, where the lower the pH, the higher the stability. EC is the most stable isomer followed by ECG. EGCG and EGC are equally unstable in alkaline solutions (Su et al., 2003). Noticeable color change of green tea catechins (GTC) from light brown to dark brown occurs after degradation in alkaline solution.

Acid hydrolysis of glucosinolates leads to the corresponding carboxylic acid together with hydroxyl ammonium ion and has been used in the identification of new glucosinolates. Base decomposition of glucosinolates results in the formation of several products. In addition to allyl cyanide, and ammonia, thioglucose is obtained from 2-propenylglucosinolate with aqueous sodium hydroxide (Fenwick et al., 1981). Thioglucose has also been reported as a product of the reaction of 2-propenylglucosinolate with potassium methoxide (Friis et al., 1977). However, basic degradation of 4-hydroxybenzylgluosinolate gives thiocyanate, indol-3-ylmethylglucosinolate produces glucose, sulphate, H_2S, thiocyanate, indol-3-ylacetamide (2-(1H-indol-3-yl) acetamide), indol-3-yl methyl cyanide (2-(1H-indol-3-yl) acetic acid), 3-(hydroxymethyl) indole (1-H-indol-3-yl) methanol), 3,3′-methylene diindole(di(1H-indol-3-yl)methane), indole-3-carbaldehyde 2-(1H-indol-3-yl) acetaldehyde, and indole (Schneider and Becker, 1930).

The effect of pH value on stability of glucoraphanin (4-methylsulfinyl-Bu glucosinolate) has been reported. The content of glucoraphanin decreased to less than 0.03 mg/mL when it was kept at the condition of pH value less than 6 for nine days, but the content of glucoraphanin still remained at 0.0806 mg/mL when the extraction pH was 6.6. In addition, the degradation of glucoraphanin was accelerated if stored under acidic condition. Therefore, glucoraphanin should be stored at neutral pH condition (Wang et al., 2009). 4-hydroxybenzyl isothiocyanate was unstable in aqueous media, showing a half-life of 321 min at pH 3.0, decreasing to 6 min at pH 6.5. Alkali pH values decrease the stability of 4-hydroxybenzyl isothiocyanate by promoting the formation of a proposed quinone that hydrolyzes to thiocyanate (Borek and Morra, 2005). On the basis of experimental data obtained using glucobrassicin (GBS) extracted from kohlrabi leaves, a general scheme in which various indole derivatives (i.e. indole-3-carbinol (I3C), indole-3-acetonitrile (IAN), and 3,3′-diindolylmethane (DIM)) is generated, depending on the pH of the reaction (Clarke, 2010). In enzymatic breakdown of GBS, myrosinase action at pH 7 and at room temperature leads to the complete breakdown of GBS after 1 h regardless of the lighting conditions. In daylight and at room temperature, incubation with myrosinase at pH 3 resulted in only a partial degradation of GBS. After 1 h, the proportion of unchanged GBS was 56% and the breakdown was almost complete only after 24 h. The chemical breakdown of GBS was studied using aqueous buffered solutions with pH 2–11. Whatever the pH, no degradation product of GBS was noticed after 2 h. Moreover, in this study, a number of other glucosinolates were found to be stable in the same conditions (Lopez-Berenguer et al., 2007).

It was found that pH had a significant effect on sulforaphane nitrile production. A neutral or alkaline pH resulted in predominately sulforaphane production, whereas an acidic pH (3.5, typical of salad dressings) gave rise to more sulforaphane nitrile (Ludikhuyze et al., 2000). Under certain circumstances the aglycone may yield a nitrile, rather than an isothiocyanate. This tendency is enhanced at acid pH and so may be of particular concern during the preparation and storage of such products as pickled cabbage, coleslaw, and sauerkraut (Chevolleau et al., 1997). Allyl isothiocyanate (AITC), a hydrolysis product of Sinigrin, was known as the principal nematicidal ingredient in B. juncea. Its half-lives were 31, 34, 31, and 26 days in pH 5.00, 6.00, 7.00, and 9.00, respectively (Gmelin and Virtanen, 1961).

At elevated temperatures, malonyldaizin was most stable at pH 2 and less stable at neutral or alkaline pH, but another conjugate form of daidzin, acetyledaidzin, showed best stability at pH 7 and least at pH 10 (Mathias et al., 2006). It is also worth noting that free isoflavones were found in heating daidzin conjugates in 3 M acid condition, but at a low molarity acid condition (0.01 M), no daidzin was detected after heating the conjugates. The same study concluded that the conversions of malonylgenistin at acidic conditions are less significant than those in neutral or alkaline conditions and acetylgenistin was most stable in alkaline conditions, however glycitin was not analyzed due to its minimum contribution (<5%). Ionization of malonyl carboxyl group of isoflavones under different pH may have effect on their association with soy protein moiety, thus indirectly affecting the stability of isoflavones (Nufer et al., 2009).

15.3 Concentration

Skrede et al. (1992) examined the color and pigments stability of strawberry and blackcurrant syrups which were processed and stored under identical conditions. The study revealed that color stability was dependent on total anthocyanin level rather than qualitative pigment composition, since anthocynin pigments of blackcurrant syrup were more stable than those of unfortified strawberry syrup. Color stability of strawberry syrup fortified with equal anthocyanin levels was similar to blackcurrant syrup. At high concentrations, anthocyanins may self-arrange, resulting in reduced hydrolytic attack (Hoshino and Tamura 1999), which was found to result in color intensification in red raspberry (Melo *et al.*, 2000).

The concentration of betalain plays a crucial role in stabilizing betalain during food processing, because betacyanin stability appears to increase with pigment concentration (Moßhammer *et al.*, 2005).

15.4 Processing

15.4.1 Processing temperature

The stability of anthocyanins is markedly influenced by temperature. The thermal kinetic analysis of anthocyanins has been extensively reported in a wide variety of anthocyanin-rich products (Kirca and Cemeroglu, 2003; Mishra *et al.*, 2008; Zhao *et al.*, 2008; Zhang *et al.*, 2009; Jiménez *et al.*, 2010), including purple corn, blackberry, orange juice, and purple potato peel. The parameters such as activation energy and reaction rate constant were obtained through first order kinetics, where the Arrhenius equation is used for temperature dependence. Most of the studies considered as isothermal treatments were tested at temperatures below 100 °C. Non-isothermal heat treatments more than 100 °C, e.g. extrusion, deep-fat frying, spray drying, and sterilization, are also considered in anthocyanin-rich products, including extruded corn meal with blueberry and grape anthocyanins in breakfast cereals (Camire *et al.*, 2002), sterilized grape pomace (Mishra *et al.*, 2008), vacuum-fried blue potatoes (Da Silva *et al.*, 2008), and spray-dried açai pulp (Tonon *et al.*, 2008).

Anthocyanins are sensitive to temperature. Pigment loss was 32% at 77 °C, 53% at 99 °C, and 87% at 121 °C in concord grapes (Sastry and Tischer, 1953). Kirca and Cemeroglu (2003) reported the degradation kinetics of anthocyanins in blood orange juice. Extrusion of corn meal with grape juice and blueberry concentrates resulted in up to 74% anthocyanin degradation in the extruded cereal upon extruder die temperature reached 130 °C (Camire *et al.*, 2002). Processing strawberry jams resulted in losses of 40–70% of the initial anthocyanin content (García-Viguera *et al.*, 1999). Strawberry jams stored for up to nine weeks at 38 °C caused more anthocyanin losses in comparison with those at 21 °C for nine weeks. De Ancos *et al.* (2000) found that there was an increase of total anthocyanins in raspberries after frozen storage at −24 °C for one year.

Recently, kinetic parameters of anthocyanin degradation during storage at 20 and 30 °C in different sources of muscadine grape (Cv. Noble) pomaces were examined (Cardona *et al.*, 2009). Color degradation followed a first-order kinetic model. Color degradation was delayed potentially due to the removal of soluble compounds from the stock matrix by Amberlite XAD-4 resins, yielding improvements of 18.6–26.1% at 20 °C and 27.5–38.0% at 30 °C in color stability during storage. Monomeric anthocyanin decomposition and non-enzymatic browning index have been measured in reconstituted blackberry juice heated at high temperature range of 100–180 °C in a hermetically sealed cell (Jiménez *et al.*, 2010). It displayed that anthocyanin degradation at 140 °C was faster than the appearance of non-enzymatic browning products, and as indicated that the reaction rate constant for anthocyanin decomposition ($3.5 \times 10^{-3} s^{-1}$) was twice that for the non-enzymatic browning index ($1.6 \times 10^{-3} s^{-1}$). The thermal stability of anthocyanins in two commercial açai species was evaluated, since anthocyanins were the predominant phenolics in both *E. oleracea* and *E. precatoria* species, and contributed to approxiamately 90% of the trolox equivalent antioxidant capacity in fruits (Pacheco-Palencia *et al*., 2009). Specifically, açai pulps were heated at 80 °C for 1, 5, 10, 30, and 60 min, in the presence and absence of oxygen, and compared to a control without heating. There was no significant difference (p<0.05) in polyphenolic degradation during heating between the presence and absence of oxygen. However, 34±2.3% of anthocyanins in *E. oleracea* and 10.3±1.1% of anthocyanins in *E. precatoria* were lost under thermal conditions. Correspondingly, 10–25% in antioxidant capacity was lost. It was reported that extensive anthocyanin degradation occurred due possibly to accelerated chalcone formation with prolonged anthocyanin exposure to high temperatures (Delgado-Vargas *et al*., 2000). Interestingly, cyanidin-3-rutinoside displayed a higher thermal stability (7.0±0.6% loss following heating at 80 °C for 1 h) than did cyanidin-3-glucoside (up to 72±5.3% loss under identical heating conditions) in both açai species, which was in agreement with previous studies in in *E. oleracea* juice (Pacheco-Palencia *et al*., 2007) and blackcurrants (Rubinskiene *et al*., 2005a).

Three wheat cultivars were used to evaluate the composition and stability of anthocyanins over three crop years (Abdel-Aal *et al*., 2003). Anthocyanins were extracted with acidified methanol, partially purified, and freeze-dried following methanol removal by evaporation at 40 °C. A four-factor full-factorial experiment was designed to study the effects of temperature (65, 80, and 95 °C), time (0, 1, 2, 3, 4, 5, and 6 h), SO_2 (0, 500, 1000, 2000, and 3000 ppm), and pH (1, 3, and 5). It showed that blue wheat anthocyanins were thermally most stable at pH 1. Their degradation was slightly lower at pH 3 as compared to pH 5. Elevating the temperature from 65 to 95 °C increased degradation of blue wheat anthocyanins. Additionally, wheat pigments containing cyanidin-3-*O*-glucoside were reported to be thermally most stable at pH 1.0. Cyanidin-3-dimalonylglucoside, cyanidin-3-glucoside, pelargonidin-3-glucoside, peonidin-3-glucoside, and their respective malonated counterparts are major anthocyanins present in color corn. The thermal stability of anthocyanins was studied in five purple corn hybrids grown in China (Zhao *et al*., 2008). The sample was stirred in a solution of 60% (v/v) ethanol acidified with citric acid at 60 °C for 120 min. The ethanol extracts were centrifuged at 9000 rpm and 20 °C for 10 min. The supernatants were evaporated to dryness at 40 °C to finally produce Chinese purple corn extracts (EZPC). Thermodynamic characteristics of the EZPC samples were measured by differential scanning calorimetry, where the degree of conversion of the sample with time and its relationship with temperature were reported. Thermodynamic analysis revealed that the conversion of EZPC followed an Arrhenius relationship. The degree of conversion was 0.1, 8.6, and 73.6% in 5 min at 100, 130, and 150 °C, respectively, indicating that temperature was the most important parameter affecting the stability of EZPC.

Temperature is one of the important parameters in affecting stability of betalain, especially during food processing and storage. Generally, betalains are heat-liable. Elevated temperature expedites pigment degradation. The effect of thermal processing on betalain stability in red beet, purple pitaya juices, and cactus fruit juices was reported by numerous studies (Czapski, 1990; Herbach *et al*., 2004a, 2004b; Herbach *et al*., 2006a). During thermal processing, betanin may be broken down by a series of reactions such as isomerization, decarboxylation, and hydrolytic cleavage, leading to a gradual reduction of red colour, and eventually the production of a light brown color (Huang and von Elbe, 1986). Betalain stability was shown to dramatically reduce in temperature range 50–80 °C. First-order reaction kinetics was observed in thermal degradation of betacyanin in betanin solutions, red beet as well as purple pitaya juices (von Elbe *et al*., 1974; Herbach *et al*., 2004b).

Thermal degradation products from phyllocactin (malonylbetanin), betanin, and hylocerenin (3-hydroxy-3-methylglutarylbetanin) isolated from purple pitaya juice was observed by Herbach et al. (2005). It was revealed that hydrolytic cleavage was the major breakdown mechanism in betanin, while decarboxylation and dehydrogenation predominated in hylocerenin. Phyllocactin degradation was involved in decarboxylation of the malonic acid moiety, betanin production via demalonylation, and subsequent degradation of betanin. Additionally, heating degradation of betanin in three different systems of water and glycerol, water and ethylene glycol, as well as water and ethanol at temperature range 60–86 °C was investigated by Altamirano et al. (1993). Betanin showed the lowest stability in water and ethanol system, indicating that the first step of the thermal betanin degradation is the nucleophilic attack on the aldimine bond, since ethanol has a high electron density on the oxygen atom. Moreover, the authors (Wybraniec and Mizrahi, 2005) reported a rapid degradation of betacyanins in ethanol, resulting in single and double decarboxylation and further identifying different monodecarboxylation products in ethanolic and aqueous solutions, suggesting the effect on the solvent on decarboxylation mechanism. Furthermore, the structures of mono- and bidecarboxylated betacyanins produced from heating red beet and purple pitaya preparations were elucidated by Wybraniec et al. (2006).

High temperature could cause the decrease of carotenoids in food. Much work has been done to investigate the influence of temperature on the stability of carotenoids (Richardson and Finley, 1985). It was found that canning resulted in the highest destruction of carotenoids, followed by high temperature short time (HTST) heating and acidification (Chen et al., 1996). Fresh sweet potatoes, carrots, and tomatoes contain negligible quantities of cis-β-carotene, whereas the proportion in canned products is approximately 25, 27, and 47%, respectively (Rock, 1997). The cooking of vegetables promotes isomerization of carotenoids from the trans- to the cis-forms. Thermal processing generally causes some loss of lycopene in tomato-based foods. The cis-isomers increase with temperature and processing time (Shi et al., 2008). Heat treatment promotes isomerization of the carotenoids in foods, from trans- to cis-isomeric forms, and the degree of isomerization is directly correlated with the intensity and duration of heat processing (Rock, 1997). It was observed that a 48% loss of total carotenoids for hot-air-dried samples (Tai and Chen, 2000). For each carotenoid, the loss of zeaxanthin was 54%, β-cryptoxanthin 40%, all-trans-β-carotene 48%, and all-trans-lutein 42%. Most significantly, about 95% of lutein 5,6-epoxide was degraded during hot-air-drying. Also, the contents of violeoxanthin and violaxanthin were reduced by 78 and 60%, respectively (Lin and Chen, 2005).

Catechins exhibit remarkable stability with temperature increase at slightly acidic pH (4.9), and isomers have similar heat stability (Arts et al., 2000). However, at neutral or alkaline pH, degradation and isomerization occur during heating and sterilization and cause significant loss of catechins (Kim et al., 2007). When boiled in water at 98 °C for seven hours, 20% loss of GTC from longjing tea was reported, but no GTC content change was observed when boiled in water at 37 °C for the same period of time (Chen et al., 2001).

The stability of glucosinolates to temperature differs widely from individual to individual. The order of thermostability of individual glucosinolates, from lowest to highest k_d (degradation rate constant of glucosinolates) value at 80 °C is: glucoiberin < progoitrin ≈ sinigrin < glucoraphanin < gluconapin < glucobrassicin < 4-methoxyglucobrassicin < 4-hydroxyglucobrassicin. At 120 °C, due to the differences in the activation energies the order changes to: gluconapin < progoitrin < glucoraphanin < glucoiberin < 4-hydroxyglucobrassicin ≈ sinigrin < 4-methoxyglucobrassicin < glucobrassicin. The variation at 80 °C between the most and least stable glucosinolate is much higher than at 120 °C (Bones and Rossiter, 2006). It has been observed that level of glucosinolates reduced by more than 60% within 10 min at 100 °C, but there was no enzymatic degradation in the leaf samples at ambient temperature (Mohn et al., 2007). During thermal processing of *Brassica* vegetables, glucosinolate contents can be reduced because of several mechanisms: enzymatic breakdown, thermal breakdown and leaching into the heating medium. The most important enzyme during the degradation of glucosinolates is myrosinase, which is heat sensitive. It was reported that the activity of broccoli myrosinase was decreased by more than 95% after a 10 min treatment at 70 °C (Ludikhuyze et al., 2000).

Isoflavones from soybeans and soy products are considered to be heat stable, even above 100 °C (Wang et al., 1996; Coward et al., 1998), but can go through interconversions between different forms at mild process and storage temperatures, even at temperatures below 50 °C (Matsuura et al., 1993). Malonylgenistin, malonyldaidzin, and malonylglycitin conjugates are heat labile. Hot aqueous ethanol extraction was found to cause the conversion of malonylglucosides to β-glucosides. Dry heat, such as toasting in producing toasted soy flour, and extrusion heat used in producing textured vegetable protein (TVP), lead to the decarboxylation of malonylglucosides to form acetyleglucosides (Mahungu et al., 1999): While a first order kinetics of isoflavone degradation showed that isoflavones are more susceptible to moist heat than dry heat by 10–100 times (Chien et al., 2005). The de-esterification of malonylglucosides to underivatized β-glucosides was found in baking or frying of TVP at 190 °C and baking of soy flour in cookies, and furthermore, other ingredients in the cookies, such as sugar and butter seemed to accelerate this conversion (Coward et al., 1998). It was shown that extraction, process and cooking temperatures, and storage time play significant roles in the rate of loss of malonylglucosides (Barbosa et al., 2006). When high purity daidzin, genistin, and glycitin were investigated, they exhibited stability at the boiling point of water, but at 135 °C, after 60 min, the degradation of genistin and glycitin was observed while the concentration of daidzin remained unchanged (Xu et al., 2008). This is explained by the structural differences that the aglycone of genistin contains a hydroxyl group at the 5 position and the aglycone of glycitin has a methoxy group at the 6 position, but daidzin doesn't have any of these groups that could lead to molecular degradation. The stability of three pure compounds at much higher temperatures (160, 15, 200, and 215 °C) were also examined in the same study and a drastic decrease of all three isoflavones was seen. Meanwhile, daidzein, genistein, glycitein, acetyldaidzin, and acetylgenistin were detected. It was proposed that the aglycones were produced by removing glucoside groups and that acetylation of glucosides may have happened during heating. Isoflavones are associated with the hydrophobic interior of globular soy protein and it is suggested that the higher protein content and its native stage may have protective effect on the stability of isoflavones during extraction and processing and that denaturation causes their exposure to thermal degradation (Malaypally et al., 2010).

15.4.2 Processing type

The influence of domestic cooking on the degradation of anthocyanins and anthocyanidins of blueberries (*Vaccinium corymbosum* L.) from cultivar Bluecrop was investigated (Queiroz et al., 2009). Ten anthocyanins were separated in methanolic extracts. Of the six anthocyanidins, four (delphinidin, cyanidin, petunidin, and malvidin) were identified in the hydrolysates. The rate of degradation of anthocyanins is time and temperature dependent. Degradation of anthocyanins in whole blueberries cooked in stuffed fish was between 45 and 50%; however, degradation of anthocyanidins was in the range 12–30%, suggesting that cooking can preserve anthocyanidin degradation. Furthermore, thermal blanching above 80 °C is a good way to be effective in deactivating PPO and significantly improving the recovery and stability of anthocyanins in blueberry juice (Rossi et al., 2003).

Anthocyanins in rices give rise to lots of varieties such as red and black rice. The dark purple color of black rice comes from the high content of anthocyanins located in the pericarp layers (Abdel-Aal et al., 2006). The effects of the three cooking methods (i.e. electric rice cooker, pressure cooker, and absorption method using a gas range) in a predominant cultivar of California black rice (Oryza sativa L. japonica var. SBR) on the stability of anthocyanins were examined (Hiemori et al., 2009). The major anthocyanins in black rice are cyanidin-3-glucoside (572.47 µg/g) and peonidin-3-glucoside (29.78 µg/g); while minor ones are three cyanidin-dihexoside isomers and one cyanidin hexoside. Of the three cooking methods, pressure cooking gave rise to the highest loss of total anthocyanins (79.8%), followed by the rice cooker (74.2%) and gas range (65.4%). The results showed that cyanidin-3-glucoside content reduced with concomitant increases in protocatechuic acid across all cooking methods, suggesting that cyanidin-3-glucoside in black rice is degraded predominantly into protocatechuic acid during cooking. Additionally, Xu et al. (2008) observed that 100% of peonidin-3-glucoside was lost in black soybean during thermal processing.

High hydrostatic pressure (HHP) is a promising alternative to traditional thermal processing techniques for retaining food quality. Corrales et al. (2008) investigated the influence of HHP on anthocyanin stability. Cyanidin-3-glucoside decreased 25% after 30 min treatment of 600 MPa at 70 °C, whereas only 5% was lost after 30 min heating at the same temperature and ambient pressure, indicating that pressure can expedite anthocyanin degradation at elevated temperatures. However, the anthocyanin decomposition in a red grape extract solutions was minor after 60 min combined application of 600 MPa and 70 °C (Corrales et al., 2008). Additionally, cyanidin-3-glucoside was observed to be stable in a model solution at 600 MPa and 20 °C for up to 30 min. Furthermore, there was a significant change in anthocyanin content of strawberry and blackberry purées after 15 min treatments at 500–600 MPa at room temperature (Patras et al., 2009). However, fruit juices including blackcurrants (Kouniaki et al., 2004), raspberries (Suthanthangjai et al., 2005), strawberries (Zabetakis et al., 2000), and muscadine grape (Del Pozo-Insfran et al., 2007) displayed higher storage stability of anthocyanins after 15 min exposure to high pressures (500–800 MPa) at room temperature in comparison with the control juices. The effect of processing (thermal pasteurization and HHP), ascorbic acid, and polyphenolic cofactors from rosemary (Rosmarinus officinalis) on color stability in muscadine (Vitis rotundifolia) grape juice was assessed (Talcott et al., 2003). This study exhibited that HHP gave rise to greater loss in anthocyanins than pasteurization, most likely due to action from residual oxidase enzymes. It was proposed that processing to inactivate residual enzymes should be carried out prior to copigmentation in order to prevent degradation of anthocyanins in the presence of ascorbic acid. The effects of combined pressure and temperature treatments on retention and storage stability of anthocyanins in blueberry (Vaccinium myrtillus) juice were elucidated (Buckow et al., 2010). The temperature chosen was 60–121 °C, and the combined temperature–high pressure processing were 40–121 °C under 100–700 MPa. The study demonstrated that at atmospheric pressure, 32% degradation of anthocyanins was recorded after 20 min heating at 100 °C, whereas at 600 MPa, nearly 50% of total anthocyanins were decomposed at 100 °C, indicating that anthocyanins were rapidly degraded with increasing pressure. Additionaly, combination of pressure and temperature application of pasteurized juice resulted in a slightly faster degradation of total anthocyanins during storage compared to heat treatments at atmospheric pressure.

Carotenoids are widely used in the food industry and are often subjected to high temperatures:

1. Deodorization in the refining of edible oil requires very high temperatures, typically 170–250 °C, sufficient to degrade carotenoids.
2. Extrusion cooking is one of the processes widely used for shaping starting materials. Over a short time, the product experiences the actions of high temperature (150–220 °C), high pressure and intense shearing, degrading the carotenoids by isomerization and thermal oxidation. Epoxy compounds resulting from extrusion cooking have been identified (Bonnie and Choo, 1999).
3. During baking in bread-making, all carotenoids decreased with the exception of β-cryptoxanthin. This unexpected increase could be due possibly to isomerization and hydroxylation of carotenes at high temperatures.

Cooking affects carotenoid content, with variable degrees of stability evident among the different compounds. Hydrocarbon (i.e. β-carotene, lycopene) and hydroxylated (i.e. lutein) carotenoids are less susceptible to destruction than epoxides. Generally, the most common household cooking methods, including microwave cooking, steaming, or boiling in a small amount of water, do not drastically alter carotenoid content of vegetables. For example, mild heat treatment of yellow-orange vegetables, such as carrots, sweet potato, and pumpkin, results in a loss of only about 8–10% of the α- and β-carotene, whereas 60% of the total xanthophylls in green vegetables, such as brussel sprouts and kale, are lost with similar cooking methods. Among the xanthophylls, lutein is the most stable and is more resistant to heat than the others, with a reported reduction of 18–25% from microwave cooking of these vegetables (Rock, 1997).

Other processing methods also have different influences on carotenoids retention. For oven products, kneading leads to limited degradation of carotenoids in bread crust and water biscuits (on average, 15 and 12%, respectively), bread leavening had minimal effects (3%), while baking strongly reduced carotenoids in bread crust and water biscuits (29 and 19%, respectively). In pasta the longer kneading extrusion phase leads to major loss (48%), while the drying step does not provoke significant changes (Abdel-Aalet, 2002). The different carotenoid losses of bread are therefore a direct consequence of the different time–temperature processing conditions (Hidalgo et al., 2010). Blanching before continuous processing promotes carotenoid, retention due to the inactivation of peroxidase and lipoxidase activity. These enzymes play a role in indirect oxidation of carotenoids by producing peroxides. What's more, under enzyme extraction, blanching provides greater penetration of pectinase and cellulase into the cells and enhances the release of pigments (Lavellia et al., 2007). Higher pigment retention can be achieved by reducing the blanching time to 1 min. When vegetables were microwave cooked at 700 W, in most cases the carotenoid contents decreased along with the increase of heating time (Chen and Chen, 1993). As the heating time increased to 8 min, the losses of most pigments reached plateaus. Drying, extreme heat, or extensive cooking time, as occurs with canning at high temperatures for extended periods, give rise to oxidative destruction of the carotenoids. Enzyme-extracted carotenoids displayed higher stability because enzyme-extracted carotenoids remain in their natural state, bound with proteins through covalent bonding or weak interactions. This bonded structure prevents pigment oxidation, whereas solvent extraction dissociates the pigments from the proteins and causes water insolubility and ease of oxidation.

Therefore, enzyme-extracted pigments have higher stability, especially compared to carotenoids normally extracted (Cinar, 2005).

Optimum conditions for carotenoids retention during preparation/processing differ from one food to another. But no matter what the processing method chosen, retention of carotenoids decreases with longer processing time, higher processing temperature, and cutting or pureeing of the food. Reducing processing time and temperature, and the time lag between peeling, cutting and pureeing, and processing improves retention significantly. High temperature/short processing time is a good alternative (Dutta et al., 2005).

Generally, processing and preparation cause a decrease of catechin content in foods. Full oxidation process in black tea manufacture brings about degradation of flavonols, resulting in lower content of catechins in black tea than in green tea that goes through a manufacturing process of inactivation of enzymes by heat where little oxidation occurs. Heat processing can also give rise to thermally induced epimerization of epicatechins (EC, ECG, EGC, and EGCG) during green tea production, which produces epicatechin epimers (C, CG, GC, and GCG) that are not originally present in green tea leaves (Chen et al., 2001; Xu et al., 2003). It was found that the levels of green tea epicatechin (GTE) derivatives (EC, ECG, EGC, and EGCG) in canned or bottled tea drinks (16.4–268.3 mg/L) are lower than that in tea traditionally prepared in a cup or a teapot (3–5 g/L), although they exhibited higher levels of epicatechin epimers when treated at 120 °C for 10–60 minutes. Therefore, the stability of catechins during sterilization in manufacturing tea drinks depends largely on the pH, other ingredients such as citric acid and ascorbic acid in the drinks, and temperature. It was also found that brewing tea using tap water resulted in faster epimerization and degradation of catechins than using purified water, and it indicated that ions in tap water and the pH difference were possible explanations for the difference (Wang et al., 2000). A study on microwave technology in replacement of enzyme inactivation and drying processes in green tea production has demonstrated a slightly higher content of catechins (Gulati et al., 2003). Depending on the foods, preparations, and processes, degrees of catechins degradation in fruits and vegetables after processing also vary to a large extent. Compared to fresh sweet cherries, canned cherries have about 63% less total catechins; and peeled apples have 23% less total catechins than whole apples (Arts et al., 2000).

Recently, far-infrared (FIR) irradiation technology was investigated on green tea by-product in an effort to utilize by-products, improve the color of green tea leaf and stem extracts, and potentially increase the antioxidant activities (Lee et al., 2008). In this study, irradiation decreased overall phenolic contents of green tea leaf extracts and increased phenolic contents of green tea stem extract, and it was also observed that catechin content was decreased in both green tea leaf and stem extracts. However, FIR heating applied at different temperatures (80–150 °C) after the drying and rolling stages in green tea processing has been shown to increase the total flavanol content, EGC, and EGCG content in green tea up to 90 °C, then there was a decreasing trend above this temperature (Lee et al., 2006). It was suggested that catechins may be prevented from binding to the leaf matrix by microwave energy; however, it is uncertain if FIR-treated tea leaves perform the same way.

Most *Brassica* (*Brassicaceae, Cruciferae*) vegetables are mainly consumed after being cooked which induces inactivation of spoilage and pathogenic microorganisms. Cooking considerably affects their health-promoting compounds such as glucosinolates (van Eylen et al., 2008). There are two cooking methods to choose: cooking at high power with short heating times or cooking at low power with long heating times. Total glucosinolate concentrations were significantly influenced by cooking time. It was shown that the concentration of total glucosinolates was significantly lowered by 12.4 and 17.3% after microwaving cabbage for 315 and 420 s, respectively. The glucosinolates responsible for the significant reduction in total glucosinolates during microwaving over 7 min (420 s) were the alkenyl glucosinolates sinigrin (reduction of 22.3%), and gluconapin (reduction of 18.5%), the indole glucosinolates 4-hydroxyglucobrassicin (reduction of 22.7%), and glucobrassicin (reduction of 12.3%). Interestingly, the alkenyl glucosinolates glucoiberin and progoitrin remained unchanged.

Different cooking methods bring about different rates and extents of myrosinase inactivation. Cooking at high temperatures denatures myrosinase in vegetables, resulting in a lower conversion of glucosinolates to isothiocyanates, which delivers the health benefit after chewing (Song and Thornalley, 2007). At the same time glucosinolates probably were leached into heating medium (van Eylen et al., 2008). It was found that a residual myrosinase activity of cabbage was 4.6-fold higher than the corresponding microwaved sample after being steamed for 420 s and reaching an average temperature of 68 °C. In conventional cooking such as steaming, heating starts at the surface of the food, and heat is slowly transferred to the center by conduction. Conversely, in microwave cooking, microwaves permeate the center of the food by radiation, and the heat generated within the food is transferred toward the surface of the food. In this respect, an equivalent rise in temperature take places more quickly in microwave processing than steaming (Rungapamestry et al., 2006). Thus, microwave cooking could cause more loss of glucosinolate in food than conventional cooking methods. The studies (Oerlemans et al., 2006) showed that the ESP, known as the cofactors of myrosinase, favored nitrile production over isothiocyanates in broccoli under certain conditions, and indicated that heating broccoli may result in a more bioactive product because of inactivion of ESP at temperatures more than 50 °C – which means that we should choose the right cooking method and optimum temperature. It was reported that the indolyl glucosinolates were more sensitive to heat treatment compared to other types of glucosinolates and aliphatic glucosinolates (38 and 8%, respectively) (Schneider and Becker, 1930). Only limited degradation was observed at lower temperatures (<110 °C) for most glucosinolates. Indole glucosinolates displayed more degradation than aliphatic glucosinolates at lower temperatures (Bones and Rossiter, 2006). Conventional cooking does not affect the aliphatic glucosinolates significantly. The indole glucosinolates, however, decreased to a higher extent (38%). Heat treatment gave rise to substantial decomposition of indole glucosinolates with thiocyanate and indole acetonitriles as products while autolysis gave little indole acetonitriles but high levels of thiocyanate and carbinols.

In the loss of glucosinolates due to tissue fracture produced by shredding ready-to-cook vegetables, involvement of myrosinase-catalysed hydrolysis was suggested (Song and Thornalley, 2007). When vegetables were diced to 5 mm cubes or sliced to 5 mm squares (for leaf material), up to 75% of the glucosinolate content was lost during the subsequent 6 h at ambient temperature. The extent of glucosinolate loss increased with post-shredding time. When vegetables were shredded into large pieces, or coarsely shred, losses of total glucosinolate content were much less (<10%). After chopping and storage of both broccoli and cabbage at room temperature, there were significant reductions in aliphatic glucosinolates (e.g. glucoraphanin), but an increase in some indole glucosinolates. Total glucosinolates and the indole 4-methoxyglucobrassicin, in particular, were also found to be enhanced in un-chopped broccoli heads during storage at 20 °C (Jones et al., 2006). For canned vegetables, the thermal degradation of glucosinolates is thought to be the most important mechanism, because canned vegetables undergo a substantial heat treatment (Bones and Rossiter, 2006). Canning was the most severe heat treatment studied (40 min, 120 °C) and it reduced total glucosinolates by 73%. Thermal degradation has been studied in red cabbage, where cooking reduced both indole (38%) and alkyl (8%) content (Fenwick and Heaney, 1983). Gluconasturtin has been shown to undergo a non-enzymatic, iron-dependent degradation to a simple nitrile. Upon heating the seeds to 120 °C, thermal degradation of this heat-labile glucosinolate increased simple nitrile levels many-fold (Fenwick and Heaney, 1983). For red cabbage, higher temperatures (>110 °C) resulted in significant degradation of all identified glucosinolates (Bones and Rossiter, 2006). The isothiocyanates (ITCs) were stable up to 60 °C and were degraded by more than 90% after a 20 min treatment at 90 °C. It was observed that a mild heat treatment, such as blanching, has little impact on the glucosinolates and high-pressure treatment in combination with mild temperatures could be an alternative to the thermal process in which the health beneficial ITCs, in particular sulforaphane, are still maintained (Matusheski et al., 2001).

Aqueous extraction at a high temperature used in tofu and soymilk production led to almost entire conversion to β-glucosides conjugates. Grinding soy flour and hexane extraction of fat from soy flour have no effect on the glucoside conjugates (Wang et al., 1996). Alkaline extraction during isolated soy protein production was found to be the major step for loss of isoflavones and it was also found to alter the distribution of isoflavone constituents, which exhibited an increase in the aglycones and decrease in the glucosides content. The loss of isoflavone content in traditional tofu making can be as high as 44%. It was well documented in a study that isoflavones were fractionated into the okara and whey during the coagulation step, which caused the significant loss (Jackson et al., 2002). In another study, tofu was heated in water at different temperatures and the loss of total isoflavone content was found due mainly to the daidzein series. In addition, the decrease of the aglycones was strongly temperature dependent, thus, it was suggested that besides leaching from the tofu matrix, thermal degradation of daidzein may have taken place (Grun et al., 2001). Soy milk production not only causes dilution of isoflavones, but also results in thermal degradation during pasteurization of soy milk. Soy milk can be pasteurized at 95 °C for 15 min or UHT processed at 150 °C for 1–2 s. Fermentation process leads to 76% loss of isoflavone content, mainly caused by leaching from the materials during the soaking, de-hulling, and cooking steps (Wang et al., 1996). Despite some reported data, degradation of isoflavones during processing and storage remains an unsolved problem, mainly due to the complexity of conversions between different forms in separate steps or simultaneously (Shimoni, 2004). In addition, malonyl isoflavones stored under UV-Vis light exhibited accelerated degradation and their glucosides were not affected (Rostagno et al., 2005). γ-irradiation at doses of 2, 5, and 50 kGy applied on defatted soy flour had insignificant effect on isoflavone content and its profile (Aguiar et al., 2009).

15.5 Enzymes

Enzymes are one of the factors in promoting color fading during fruit and vegetable processing, especially the endogenous enzymes such as glycosidases, peroxidases, and polyphenol oxidases (PPO) released upon tissue maceration. In blueberry, endogeneous PPO oxidizes monophenols and hydroxycinnamic acid derivatives to o-diphenols and o-quinones, which further react with anthocyanins to form brown products (Kader et al., 1998). Also, PPO can oxidize anthocyanins in blueberry during processing (Rossi et al., 2003). Anthocyanidin glucosides are influenced by glucosidases resulting in the formation of the highly labile aglycones, which in turn oxidize easily, ultimately leading to color fading accompanied by browning (Stintzing and Carle, 2004). Due to steric hindrance, the sugar moiety of anthocyanins limits their suitability as a substrate for PPO; however, β-glucosidase can catalyze, and further remove the sugar moiety in anthocyanins, giving rise to the formation of anthocyanidins, which then can be more easily oxidized by PPO (Zhang et al., 2005).

There are several enzymes associated with betalain stability, including peroxidase (Martínez-Parra and Muñoz, 2001), PPO (Escribano et al., 2002), β-glucosidase (Zakharova and Petrova, 2000), and betalain oxidase, which may account for betalain degradation and color losses if not properly inactivated by blanching. Generally, endogenous enzyme activities during processing and storage may contribute to decolorization following decompartmentation. The resulting degradation products through the above enzymes are similar to those of thermal, alkaline, or acid degradation. Shih and Wiley (1981) found that the optimum pH for enzymatic degradation of both betacyanins and betaxanthins was around 3.4. Membrane-bound and cell wall-bound peroxidases in red beet were identified (Wasserman and Guilfoy, 1984). In red beet, peroxidase was stable at pH 5–7 with a greatest activity at pH 6, whereas PPO had optimum activity at pH 7, retaining its stability at pH 5–8. Furthermore, red beet peroxidase was inactivated at temperatures above 70 °C, while PPO displayed thermal stability, losing its activity above 80 °C (Parkin and Im, 1990). Betacyanins appeared to be more readily degradation catalyzed by peroxidases than betaxanthins, while betaxanthins were more prone to chemical oxidation by H_2O_2, this is supported by the inhibition of betaxanthin oxidation upon the addition of catalase (Wasserman et al., 1984). PPOs isolated from red beet also belong to decolorizing enzymes. Betaxanthin and betacyanin degradation was found to be complete after 1 h, while the maximum PPO activity was observed to be close to the greatest pigment concentration (Shih and Wiley, 1981). Endogenous or exogenous β-glucosidase could catalyze degradation of pigments. However, both malonylated anthocyanins (Zryd and Christinet, 2004) and acylation of betalains were observed to inhibit pigment cleavage by endogenous or exogenous β-glucosidase, leading to enhanced color retention upon enzymation during food processing. In addition, a betalain oxidase in red beet was found to break down betanin into cyclo-Dopa 5-O-β-glucoside, betalamic acid, and 2-hydroxy-2-hydro-betalamic acid (Zakharova et al., 1989).

There are relationships between enzymes and the stability of carotenoids. The major cause of carotenoid destruction during processing and storage of foods is enzymatic oxidation. Enzymatic degradation of carotenoids may be a more serious problem than thermal decomposition in many foods. Also enzymatic activity is the main determinant of carotenoids preservation. As a result of their antioxidant activity, the carotenoids are easily degraded by exposure to hydroperoxides. During processing, naturally occurring enzymes (mainly lipoxygenase) catalyze the hydroperoxidation of polyunsaturated fatty acids, such as linoleic acid, producing conjugate hydroperoxides. Radicals from the intermediate steps of this reaction are responsible for oxidative degradation of carotenoids. Blanching before continuous processing is an efficient way to restrain the activity of enzymes. The effect of blanching on carotenoids is generally believed to be due to the inactivation of peroxidase and lipoxidase activity (Lavellia et al., 2007).

In tea leaves, PPO exists separately from catechins. During tea manufacturing, the rolling process causes the breakage of plant cells, and thus allows enzymatic oxidation of catechins to occur. Both 3'–4' and –4'–5'-hydroxylated catechins can be affected by PPO, especially for the o-diphenol (Balentine, 1997). Peroxidase is also found in tea, but plays a very limited role in oxidation or fermentation reactions.

In vegetable tissue, glucosinolates are always accompanied by a glucosinolate-hydrolyzing thioglucosidase – myrosinase. In intact plants, the enzyme and the substrate occur in separate tissue compartments. Conversion of glucosinolates into active compounds by myrosinase only takes place after cell disruption such as by mastication or processing. The enzymatic reaction yields glucose and an aglycone, which spontaneously decomposes into a wide range of products depending on the reaction conditions, such as pH, substrate, and the cofactors including ascorbic acid, ESP, and ferrous ions. For example, in the presence of residual ESP activity at these stages (microwaved up to 45 s or steamed up to 210 s) of cooking cabbage, despite possessing the highest myrosinase activity, the yield of allyl isothiocyanate (AITC) from sinigrin was minimal. But with further cooking (microwaved for 120 s or steamed for 420 s) and denaturation of ESP (ESP activity was significantly reduced at a temperature of 50 °C and above) the production of AITC from the hydrolysis of cooked cabbage increased with a proportionate reduction in formation of the cyanoepithioalkane. The high yield of AITC on hydrolysis was due to the denaturation of ESP despite the low myrosinase activity as cooking time extended. It has been suggested that ESP may cause allosteric inhibition of myrosinase and 1-cyano-2,3-epithiopropane (CEP) from sinigrin in cabbage in the presence of residual ESP activity (Rungapamestry et al., 2006). Some of the hydrolysis products have an undesired effect on odor and taste. For instance, bitterness can be caused by gluconapin, sinigrin, and 5-vinyloxazolidine-2-thione. AITC produces a pungent and lachrymatory response upon chewing and cutting of *Brassica*. Since odor and taste are important to consumers in selecting the health-promoting vegetables, it is of interest to study the activity of myrosinase (Ludikhuyze et al., 2000; Verkerk and Dekker, 2004).

Myrosinase is heat sensitive. Broccoli myrosinase was stable until 45 °C, and its activity was reduced by more than 95% after a 10 min treatment at 70 °C. The stability of myrosinase to temperature differs widely from vegetable to vegetable. Myrosinase in a crude red cabbage extract was stable up to 60 °C, while myrosinase in a crude white cabbage extract was only stable up to 50 °C (Verkerk and Dekker, 2004). During processing, the myrosinase activity and glucosinolate concentrations are dependent on the method and duration of processing (Rungapamestry et al., 2006). The activity of myrosinase in cooked cabbage was significantly influenced by cooking treatment, cooking time, and an interaction between the two factors. Myrosinase in microwaved cabbage showed an initial significant decrease of 27.4% in activity after being cooked for 45 s and an abrupt reduction of 96.7% after cooking for 120 s, as compared to raw cabbage. The activity then remained stable in cabbage cooked for up to 420 s. Myrosinase is very stable under pressure. Moderate pressure (50–250 MPa) only had a limited effect on its activity (van Eylen et al., 2008). Pressure treatment at 700 Mpa for 50 min at 50 °C resulted in approximately 5% of enzyme inactivation, while pressure treatment at 750 MPa for 50 min at 50 °C gave rise to an approximately 20% of enzyme inactivation during the isothermal/isobaric conditions, suggesting that the reaction rate increases with elevating pressure. This implies that a high myrosinase activity can still be retained after pressure treatment. As myrosinase is heat sensitive, blanching will lead to myrosinase inactivation. Pressure blanching can be a good alternative to thermal blanching because undesired quality related enzymes such as lipoxygenase (i.e. blanching indicator) can be inactivated while desired nutrition related enzymes can be maintained (Verkerk and Dekker, 2004). On the other hand, cooking methods that retain some of the endogenous myrosinase activity may also be beneficial by increasing the conversion of glucosinolates to isothiocyanates that have great benefit to human beings during chewing (Song and Thornalley, 2007).

15.6 Structure

The stability of anthocyanins is structure dependent. The modification of anthocyanin structure has an influence on the stability of the natural colorant throughout the food processing and storage. Furthermore, structure modifications also affect anthocyanin bioavailability, metabolism, and biological properties. Acylation in the anthocyanin molecules confers stability as deacylated pigments were found to be less stable, which was reported in sweet potato (Bassa and Francis, 1987), morning glory (Teh and Francis, 1988), and *Tradescanina paccida* (Malien-Aubert et al., 2001). Further study showed that acylation in anthocyanins boosts both heat and light stabilities, whereas glucosidation stablizes anthocyanins only in the presence of light (Inami et al., 1996). It was proposed that acylation through intramolecular copigmentation (Garzón and Wrolstad, 2001; Baublis et al., 1994) and the sterical conditions given by the glycosylation pattern (Eiro and Heinonen, 2002) stabilize the anthocyanin molecules. Aromatic residues of the acyl groups stack hydrophobically with the pyrylium ring of the flavylium cation and dramatically reduce the susceptibility of nucleophillic attack of water. In general, 5-glycosylated structures break down more easily than 3-glycosides followed by aliphatic acyl-anthocyanins and aromatic acyl derivatives. It was evident that the stability of pelargonidin 3-glucoside was stronger than that of acylated pelargonidin 3-sophoroside-5-glucoside. The copigmentation increases with the degree of methoxylation and glycosylation of the anthocyanin chromophore. The stability of the chromophore is enhanced by preventing the formation of a pseudobase or chalcone. Pelargonidin 3-glucoside and cyanidin 3-glucoside lost 80 and 75% of their color, respectively, after six months of storage; however, cyanidin 3-(2"-xylosyl-6"-glucosyl)-galactoside had 40% of its color left, suggesting that trisaccharidic anthocyanins retained their color better than the monoglucosidic ones due to the sterically compact structure, which inhibits the copigments from intervening with the anthocyanin chromophore to form intermolecular complexes (Mazza and Brouillard, 1990; Eiro and Heinonen, 2002). Polyacylated anthocyanins including *Tradescantia pallida*, *Ipoema tricolor* cv Heavenly Blue, red cabbage, and *Zebrina pendula* have demonstrated great stability during processing, storage, and pH changes (Teh and Francis, 1988; Dangles et al., 1993). Anthocyanins from *Zebrina pendula* and *Ipomoea tricolor* Cav. (cultivar Heavenly Blue) have been reported to be very stable (Asen et al., 1977). The pigments in *Zebrina pendula* have been identified as tricaffeoylcyanidin-3,7,3'-triglucoside and caffeoylferuloylcyanidin-3,7,3'-triglucoside, which exhibited exceptional stability to pH change, being ascribed to the acyl groups which prevented the formation of a pseudobase or chalcone.

Anthocyanin stabilization may be ascribed to acylation by the nonaromatic malonic acid, which is facilitated by hydrogen bonding between the carboxylate group and the core aglycone (Dangles, 1997). For instance, Saito et al. (1988) reported that it boosted stability of both malonynated and glucuronosylated anthocyanins from flowers in the red daisy (*Bellis perennis*). Glucuronosylated anthocyanins isolated from red daisy (*Bellis perennis*) flower petals or obtained by enzymatic *in vitro* synthesis of red daisy glucuronosyltransferase *Bp*UGT94B1 were used to evaluate the effect of glucuronosylation on the color stability of anthocyanins toward light and heat stress (Osmani et al., 2009). Cyanidin-3-O-2"-O-glucuronosylglucoside displayed enhanced color stability under light in comparison with both cyanidin 3-O-glucoside and cyanidin 3-O-2"-O-diglucoside. However, there was no difference in heat stability observed among monoglucosylated, diglucosylated, and glucuronosylated cyanidin derivatives, whereas the glucuronosylated elderberry extract exhibited increased heat stability. Glucuronosylation of around 50% of total anthocyanins in elderberry extract led to increased color stability in response to both heat and light, suggesting that enzymatic glucuronosylation may be utilized to stabilize natural colorants in industry. Some studies indicated that acylated anthocyanins demonstrate increased resistance to heat, light, and SO_2, which may have contributed to glycosidic residues as spacers in folding, assuring the correct positioning of the aromatic rings (Figueiredo et al., 1996). Rommel et al. (1992) reported that acylated anthocyanins are less prone to color changes via endogenous β-glucosidase. Additionally, acylated structures had higher resistance to heat and light degradation (Inami et al., 1996).

Structurally, condensation of betalamic acid with amino compounds or cyclo-Dopa leads to the different stability of betazanthins and betacyanins. Betacyanins were much more stable than betaxanthins at ambient temperature (Sapers and Hornstein, 1979) and heat (Herbach et al., 2004a). The stability of betacyanin could be explained by substitution with aromatic acids because of intramolecular stacking, which is ascribed to the U-shape folding of the molecule in prevention of the aldimine bond from hydrolysis. Among different betacyanins, glycosylated structures are more stable than aglycones, due probably to the higher oxidation-reduction potentials of the former (von Elbe and Attoe, 1985). In addition, the sterical position of the aromatic acid was assumed to affect stability, with 6-O-substitution being more effective than 5-O-substitution (Schliemann and Strack, 1998). The esterification of betacyanins with aliphatic acids was also found to result in enhanced pigment stability (Barrera et al., 1998). The study showed that pigment solutions of *Myrtillocactus geometrizans* (Martius) Console appear to be more stable than the respective solutions in red beet, due partly to the existence of betanidin 5-O-(6'-O-malonyl)-β-glucoside in *Myrtillocactus geometrizans* solution. Herbach et al. (2005) reported that phyllocactin and hylocerenin were less susceptible to hydrolytic cleavage than betanin, implying the protection of the aldimine bond by aliphatic acid moieties. Additionally, phyllocactin and hylocerenin appear to be decarboxylated upon thermal treatment. The decarboxylated betacyanins show absorption maxima identical to their precursors, and possess boosted pigment integrity. Therefore, hylocerenin solutions have higher chromatic and tinctorial stability towards thermal degradation than betanin-based solutions, although the half-life of heated betanin was 11-fold higher than that of vulgaxanthin I. Phyllocactin solutions were found to be less stable because of substitution with malonic acid, which is prone both to cleavage of the carboxyl group at the β-position and to deacylation. Moreover, betanidin has 17-fold higher half-life than isobetanidin upon degradation by active oxygen species, which has been ascribed to the glycosylation and the lower oxidation–reduction potential of betanidin when compared to betani (von Elbe and Attoe, 1985).

15.7 Copigments

Copigmentation is a natural phenomenon and occurs in fruit and vegetable-derived products such as juices and wines. It is considered as an interaction in which pigments and other non-colored organic components form molecular associations or complexes, leading to an enhancement in the absorbance and/or a shift in the wavelength of the maximum absorbance of the pigment by being detected both as a hyperchromic effect and as a bathochromic shift (Baranac et al., 1996). As one of the color stabilizing mechanisms, intermolecular copigmentation by the formation of anthocyanins and copigments (mostly polyphenolics) can explain the hyperchromic and bathochromic effects. In principle, it is formed by stacking the copigment molecule on the planar polarizable nuclei of the anthocyanin-colored forms. Therefore, the nucleophilic attack from water at position 2 of the pyrylium nucleus, which results in colorless hemiketal and chalcone forms, is partially prevented. Additionally the copigment molecule interacts with the excited state anthocyanin more strongly than with the ground state anthocyanin, exhibiting bathochromic shift (Alluis et al., 2000). In copigmentation, an anthocyanin chromophore is covalently linked to an organic acid, a simple phenolic acid, an aromatic acyl group, or a flavonoid (Bloor and Falshae, 2000). It would be most efficient if anthocyanin and copigment moieties are covalently linked. For example, the sugar residues of anthocyanins are acylated by phenolic acids. Copigmentation provides brighter, stronger, and more stable colors than those produced by anthocyanin alone. On the other hand, anthocyanins make reactions with other phenolics via weak hydrophobic forces to form loose intermolecular interactions. Intermolecular copigmentation reactions have long been elucidated in wines (Boulton, 2001). Simple anthocyanin molecules copigmentation in fruits, vegetables, and beverages has also been investigated. It was observed that the addition of copigments enhanced anthocyanin color stability during storage (Eiro and Heinonen, 2002). For example, ferulic and caffeic acid addition dramatically boosted the color of pelargonidin 3-glucoside throughout a six month storage period, being 220 and 190% of the original color intensity, respectively, at the end of storage. On the other hand, the addition of gallic, ferulic, and caffeic acids lowered the color stability of the acylated anthocyanin during storage, indicating that these phenolic acids reduce the protective intramolecular mechanism of the acylated anthocyanins. Generally, acylation of the anthocyanindin causes an increase in the relative proportion of the flavylium cation, therefore protecting the red color at higher pH, enhancing the stability of anthocyanin. Also, flavones enhance the color of anthocyanins by stabilizing the quinoidal bases due to intermolecular copigmentation phenomena in the presence of colorless matrix compounds (Mistry et al., 1991). These copigments may also protect the flavylium ion from hydration.

The stability of the anthocyanins is in direct proportion to phenolic concentration (Bakowska et al., 2003), suggesting that copigmentation confers the increase in stability of anthocyanins. Additionally, the copigmentation of caffeic acid enhanced the stability of the grape anthocyanins in a yoghurt system (Gris et al., 2007). Furthermore, polysaccharide copigmentation boosted anthocyanin stability in Hibiscus sabdariffa L. in the solid state (Gradinaru et al., 2003). Eiro and Heinonen (2002) studied intermolecular copigmentation with five anthocyanins (pelargonidin 3-glucoside, cyanidin 3-glucoside, malvidin 3-glucoside, acylated cyanidin trisaccharide, cyanidin trisaccharide) and five phenolic acids (gallic, ferulic, caffeic, rosmarinic, chlorogenic) acting as copigments. A UV-visible spectrophotometer was used to monitor the hyperchromic effect and the bathochromic shift of the pigment–copigment complexes. The stability of the complexes with different molar ratios of the anthocyanin-copigment was examined during a storage period of six months. During the storage period, cyanidin 3-(2″-xylosyl-6″-(coumaroyl-glucosyl))-galactoside exhibited the maximum color stability among five anthocyanins. The strongest copigmentation reactions occurred in malvidin 3-glucoside solutions. Malvidin 3-glucoside lost its color quickly, disappearing after 55 days. The greatest copigments for all anthocyanins were ferulic and rosmarinic acids.

Baublis et al. (1994) investigated stabilities of anthocyanins from Concord grapes, red cabbage, tradescantia, and ajuga by using RP-HPLC analysis. Except tradescantia, the other three pigments exhibited approximately 90% degradation after being stored for 15 days. Further analysis found that copigments such as rutin, chlorogenic acid, and caffeic acid could assist in intermolecular stabilization of anthocyanins in tradescantia. Such increased stability in tradescantia was attributed to B ring substitution of the chromophore, intramolecular copigmentation, and the high degree of acylation, enhancing in stability by inhibiting hydration and fading. The study of colorants' stability has been conducted in sugar and non-sugar drink models at three pH values (3, 4, and 5) under thermal and light conditions mimicking rapid food aging (Malien-Aubert et al., 2001). It was concluded that colorants rich in acylated anthocyanins (purple carrot, red radish, and red cabbage) show great stability due to intramolecular copigmentation. For colorants without acylated anthocyanins (grape-marc, elderberry, black currant, and chokeberry), intermolecular copigmentation confers a major color protection. Colorants with high amount of flavonols and with the highest copigment–pigment ratio display a remarkable stability. No singinficant color change was found by the addition of sugar. The effect of the copigmentation on the kinetics and thermodynamics of purple potato peel anthocyanins was measured by the shift of the exothermic peaks with the differential scanning calorimetry (DSC) analysis in both liquid and solid states (Zhang et al., 2009). It was observed that citric acid monohydrate and glucose increased the stability of purple potato peel, while ascorbic acid lowered the stability of purple potato peel by the activation–energy evaluation in the liquid state. On the other hand, the copigmentation with the ascorbic acid, citric acid monohydrate, and glucose improved the stability of purple potato peel by the transition–temperature evaluation in the solid state.

An increase in color intensity in strawberry beverages has been reported after addition of phenolic acids serving as copigments (Rein and Heinonen, 2004). Heat stability and copigmentation behavior of purified strawberry anthocyanins (PSA) in the presence of rose petal polyphenolics (RPP), along with strawberry beverage color change during thermal treatment upon the addition of polyphenolic copigments in RPP has been examined (Mollov et al., 2007). The copigments are sinapic acid, chlorogenic acid, and caffeic acid. The results revealed that the addition of polyphenolic copigments extracted from distilled RPP lowers the thermal degradation of anthocyanins, permiting enhanced color stability of the processed strawberries. For instance, after 1 h heating at 85 °C, the non-copigmented and copigmented strawberry anthocyanins retained 71 and 79% of relative concentration, respectively. High concentrations of copigments such as flavonols (quercetin and kaempferol) have been reported (Henning, 1981). Anthocyanin polymerization mediated by flavan 3-ols has been found in strawberry jam, red raspberry jams (García-Viguera et al., 1999), and wines (Rivas-Gonzalo et al., 1995) during storage. Aside from intramolecular stabilization mechanisms, flavonoids as copigments can contribute to anthocyanin stability and color tonation, due mainly to intermolecular interactions by aromatic ring stacking and H-bonding with the anthocyanins (Malien-Aubert et al., 2001; Eiro et al., 2002). A pigment was formed in the acetaldehyde-mediated condensation between malvidin 3-O-glucoside and catechin (Escribano-Bailón et al., 2001). The stability of this pigment in relation to pH, discoloration by SO_2, and storage in aqueous solution was examined. The color of the pigment exhibited more stability with regard to bleaching by SO_2 than that of malvidin 3-O-glucoside. When the pH was raised from 2.2 to 5.5, the pigment solution became more violet, whereas anthocyanin solutions were almost colorless at pH 4.0, indicating that the anthocyanin moiety of the pigment was protected against water attack, and thus the formation of its quinonoidal forms was favored.

Catechins can react with anthocyanins to form copigments though non-covalent bonds, which results in the red wine color changes during maturation and aging (Mazza et al., 1993; Gonzalez-Manzano et al., 2008). As a copigment, catechins were reported to stabilize anthocyanin and inhibit its degradation during UV treatment (Parisa et al., 2007).

Isoflavones extracted from red clove, including formononetin, biochanin A, and prunetin, were reported to form copigments with anthocyanin in muscadine grape juice and wine and to improve color stability (Talcott et al., 2005).

Dangles et al. (1992) have studied anti-copigmentation. It was shown that cyclodextrins expedited color loss of the anthocyanin solution by forming inclusion complexes with the colorless carbinol pseudobase and the yellowish chalcones. It was proposed that starch-rich foods might also be susceptible to a fading reaction upon processing.

15.8 Matrix

15.8.1 Presence of SO_2

Sulfite has an effect on anthocyanin stability (Berké et al., 1998). It has been shown to lose color, but, in an almost reversible reaction. Addition of SO_2 and maintaining a given pH during processing of blue wheat whole meals or of the isolated anthocyanins play a crucial role in stabilizing pigments (Abdel-Aal et al., 2003). The optimal SO_2 concentrations were 500–1000 ppm for whole wheat meals and 1000–3000 ppm for isolated anthocyanins.

The addition of sodium hydrogen sulfite has been reported to possess antioxidant activity (Lindsay, 1996). Although the reason for antioxidant activity is complex, the possible reaction of sulfite with carbonyl groups, reducing sugars and disulfide bonds in proteins may result in a protective effect. In addition, the application of sodium hydrogen sulfite prevents the growth of microorganisms such as fungi (Cinar, 2005).

The advantage of using sodium hydrogen sulfite was undoubtful in retaining a high amount of carotenoids; however, using of sulfite salts in foods has been severely regulated because of possible adverse reactions to these compounds by asthmatic patients. Finding a proper substitute may be a solution to this problem in the future.

It has been shown that some kinds of glucosinolates could react with sulphur dioxide. The pungency of mustard, cress, and radish is dependent upon the hydrolysis of glucosinolates and on the nature and amount of the products thus formed. Condiment mustard is prone to oxidation and certain continental type preparations contain SO_2 to inhibit both this process and subsequent discoloration of the product. It is possible for the pungent 2-propenyl isothiocyanate to react with SO_2 to form 2-propenyl aminothiocarbonyl sulphonate. Not only does this reaction lead to a loss of pungency, but decomposition of the involatile sulphonate to 2-propenyl thiol and related sulphides is linked to the development of a garlic-like off-odor on storage (Austin et al., 1968b). The addition of SO_2 to horseradish powder has a detrimental effect on flavor. Preliminary experiments have shown that the reaction of 2-propenyl isothiocyanate with SO_2 is much faster than that of 2-phenylethyl isothiocyanate, the other major component of horseradish essence. This presumably leads to the reduction in odor already mentioned (Chevolleau et al., 1997).

In order to reduce the degradation of carotenoids, there are many treatments prior to the dehydration process. One of the common treatments is pre-soaking (Tai and Chen, 2000). Being soaked in 1% sodium hydrogen sulfite solution prior to dehydration can reduce the loss of carotenoids in daylily flowers. Comparing with the daylily flowers, which are dried by hot air directly, the loss of zeaxanthin was reduced by 54%, β-cryptoxanthin 44%, all-*trans*-lutein 28%, 13-*cis*-lutein 39%, and all-*trans*-β-carotene 44%. Obviously, the application of sodium hydrogen sulfite solution could protect carotenoids from undergoing oxidative degradation during hot-air-drying.

15.8.2 Presence of ascorbic acids and other organic acids

Kaack and Austed (1998) investigated the effect of vitamin C on flavonoid degradation in 13 European elderberry cultivars. They revealed that ascorbic acid protected the anthocyanins, but not quercetin, from oxidative degradation. Under condition of high concentration of oxygen, anthocyanin level reduced dramatically during juice processing and storage. However, purging of the elderberry juice with N_2 and/or addition of ascorbic acid decreased the oxidative degradation rate of the cyanidin-3-glucoside, cyanidin-3-sambubioside, and quercetin. On the other hand, ascorbic acid impairs anthocyanin stability, although betalains can effectively be stabilized by ascorbic acid (Shenoy, 1993). Nonetheless, anthocyanins after chelating metal ions were reported to be capable of preventing ascorbic acid oxidation by complex formation (Sarma et al., 1997). The addition of ascorbic acid in reduction of color stability of strawberry syrup was reported. Concurrent loss of anthocyanin pigments and ascorbic acid has been observed in different fruit juices such as strawberry and blackcurrant juices (Skrede et al., 1992), as well as cranberry juice (Shrikhande and Francis, 1974), which gave rise to both anthocyanin and ascorbic acid degradation by either H_2O_2 formed through oxidation or condensation of ascorbic acid directly with anthocyanins (Markakis, 1982). The study from Choi et al. (2002) characterized the change in pigment composition with two different levels of ascorbic acid content in blood orange juice placed in HDPE plastic bottles, pasteurized, and stored at 4.5 °C. The result revealed that ascorbic acid degradation was highly correlated ($r>0.93$) to anthocyanin pigment degradation, suggesting a possible interaction between the two compounds in stored blood orange juice.

Açai anthocyanin stability against hydrogen peroxide (0 and 30 mmol/L) over a range of temperatures (10–30 °C) was evaluated, and further compared to other anthocyanin sources including black carrot, red cabbage, red grape, purple sweet potato, and a noncommercial extract from red hibiscus flowers (*Hibiscus sabdariffa*) (Del Pozo-Insfran et al., 2004). Also, pigment stability in a model beverage system was measured in the presence of ascorbic acid and naturally occurring polyphenolic cofactors. It was found that anthocyanins were the most stable in the presence of hydrogen peroxide in red grape, while açai and other pigments rich in acylated anthocyanins exhibited lower color stability in a temperature-dependent pattern. In the presence of ascorbic acid, acylated anthocyanin sources usually possessed improved color stability. Improvement of color stability in blood orange juice by the addition of tartaric acid, tannic acid, and antioxidant agents was reported (Maccarone et al., 1985). There was a positive correlation between anthocyanin content and juice stability in the conditions of three different storage temperatures with aseptic glass bottles (Bonaventura and Russo, 1993).

Potential interactions of betalains in red beet with their matrix compounds have been examined in detail (Delgado-Vargas et al., 2000). The presence of matrix compounds such as organic acid supplements efficiently boosted betacyanin stabilization. The effective result was reported in pitaya juice heated at pH 4 when adding 1.0% ascorbic acid (Herbach et al., 2006a). Addition with some compounds, especially antioxidants, stabilizes betalain. For instance, the supplementation of ascorbic and isoascorbic acids was found to enhance betalain stability by O_2 removal, although different optimum ascorbic acid concentration ranges of 0.003–1.0% were reported (Attoe and von Elbe, 1984). The addition of ascorbic, isoascorbic, and citric acids to purple pitaya juice and purified pigment preparation can stabilize betacyanins (Herbach et al., 2006a). On the contrary, Pasch and von Elbe (1979) revealed that 1000 ppm ascorbic acid as a pro-oxidant decreased half-life time of betanin, being attributed to hydrogen peroxide bleaching effect during ascorbic acid degradation.

There exist some discrepancies between ascorbic and isoascorbic acid in enhancement of betalain stability (Barrera et al., 1998; Herbach et al., 2006a), even though isoascorbic acid displayed superior oxygen conversion due to its higher redox potential. For example, ascorbic acid exhibited greater betacyanin retention than isoascorbic acid upon utilization at identical concentrations in purple pitaya. The ineffectiveness of phenolics in betalain stability suggests that betanin oxidation does not involve a free radical chain mechanism (Attoe and von Elbe, 1985). Chen et al. (2007) researched on different drying treatments to the stability of Taiwanese mango. It was reported that the pre-soaking step may prevent the epoxy-containing carotenoids from degradation during the subsequent extraction procedure. They also compare the protective effect to different kinds of carotenoids (violaxanthin, neochrome, neoxanthin, and so on) by soaking in ascorbic acid with soaking in sodium hydrogen. Luteoxanthin can be formed from violaxanthin under acidic conditions. Neochrome can be attributed to conversion of neoxanthin under acidic conditions too. It was showed that different kinds of carotenoids fit to different soaking conditions.

Ascorbic acid has been reported to significantly increase the stability of catechins at a neutral or alkaline pH, maybe due to its antioxidant properties or its ability to lower oxygen concentration dissolved in solutions (Chen et al., 2001), however, it can also act as prooxidant and accelerate the degradation of catechins. In addition to ascorbic acid, other reducing agents, such as dithiothreitol (DTT), tris (2-carboxyethyl) phosphine (TCEP), as well as encapsulation in chitosan–tripolyphosphate nanoparticles, were also found to improve the stability of catechins in an alkaline solution .

In the presence of ascorbic acid, indolylglucosinolates give rise to a variety of products on hydrolysis by myrosinase. A feature of plant myrosinase is its activation by ascorbate. Early work had shown that ascorbate created an allosteric effect on the activity of the enzyme. Subsequently the mechanism of ascorbate activation has been worked out where it was shown that ascorbate acts as a catalytic base (Schneider and Becker, 1930).

It was shown that the content of glucoraphanin was 0.0815 mg/mL when ascorbic acid was added to the extraction, and it remained at 0.0925 mg/mL without this addition. So removing ascorbic acid from extraction upon stored has been suggested (Wang et al., 2009).

15.8.3 Presence of metallic ions

Some metal cations including Al^{3+}, Cr^{3+}, Cu^{2+}, Fe^{2+}, Fe^{3+}, and Sn^{2+} were found to expedite betanin degradation (Czapski, 1990). For instance, metal ions such as Cu^+, Cu^{2+}, and Hg^{2+} combine with betanin to form metal-pigment complexes accompanied by bathochromic and hypochromic shifts. However, the spectra could be reversible by the addition of EDTA (Attoe and von Elbe, 1984), where EDTA inhibits metal-catalyzed betanin degradation by pigment stabilization and metal-pigment complex formation. EDTA improved betanin's half-life time by 1.5-fold (Pasch and von Elbe, 1979). As a chelating agent, citric acid boosted betacyanin stability (Herbach et al., 2006a), explained by partially neutralizing the electrophilic center of betanin via association with the positively charged amino nitrogen. Interestingly, the addition of EDTA didn't enhance vulgaxanthin I stability (Savolainen and Kuusi, 1978).

The effect of metal ions on stability of carotenoids was reported. Fe^{3+}, Fe^{2+}, and Al^{3+} had the negative effect on the stability of carotenoids, whereas Mn^{2+} had the smaller effect (Yao and Han, 2008). Astaxanthin is one of the few carotenoids containing four oxygen donors. Usually, these oxygen donors can coordinate with heavy metal ions such as Cu(II) and Fe(III). It was found that Cu(II) markedly induces the conversion of *trans*-astaxanthin to its *cis*-forms, which mainly consist of 9-*cis*-astaxanthin and 13-*cis*-astaxanthin as suggested by UV-visible spectra and HPLC measurements. Increasing either incubation time of Cu(II) and *trans*-astaxanthin in ethanol or the Cu(II)/astaxanthin ratio gave rise to an increased percentage of *cis*-isomers derived from *trans*-astaxanthin. These results provide important information on the effects of dietary factors on the bioavailability and bioactivity of *trans*-astaxanthin (Zhao *et al.*, 2005).

Catechins can chelate metal ions such as iron and copper, and this may be due to their antioxidant activities by inhibiting transition metal-catalyzed free radical formation. The *O*-3'4'-dihydroxyl group on the B ring is likely to be the metal binding site (Hider *et al.*, 2001). However, it was suggested that the gallate moiety of the gallocatechins also binds metals (Midori *et al.*, 2001). The same study found that Cu^{2+} strongly increased the antioxidant activity of EGCG during 2,2'-azobis(2,4-dimethylvaleronitrile) (AMVN)-initiated lipid peroxidation, but Fe^{2+} adversely affected the antioxidant activity of EGCG (Midori *et al.*, 2001). Although potassium, calcium, magnesium and aluminum are the predominant minerals found in tea (Eden, 1976), there may be weak interaction between catechins and aluminum or magnesium, and catechins do not chelate potassium, calcium (Hider *et al.*, 2001).

(S)-2-Hydroxybut-3-enylglucosinolate of Crambe abyssinica meal has been found to be decomposed by heating in the presence of metal salts including $FeSO_4$, $Fe(NO_3)_3$, $CuCl_2$, and $CuSO_4$ with Fe^{2+} and Cu^+ being the most active species. N-methylindol-3-methylglucosinolate was self decomposed by Fe^{2+} (aq) to the nitrile (Zabala *et al.*, 2005). Glucosinolates with a hydroxyl group in the 2 position of the side chain (e.g. 2-hydroxybut-3-enylglucosinolate) have been shown to give thionamides with Fe^{2+}, although an eight-fold excess of Fe^{2+} is necessary, but it seems unlikely that these compounds would be generated during food processing (Austin *et al.*, 1968a).

15.8.4 Others

The effect of sugar addition on the anthocyanin stability is influenced by anthocyanins' structure and concentration as well as type of sugar. Anthocyanin stability by linkage of sugar residues could take place by formation of hydrogen bonds between glycosyl groups and the aglycone, which is influenced by the site of glycosylation and the type of sugar added (Delgado-Vargas *et al.*, 2000; Stintzing *et al.*, 2002). At low concentration of sucrose (86 g/L), anthocyanin degradation in blackcurrant, elderberry, and red cabbage extracts was higher in soft drinks compared to buffer systems both at pH 3 (Dyrby *et al.*, 2001). Conversely, the anthocyanins' stability in strawberry increased after the sucrose concentration increased by 20% (Wrolstad *et al.*, 1990). Additionally, although some anthocyanic extracts displayed lower stability in sugar-added systems, the addition of 20 g/L of sucrose to a pH 3 drink system containing red cabbage and grape extracts did not affect light and thermal stability of the anthocyanins (Duhard *et al.*, 1997). In the study from Queiroz *et al.* (2009), 200 g of blueberries were mixed with 200 g of sugar to make jams. A 5 g measure of blueberry samples ground to paste with mortar and mixed with water was used to evaluate the degradation of anthocyanins and anthocyanidins at 100 °C. °Brix in jams play an important role in degradation of anthocyanins and anthocyanidins, where 64–76 °Brix caused 20–30% degradation, whereas 80 °Brix led to degradation of 50–60%, suggesting that anthocyanin degradation is realted to °Brix of the samples as well as the rate of sugar degradation resulted from Maillard reaction (Duhard *et al.*, 1997). Anthocyanin thermostability was affected by sugar addition such as fructose and glucose (Rubinskiene *et al.*,

2005b). Additionally, anthocyanidins' degradation in blueberry jam displayed similar patterns to that mintored for anthocyanin degradation in jam. Jams with high 80 °Brix showed degradation approximately 60%. The degradation rate constant of anthocyanins in an isotonic soft drink system in the dark was $6.0\times10^{-2}\,h^{-1}$ for acerola and $7.3\times10^{-4}\,h^{-1}$ for açai, respectively, indicating that the addition of sugars (55 g of sucrose, 5.5 g of fructose, 5.5 g of glucose in 1 l) and salts (0.15 g of sodium benzoate, 3.0 g of citric acid, 0.14 g of sodium citrate, 0.5 g of sodium chloride, 0.5 g of potassium chloride, and 0.4 g of potassium phosphate monobasic in 1 l) had a negative effect on the anthocyanin stability (Vera de Rossoa and Mercadante, 2007).

15.9 Storage conditions

15.9.1 Light

The effect of fluorescent light on the degradation rates of the major cranberry anthocyanins was assayed in model systems in the presence of oxygen in the temperature range 25–55°C (Attoe and Von Elbe, 1981). The study revealed that light degraded most of anthocyanins at 40 °C. Light exhibited a significant effect on anthocyanin degradation in the presence of molecular oxygen. Conversely, light-induced *trans-cis*-isomerization of coumaric acid substituents in anthocyanins offers a way to stabilize color (George *et al.*, 2001). Acerola is a good source of ascorbic acid, carotenoids, as well as cyanidin-3-α-*O*-rhamnoside and pelargonidin-3-α-*O*-rhamnoside. Açai is rich in the anthocyanins cyanidin-3-glucoside and cyanidin-3-rutinoside. The addition of anthocyanic extracts from acerola (*Malpighia emarginata* DC.) and açai (*Euterpe oleracea* Mart.) as a colorant and functional ingredient in isotonic soft drinks and in buffer solution was evaluated (Vera de Rossoa and Mercadante, 2007). The study revealed that the degradation of anthocyanins from both tropical fruit sources followed first-order kinetics in all the systems under air, either in the presence or absence of light. Light exerted a significantly negative influence on anthocyanin stability in both açai added systems, isotonic soft drink ($p<0.001$) and buffer ($p<0.001$). The degradation rate of açai anthocyanins extract in the buffer system was 7.1 times faster under light than in the dark. Additionally, in the presence of light, the anthocyanin degradation was 1.2 times quicker for acerola and 1.6 times quicker for açai in soft drink isotonic systems, as compared to their respective buffer solutions.

Light was found to affect betalain stability (Cai *et al.*, 1998; Herbach *et al.*, 2006a), which can be attributed to betalain absorption of light in the UV and visible range resulting in excitation of electrons of chromophore to a more energetic state, thus bringing about higher reactivity or lowered activation energy of the molecule (Jackman and Smith, 1996). An inverse relationship between betalain stability and light intensity in the range 2200–4400 lux was reported by Attoe and von Elbe (1981). Deterioration of betalain stability influenced by light was observed at temperatures below 25 °C, while there was no impact of light on stability at temperatures above 40 °C (Huang and von Elbe, 1986). Additionally, light-induced degradation was found to be oxygen dependent. On the other hand, the addition of 0.1% isoascorbic acid and 1.0% ascorbic acid, respectively, into red beet and purple pitaya juices was shown to inhibit light-induced betacyanin degradation during juice storage.

β-carotene effectively protects oil against light deterioration by quenching singlet oxygen even at concentration below 20 ppm. But light can make the carotenoids unstable. Carotenoids are known to exist in different geometric forms (*cis*- and *trans*-isomers). These forms may be interconverted by light (Rock, 1997). Stability of biomass of two microalgae species was predominantly affected by light, followed by oxygen content, whereas the storage temperature is only important to a lesser extent (Gouveia and Empis, 2003). A similar conclusion has been made by Lin and Chen (2005), that light has a greater influence on isomerization and degradation of *cis*-isomers of β-carotene than temperature. For lycopene pigments, light effects were more destructive than those of the high temperature (Nachtigall *et al.*, 2009). The sensitivity of carotenoids to non-sensitized direct light is dependent on the wavelength of irradiation. Under fluorescent light, which means the involvement of singlet oxygen was ruled out, the higher the unsaturation, the slower is the rate of carotenoids autoxidation. This reveals that a higher degree of unsaturation offers a greater protection to β-carotene against autoxidation. Also it has been reported that the deterioration of the carotene was probably due to absorption of light in the visible region. The photocatalyzed oxidation of β-carotene is also more severe in ultraviolet than in visible light (Bonnie and Choo, 1999). Exposure to light, especially direct sunlight or ultraviolet light, induces *trans-cis* photoisomerization and photodestruction of carotenoids. Thus, work on carotenoids must be performed under subdued light; for example, all the extraction procedures were conducted under dimmed light to avoid isomerization or degradation loss of carotenoids (Chen *et al.*, 2007). As compared with pure carotenoids, carotenoid-arabinogalactan complexes exhibit an enhanced stability toward photodegradation (Polyakov *et al.*, 2010). Open columns and vessels containing carotenoids should be wrapped with aluminum foil, and thin-layer chromatography development tanks should be kept in the dark or covered with dark material or aluminum foil. Polycarbonate shields are available for fluorescent lights, which are notorious for emission of high-energy, short-wavelength radiation (Sajilata *et al.*, 2008).

In darkness, I3C was the unique compound formed that accounted for the total radioactivity after 1 h. A very weak proportion of 3,3′-diindolylmethane (DIM) (4%) was observed only after a 24 h incubation period. This indicates that the enzymatic activity of myrosinase is not influenced by the lighting conditions but indole derivatives are more reactive in the presence of light and that condensation must be photochemically initiated (Lopez-Berenguer *et al.*, 2007). It has been reported that the content of glucoraphanin was 0.0548 mg/mL when the extraction was illuminated for nine days. If stored in darkness for nine days, it remained at 0.0775 mg/mL. So glucoraphanin should be stored away from light and packaged (Wang *et al.*, 2009).

Light storage can be more destructive to each carotenoid and vitamin A than dark storage during the storage of carrot juice. The 9-*cis* isomers were the major carotenoid isomers formed in carrot juice under light storage, while 13-*cis* was favored under dark storage. In canned tomato juice, because it is in a dark environment and the exposure of juice to atmospheric oxygen is excluded, the degradation of all-*trans* plus *cis* forms of lutein preceded slowly. Compared to dark storage, most *cis*-isomers of β-carotene were at lower levels in canned tomato juice, which can be due to the fact that the canned juice is in a dark environment and the exposure to air is excluded (Xu *et al.*, 2006). Generally, lower storage temperature, lower oxygen pressure, comparatively dry, under dark condition, addition of antioxidant like BHT and so on, could make carotenoids relatively stable in storage process.

15.9.2 Temperature

The aim of post-harvest treatment is to manipulate metabolism in fruits and vegetables during storage in order to extend shelf life. Ferreres *et al.* (1996) reported on the stability of the anthocyanin pigments of Spanish red onion (cultivar 'Morada de Amposta') stored in perforated films for seven days at 8 °C. A small increase in anthocyanins was found after one day of storage, followed by a decrease after seven days of storage. It was observed that there was a big difference in the stability of the individual anthocyanins. The glucosides were more stable than the corresponding arabinosides. The malonated anthocyanins were more stable than the corresponding non-acylated pigments, suggesting that anthocyanin acylation is one of the major structural factors influencing pigment stability, which is in agreement with previous reports (Mazza and Miniati, 1993). Rodríguez-Saona et al. (1999) have evaluated two acylated pelargonidin-based anthocyanins from red-fleshed potatoes (*Solanum tuberosum*) and red radishes (*Raphanus sativus*) and two extraction methods (C-18 resin and juice processing) during 65 weeks of storage at 25 °C and 2 °C in the dark. It was shown that higher stability was obtained in juices with C-18 purified radish anthocyanins (22 week half-life) and lowest stability with potato juice concentrate (ten week half-life). Anthocyanin degradation greatly depended on storage temperature, with degradation kinetics following a quadratic model at 25 °C and a linear model at 2 °C. The addition of 10, 20, and 40% of sucrose by weight to IQF strawberries prior to freezing displayed a protective effect on the anthocyanin degradation after storage of −15 °C for three years (Wrolstad et al., 1990). The sucrose addition also delayed browning and polymeric color formation. Additionally, the effect of thawing on anthocyanin stability was examined by maintaining strawberry samples at 20 °C for 24 h. The thawed samples were then refrozen at −80 °C and powdered for analysis. The result revealed that thawing accelerated the color decomposition rate.

The effect of cultivars (Chandler, Tudla, and Oso Grande) and storage temperature on the color stability of strawberry (Fragaria×ananassa) jam was investigated (García-Viguera et al., 1999). Strawberry fruit and sugar were mixed with pectin and citric acid, manufactured for 15 min at 78 °C under 500 mm Hg vacuum, heated at 92 °C and allowed to cool to 88 °C before filling into glass jars. Jams were stored at 20, 30, or 37 °C in the dark for 200 days. The result indicated that Oso Grande cultivar showed the highest anthocyanin degradation (35.30% cyanidin 3-glucoside, 44.71% pelargonidin 3-glucoside, and 33.02% pelargonidin 3-rutinose) during processing. There were no differences in degradation kinetics of anthocyanins among cultivars at the same temperature. However, differences were found in anthocyanin stability under storage temperatures, being much more stable at the lower temperature, which was in agreement with previous studies (Jackman and Smith, 1996). Patras et al. (2009) studied the effect of storage time and temperature on degradation of anthocyanins in strawberry jam. The data exhibited that lightness value significantly reduced ($p<0.05$) over 28 days of storage at 4 and 15 °C. The reaction rate constant for anthocyanins increased from 0.95×10^{-2} day^{-1} to 1.71×10^{-2} day^{-1} at 4 and 15 °C. The effect of storage time, temperature, and light on the degradation of monomeric anthocyanin pigments extracted from skins of grape (*Vitis vinifera* var. Red globe) was evaluated through stepwise regression analysis (Morais et al., 2002). The extract of pigments redissolved in distilled water containing 0.01% HCl were stored in the air at 24, 32, and 40 °C, and analyzed after 1, 3, 6, 8, and 14 days of storage both in light, using a lamp of 1.5 W, and in the dark. It was concluded that the overall decomposition rate of peonidin-3-glucoside and malvidin-3-glucoside was significantly dependent on storage time and temperature. However, light exerted a negligible impact on the decomposition rate.

During the storage of carrot juice, the concentration changes of lutein, α-carotene, β-carotene, and vitamin A in the carrot juice decreased with increasing storage temperature (Chen et al., 1996). Saldana et al. (1976) studied the effect of storage on quality of carrot juice and found that there was no effect on color or β-carotene when carrot juice was stored at 20 °C for nine months. Both isomerization and degradation of β-carotene may proceed simultaneously during the storage of tomato juice. The contents of all-*trans* plus *cis* forms of lutein were found to decrease following the increase of storage temperature, implying that degradation may still proceed even in the absence of light. All-*trans*-lutein was not detected after storage at 4 and 25 °C for five weeks. However, the same phenomenon occurred after storage for four weeks at 35 °C. This result showed that the higher the storage temperature, the faster was the degradation of all-*trans*-lutein. Similar trends also applied to 9-*cis*- and 13-*cis*-lutein, under light the degradation tends to be faster than under dark (Lin and Chen, 2005). Generally, the higher the storage temperature, the greater are the losses of all-*trans* plus *cis*- forms of carotenoids and more *cis*-isomers of carotenoids were formed during storage.

Refrigeration at 4 °C and freezing might be the best preservation processes for maintaining high level of glucosinolates in broccoli (Rodrigues and Rosa, 1999). The frozen vegetables had, however, been blanched by steaming prior to freezing. When stored in a domestic refrigerator (4–8 °C), the contents of individual and total glucosinolates of vegetables decreased during storage for seven days. There were slight changes in the first three days of storage. After storage for seven days the total glucosinolate analyte content was decreased 11–27%. For individual glucosinolates, the losses of glucoiberin, glucoraphanin, and glucoalyssin were higher than of sinigrin, gluconapin, and progoitrin. The loss of the glucoiberin in broccoli was 40–50%, whereas the loss for gluconapin was 5–10% in all vegetables studied. When stored at ambient temperature (12–22 °C), there was no significant decrease in glucosinolate content of the *Brassica* vegetables studied. But at that time, the vegetables had visibly started to decay. Storage at –85 °C may cause significant loss of glucosinolates due to freeze–thaw fracture of plant cells and accessibility of myrosinase to glucosinolates with subsequent enzymatic conversion of glucosinolates to isothiocyanates during thawing. The loss of individual glucosinolates was 10–53% and the loss of total glucosinolates was 33% – much higher than for storage in a refrigerator at 4–8 °C (Song and Thornalley, 2007).

Lower storage temperature can extend the shelf life of catechins and for ready-to-drink tea beverages. Low temperature (4 °C) and acidic pH (4.0) were found to be the optimal storage conditions for catechin preservation (Bazinet *et al.*, 2010). Addition of butylated hydroxytoluene (BHT) at a level of 0.1% was reported to have a significant effect on longer stability of catechins with over 90% EGCG remaining on day 130 stored at 37 °C (Demeule *et al.*, 2002). BHT in glycerin was also found to improve the t_{90} (time for 10% degradation to occur) to up to 76 days at 50 °C, which offers a potential for glycerin based vehicles to stabilize EGCG (Proniuk *et al.*, 2002).

At lower storage temperature (10 °C) up to seven days, individual isoflavone concentrations in ethanol extraction solutions remained constant, but at higher temperatures (25 and 40 °C) malonylglucosides degraded to glucosyl forms and aglycones are unchanged (Rostagno *et al.*, 2005). Interestingly, genistein was reported to be able to react with itself or with lysine in the non-enzymatic Maillard browning reaction, thus, leading to loss of isoflavones during storage of soy protein isolates at mild conditions, therefore, a storage condition was suggested at a_w less than 0.3 and temperature greater than 4 °C to prevent Maillard reaction (Davies *et al.*, 1998).

15.9.3 Relative humidity (RH)

A high RH of 98–100% is recommended to maintain post-harvest quality in broccoli (Rodrigues and Rosa, 1999). RH only appears to be a critical factor in glucosinolate retention when post-harvest temperatures rise above approximately 4 °C. For example, glucoraphanin content declined by more than 80% in broccoli heads left at low RH and 20 °C for five days. Similarly, broccoli heads stored in open boxes with low RH at 20 °C showed a 50% decrease in glucoraphanin content during the first three days of storage, whereas heads stored in plastic bags with high RH more than 90% displayed no significant loss at the same temperature. The decrease in glucoraphanin coincided with a marked loss of visual quality (i.e. yellowing), indicating probable loss of membrane integrity and mixing of glucosinolates with myrosinase at low RH (Toivonen and Forney, 2004).

15.9.4 Water activity (a_w)

a_w plays a crucial role in betanin susceptibility to aldimine bond cleavage because of the water-dependent hydrolytic reaction, a reduced mobility of reactants and limited oxygen solubility. The improvement of betanin stability with lower a_w was reported by Kearsley and Katsaborakis (1981), where a_w reduction was most effective below 0.63. An increase of around one order of magnitude in betalain degradation rates was observed when a_w increased from 0.32 to 0.75 (Cohen and Saguy, 1983). Serris and Biliaderis (2001) revealed that betanin had the greatest degradation at a_w of 0.64 in encapsulated beet root pigments, being explained by lowering mobility of the reactants at reduced a_w values. *Amaranthus* pigment powders showed higher stability than the respective aqueous solutions, being ascribed to varying a_w values (Cai *et al.*, 1998). Additionally, some stabilizers like pectin, guar gum, and locust bean gum appeared to improve storage stability of red beet solutions by lowering the a_w value. Betacyanins' stability was reported to increase after reduction of a_w by spray-drying (Cai and Corke, 2000) and by concentration (Castellar *et al.*, 2006).

The study from Lavellia *et al.* (2007) gives rise to some practical points about processing and storage conditions required to maintain high carotenoid contents in dehydrated carrots. Partial dehydration of carrots to intermediate moisture levels could be proposed instead of removing water completely, according to the following protocols: (1) reduction of a_w values to 0.31–0.54, corresponding to 6–11% of moisture (on wet weight basis) – in this a_w range microbial growth is arrested, enzymatic activity and non-enzymatic browning are at minimum, and our data indicate maximum carotenoid stability; and (2) reduction of a_w values to 0.54–0.75, corresponding to 11–22% of moisture – In this a_w range the microbial growth rate and the enzymatic activity are still at minimum; however, the most effective factors which account for carotenoid stability are still to be investigated. Furthermore, the occurrence of non-enzymatic browning cannot be ruled out. Both criteria should be combined with optimized packaging conditions, which reduce exposure of product to air and light during storage.

There is no direct effect of a_w on the degradation of isoflavones in soy protein products, but high a_w (above 0.6) may favor the activities of endogenous β-glucosidase, which hydrolyze glucosides to their respective aglycones (Huang *et al.*, 2009).

15.9.5 Atmosphere

In the presence of O_2, both betanidin and betanin were found to be unstable. The stability of betanin was negatively correlated with oxygen concentration (Czapski, 1990), indicating the involvement of O_2 in betanin degradation. The degradation kinetics of betanin influenced by O_2 was reported (Attoe and von Elbe, 1984). Conversely, betanin stability was observed to be improved in a N_2 environment (Drunkler *et al.*, 2006). It was reported that amaranthin was less stable than betanin under anaerobic conditions (Huang and von Elbe, 1986).

Oxygen was a critical factor in β-carotene degradation. It was also found that oxidation was the major cause of β-carotene destruction. Exclusion of oxygen during storage of powders would extend their shelf life (Wagner and Warthesen, 1995). Carotenoids, even in the crystalline state, are susceptible to oxidation and may be broken down rapidly if samples are stored in the presence of even traces of oxygen. The structures are broken down when attacked by free radicals, such as singlet molecular oxygen and other reactive species. The mechanism of carotenoids oxidative degradation can be described as auto-oxidation:

$$BC\bullet + ROO\bullet \rightarrow BROO\text{-}BC\bullet \tag{a}$$

$$ROO\text{-}BC\bullet + ROO\bullet \rightarrow \text{non-radical products} \tag{b}$$

$$ROO\text{-}BC\bullet + O_2 \rightarrow ROO\text{-}BC\text{-}OO\bullet \tag{c}$$

(BC•: β-carotene; ROO•: peroxyl radical)

When the reaction follows equation (a) then (c), β-carotene is degraded without any free radical being trapped. This is auto oxidation. In contrast, the antioxidant reaction follows equation (a) then (b), with consumption of radicals. Both reactions depend on the oxygen partial pressure (PO_2). At high PO_2, oxygen addition to the β-carotene-radical adduct is favored, β-carotene-derived peroxyl radicals are formed, and most of the β-carotene are consumed by auto oxidation. However, at low PO_2, addition of oxygen to the β-carotene-radical adduct is less favored, and the adduct may trap a second peroxyl radical to produce an antioxidant effect.

In photosensitized oxidation, carotenoids, especially lycopene and β-carotene, are effective quenchers of singlet oxygen ($^1O_2$). The quenching of $^1O_2$ by β-carotene can be the physical quenching of: (1) excited sensitizer molecules or (2) singlet oxygen. Physical quenching proceeds by energy transfer from $^1O_2$ to the carotenoids molecule. A similar process can occur between a carotenoid and an excited sensitizer. If truly catalytic, the carotenoid should remain intact. However, usually a chemical reaction sets in, destroying the carotenoid molecules (Bonnie and Choo, 1999).

$$^1O_2 + CAROTENOID \rightarrow {}^3O_2 + {}^3CAROTENOID \tag{d}$$

$$^3CAROTENOID \rightarrow CAROTENOID + HEAT \tag{e}$$

Pérez and Mínguez (2000) indicated that a more convenient strategy to prevent carotenoids degradation in oily food additives such as paprika oleoresins should be to minimize contact of food with oxygen. This could be achieved by applying vacuum, use of novel packaging materials, or by encapsulation techniques, thus avoiding oxygen-mediated auto-oxidation reactions.

Storage under regular atmosphere revealed that the keeping quality for all four glucosinolates is between four and seven days, while storage under 1.5 kPa O_2 and 15 kPa CO_2 showed a keeping quality of at least 14 days, indicating that to preserve the nutritional quality of broccoli, modified atmosphere packing (MAP) is a viable option (Rangkadilok et al., 2002). Controlled atmosphere (CA) storage is very effective in maintaining broccoli quality, and can double post-harvest life (Rodrigues and Rosa, 1999). Ideal atmospheres to maintain quality were 1–2% O_2; 5–10% CO_2 when temperatures were kept between 0 and 5°C (Schouten et al., 2009). Care needs to be taken that O_2 does not drop below 1% as this can cause the development of off-odors (Cantwell and Suslow, 1999). 'Marathon' broccoli heads stored for 25 days at 4°C, under a CA atmosphere of 1.5% O_2, 6% CO_2 contained significantly higher glucoraphanin levels than heads stored in air at the same temperature.

Glucoraphanin and glucoiberin contents reflected the rises in total glucosinolates. Transferring heads to air after storage had no effect on glucosinolate content. The reported increase in glucoraphanin under air is puzzling, as it contradicts much of the storage work showing a decline in glucosinolates in broccoli heads held in air at 4 and 20 °C (Forney *et al.*, 1991). As it is often difficult to maintain low temperatures throughout the broccoli distribution and marketing phase and in fluctuating temperatures, MAP can help extend shelf life. Optimum broccoli quality was obtained when atmospheres within MAP reached 1–2% O_2 and 5–10% CO_2. At 20 °C, however, broccoli in air lost 50% of its glucoraphanin in seven days. In contrast, under MAP there was no significant decrease in glucoraphanin over ten days.

Catechins are effective scavengers and more superior than vitamin C and E with respect to some active oxygen radicals, but they have less prominent effect on hydroxyl free radicals. It was shown that the rate of catechin oxidation increases with pH and concentration of oxygen (Chen *et al.*, 2001; Mochizuki *et al.*, 2002). Another study indicated that as the number of phenolic hydroxyl groups increases, the more rapid they can scavenge peroxyl radicals, thus, EGCG can scavenge peroxyl radicals more quickly than EC and EGC; and, furthermore, the pyrogallol structure in the B ring plays an important role in rapid scavenging ability of catechins, therefore, EGC can scavenge peroxyl radicals more quickly than EC (Kondo *et al.*, 2001). The oxidation mechanism was proposed to undergo sequential steps, involving the loss of hydrogen atoms, generation of semiquinone radical intermediate (reversible), and formation of quinine oxidized product (irreversible).

15.10 Conclusion

Phytochemicals are plant-derived secondary compounds that offer functions to improve the overall appearance in foods, and may exert physiological effects beyond nutrition promoting human health and well-being. In order to maintain phytochemicals' functionality and better physiological and biochemical activities, new technologies and measurements need to be taken to minimize phytochemicals' degradation. For instance, anthocyanins as colorants commercially have been limited due to their lack of stability and difficult purification. Carotenoids are the natural yellow-orange color range pigments present in a selection of foods. However, they are poorly soluble in water. Betalain sources such as betaxanthin may substitute carotenoids' utilization in the range of yellow-orange as food colorants. Anthocyanins are considered as the most widely applied natural colors. Nevertheless, the instability of anthocyanins at pH values above 3 makes betacyanins the natural color of selection to offer red-purple color utilized in low acid foods. Additionally, intensified research is needed to fill the gaps of intraspecific variability, horticultural, and technological measures targeting at optimizing phytochemicals' quality and quantity. Moreover, phytochemical loss can be minimized during processing and storage by selecting the respective temperature and pH regimes as well as minimizing oxygen and light access.

It is becoming evident that beneficial aspects of phytochemicals supporting human defense mechanisms are of increasing interest. However, the biological effects of phytochemicals are far from being clear. For instance, there is little information published with respect to the physiological role of betalains in mammals. Less data were found to be associated with changes of health benefits such as antioxidant capacity of phytochemicals after being subjected to heat, light, pH, and processing treatments. It remains to be clarified if the whole structures or rather their degraded compounds are responsible for the reported bioactivities.

The relationship between phytochemicals' stability changes and health benefit would be the next urgent topic. Future studies including pharmacokinetics, bioavailability, and tissue distribution of ingested compounds, and interference with signal transduction pathways need to be conducted. Structural transformations with altering pH, charge changes, copigmentation, and the formation of degradation products will also need to be considered to assess the benefit of phytochemicals.

References

Abdel-Aal, E. –S. M., and Hucl, P. (2003) Composition and stability of anthocyanins in blue-grained wheat. *Journal of Agricultural and Food Chemistry*, 51, 2174–2180.

Abdel-Aal, E. –S, Young, J.S.M, and Rabalski, I. (2006) Anthocyanin Composition in Black, Blue, Pink, Purple, and Red Cereal Grains. *Journal of Agricultural and Food Chemistry*, 54, 4696–4704.

Abdel-Aal, E.-S.M., Young, J.C., Wood, P.J., Rabalski, I., Hucl, P., Falk, D. *et al.* (2002) Einkorn: A potential candidate for developing high lutein wheat. *Cereal Chemistry*, 79, 455–457.

Aguiar, C.L., Baptista, A.S., Walder, J.M.M., Tsai S.M., Carra-O-Panizzi, M.C., and Kitajima, E.W. (2009) Changes in isoflavone profiles of soybean treated with gamma irradiation. *International Journal of Food Science and Nutrition*, 60, 387–394.

Alluis, B., Perol, N., Elhajji, H., and Dangles, O. (2000) Water-soluble flavonol (3-hydroxy-2-phenyl-4H-1-benzopyran-4-one) derivatives: Chemical synthesis, colouring, and antioxidant properties. *Helvetica Chimica Acta*, 83, 428–443.

Altamirano, R.C., Drdá k, M., Simon, P., Rajniakova, A., Karovicová, J. and Preclík, L. (1993) Thermal degradation of betanine in various water alcohol model systems. *Food Chemistry*, 46, 73–75.

Arts, I.C.W., Putte, B., and Hollman, P.C.H. (2000) Catechin Contents of Foods Commonly Consumed in The Netherlands. 1. Fruits, Vegetables, Staple Foods, and Processed Foods. *Journal of Agricultural and Food Chemistry*, 48, 1746–1751.

Asen, S., Stewart, R.N., and Norris, K.H. (1977) Anthocyanin and pH involved in the color of Heavenly Blue morning glory. *Phytochemistry*, 16, 1118.

Attoe, E.L. and von Elbe, J.H. (1981) Photochemial Degradation of Betanine and Selected Anthocyanins. *Journal of Food Sciences*, 46, 1934–1937.

Attoe, E.L. and von Elbe, J.H. (1984) Oxygen involvement in betanin degradation. *Z Lebensm Unters Forsch*, 179, 232–236.

Attoe, E.L. and von Elbe, J.H. (1985) Oxygen involvement in betanine degradation: effect of antioxidants. *Journal of Agricultural and Food Chemistry*, 50, 106–110.

Austin, F.L., Gent, C.A., and Wolff, I.A. (1968a) Degradation of natural thioglucosides with ferrous salts. *Journal of Agricultural and Food Chemistry*, 16, 752–755.

Austin, F.L., Gent, C.A., and Wolff, I.A. (1968b) Enantiomeric 3-hydroxy-pent-4- enethionamides from thioglucosides of Crambe and Brassica seeds by action of ferrous salts. *Canadian Journal of Chemistry*, 46, 1507–1512.

Bakowska, A., Kucharska, A.Z., and Oszmianski, J. (2003) The effects of heating, UV irradiation, and storage on stability of the anthocyanin-polyphenol copigment complex. *Food Chemistry*, 81, 349–355.

Balentine, D.A. (1997) Tea. In: Kirk-Othmer Encyclopedia of Chemical Technology 4th ed. New York: John Wiley & Sons, Inc.

Bao, J.S., Cai, Y.Z., Sun, M., Wang, G.Y., and Corke, H. (2005) Anthocyanins, flavonols, and free radical scavenging activity of Chinese bayberry (Myrica rubra) extracts and their color properties and stability. *Journal of Agricultural and Food Chemistry*, 53, 2327–2332.

Baranac, J.M., Petranović, N.A., and Dimitrić-Marković, J.M. (1996) Spectrophotometric study of anthocyan copigmentation reactions. *Journal of Agricultural and Food Chemistry*, 44, 1333–1336.

Barbosa, A.C.L., Lajolo, F.M., and Genovese, M.I. (2006) Influence of temperature, pH and ionic strength on the production of isoflavone-rich soy protein isolates. *Food Chemistry*, 98, 757–766.

Barrera, F.A., Reynoso, C.R., and Gonzáles de Mejía, E. (1998) Estabilidad de las betala´ınas extra ´ıdas del garambullo (*Myrtillocactus geometrizans*). *Food Science Technology International*, 4, 115–120.

Bassa, I.A. and Francis, F.J. (1997) Stability of anthocyanins from sweet potatoes in a model beverage. *Journal of Food Science*, 52, 1753–1755.

Baublis, A., Spomer, A., and Berber-Jiménez, M.D. (1994) Anthocyanin pigments: comparison of extract stability. *Journal of Food Science*, 59, 1219–1221.

Bazinet, L., Araya-Farias, M., Doyen, A., Trudel, D., and Têtu, B. (2010) Effect of process unit operations and long-term storage on catechin contents in EGCG-enriched tea drink. *Food Research International*, 43, 1692–1701.

Berké, B., Chéze, C., Vercauteren, J., and Deffieux, G. (1998) Bisulfite addition to anthocyanins: revisited structures of colourless adducts. *Tetrahedron Letters*, 39, 5771–5774.

Bloor, S.J. and Falshaw, R. (2000) Covalently linked anthocyanin-flavonol pigments from blue *Agapanthus* flowers. *Phytochemistry*, 53, 575–579.

Bonaventura, S. and Russo, C. (1993) Refrigeration of blood oranges destined for transformation. *Fruit Processing*, 10, 284–289.

Bones, A.M. and Rossiter, J.T. (2006) The enzymic and chemically induced decomposition of glucosinolates. *Phytochemistry*, 67, 1053–1067.

Bonnie, T.Y. and Choo, Y.M. (1999) Oxidation and thermal degradation of carotenoids. *Journal of Oil Palm Research*, 2, 62–78.

Boon, C.S., McClements, D.J., Weiss, J., and Decker, E.A. (2010) Factors influencing the chemical stability of carotenoids in foods. *Critical Reviews in Food Science and Nutrition*, 515–532.

Borek, V. and Morra, M.J. (2005) Ionic thiocyanate (scn-) production from 4-hydroxybenzyl glucosinolate contained in sinapis alba seed meal. *Journal of Agricultural and Food Chemistry*, 53, 8650–8654.

Boulton, R. (2001) The copigmentation of anthocyanins and its role in the color of red wine: a critical review. *American Journal of Enology and Viticulture*, 52, 67–87.

Buckow, R., Kastell, A., Terefe, N.S., and Versteeg, C. (2010) Pressure and temperature effects on degradation kinetics and storage stability of total anthocyanins in blueberry juice. *Journal of Agricultural and Food Chemistry*, 58, 10076–10084.

Cabrita, L., Fossen, T., and Andersen, M. (2000) Colour and stability of the six common anthocyanidin 3-glucosides in aqueous solutions. *Food Chemistry*, 68, 101–107.

Cai, Y. and Corke, H. (2000) Production and properties of spray-dried *Amaranthus* betacyanin pigments. *Journal of Food Science*, 65, 1248–1252.

Cai, Y., Sun, M., and Corke, H. (1998) Colorant properties and stability of *Amaranthus* betacyanin pigments. *Journal of Agricultural and Food Chemistry*, 46, 4491–4495.

Cai, Y., Sun, M., and Corke, H. (2001) Identification and Distribution of Simple and Acylated Betacyanins in the Amaranthaceae. *Journal of Agricultural and Food Chemistry*, 49, 1971–1978.

Cai, Y. and Corke, H. (2001) Effect of postharvest treatments on *Amaranthus* betacyanin degradation evaluated by visible/near-infrared spectroscopy. *Journal of Food Science*, 66, 1112–1118.

Cai, Y., Sun, M., and Corke, H. (2005) Characterization and application of betalain pigments from plants of the *Amaranthaceae*. *Trends in Food Science Technology*, 16, 370–376.

Camire, M.E., Chaovanalikit, A., Dougherty, M.P., and Briggs, J. (2002) Blueberry and grape anthocyanins as breakfast cereal colorants. *Journal of Food Science*, 67, 438–441.

Canfield, L.M., Krinski, N.I., and James, A.O. (eds) (1993) *Carotenoids in Human Health*, New York, Academy of Science.

Cantwell, M. and Suslow, T. (1999) Broccoli: recommendations for maintainin postharvest quality. http://www.postharvest.ucdavis.edu/produce/producefacts/ veg/broccoli.shtml.

Cardona, J.A., Lee, J-H., and Talcott, S.T. (2009) Color and Polyphenolic Stability in Extracts Produced from Muscadine Grape (Vitis rotundifolia) Pomace. *Journal of Agricultural and Food Chemistry*, 57, 8421–8425.

Castellar, M.R., Obón, J.M., and Fernández-López, J.A. (2006) The isolation and properties of a concentrated red-purple betacyanin food colourant from Opuntia stricta fruits. *Journal of the Science of Food and Agriculture*, 86, 122–128.

Chen, B.H. and Chen, Y.Y. (1993) Stability of chlorophylls and carotenoids in sweet potato leaves during microwave cooking. *Journal of Agricultural and Food Chemistry*, 41, 1315–1320.

Chen, H.E., Peng, H.Y., and Chen, B.H. (1996) Stability of carotenoids and vitamin A during storage of carrot juice. *Food Chemistry*, 57, 497–503.

Chen, J.P., Tai, C.Y., and Chen, B.H. (2007) Effects of different drying treatments on the stability of carotenoids in Taiwanese mango (*Mangifera indica* L.). *Food Chemistry*, 100, 1005–1010.

Chen, P.N., Kuo, W.H., Chiang, C.L., Chiou, H.L. *et al.* (2006) Black rice anthocyanins inhibit cancer cells invasion via repressions of MMPs and u-PA expression. *Chemico–Biological Interacttions*, 163, 218–229.

Chen, Z., Zhu, Q.Y., Tsang, D., and Huang, Y. (2001) Degradation of Green Tea Catechins in Tea Drinks. *Journal of Agricultural and Food Chemistry*, 49, 477–482.

Chevolleau, S., Gasc, N., Rollin, P., and Tulliez, J. (1997) Enzymatic, chemical, and thermal breakdown of ^{3}H-labeled glucobrassicin, the parent indole glucosinolate. *Journal of Agricultural and Food Chemistry*, 45, 4290–4296.

Chien, J.T., Hsieh, H.C., Kao, T.H., and Chen, B.H. (2005) Kinetic model for studying the conversion and degradation of isoflavones during heating. *Food Chemistry*, 91, 425–434.

Choi, M.H., Kima, G.H., and Lee, H.S. (2002) Effects of ascorbic acid retention on juice color and pigment stability in blood orange (Citrus sinensis) juice during refrigerated storage. *Food Research International*, 35, 753–759.

Cinar, I. (2005) Stability studies on the enzyme extracted sweet potato carotenoproteins. *Food Chemistry*, 89, 397–401.

Clarke, D.B. (2010) Glucosinolates, structures and analysis in food. *Food Analytical Methods*, 2, 310–325.

Cohen, E. and Saguy, I. (1983) Effect of water activity and moisture content on the stability of beet powder pigments. *Journal of Food Science*, 48, 703–707.

Corrales, M., Lindauer, R., Butz, P., and Tauscher, B. (2008) Effect of heat/pressure on cyanidin-3-glucoside ethanol model solutions. *Journal of Physics: Conference Series*, 121, 142003.

Coward, L., Smith, M., Kirk, M., and Barnes, S. (1998) Chemical modification of isoflavones in soyfoods during cooking and processing. *American Journal of Clinical Nutrition*, 68, 1486S–1491S.

Czapski, J. (1990) Heat stability of betacyanins in redbeet juiceandin betanine solutions. *Z Lebensm Unters Forsch*, 191, 275–278.

Da Silva, P.F. and Moreira, R.G. (2008) Vacuum frying of high-quality fruit and vegetable-based snacks. *LWT–Food Science Technolnology*, 41, 1758–1767.

Dangles, O., Saito, N., and Brouillard, R. (1993) Anthocyanin intramolecular copigment effect. *Phytochemistry*, 34, 119–124.

Dangles, O., Stoeckel, C., Wigand, M.C., and Brouillard, R. (1992) Two very distinct types of anthocyanin complexation: copigmentation and inclusion. *Tetrahedron Letters*, 33, 5227–5230.

Davies, C.G.A., Netto, F.M., Glassenap, N., Gallaher, C.M., Labuza, T.P., and Gallaher, D.D. (1998) Indication of the maillard reaction during storage of protein isolates. *Journal of Agricultural and Food Chemistry*, 46, 2485–2489.

De Ancos, B., Ibanez, E., Reglero, G. et al. (2000) Frozen storage effects on anthocyanins and volatile compounds of raspberry fruit. *Journal of Agricultural and Food Chemistry*, 48, 873–879.

Del Pozo-Insfran, D., Del Follo-Martinez, A., Talcott, S.T., and Brenes, C.H. (2007) Stability of copigmented anthocyanins and ascorbic acid in muscadine grape juice processed by high hydrostatic pressure. *Journal of Food Science*, 72, S247–S253.

Delgado-Vargas, F., Jimenez, A.R., and Paredes-Lopez, O. (2000) Natural pigments: carotenoids, anthocyanins, and betalains: characteristics, biosynthesis, processing, and stability. *Critical Reviews in Food Science Nutrition*, 40, 173–289.

Del Pozo-Insfran, D., Brenes, C.H., and Talcott, S.T. (2004) Phytochemical Composition and Pigment Stability of Açai (*Euterpe oleracea* Mart.). *Journal of Agricultural and Food Chemistry*, 52, 1539–1545.

Demeule, M., et al. (2002) Green tea catechins as novel antitumor and antiangiogenic compounds. *Current Medicinal Chemistry*, 2, 441–463.

Drunkler, D.A., Fett, R., and Bordignon-Luiz, M.T. (2006) Avaliacão da estabilidade de betalaı́nas em extrato de beterraba (Beta vulgaris L.) com α-, β-, e, γ-ciclodextrinas. *Boletim Centro de Pesquisa de Processamento de Alimentos*, 24, 259–276.

Dutta, D., Chaudhuri, U.R., and Chakraborty, R. (2005) Structure, health benefits, antioxidant property and processing and storage of carotenoids. *African Journal of Biotechnology*, 4, 1510–1520.

Duhard, V., Garnier, J.C., and Megard, D. (1997) Comparison of the stability of selected anthocyanins colorants in drink model systems. *Agro Food Industry Hi-Tech*, 8, 28–34.

Dyrby, M., Westergaard, N., and Stapelfeldt, H. (2001) Light and heat sensitivity of red cabbage extract in soft drink model systems. *Food Chemistry*, 72, 431–437.

Eden, T. (1976) The chemistry of tea leaf and of its manufacture. In: T. Eden, *Tea* (3rd edition) London: Longman Group Limited.

Eiro, M.J. and Heinonen, M. (2002) Anthocyanin color behavior and stability during storage: effect of intermolecular copigmentation. *Journal of Agricultural and Food Chemistry*, 50, 7461–7466.

Escribano-Bailón, T., Älvarez-García, M., Rivas-Gonzalo, J.C., Heredia, F.J., and Santos-Buelga, C. (2001) Color and stability of pigments derived from the acetaldehyde-mediated condensation between malvidin 3-*o*-glucoside and (+)-catechin. *Journal of Agricultural and Food Chemistry*, 49, 1213–1217.

Escribano, J., Gandía-Herrero, F., Caballero, N., and Pedreno, M.A. (2002) Subcellular localization and isoenzyme pattern of peroxidase and polyphenol oxidase in beet root (*Beta vulgaris* L.). *Journal of Agricultural and Food Chemistry*, 50, 6123–6129.

Fenwick G.R. and Heaney R.K. (1983) Glucosinolates and their breakdown products in cruciferous crops, foods and feeding stuffs. *Food Chemistry*, 11, 249–271.

Fenwick, G.R., Heaney, R.K., Gmelin, R., Rakow, D., and Thies, W. (1981) Glucosinalbin in *Brassica napus* – a re-evaluation. *Zeitschrift fuer Pflanzenzuechtung*, 87, 254–259.

Ferreres, F., Gil, M.I., and Tomas-Barberan, F.A. (1996) Anthocyanins and flavonoids from shredded red onion and changes during storage in perforated films. *Food Research International*, 29, 389–395.

Figueiredo, P., Elhabiri, M., Toki, K., Saito, N., Dangles, O., and Brouillard, R. (1996) New aspects of anthocyanin complexation. Intramolecular copigmentation as a means for colour loss. *Phytochemistry*, 41, 301–308.

Forney, C.F., Mattheis, J.P., and Austin, R.K. (1991) Volatile compounds produced by broccoli under anaerobic conditions. *Journal of Agriculrural and Food Chemistry*, 39, 2257–2259.

Fossen, T., Cabrita, L., and Andersen, Ø.M. (1998) Color and stability of pure anthocyanins influenced by pH including the alkaline region. *Food Chemistry*, 63, 435–440.

Francis, F.J. (2000) Anthocyanins and betalains: composition and applications. *Cereal Foods World*, 45, 208–213.

Friis, P., Larsen, P.O., and Olsen, C.E. (1977) Base-catalyzed Neber-typ rearrangement of glucosinolates [1-(b-D-glucosylthio)-N-(sulfonatooxy)alkylideneamines]. *Journal of the Chemistry Society, Perki Trans I: Org Bioorg Chem (1972–1999)*, 1977, 661–665.

García-Viguera, C., Zafrilla, P., Romero, F., and Abellán, P. *et al.* (1999) Color stability of strawberry jam as affected by cultivar and storage temperature. *Journal if Food Science*, 64, 243–247.

Garzón, G.A. and Wrolstad, R.E. (2001) The stability of pelargonidinbased anthocyanins at varying water activity. *Food Chemistry*, 75, 185–196.

Gentile, C., Tesoriere, L., Allegra, M., Livrea, M.A., and D'Alessio, P. (2004) Antioxidant betalains fromcactus pear (*Opuntia ficus-indica*) inhibits endothelial ICAM-1expression. *Annals of the New York Academy of Science*, 1028, 481–486.

George, F., Figueiredo, P., Toki, K., Tatsuzawa, F., Saito, N., and Brouillard, R. (2001) Influence of *trans-cis*-isomerization of coumaric acid substituents on colour variance and stabilisation in anthocyanins. *Phytochemistry*, 57, 791–795.

Gmelin, R. and Virtanen, A.I. (1961) Glucobrassicin der precursor von SCN-, 3-indolylacetonitril and ascorbigen in *Brassica oleracea* species. *Annals of the New York Academy of Science*, 107, 1–25.

Gonzalez-Manzano, S., Mateus, N., de Freitas, V., and Santos-Buelga, C. (2008) Influence of the Degree of Polymerisation in the Ability of Catechins to Act as Anthocyanin Copigments. *European Food Research and Technology*, 227, 83–92.

Gouveia, L. and Empis, J. (2003) Relative stabilities of microalgal carotenoids in microalgal extracts, biomass and fish feed: effect of storage conditions. *Innovative Food Science and Emerging Technologies*, 4, 227–233.

Gradinaru, G., Biliaderis, C.G., Kallithraka, S., Kefalas, P., and Garcia-Viguera, C. (2003) Thermal stability of *Hibiscus sabdariffa* L. anthocyanins in solution and in solid state: effects of copigmentation and glass transition. *Food Chemistry*, 83, 423–436.

Gris, E.F., Ferreira, E.A., Falcao, L.D., and Bordignon-Luiz, M.T. (2007) Caffeic acid copigmentation of anthocyanins from Cabernet Sauvignon grape extracts in model systems. *Food Chemistry*, 100, 1289–1296.

Grun, I.U. *et al.* (2001) Changes in the profile of genistein, daidzein, and their conjugates during thermal processing of tofu. *Journal of Agricultural and Food Chemistry*, 49, 2839–2843.

Gulati, A., Rawat, R., Singh, B., and Ravindranath, S.D. (2003) Application of Microwave Energy in the Manufacture of Enhanced-Quality Green Tea. *Journal of Agricultural and Food Chemistry*, 51, 4764–4768.

Henning, W. (1981) Flavonolglycoside der Erdbeered (*Fragaria x ananassa* Duch.), Himbeeren (*Rubus idaeus* L.) und Brombeeren (*Rubus fruticosus* L.). *Z. Lebensm Unters Forsch.*, 173, 180–187.

Herbach, K.M., Stintzing, F.C., and Carle, R. (2004a) Impact of thermal treatment on color and pigment pattern of red beet (*Beta vulgaris* L.) preparations. *Journal of Food Science*, 69, C491–C498.

Herbach, K.M., Stintzing, F.C., and Carle, R. (2004b) Thermal degradation of betacyanins in juices from purple pitaya (*Hylocereus polyrhizus* [Weber] Britton and Rose) monitored by high-performance liquid chromatography-tandemmassspectrometric analyses. *European Food Research and Technology*, 219, 377–385.

Herbach, K.M., Stintzing, F.C., and Carle, R. (2005) Identification of heat-induced degradation products from purified betanin, phyllocactin and hylocerenin by high-performance liquid chromatography/ electrospray ionization mass spectrometry. *Rapid Comm Mass Spectrom.*, 19, 2603–2616.

Herbach, K.M., Stintzing, F.C., and Carle, R. (2006a) Stability and color changes of thermally treated betanin, phyllocactin, and hylocerenin solutions. *Journal of Agricultural and Food Chemistry*, 54, 390–398.

Herbach, K.M., Rohe, M., Stintzing, F.C., and Carle, R. (2006b) Structural and chromatic stability of purplepitaya (*Hylocereus polyrhizus* [Weber] Britton and Rose) betacyanins as affected by the juice matrix and selected additives. *Food Research International*, 39, 667–677.

Hidalgo, A., Brandolini, A., and Pompei, C. (2010) Carotenoids evolution during pasta, bread and water biscuit preparation from wheat flours. *Food Chemistry*, 121, 746–751.

Hider, R.C., Liu, Z.D., and Khodr, H.H. (2001) Metal chelation of polyphenols. *Methods in Enzymology*, 335, 190–203.

Hiemori, M., Koh, E., and Mitchell, A.E. (2009) Influence of Cooking on Anthocyanins in Black Rice (*Oryza sativa* L. japonica var. SBR). *Journal of Agricultural and Food Chemistry*, 57, 1908–1914.

Hoshino, T. and Tamura, H. (1999) Anthocyanidin glycosides: color variation and color stability. In: R. Ikan (ed.) *Naturally Occurring Glycosides*, New York/Weinheim: John Wiley & Sons, pp. 43–82.

Huang, A.S. and von Elbe, J.H. (1986) Stability comparison of two betacyanine pigments-Amaranthine and betanine. *Journal of Food Science*, 51, 670–674.

Huang, R., Chou, C. Stability of Isoflavone Isomers in Steamed Black Soybeans and Black Soybean Koji Stored under Different Conditions. *Journal of Agricultural and Food Chemistry*, 57, 1927–1932.

Inami, O., Tamura, I., Kikuzaki, H., and Nakatani, N. (1996) Stability of anthocyanins of *Sambucus canadensis* and *Sambucus nigra*. *Journal of Agricultural and Food Chemistry*, 44, 3090–3096.

Jackman, R.L. and Smith, J.L. (1996) Anthocyanins and betalains. In: G.F. Hendry, and J.D. Houghton (eds) *Natural Food Colorants*, London: Blackie Academic and Professional, pp. 244–309.

Jackson, C.-J.C. et al. (2002) Effects of processing on the content and composition of isoflavones during manufacturing of soy beverage and tofu. *Process Biochemistry*, 37, 1117–1123.

Jiménez, N., Bohuon, P., Lima, J., Dornier, M., Vaillant, F., and Perez, A.M. (2010) Kinetics of anthocyanin degradation and browning in reconstituted blackberry juice treated at high temperatures (100–180 °C). *Journal of Agricultural and Food Chemistry*, 58, 2314–2322.

Jones, R. B., Faragher, J, B., Winkler, S. A review of the influence of postharvest treatments on quality and glucosinolate content in broccoli (*Brassica oleracea* var. italica) heads. *Postharv Biol Technol.*, **2006**, 41, 1–8.

Jørjensen, K. and Skibsted, L.H. (1993) Carotenoid scavenging of radicals. Effect of carotenoid structure and oxygen partial pressure on antioxidative activity. *Z. Lebensm.-Unters.-Forsch.*, 196, 423–429.

Kaack, K. and Austed, T. (1998) Interaction of vitamin C and flavonoids in elderberry (*Sambucus nigra* L.) during juice processing. *Plant Foods and Human Nutrition*, 52, 187–198.

Kader, F., Haluk, J.-P., Nicolas, J.-P., and Metche, M. (1998) Degradation of cyanidin 3–glucoside by blueberry polyphenol oxidase: kinetic studies and mechanisms. *Journal of Agricultural and Food Chemistry*, 46, 3060–3065.

Kanner, J., Harel, S., and Granit, R. (2001) Betalains – A new class of dietary cationized antioxidants. *Journal of Agricultural and Food Chemistry*, 49, 5178–5185.

Kearsley M.W. and Rodriguez N. (1981) The stability and use of natural colours in foods: anthocyanin, p-carotene and riboflavin. *Journal of Food Technology*, 16, 421–431.

Kelly, P.J., Bones, A., and Rossiter, J.T. (1998) Sub-cellular immunolocalization of the glucosinolate sinigrin in seedlings of *Brassica juncea*. *Planta*, 206, 370–377.

Kim, E.S. et al. (2007) Impact of heating on chemical compositions of green tea liquor. *Food Chemistry*, 103, 1263–1267.

Kirca, A. and Cemeroglu, B. (2003) Degradation kinetics of anthocyanins in blood orange juice and concentrate. *Food Chemistry*, 81, 583–587.

Kondo, K., Kurihara, M., and Fukuhara, K. (2001) Mechanism of Antioxidant Effect of Catechins. *Methods of Enzymology*, 335, 203–217.

Kouniaki, S., Kajda, P., and Zabetakis, I. (2004) The effect of high hydrostatic pressure on anthocyanins and ascorbic acid in blackcurrants (*Ribes nigrum*). *Flavour Fragrance Journal*, 19, 281–286.

Kwon, S.H., Ahn, I.S., Kim, S.O., Kong, C.S., Chung, H.Y., Do, M.S., and Park, K.Y. (2007) Anti-obesity and hypolipidemic effects of black soybean anthocyanins. *Journal of Medicinal Food*, 10, 552–556.

Lavellia, V., Zanonib, B., and Zaniboni, A. (2007) Effect of water activity on carotenoid degradation in dehydrated carrots. *Food Chemistry*, 104, 1705–1711.

Lee, S., Kim, S., Jeong, S., and Park, J. (2006) Effect of far-infrared irradiation on green tea. *Journal of agricultural and Food Chemistry*, 54, 399–403.

Lee, S.C., Jeong, S.M., Lee, J.M., Jang, A., Kim, D.H., and Jo, C. (2008) Effect of irradiation on total phenol and catechins contents and radical scavenging activity of green tea leaf and stem extract. *Journal of Food Biochemistry*, 32, 782–794.

Lin, C.H. and Chen, B.H. (2005) Stability of carotenoids in tomato juice during storage. *Food Chemistry*, 90, 837–846.

Lindsay, R.C. (1996) Food additives. In: O.R. Fennema (ed.) *Food Chemistry* Marcel Dekker: New York, pp. 768–823.

Lopez-Berenguer, C., Carvajal, M., and Moreno, D.A., Garcia-Viguera, C. (2007) Effects of microwave cooking conditions on bioactive compounds present in broccoli inflorescences. *Journal of Agricultural and Food Chemistry*, 55, 10001–10007.

Ludikhuyze, L., Rodrigo, L., and Hendrickx, M. (2000) The activity of myrosinase from broccoli (*Brassica oleracea* L. cv. Italica): Influence of intrinsic and extrinsic factors. *Journal of Food Protection*, 63, 400–403.

Maccarone, E., Maccarrone, A., and Rapisarda, P. (1985) Stabilization of anthocyanins of blood orange fruit juice. *Journal of Food Science*, 50, 901–904.

Mahungu, S.M., Diaz–Mercado, S., Li, J., Schwenk, M., Singletary, K., and Faller, J. (1999) Stability of Isoflavones during Extrusion Processing of Corn/Soy Mixture. *Journal of Agricultural and Food Chemistry*, 47, 279–284.

Malaypally, S.P. and Ismail, B. (2010) Effect of protein content and denaturation on the extractability and stability of isoflavones in different soy systems. *Journal of Agricultural and Food Chemistry*, 58, 8958–8965.

Malien-Aubert, C., Dangles, O., and Amiot, M.J. (2001) Color stability of commercial anthocyanin-based extracts in relation to the phenolic composition. Protective effects by intra- and intermolecular copigmentation. *Journal of Agricultural and Food Chemistry*, 49, 170–176.

Markakis, P. (1982) Stability of anthocyanins in foods. In: P. Markakis (ed.) *Anthocyaninsas Food Colors*, London: Academic Press Inc., pp. 163–180.

Martínez-Parra, J. and Muñoz, R. (2001) Characterization of betacyanin oxidation catalyzed by a peroxidase from Beta vulgaris L. roots. *Journal of Agricultural and Food Chemistry*, 49, 4064–4068.

Mathias, K., Ismail, B., Corvalan, C.M., andv Hayes, K.D. (2006) Heat and pH effects on the conjugated forms of genistin and daidzin isoflavones. *Journal of Agricultural and Food Chemistry*, 54, 7495–7502.

Matsuura, M. and Obata, A. (1993) β-glucosidases from soybeans hydrolyze daidzin and genistin. *Journal of Food Science*, 58, 144–147.

Matusheski, N.V., Wallig, M.A., Juvik, J.A., Klein, B.P., Kushad, M.M., and Jeffery, E. (2001) H. Preparative HPLC method for the purification of sulforaphane and sulforaphane nitrile from Brassica oleracea. *Journal of Agricultural and Food Chemistry*, 49, 1867–1872.

Mazza, G. and Brouillard, R. (1990) The mechanism of co-pigmentation of anthocyanins in aqueous solutions. *Phytochemistry*, 29, 1097–1102.

Mazza, G. and Miniati, E. (1993) *Anthocyanins in Fruits, Vegetables and Gains*, Boca Raton, Florida: CRC Press.

Melo, M.J., Moncada, M.C., and Pina, F. (2001) On the red colour of raspberry (*Rubus idaeus*). *Tetrahedron Leters*, 41, 1987–1991.

Midori, K., Tamiyoshi, S., Kinuyo, N., and Masaaki, T. (2001) Effects of pH and Metal Ions on Antioxidative Activities of Catechins. *Bioscience, Biotechnology and Biochemistry*, 65, 126–132.

Mishra, D.K., Dolan, K.D., and Yang, L. (2008) Confidence intervals for modeling anthocyanin retention in grape pomace during nonisothermal heating. *Journal of Food Science*, 73, E9–E15.

Mistry, T.V., Cai, Y., Lilley, T.H., and Haslam, E. (1991) Polyphenol interactions. Part 5: anthocyanin co-pigmentation. *Journal of the Chemistry Society Perkin Transactions*, 2, 1287–1296.

Moßhammer, M.R., Stintzing, F.C., and Carle, R. (2005) Development of a process for the production of a betalain-based colouring foodstuff from cactus pear. *Innovative Food Science and Emerging Technologies*, 6, 221–231.

Mochizuki, M., Yamazaki, S., Kano, K., and Ikeda, T. (2002) Kinetic analysis and mechanistic aspects of autoxidation of catechins. *Biochimica et Biophysica Acta*, 1569, 35–44.

Mohn, T., Cutting, B., Ernst, B., and Hamburger, M. (2007) Extraction and analysis of intact glucosinolates-A validated pressurized liquid extraction/liquid chromatography-mass spectrometry protocol for Isatis tinctoria, and qualitative analysis of other cruciferous plants. *Journal of Chromatology A*, 1166, 142–151.

Mollov, P., Mihalev, K., Shikov, V., Yoncheva, N., and Karagyozov, V. (2007) Colour stability improvement of strawberry beverage by fortification with polyphenolic copigments naturally occurring in rose petals. *Innovative Food Science and Emerging Technolologies*, 8, 318–321.

Morais, H., Ramos, C., Forgács, E., Cserháti, T., and Oliviera, J. (2002) Influence of storage conditions on the stability of monomeric anthocyanins studied by reversed-phase high-performance liquid chromatography. *Journal of Chromatology B*, 770, 297–301.

Murphy, P.A. et al. (1999) Isoflavones in retail and institutional soy foods. *Journal of Agricultural and Food Chemistry*, 47, 2697–2704.

Nachtigall, A.M., Da Silva, A.G., Stringheta, P.C., Silva, P.I., and Bertoldi, M.C. (2009) Correlation between spectrophotometric and colorimetric methods for the determination of photosensitivity and thermosensitivity of tomato carotenoids, *Boletim do Centro de Pesquisa e Processamento de Alimentos*, 27, 11–18.

Nufer, K.R., Ismail, B., and Hayes, K.D. (2009) The Effects of Processing and Extraction Conditions on Content, Profile, and Stability of Isoflavones in a Soymilk System. *Journal of Agricultural and Food Chemistry*, 57, 1213–1218.

Oerlemans, K., Barrett, D.M., Suades, C.B., Verkerk, R., and Dekker, M. (2006) Thermal degradation of glucosinolates in red cabbage. *Food Chemistry*, 95, 19–29.

Osmani, S.A., Hansen, E.H., Malien-Aubert, C.L., Olsen, C-E., Bak, S., and Møller, B.L. (2009) Effect of Glucuronosylation on Anthocyanin Color Stability. *Journal of Agricultural and Food Chemistry*, 57, 3149–3155.

Pacheco-Palencia, L.A., Duncan, C.E., and Talcott, S.E. (2009) Phytochemical composition and thermal stability of two commercial açai species, *Euterpe oleracea* and *Euterpe precatoria*. *Food Chemistry*, 115, 1199–1205.

Pacheco-Palencia, L., Hawken, P., and Talcott, S. (2007) Juice matrix composition and ascorbic acid fortification effects on the phytochemical, antioxidant, and pigment stability of açai (*Euterpe oleracea* Mart.). *Food Chemistry*, 105, 28–35.

Palozza, P. and Krinsky, N.I. (1992) Antioxidant effects of carotenoids in vivo and in vitro: An overview. *Methods Enzymology*, 213, 403–420.

Parisa, S., Reza, H., Elham, G., and Rashid, J. (2007) Effect of Heating, UV irradiation and pH on the Stability of the Anthocyanin Copigment Complex. *Pakistan Journal of Biological Science*, 267–272.

Parkin, K.L. and Im, J-S. (1990) Chemical and physical changes in beet (*Beta vulgaris* L.) root tissue during simulated processing—Relevance to the "black ring" defect in canned beets. *Journal of Food Science*, 55, 1039–1041.

Pasch, J.H. and von Elbe, J.H. (1979) Betanine stability in buffered solutions containing organic acids, metal cations, antioxidants, or sequestrants. *Journal of Food Science*, 44, 72–74.

Patras, A., Brunton, N.P., Da Pieve, S., and Butler, F. (2009) Impact of high pressure processing on total antioxidant activity, phenolic, ascorbic acid, anthocyanin content and colour of strawberry and blackberry purées. *Innovative Food Science and Emerging Technologies*, 10, 308–313.

Pérez, A., Jarén, M.,and Mínguez, M.I. (2000) Effect of high-temperature degradative processes on ketocarotenoids present in paprika oleoresins. *Journal of Agricultural and Food Chemistry*, 48, 2966–2971.

Polyakov, N.E., Leshina, T.V., Meteleva, E.S., Dushkin, A.V., Konovalova, T.A., and Lowell, D. (2010) Enhancement of the photocatalytic activity of tio2 nanoparticles by water-soluble complexes of carotenoids. *Journal of Physical Chemistry*, 114, 14200–14204.

Price, C.L. and Wrolstad, R.E. (1995) Anthocyanin pigments of Royal Okanogan Huckleberry Juice. *Journal of Food Science*, 60, 369–374.

Proniuk, S., Liederer, B.M., and Blanchard, J. (2002) Preformulation study of epigallocatechin gallate, a promising antioxidant for topical skin cancer prevention. *Journal of Pharmceutical Science*, 91, 111–116.

Queiroz, F., Oliveira, C., Pinho, O., (2009) Ferreira, I.M.P.L.V.O. Degradation of anthocyanins and anthocyanidins in blueberry jams/stuffed fish. *Journal of Agricultural and Food Chemistry*, 57, 10712–10717.

Rangkadilok, N., Tomkins, B., Nicolas, M.E., Premier, R.R., Bennett, R.N., Eagling, D.R., and Taylor, P.W.J. (2002) The effect of post-harvest and packaging treatments on glucoraphanin concentration in broccoli (*Brassica oleracea* var. italica). *Journal of Agriclutrural and Food Chemistry*, 50, 7386–7391.

Rao, A.V. and Rao L. G. (2007) Carotenoids and human health. *Pharmacology Research*, 55, 207–216.

Rascón, M.P., Beristain, C.I., García, H.S., and Salgado, M.A. (2011) Carotenoid retention and storage stability of spray-dried encapsulated paprika oleoresin using gum Arabic and Soy protein isolate as wall materials. *LWT-- Food Science Technology*, 44, 549–557.

Rein, M.J. and Heinonen, M. (2004) Stability and enhancement of berry juice color. *Journal of Agricultural and Food Chemistry*, 52, 3106–3114.

Richardson, T. and Finley, J.W. (1985) Chemical changes in natural food pigments. In: T. Richardson and J.W. Finley (eds) *Chemical Chances in Food during Processing*, London: Springer Verlag, p. 431.

Rivas-Gonzalo, J.C., Bravo-Haro, S., and Santos-Buelga, C. (1995) Detection of compounds formed through the reaction of malvidin 3- monoglucoside and catechin in the presence of acetaldehyde. *Journal of Agricultural and Food Chemistry*, 43, 1444–1449.

Rock, C.L. (1997) Carotenoids: biology and treatment. *Pharmacology and Therapeutics*, 75, 185–197.

Rodrigues, A.S. and Rosa, E.A.S. (1999) Effect of post-harvest treatments on the level of glucosinolates in broccoli. *Journal of Science and Food Agriculture*, 79, 1028–1032.

Rodríguez-Saona, L.E., Giusti, M.M., and Wrolstad, R.E. (1999) Color and Pigment Stability of Red Radish and Red-Fleshed Potato Anthocyanins in Juice Model Systems. *Journal of Food Science*, 64, 451–456.

Rommel, A., Wrolstad, R.E., and Heatherbell, D.A. (1992) Blackberry juice and wine: processing and storage effects on anthocyanin composition, color and appearance. *Journal of Food Science*, 57, 385–391.

Rossi, M., Giussani, E., Morelli, R., Lo Scalzo, R., Nani, R.C., and Torreggiani, D. (2003) Effect of fruit blanching on phenolics and radical scavenging activity of highbush blueberry juice. *Food Research International*, 36, 999–1005.

Rostagno, M.A., Palma, M., and Barroso, C.G. (2005) Short-term stability of soy isoflavones extracts: Sample conservation aspects. *Food Chemistry*, 93, 557–564.

Rubinskiene, M., Jasutiene, I., Venskutonis, P., and Viskelis, P. (2005a) HPLC determination of the composition and stability of blackcurrant anthocyanins. *Journal of Chromatology Sciemce*, 43, 478–482.

Rubinskiene, M., Viskelis, P., Jasutiene, I., Viskeliene, R., and Bobinas, C. (2005b) Impact of various factors on the composition and stability of black currant anthocyanins. *Food Research International*, 38, 867–871.

Rungapamestry, V., Duncan, A.J., Fuller, Z., and Ratcliffe, B. (2006) Changes in glucosinolate concentrations, myrosinase activity, and production of metabolites of glucosinolates in cabbage (*Brassica oleracea* Var. capitata) cooked for different durations. *Journal of Agricultrural and Food Chemistry*, 54, 7628–7634.

Saito, N., Toki, K., Honda, T., and Kawase, K. (1998) Cyanidin 3-malonylglucuronylglucoside in *Bellis* and cyanidin 3-malonylglucoside in *Dendranthema*. *Phytochemistry*, 27, 2963–2966.

Sajilata, M.G., Singhal R.S., and Kamat, M.Y. (2008) The carotenoid pigment zeaxanthin-a review. *Comprehensive Reviews in Food Science and Food Safety*, 7, 29–49.

Saldana, G., Stephens, T.S., Lime, and B.J. (1976) Carrot beverages and lots of padding. *Journal of Food Science*, 41, 1243–1244.

Sapers, G.M. and Hornstein, J.S. (1979) Varietal differences in colorant properties and stability of red beet pigments. *Journal of Food Science*, 44, 1245–1248.

Sarma, A.D., Sreelakshmi, Y., and Sharma, R. (1997) Antioxidant ability of anthocyanins against ascorbic acid oxidation. *Phytochemistry*, 45, 671–674.

Sasaki, R., Nishimura, N., Hoshino, H., Isa, Y., Kadowaki, M., Ichi, T. *et al.* (2007) Cyanidin 3-glucoside ameliorates hyperglycemia and insulin sensitivity due to down regulation of retinol binding protein 4 expression in diabetic mice. *Biochemical Pharmacology*, 74, 1619–1627.

Sastry, L. and Tischer, R.G. (1953) Behavior of the anthocyanin pigments in concord grapes during heat processing and storage. *Food Technology*, 6, 82.

Satue-Gracia, M., Heinonen, I.M. and Frankel, E.N. (1997) Anthocyanins as antioxidants on human low-density lipoprotein and lecithinliposome systems. *Journal of Agricultural and Food Chemistry*, 45, 3362–3367.

Savolainen, K. and Kuusi, T. (1978) The stability properties of golden beet and red beet pigments: influence of pH, temperature, and some stabilizers. *Z Lebensm Unters Forsch*, 166, 19–22.

Schliemann, W. and Strack, D. (1998) Intramolecular stabilization of acylated betacyanins. *Phytochemistry*, 49, 585–588.

Schneider, W. and Becker, M.A. (1930) Case of Walden inversion in glucoside cleavage. *Naturwissenschaften*, 18, 133.

Schouten, R.E., Zhang, X.B., Verkerk, R., Verschoor, J.A., Otma, E.C. *et al.* (2009) Modeling the level of the major glucosinolates in broccoli as affected by controlled atmosphere and temperature. *Postharvest Biology and Technology*, 53, 1–10.

Schwartz, S.J. and von Elbe, J.H. (1983) Identification of betanin degradation products. *Zeitschrift für Lebensmittel-Untersuchung und -Forschung*, 176, 448–453.

Serris, G.S. and Biliaderis, C.G. (2001) Degradation kinetics of beetroot pigment encapsulated in polymeric matrices. *Journal of Science and Food Agriculture*, 81, 691–700.

Shahidi, F. and Naczk, M. (2004) *Phenolics in Food and Neutraceuticals*. Boca Raton, Florida: CRC Press.
Shenoy, V.R. (1993) Anthocyanins – prospective food colours. *Current Science*, 64, 575–579.
Shrikhande, A.J. and Francis, F.J. (1974) Effect of flavonols on ascorbic acid and anthocyanin stability in model system. *Journal of Food Science*, 39, 904–906.
Shi, J. and Sophia, J. (2008) Stability of lycopene during food processing and storage. *Agriculture and Agri-Food Canada, Lycopene*, 17–36.
Shi, Z.L., Lin, M. and Francis, F.J. (1992) Anthocyanins of *Tradescantia pallida* potential food colorants. *Journal of Food Science*, 57, 761–765.
Shih, C.C. and Wiley, R.C. (1981) Betacyanine and betaxanthine decolorizing enzymes in the beet (*Beta vulgaris* L.) root. *Journal of Food Science*, 47, 164–172.
Shimoni, E. (2004) Stability and shelf life of bioactive compounds during food processing and storage: soy isoflavones. *Journal of Food Science*, 69, R160–R166.
Sims, C.A., Balaban, M.O., and Mathews R.F. (1993) Optimization of carrot juice color and cloud stability. *Journal of Food Science*, 58, 1129–1131.
Skrede, G., Wrolstad, R.E., Lea, P., and Enersen, G. (1992) Color stability of strawberry and blackcurrant syrups. *Journal of Food Science*, 57, 172–177.
Song, L.J. and Thornalley, P.J. (2007) Effect of storage, processing and cooking on glucosinolate content of Brassica vegetables. *Food Chemistry and Toxicology*, 45, 216–224.
Stintzing, F.C. and Carle, R. (2004) Functional properties of anthocyanins and betalains in plants, food, and in human nutrition. *Trends in Food Science and Technology*, 15, 19–38.
Stintzing, F.C., Herbach, K.M., Moßhammer, M.R., Carle, R., et al. (2005) Color, betalain pattern, and antioxidant properties of cactus pear (*Opuntia* ssp.) clones. *Journal of Agricultural Food Chemistry*, 53, 442–451.
Stintzing, F.C., Schieber, A., and Carle, R. (2001) Phytochemical and nutritional significance of cactus pear. *European Food Research and Technology*, 212, 396–407.
Stintzing, F.C., Stintzing, A.S., Carle, R., Frei, B., and Wrolstad, R.E. (2002) Color and antioxidant properties of cyanidin-based anthocyanin pigments. *Journal of Agricultural and Food Chemistry*, 50, 6172–6181.
Strack, D., Steglich, W., and Wray, V. (1993) Betalains. In: P.M. Dey and J. Harborne (eds) *Methods in Plant Biochemistry*. London: Academic Press Limited, pp: 421–450.
Su, Y.L., Leung, L.K., Huang, Y., and Chen, Z. (2003) Stability of tea theaflavins and catechins. *Food Chemistry*, 83, 189–195.
Suthanthangjai, W., Kajda, P., and Zabetakis, I. (2005) The effect of high hydrostatic pressure on the anthocyanins of raspberry (*Rubus idaeus*). *Food Chemistry*, 90, 193–197.
Tai, C.Y. and Chen B.H. (2000) Analysis and stability of carotenoids in the flowers of daylily (*Hemerocallis disticha*) as affected by various treatments. *Journal of Agricultural and Food Chemistry*, 48, 5962–5968.
Talcott, S.T., Brenes, C.H., Pires, D.M., and Del Pozo-Insfran, D. (2003) Phytochemical stability and color retention of copigmented and processed muscadine grape juice. *Journal of Agricultural and Food Chemistry*, 51, 957–963.
Teh, L.S. and Francis, F.J. (1988) Stability of anthocyanins from *Zebrina pendula* and *Ipomoea tricolor* in a model beverage. *Journal of Food Science*, 53, 1580–1581.
Toivonen, P.M.A. and Forney, C. (2004) Broccoli. In: *The Commercial Storage of Fruits, Vegetables and Florist and Nursery Stock*. USDA, ARS Agriculture Handbook #66.
Tonon, R.V., Brabet, C., and Hubinger, M.D. (2008) Influence of process conditions on the physicochemical properties of acai (*Euterpe oleraceae* Mart.) powder produced by spray drying. *Journal of Food Engineering*, 88, 411–418.
Tsai, P-J., Sheu, C-H., Wu, P-H., and Sun, Y-F. (2010) Thermal and pH stability of betacyanin pigment of djulis (*Chenopodium formosanum*) in Taiwan and their relation to antioxidant activity. *Journal of Agricultural and Food Chemistry*, 58, 1020–1025.
Tsuda, T., Horio, F., Osawa, and T. Cyanidin 3-*O*-â-glucoside suppresses nitric oxide production during a zymosan treatment in rats. *Journal of Nutritional Science and Vitaminology*, 48, 305–310.
van Eylen, D., Oey, I., Hendrickx, M., and van Loey, A. (2008) Effects of pressure/temperature treatments on stability and activity of endogenous broccoli (*Brassica oleracea* L. cv. Italica) myrosinase and on cell permeability. *Journal of Food Engineering*, 89, 178–186.
van Poppel, G., Verhoeven, D.T.H., Verhagen, H., and Goldbohm, R.A. (1999) Brassica vegetables and cancer preventions epidemiology and mechanisms. *Advances in Experimental Medicine and Biology*, 472, 159–68.

Vera de Rossoa, V. and Mercadante, A.Z. (2007) Evaluation of colour and stability of anthocyanins from tropical fruits in an isotonic soft drink system. *Innovative Food Science and Emerging Technologies*, 8, 347–352.

Verkerk, R. and Dekker, M. (2004) Glucosinolates and myrosinase activity in red cabbage *Brassica oleracea* L. Var. Capitata f. rubra DC. after various microwave treatments. *Journal of Agricultural and Food Chemistry*, 52, 7318–7323.

von Elbe, J.H. and Attoe, E.L. (1985) Oxygen involvement in betanine degradation-measurement of active oxygen species and oxidation reduction potentials. *Food Chemistry*, 16, 49–67.

von Elbe, J.H., Maing, I.Y., and Amundson, C.H. (1974) Colour stability of betanin. *Journal of Food Science*, 39, 334–337.

Wagner, L.A. and Warthesen, J.J. (1995) Stability of spray-dried encapsulated carrot carotenes. *Journal of Food Science*, 60, 1048–1053.

Wang, H. and Helliwell, K. (2000) Epimerisation of catechins in green tea infusions. *Food Chemistry*, 70, 337–344.

Wang H. and Murphy P.A. (1996) Mass Balance Study of Isoflavones during Soybean Processing. *Journal of Agricultural and Food Chemistry*, 44, 2377–2383.

Wang, X.Y., Zhou, R., and Jiang, L.J. (2000) Study on the stability of glucoraphanin extracted from broccoli. *Shipin Keji*, 34, 250–252, 257.

Wasserman, B.P., Eiberger, L.L., and Guilfoy, M.P. (1984) Effect of hydrogen peroxide and phenolic compounds on horseradish peroxidase-catalysed decolorization of betalain pigments. *Journal of Food Science*, 49, 536–538.

Wasserman, B.P. and Guilfoy, M.P. (1984) Solubilization of the red beet cell wall betanin decolorizing enzyme. *Journal of Food Science*, 49, 1075–1077.

Wrolstad, R.E., Skrede, G., Lea, P., and Enersen, G. (1990) Influence of sugar on anthocyanin pigment stability in frozen strawberries. *Journal of Food Science*, 55, 1064–1065, 1072.

Wu, X., Beecher, G.R., Holden, J.M., Haytowitz, D.B., Gebhardt, S.E., and Prior, R.L. (2006) Concentration of anthocyanins in common foods in the United States and estimation of normal consumption. *Journal of Agricultural and Food Chemistry*, 54, 4069–4075.

Wybraniec, S. and Mizrahi, Y. (2005) Generation of decarboxylated and dehydrogenated betacyanins in thermally treated purified fruit extract from purple pitaya (*Hylocereus polyrhizus*) monitored by LC-MS/MS. *Journal of Agricultural and Food Chemistry*, 53, 6704–6712.

Wybraniec, S., Nowak-Wydra, B., and Mizrahi, Y. (2006) ^{1}H and ^{13}CNMR spectroscopic structural elucidation of new decarboxylated betacyanins. *Tetrahedron Letters*, 47, 1725–1728.

Wyler, H. and Dreiding, A.S. (1962) Constitution of the beet pigment betanin. IV. Degradation products of betanidin. *Helv Chim Acta*, 45, 638–640.

Wyler, H. and Dreiding, A.S. (1984) Deuteration of betanidin and indicaxanthin. (*E*/*Z*)-Stereoisomerism in betalaines. *Helv Chim Acta*, 67, 1793–1800.

Xia, X., Ling, W., Ma, J., Xia, M. *et al.* (2006) An anthocyanin-rich extract from black rice enhances atherosclerotic plaque stabilization in apolipoprotein E-deficient mice. *Journal of Nutrition*, 136, 2220–2225.

Xu, B. and Chang, S.K.C. (2008) Total phenolics, phenolics acids, isoflavones, and anthocyanins and antioxidant properties of yellow and black soybeans as affected by thermal processing. *Journal of Agricultural and Food Chemistry*, 56, 7165–7175.

Xu, J.Z., Leung, L.K., Huang, Y., and Chen, Z.Y. (2003) Epimerisation of tea polyphenols in tea drinks. *Journal of Food Science and Agriculture*, 83, 1617–1621.

Xu, X.M., Chen, X.M., and Jin, Z.Y. (2006) Stability of astaxanthin from Xanthophyllomyces dendrorhous during storage. *Shipin Yu Shengwu Jishu Xuebao*, 25, 29–36.

Yang, J., Martinson, T.E., and Liu, R.H. (2009) Phytochemical profiles and antioxidant activities of wine grapes. *Food Chemistry*, 116, 332–339.

Yao, L. and Han, J.R. (2008) Effects of environmental factors on the stability of carotenoid in sclerotia of Penicillium sp. PT95. *Shanxi Nongye Daxue Xuebao*, 28, 73–76.

Zabala, M.T., Grant, M., Bones, A.M., Bennett, R., Lim, Y.S., Kissen, R., and Rossiter, J.T. (2005) Characterisation of recombinant epithiospecifier protein and its over-expression in Arabidopsis thaliana. *Phytochemistry*, 66, 859–867.

Zabetakis, I., Leclerc, D., and Kajda, P. (2000) The effect of high hydrostatic pressure on the strawberry anthocyanins. *Journal of Agricultural and Food Chemistry*, 48, 2749–2754.

Zakharova, N.S., Petrova, T.A., and Bokuchava, M.A. (1989) Betalain oxidase and betalain pigments in table beet seedlings. *Soviet Hournal of Plant Physiology*, 36, 273–277.

Zakharova, N.S. and Petrova, T.A. (2000) β-Glucosidases from leaves and roots of common beet, Beta vulgaris. *Applied Biochemistry and Microbiology*, 36, 458–461.

Zhang, C., Ma, Y., hao, X.Y., and Mu, J. (2009) Influence of copigmentation on stability of anthocyanins from purple potato peel in both liquid state and solid state. *Journal of Agricultural and Food Chemistry*, 57, 9503–9508.

Zhang, Z., Pang, X., Xuewu, D., Ji, Z., and Jiang, Y. (2005) Role of peroxidase in anthocyanin degradation in litchi fruit pericarp. *Food Chemistry*, 90, 47–52.

Zhao, L.Y., Chen, F., Zhao, G.H., Wang, Z.F., Liao, X.J., and Hu, X.S. (2005) Isomerization of trans-astaxanthin induced by copper(II) ion in ethanol. *Journal of Agricultural and Food Chemistry*, 53, 9620–9623.

Zhao, X.Y., Corrales, M., Zhang, C., Hu, X. S., Ma, Y., and Tauscher, B. (2008) Composition and thermal stability of anthocyanins from Chinese purple corn (*Zea mays* L.). *Journal of Agricultural and Food Chemistry*, 56, 10761–10766.

Zryd, J-P. and Christinet, L. (2004) Betalains. In: K.M. Davies (ed.) *Plant Pigments and their Manipulation*, Boca Raton, Florida: CRC Press, Pp. 185–247.

16 Stability of phytochemicals at the point of sale

Pradeep Singh Negi

Human Resource Development
Central Food Technological Research Institute (CSIR), Mysore, India

16.1 Introduction

Phytochemicals are a heterogeneous group of substances found in all plant products, therefore constituting an important component of human diets. It is estimated that there are 250 000–500 000 plant species on Earth (Borris, 1996) and approximately 1–10% of these are used as food by human or other animals. Besides food, plants have also been used for centuries as remedies for human diseases because they contain components (phytochemicals) with therapeutic values. Phytochemicals have been isolated from several herbs and plants, and they have shown potent biological activities (Cowan, 1999; Negi et al., 1999; Beuchat, 2001; Burt, 2004; Negi and Jayaprakasha, 2004; Jayaprakasha et al., 2007; Tiwari et al., 2009; Raybaudi-Massilia et al., 2009; Negi et al., 2010; Negi, 2012). Food can be used as a vehicle for the delivery of phytochemicals that provide health benefits for increased well-being. Consumption of foods rich in phytochemicals has been reported to protect against various degenerative diseases, but their beneficial properties may alter as food products undergo processing and subsequent storage prior to consumption, which can affect stability of phytochemicals. Given the recent trend of health promotion through diet, understanding processing and storage effects is critical for conserving active phytochemicals. Simultaneously, there should not be any compromise on maintaining quality and enhancing shelf life of the food after addition of phytochemicals.

16.2 Stability of phytochemicals during storage

Phytochemical content in plants can vary greatly with variety, maturity, growing conditions, and agro climatic factors; therefore it is difficult to pinpoint whether the differences in composition of commercial samples are due to agronomic reasons or degradation during storage and processing methods followed prior to storage. In general, carotenoids are very susceptible to degradation, and the major mechanism of degradation is oxidation. The rate of carotenoid oxidation depends on carotenoid structure, oxygen, temperature, light, water activity, pH, enzymes, presence of metals and unsaturated lipids, type and physical state of the carotenoids present, severity and duration of processing, packaging material, storage conditions, and the presence of pro- and anti-oxidants (Namitha and Negi, 2010). Anthocyanins are destabilized by heat, high pH, light exposure, dissolved oxygen, and enzymes such as PPO, whereas copigmentation with acids or other flavonoids and metals enhances their color during storage (Fennema and Tannenbaum, 1996). Anthocyanin stability during storage follows first order kinetics (Garzon and Wrolstad, 2002; Turker et al., 2004; Brenes et al., 2005). Anthocyanins are found in their monomeric form in fresh fruit and juices, and the monomeric anthocyanins may undergo a condensation reaction to form polymeric pigments during storage (Bishop and Nagel, 1984; Dallas et al., 1996; Es-Safi et al., 2000; Hillebrand et al., 2004), which results in greater color stability of the matrix during further processing and storage (Gutierrez et al., 2004). Light exposure reduces betalain stability and metal cations are capable of accelerating betalain degradation (Herbach et al., 2006). Both betanidin and betanin were reported to be unstable in the presence of oxygen (Pasch and von Elbe, 1978; Schwartz et al., 2008).

16.2.1 Effect of water activity

The phytochemical stability of freeze-dried apple products during storage (up to 45 days) at 30 °C was most affected by a_w and highest losses were observed at highest moisture activity (Corey et al., 2011). Phytochemical degradation for added green tea extracts also occurred more rapidly at higher moisture contents, except for caffeine, which was stable throughout the storage period irrespective of moisture content. Similarly, absorption of moisture by the acerola concentrate affected the stability of phytochemicals and almost half of the phenolics were lost within the first hour of storage at 65% RH (Ikonte et al., 2003). Superior stability of *Amaranthus* pigment powders as compared to the respective aqueous solutions may be attributed to lower a_w values. Spray drying was found to increase stability of betacyanins, probably by increasing dry matter. Therefore, it was recommended that moisture content of pigment should be kept below 5% for enhancing their stability (Cai et al., 2005). Some matrix compounds like pectin, guar gum, and locust bean gum were shown to enhance storage stability of red beet solutions, probably by lowering the a_w value (Herbach et al., 2006).

16.2.2 Effect of temperature

The stability of a phytochemical depends on storage environment and length of storage. It has been stated that temperature is the most important factor to affect the overall stability of various phytochemicals during storage (Rodrigues et al., 1991; Su et al., 2003). In Hawthorn fruit, Procyanidin B2 (PC-B2), (-)Epicatechin (EC), Chlorogenic acid (ChA), Hyperoside (HP), and Isoquercitrin (IQ) were stable for six months at a storage temperature of 4 °C. At room temperature their stability varied from relatively stable (HP and IQ, 8% loss), to intermediate stable (ChA, 30% loss), to quite unstable (EC and PC-B2, 50% loss), however, these compounds were unstable at the higher temperature of 40 °C (Chang et al., 2006). In Hawthorn drink also these compounds were stable for six months at lower temperatures (4 °C) and relatively unstable at higher temperatures (23 and/or 40 °C). Wang and Stretch (2001) observed that in cranberries storage temperature has profound effect on anthocyanins content as storage at 15 °C promoted anthocyanins biosynthesis as compared to storage at higher or lower temperature. Fresh cactus fruit of Yellow spineless also showed an increase in the concentration of active substances during three to four weeks storage at 5–8 °C (Nazareno *et al.*, 2009). In milled rice also there was a consistent decrease in phenolic acid content during storage and the decline was greater at 37 °C than at 25 °C (Thanajiruschaya *et al.*, 2010).

Uddin *et al.* (2002) studied the effects of degradation of ascorbic acid in dried guava during storage and observed that as the storage time and temperature increased, there was a progressive decrease in ascorbic acid content. Similarly, total vitamin C was found to decrease with increase in storage temperature and duration in green leaves (Negi and Roy, 2001a, 2001b) and carrots (Negi and Roy, 2000). Total vitamin C in grapefruit juice was retained higher at the lowest temperature (10 °C) at the end of 12 weeks of storage (Smoot and Nagy, 1990). However, Rodriguez *et al.* (1991) observed that ascorbic acid degradation of an alcoholic orange juice beverage was not influenced by temperature while other quality parameters such as degree of browning, accumulation of furfural, and limonene content were highly correlated with temperature of storage.

Fruit juices should be kept refrigerated to increase the stability of phytochemicals as the ascorbic acid present in most of the fruit combines with anthocyanins, which may be mutually destructive, more so at higher temperatures (Brenes *et al.*, 2005; Choi *et al.*, 2001). This destruction has been linked to the formation of dehydroascorbic acid breakdown products, but the exact mechanism for their adverse interaction has not been completely elucidated. Ascorbic acid stability is important in anthocyanin containing juice blends since the degradation products of ascorbic acid can degrade anthocyanins (Es-Safi *et al.*, 1999, 2002; Brenes *et al.*, 2005). Brenes *et al.* (2005) reported a 12% decrease in total anthocyanins in a grape juice (*Vitis vinifera*) model system with and without added ascorbic acid. Pozo-Insfran *et al.* (2007) reported that the added ascorbic acid decreased the anthocyanin stability, as opposed to the wine where polyphenolics helps to stabilize anthocyanins (Gutierrezz *et al.*, 2004).

Juice containing cyanidin-3-glucoside retained 90% of its color for 60 days at 10 °C (Fossen *et al.*, 1998), whereas at 25 °C storage color loss was 10–17%, which accelerated further to 35–49% at 38 °C. In thermally processed blueberry (*Vaccinium myrtillus*) juices degradation of anthocyanins was also significantly accelerated with increasing storage temperatures. Combined pressure temperature treatment (100–700 MPa, 40–121 °C) of pasteurized juice led to a slightly faster degradation of total anthocyanins during storage compared to heat treatments at ambient pressure (Buckow *et al.*, 2010). Phytochemicals were stable after high hydrostatic pressure processing (400 and 550 MPa for 15 min) in ascorbic acid-fortified muscadine grape juice at 25 °C for 21 days, and addition of rosemary and thyme polyphenolic extracts increased muscadine grape juice color, antioxidant activity, and also reduced phytochemical losses during storage (Pozo Insfran *et al.*, 2007). Addition of rosemary extract readily forms copigment complexes with anthocyanins in concentration-dependent manner and increases its antioxidant activity (Talcott *et al.*, 2003a). Condensation of the other polyphenolic compounds with the anthocyanins may cause higher retention of polyphenolics, as observed in blood orange juice (Hillebrand *et al.*, 2004). Guava juice showed a protecting effect on several of the juice blends, which is attributed to the guava polyphenolics forming more stable polymeric compounds with the anthocyanins (Bishop and Nagel, 1984; Dallas *et al.*, 1996).

Storage of commercial tea leaves at 20 °C for six months resulted in a progressive decrease in the total phytochemical content, most of which were attributed to losses in the epigallocatechin 3-gallate and epicatechin 3-gallate (Friedman *et al.*, 2009). Epigallocatechins were shown to be isomerized into (-)-catechin during storage at 40 °C after a few days (Komatsu *et al.*, 1993; Wang and Helliwell, 2000). Changes in the antioxidant capacity of a green tea infusion were directly related to the changes in catechins that showed considerably higher stability at lower pH. Green tea catechins used in oil/water emulsions were found to decrease to 70% of the initial content at room temperature and almost negligible amount remained at 40 °C after six months (Frauen *et al.*, 2000). Spanos *et al.* (1990) also observed a complete degradation of procyanidins, including catechin, epicatechin, and procyanidins B1, B2, B3, and B4, after the storage of concentrated apple juice at 25 °C for nine months. Similarly, isoquercetin and kaempferol 3-glucoside present in red raspberry jam decreased slightly after six months of storage (Zafrilla *et al.*, 2001). In a model system, catechin was more stable in aqueous solutions stored in the dark as compared to illuminated storage, and its stability was further enhanced when stored at refrigeration temperature (4 °C) over a storage period of 80 days (Callemien and Collin, 2007). Degradation of catechins in fruit juice was greater at high storage temperature (23 °C) than low storage temperature (4 °C), and the degradation pathway was related to oxidative processes (Chang *et al.*, 2006). Although, oxidation can occur under a variety of temperature conditions, reaction rates are generally faster at higher temperatures and depend on the dissolved oxygen in aqueous solution (Devlin and Harris, 1984; Alnaizy and Akgerman, 2000).

Individually quick frozen black raspberries retained anthocyanins during long-term storage at −20 °C, but heating followed by storage for six months resulted in dramatic losses in total anthocyanins ranging from 49–75% (Hager *et al.*, 2008). Blueberries stored at −25 °C for six months after initial heat processing showed 62–85% losses in total anthocyanins (Brownmiller *et al.*, 2008). Heat processing for different durations at 95 °C did not have an effect on the initial concentration of tea catechins, but it significantly influenced the stability of these compounds during storage. The heat treatment decreased the storage stability of all tea catechins, and the duration of heating was not a factor in polyphenolic stability. In green tea, mild heat pasteurization (85 °C) retains characteristic color and flavor better for longer storage duration than higher temperatures treatments (Kim *et al.*, 2007). High temperature also induces a negative effect by lowering polyphenolic stability during storage regardless of time duration, which may be attributed to the loss of ascorbic acid by heat treatment.

A reduction in the carotene content of fresh vegetables irrespective of storage conditions has been reported. Unfavorable relative humidity and temperature has been shown to hasten the loss of carotenes during storage of fresh produce, wherein spinach lost almost 63.5% of the original carotenoids after wilting (Akpapunam, 1984). Negi and Roy (2003) reported up to 85% losses in β-carotene in fresh green leaves depending on duration and storage conditions with packaging helping in retaining higher β-carotene. During storage of fresh carrots, a steady decrease (Negi and Roy, 2000) and a slight increase followed by decrease (Lee, 1986) in β-carotene content have been reported. Total carotenoids in hand peeled carrot disks were significantly higher than fine or coarse carborundum plates abrasion peeled carrot disks throughout eight days of storage at 4 °C (Kenny and O'Beirne, 2010). During storage of tomatoes and their processed products all-*trans*-lycopene was more stable at 20 °C compared to −10, 2, and 37 °C, as re-isomerization from *cis*- to *trans*- is favored at this temperature (Lovric *et al.*, 1970).

The Phytochemical composition of five varieties of black soybeans (*Glycine max*) and their stability at room temperature, 4 and −80 °C over 14 months were determined by Correa *et al.* (2010). No significant decrease was found in total phenols of black soybeans during storage for 14 months. On the other hand, lutein and γ-tocopherol degraded significantly within a month of storage at room temperature, whereas they remained stable up to six months at 4 °C and up to 14 months at −80 °C. Storage at low temperature can reduce the loss of fat-soluble phytochemicals in black soybeans over an extended period of time; however no significant decrease occurs in total phenols even at room temperature for 14 months.

Koski et al. (2002) found that the content of α-tocopherol in cold pressed rape seed oil declined to nil from the initial value of about 200 mg/kg in fresh oil within 7–11 days of storage at a temperature of 60 °C, while γ-tocopherol was retained to 5–10% of the initial value (600 mg/kg) after two weeks of storage. Morello et al. (2004) also found that α-tocopherol was totally absent in olive oil after 12 months of storage at room temperature, but at lower temperatures a slower rate of reduction of α-tocopherol (60% loss after 12 months) in virgin olive oil was observed (Okogeri and Tasioula-Margari, 2002).

Kopelman and Augsburger (2002) determined the influence of capsule shell composition and sealing on the stability of the phytochemical in fresh and formulated *Hypericum perforatum* extract capsules stored at 25 °C/60% RH for 60 days. Phytochemicals had varying stability towards capsule shell composition and sealing. Except with gelatin capsules of neat *Hypericum perforatum* extract, sealing of the capsules did not offer much protection. Neat *Hypericum perforatum* extract was typically more sensitive to the effects of shell composition and sealing relative to formulated *Hypericum perforatum* extract.

Long-term storage at −24 °C of raw carrot cubes reduced the falcarinol content by almost 35%. Blanching before storage reduces almost one-third of the falcarinol content of carrot, although no further reduction in the falcarinol content was reported after steam blanching during long-term storage (Hansen et al., 2003). Studies testing the stability of Policosanol (PC) supplement under environments that favor acid hydrolysis, basic hydrolysis, oxidation, photolytic degradation, and thermolysis indicated a shelf life of five years for the original PC supplement (Mas, 2000; Cabrera et al., 2002; Castano et al., 2002; Cabrera et al., 2003). Storage temperature has no effect on the degradation of bixin during initial storage period, but at later stages the degradation accelerated with temperature, and it followed second-order rate kinetics. The reaction rates increased by increasing the interaction between oxygen molecules and bixin, and temperature had a positive effect on reaction rate causing faster degradation (PrabhakaraRao et al., 2005).

16.2.3 Effect of light and oxidation

Besides water activity and temperature, light exposure is also an important factor to influence the stability of phytochemicals during storage (Schwartz et al., 2008). It is known that light induced oxidation of carotenoids, proteins, lipids, and vitamins are common in many food systems (Wishner, 1964; Pesek and Warthesen, 1987; Solomon et al., 1995). Carotenoids are very susceptible to degradation, particularly once they have been extracted from biological tissues. The major cause of carotenoid destruction during the storage of food is oxidation (Zanoni et al., 1998), and they are susceptible to oxidation when exposed to light (Saguy et al., 1985) and enzymes (Gregory, 1996), but reduced water activity of the medium has a protective role (Minguez and Galan, 1995). Oxidation of carotenoids occurs as a result of either auto-oxidation in the presence of oxygen, or by photo-oxidation in the presence of light (MacDougall, 2002). Overall, the rate of carotenoid oxidation depends on carotenoid structure, oxygen, temperature, light, water activity, pH, metals, enzymes, presence of unsaturated lipids, type and physical state of the carotenoids present, severity and duration of processing, packaging material, storage conditions, and the presence of pro- and anti-oxidants (Rodriguez-Amaya, 2003; Namitha and Negi, 2010). Oxidation of carotenoids results in the formation of colorless end products such as compounds with epoxy, hydroxyl, and carbonyl groups (MacDougall, 2002). Therefore, while designing the delivery system for their use as functional food ingredients, appropriate measures should be taken to protect them. Addition of carotenoids to functional foods should be done by incorporating into edible oil as it makes them more bioavailable than carotenoids in a plant cellular matrix (Lakshminarayana et al., 2007).

Lycopene stability in products such as guava nectar is also a function of light, water activity, oxygen, pH, temperature, and the presence of pro-oxidants or antioxidants (Chou and Breene, 1972). Oxygen-independent reactions affect Yellow passion fruit juice (*Passiflora edulis*) color and antioxidant activity, and ascorbic acid and sucrose fortification increases stability of carotenoids (Talcott et al., 2003b). Dehydrated vegetables lose color due to the oxidation of highly unsaturated molecules upon exposure to air during storage and β-carotene degradation is associated with the development of an off flavor in dehydrated carrots (Ayer et al., 1964).

The three most predominant phenolic compounds in tea (ECG, EGCG, and EGC) showed higher stability at lower temperature in the dark, indicating that the two storage conditions (temperature and light) were a significant factor to influence phenolic stability during green tea storage (Callemien and Collin, 2007). During storage for six months of virgin olive oil under diffused light in the temperature range of 6–18 °C, an almost 60% decrease in the total phenols occured, whereas storage in darkness resulted in a decrease of 50% of total phenols after 12 months (Okogeri and Tasioula-Margari, 2002). Tsimidou et al. (1992) also found significant losses of phenolic compounds in virgin olive oil stored in the dark at 20 °C in closed bottles. A significant decrease of phenol content in virgin olive oil after 12 months of storage in darkness at room temperature with subsequent loss of oxidative stability was also established in the study by Morello et al. (2004). The total anthocyanins contents in colored rice were retained under low O_2 concentrations (0, 5, and 10%). Polyphenol contents significantly declined during four months of storage with free and soluble conjugated phenolic contents showing minimum losses at 0% O_2 storage, whereas minimum loss of insoluble bound phenolics was detected in samples stored at 5% O_2 (Htwe et al., 2010).

Ascorbic acid was not affected by light exposure in juice stored in air-tight containers for 52 days at 8 °C (Solomon et al., 1995); however, commercial juice in foil-covered bottles retained higher ascorbic acid than clear bottles during 18 days of storage at 3 °C (Andrews and Driscoll, 1977). Similarly, green tea stored in lightproof packaging retained higher ascorbic acid (Yaminish, 1996). Light exposure reduced betalain stability (von Elbe et al., 1974; Bilyk et al., 1981; Cai et al., 2005, Herbach et al., 2007) and detrimental effects of light were observed at temperatures below 25 °C, but no effect of light was observed at storage temperatures above 40 °C (Attoe and von Elbe, 1981; Huang and von Elbe, 1986). High ascorbic acid concentrations were capable of reducing betalain degradation (Pasch and von Elbe, 1978) and supplementation with ascorbic and isoascorbic acids was reported to enhance betalain stability by oxygen removal (Attoe and von Elbe, 1982).

Additive effects of light and oxygen were observed as light alone caused 15.6% betanin degradation, and oxygen alone caused 14.6% betanin degradation, whereas their simultaneous presence was responsible for 28.6% betanin decomposition (von Elbe et al., 1974). Attoe and von Elbe (1981) reported that light-induced degradation was oxygen dependent as the detrimental effects of light were found to be negligible under anaerobic conditions (Huang and von Elbe, 1986). Supplementation of red beet and purple pitaya juices with acids has been shown to inhibit light-induced betacyanin degradation during juice storage (Bilyk et al., 1981; Herbach et al., 2007). The exposure to light during storage showed a more pronounced decrease (15% at 38 °C) in color than those kept in the dark in rose extracts (PrabhakaraRao et al., 2005). Similarly, the effect of light on degradation of bixin was seen from the initial days of the storage period in both oleoresin and dye (Balaswamy et al., 2006). Betanin stability decreases linearly with increasing oxygen concentration (Czapski, 1985), and storage in a nitrogen atmosphere significantly increased its stability (Attoe and von Elbe, 1982; von Elbe and Attoe, 1985). In addition to oxygen, hydrogen peroxide was also reported to accelerate betanin degradation (Wasserman et al., 1984).

16.2.4 Effect of pH

The stability of phenolic compounds is highly pH dependent and varies depending on the structural conformation. Flavan-3-ols show high storage stability under acidic conditions but are unstable in neutral pH (Komatsu et al., 1991; Suematsu et al., 1992; Zhu et al., 1997; Chen et al., 1998; Xu et al., 2003). Addition of acids confer stability to tea beverages since a lower pH is more effective for stabilizing tea catechins during storage (Chen et al., 1998), while neutral pH degraded tea catechins faster. Adding ascorbic acid or organic acids (citric and malic acid) to mimic citrus flavor improves storage stability and flavor of green tea (Aoshima and Ayabe, 2007), probably by lowering the pH. Moreover, ascorbic acid is more stable at lower pH, indicating that the protective effect on polyphenolics is higher at lower pH (Gallarate et al., 1999). Lowering pH is effective in slowing down the reduction of predominant compounds (chlorogenic acid and its isomers) formed during oxidative degradation (Schmalko and Alzamora, 2001). In green tea, changing pH affects the rate of hydrogen peroxide production, and when the pH of green tea infusion was lowered, the production rate of hydrogen peroxide and superoxide was significantly reduced (Akagawa et al., 2003). Presence of 3-deoxyanthocyanins, which lacks the hydroxyl group at 3 position of C ring in sorghum, increases the anthocyanin stability at high pH making it a good colorant for food use (Awika et al., 2004).

Polyphenols are readily oxidized during storage, which results in the production of H_2O_2 (Akagawa et al., 2003; Chai et al., 2003; Aoshima and Ayabe, 2007). The H_2O_2 produced during storage can degrade a polyphenol-rich product (Long et al., 1999), and ascorbic acid may be effective in reducing the rate of oxidative degradation during storage by quenching of H_2O_2. Ascorbic acid fortification may reduce free radical production in polyphenol-rich beverages by lowering pH, while no protective effect was observed when pH of the tea beverage was neutral (Aoshima and Ayabe, 2007). Ascorbic acid present in guava juice is known to stabilize lycopene (Mortensen et al., 2001) by a radical scavenging mechanism, although this protecting effect was not observed during heating probably due to degradation of ascorbic acid.

16.3 Food application and stability of phytochemicals

In general, it has been found that a higher concentration of bioactive compounds is required to achieve similar efficacy in foods as demonstrated in *in vitro* experiments (Shelef, 1983; Tassou et al., 1995; Smid and Gorris, 1999; Burt, 2004; Holley and Patel, 2005; Negi, 2012), and experiments have proved that the concentration of essential oils to achieve the desired antibacterial effect should be approximately two-fold in semi-skimmed milk (Karatzas et al., 2001), ten-fold in pork liver sausage (Pandit and Shelef, 1994), 50-fold in soup (Ultee and Smid, 2001), and up to100-fold in soft cheese (Mendoza-Yepes et al., 1997). Most studies on food application of phytochamicals are limited to examining their bioactive efficacy rather than their stability during storage (Burt, 2004; Fisher and Phillips, 2008; Negi, 2012). Stability of phytochemicals have been discussed in detail elsewhere in this book (Chapters 14 and 15).

16.4 Edible coatings for enhancement of phytochemical stability

The use of edible coatings to extend the shelf life and improve the quality of fruits and vegetables has been studied extensively due to their ecofriendly and biodegradable nature. Edible coatings can provide a supplementary and sometimes essential means of controlling physiological, morphological, and physicochemical changes in fruit and preserves the phytochemicals present in them.

The functionality of edible coatings can be improved by incorporating natural or synthetic antimicrobial agents, antioxidants, and functional ingredients such as minerals and vitamins. The addition of preservatives is of special interest for minimally processed fruit and vegetables, which have an extremely short shelf life because of microbiological concerns as well as sensory and nutritional losses that occur during their distribution and storage. Antioxidants are added to edible coatings to protect fruit against oxidative rancidity and discoloration (Baldwin et al., 1995). The antioxidants were also added in edible coatings to control oxygen permeability and reduce vitamin C losses in apricots during storage (Ayranci and Tunc, 2004). Anti-browning agents (McHugh and Senesi, 2000; Baldwin et al., 1996; Lee et al., 2003; Perez-Gago et al., 2006) and texture enhancers like $CaCl_2$ (Wong et al., 1994) and milk proteins (Le Tien et al., 2001) have also been used in edible coatings for preservation purpose. Eswaranandam et al. (2006) extended the shelf life of fresh-cut cantaloupe melon by incorporating malic and lactic acid into soy protein coatings. Antimicrobial and antioxidant coatings have advantages over direct incorporation of the antimicrobial or antioxidant agents because they can be designed for slow release of the active compounds from the surface of the coated commodity. By slowing their diffusion into coated foods, the preservative activity at the surface of the food is maintained for a longer storage period, and a smaller amount of antimicrobials/antioxidants would come into contact with the food compared to dipping, dusting, or spraying the preservatives onto the surface of the food to achieve a target shelf life (Min and Krochta, 2005).

Use of natural antimicrobials in the development of coatings, which use inherently antimicrobial polymers as a support matrix has been studied in detail using chitosan, which is mainly obtained from the deacetylation of crustacean chitin and is one of the most effective antimicrobial film forming biopolymers (it is out of purview of this topic but readers can refer to Vargas et al., 2006; El Gaouth et al., 1991; Zhang and Quantick, 1997, 1998; Romanazzi et al., 2003; Devlieghere et al., 2004; Park et al., 2005). Chitosan-based edible coatings can be also used to carry other antimicrobials compounds such as organic acids (Outtara et al., 2000), essential oils (Zivanovich et al., 2005), spice extracts (Pranoto et al., 2005), lysozyme (Park et al., 2004), and nisin (Pranoto et al., 2005; Cha et al., 2003). Natural antimicrobial compounds have been incorporated into protein or polysaccharide-based matrices, thereby obtaining a great variety of multi-component antimicrobial coatings by adding oregano, rosemary, and garlic essential oils (Seydim and Sarykus, 2006). Rojas-Grau et al. (2006) used apple puree and high methoxyl pectin combined with oregano, lemon grass, or cinnamon oil at different concentrations as coatings for enhancing phytochemical stability. Greater details about effect of phytochemicals on minimally processed fruit and vegetables can be found in Chapter 10 of this book.

Although several coatings have shown their efficacy in *in vitro* tests against a range of microorganisms, they were not tested in food systems and therefore information about their possible impact on the aroma and flavor of the coated products is not available. The influence of the incorporation of antimicrobial phytochemicals into edible films and coatings on sensory properties of coated commodities needs much deeper investigation.

16.5 Modified atmosphere storage for enhanced phytochemical stability

The use of elevated CO_2 as the packaging gas reduced the overall antioxidative capacity of cranberries during the initial storage period. The antioxidant status of air packaged fruit decreased initially but increased on further storage. Berries stored under elevated O_2 exhibited good antioxidative capacity over the first four days of storage but this declined with prolonged storage, possibly due to O_2 promoted oxidation of the constitutive anthocyanins and phenolics. However, during the first four days of storage the effect of elevated O_2 on antioxidative status was minimal. High levels of oxygen in controlled atmosphere storage had little effect on post-harvest anthocyanins development and total phenolics in cranberries (Gunes et al., 2002).

Modified atmospheres with controlled concentrations of CO_2 and O_2 have been used to maintain the quality of fresh-cut spinach. The total flavonoid content remained constant during storage in both air and MAP atmospheres, while vitamin C was better preserved in MAP stored spinach. Ascorbic acid was transformed to dehydroascorbic acid during storage, and its concentration was higher in MAP-stored tissues. A decrease in the total antioxidant activity was observed during storage in MAP-stored spinach, which may be due to higher content of dehydroascorbic acid and lower content of both ascorbic acid and antioxidant flavonoids in the MAP-stored samples (McGill et al., 1966; Izumi et al., 1997). Neither controlled atmospheric nor cold storage had any adverse effect on antioxidant activity in apples. After 25 weeks of cold storage there was no decrease in chlorogenic acid, but catechin content decreased slightly. Storage at 0 °C for nine months had little effect on phenolic content of apple peel (Goulding et al., 2001). Lattanzio et al. (2001) also found that after 60 days of cold storage the concentration of total phenolics in the skin of Golden Delicious apples increased. Quercetin glycosides, phloridzin, and anthocyanin content of various apple cultivars were not affected by 52 weeks of storage in controlled atmospheric conditions, although chlorogenic acid and total catechins decreased slightly in Jonagold apples; total catechin concentration decreased slightly in Golden Delicious; and chlorogenic acid concentrations remained stable during storage period (van der Sluis et al., 2001) indicating stability of phytochemicals under modified atmosphere is a function of crop, variety, and phytochemical in question.

Fiber content in asparagus increases significantly during storage of up to 13 days irrespective of storage temperatures (10 and 15 °C), but a slow increase in fiber content was observed during MAP-stored asparagus (at 4 °C) up to 30 days of storage (Sothornvit and Kiatchanapaibul, 2009). Asparagus stored in different packaging conditions at 10 °C for four days showed an increase in lignin content also (Huyskens-Keil and Kadau, 2003).

Total phenolic content decreased during storage of fresh-cut jackfruit bulbs during 35 days of storage at 6 °C. Bulbs dipped in a solution containing $CaCl_2$, ascorbic acid, citric acid, and sodium benzoate coupled with MAP resulted in significantly lower loss in phenolics (Saxena et al., 2009). Mateos et al. (1993) also found inhibition of enzyme mediated phenolic metabolism in fresh-cut lettuce stored under low O_2 and high CO_2 atmosphere. Alasalvar et al. (2001) reported that storage under low O_2 conditions reduces the accumulation of total phenols in shredded oranges and purple carrots as compared to air or high O_2 storage.

16.6 Bioactive packaging and micro encapsulation for enhanced phytochemical stability

Bioactive packaging is a process in which a food package or coating plays the unique role of enhancing impact of food over the consumer's health. The bioactive packaging material should be capable of withholding desired bioactive principle in optimum conditions until their eventual release into the food product either during storage or just before consumption. Bioactive packaging can be achieved by integration and controlled release of bioactive components or nanocomponents from a biodegradable packaging system, micro or nano encapsulation of active substances in the packaging, and packaging with active enzymes exerting a health-promoting benefit through transformation of specific food components (Lagaron, 2005). Method of fabrication of the films, the optimal time temperature conditions for mixing the biomaterial with phytochemical, and the suitable mechanism to attain the desired release rate just upon packaged food opening and before consumption are important for phytochemical based bioactive packaging systems. An antimicrobial agent releasing plastic film for cheese packaging has been developed (Han, 2002) that has the potential to incorporate other phytochemicals using a similar system. The edible films can be modified using polysaccharide (starch, alginates, etc.), protein (gelatin, soy protein, wheat gluten, etc.), and lipids (waxes, triglycerides, fatty acids, etc.) to contain phytochemicals for food use.

Microencapsulation is defined as "the technology of packaging solid, liquid and gaseous materials in small capsules that release their contents at controlled rates at specific conditions over prolonged periods of time" (Champagne and Fustier, 2007). Release can be solvent activated or signaled by changes in pH, temperature, irradiation, or osmotic shock. As the encapsulated materials are protected from moisture, heat, or other extreme conditions, their stability is enhanced and they maintain viability for longer durations. Lopez-Rubio (2006) observed that microencapsulation is suitable for incorporating functional ingredients that are very susceptible to lipid oxidation or to mask off odors or tastes expected in foods after addition of phytochemicals. Microencapsulation promotes the delivery of active ingredients without their interaction with food components. As it is used to provide barriers between the sensitive bioactive materials and the environment (food or oxygen), it can also be used to mask unpleasant flavors and odors, or to modify texture or preservation properties (Fang and Bhandari, 2010). Omega-3 and omega-6 fatty acids are used for food fortification, but the taste and smell of these oils and their tendency to oxidize rapidly is a problem in their food application (Augustin and Sanguansri, 2003). It was demonstrated that the consumption of food enriched with microencapsulated fish oil obtained by emulsion spray-drying was as effective as the daily intake of fish oil gelatine capsules in meeting the dietary requirements of omega-3 fatty acid (Wallace et al., 2000).

Several technologies have been used for microencapsulation of bioactive ingredients, which basically include three steps; formation of a wall around the material, prevention of undesirable leakage, and leaving undesirables out of encapsulated material (Gibbs et al., 1999; Mozafari et al., 2008). The current encapsulation techniques include spray-drying, spray-chilling, fluidized-bed coating, extrusion, liposome entrapment, coacervation, and nanoemulsions (Arneado, 1996; Gibbs et al., 1999; Tan and Nakajima, 2005; Garti et al., 2005; Weiss et al., 2006; Flanagan and Singh, 2006; Augustin and Hemar, 2009).

Spray-drying has been traditionally used for the encapsulation of oil-based vitamins and fatty acids. For many emulsions, spray-chilling and liposome techniques have shown potential for the controlled release of bioactive compounds. Spray-chilling and fluidized-bed coatings are the most popular methods for encapsulating water-soluble vitamins, whereas spray-drying of emulsions is generally recommended for the encapsulation of lipid-soluble vitamins (Kirby et al., 1991; Arnaud, 1995; Reineccius, 1995; Augustin et al., 2001; Guimberteau et al., 2001; McClements, 2005; Goula and Adamopoulos, 2012; Wang et al., 2012). In spray-chilling and spray-cooling, the core and wall mixtures are atomized into the cooled or chilled air, which causes the wall to solidify around the core. The coating materials used are vegetable oils or their derivatives, fats and stearin, and mono- and di-acylglycerols (Cho et al., 2000; Taylor, 1993). Atomization causes quick and intimate mixing of droplets with the cooling medium and evaporation does not occur due to low temperatures, therefore it yields droplets of almost perfect spheres to give free-flowing powders. Microcapsules are insoluble in water as oils are used as a coating material, therefore this technique can be utilized for encapsulating water-soluble core materials such as minerals, water-soluble vitamins, enzymes, acidulants, and flavors (Lamb, 1987). Fluidized-bed coating involves fluidization of the solid particles in a temperature- and humidity-controlled chamber of high velocity air where the coating material is atomized (Balassa and Fanger, 1971; Zhao et al., 2004). Wall materials used in this technique include cellulose derivatives, dextrins, emulsifiers, lipids, protein derivatives, and starch derivatives, which may be used in a molten state or dissolved in an evaporable solvent either by top-spray, bottom-spray, or tangential spray (Jackson and Lee, 1991). Top spray method was used to obtain microencapsulated ascorbic acid after fluidization with hydrophobic coating materials (Knezevic et al., 1998). Microfluidization involves high pressure homogenization to produce fine emulsion, which can be further evaporated to obtain nano particles (O'Donnell and McGinity, 1997; Couvreur et al., 1997), and Salvia-Trujilo et al. (2013) showed that the microfluidization has the potential for obtaining nano-emulsions of essential oils.

The Liposome entrapment technique utilizes liposomes, which consist of an aqueous phase that is completely surrounded by a phospholipid-based membrane and both the aqueous and lipid-soluble materials can be enclosed in the liposome. Permeability, stability, surface activity, and affinity of liposomes can be varied through size and lipid composition variations (Gregoriadis, 1984; Kirby and Gregoriadis, 1984). Encapsulation of ascorbic acid in a liposome together with vitamin E produces a synergistic antioxidant effect (Reineccius, 1995).

Encapsulation by extrusion involves forcing a core material in a molten carbohydrate mass through a series of dyes into a bath of dehydrating liquid. In this encapsulation method, the pressure is kept around 100 psi and temperature rarely goes beyond 115 °C (Reineccius, 1989). The coating material hardens on contacting the liquids, forming an encapsulating matrix to entrap the core material. The extruded filaments are then separated from the liquid bath, dried, and sized (Shahidi and Han, 1993). The carrier used may be composed of more than one ingredient, such as sucrose, maltodextrin, glucose syrup, glycerine, and glucose (Arshady, 1993). Several polyphenolic antioxidants from medicinal plants were encapsulated using extrusion procedure by Belscak-Cvitanovic et al. (2011).

Centrifugal suspension separation involves mixing the core and wall materials and then adding to a rotating disk. The core material then leaves the disk with a coating of residual liquid, which is then dried or chilled (Sparks, 1989), whereas centrifugal extrusion is a liquid co-extrusion process that uses nozzles consisting of concentric orifice located on the outer circumference of a rotating cylinder through which coating and core materials are pumped separately on the outer surface of the device. While the core material passes through the center tube, coating material flows through the outer tube. As the cylinder rotates, the core and coating materials are co-extruded and the coating material envelops the core material. The wall materials used include gelatin, sodium alginate, carrageenan, starches, cellulose derivatives, gum acacia, fatty acids, waxes, and polyethylene glycol (Schlameus, 1995). Using alginate or alginate- hydroxy propyl methyl cellulose combinations, stability of avocado oil encapsulated by co-extrusion process was maintained for 90 days at 37 degree C (Sun-Waterhouse et al., 2011).

Many emulsion and coating technologies offer significant opportunities for the co-encapsulation of various hydrophobic and hydrophilic bioactives (Champagne and Fustier, 2007). Co-crystallization utilizes sucrose syrup as a wall material, which is concentrated to the supersaturated state and maintained at a temperature high enough to prevent crystallization. A predetermined amount of core material is then added to the concentrated syrup with vigorous mechanical agitation until the agglomerates are discharged from the vessel. The encapsulated products are then dried to the desired moisture and screened to a uniform size (Rizzuto et al., 1984). Yerba mate (*Ilex paraguariensis*) extract was encapsulated by co-crystallization in a super saturated sucrose solution (Lorena et al., 2007). Coacervation involves the separation of a liquid phase of coating material from a polymeric solution followed by the coating of that liquid phase around suspended core particles followed by solidification of the coating. The coacervation process consists of three steps, which involves formation of a three-immiscible chemical phase consisting of a liquid vehicle phase, a core material phase, and a coating material phase; deposition of the coating by controlled physical mixing; and solidification of the coating by thermal, cross-linking, or desolventization techniques to form a self-sustaining microcapsule. The coating materials used for coacervation microencapsulation include gelatin-gum acacia, gliadin, heparin-gelatin, carrageenan, chitosan, soy protein, polyvinyl alcohol, gelatin-carboxymethylcellulose, β-lactoglobulin-gum acacia, and guar gum-dextran (Gouin, 2004). Using coacervation of gelatin A with sodium carboxy methyl cellulose, neem seed oil was encapsulated (Devi and Maji, 2011), whereas gelatin and gum arabic were used for obtaining microcapsules of peppermint oil by complex coacervation process (Dong et al., 2011). Molecular inclusion (Inclusion Complexation) achieves encapsulation at a molecular level, which typically employs β-cyclodextrin as an encapsulating medium. The external part of the cyclodextrin molecule is hydrophilic, whereas the internal part is hydrophobic, therefore the apolar flavor compounds can be entrapped into the apolar internal cavity through a hydrophobic interaction and are entrapped inside the hollow center of a β-cyclodextrin molecule (Pagington, 1986). Nunes and Mercadante (2007) encapsulated lycopene using beta-cyclodextrin as an encapsulating medium by molecular inclusion process, but reported a slight decrease in its purity after encapsulation.

The solubility of functional ingredients in food formulations is a major consideration as the bioavailability of water insoluble or low-water-soluble ingredients gets reduced in many foods. Nano sized particles have shown substantial increase in solubility in water, which improves bioavailability (Grau et al., 2000; Muller et al., 1999; Trotta et al., 2001). Tan and Nakajima (2005) evaluated stability of β-carotene nanodispersions prepared by emulsification evaporation technique. They observed significant effect of homogenization pressure and homogenization cycle on the size of particles, which in-turn was responsible for β-carotene stability during storage. High homogenization pressure ensured a good emulsification and led to the formation of smaller sized particles, but had an adverse influence on the stability of β-carotene. The smaller particles showed higher losses during 12 weeks of storage, probably on account of increase in surface area in comparison to higher diameter particles.

16.7 Conclusions

Phytochemicals are effective in promoting health and reducing the disease risk to human beings. Phytochemicals are often lost during many of the commonly practiced processes and subsequent storage and food preparation. Modified atmospheres with controlled concentrations of CO_2 and O_2 have been used to maintain the quality of several fruits and vegetables to extend shelf life and preserve the phytochemicals present in them. Edible coating technology is a promising method for preserving the quality of fresh and minimally processed fruit, and research efforts have resulted in an improvement of the functional characteristics of the coatings. Microencapsulation and nanoencapsulation are promising techniques that can be potentially used to incorporate phytochemicals into edible coatings. Investigations related to the additional benefits of microencapsulation on the stability of bioactive ingredients in the gastric environment and on the release of bioactive ingredients into the GI tract needs attention.

Processing is a critical aspect of phytochemical production, especially due to the low yield of extracts. Processing methods are usually based on traditional methods such as water or solvent extraction. New innovative methods such as microwave and ultrasound assisted techniques or supercritical fluid extraction to obtain phytochemicals need to be explored for production of a higher yield of phytochemicals at lower operating costs and faster production times. There are a number of stability issues that must be overcome before phytochemicals can be successfully used as functional food ingredients since extracted phytochemicals can be less stable than naturally occurring phytochemicals in tissues.

Packaging can affect quality of phytochemicals and phytochemical-containing foods by influencing browning, flavor, and nutrient losses during storage, but studies reporting the phytochemical stability as affected by various packaging materials are lacking. Bioactive packaging, a novel technology for enhanced delivery of phytochemicals, is being investigated, but the issues related to feasibility, stability, and bioactivity of phytochemicals for food industry are yet to be studied.

References

Akagawa, M., Shigemitsu, T., and Suyama, K. (2003) Production of hydrogen peroxide by polyphenols and polyphenol-rich beverages under quasi-physiological conditions. *Bioscience Biotechnology and Biochemistry*, 67, 2632–2640.

Akpapunam, M.A. (1984) Effect of wilting, blanching and storage temperatures on ascorbic acid and total carotenoids content of some Nigerian fresh vegetables. *Qualitas Plantarum Plant Foods for Human Nutrition*, 34, 177–180.

Alasalvar, C., Grigor, J.M., Zhang, D., Quantick, P.C., and Shahidi, F. (2001) Comparison of volatiles, phenolics, sugars, antioxidant vitamins, and sensory quality of different colored carrot varieties. *Journal of Agricultural and Food Chemistry*, 49, 1410–1416.

Alnaizy, R. and Akgerman, A. (2000) Advanced oxidation of phenolic compounds. *Advance Environmental Research*, 4, 233–244.

Andrews, F.E. and Driscoll, P.J. (1977) Stability of ascorbic acid in orange juice exposed to light and air during storage. *Journal of The American Dietic Association*, 71, 140–142.

Aoshima, H. and Ayabe, S. (2007) Prevention of the deterioration of polyphenol-rich beverages. *Food Chemistry*, 100, 350–355.

Arnaud, J.P. (1995) Pro-liposomes for the food industry. *Food Technology Europe*, 2, 30–34.

Arneado, C.J.F. (1996) Microencapsulation by complex coacervation at ambient temperature. Patent No. FR 2732240 A1.

Arshady, R. (1993) Microcapsules for food. *Journal of Microencapsulation*, 10, 413–435.

Attoe, E.L. and von Elbe, J.H. (1981) Photochemical degradation of betanine and selected anthocyanins. *Journal of Food Science*, 46, 1934–1937.

Attoe, E.L. and vonElbe, J.H. (1982) Degradation kinetics of betanine in solutions as influenced by oxygen. *Journal of Agricultural and Food Chemistry*, 30, 708–712.

Augustin, M.A. and Hemar, Y. (2009) Nano- and micro-structured assemblies for encapsulation of food ingredients. *Chemical Society Reviews*, 38, 902–912.

Augustin, M.A., Sanguansri, L., Margetts, C., and Young, B. (2001) Microencapsulation of food ingredients. *Food Australia*, 53, 220–223.

Augustin, M.A. and Sanguansri, L. (2003) Polyunsaturated fatty acids, delivery, innovation and incorporation in foods. *Food Australia*, 55, 294–296.

Awika, J.M., Rooney, L.W., and Waniska, R.D. (2004) Properties of 3-deoxyanthocyanins from sorghum. *Journal of Agricultural and Food Chemistry*, 52, 4388–4394.

Ayer, J.E., Fishwick, M.J., Land, D.G., and Swain, T. (1964) Off-flavour of dehydrated carrot stored in oxygen. *Nature*, 203, 81.

Ayranci, E. and Tunc, S. (2004) The effect of edible coatings on water and vitamin C loss of apricots (*Armeniaca vulgaris* Lam.) and green peppers (*Capsicum annuum*, L.). *Food Chemistry*, 87, 339–342.

Balassa, L.L. and Fanger, G.O. (1971) Microencapsulation in the food industry. *CRC Reviews in Food Technology*, 2, 245–263.

Balaswamy, K., PrabhakaraRao, P.G., Satyanarayana, A., and Rao, D.G. (2006) Stability of bixin in annatto oleoresin and dye powder during storage. *LWT- Food Science and Technology*, 39, 952–956.

Baldwin, E.A., Nisperos-Carriedo, M.O., and Baker, R.A. (1995) Edible coatings for lightly processed fruits and vegetables. *HortScience*, 30, 35–38.

Baldwin, E.A., Nisperos-Carriedo, M.O., Chen, X., and Hagenmaier, R.D. (1996) Improving storage life of cut apple and potato with edible coating. *Postharvest Biology and Technology*, 9, 151–163.

Belscak-Cvitanovic, A., Stojanovic, R., Manojlovic, V., Komes, D., Cindric, I.J., Nedovic, V., and Bugarski, B. (2011) Encapsulation of polyphenolic antioxidants from medicinal plant extracts in alginate-chitosan system enhanced with ascorbic acid by electrostatic extrusion. *Food Research International*, 44, 1094–1101.

Beuchat, L.R. (2001) Control of foodborne pathogens and spoilage microorganisms by naturally occurring microorganisms. In: C.L. Wilson and S. Droby (eds) *Microbial Food Contamination*, London, UK: CRC Press, 149–169.

Bilyk, A., Kolodij, M.A., and Sapers, G.M. (1981) Stabilization of red beet pigments with isoascorbic acid. *Journal of Food Science*, 46, 1616–1617.

Bishop, P.D. and Nagel, C.W. (1984) Characterization of the condensation product of malvidin 3, 5-diglucoside and catechin. *Journal of Agricultural and Food Chemistry*, 32, 1022–1026.

Borris, R.P. (1996) Natural product research: Perspectives from a major pharmaceutical company. *Journal of Ethanopharmacology*, 51, 29–38.

Brenes, C., Del Pozo-Insfran, D., and Talcott, S. (2005) Stability of copigmented anthocyanins and ascorbic acid in a grape juice model system. *Journal of Agricultural and Food Chemistry*, 53, 49–56.

Brownmiller, C., Howard, L.R., and Prior, R.L. (2008) Processing and storage effects on monomeric anthocyanins, percent polymeric color, and antioxidant capacity of processed blueberry products. *Journal of Food Science*, 73, H72–H79.

Buckow, R., Kastell, A., Terefe, N.S., and Versteeg, S. (2010) Pressure and temperature effects on degradation kinetics and storage stability of total anthocyanins in blueberry Juice. *Journal of Agricultural and Food Chemistry*, 58, 10076–10084.

Burt, S. (2004) Essential oils: Their antibacterial properties and potential application in foods: A review. *International Journal of Food Microbiology*, 94, 223–253.

Cabrera, L., Gonzalez, V., Uribarri, E., Sierra, R., Laguna, A., Magraner, J., Mederos, D., and Velazquez, C. (2002) Study of the stability of tablets containing 10 mg of policosanol as active principle. *Boll Chim Farm*, 141, 223–229.

Cabrera, L., Rivero, B., Magraner, J., Sierra, R., Gonzalez, V., Uribarri, E., Laguna, A., Cora, M., Tejeda, Y., Rodriguez, E., and Velazquez, C. (2003) Stability studies of tablets containing 5 mg of policosanol. *Boll Chim Farm*, 142, 277–284.

Cai, Y., Sun, M., and Corke, H. (2005) Characterization and application of betalain pigments from plants of the Amaranthaceae. *Trends in Food Science and Technology*, 16, 370–376.

Callemien, D. and Collin, S. (2007) Involvement of flavonoids in beer color instability during storage. *Journal of Agricultural and Food Chemistry*, 55, 9066–9073.

Castano, G., Fernandez, L., Mas, R., Illnait, J., Fernandez, J., Mesa, M., Alvarez, E., and Lezcay, M. (2002) Comparison of the efficacy, safety and tolerability of original policosanol versus other mixtures of higher aliphatic primary alcohols in patients with type II hypercholesterolemia. *International Journal of Clinical Pharmacological Research*, 22, 55–66.

Cha, D.S., Cooksey, K., Chinnan, M.S., and Park, H.J. (2003) Release of nisin from various heat-pressed and cast films. *LWT – Food Science and Technology*, 36, 209–213.

Chai, P.C., Long, L.H., and Halliwell, B. (2003) Contribution of hydrogen peroxide to the cytotoxicity of green tea and red wines. *Biochemical Biophysical Research Communication*, 304, 650–654.

Champagne, C.P. and Fustier, P. (2007) Microencapsulation for the improved delivery of bioactive compounds into foods. *Current Opinion in Biotechnology*, 18, 184–190.

Chang, Q., Zuo, Z., Chow, M.S.S., and Ho, W.K.K. (2006) Effect of storage temperature on phenolics stability in hawthorn (*Crataegus pinnatifida* var. Major) fruits and a hawthorn drink. *Food Chemistry*, 98, 426–430.

Chen, Z., Zhu, Q.Y., Wong, Y.F., Zhang, Z., and Chung, H.Y. (1998) Stabilizing effect of ascorbic acid on green tea catechins. *Journal of Agricultural and Food Chemistry*, 46, 2512–2516.

Cho, Y.H., Shin, D.S., and Park, J. (2000) Optimization of emulsification and spray drying processes for the microencapsulation of flavor compounds. *Korean Journal of Food Science and Technology*, 32, 132–139.

Choi, M.H., Kim, G.H., and Lee, S.H. (2001) Effects of ascorbic acid retention on juice color and pigment stability in blood orange juice during refrigerated storage. *Food Research International*, 35, 753–759.

Chou, H.E. and Breene, W.M. (1972) Oxidative decoloration of β-carotene in low moisture model systems. *Journal of Food Science*, 37, 66–68.

Corey, M.E., William, L.K., Jake, H.M., and Vera, L. (2011) Phytochemical stability in dried apple and green tea functional products as related to moisture properties. *LWT – Food Science and Technology*, 44, 67–74.

Correa, C.R., Lei, L., Giancarlo, A., Marina, C., Chen, C.Y.O., Hye-Kyung, C., Soo-Muk, C., Ki-Moon, P., Robert, M.R., Jeffrey, B.B., and Kyung-Jin, Y. (2010) Composition and stability of phytochemicals in five varieties of black soybeans (*Glycine max*). *Food Chemistry*, 123, 1176–1184.

Couvreur, P., Blanco-Prieto, M.J., Puisieux, F., Roques, B., and Fattal, E. (1997) Multiple emulsion technology for the design of microspheres containing peptides and oligopeptides. *Advanced Drug Delivery Reviews*, 28, 85–96.

Cowan, M.M. (1999) Plant products as antimicrobial agents. *Clinical Microbiology Reviews*, 12, 564–582.

Czapski, J. (1985) The effect of heating conditions on losses and regeneration of betacyanins. *Z Lebensm Unters Forsch*, 180, 21–25.

Dallas, C., Ricoardo-da-Silva, J.M., and Laureano, O. (1996) Products formed in model wine solutions involving anthocyanins, procyanidin B2, and acetaldehyde. *Journal of Agricultural and Food Chemistry*, 44, 2402–2407.

Devi, N. and Maji, T.K. (2011) Study of complex coacervation of gelatin A with sodium carboxymethyl cellulose: Microencapsulation of Neem (*Azadirachta indica* A. Juss.) seed oil (NSO). *International Journal of Polymeric Materials*, 60, 1091–1105.

Devlieghere, F., Vermeulen, A., and Debevere, J. (2004) Chitosan, Antimicrobial activity, interactions with food components and applicability as a coating on fruit and vegetables. *Food Microbiology*, 21, 703–714.

Devlin, H.R. and Harris, I.J. (1984) Mechanism of the oxidation of aqueous phenol with dissolved oxygen. *Ind. Eng. Chem. Fundam*, 23, 387–392.

Dong, Z., Ma, Y., Hayat, K., Jia, C., Xia, S., and Zhang, X. (2011) Morphology and release profile of microcapsules encapsulating peppermint oil by complex coacervation. *Journal of Food Engineering*, 104, 455–460.

El Gaouth, A., Arul, J., Ponnampalam, R., and Boulet, M. (1991) Chitosan coating effect on storability and quality of fresh strawberries. *Journal of Food Science*, 12, 1618–1632.

Es-Safi, N., Fulcrand, H., Cheynier, V., and Moutounet, M. (1999) Studies on the acetaldehyde-induced condensation of (−)-epicatechin and malvidin 3-O-glucoside in a model solution system. *Journal of Agricultural and Food Chemistry*, 47, 2096–2102.

Es-Safi, N., Cheynier, V., and Moutounet, M. (2000) Study of the reaction between (+)- catechin and furfural derivatives in the presence of anthocyanins and their implication in food color change. *Journal of Agricultural and Food Chemistry*, 48, 5946–5954.

Es-Safi, N., Cheynier, V.R., and Moutounet, M. (2002) Role of aldehydic deriviatives in the condensation of phenolic compounds with emphasis on the sensorial properties of fruit-derived foods. *Journal of Agricultural and Food Chemistry*, 50, 5571–5585.

Eswaranandam, S., Hettiarachchy, N.S., and Meullenet, J.F. (2006) Effect of malic and lactic acid incorporated soy protein coatings on the sensory attributes of whole apple and fresh-cut cantaloupe. *Journal of Food Science*, 71, 307–313.

Fang, Z. and Bhandari, B. (2010) Encapsulation of polyphenols – A review. *Trends in Food Science and Technology*, 21, 510–523.

Fennema, O.R. and Tannenbaum, S.R. (1996) Introduction to Food Chemistry. In: O.R. Fennema (ed.) *Food Chemistry*, New York: Marcel Dekker, Inc, pp. 1–16.

Fisher, K. and Phillips, C. (2008) Potential antimicrobial uses of essential oils in food, is citrus the answer? *Trends in Food Science and Technology*, 19, 156–164.

Flanagan, J. and Singh, H. (2006) Microemulsions: a potential delivery system for bioactives in foods. *Critical Reviews in Food Science and Nutrition*, 46, 221–237.

Fossen, T., Cabrita, L., and Andersen, O.M. (1998) Colour and stability of pure anthocyanins influenced by pH including the alkaline region. *Food Chemistry*, 63, 435–440.

Frauen, M., Rode, T., Steinhart, H., and Rapp, C. (2000) Determination of green tea catechins by HPLC/EIS-MS- A method for stability analysis of green tea extracts in cosmetic formulations. *Lebensmittel Chemie*, 54, 141–142.

Friedman, M., Levin, C.E., Lee, S.U., and Kozukue, O. (2009) Stability of green tea catechins in commercial tea leaves during storage for 6 months. *Journal of Food Science*, 74, H47–H51.

Gallarate, M., Carlotti, M.E., Trotta, M., and Bovo, S. (1999) On the stability of ascorbic acid in emulsified systems for topical and cosmetic use. *International Journal of Pharmacology*, 188, 233–241.

Garti, N., Spernath, A., Aserin, A., and Lutz, R. (2005) Nano-sized self assemblies of non-ionic surfactants as solubilization reservoirs and microreactors for food systems. *Soft Materials*, 1, 206–218.

Garzon, G.A. and Wrolstad, R.E. (2002) Comparison of the stability of pelagronidin-based anthocyanins in strawberry juice and concentrate. *Journal of Food Science*, 67, 1288–1299.

Gibbs, B.F., Kermasha, S., Alli, I., and Mulligan, C.N. (1999) Encapsulation in the food industry: a review. *International Journal of Food Science and Nutrition*, 50, 213–224.

Goula, A.M. and Adamopoulos, K.G. (2012) A new technique for spray-dried encapsulation of lycopene. *Drying Technology*, 36, 641–652.

Goulding, J., McGlasson, B., Wyllie, S., and Leach, D. (2001) Fate of apple phenolics during cold storage. *Journal of Agricultural and Food Chemistry*, 49, 2283–2289.

Gouin, S. (2004) Microencapsulation: Industrial appraisal of existing technologies and trends. *Trends in Food Science Technology*, 15, 330–347.

Grau, M.J., Kayser, O., and Muller, R.H. (2000) Nanosuspensions of poorly soluble drugs – reproducibility of small scale production. *International Journal of Pharmaceutics*, 196, 155–157.

Gregoriadis, G. (1984) *Liposome Technology*, Vol. 1–3. Boca Raton, FL, CRC Press.

Gregory, J.F. (1996) Vitamins. In: O.R. Fennema (ed.) *Food Chemistry*, New York, Marcel Dekker, pp. 532–616.

Guimberteau, F., Dagleish, D., and Bibette, J.M.B. (2001) Multiple food emulsion composed of an inversed primary emulsion dispersed within an aqueous phase. Patent # FR 828,378–A1.

Gunes, G., Liu, R.H., and Watkins, C.B. (2002) Controlled atmosphere effects on postharvest quality and antioxidant activity of cranberry fruits. *Journal of Agricultural and Food Chemistry*, 50, 5932–5938.

Gutierrez IH, Lorenzo E, Espinosa EV. (2004) Phenolic composition and magnitude of copigmentation in young and shortly aged red wines made from the cultivars, cabernet sauvignon, cencibel, and syrah. *Food Chemistry*, 92, 269–283.

Hager, A., Howard, L.R., Prior, R.L., and Brownmiller, C. (2008) Processing and storage effects on monomeric anthocyanins, percent polymeric color, and antioxidant capacity of processed black raspberry products. *Journal of Food Science*, 73, H134–H140.

Han, J.H. (2002) Protein-based edible films and coatings carrying antimicrobial agents. In: A. Gennadios (ed.) *Protein-based Films and Coatings*, Boca Raton, CRC Press, pp. 485–499.

Hansen, S.L., Purup, S., and Christensen, L.P. (2003) Bioactivity of falcarinol and the influenceof processing and storage on its content in carrots (*Daucus carota* L). *Journal of the Science of Food and Agriculture*, 83, 1010–1017.

Herbach, K.M., Stintzing, F.C., and Carle, R. (2006) Betalain Stability and Degradation-Structural and Chromatic Aspects. *Journal of Food Science*, 71, R41–R50.

Herbach, K.M., Maier, C., Stintzing, F.C., and Carle, R. (2007) Effects of processing and storage on juice color and betacyanin stability of purple pitaya (*Hylocereus polyrhizus*) juice. *European Food Research and Technology*, 224, 649–658.

Hillebrand, S., Schwarz, M., and Winterhalter, P. (2004) Characterization of anthocyanins and pyranoanthocyanins from blood orange [*Citrus sinensis* (L.) osbeck] juice. *Journal of Agricultural and Food Chemistry*, 52, 7331–7338.

Holley, R.A. and Patel, D. (2005) Improvement in shelf-life and safety of perishable foods by plant essential oils and smoke antimicrobials. *Food Microbiology*, 22, 273–292.

Htwe, N.N., Srilaong, V., Tanprasert, K., Photchanachai, S., Kanlayanarat, S., and Uthairatanakij, A. (2010) Low oxygen concentrations affecting antioxidant activity and bioactive compounds in coloured rice. *Asian Journal of Food and Agro-Industry*, 3(2), 269–281.

Huang, A.S. and von Elbe, J.H. (1986) Stability comparison of two betacyanine pigments-Amaranthine and betanine. *Journal of Food Science*, 51, 670–674.

Huyskens-Keil, S. and Kadau, R. (2003) Quality and shelf life of fresh-cut asparagus (*Asparagus officinalis* L.) in different packaging films. *Acta Horticulture*, 600, 787–790.

Ikonte, C.J., Franco, E., David, P., and Murray, M. (2003) Effect of manufacturing conditions on the phytochemical stability of acerola concentrates. Access Business Group. http://www.aapsj.org/abstracts/Am_2003/AAPS2003–002235.PDF

Izumi, H., Nonaka, T., and Muraoka, T. (1997) Physiology and quality of fresh-cut spinach stored in low O_2 controlled atmospheres at various temperatures. In: J.R. Gorny (ed.) *CA'97 Proceedings Volume 5: Fresh-cut Fruits and Vegetables and MAP*, Davis, University of California, pp. 130–133.

Jackson, L.S. and Lee, K. (1991) Microencapsulation and the food industry. *Lebensmittel- Wiss. U.-Technology*, 24, 289–297.

Jayaprakasha, G.K., Negi, P.S., Jena, B.S., and Rao, L.J.M. (2007) Antioxidant and antimutagenic activities of *Cinnamomum zeylanicum* fruit extracts. *Journal of Food Composition and Analysis*, 20, 330–336.

Karatzas, A.K., Kets, E.P.W., Smid, E.J., and Bennik, M.H.J. (2001) The combined action of carvacrol and high hydrostatic pressure on *Listeria monocytogenes* Scott A. *Journal of Applied Microbiology*, 90, 463– 469.

Kenny, O. and O'Beirne, D. (2010) Antioxidant phytochemicals in fresh-cut carrot disks as affected by peeling method. *Postharvest Biology and Technology*, 58, 247–253.

Kim, E.S., Liang, Y.R., Jin, J., Sun, Q.F., Lu, J.L., Du, Y.Y., and Lin, D.C. (2007) Impact of heating on chemical compositions of green tea liquor. *Food Chemistry*, 103, 1263–1267.

Kirby, C.J. and Gregoriadis, G. (1984) A simple procedure for preparing liposomes capable of high encapsulation efficiency under mild conditions. In: G. Gregoriadis (ed.) *Liposome Technology*, Vol 1, Boca Raton, Florida, CRC Press, pp. 19–28.

Kirby, C.J., Whittle, C., Rigby, N., Coxon, D.T., and Law, B.A. (1991) Stabilization of ascorbic acid by microencapsulation. *International Journal of Food Science and Technology*, 26, 437–444.

Knezevic, Z., Gosak, D., Hraste, M., and Jalsenjako, I. (1998) Fluid-bed microencapsulation of ascorbic acid. *Journal of Microencapsulation*, 15, 237–252.

Komatsu, Y., Hisanobu, Y., Suematsu, S., Matsuda, R., Saigo, H., and Hara, K. (1991) Stability of functional constituents in canned tea drinks during extraction, processing and storage. In: *Proceedings of International Symposium on Tea Science*, Shizuoka, Japan, pp. 571–575.

Komatsu, Y., Suematsu, S., Hisanobu, Y., Saigo, H., Matsuda, R., and Hara, K. (1993) Studies on preservation of constituents in canned drinks. Part II. Effects of pH and temperature on reaction kinetics of catechins in green tea infusion. *Bioscience, Biotechnology and Biochemistry*, 57, 907–910.

Kopelman, S.H. and Augsburger, L.L. (2002) Capsule Shell Composition and Sealing Effects on Phytochemical Profile Stability of *Hypericum perforatum* Extract. http://www.aapsj.org/abstracts/AM_2002/AAPS2002–003199.pdf

Koski, A., Psomiadou, E., Tsimidou, M., Hopia, A., Kefalas, P. et al. (2002) Oxidative stability and minor constituents of virgin olive oil and cold-pressed rapeseed oil. *European Food Research and Technology*, 214, 294–298.

Lagaron, J.M. (2005) Bioactive packaging: A novel route to generate healthier foods. In: *Second Conference in Food Packaging Interactions*, Chipping Campden (UK, CAMPDEM (CCFRA).

Lakshminarayana, R., Raju, M., Krishnakantha, T.P., and Baskaran, V. (2007) Lutein and zeaxanthin in leafy greens and their bioavailability: Olive oil influences the absorption of dietary lutein and its accumulation in adult rats. *Journal of Agricultural and Food Chemistry*, 55, 6395–6400.

Lamb, R. (1987) Spray chilling. *Food Flavor Ingredients Packaging and Processing*, 9 (12), 39–42.

Lattanzio, V., Di, Vinere, D., Linsalata, V., Bertolini, P., Ippolito, A., and Salerno, M. (2001) Low temperature metabolism of apple phenolics and quiescence of *Phlyctaena vagabunda*. *Journal of Agricultural and Food Chemistry*, 49, 5817–5821.

Le Tien, C., Vachon, C., Mateescu, M.A., and Lacroix, M. (2001) Milk protein coatings prevent oxidative browning of apples and potatoes. *Journal of Food Science*, 66, 512–516.

Lee, J.Y., Park, H.J., Lee, C.Y., and Choi, W.Y. (2003) Extending shelf-life of minimally processed apples with edible coatings and antibrowning agents. *LWT – Food Science and Technology*, 36, 323–329.

Lee, C.Y. (1986) Changes in carotenoid content of carrots during growth and post harvest storage. *Food Chemistry*, 20, 285–293.

Long, L.H., Lan, A.N.B,, Hsuan, F.T.Y., and Halliwell, B. (1999) Generation of hydrogen peroxide by "antioxidant" beverages and the effect of milk addition: Is cocoa the best beverage? *Free Radical Research*, 31, 67–71.

Lopez-Rubio, A., Gavara, R., and Lagaron, J.M. (2006) Bioactive packaging, turning foods into healthier foods through biomaterials. *Trends in Food Science and Technology*, 17, 567–575.

Lorena, D., Anbinder, P.S., Navarro, A.S., and Martino, M.N. (2007) Co-crystallization of Yerba mate extract (*Ilex paraguariensis*) and mineral salts within a sucrose matrix. *Journal of Food Engineering*, 80, 573–580.

Lovric, T., Sablek, Z., and Boskovic, M. (1970) Cis-trans isomerization of lycopene and color stability foam-mat dried tomato powder during storage. *Journal of the Science of Food and Agriculture*, 21, 641–647.

Macdougall, D. (2002) *Colour in Food*. Cambridge, Woodhead Publishing.

Mas, R. (2000) Policosanol: Hypolipidemic, antioxidant, treatment of atherosclerosis. *Drugs Future*, 25, 569–586.

Mateos, M., Ke, D., Cantwell, M., and Kader, A.A. (1993) Phenolic metabolism and ethanolic fermentation of intact and cut lettuce exposed to CO_2-enriched atmospheres. *Postharvest Biology and Technology*, 3, 225–233.

McClements, D.J. (2005) Emulsion stability. In: D.J. McClements (ed.) *Food Emulsions: Principles, Practices and Techniques*, 2nd edition, Washington, DC, CRC Press, pp. 185–233.

McGill, J.N., Nelson, A.I., and Steinberg, M.P. (1966) Effect of modified atmosphere on ascorbic acid and other quality characteristics of spinach. *Journal of Food Science*, 31, 510–517.

McHugh, T.H. and Senesi, E. (2000) Apple wraps: A novel method to improve the quality and extend the self life of fresh-cut apples. *Journal of Food Science*, 65, 480–485.

Mendoza-Yepes, M.J., Sanchez-Hidalgo, L.E., Maertensm G., and Marin-Iniesta, F. (1997) Inhibition of *Listeria monocytogenes* and other bacteria by a plant essential oil (DMC) on Spanish soft cheese. *Journal of Food Safety*, 17, 47–55.

Min, S. and Krochta, J.M. (2005) Antimicrobial films and coatings for fresh fruit and vegetables. In: W. Jongen (ed.) *Improving the Safety of Fresh Fruit and Vegetables*, New York, CRC Press, pp. 354–492.

Minguez, M.I. and Galan, J.M. (1995) Kinetics of decoloring of carotenoid pigments. *Journal of Science of Food and Agriculture*, 67, 153–161.

Morello, J.R., Motiva, M.J., Tovar, M.J., and Romero, M.P. (2004) Changes in commercial virgin olive oil (cv. Arbequina) durino storage,with special emphasis on the phenolic fraction. *Food Chemistry*, 85, 357–364.

Mortensen, A., Skibsted, L.H., and Truscott, T.G. (2001) The interaction of dietary carotenoids with radical species. *Archives of Biochemistry and Biophysics*, 385, 13–19.

Mozafari, M.R., Khosravi-Darani, K., Borazan, G.G., Cui, J., Pardakhty, A., and Yurdugul, S. (2008) Encapsulation of food ingredients using nonoliposome technology. *International Journal of Food Properties*, 11, 833–844.

Muller, R.H., Beker, R., Kruss, B., and Peters, K. (1999) Pharmaceutical nanosuspensions for medicament administration as systems with increased saturation solubility and rate of solution. US Patent No. 5,858,410.

Namitha, K.K. and Negi, P.S. (2010) Chemistry and biotechnology of carotenoids. *Critical Reviews in Food Science and Nutrition*, 50, 728–260.

Nazareno, M.A., Coria Cayupan, Y., Targa, G., and Ochoa, J. (2009) Bioactive substance content and antioxidant activity changes during refrigerated storage of yellow without spines cactus pears. In: F.A.P. Campos, J.C.B. Dubeux Jr., J. de Melo Silva (eds) *VI International Congress on Cactus Pear and Cochineal*, ISHS Acta Horticulturae 811, Joao Pessoa Paraiba, Brazil.

Negi, P.S. (2012) Plant extracts for the control of bacterial growth: Efficacy, stability and safety issues for food application. *International journal of Food Microbiology*, 156, 7–17.

Negi, P.S., Jayaprakasha, G.K., Rao, L.J.M., and Sakariah, K.K. (1999) Antibacterial activity of turmeric oil – A byproduct from curcumin manufacture. *Journal of Agricultural and Food Chemistry*, 47, 4297–4300.

Negi, P.S. and Jayaprakasha, G.K. (2004) Control of Food borne pathogenic bacteria by garcinol and *Garcinia indica* extracts and their antioxidant activity. *Journal of Food Science*, 69, FMS 61–FMS 65.

Negi, P.S., Jayaprakasha, G.K., and Jena, B.S. (2010) Evaluation of antioxidant and antimutagenic activities of the extracts from the fruit rinds of *Garcinia cowa*. *International Journal of Food Properties*, 13, 1256–1265.

Negi, P.S. and Roy, S.K. (2000) Effect of low cost storage and packaging on quality and nutritive value of fresh and dehydrated carrots. *Journal of Science of Food and Agriculture*, 80, 2169–2175.

Negi, P.S. and Roy, S.K. (2001a) Retention of quality characteristics of dehydrated green leaves during storage. *Plant Foods for Human Nutrition*, 56, 285–295.

Negi, P.S. and Roy, S.K. (2001b) Effect of drying conditions on quality of green leaves during long term storage. *Food Research International*, 34, 283–287.

Negi, P.S. and Roy, S.K. (2003) Changes in β-carotene and ascorbic acid content of amaranth and fenugreek leaves during storage by low cost technique. *Plant Foods for Human Nutrition*, 58, 1–8.

Nunes, I.L. and Mercadante, A.Z. (2007) Encapsulation of lycopene using spray drying and molecular inclusion processes. *Brazilian Archives of Biology and Technology*, 50, 893–900.

O'Donnell, P.B. and McGinity, J.W. (1997) Preparation of microspheres by the solvent evaporation technique. *Advanced Drug Delivery Reviews*, 28, 25–42.

Okogeri, O. and Tasioula-Margari, M. (2002) Changes occurring in phenolic compounds and α–tocopherol of virgin olive oil during storage. *Journal of Agricultural and Food Chemistry*, 50, 1077–1080.

Outtara, B., Simard, R.E., Piette, G., Begin, A., and Holley, R.A. (2000) Diffusion of acetic and propionic acids from chitosan-based antimicrobial packaging films. *Journal of Food Science*, 65, 768–773.

Pagington, J.S. (1986) β-Cyclodextrin and its uses in the flavour industry. In: G.G. Birch and M.G. Lindley (eds) *Developments in Food Flavours*, London, Elsevier Applied Science, pp. 131–150.

Pandit, V.A. and Shelef, L.A. (1994) Sensitivity of *Listeria monocytogenes* to rosemary (*Rosmarinus officinalis* L.). *Food Microbiology*, 11, 57–63.

Park, S.I., Daeschel, M.A., and Zhao, Y. (2004) Functional properties of antimicrobial lysozyme-chitosan composite films. *Journal of Food Science*, 69, 215–221.

Park, S.I., Stan, S.D., Daeschel, M.A., and Zhao, Y. (2005) Antifungal coatings on fresh strawberries (*Fragaria ananassa*) to control mold growth during cold storage. *Journal of Food Science*, 70, 202–207.

Pasch, J.H. and von Elbe. J.H. (1978) Sensory evaluation of betanine and concentrated beet juice. *Journal of Food Science*, 43, 1624–1625.

Perez-Gago, M.B., Serra, M., Alonso, M., Mateos, M., and delRıo, M.A. (2006) Color change of fresh-cut apples coated with whey protein concentrate-based edible coatings. *Postharvest Biology and Technology*, 39, 84–92.

Pesek, C.A. and Warthesen, J.J. (1987) Photodegradation of carotenoids in a vegetable juice system. *Journal of Food Science*, 52, 744–746.

Pozo-Insfran, D.D., Follo-Martinez, A.D., Talcott, S.T., and Brenes, C.H. (2007) Stability of copigmented anthocyanins and ascorbic acid in muscadine grape juice processed by high hydrostatic pressure. *Journal of Food Science*, 72, S247–S253.

PrabhakaraRao, P.G., Jyothirmayi, T., Balaswamy, K., Satyanarayana, A., and Rao, D.G. (2005) Effect of processing conditions on the stability of annatto (*Bixa orellana* L.) dye incorporated into some foods. *LWT – Food Science and Technology*, 38, 779–784.

Pranoto, Y., Rakshit, S.K., and Salokhe, V.M. (2005) Enhancing antimicrobial activity of chitosan films by incorporating garlic acid, potassium sorbate and nisin. *LWT – Food Science and Technology*, 38, 859–865.

Raybaudi-Massilia, R.M., Mosqueda-Melgar, J., Soliva-Fortuny, R., and Martin-Belloso, O. (2009) Control of pathogenic and spoilage microorganisms in fresh cut fruits and fruit juices by traditional and alternative natural antimicrobials. *Comprehensive Reviews in Food Science and Food Safety*, 8, 157–180.

Reineccius, G.A. (1989) Flavor encapsulation. *Food Reviews International*, 5, 147–150.

Reineccius, G.A. (1995) Liposomes for controlled release in the food industries. In encapsulation and controlled release of food ingredients. *ACS Symposium Series*, 590, 113–131.

Rizzuto, A.B., Chen, A.C., and Veiga, M.F. (1984) Modification of the sucrose crystal structure to enhance pharmaceutical properties of excipient and drug substances. *Pharmaceutical Technology*, 8 (9), 32–35.

Rodriguez, M., Sadler, G.D., Sims, C.A., and Braddock, R.J. (1991) Chemical changes during storage of an alcoholic orange juice beverage. *Journal of Food Science*, 56, 475–479.

Rodriguez-Amaya, D.B. (2003) Food carotenoids: Analysis, composition and alterations during storage and processing of foods. *Forum Nutrition*, 56, 35–37.

Rojas-Grau, M.A., Avena-Bustillos, R.J., Friedman, M., Henika, P.R., Martin-Belloso, O., and McHugh, T.H. (2006) Mechanical, barrier, and antimicrobial properties of apple puree edible films containing plant essential oils. *Journal of Agricultural and Food Chemistry*, 54, 9262–9267.

Romanazzi, G., Nigro, F., and Ippolito, A. (2003) Short hypobaric treatments potentiate the effect of chitosan in reducing storage decay of sweet cherries. *Postharvest Biology and Technology*, 29, 73–80.

Saguy, I., Goldman, M., and Karel, M. (1985) Prediction of beta-carotene decolorization in model system under static and dynamic conditions of reduced oxygen environment. *Journal of Food Science*, 50, 526–530.

Salvia-Trujillo, L., Rojas-Grau, M.A., Soliva-Fortuny, R., and Martin-Belloso, O. (2013) Effect of processing parameters on physicochemical characteristics of microfluidized lemongrass essential oil-alginate nanoemulsions. *Food Hydrocolloids*, 30, 401–407.

Saxena, A., Bawa, A.S., and Rajum P.S. (2009) Phytochemical changes in fresh-cut jackfruit (*Artocarpus heterophyllus* L.) bulbs during modified atmosphere storage. *Food Chemistry*, 115, 1443–1449.

Schlameus, W. (1995) Centrifugal extrusion encapsulation. In: S.J. Risch and G.A. Reinessius (eds) *Encapsulation and Controlled Release of Food Ingredients*, Washington, DC, American Chemical Society, pp. 96–103.

Schmalko, M.E. and Alzamora, S.M. (2001) Color, chlorophyll, caffeine, and water content variation during processing. *Drying Technology*, 19, 599–610.

Schwartz, S.J., von-Elbe, J.H., and Giusti, M.M. (2008) Colorants. In: S. Damodaran, K.L. Parkin and O.R. Fennema (eds) *Fennema's Food Chemistry*, Boca Raton, Taylor and Francis, pp. 571–638.

Seydim, A.C. and Sarikus, G. (2006) Antimicrobial activity of whey protein based edible films incorporated with oregano, rosemary and garlic essential oils. *Food Research International*, 39, 639–644.

Shahidi, F. and Han, X.Q. (1993) Encapsulation of food ingredients. *Critical Reviews in Food Technology*, 33, 501–504.

Shelef, L.A. (1983) Antimicrobial effects of spices. *Journal of Food Safety*, 6, 29–44.

Smid, E.J. and Gorris, L.G.M. (1999) Natural antimicrobials for food preservation. In: M.S. Rahman (ed.) *Handbook of Food Preservation*, New York, Marcel Dekker, pp. 285–308.

Smoot, J.M. and Nagy, S. (1990) Effects of storage temperature and duration on total vitamin C content of canned single-strength grapefruit juice. *Journal of Agricultural and Food Chemistry*, 28, 417–424.

Solomon, O., Svanberg, U., and Sahlstrom, A. (1995) Effect of oxygen and fluorescent light on the quality of orange juice during storage at 8 °C. *Food Chemistry*, 53, 363–366.

Sothornvit, R. and Kiatchanapaibul, P. (2009) Quality and shelf-life of washed fresh–cut asparagus in modified atmosphere packaging. *LWT – Food Science and Technology*, 42,1484–1490.

Spanos, G.A., Wrolstad, R.E., and Heatherbell, D.A. (1990) Influence of processing and storage on the phenolic composition of apple juice. *Journal of Agricultural and Food Chemistry*, 38, 1572–1579.

Sparks, R.E. (1989) Microencapsulation. In: J.J. McKetta (ed.) *Encyclopedia of Chemical Processing and Design*, New York, Marcel Dekker, pp. 162–180.

Su, Y.L., Leung, L.K., Huang, Y., and Chen, Z. (2003) Stability of tea theaflavins and catechins. *Food Chemistry*, 83, 189–195.

Suematsu, S., Hisanobu, Y., Saigo, H., Matsuda, R., Hara, K., and Komatsu, Y. (1992) Effect of pH on stability of constituents in canned tea drinks. Studies on preservation of constituents in canned tea drink. *Journal of Japan Society of Food Science and Technology*, 39, 178–182.

Sun-Waterhouse, D., Penin-Peyta, L., Wadhwa, S.S., and Waterhouse, G.I.N. (2011) Storage stability of phenolic-fortified avocado oil encapsulated using different polymer formulations and co-extrusion technology. *Food Bioprocess Technology*, DOI 10.1007/s11947-011-0591-x

Talcott, S.T., Brenes, C.H., Pires, D.M., and Pozo-Insfran, D.D. (2003a) Phytochemical stability and color retention of copigmented and processed muscadine grape juice. *Journal of Agricultural and Food Chemistry*, 51, 957–963.

Talcott, S.T., Percival, S.S., Pittet-Moore, J., and Celoria, C. (2003b) Phytochemical composition and antioxidant stability of fortified yellow passion fruit (*Passiflora edulis*). *Journal of Agricultural and Food Chemistry*, 51, 935–941.

Tan, C.P. and Nakajima, M. (2005) β-carotene nanodispersions, Preparation, characterization and stability evaluation. *Food Chemistry*, 92, 661–671.

Tassou, C., Drosinos, E.H., and Nychas, G.J.E. (1995) Effects of essential oil from mint (*Mentha piperita*) on *Salmonella enteritidis* and *Listeria monocytogenes* in model food systems at 4 °C and 10 °C. *Journal of Applied Bacteriology*, 78, 593–600.

Taylor, A.H. (1983) Encapsulation systems and their applications in the flavor industry. *Food Flavor Ingredients Packaging and Processing*, 5 (9), 48–51.

Thanajiruschaya, P., Doksaku, W., Rattanachaisit, P., and Kongkiattikajorn, J. (2010) Effect of storage time and temperature on antioxidant components and properties of milled rice. *KKU Research Journal*, 15, 843–851.

Tiwari, B.K., Valdramidis, V.P., O'Donnell, C.P., Muthukumarappan, K., Bourkes, P., and Cullen, P.J. (2009) Application of Natural Antimicrobials for Food Preservation. *Journal of Agricultural and Food Chemistry*, 57, 5987–6000.

Trotta, M., Gallarate, M., Pattarino, F., and Morel, S. (2001) Emulsions containing partially water-miscible solvents for the preparation of drug nanosuspensions. *Journal of Controlled Release*, 76, 119–128.

Tsimidou, M., Papadopoulos, G., and Boskou, D. (1992) Phenolic compounds and stability of virgin olive oil-Part 1. *Food Chemistry*, 45, 141–144.

Turker, N., Aksay, S., Ekiz, H.I. (2004) Effect of storage temperature on the stability of anthocyanins of a fermented black carrot beverage, shalgam. *Journal of Agricultural and Food Chemistry*, 52, 3807–3813.

Uddin, M.S., Hawlader, M.N.A., Ding, L., and Mujumdar, A.S. (2002) Degradation of ascorbic acid in dried guava during storage. *Journal of Food Engineering*, 51,21–26.

Ultee, A. and Smid, E.J. (2001) Influence of carvacrol on growth and toxin production by *Bacillus cereus*. *International Journal of Food Microbiology*, 64, 373–378.

Van der Sluis, A., Dekker, M., de Jager, A., and Jongen, W. (2001) Activity and concentration of polyphenolic antioxidants in apple: effect of cultivar, harvest year, and storage conditions. *Journal of Agricultural and Food Chemistry*, 49, 3606–3613.

Vargas, M., Albors, A., Chiralt, A., and Gonzalez-Martınez, C. (2006) Quality of cold-stored strawberries as affected by chitosan-oleic acid edible coatings. *Postharvest Biology and Technology*, 41, 164–171.

von Elbe, J.H. and Attoe, E.L. (1985) Oxygen involvement in betanine degradation-Measurement of active oxygen species and oxidation reduction potentials. *Food Chemistry*, 16, 49–67.

von Elbe, J.H., Maing, I.Y., and Amundson, C.H. (1974) Colour stability of betanin. *Journal of Food Science*, 39, 334–337.

Wallace, J.M.W., McCabe, A.J., Robson, P.J., Keogh, M.K., Murray, C.A., Kelly, P.M., Marquez-Ruiz, G., McGlyn, H., Gilmore, W.S., and Strain, J.J. (2000) Bioavailability of n-3 polyunsaturated fatty acids in foods enriched with microencapsulated oils fish oil. *Ann Nutr Metab*, 44, 157–162.

Wang, H. and Helliwell, K. (2000) Epimerisation of catechins in green infusions. *Food Chemistry*, 70, 337–344.

Wang, S.Y. and Stretch, A.W. (2001) Antioxidant capacity in cranberry is influenced by cultivar and storage temperature. *Journal of Agricultural and Food Chemistry*, 49, 969–974.

Wang, Y., Ye, H., Zhou, C., Lv, F., Bie, X., and Lu, Z. (2012) Study on the spray-drying encapsulation of lutein in the porous starch and gelatin mixture. *European Food Research and Technology*, 234, 157–163.

Wasserman, B.P., Eiberger, L.L., and Guilfoy, M.P. (1984) Effect of hydrogen peroxide and phenolic compounds on horseradish peroxidase-catalysed decolorization of betalain pigments. *Journal of Food Science*, 49, 536–538.

Weiss, J., Takhistov, P., and McClements, J. (2006) Functional materials in food nanotechnology. *Journal of Food Science*, 71, 107–116.

Wishner, L.A. (1964) Light-induced oxidation in milk. *Journal of Dairy Science*, 47, 216–221.

Wong, D.W.S., Tillin, S.J., Hudson, J.S., and Pavlath, A.E. (1994) Gas exchange in cut apples with bilayer coatings. *Journal of Agricultural and Food Chemistry*, 42, 2278–2285.

Xu, J.Z., Leung, L.K., Huang, Y., and Chen, Z. (2003) Epimerization of tea polyphenols in tea drinks. *Journal of Agricultural and Food Chemistry*, 83, 1617–1621.

Yaminish, T. (1996) Chemical changes during storage of tea. In: G. Charalambous (ed.) *Handbook of Food and Beverage Stability*, Orlando, Florida, Academic Press, pp. 1–840.

Zafrilla, P., Ferreres, F., and Tomas-Barberan, F.A. (2001) Effect of processing and storage on the antioxidant ellagic acid derivatives and flavonoids of red raspberry (*Rubus idaeus*) jams. *Journal of Agricultural and Food Chemistry*, 49, 3651–3655.

Zanoni, B., Peri, C., Nani, R., and Lavelli, V. (1998) Oxidative heat damage of tomato halves as affected by drying. *Food Research International*, 31, 395–401.

Zhang, D. and Quantick, P.C. (1997) Effects of chitosan coating on enzymatic browning and decay during postharvest storage of litchi (*Litchi Chinensis* Sonn.) fruit. *Postharvest Biology and Technology*, 12, 195–202.

Zhang, D. and Quantick, P.C. (1998) Antifungal effect of chitosan coating on fresh strawberries and raspberries during storage. *Journal of Horticulture Science and Biotechnology*, 73, 763–767.

Zhao, L., Pan, Y., Li, J., Chen, G., and Mujumdar, A.S. (2004) Drying of a dilute suspension in a revolving flow fluidized bed of inert particles. *Drying Technology*, 22,, 363–376.

Zhu, Q.Y., Zhang, A., Tsang, D., Huang, Y., and Chen, Z.Y. (1997) Stability of green tea catechins. *Journal of Agricultural and Food Chemistry*, 45, 4624–4628.

Zivanovic, S., Chi, S., and Draughon, F. (2005) Antimicrobial activity of chitosan films enriched with essential oils. *Journal of Food Science*, 70, 45–51.

Part V
Analysis and Application

17 Conventional extraction techniques for phytochemicals

Niamh Harbourne, Eunice Marete, Jean Christophe Jacquier and Dolores O'Riordan

Institute of Food and Health, University College Dublin

17.1 Introduction

Phytochemicals are a diverse group of plant derived chemicals which have received much attention in recent years due to their many health benefits including antioxidant, anticarcinogenic and anti-inflammatory activity (Dillard and German, 2000; Schreiner and Huyskens-Keil, 2006). They can be classified into sub-groups according to their chemical structure, which include terpenoids (e.g. carotenoids), phytosterols, polyphenols (e.g. tannins, flavonoids, phenolic acids) and glucosinolates (Chapter 4).

Phytochemicals make up less than 10% of the plant matrix (Harjo, Wibowo and NG, 2004) therefore to prepare phytochemical rich foods they may first need to be extracted from the plant matrix. It is important to note that the phytochemical content of plants used may vary depending on the species or organ (e.g. roots, leaves, flowers, fruits), therefore extraction conditions used may also vary. Some phytochemicals are limited to specific taxonomic groups, for example, glucosinates are specific to cruciferous vegetable crops, while others (e.g. polyphenols) are present in a wide range of plants (Schreiner and Huyskens-Keil, 2006).

Firstly, this chapter will focus on the principles of conventional methods used to extract phytochemicals from plants with a view to incorporating them into foods and beverages, including the various extraction methods used, the factors affecting extraction of these bioactive compounds, and limitations to the use of these methods. The final section of this chapter will give an account of conventional extraction techniques used to extract phytochemicals from various plant species and plant organs, from roots to fruits, reported in the literature.

17.2 Theory and principles of extraction

The aim of extraction is to maximise the yield of compounds of interest, while minimising the extraction of undesirable compounds. Traditionally, fresh plant material was used for the extraction of plants. However, nowadays this is not as common, as it requires very rapid post-harvest processing to avoid degradation of the plant. The extraction of phytochemicals from plants is now mostly done using dried plant as the starting material in order to inhibit the metabolic processes which can cause degradation of the active compounds, therefore extending the shelf life of the plant material.

Traditionally, phytochemicals have been extracted from plants using solid-liquid extraction techniques. This section will first cover conventional methods used for the extraction of phytochemicals from plants and it will then focus on the factors that influence the quality of the resulting extract.

17.2.1 Conventional extraction methods

Solid-liquid extraction methods used for the extraction of phytochemicals from plants include maceration, infusion and Soxhlet extraction. These extraction processes involve firstly the diffusion of the solvent into the plants cells, solubilisation of the phytochemical compounds within the plant matrix and finally diffusion of the phytochemical-rich solvent out of the plant cells. This section will also cover the extraction of essential oils from plants using steam and hydrodistillation.

17.2.1.1 Maceration

Macerations are produced by steeping the plant material in a liquid, which is generally an organic solvent, at room temperature. For this extraction process the plant material is soaked in the solvent in a closed container. The solution can be stirred to increase the rate of extraction of the phytochemicals from the plant material. After extraction is complete the plant material is separated from the solvent by filtration. The plant material can then undergo another extraction step by adding fresh solvent to the material and letting it soak. This step can be repeated several times to ensure complete extraction of phytochemicals from the plant material, however it is a very time and solvent consuming process. Maceration can take from hours to days for a single extraction, and can take weeks for repeated maceration of the plant material (Seidel, 2006). Although it is time consuming it is a useful extraction method for heat labile compounds as it is carried out at room temperature.

17.2.1.2 Infusions

Infusion is a similar process to maceration but the extraction is carried out at a set temperature (normally higher than room temperature and up to 100 °C) for a set period of time (from minutes to hours) and water is generally used as the extraction solvent. As for maceration, after extraction is complete the mixture is filtered. Traditionally, infusions were made by using boiling water as the extracting solvent, for example, making a cup of tea. Following immersion in boiling water the plant material was left to steep and finally filtered to remove the plant material from the extract.

17.2.1.3 Soxhlet extraction

Soxhlet extraction has been used for many years in the extraction of phytochemicals from plants, and is often used as a reference for evaluating other solid-liquid extraction methods or new non-conventional extraction methods (Wang and Weller, 2006). In a Soxhlet extraction system the plant material is put in a thimble-holder, which has perforated sides and bottom so liquid can fall through. There is a collection flask below the thimble and a reflux condenser above it. Heat is applied to the flask containing solvent; the solvent evaporates and travels to the condenser. Condensed solvent then falls into the thimble containing the plant material, when it reaches a certain level it is unloaded back into the solvent flask. The solute is separated from the solvent by distillation, as the solute is left in the flask and fresh solvent passes into the plant material. This procedure is repeated until complete extraction of plant material is achieved (Wang and Weller, 2006). Solvent and particle size will need to be selected depending on the phytochemicals which need to be extracted; this will be discussed in more detail in section 17.2.1.4.

17.2.1.4 Steam and hydrodistillation

Steam and hydrodistillation are extraction techniques used to extract water-insoluble volatile constituents from various matrices, including the extraction of essential oils from plants and are widely used in the perfume industry for extraction of essential oils. For steam distillation the steam is percolated through the plant material. The steam dissolves the essential oil in the plant material and then enters a condenser. The mixture of condensed water and oil is collected and finally separated by decanting. For hydrodistillation the only difference is that the plant material is submerged in the water, which is then heated until it boils. The extraction conditions can be optimised by modifying the distillation time and temperature. The conditions may also need to be modified depending on the material being extracted, for example, for the extraction of tough material (roots or bark) glycerol may be added to the water to assist extraction (Seidel, 2006). However, for many medicinal plants the conditions used to extract essential oil are well defined in the *European Pharmacopeia* (2004).

17.2.2 Factors affecting extraction methods

The efficiency of the solid-liquid extraction methods are affected by factors such as solvent type, ratio of solvent to plant material, temperature, time and structure of the matrix (e.g. particle size, plant organ).

17.2.2.1 Solvent

As previously mentioned, the extraction of phytochemicals is dependent on the dissolution of each compound in the plant material matrix and their diffusion into the external solvent (Shi, Nawaz, Pohorly, Mittal, Kakuda and Jiang, 2005), therefore the choice of extraction solvent is one of the most important matters to consider for solid-liquid extraction. The factors that need to be considered when choosing the solvent or solvent system for extraction of phytochemicals are safety of the solvent and potential for formation or extraction of undesirable compounds and finally solubility of the target compounds (Seidel, 2006).

In recent years, organic solvents (e.g. methanol) have been used to extract phytochemicals from plant material (Naczk and Shahidi, 2004). These extraction procedures were efficient and resulted in high yields of phytochemicals, but the solvents may be harmful to human health if ingested and therefore would not be desirable for inclusion in a food or beverage. To produce phytochemical rich extracts for incorporation into foods and beverages it is necessary to use food grade solvents (e.g. water, ethanol or mixtures of these).

Water is a polar solvent that has been used for many years to extract phytochemicals from plant materials, for example, infusions of medicinal herbs or teas have been used traditionally to treat many conditions including inflammation. Ethanol may also be used as an extraction solvent, because even if it is found in the final extract it is safe for human consumption. However, under EU legislation if a food contains more than 1.2% ethanol, no health claims on the efficacy of the resulting extract can be made (European Parliament and Council of Europe, 2006). Therefore, if the extraction solvent contains ethanol it must be removed before inclusion of the extract into a functional food or beverage. It should be noted that if any other organic solvents are used for the extraction of phytochemicals all solvent residues must be totally removed from the extracts before they can be incorporated into foods or beverages. Depending on the polarity of the compounds to be extracted mixtures of ethanol and water may need to be used, so the water can extract the more polar compounds and the ethanol the more hydrophobic compounds. An example of the extraction of various bioactives of with mixtures of ethanol and water is shown in Figure 17.1. It is clear that the maximum level of the bioactive marker from chamomile flowers (apigenin-7-glucoside) is extracted at an ethanol content of 50%, while the maximum bioactive marker is extracted from feverfew (parthenolide) at 90% ethanol.

Finally, the pH of the extraction solvent can be changed to selectively extract or improve the extraction of certain plant bioactives. For example, anthocyanins are unstable at neutral or alkaline pH and as a result acidic aqueous solvents are often used for the extraction of these compounds (Mateus and de Freitas, 2009).

17.2.2.2 Temperature

Many studies have investigated the effect of temperature on the extraction of polyphenolics from plant material (Joubert, 1990; Price and Spitzer, 1994; Labbe, Tremblay and Bazinet, 2006; Lim and Murtijaya, 2007). In general, a higher extraction temperature causes an increase in the rate of diffusion of the soluble plant phytochemicals into the extraction solvent, thereby reducing extraction time. An increase in temperature can cause an increase in the concentration of some phytochemicals, which is possibly due to an increase in the solubility of many of these bioactive compounds, or to the breakdown of cellular constituents resulting in the release of the phytochemicals (Lim and Murtijaya, 2007). In addition an increase in the extraction temperature may also inhibit enzymatic activities thus resulting in an increase in the yield of the bioactive compounds. Marete, Jacquier and O'Riordan (2009) reported that extraction temperature of 70 °C and above resulted in a significant increase of total phenols from feverfew due to inactivation of polyphenol oxidase. The temperature used for extraction will be limited depending on the extraction solvent chosen as they all have different boiling points, for example, the boiling point of acetone is 56–57 °C whereas the boiling point of water is 100 °C.

17.2.2.3 Time

The time given to the extraction of phytochemicals from plant material by a food manufacturer may be a compromise between complete extraction of these components and having an extraction process which is both time and cost effective. The time it takes for extraction of phytochemicals will vary depending on the plant species to be extracted, the particle size of the material and the plant organ. For example, the extraction of phytochemicals from leafy material will be faster than the extraction from harder material such as roots or bark (Whitehead, 2005). To produce extracts high in desirable and low in undesirable compounds, the extraction kinetics of both the wanted and unwanted compounds may need to be studied.

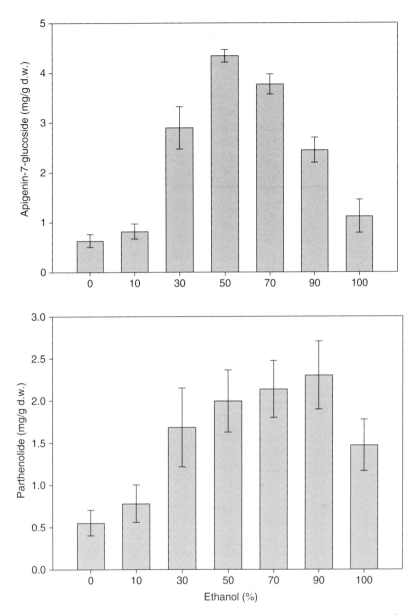

Figure 17.1 Effect of content of ethanol (%) in water on extraction of apigenin-7-glucoside from chamomile flowers and parthenolide from feverfew leaves.

17.2.2.4 Particle size

Plant material can undergo grinding or milling before extraction to reduce the particle size. The smaller the particle size of the material the shorter the path that the solvent has to travel, which decreases the time for maximum phytochemical content to be extracted (Shi et al., 2005). Also, grinding or milling the plant material to reduce the particle size damages the plant cells which can also lead to increased extraction of phytochemical compounds. For example, Fonseca, Rushing, Thomas, Riley and Rajapakse (2006) explained the suitability of using finely ground samples for the maximum extraction yield of parthenolide in feverfew, which may be localised in trichomes, in small oil glands or may be bound in other tissues. The disadvantage of grinding or milling the plant before extraction is that plant material of small particle size may block filters quicker than bigger particles and this could possibly result in wastage of the extract and extended extraction times (Whitehead, 2005).

17.2.3 Limitations of extraction techniques

In general, the disadvantages of using conventional extraction techniques that are commonly cited in the literature include long extraction times, requirement of large quantities of solvent and the degradation of heat labile phytochemicals by using high temperatures for extraction. The disadvantages of conventional methods as a result of solvent and temperature will be discussed in more detail.

17.2.3.1 Solvent

As mentioned earlier, for conventional extraction methods solvent choice is very important. The extraction of phytochemicals with a view to incorporating these compounds into foods means that the solvent choice is limited as it must be food grade. The small selection of solvents available for use may mean that some bioactive compounds may not be soluble in the solvent system chosen. Furthermore, unlike extracts made from water, those made by the extraction of phytochemicals from plants with any organic solvent must also undergo an evaporation step to remove the solvent. It is possible that this evaporation step may result in the formation of undesirable compounds or degradation of the bioactive(s) of interest. This evaporation step may also be costly, as companies have to ensure for health and safety reasons that the solvent is completely removed from the end product. Dried extracts can easily be added into foods, however a liquid extract (e.g. infusions) is preferred for the majority of functional beverages as it is the easiest to mix into the end product (Whitehead, 2005). Lastly, some conventional methods use large amounts of organic solvents during extraction (e.g. maceration) which is neither environmentally friendly nor cost effective.

17.2.3.2 Temperature

Many phytochemicals are heat stable and extraction at high temperature has no adverse effects; however there are some phytochemicals which are heat labile. A recent study showed that in acai fruit extracts heat had no effect on its phenolic content, which included flavone glycosides, flavonol derivatives and phenolic acids, but resulted in a significant reduction in the anthocyanin content possibly due to accelerated chalcone formation on exposure to high temperatures (Pacheco-Palencia, Duncan and Talcott, 2009). The effect of heat on anthocyanins degradation is well documented in literature (Harbourne, Jacquier, Morgan and Lyng, 2008; Patras, Brunton, O'Donnell and Tiwari, 2010) therefore high temperatures for long extraction times would not be suitable for extraction of plants containing these compounds. Another phytochemical which has been shown to be heat labile is parthenolide in feverfew extracts (Marete, Jacquier and O'Riordan, 2011). Steam distillation can also result in degradation of bioactive volatiles during the extraction of essential oils. For example, chamomile essential oil extracted by steam distillation had a low content of matricine as it degraded to its breakdown product chamazulene, whereas during supercritical fluid extraction (SFE) there were higher levels of matricine and very little chamazulene present in the essential oil (Kotnik, Skerget and Knez, 2007). Before choosing extraction conditions the bioactive of interest needs to be assessed for heat stability. If the bioactive is heat sensitive it may need to be extracted at lower temperatures for a longer time period or alternatively a non conventional method, which does not use high temperatures, could be used.

17.3 Examples of conventional techniques

As already mentioned extraction conditions can vary greatly depending on the part of the plant used for extraction. Therefore, in this section the extraction of phytochemicals using conventional techniques from various plant species and from different plant organs from roots to fruits will be reviewed.

17.3.1 Roots

Roots are the underground part of the plant and include vegetables such as carrots, potatoes and sweet potatoes. Purple sweet potatoes are a good source of anthocyanins and the extraction of these compounds has been studied as they show potential as natural food colorants and also have many health benefits (Fan, Han, Gu and Chen, 2008). Response surface methodology was used to optimise the extraction of fresh pulverised purple sweet potato using acid-ethanol (1.5M HCl) at various extraction temperatures (40–80 °C), times (60–120 min) and solvent to solid ratios (15:1–35:1) (Fan et al., 2008). The factors that had the most significant effect on the extraction of the anthocyanins from sweet potato were temperature and solvent to solid ratio. The optimum conditions for extraction were extraction temperature of 80 °C for 60 min at a solvent to solid ratio of 32:1 to yield 1.58 mg/g d.w. of purple sweet potato anthocyanins.

Black carrot is another root vegetable which is also a potential source of anthocyanins that can be used as a functional ingredient, especially as it contains acylated cyanidin derivatives which demonstrate superior heat and pH stability in comparison with other anthocyanins (Turker and Erdogdu, 2006). The effect of extraction temperature (25–50 °C) and pH (2–4) on the diffusion coefficient of anthocyanins from black carrot slices were studied. The diffusion coefficient increased with an increase in temperature, due to an increase in the solubility of the anthocyanins. For example, in the extract at pH 2 the diffusion coefficient increased from 3.73 to 7.37 m^2/s at 25 and 50 °C, respectively (Turker et al., 2006). However, as anthocyanins are heat labile the extraction temperature cannot be increased indefinitely or it may result in degradation. The diffusion coefficient also increased with a decrease in pH; at an extraction temperature of 37.5 °C the diffusion coefficient increased from 0.25 to 5.00 m^2/s as the pH decreased from 4 to 2. Overall, extraction of black carrot anthocyanins at high temperatures and low pH results in a more time effective process.

17.3.2 Leaves and stems

In recent years the extraction of the leaves and stems of many plants has been studied with a view to using the extracts as functional ingredients for incorporation into foods (Harbourne, Marete, Jacquier and O'Riordan, 2011) including meadowsweet (Harbourne, Jacquier and O'Riordan, 2009) and feverfew (Marete et al., 2009). Meadowsweet contains tannins, phenolic acids, flavonoids and salicylates (Blumental, Goldberg and Brinckmann, 2000). Flavonoids have been extracted from meadowsweet leaves using hot aqueous ethanol (70%)

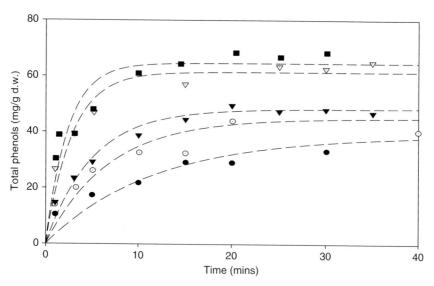

Figure 17.2 Extraction kinetics of total phenols from meadowsweet at temperatures 60(●), 70(○), 80(▼), 90(▽) & 100(■) °C.

(Krasnov, Raldugin, Shilova and Avdeeva, 2006) and also using methanol in a Soxhlet apparatus (Papp et al., 2004), however this was with a view of analysing the compounds present and not for incorporation into foods. Traditionally meadowsweet has been extracted by infusion or maceration (Mills and Bone, 2000). The aqueous extraction kinetics of phenolic compounds from meadowsweet (Harbourne et al., 2009) has recently been examined with a view to maximising the total phenolic content and minimising the content of tannins, which impart a bitter and astringent taste to the extract, and was found to follow a pseudo-first order kinetic model (Figure 17.2). The concentration of phenolic compounds, which included salicylic acid and quercetin (bioactives thought to be responsible for meadowsweet's health benefits), rose rapidly initially before reaching an equilibrium concentration (Figure 17.2). The rate constant (k) increased from 0.09 to 0.44 min^{-1} as the temperature increased from 60 to 100 °C. In addition, increasing the temperature from 60 to 90 °C resulted in an increase in the total phenols from 39 ± 2 to 61 ± 2 mg/g d.w, however increasing the temperature to 100 °C showed no further increase (Figure 17.2). Increasing the temperature of the extraction solvent did not have a significant effect on the proportion of tannins and non-tannins extracted. Interestingly, the non-tannin fraction was extracted faster than both the tannins and total phenols from meadowsweet. Therefore, the extraction of tannins is likely to be the rate limiting step in the extraction of total phenols from meadowsweet. The optimum temperature for the extraction of phenolic compounds (including salicylic acid and quercetin) without having any adverse effects on the tannin concentration of meadowsweet was ≥ 90 °C for 15 min (Harbourne et al., 2009).

The pH of the extraction solvent can be changed to improve the extraction of some bioactives from the plant matrix. However, it tends to be different depending on the plant species and bioactives of interest and therefore should be optimised specifically for each plant. For example, the maximum total phenols extracted from meadowsweet increased from 43 ± 2 to 57 ± 2 mg/g d.w. with an increase in pH from 3.9 to 6.4, however increasing the pH had no significant effect on the salicylic acid or quercetin content in the extracts (Harbourne et al., 2009). Unlike meadowsweet, the extraction of total catechins from green tea leaves did not significantly change when the green tea was extracted at pH 4, 5 or 6 but when the extraction pH was increased to 7 the total catechin content decreased significantly (Kim, Park, Lee and Han, 1999). Also, Spiro and Price (1987) studied the effect of pH on theaflavins in black tea; in this case increasing the pH to 6.8 had no significant effect on the theaflavin content but decreasing the pH caused an increase in theaflavin content. Likewise, Liang and Xu (2001) found that more acidic conditions favoured the extraction of polyphenols in black tea, while increasing the pH from 4.9 to 9.45 caused a decrease in some of the tea phenols, including thearubigins, theaflavins and catechins.

17.3.3 Flowers

The flowers of many plants also have a high content of phytochemicals. Chamomile (*Matricaria chamomilla* L.) flowers are a popular ingredient in many foods and beverages due to their many health benefits. They have a high content of phenolic compounds, including flavonoids such as flavone glycosides (e.g. apigenin-7-glucoside), flavonols (e.g. quercetin glycosides, luteolin glucosides) and caffeic and ferulic acid derivatives. Chamomile also contains an essential oil, the main components of it being α-bisabolol and chamazulene (which is responsible for its blue colour) (Harbourne *et al.*, 2011). The essential oil of chamomile can be extracted using steam or hydrodistillation according to the pharmacopeia standards (4 h distillation period).

Traditionally, chamomile flowers were extracted by infusion or maceration. The effect of extraction temperature and time on the aqueous extraction of total phenols and apigenin-7-glucoside from whole chamomile flowers was optimised (Harbourne, Jacquier and O'Riordan, 2009a). The extraction of phenolic compounds from chamomile flowers followed a pseudo-first order kinetic model. As the extraction temperature increased from 57 to 100 °C the rate of extraction (k) of total phenols increased from 0.028 ± 0.004 to 0.31 ± 0.03 min^{-1}. Also, the total phenolic content increased from 14.6 to 24.5 mg/g d.w with an increase in temperature from 57 to 100 °C, while the apigenin-7-glucoside content reached a maximum at 90 °C (0.29 mg/g d.w.). It should be noted that between 90 and 100 °C there was a significant increase in turbidity of the extract possibly due to tissue degradation of the chamomile flowers during boiling. Therefore, aqueous extraction at 90 °C × 20 min were the optimum conditions for preparing an extract with a high phenolic content and low turbidity which would be ideal for incorporation into foods or beverages.

The content of ethanol (%) in water on extraction of polyphenols from whole chamomile flowers after steeping at room temperature for 24 h has been examined in our laboratory. An ethanol content of 50% yielded extracts with the maximum apigenin-7-glucoside (Figure 17.1) and total phenol content (unpublished results).

17.3.4 Fruits

Apple pomace is a by-product from apple juice and apple cider and grape pomace is a by-product from wine making, which have been investigated as a source of phenolic compounds due to their abundance. Extraction of phytochemicals from waste products, such as apple pomace and grape pomace, has received much interest in recent years due to the interest in using natural and low cost sources of phytochemicals for incorporation into foods or beverages.

Apple pomace consists of the peel, core, seed, calyx, stem and soft tissue. It contains many polyphenols including chlorogenic acid, catechins, procyanididns and quercetin glycosides (Lu and Foo, 1997; Cam and Aaby, 2010). Mostly, the extraction of these compounds from apple pomace has been done using organic solvents such as methanol (Cam and Aaby, 2010), however recently the extraction of polyphenols using food grade solvents has also been investigated. Response surface methodology was used to optimise the extraction of apple pomace phenolics with water (Cam and Aaby, 2010; Wijngaard and Brunton, 2010), ethanol and acetone (Wijngaard and Brunton, 2010). In both studies the apple pomace was freeze-dried and milled to a fine powder before extraction. Cam and Aaby (2010) studied the effect of temperature, extraction time and solvent to solid ratio on the extraction of total phenols and 5-Hydroxymethylfurfural (HMF), high levels of which are undesirable in foods and beverages, from apple pomace. Aqueous extraction at 100 °C for 37 min at a solvent to solid ratio of 100 mL/g were the optimum to yield extracts rich in phenolic compounds (8.3 mg/g d.w.) with limited quantity of HMF (42 mg/L). At times greater than 37 min the total phenol concentration increased but so did the HMF content and at solvent to solid ratios above 100 ml/g there was no further increase in total phenolics. To maximise the extraction of polyphenols from apple pomace using acetone and ethanol, the extraction time, temperature and content of water in the solvent was optimised to give the highest antioxidant activity (Wijngaard and Brunton, 2010). The conditions using ethanol as the solvent were 56% ethanol at 80 °C for 31 min, which resulted in extracts with an antioxidant value of 4.44 mg Trolox/g d.w. and a total phenolic content of 10.92 mg/g d.w. Optimum conditions using acetone as the extraction solvent were 65% acetone at 25 °C for 60 min, which yielded an antioxidant value of 5.29 mg Trolox/g d.w. and total phenol content of 14.15 mg/g d.w. Depending on the solvent used the content of individual polyphenols changed. For example, the content of procyanidins, catechin, epicatechin and caffeoylquinic acids were higher in extracts made using water as the solvent, while quercetin glycosides were much higher when acetone was used as the extraction solvent (Cam and Aaby, 2010). Therefore, as mentioned earlier, depending on the compounds of interest the extraction solvent can be modified. Although in contrast to using water all traces of organic solvent used to extract polyphenols must be removed prior to incorporation in foods.

Grape pomace consists of pressed skins and seeds of grapes. They are rich in phenolic compounds including anthocyanins, catechins, cinnamic acids and proanthocyanidins. Anthocyanin-rich extracts from grape pomace are widely used as natural food colorants in many foods and beverages including yogurts, soft drinks and confectionary (Mateus and de Freitas, 2009). The extraction of phenolic compounds from dried grape pomace has been studied using ethanol, water and mixtures of these solvents (Pinelo, Rubilar, Jerez, Sineiro and Nunez, 2005; Spigno, Toramelli and De Faveri, 2007; Lapornik, Prosek and Wondra, 2005). Spigno et al. (2007) studied the extraction of phenolics from dried and milled grape pomace using ethanol at two temperatures (45 and 60 °C) over a time period of 1–24 h. In general, the extraction of all phenolic compounds increased with an increase in extraction temperature and time. However, after extraction for 20 h at 60 °C there was a reduction in total phenols, particularly for anthocyanins and tannins due to thermal degradation or polymerisation. As this process may be used for industrial applications, cost and time were considered and as a result extraction conditions of 60 °C for 5 h were selected even though they did not yield the highest phenolic content. Using these conditions the grape pomace was extracted using mixtures of ethanol and water (10–60%). There was an increase in the yield of total phenols with an increase in water content in the extraction solvent (ethanol) from 10% (24.5 mg/g) to 30% (41.3 mg/g); however above this there was no further significant increase. In another study the effect of using water or aqueous ethanol (70%) on the extraction of anthocyanins from grape pomace was compared, and it was found that the ethanol extracts contained seven times more anthocyanins than those extracted with water (Lapornik *et al.*, 2005). The extraction kinetics of dried and milled grape pomace phenolics have been studied in aqueous ethanol (60%) at 60 °C over a period of 5 h. Similar to chamomile flowers and meadowsweet leaves the extraction of grape pomace phenolics followed a first order kinetic model. The equilibrium concentration of phenolics was extracted after 120 min (Amendola, Faveri and Spigno, 2010).

17.4 Conclusion

By varying the factors affecting extraction, including solvent, temperature, particle size, solvent to solid ratio and time, a wide range of phytochemicals can be extracted using conventional methods. However, the extraction parameters must be optimised depending on the bioactives of interest, the plant species and the plant organ, as the extraction conditions can vary greatly.

References

Amendola, D., De Faveri, D.M. and Spigno, G. (2010) Grape marc phenolics: Extraction kinetics, quality and stability of extracts. *Journal of Food Engineering*, 97, 384–392.

Blumental, M., Goldberg, A. and Brinckmann, J. (2000) *Herbal Medicine: Expanded Commission E Monographs*. Massachusetts: Integrative Medicine Communications.

Cam, M. and Aaby, K. (2010) Optimization of extraction of apple pomace phenolics with water by response surface methodology. *Journal of Agriculture and Food Chemistry*, 58, 9103–9111.

Dillard, C.J. and German, J.B. (2000) Phytochemicals: Nutraceuticals and human health. *Journal of the Science of Food and Agriculture*, 80, 1744–1756.

European Parliament and Council of Europe (2006) Directive 2006/1924/EC, relating to the nutrition and health claims made on foods. *Official Journal, L 404*. Luxembourg: Office for Official Publications of the European Communities.

European Pharmacopoeia (2004) Directorate for the quality of medicines, Council of Europe: Strasbourg.

Fan, G., Han, Y., Gu, Z. and Chen, D. (2008) Optimizing conditions for anthocyanins extraction from purple sweet potato using response surface methodology (RSM). *LWT – Food Science and Technology*, 41,155–160.

Fonseca, J.M., Rushing, J.W., Thomas, R.L., Riley, M.B. and Rajapakse, N.C. (2006) Post-production stability of parthenolide in Feverfew (*Tanacetum parthenium*). *Journal of Herbs, Spices and Medicinal Plants*, 12, 139–152.

Harbourne, N., Jacquier, J.C., Morgan D. and Lyng, J. (2008) Determination of the degradation kinetics of anthocyanins in a model juice system using isothermal and non-isothermal methods, *Food Chemistry*, 111, 204–208.

Harbourne, N., Jacquier, J.C. and O'Riordan, D. (2009) Optimisation of the aqueous conditions of phenols from meadowsweet (*Filipendula ulmaria*) for incorporation into beverages. *Food Chemistry*, 116, 722–727.

Harbourne, N., Jacquier, J.C. and O'Riordan, D. (2009a) Optimisation of the extraction and processing conditions of chamomile (*Matricaria chamomilla* L.) for incorporation into a beverage. *Food Chemistry*, 115, 15–19.

Harbourne, N., Marete, E., Jacquier, J.C. and O'Riordan, D. (2011) Stability of phytochemicals as sources of anti-inflammatory nutraceuticals in beverages – A review. *Food Research International*, doi:10.1016/j.foodres.2011.03.009.

Harjo, B., Wibowo, C. and Ng, K.M. (2004) Development of natural product manufacturing processes: Phytochemicals. *Chemical Engineering Research and Design*, 82, 1010–1028.

Joubert, E. (1990) Effect of batch extraction conditions on extraction of polyphenols from rooibos tea (Aspalathus-Linearis). *International Journal of Food Science and Technology*, 25(3), 339–343.

Kim, S.H., Park, J.D., Lee, L.S. and Han, D.S. (1999) Effect of pH on green tea extraction. *Korean Journal of Food Science and Technology*, 31(4), 1024–1028.

Kotnik, P., Skerget, M. and Knez, Z. (2007) Supercritical fluid extraction of chamomile flower heads: Comparison with conventional extraction, kinetics and scale up. *Journal of Supercritical Fluids*, 43, 192–198.

Krasnov, E.A., Raldugin, V.A., Shilova, I.V. and Avdeeva, E.Y. (2006) Phenolic compounds from *Filipendula ulmaria*. *Chemistry of Natural Compounds*, 42(2), 148–151.

Labbe, D., Tremblay, A. and Bazinet, L. (2006) Effect of brewing temperature and duration on green tea catechin solubilisation: basis for production of EGC and EGCG-enriched fractions. *Separation and Purification Technology*, 49(1), 1–9.

Lapornik, B., Prosek, M. and Wondra, A.G. (2005) Comparison of extracts prepared from plant by-products using different solvents and extraction time. *Journal of Food Engineering*, 71, 214–222.

Liang, Y.R. and Xu, Y.R. (2001) Effect of pH on cream particle formation and solids extraction yield of black tea. *Food Chemistry*, 74(2), 155–160.

Lim, Y.Y. and Murtijaya, J. (2007) Antioxidant properties of *Phyllanthus amarus* extracts as affected by different drying methods. *LWT – Food Science and Technology*, 40(9), 1664–1669.

Lu, Y. and Foo, L.Y. (1997) Identification and quantification of major polyphenols in apple pomace, *Food Chemistry*, 59, 187–194.

Marete, E., Jacquier, J.C. and O' Riordan, D. (2009) Effects of extraction temperature on the phenolic and parthenolide contents, and colour of aqueous feverfew (*Tanacetum parthenium*) extracts. *Food Chemistry*, 117, 226–231.

Marete, E., Jacquier, J.C. and O'Riordan, D. (2011) Effect of processing temperature on the stability of parthenolide in acidified feverfew infusions. *Food Research International*, doi:10.1016/j.foodres.2011.03.042.

Mateus, N. and de Freitas, V. (2009) Anthocyanins as food colorants. In: K. Gould, K. Davies and C. Winefield (eds.) *Anthocyanins: Biosynthesis, Functions, and Applications*, New York: Springer Verlag, pp. 283–304.

Mills, S. and Bone, K. (2000) *Principles and Practice of Phytotherapy:Modern Herbal Medicine*, London: Churchill Livingstone.

Naczk, M. and Shahidi, F. (2004) Extraction and analysis of phenolics in food. *Journal of Chromatography A*, 1054, 95–111.

Pacheco-Palencia, L.A., Duncan, C.E. and Talcott, S.T. (2009) Phytochemical composition and thermal stability of two commercial acai species, *Euterpe oleracea* and *Euterpe precatoria*. *Food Chemistry*, 115, 1199–1205.

Papp, I., Apati, P., Andrasek, V., Blazovics, A., Balazs, A., Kursinszki, L., Kite, G.C., Houghton, P.J., and Kery, A. (2004) LC-MS analysis of antioxidant plant phenoloids. *Chromatographia*, 60, S93–S100.

Patras, A., Brunton, N., O'Donnell, C. and Tiwari, B. (2010) Effect of thermal processing on anthocyanin stability in foods: mechanisms and kinetics of degradation. *Trends in Food Science and Technology*, 21, 3–11.

Pinelo, M., Rubilar, M., Jerez, M., Sineiro, J. and Nunez, M.J. (2005) Effect of solvent, temperature, and solvent-to-solid ratio on the total phenolic content and antiradical activity of extracts from different components of grape pomace. *Journal of Agricultural and Food Chemistry*, 53, 2111–2117.

Price, W.E. and Spitzer, J.C. (1994) The kinetics of extraction of individual flavanols and caffeine from a Japanese green tea (Sen Cha Uji Tsuyu) as a function of temperature. *Food Chemistry*, 50, 19–23.

Seidel, V. (2006) Initial and bulk extraction. In: S. D. Sarker, Z. Latif, A. I. Gray (eds.) *Natural Products Isolation*, New Jersey: Humana Press Inc., pp. 27–46.

Schreiner, M. and Huyskens-Keil, S. (2006) Phytochemicals in Fruit and Vegetables: Health Promotion and Postharvest Elicitors. *Critical Reviews in Plant Sciences*, 25, 267–278.

Shi, J., Nawaz, H., Pohorly, J., Mittal, G., Kakuda, Y. and Jiang, Y. (2005) Extraction of polyphenols from plant material for functional foods - Engineering and Technology. *Food Reviews International*, 21, 139–166.

Spigno, G., Toramelli, L. and De Faveri, D.M. (2007) Effects of extraction time, temperature and solvent on concentration and antioxidant activity of grape marc phenolics. *Journal of Food Engineering*, 81, 200–208.

Spiro, M. and Price, W.E. (1987) Kinetics and equilibria of tea infusion .6. The effects of salts and of pH on the concentrations and partition constants of theaflavins and caffeine in kapchorua pekoe fannings. *Food Chemistry*, 24, 51–61.

Turker, N. and Erdogdu, F. (2006) Effects of pH and temperature of extraction medium on effective diffusion coefficient of anthocyanin pigments of black carrot. *Journal of Food Engineering*, 76, 579–583.

Wang, L. and Weller, C.L. (2006) Recent advances in extraction of nutraceuticals from plants. *Trends in Food Science and Technology*, 17, 300–312.

Whitehead, J. (2005) Functional drinks containing herbal extracts. In: P.R. Ashurst (ed.) *Chemistry and Technology of Soft Drinks and Fruit Juices*, Oxford: Blackwell Publishing Ltd., pp. 300–335.

Wijngaard, H.H. and Brunton, N. (2010) The optimisation of solid-liquid extraction of antioxidants from apple pomace by response surface methodology. *Journal of Food Engineering*, 96, 134–140.

18 Novel extraction techniques for phytochemicals

Hilde H. Wijngaard,[1] Olivera Trifunovic[2] and Peter Bongers[2]*

[1] Dutch Separation Technology Institute, Amersfoort, The Netherlands
[2] Structured Materials and Process Science, Unilever Research and Development Vlaardingen, Vlaardingen, The Netherlands

*Sadly, Peter passed away in 2012

18.1 Introduction

Traditionally organic solvents are used to extract phytochemicals from plant materials. Hexane is generally used in oil extraction (Wakelyn and Wan, 2001), while polyphenols can be extracted with various solvents including methanol and ethylacetate (Shi et al., 2005). Sage and rosemary extracts have been extracted with hexane, benzene, methanol, ethyl ether, chloroform, ethylene dichloride and dioxane (Chang et al., 1977). Although solvents are generally removed by ultrafiltration or evaporation (Wakelyn and Wan, 2001), there is a higher risk that unsafe solvents are still present in the final product than when less harmful solvents are used, such as ethanol or CO_2.

In addition, interest in more sustainable and non-toxic routes of phytochemical extraction has increased. It has become important to enhance the naturalness of food ingredients from a customer-oriented point of view. In addition, the negative impact on the environment can be reduced by using more environmentally friendly extraction methods. Conventional solid-liquid extraction techniques such as Soxhlet extraction and maceration, are time consuming and use high amounts of solvents (Wang and Weller, 2006). This has highlighted the necessity for more sustainable techniques.

In most cases organic solvents show better solubilities for phytochemicals than environmentally safe solvents such as water and CO_2 due to their chemical characteristics. There are various ways of avoiding organic solvents, which are generally not environmentally friendly and seen as unnatural. An additional challenge of extracting phytochemicals from food materials in a sustainable way is the fact that phytochemicals usually are embedded within the plant matrix. Solid-liquid extraction of plants is therefore not as straightforward as general chemical extractions. Besides parameters, such as diffusion coefficients, solvent choice, temperature and concentration difference, the matrix itself plays a large role. Phytochemicals are usually embedded within the plant cellular matrix, and are not readily available. By applying a 'pretreatment' to the sample, the matrix can become more accessible to the solvent. Enzymes can be added as described in Kim *et al.* (2005) and Pinelo *et al.* (2008) in order to breakdown plant cell walls. Pulsed electric fields can be applied, which entails poration of cell membranes by a field of electrical pulses (Soliva-Fortuny *et al.*, 2009) or ultrasound waves, which can enhance extraction by the effects of bubble cavitation (Luque-García and Luque de Castro, 2003). Another approach is to use other pressure and temperature conditions with environmentally safe and food-grade solvents, such as water, CO_2 and ethanol. Hereby the properties of these solvents can be adapted, for example the dielectric constant will alter and therefore solubility parameters will change, which can lead to enhanced extractability of phytochemicals. Pressurised fluids, which are discussed in section 18.2, are an example of this approach.

18.2 Pressurised solvents

As the name suggests, pressurised solvents apply pressure to a solvent system, which affects the target molecule's specificity and speed. By applying certain pressure and temperature conditions, the physicochemical properties of the solvents, including density, diffusivity, viscosity and dielectric constant, can be controlled. By using high pressures and temperatures the extraction of phytochemicals is generally enhanced and the environmentally friendly solvents such as water can obtain similar physicochemical properties as organic solvents. Another advantage may be that through the high pressures used, the cellular matrix is more penetrable.

The two techniques that fall under this category are:

1. Supercritical fluid extraction, which is called supercritical CO_2 extraction (SC-CO_2) when carbondioxide is used.
2. Pressurised liquid extraction (PLE) or when 100% water is used, this technique is called subcritical water extraction (SWE) or superheated water extraction (Pronyk and Mazza 2009).

The application of supercritical fluid extraction in industry has been proved in the late sixties, by Zosel and his patent on caffeine removal from coffee with SC-CO_2 (Zosel, 1981). SC-CO_2 is an alternative process to coffee decaffeination by extraction with ethylacetate, dichloromethane or benzene. Another example of an industrial application is the production of hop oils. More recently SC-CO_2 has also been considered as a technique to extract phytochemicals, mainly from by-product. By-products are an inexpensive source of phytochemicals that in many cases is underutilised. PLE is a newer technique and has mainly been used in analytical chemistry as a sample preparation system. Both SC-CO_2 and PLE are perceived as environmentally friendly and sustainable technologies, since both consume relatively less organic solvents and can show higher extraction efficiencies than conventional techniques.

18.2.1 Supercritical fluid extraction

Supercritical fluid extractions are taking place above the critical temperature and critical pressure of the applied solvent. The critical temperature is the highest temperature at which an increase in pressure can convert a gas to a liquid phase and the critical pressure is the highest pressure at which a liquid can be converted into a gas by an increase in temperature.

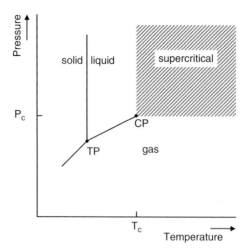

Figure 18.1 Pressure–temperature diagram for a pure component.

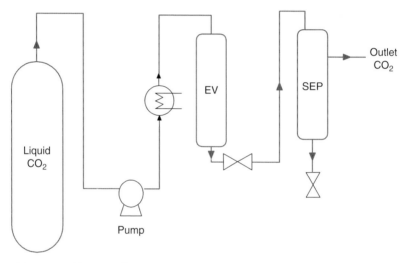

Figure 18.2 A possible setup of a SC-CO_2 system.
The sample is added to the extraction vessel (EV). Liquid CO_2 is pumped and heated while the backpressure is regulated. The SC-CO_2 is pumped through the extraction vessel and the targeted solute will be solubilised. By releasing the pressure the solutes will come out of solution in the separation vessel (SEP) and can be collected.

If a solvent system is set in a temperature higher than the critical temperature and a pressure higher than the critical pressure, the solvent will be in the so-called 'supercritical region' as is shown in Figure 18.1 (Taylor, 1996). In the supercritical region, the physico-chemical properties of the solvent can be advantageous. Supercritical fluids possess a relatively high density (more comparable to liquids) and a relatively low viscosity (more comparable to gases) (Lang and Wai, 2001). Although other solvents, such as nitrous oxide, ammonia and propane can be used, the main medium applied is CO_2. A typical setup of a SC-CO_2 system is shown in Figure 18.2. The main reason to use SC-CO_2 in extraction of phytochemicals is that CO_2 has advantageous physical characteristics with its relatively low critical points (McHugh and Krukonis, 1994). CO_2 has a critical temperature of 31.3 °C, which is low enough to retain thermo-labile phytochemicals and a critical pressure of 7.3 MPa (=73 bars). In addition, CO_2 is environmentally safe, inexpensive, can easily be obtained at a high purity and is food-grade: when used to process foods it does not need to be declared on the food label (Brunner, 2005; Herrero et al., 2010). Another advantage is that by manipulating pressure and temperature, SC-CO_2 can be very selective. This makes the application very useful for plant matrices, in which the targeted phytochemicals are in general present at low concentrations and have complex compositions (Lang et al., 2001). Finally, by using SC-CO_2, the phytochemical can be easily separated by depressurising, which can eliminate the process of solvent evaporation and phytochemical concentration. These processes are in general very costly and time-consuming (Brunner, 2005).

Various studies and books have described the mass transfer process during SC-CO_2 extraction. During solid-liquid extraction the following steps usually take place: (1) entering of the solvent into the solid plant matrix, (2) solubilisation/breakdown of components, (3) transport of solute to the exterior of the plant matrix and (4) migration of the solute from the plant surface layer into the bulk solution (external diffusion) (Aguilera and Stanley, 1999). The mass transfer process can be limited by different parts of the extraction process. When the transport of the solute through the matrix or its pores is limiting, mass transfer is controlled by intra-particle diffusion. The properties of the matrix and solute, and not the flow rate play a large role when this is the case. On the other hand, when the process is controlled by external diffusion, the flow rate has a large effect and enhances mass transfer rates when the flow rate is increased. If the limiting step is to solubilise the components, it is a thermodynamic restraint, but solvent rate will still have an effect (Anekpankul et al., 2007). When developing extraction methods for phytochemical ingredients these steps need to be taken into consideration. When SC-CO_2 is used the same principles apply and therefore various studies have been carried out to model mass transfer of phytochemicals into SC-CO_2. For example, the SC-CO_2 extraction of the triterpenoid nimbin from neem seeds has been modelled successfully by determining the intraparticle diffusion coefficient and the external mass transfer parameter (Mongkholkhajornsilp et al., 2005). To determine if effects are external or internal flow rates can be varied and characteristics of the matrix, such as porosity, can be changed. Determining thermodynamic properties and solubility parameters of solutes in SC-CO_2 is another approach researchers have taken. The properties can be described by various models, such as Peng-Robinson equation or group contribution methods (Fornari et al., 2005; Murga et al., 2002).

Sovová (2005) used the concept of broken and intact cells to model the mass transfer of natural products with supercritical fluids. The author divided the extraction into two phases, the first phase controlled by phase equilibrium and the second by internal diffusion. Four types of extraction curves were defined based on composition of solid and fluid phases. Reverchon (1996) used models with different particle shapes when extracting essential oils from sage with SC-CO_2. The particle shape of the ground material, proved to be an important factor when fitting the results. Using a slab as a particle shape resulted in better models than when using a more conventional sphere.

The importance of the matrix effect was also emphasised by Björkland et al. (1998). They analysed solubility of solutes such as clevidipine and various oils in SC-CO_2 when applied to different matrices, for example filter paper and stainless steel beads. The various matrices showed a large effect on the extraction of the solute into SC-CO_2. Especially cellulose based materials inhibited extraction. Most investigators add modifiers in order to increase the polarity of the SC-CO$_2$ system, but adding modifiers has another advantageous effect, namely that interactions between analyte and matrix can be broken. For example, adding 4% methanol to a filter paper matrix in order to extract added clevidipine with SC-CO$_2$ enhanced the recovery of clevidipine 15 times (Björklund et al., 1998). Another way to release the solute from the matrix is to increase the temperature, which can also help in breaking the solute–matrix bond (Langenfeld et al., 1995). An additional important parameter is the particle size of the matrix: the smaller the particle size, the higher the extraction rate, up to a certain point. A smaller particle has a relatively large surface area, which would enhance the extraction. But when the particles are too small, channelling can start to take place and the extraction rate will decrease again (Reverchon and De Marco, 2006). For example, lycopene was extracted from tomato waste by SC-CO$_2$. When a particle size of 0.080 mm was used the extraction rate decreased and the matrix was inhomogenous after extraction, which pointed at the fact that channelling had occurred in the matrix (Sabio et al., 2003).

Another approach is to use empirical models in order to optimise extraction conditions, such as response surface methodology (RSM). Empirical models are based on various conditions such as temperature and pressure and have no need for other facts, such as density change.

Various groups of bioactive compounds can be extracted by SC-CO$_2$. For example, alkaloids, such as caffeine, are soluble in SC-CO$_2$, especially at high densities. Decaffeination of coffee is probably the best known example of SC-CO$_2$, and is largely applied on industrial scale. More recently many studies have focussed on decaffeination of tea (Içen and Gürü, 2009; Kim et al., 2008; Park et al., 2007a; Park et al., 2007b). Since caffeine is not a phytochemical it will not be discussed further here.

Carotenoids are a group of phytochemicals that has often been extracted by SC-CO$_2$. To extract carotenoids a modifier can be added, but this is not essential. For example, in order to extract trans-lycopene from tomato skin, no addition of modifier was necessary (Kassama et al., 2008). Saldaña et al. (2006) did not use modifiers either to study the solubility of β-carotene with SC-CO$_2$. Solubilising free β-carotene was carried out by the quartz crystal microbalance technique and β-carotene that was present in the carrot matrix was extracted by a dynamic SC-CO$_2$ extraction. Although carrots were freeze-dried and ground to a particle size distribution of 0.5–1.0 mm, the solubility of β-carotene from the carrot matrix was five to ten times lower than when free β-carotene was extracted. The authors concluded that the cellular structure of carrots and the presence of carbohydrates to which the carotenoids can be bound interfered with solubilising β-carotene in SC-CO$_2$ (Saldaña et al., 2006). When necessary, a modifier that is often used is ethanol, which can enhance the polarity of SC-CO$_2$ and also help to desorb the solute from the plant matrix (Sanal et al., 2004). Vega et al. (1996) extracted β-carotene from carrot press cake. They optimised the extraction by response surface methodology (RSM). They reported that an ethanol concentration of 10% was optimal due to enhanced solubility of β-carotene in the CO$_2$ at this condition. Another modifier that can be used for extraction of carotenoids from carrots is canola oil. By adding 5% canola oil as a co-solvent, the extraction of α- and β-carotene was enhanced twice, while the extraction of lutein was enhanced four times (Sun and Temelli, 2006). Using vegetable oils as modifier is an interesting application since no evaporation process is needed as when ethanol is added. In order to extract lycopene from tomato, the addition of sunflower seed, peanut, almond and hazelnut oil were tested. Only the addition of hazelnut oil had an enhancing effect, possibly due to the lower acidity. The recovery of lycopene was 60% when hazelnut oil was added in 10%, a pressure of 450 bars was used, a temperature of 65–70 °C and a flow rate of 18–20 kg CO$_2$/h and an average particle size of 1 mm (Vasapollo et al., 2004). An even higher recovery of 75% lycopene was reported when a combination of olive oil and ethanol has been used to enhance the extraction from tomato skins. The maximum recovery of lycopene of 75% was reached when 10% ethanol and 10% olive oil were added to the SC-CO_2 system. In this report it was also mentioned that addition of any of the modifiers ethanol, olive oil and water increased lycopene extraction with SC-CO_2 (Shi et al., 2009).

In addition to tomato, fruits such as pitanga fruit and watermelon have been extracted with SC-CO_2 in order to extract carotenoids (Filho et al., 2008; Vaughn Katherine et al., 2008). But the most extracted fruit is the tomato and predominantly by-products are used. Various studies exist, which are focussed on the optimisation of conditions to maximise the recovery of carotenoids with SC-CO_2. Especially pressure (density), temperature, CO_2 flow rate, modifier percentage, moisture content and particle size have been tested and reported to affect the extraction. The main phytochemicals recovered from tomato by-product are lycopene and β-carotene, but γ-carotene, lutein, chryptoxanthines and lycoxanthines were also extracted (Vági et al., 2007).

To supercritically extract maximum amounts of lycopene and β-carotene from tomato by-products a minimum pressure of 30 MPa is required. In addition, relatively high temperatures are generally optimal. When a temperature of 80 °C and a pressure of 30 MPa was applied, 88% of the present lycopene was extracted from skins and seeds and 80% of the total present β-carotene (Sabio et al., 2003). In other studies it was noted that the chemical form of the solute mattered. To maximally extract *trans*-lycopene from tomato skins a temperature of 60 °C instead of 80 °C was beneficial (Nobre et al., 2009). These results were confirmed by Kassama et al. (2008). With regard to flow rate, a lower flow rate of 0.59 g CO_2/min was preferred over a flow rate of 1.14 g CO_2/min. It is possible that channelling was taking place at the higher flow rate (Nobre et al., 2009). These results agreed with results from Rozzi et al. (2002), which showed that an increase in flow rate decreased the extraction of lycopene from tomato seeds and skins. Various studies show that very wet samples are not suitable for extraction. Vasopollo et al. (2004) showed that drying was needed to obtain quantifiable amounts of lycopene from sun-dried tomatoes, which had an initial moisture content of ca. 60%. A similar effect was detected when wet tomato pomace (82% moisture) was extracted and no traceable amounts of lycopene were recovered. In addition, drying to 58 and 23% moisture did not have a large effect. It was only at a moisture percentage of 5% that the extraction was sufficient (Nobre et al., 2009). In other fruit pomaces, in this case apricot bagasse, the moisture content was also a major factor when extracting β-carotene with SC-CO_2. When the moisture content of the freeze-dried apricot pomace was reduced from 14 to 10%, the extraction yield of β-carotene was increased five-fold (Sanal et al., 2004).

Oils can be easily extracted with SC-CO_2. It is recommended to use pressures higher than 60 MPa and temperatures may vary from 40 to 80 °C. In general, pressures higher than 300 MPa are not needed and modifiers do not need to be applied. It is out of the scope of this book chapter to discuss the possibilities of oil extraction by SC-CO_2. Therefore the authors would like to refer to extensive reviews that are present in the literature on oil and lipid extraction with supercritical fluids (Herrero et al., 2010; Sahena et al., 2009; Temelli, 2009). The only thing that should be noted is that terpenes, which are a group of apolar phytochemicals, are often extracted within the oil fraction. Terpenes are usually present in high amounts in essential oils obtained from citrus peels. Because they are highly reactive they can cause off-flavours in oils. Therefore citrus peels are usually deterpenated before use and SC-CO_2 is an appropriate method to accomplish this (Diaz et al., 2005; Jeong et al., 2004).

Rosemary phytochemicals were extracted with SC-CO_2 without modifier. SC-CO_2 extracted carnosic acid 136% better than in the conventional extraction, which was an acetonic extraction with ultrasound (Tena et al., 1997). Bioactive extracts were produced from sweet cherries under various conditions with SC-CO_2. When ethanol was used as a modifier at 10%, the extracts show the highest antioxidant and anticarcinogenic activity. In the optimal extract sakuranetin and sakuranin were the polyphenols responsible for the major antioxidant activity, while perillyl alcohol was the major compound contributing to anticarcinogenic activity. To obtain sufficiently concentrated extracts with antiproliferative activity a pre-treatment with SC-CO_2 was recommended (Serra et al., 2010).

Phenolics have also been extracted from guava seeds. When the seeds were extracted with SC-CO_2, ethanol was found to be a better modifier than ethyl acetate. Best results in extracting phenolics were obtained when ethanol was added at 10% and a pressure of 30 MPa and a temperature of 50 °C was applied (Castro-Vargas et al., 2010). SC-CO_2 extraction of polyphenols from cocoa seeds was optimised by RSM, showing that a high level of ethanol added to sample in the extractor (200% m/m) was optimal for an extraction at 40 °C. In this way 40% of total present polyphenols could be extracted when compared with traditional organic solvent extraction (Sarmento et al., 2008). Another source of polyphenols that is often used is soy and its isoflavones. When extracting isoflavones from soybean pressed cake, the highest amount of isoflavones was extracted by SC-CO_2 using a pressure of 350 bars and 60 °C. By adding manually 16% (m/m) of 70% ethanol/water solution as modifier ((3279/5010)*100 µg/g =) 65% of total present isoflavones were collected. They analysed 12 different isoflavones. It was noted that malonyl glucosides and glucosides were optimally extracted at 350 bars and 60 °C, while acetyl glucosides and aglycones were best extracted at 350 bars and 80 °C (Kao et al., 2008). Others, such as Rostagno et al. (2002) only recovered 40% of total isoflavones present in soy flour with SC-CO_2 when compared with a traditional Soxhlet extraction. These authors only measured three isoflavones of which two were aglycones. Optimal conditions determined were 50 °C and 360 bars and 10 mol% of methanol. In other studies methanol was also used as a modifier to extract isoflavones from defatted soybean meal. A maximum of 87.3% isoflavones could be extracted when 80% methanol was added at 7.8 mass%, a pressure of 500 bars was applied, a temperature of 40 °C and a flow rate of 9.80 kg/h (Zuo et al., 2008). Also extraction with SC-CO_2 of soy isoflavones daidzein, genistein, glycitein and their glycosides was compared. Ethanol was required as a modifier to extract the isoflavones with SC-CO_2. By adding ten times more ethanol, daidzin could be extracted even 1000 times better. Daidzin was easier to extract than daidzein, which wasn't expected based on the K_{ow} values. By using thermodynamic modelling, it was concluded that ethanol facilitates cluster formation of CO_2 molecules, which enhances isoflavone solubility (Nakada et al., 2009).

Catechins have also been extracted with SC-CO_2 from green tea leaves. By using 95% (v/v) ethanol as co-solvent at a mass percentage of 4.6%, Park et al. (2007a) managed to extract 82% of the present epigallo-catechin, 71% of epicatechin-gallate, 70% of epigalloctechin-gallate and 50% of epicatechin. The conditions they used were 300 bars and 80 °C. Catechin, epicatechin and gallic acid have also been extracted from grape seed concentrates. Some phenols, such as protocatechuic acid aldehyde, could be extracted with yields higher than 90%, but it was mentioned that at the moment SC-CO_2 could probably not compete with traditional grape seed extraction (Murga et al., 2000). The solubility of catechins in SC-CO_2 has been modelled by using Peng-Robinson equation of state and the Chrastil model. These models can assist in developing methods to extract polyphenols from food matrices (Murga et al., 2002).

18.2.2 Pressurised liquid extraction (PLE)

Pressurised liquid extraction is another technique that has potential to be used in the extraction of phytochemicals. It is a technique that makes use of pressurised fluids. Other names for the technique are accelerated solvent extraction, pressurised solvent extraction and subcritical solvent extraction. Any solvent that is normally applied in extractions can be used. Most applications are in the development of analytical methods and the extraction of many groups of phytochemicals has been optimised using organic solvents: capsaicinoids from peppers (Barbero *et al.*, 2006), polyphenols from apples (Alonso-Salces *et al.*, 2001), carotenoids from microalgae (Rodríguez-Meizoso *et al.*, 2008) and polyacetylenes from carrots (Pferschy-Wenzig *et al.*, 2009). If natural and sustainable processing is a requirement, usually ethanol and/or water are used. When extractions are carried out with 100% water, the technique is also called superheated water extraction (SWE), subcritical water extraction, pressurised low polarity water extraction or pressurised hot water extraction (Pronyk and Mazza, 2009). Also in PLE it is important how the phytochemicals are embedded in the matrix, depending on their physicochemical properties and the properties of the matrix itself (Runnqvist *et al.*, 2010). PLE is a technique in which pressure is applied during extraction, which allows the use of temperatures above the boiling point of the solvent. Extracting at elevated temperatures has various advantages on mass transfer and surface equilibria. In general, a higher temperature increases mass transfer by a higher capacity of the solvent to solubilise the phytochemical, and a decrease in viscosity of the solvent. Higher temperatures can also have a beneficial effect on releasing solutes from the matrix as described earlier in the paragraph on supercritical fluid extraction. A higher accessibility of the matrix has also been mentioned due to the applied pressure (Richter *et al.*, 1996), but scientific proof is still ambiguous.

PLE requires smaller amounts of solvent use than traditional extraction and a shorter extraction time. Therefore PLE is generally perceived as a green and sustainable extraction technique (Mendiola *et al.*, 2007). Over recent years many applications have been mentioned and PLE is gaining popularity. Especially since the publication of a patent on extracting polyphenolic compounds from fruits and vegetables with subcritical water (King and Grabiel, 2007), PLE procedures have been further developed. Still most applications are analytical and have used laboratory systems.

During PLE, various conditions are important, such as particle size, temperature and solid to solvent ratio (Luthria and Natarajan, 2010; Mukhopadhyay *et al.*, 2010). Pressure is important in order to be able to reach the required temperature and maintain the water (or other solvent) in the liquid form and not to produce steam. Steam has a much lower dielectric constant, and gas-like diffusion rates and viscosity properties (Smith, 2002). On the other hand, unlike SC-CO_2 extractions, changes in pressure do not have large effects on PLE (Mendiola *et al.*, 2007). This is one of the main differences between SC-CO_2 and PLE.

As in any extraction the choice of solvent is important as well. In addition, like in SC-CO_2, modifiers can play a large role in breaking solute–matrix interactions (Björklund *et al.*, 1998). The literature discussed here will focus on the use of food-grade and more sustainable solvents, such as ethanol and water.

Phenols have been extracted from parsley with PLE. The smallest particle size used (<0.425mm) resulted in an optimal extraction of phenols. The solid to liquid ratio also affected phenol extraction. Temperature had a small effect on the total level of phenols extracted, but a large effect on the profile of phenols extracted from parsley. When heated from 40 to 160°C, apiin and acetyl-apiin were increased, but the amount of malonyl apiin decreased. This indicated that malonyl-apiin is unstable and is partially converted to apiin and acetyl-apiin (Luthria, 2008). Other herbs have been successfully extracted with PLE as well. Phytochemicals from rosemary were extracted successfully with similar results as with SC-CO_2. The optimal temperature was dependent on the phytochemical compound, especially its polarity. For example, rosamanol, which is the most polar phytochemical compound present in rosemary, was best extracted at 25 °C (so not at subcritical conditions), while carnosic acid, which is an apolar compound, was best extracted at 200 °C with subcritical water (Ibáñez et al., 2002). The dielectric constant of water changes with increased temperatures. At room temperature and atmospheric pressure water has a high polarity with an ε of 80 (Kim and Mazza, 2006), while at a temperature of 220 °C, water has a lower polarity with an ε of 30. In comparison, methanol has an ε of 33 at room temperature (Smith, 2002). Therefore apolar compounds, such as carnosic acid are better extracted at higher temperatures. When American skullcap, another herb, was extracted, SWE was also a good alternative. The level of total flavonoids extracted was similar to conventional extraction (Bergeron et al., 2005). Saponins were extracted by PLE from cockle seeds, another North American herb. The amount extracted was optimal at 80% ethanol and 125 °C. The procedure was affected by extraction solvent and method, optimal results were obtained with whole cockle seeds (Güçlü-Üstündag et al., 2007).

One of the main tested matrices is red grape pomace; both table grape pomace and wine grape pomace have been optimised for their polyphenol extraction with PLE. When ethanol and water combinations were used of 50 or 70% ethanol, similar amounts of anthocyanins were extracted as with a conventional extraction with methanol/water/formic acid (60:37:3) (Monrad et al., 2010a). When the PLE procedure was optimised for the extraction of procyanidins from the same red grape pomace, 115% of procyanidins could be extracted when an ethanol percentage of 50% was used in comparison to a conventional acetone based extraction. Epicatechin and catechin levels were even increased to 205 and 221%, respectively, when 50% ethanol was used as solvent (Monrad et al., 2010b). The increased extraction of flavanols was also noted by García-Marino et al. (2006). They also noted a higher recovery of catechins and procyanidins from grape seeds with superheated water (thus without addition of ethanol), than with a conventional methanolic extraction. When wine grape skins were extracted with superheated water, an increase in antioxidant activity was noted with an increase in temperature. But the level of total phenols and anthocyanins decreased at temperatures higher than 110 °C. Using subcritical water extraction at 110 °C seemed an excellent alternative to traditional extractions, since the level of extracted total phenols and anthocyanins was the same or higher than conventional hot aqueous and methanolic extractions (Ju and Howard, 2003).

Wijngaard and Brunton (2009) measured levels of total flavonols, chlorogenic acid and phloretin glycoside at various ethanol concentrations using PLE. Optimal extraction conditions were estimated with response surface methodology (RSM). It was reported that the level of the phenols was largely affected by ethanol concentration and little by temperature. Temperatures between 75 and 125 °C were recommended. Temperatures higher than 150 °C formed hydroxymethylfurfural and increased browning components, and hence antioxidant activity (Wijngaard and Brunton, 2009). Similar results were reported when flavonoids were extracted from spinach by PLE. They also noted that at temperatures higher than 150 °C, browning components were formed, which correlated well with antioxidant activities as measured by ORAC. A temperature below 150 °C was advised for PLE with aqueous ethanol, and a temperature below 130 °C for subcritical water extractions (Howard and Pandjaitan, 2008).

Soybeans were extracted with PLE to optimise the level of isoflavones. The best solvent composition was 70% ethanol and a temperature of 100 °C. It was found that malonyl glycosides were the most heat labile and were degraded at 100 °C, while acetyl glycosides were broken down at 135 °C (Rostagno et al., 2004). Similar results were reported when de-fatted soybean flakes were extracted with PLE. A percentage of 80% ethanol and a temperature of 110 °C were optimal. At these conditions 95% of isoflavones and 75% of soysaponins could be recovered (Chang and Chang, 2007). Several cereal sources have undergone pressurised liquid extraction. The main aim was to obtain ferulic acid, which is a phenolic acid that is mainly present in bound form in many cereals and a precursor for the production of vanillin, a costly aroma. PLE did not increase the levels of phenolic acids present when wheat or corn bran were extracted, but by using PLE vanillin was formed from the bound ferulic acid. Buranov and Mazza (2009) reported that both aqueous ethanol and SWE promoted the formation of vanillin.

Srinivas et al. (2010) modelled solubility of gallic acid, catechin and protocatechuic acid in subcritical water. All compounds were better solubilised when the temperature was increased. For example, the solubility of catechin hydrate increased from 2 to 576 g/l when the temperature was increased from 25 to 143 °C. Pressure was applied when needed to keep the water in the liquid state. Thermodynamic properties were calculated from the solubility data.

Subcritical water extraction of mannitol was modelled by measuring the effects of temperature, pressure and flow rate. External mass transfer coefficients and equilibrium coefficients were calculated assuming a fixed-bed system (Ghoreishi and Shahrestani, 2009). Kim and Mazza (2007) modelled the mass transfer of free phenolic acids, such as vanilic acid and syringic acid, from flax shives. They determined by using kinetic and thermodynamic models that the extraction process was controlled by both internal diffusion and external elusion. In addition they noted that the internal diffusion process could also be influenced by the flow rate. In another study the solute–subcritical solvent interactions were studied by modelling of Hansen solubility parameters. It was concluded that by using group contribution methods and computerised algorithms, solubility and extraction conditions can be estimated (Srinivas et al., 2009).

18.3 Enzyme assisted extraction

As mentioned earlier phytochemicals are often embedded in the plant matrix, which can cause problems when extracting phytochemicals. For example, phenols can be associated with polysaccharides. It is suggested that they are bound by hydrogen bonds between the hydroxyl groups of phenols and the cross-linking oxygen atoms of polysaccharides or that polysaccharides form secondary hydrophobic structures, such as nanotubes in which complex phenols may be encapsulated (Pinelo et al., 2006). Exogenous enzymes can be added, in order to enhance the extraction of phytochemical compounds. The exact effect of enzymes is still somewhat unclear though. The leading theory suggests that the enzymes degrade the cell walls partially, which enhances porosity and pore size and therefore increases extractability of polyphenols. Another theory is that phytochemicals are chemically bound to a cell structure and can be released by adding exogenous enzymes (Landbo and Meyer, 2001). The composition of cell walls depends on the source, but cell walls mainly consist of polysaccharides, such as cellulose and pectins. Therefore, for extraction purposes, many applications of cellulases and pectinases have been reported.

One popular application is apple peels. Apple peels mainly comprise of hemicellulose, cellulose and pectin. Pectins are known to be able to encapsulate procyanidins, one group of phenols present in apples (Le Bourvellec et al., 2005). By applying exogenous enzymes, which are targeting the polysaccharides, the polysaccharides are partially degraded and dissolved. This also affects the availability of other compounds present, such as polyphenols (Dongowski and Sembries, 2001). Kim et al. (2005) successfully enhanced the extraction of phenolics from apple peels when they added cellulases from *Thermobifida fusca*. The increase in phenol content went parallel with an increase in degraded polysaccharides. The authors also reported a correlation between the level of phenols and enzyme activity (Kim et al., 2005). The mass transfer of phenols from apple peels by enzyme-assisted extraction has also been described. The mass transfer was modelled by Fick's law. Adding cellulolytic, pectinolytic and proteolytic enzymes enhanced the diffusion factors in addition to the mass transfer of phenols (Pinelo et al., 2008).

Black currant pomace is another studied matrix, since it is known that many polyphenols in this pomace are bound (Kapasakalidis et al., 2009). Black currant pomace was treated with four different commercial pectinases and one protease. Except for one pectinase, all enzymes enhanced the extraction of phenols. The same pectinase did not enhance the extraction of phenols from red wine pomace. A reduction in particle size also increased the phenol level (Landbo and Meyer, 2001). In a later study black currant pomace was treated with cellulase from *Trichoderma* and individual phenol release was measured. Enzyme activity tests demonstrated that besides endocellulase activity the mixture also contained cellobiohydrase and β-glucosidase activities, which is often the case in commercial enzyme preparations. The effects of enzyme concentration, hydrolysis time and temperature were investigated. At 50 °C and a hydrolysis time of 1.5 h the anthocyanin content could be increased 60%. It was concluded that enzyme addition can significantly enhance the extraction of phenols, especially anthocyanins, from the black currant matrix (Kapasakalidis et al., 2009). On the other hand, when grape pomace was treated with pectinases and cellulases the level of anthocyanins was not correlated with the degradation of polysaccharides. Also, rutin extraction was not enhanced by enzymatic treatment, but the level of extracted phenolic acids was well correlated with the degradation of polysaccharides, in particular pectin. In addition, the various polyphenol groups reacted differently biochemically to the enzymatic treatment. Anthocyanins were extracted in the early stage of the enzymatic hydrolysis and were degraded after prolongation. Flavonols were hydrolysed to their aglycones, for example rutin was degraded to quercetin by enzymatic treatment. The study proved the difference in the behaviour of various polyphenols to enzymatic treatment, and hence the importance of monitoring the various groups of polyphenols (Arnous and Meyer, 2010). Attempts to upscale the enzyme-assisted extraction of polyphenols from grape pomace to pilot-scale size have been reported, since this stream is an abundant and polyphenol-rich source. Usually polyphenols are extracted with acidified alcohols or sulfited water. To enhance naturalness of the polyphenol ingredients alternative processes such as enzyme assisted extraction can be applied: but first attempts to upscale resulted in very low recoveries of anthocyanins of 8% (Kammerer et al., 2005). In later studies though, the recovery of anthocyanins could be enhanced until an impressive 64%. The described optimised process existed of a pre-extraction in order to inhibit polyphenol degrading enzymes and an enzymatic extraction: water of 80 °C was added to the pomace, the pomace was ground and left for 1 min at 90 °C to inactivate polyphenol oxidases (pasteurisation), the slurry was then milled with a colloid mill, and pressed with a rack and cloth press. The liquid fraction was collected and the solid fraction was further treated with enzymes. Pectinolytic and cellulolytic enzymes were added in a ratio of 2:1, at pH 4, a temperature of 40 °C and an incubation time of 2 h. After the incubation, the slurry was again pasteurised. The resulting extract was pressed again and the collected liquid fractions were added together and spray-dried into a powder. By using this process, 92% phenolic acids, 92% non-anthocyanin flavonoids and 64% anthocyanins could be recovered (Maier et al., 2008).

In addition to apple peels, black currant pomace and grape pomace, the extraction of various bioactive compounds from many other fruits and vegetables have been studied for enzyme-assisted extraction. For instance, terpenes have been extracted from celery seeds (Sowbhagya et al., 2010), carotenoids from orange peel, sweet potato and carrot (Çinar, 2005), capsaicinoids and carotenoids from peppers (Santamaria et al., 2000), flavones from pigeonpea leaves (Fu et al., 2008) and phenolics from citrus peels (Li et al., 2006). In general, cellulases, pectinases and proteases are added. With some exceptions all enzymes increase the yield of polyphenols. The optimisation and improvement in yield depends on the source and should be considered when enzymes are applied. The added cost should be earned back in the higher yields of phytochemicals and considered per individual case.

18.4 Non-thermal processing assisted extraction

Plant phytochemicals are usually entrapped in insoluble cell structures, such as vacuoles or lipoprotein bilayers, which offer significant diffusional resistance to the extraction. Furthermore, the ability of some phytochemicals to form hydrogen bounds with bulk constituents of the cell matrix additionally limits the yield of the extraction process. To overcome diffusion limitations a pre-treatment that will allow larger solute-solvent contact area, that is, release of intracellular compounds into the appropriate solvent, can be used. This chapter will mainly focus on two emerging technologies that increase the plant material porosity by cell disruption: ultrasound and pulsed electric fields (PEF).

18.4.1 Ultrasound

Ultrasound is a technique in which soundwaves that are higher in frequency than the human hearing (ca. 16 kHz) are applied to a medium. The lowest frequency generally applied is 20 kHz. If the ultrasound is strong enough, bubbles are formed in the liquid. Eventually the formed bubbles cannot take up the energy any longer and will collapse; this implosion is called 'cavitation'. This collapse generates the energy for chemical reactions by a change in temperature and pressure within the bubble. Extremely high temperatures of 5000 °C and pressures of 1000 bars have been measured. When a solid matrix is present, it is affected by mechanical forces surrounding the bubble (Luque-García and Luque de Castro, 2003). Ultrasound can result in a higher swelling of the plant material that increases extraction (Vinatoru, 2001). The choice of solvent is important, since solvent properties, including vapour pressure, surface tension, viscosity and density play important roles in cavitational activity (Xu et al., 2007). The authors found that 50% ethanol extracted isoflavones from Ohwi roots much better in comparison to 95% ethanol at an ultrasound power of 20 kHz. When the electrical power was increased from 0 to 650 W the extraction rate of isoflavones was increased as well. The electrical power was an important factor when modelling the mass transfer from isoflavones of the Ohwi root (Xu et al., 2007). Isoflavones have also been extracted from soy products by the assistance of ultrasound. Optimal results were obtained after 20 min at 60 °C and by using 50% ethanol. A percentage of 0–20% higher extraction was achieved at a power of 200 W and a frequency of 24 kHz than when extracted by mixing and stirring (Rostagno et al., 2003). Other flavanoids have also been extracted with assistance of ultrasound. Orange peel particles with a surface area of 2.0 cm^2 were optimal. Smaller particles decreased extraction rates, since at smaller sizes particles started floating. Sonication power, temperature and ethanol to water ratio were optimised using a central composite design. An optimal flavonone concentration of 70.3 mg naringin and 205.2 mg hesperidin per 100 g fresh peels was reached at a temperature of 40 °C, a 4:1 (v/v) ethanol to water ratio and a sonication power of 150 W. At these conditions sonication led to an increased extraction percentage of approximately 40% (Khan et al., 2010). Mandarin peels have been subjected to the study of ultrasonic extraction of phenolic acids. The optimal conditions of ultrasonic extraction depended on the type of polyphenol. Phenolic acids were best extracted at 20 min, 30 °C and ultrasonic power of 8 W. For ultrasonic extraction of flavanone glycosides an extraction time of 60 min, a temperature of 40° C and an ultrasonic power of 8 W were found to be optimal. Cinnamic acids were more susceptible to degradation than benzoic acids, when an intensive ultrasound procedure was applied. Depending on the type of phenolic extracted, extraction rates could be enhanced maximally by approximately 50% (Ma et al., 2008). Ghafoor et al. (2009) used response surface methodology to optimise ultrasonic extraction of polyphenols from grape seeds by ethanol concentration, temperature and time. The optimal predicted conditions for ethanol extraction by US were 53% ethanol, 56° C and 29 min, which were very close to the final optimal experimental values. The frequency and power were constant at 40 kHz and 250 W (Ghafoor et al., 2009). Catechins, a group of flavanoids, were 20% better extracted from apple pomace, a by-product of the cider industry, when ultrasound was used in comparison to conventional extraction. In addition, the ultrasonic process was upscaled successfully to a volume of 30 l (Virot et al., 2010).

Capsaicinoids have been extracted from peppers and the procedures optimised. The optimal extraction was carried out with 100% methanol and a temperature of 50° C and 10 min. A constant power of 360 W was used (Barbero et al., 2008). In a later study the extraction of capsaicinoids from peppers was done with the more environmental friendly solvent ethanol in order to upscale the process. First lab trials were conducted with a fixed frequency of 35 kHz and a power of 600 W, which showed optimal results of a recovery of 83% at 1:5 solid to liquid ratio, an ethanol percentage of 95% and a temperature of 45 °C. At the 20 l pilot scale a 76% recovery was reached at a frequency of 26 kHz and 1.08 kW. This recovery was 7% lower than of an industrial maceration process at which peppers are soaked overnight and extracted at 78° C for 3 h. Although the recovery of the UAE was slightly lower, industrial potential may exist, because of possible lower operational costs (Boonkird et al., 2008). Ultrasound as an extraction technique has the potential to be upscaled, at low costs. It has already been used for alcohol beverage maceration at volumes of 100–1000 l (Virot et al., 2010).

18.4.2 Pulsed electric fields

The first records on the research on the influence of electric current on biological cells date almost as early as the end of the nineteenth century (Töpfl, 2006). Technological application of pulsed electric fields (PEF) on food production was first explored almost 50 years ago, mainly as a non-thermal alternative to pasteurisation. Numerous research groups are working on different PEF applications in food production, but the number of current successful industrial applications is limited at best.

PEF utilises the influence of a strong electrical field on material located between two electrodes, which leads to cell membrane disruption, thus increasing cell permeability. The exact mechanism of this occurrence is still under debate, but the most accepted theory is the electromechanical model developed by Zimmermann et al. (1974). In essence, cells are highly complicated structures that consist of an intracellular space, which is filled with different organelles and is surrounded by a cell membrane. The cell membrane separates the intracellular and extracellular space and is essentially electroneutral (equivalent to a capacitor in electrical circuits), while free charges of opposite polarities are present on both sides of the membrane. This creates a naturally occurring transmembrane potential. When an external electrical field is applied the additional transmembrane potential is formed, which increases attraction between opposite charges on both sides of the membranes and compresses the membrane. If the stress on the membrane is large enough pore formation occurs. Depending on the treatment applied (electric field strength, pulse duration, number of pulses) the pore formation can be reversible or irreversible; in the latter case cells are destroyed. According to Angersbach *et al.* (2000), the pore formation is reversible only if the formed pores are small in comparison to the membrane area (low intensity treatment), while if the field strength is large enough irreversible breakdown occurs. The critical field strength where electroporation occurs depends on cell diameter and it typically is in the range of 1–2 kV/cm for plant cells (diameter 40–200 μm) and 12–20 kV/cm for microorganisms (diameter 1–10 μm) (Heinz *et al.*, 2002; Soliva-Fortuny *et al.*, 2009). In the case of plant tissues the cell wall perforation by PEF can potentially lead to higher extractability of cell contents. According to (2010) the electroporation of plant cells and extraction of materials from solid foods typically require 0.1–5 kV/cm field strengths and pulses of 10–000 μs. The reason for higher extractability could be two-fold:

1. by applying a severe treatment irreversible permeabilisation leads to release of bioactive compounds that were previously enclosed in cells, i.e. the extraction can be enhanced due to a decrease in diffusional resistance; and
2. by applying moderate treatment conditions electroporation is reversible and the viability of cell is preserved. In this case PEF induces a stress reaction in the plant tissue, which leads to additional synthesis of secondary metabolites. The majority of bioactive compounds, like polyphenols, are secondary cell metabolites.

The influence of PEF-induced plant tissue perforation (reversible and irreversible) was explored for various applications in the food industry, such as improving extraction yields after cold pressing of different raw materials, enhancement of drying efficiency or increasing the content of secondary metabolites in cells by stressing plant tissues. Several of the findings will be briefly discussed. An extensive review on the research that has been undertaken so far in this field can be found elsewhere in the literature (Soliva-Fortuny *et al.*, 2009; Vorobiev and Lebovka, 2010).

Several authors studied PEF assisted mechanical pressing of apple mash (Bazhal and Vorobiev, 2000; Schilling *et al.*, 2007; Schilling *et al.*, 2008; Wang and Sastry, 2002). The reported improvements of the juice yields were low to relatively high, depending on process conditions employed (particle size, PEF parameters, etc.), which makes it difficult to clearly conclude whether a higher yield was caused by PEF treatment. For instance, particle size and the type of size reduction method (slicing, milling, grinding) can be essential for an improvement in juice yield (Vorobiev and Lebovka, 2010). Next to the improvement of juice yield, several studies monitored the changes in antioxidant activity and phenolic content of pressed juices after PEF treatment. Schilling *et al.* (2007) have applied field strengths in the range of 1–5 kV/cm (30 pulses) on apple mash, but could not find any significant improvement in phenolic contents. However, in a subsequent study (Schilling *et al.*, 2008), where field strengths of 3 kV/cm were used, although the juice yield was not increased, there was almost the double amount of the main apple juice phenolic compounds chorogenic acid and phloridizin present in the juice where PEF was used as a pre-treatment.

Guderjan and co-authors (Guderjan *et al.*, 2005; Guderjan *et al.*, 2007) have investigated PEF induced recovery of plant oils and additional content of secondary metabolites in maize, olives, soybeans and rape seed. Guderjan *et al.* (2005) reported that although there was no significant oil yield improvement as a result of PEF treatment, there was almost a 32% higher content of phytosterols in maize germs and around 20% more soy isoflavonoids present in soybeans when low intensity treatment (reversible electroporation) was used (0.6–1.3 kV/cm, 20–50 pulses). On the other hand, when conditions for irreversible electroporation were used (7.3 kV/cm and 120 pulses) these high yields of secondary metabolites could not be observed. In the subsequent study (Guderjan *et al.*, 2007) they investigated the influence of higher intensity PEF treatment (5 kV/cm with 60 pulses and 7 kV/cm with 120 pulses) on rape seed oil production and bioactive compounds content in the oil obtained both with mechanical pressing and solvent extraction. Although the influence of the irreversible electroporation on increased oil yield was negligible, there was a significant increase in antioxidant capacity of extracted hulled and non-hulled rapeseed. The antioxidant capacity in rape seed is mainly attributed to tocopherols and polyphenols.

Several studies were performed on the influence of different PEF parameters on the evolution of different phenolic compounds in wine production (López *et al.*, 2008; López *et al.*, 2009; Puértolas *et al.*, 2010a) and aging (Puértolas *et al.*, 2010b). In the wine making process all authors observed higher amounts of phenolic compounds present in wine samples after PEF treatment in comparison with the control sample. López *et al.* (2008) employed a field of 5 and 10 kV/cm and observed that total polyphenol content in wine increased with field intensity, while total content and colour intensity of anthocyanins was the highest at 5 kV/cm. A more intense treatment did not improve the afore mentioned characteristics. In a subsequent study (López *et al.*, 2009) the influence of maceration time was explored together with PEF treatment (5 kV/cm) and it was found that irrespective of maceration times all PEF-treated fresh wines had a higher colour intensity and total polyphenol index. This was confirmed by Puértolas *et al.* (2010a) with additional information that after four months of wine aging in bottles there was 11% more polyphenolic compounds present in PEF treated wine than in the control sample.

Corrales *et al.* (2008) have investigated the influence of PEF, ultrasonic (US) and high hydrostatic pressure (HHP) on the extraction of anthocyanins from grape by-products (skin, stems and seeds). Although all applied techniques increased anthocyanin extraction compared to the control sample, PEF treatment (3 kV/cm) and US treatment (35 kHz) showed almost 75% higher yields. On the other hand total antioxidant capacity was the highest in PEF treated samples (4.4 times increase compared to the control sample).

In general the amount of information on influence of PEF treatment on recovery and stimulation of production of secondary metabolites is relatively low compared to results on food pasteurisation. Results so far are sometimes conflicting and more research is required to prove definite advantage of PEF as a pre-treatment for enhanced bioactives recovery.

18.5 Challenges and future of novel extraction techniques

First of all it is important to know with what matrix and what phytochemicals you are dealing. As mentioned before, phenols, for example, can often be bound to polysaccharides that are part of plant cell walls. But phenols have also been found within the cell cytoplasm, in cell vacuoles or near the cell nucleus (Pinelo *et al.*, 2006). When the phenol is located in the vacuole, it will be more easily available than when it is embedded in the cellular matrix. Thermodynamics and solubility conditions are in this case more important than the breakdown of the cell structure. On the other hand, if the targeted polyphenol molecule is embedded in a cellular structure, the cells need to be broken open, which will make the polyphenols more available. In this case the use of exogenous enzymes or ultrasound would be a better choice. Many studies were carried out on the optimisation of extraction conditions of bioactive compounds from plant matrices, but surprisingly little research considers the matrix effects and the location of the bioactive compound. More research should be focussed on these subjects.

Of the techniques discussed, the application of pulsed electric fields still needs proof of a significant advantage as an extraction aid. More research needs to be carried out if installation of PEF actually increases the yield of the extracted bioactive compounds. As already discussed, the papers reported on PEF are ambiguous on the extraction yields and no clear advantage is apparent.

One of the main challenges in the area of novel extraction techniques is to evaluate the ability of the techniques to be scaled up and implemented at an industrial level. Tests need to be carried out to estimate the capital and production costs in order to make the advantage of the process economically viable. Ultrasound as an industrial technique to enhance extraction has a large potential. Since the relevant parameters energy (energy input per volume of treated material in kWh/L) and intensity (actual power output per surface area of the sonotrode in W/cm^2) do not depend on the size of the scale, industrial ultrasonic extractions are promising. An example is the business case in which ultrasound that was implemented to enhance the yield of extraction could be paid back in four months (Patist and Bates, 2008). In addition, several suppliers offer large scale ultrasound extractors and it has already been used in the beverage industry (Virot et al., 2010).

On the other hand, the economic benefits of the use of pressurised fluids in order to enhance extraction need to be further evaluated. Implementing pressurised fluid systems entails a high investment cost since the equipment needs to be adapted to handle high pressures. In addition, the energy to apply the necessary heat and pressure may not be economically viable when compared to traditional extraction processes, such as maceration. The estimation of implementation costs may not be straightforward since PLE systems, for example, do not even exist on a large scale to our knowledge. Therefore in the case of PLE, large scale PLE systems first need to be designed. In order to achieve this, modelling of mass transfer processes is a key to gaining more understanding of the PLE process. Besides, modelling the thermodynamic behaviour of the targeted bioactive molecules under pressurised conditions may result in useful information (Pronyk and Mazza, 2009). In the case of SC-CO_2, large scale systems exist, but per specific business case it should be evaluated if the investment in the expensive equipment is profitable.

In addition to upscaling, trends towards more continuous systems have been proposed and need to be designed. This applies to pressurised fluids using multiple units with subcritical and supercritical fluids (King and Shrivinas, 2009), but also to ultrasound systems (Patist and Bates, 2008). In addition, combinations of various novel extraction techniques could make the complete process viable.

One area that has not yet been discussed but may be more developed over the coming years is the use of ionic liquids instead of solvents in the food industry. Ionic liquids consist entirely of ions (Tao et al., 2006). They are generally seen as sustainable since they have a negligible vapour pressure and hence cannot be inhaled or emitted into the environment. Although this does not necessarily mean they are to be considered environmentally safe, since they still could pose threats to ecosystems (Thuy Pham *et al.*, 2010). Food grade ionic liquids exist, and are generally derived from α-amino acids and their ester salts (Tao *et al.*, 2005; Tao *et al.*, 2006). One advantage of ionic liquids is that they are tuneable and can be designed for various purposes (Kroon *et al.*, 2008). In the future, food grade ionic liquids may be designed with a specific function that can replace organic solvents as an extraction medium.

In conclusion, novel extraction technologies of phytochemicals is an area in which many developments are taking place at the moment. Some techniques, such as ultrasound show more potential than others, such as PEF. Especially the development of industrial applicable systems that are economically viable will be a challenge for the future.

References

Aguilera, J.M., and Stanley, D.W. (1999) Microstructure and mass transfer: solid-liquid extraction. In: J.M. Aguilera *Microstructural Principles of Food Processing and Engineering*, Gaithersburg, Aspen Publishers, pp. 325–373.

Alonso-Salces, R.M., Korta, E., Barranco, A., Berrueta, L.A., Gallo, B. and Vicente, F. (2001) Pressurized liquid extraction for the determination of polyphenols in apple. *Journal of Chromatography A*, 933(1–2), 37–43.

Anekpankul, T., Goto, M. Sasaki, M., Pavasant, P. and Shotipruk, A. (2007) Extraction of anti-cancer damnacanthal from roots of Morinda citrifolia by subcritical water. *Separation and Purification Technology*, 55(3), 343–349.

Angersbach, A., Heinz, V. and Knorr, D. (2000) Effects of pulsed electric fields on cell membranes in real food systems. *Innovative Food Science and Emerging Technologies*, 1(2), 135–149.

Arnous, A., and Meyer, A.S. (2010) Discriminated release of phenolic substances from red wine grape skins (Vitis vinifera L.) by multicomponent enzymes treatment. *Biochemical Engineering Journal*, 49(1), 68–77.

Barbero, G.F., Liazid, A., Palma, M. and Barroso, C.G. (2008) Ultrasound-assisted extraction of capsaicinoids from peppers. *Talanta*, 75(5),1332–1337.

Barbero, G.F., Palma, M. and Barroso, C.G. (2006) Pressurized Liquid Extraction of Capsaicinoids from Peppers. *Journal of Agricultural and Food Chemistry*, 54(9), 3231–3236.

Bazhal, M. and Vorobiev, E. (2000) Electrical treatment of apple cossetes for intensifying juice pressing. *Journal of the Science of Food and Agriculture*, 80(11), 1668–1674.

Bergeron, C., Gafner, S., Clausen, E. and Carrier, D.J. (2005) Comparison of the chemical composition of extracts from scutellaria lateriflora using accelerated solvent extraction and supercritical fluid extraction versus standard hot water or 70% ethanol extraction. *Journal of Agricultural and Food Chemistry*, 53(8), 3076–3080.

Björklund, E., Järemo, M., Mathiasson, L., Jönsson, J. and Karlsson, L. (1998) Illustration of important mechanisms controlling mass transfer in supercritical fluid extraction. *Analytica Chimica Acta*, 368(1–2), 117–128.

Boonkird, S., Phisalaphong, C. and Phisalaphong, M. (2008) Ultrasound-assisted extraction of capsaicinoids from Capsicum frutescens on a lab- and pilot-plant scale. *Ultrasonics Sonochemistry*, 15(6), 1075–1079.

Brunner, G. (2005) Supercritical fluids: technology and application to food processing. *Journal of Food Engineering*, 67(1–2), 21–33.

Buranov, A.U. and Mazza, G. (2009) Extraction and purification of ferulic acid from flax shives, wheat and corn bran by alkaline hydrolysis and pressurised solvents. *Food Chemistry*, 115(4), 1542–1548.

Castro-Vargas, H.I., Rodríguez-Varela, L.I., Ferreira, S.R.S. and Parada-Alfonso, F. (2010) Extraction of phenolic fraction frem guava seeds (Psidium guajava L.) using supercritical carbon dioxide and co-solvents. *The Journal of Supercritical Fluids*, 51(3), 319–324.

Chang, L.H. and Chang, C.M. (2007) Continuous hot pressurized fluids extraction of isoflavones and soyasaponins from defatted soybean flakes. *Journal of the Chinese Institute of Chemical Engineers*, 38(3–4), 313–319.

Chang, S.S., Ostric-Matijasevic, B., Hsieh, O. and Huang, C.-L. (1977) Natural antioxidants from rosemary and sage. *Journal of Food Science*, 42(4),1102–1106.

Çinar, I. (2005) Effects of cellulase and pectinase concentrations on the colour yield of enzyme extracted plant carotenoids. *Process Biochemistry*, 40(2), 945–949.

Corrales, M., Toepfl, S., Butz, P., Knorr, D. and Tauscher, B. (2008) Extraction of anthocyanins from grape by-products assisted by ultrasonics, high hydrostatic pressure or pulsed electric fields: A comparison. *Innovative Food Science and Emerging Technologies*, 9(1), 85–91.

Diaz, S., Espinosa, S. and Brignole, E.A. (2005) Citrus peel oil deterpenation with supercritical fluids: Optimal process and solvent cycle design. *The Journal of Supercritical Fluids*, 35(1), 49–61.

Dongowski, G. and Sembries, S. (2001) Effects of Commercial Pectolytic and Cellulolytic Enzyme Preparations on the Apple Cell Wall. *Journal of Agricultural and Food Chemistry*, 49(9), 4236–4242.

Filho, G.L., De Rosso, V.V., Meireles, M.A., Rosa, P.T.V., Oliveira, A.L., Mercadante, A.Z. and Cabral, F.A. (2008) Supercritical CO_2 extraction of carotenoids from pitanga fruits (Eugenia uniflora L.). *The Journal of Supercritical Fluids*, 46(1), 33–39.

Fornari, T., Chafer, A., Stateva, R.P. and Reglero, G. (2005) A New Development in the Application of the Group Contribution Associating Equation of State To Model Solid Solubilities of Phenolic Compounds in SC-CO_2. *Industrial and Engineering Chemistry Research*, 44(21), 8147–8156.

Fu, Y.J., Liu, W., Zu, Y.C., Tong, M.H., Li, S.M., Yan, M.M., Efferth, T. and Luo, H. (2008) Enzyme assisted extraction of luteolin and apigenin from pigeonpea [Cajanus cajan (L.) Millsp.] leaves. *Food Chemistry*, 111(2), 508–512.

García-Marino, M., Rivas-Gonzalo, J.C., Ibáñez, E. and García-Moreno, C. (2006) Recovery of catechins and proanthocyanidins from winery by-products using subcritical water extraction. *Analytica Chimica Acta*, 563(1–2), 44–50.

Ghafoor, K., Choi, Y.H., Jeon, J.Y. and Jo, I.H. (2009) Optimization of Ultrasound-Assisted Extraction of Phenolic Compounds, Antioxidants, and Anthocyanins from Grape (Vitis vinifera) Seeds. *Journal of Agricultural and Food Chemistry*, 57(11), 4988–4994.

Ghoreishi, S.M. and Shahrestani, R.G. (2009) Subcritical water extraction of mannitol from olive leaves. *Journal of Food Engineering*, 93(4), 474–481.

Güçlü-Üstündag, O., Balsevich, J. and Mazza, G. (2007) Pressurized low polarity water extraction of saponins from cow cockle seed. *Journal of Food Engineering*, 80(2), 619–630.

Guderjan, M., Töpfl, S., Angersbach, A. and Knorr, D. (2005) Impact of pulsed electric field treatment on the recovery and quality of plant oils. *Journal of Food Engineering*, 67(3), 281–287.

Guderjan, M., Elez-MartØnez, P. and Knorr, D. (2007) Application of pulsed electric fields at oil yield and content of functional food ingredients at the production of rapeseed oil. *Innovative Food Science and Emerging Technologies*, 8(1), 55–62.

Heinz, V., Alvarez, I., Angersbach, A. and Knorr, D. (2002) Preservation of liquid foods by high intensity pulsed electric fields-basic concepts for process design. *Trends in Food Science and Technology*, 12(3–4), 103–111.

Herrero, M., Mendiola, J.A., Cifuentes, A. and Ibáñez, E. (2010) Supercritical fluid extraction: Recent advances and applications. *Journal of Chromatography A*, 1217(16), 2495–2511.

Howard, L.R. and Pandjaitan, N. (2008) Pressurized Liquid Extraction of Flavonoids from Spinach. *Journal of Food Science*, 73(3), C151–C157.

Ibáñez, E., Kubátová, A., Señoráns, F.J., Cavero, S., Reglero, G. and Hawthorne, S.B. (2002) Subcritical Water Extraction of Antioxidant Compounds from Rosemary Plants. *Journal of Agricultural and Food Chemistry*, 51(2), 375–382.

Içen, H. and Gürü, M. (2009) Extraction of caffeine from tea stalk and fiber wastes using supercritical carbon dioxide. *The Journal of Supercritical Fluids*, 50(3), 225–228.

Jeong, S.M., Kim, S.Y., Kim, D.R., Jo, S.C., Nam, K.C., Ahn, D.U. and Lee, S.C. (2004) Effect of Heat Treatment on the Antioxidant Activity of Extracts from Citrus Peels. *Journal of Agricultural and Food Chemistry*, 52(11), 3389–3393.

Ju, Z.Y. and Howard, L.R. (2003) Effects of Solvent and Temperature on Pressurized Liquid Extraction of Anthocyanins and Total Phenolics from Dried Red Grape Skin. *Journal of Agricultural and Food Chemistry*, 51(18), 5207–5213.

Kammerer, D.R., Claus, A., Schieber, A. and Carle, R. (2005) A Novel Process for the Recovery of Polyphenols from Grape (Vitis vinifera L.) Pomace. *Journal of Food Science*, 70(2), C157–C163.

Kao, T.H., Chien, J.T. and Chen, B.H. (2008) Extraction yield of isoflavones from soybean cake as affected by solvent and supercritical carbon dioxide. *Food Chemistry*, 107(4), 1728–1736.

Kapasakalidis, P.G., Rastall, R.A. and Gordon, M.H. (2009) Effect of a Cellulase Treatment on Extraction of Antioxidant Phenols from Black Currant (Ribes nigrum L.) Pomace. *Journal of Agricultural and Food Chemistry*, 57(10), 4342–4351.

Kassama, L.S., Shi, J. and Mittal, G.S. (2008) Optimization of supercritical fluid extraction of lycopene from tomato skin with central composite rotatable design model. *Separation and Purification Technology*, 60(3), 278–284.

Khan, M.K., bert-Vian, M., Fabiano-Tixier, A.S., Dangles, O. and Chemat, F. (2010) Ultrasound-assisted extraction of polyphenols (flavanone glycosides) from orange (Citrus sinensis L.) peel. *Food Chemistry*, 119(2), 851–858.

Kim, J.W. and Mazza, G. (2006) Optimization of Extraction of Phenolic Compounds from Flax Shives by Pressurized Low-Polarity Water. *Journal of Agricultural and Food Chemistry*, 54(20), 7575–7584.

Kim, J.W. and Mazza, G. (2007) Mass Transfer during Pressurized Low-Polarity Water Extraction of Phenolics and Carbohydrates from Flax Shives. *Industrial and Engineering Chemistry Research*, 46(22), 7221–7230.

Kim, W.J., Kim, J.D., Kim, J., Oh, S.G. and Lee, Y.W. (2008) Selective caffeine removal from green tea using supercritical carbon dioxide extraction. *Journal of Food Engineering*, 89(3), 303–309.

Kim, Y.J., Kim, D.O., Chun, O.K., Shin, D.H., Jung, H., Lee, C.Y. and Wilson, D.B. (2005) Phenolic Extraction from Apple Peel by Cellulases from Thermobifida fusca. *Journal of Agricultural and Food Chemistry*, 53(24), 9560–9565.

King, J.W. and Grabiel, R. (inventors) (2007) Isolation of polyphenolic compounds from fruits or vegetables utilizing sub-critical water extraction.

King, J.W. and Srinivas, K. (2009) Multiple unit processing using sub- and supercritical fluids. *The Journal of Supercritical Fluids*, 47(3), 598–610.

Kroon, M.C., Hartmann, D. and Berkhout, A.J. (2008) Toward a Sustainable Chemical Industry: Cyclic Innovation Applied to Ionic Liquid–Based Technology. *Industrial and Engineering Chemistry Research*, 47(22), 8517–8525.

Landbo, A.K. and Meyer, A.S. (2001) Enzyme–Assisted Extraction of Antioxidative Phenols from Black Currant Juice Press Residues (Ribes nigrum). *Journal of Agricultural and Food Chemistry*, 49(7), 3169–3177.

Lang, Q. and Wai, C.M. (2001) Supercritical fluid extraction in herbal and natural product studies – a practical review. *Talanta*, 53(4), 771–782.

Langenfeld, J.J., Hawthorne, S.B., Miller, D.J. and Pawliszyn, J. (1995) Kinetic Study of Supercritical Fluid Extraction of Organic Contaminants from Heterogeneous Environmental Samples with Carbon Dioxide and Elevated Temperatures. *Analytical Chemistry*, 67(10), 1727–1736.

Le Bourvellec, C., Bouchet, B. and Renard, C.M.G.C. (2005) Non-covalent interaction between procyanidins and apple cell wall material. Part III: Study on model polysaccharides. *Biochimica et Biophysica Acta (BBA) – General Subjects*, 1725(1), 10–18.

Li, B.B., Smith, B. and Hossain, M. (2006) Extraction of phenolics from citrus peels: II. Enzyme-assisted extraction method. *Separation and Purification Technology*, 48(2), 189–196.

López, N., Puértolas, E., Condón, S., Álvarez, I. and Raso, J. (2008) Effects of pulsed electric fields on the extraction of phenolic compounds during the fermentation of must of Tempranillo grapes. *Innovative Food Science and Emerging Technologies*, 9(4), 477–482.

López, N., Puértolas, E., Hernández-Orte, P., Álvarez, I. and Raso, J. (2009) Effect of a pulsed electric field treatment on the anthocyanins composition and other quality parameters of Cabernet Sauvignon freshly fermented model wines obtained after different maceration times. *LWT – Food Science and Technology*, 42(7), 1225–1231.

Luque-García, J.L. and Luque de Castro, M.D. (2003) Ultrasound: a powerful tool for leaching. *(TrAC) Trends in Analytical Chemistry*, 22(1), 41–47.

Luthria, D.L. and Natarajan, S.S. (2010) Influence of Sample Preparation on the Assay of Isoflavones. *Planta Medicina*.

Luthria, D.L. (2008) Influence of experimental conditions on the extraction of phenolic compounds from parsley (Petroselinum crispum) flakes using a pressurized liquid extractor. *Food Chemistry*, 107(2), 745–752.

Ma, Y.Q., Ye, QX.Q., Fang, Z.X., Chen, J.C., Xu, G.H. and Liu, D.H. (2008) Phenolic Compounds and Antioxidant Activity of Extracts from Ultrasonic Treatment of Satsuma Mandarin (Citrus unshiu Marc.) Peels. *Journal of Agricultural and Food Chemistry*, 56(14), 5682–5690.

Maier, T., Göppert, A., Kammerer, D., Schieber, A. and Carle, R. (2008) Optimization of a process for enzyme-assisted pigment extraction from grape (Vitis vinifera L.) pomace. *European Food Research and Technology*, 227(1), 267–275.

McHugh, M.A. and Krukonis, V.J. (1994) Introduction. In: Taylor, L.T. *Supercritical Fluid Extraction*, Stoneham, Butterworth-Heinemann, pp. 1–16.

Mendiola, J.A., Herrero, M., Cifuentes, A. and Ibáñez, E. (2007) Use of compressed fluids for sample preparation, Food applications. *Journal of Chromatography A*, 1152(1–2), 234–246.

Mongkholkhajornsilp, D., Douglas, S., Douglas, P.L., Elkamel, A., Teppaitoon, W. and Pongamphai, S. (2005) Supercritical CO_2 extraction of nimbin from neem seeds—a modelling study. *Journal of Food Engineering*, 71(4), 331–340.

Monrad, J.K., Howard, L.R., King, J.W., Srinivas, K. and Mauromoustakos, A. (2010a) Subcritical Solvent Extraction of Anthocyanins from Dried Red Grape Pomace. *Journal of Agricultural and Food Chemistry*, 58(5), 2862–2868.

Monrad, J.K., Howard, L.R., King, J.W., Srinivas, K. and Mauromoustakos, A. (2010b) Subcritical solvent extraction of procyanidins from dried red grape pomace. *Journal of Agricultural and Food Chemistry*, 58, 4014–4021.

Mukhopadhyay, M., Luthria, D.L. and Robbins, R.J. (2010) Optimization of extraction process for phenolic acids from black cohosh (*Cimicifuga racemosa*) by pressurized liquid extraction. *Journal of the Science of Food and Agriculture*, 86(1), 156–162.

Murga, R., Ruiz, R., Beltrán, S. and Cabezas, J.L. (2000) Extraction of Natural complex phenols and tannins from grape seeds by using supercritical mixtures of carbon dioxide and alcohol. *Journal of Agricultural and Food Chemistry*, 48(8), 3408–3412.

Murga, R., Sanz, M.T., Beltrán, S. and Cabezas, J.L. (2002) Solubility of some phenolic compounds contained in grape seeds, in supercritical carbon dioxide. *The Journal of Supercritical Fluids*, 23(2), 113–121.

Nakada, M., Imai, M. and Suzuki, I. (2009) Impact of ethanol addition on the solubility of various soybean isoflavones in supercritical carbon dioxide and the effect of glycoside chain in isoflavones. *Journal of Food Engineering*, 95(4), 564–571.

Nobre, B.P., Palavra, A.n.F., Pessoa, F.L.P. and Mendes, R.L. (2009) Supercritical CO_2 extraction of trans-lycopene from Portuguese tomato industrial waste. *Food Chemistry*, 116(3), 680–685.

Park, H.S., Choi, H.K., Lee, S.J., Park, K.W., Choi, S.G. and Kim, K.H. (2007a) Effect of mass transfer on the removal of caffeine from green tea by supercritical carbon dioxide. *The Journal of Supercritical Fluids*, 42(2), 205–211.

Park, H.S., Lee, H.J., Shin, M.H., Lee, K.W., Lee, H., Kim, Y.S., Kim, K.O. and Kim, K.H. (2007b) Effects of cosolvents on the decaffeination of green tea by supercritical carbon dioxide. *Food Chemistry*, 105(3), 1011–1017.

Patist, A. and Bates, D. (2008) Ultrasonic innovations in the food industry: From the laboratory to commercial production. *Innovative Food Science and Emerging Technologies*, 9(2), 147–154.

Pferschy-Wenzig, E.M., Getzinger, V., Kunert, O., Woelkart, K., Zahrl, J. and Bauer, R. (2009) Determination of falcarinol in carrot (Daucus carota L.) genotypes using liquid chromatography/mass spectrometry. *Food Chemistry*, 114(3), 1083–1090.

Pinelo, M., Arnous, A. and Meyer, A.S. (2006) Upgrading of grape skins: Significance of plant cell-wall structural components and extraction techniques for phenol release. *Trends in Food Science and Technology*, 17(11), 579–590.

Pinelo, M., Zornoza, B. and Meyer, A.S. (2008) Selective release of phenols from apple skin: Mass transfer kinetics during solvent and enzyme-assisted extraction. *Separation and Purification Technology*, 63(3), 620–627.

Pronyk, C. and Mazza, G. (2009) Design and scale-up of pressurized fluid extractors for food and bioproducts. *Journal of Food Engineering*, 95(2), 215–226.

Puértolas, E., Hernández-Orte, P., Saldaña, P.G., Álvarez, I. and Raso, J. (2010a) Improvement of winemaking process using pulsed electric fields at pilot-plant scale. Evolution of chromatic parameters and phenolic content of Cabernet Sauvignon red wines. *Food Research International*, 43(3), 761–766.

Puértolas, E., Saldaña, G., Condón, S., Álvarez, I. and Raso, J. (2010b) Evolution of polyphenolic compounds in red wine from Cabernet Sauvignon grapes processed by pulsed electric fields during aging in bottle. *Food Chemistry*, 119(3), 1063–1070.

Reverchon, E. (1996) Mathematical modeling of supercritical extraction of sage oil. *AIChE Journal*, 42(6), 1765–1771.

Reverchon, E. and De Marco, I. (2006) Supercritical fluid extraction and fractionation of natural matter. *The Journal of Supercritical Fluids*, 38(2), 146–166.

Richter, B.E., Jones, B.A., Ezzell, J.L., Porter, N.L., Avdalovic, N. and Pohl, C. (1996) Accelerated solvent extraction (ASE®): a technique for sample preparation. *Analytical Chemistry*, 68(6), 1033–1039.

Rodríguez-Meizoso, I., Jaime, L., Santoyo, S., Cifuentes, A., García-Blairsy Reina, Señoráns, F.J. and Ibáñez, E. (2008) Pressurized fluid extraction of bioactive compounds from phormidium species. *Journal of Agricultural and Food Chemistry*, 56(10), 3517–3523.

Rostagno, M.A., Palma, M. and Barroso, C.G. (2004) Pressurized liquid extraction of isoflavones from soybeans. *Analytica Chimica Acta*, 522(2), 169–177.

Rostagno, M.A., Araújo, J.M.A. and Sandi, D. (2002) Supercritical fluid extraction of isoflavones from soybean flour. *Food Chemistry*, 78(1), 111–117.

Rostagno, M.A., Palma, M. and Barroso, C.G. (2003) Ultrasound-assisted extraction of soy isoflavones. *Journal of Chromatography A*, 1012(2), 119–128.

Rozzi, N.L., Singh, R.K., Vierling, R.A. and Watkins, B.A. (2002) Supercritical fluid extraction of lycopene from tomato processing byproducts. *Journal of Agricultural and Food Chemistry*, 50(9), 2638–2643.

Runnqvist, H., Søren, A.B., Hansen, M., Styrishave, B., Halling-Sørensen, B. and Björklund, E. (2010) Determination of pharmaceuticals in environmental and biological matrices using pressurised liquid extraction–Are we developing sound extraction methods? *Journal of Chromatography A*, 1217(16), 2447–2470.

Sabio, E., Lozano, M., Montero de Espinosa, V., Mendes, R.L., Pereira, A.P., Palavra, P.F. and Coelho, J.A. (2003) Lycopene and β-carotene extraction from tomato processing waste using supercritical CO_2. *Industrial and Engineering Chemistry Research*, 42(25), 6641–6646.

Sahena, F., Zaidul, I.S.M., Jinap, S., Karim, A.A., Abbas, K.A., Norulaini, N.A.N/ and Omar, A.K.M. (2009) Application of supercritical CO2 in lipid extraction – A review. *Journal of Food Engineering*, 95(2), 240–253.

Saldaña, M.D.A., Sun, L., Guigard, S.E. and Temelli, F. (2006) Comparison of the solubility of [beta]-carotene in supercritical CO2 based on a binary and a multicomponent complex system. *The Journal of Supercritical Fluids*, 37(3), 342–349.

Sanal, I.S., Gnvenτ, A., SalgIn, Mehmetoglu, U. and !alImlI, A. (2004) Recycling of apricot pomace by supercritical CO2 extraction. *The Journal of Supercritical Fluids*, 32(1–3), 221–230.

Santamaria, R.I., Reyes-Duarte, M.D., Barzana, E., Fernando, D., Gama, F.M., Mota, M. and Lopez-Munguia, A. (2000) Selective enzyme-mediated extraction of capsaicinoids and carotenoids from chili guajillo puya (Capsicum annuum L.) using ethanol as solvent. *Journal of Agricultural and Food Chemistry*, 48(7), 3063–3067.

Sarmento, L.A.V., Machado, R.A.F., Petrus, J.C.C., Tamanini, T.R. and Bolzan, A. (2008) Extraction of polyphenols from cocoa seeds and concentration through polymeric membranes. *The Journal of Supercritical Fluids*, 45(1), 64–69.

Schilling, S., Alber, T., Toepfl, S., Neidhart, S., Knorr, D., Schieber, A. and Carle, R. (2007) Effects of pulsed electric field treatment of apple mash on juice yield and quality attributes of apple juices. *Innovative Food Science and Emerging Technologies*, 8(1), 127–134.

Schilling, S., Toepfl, S., Ludwig, M. Dietrich, H., Knorr, D., Neidhart, S., Schieber, A. and Carle, R. (2008) Comparative study of juice production by pulsed electric field treatment and enzymatic maceration of apple mash. *European Food Research and Technology*, 226(6), 1389–1398.

Serra, A.T., Seabra, I.J., Braga, M.E.M., Bronze, M.R., de Sousa, H.C. and Duarte, C.M.M. (2010) Processing cherries (Prunus avium) using supercritical fluid technology. Part 1– Recovery of extract fractions rich in bioactive compounds. *The Journal of Supercritical Fluids*, 55(1), 184–191.

Shi, J., Nawaz, H., Pohorly, J., Mittal, G., Kakuda, Y. and Jiang, Y. (2005) Extraction of Polyphenolics from Plant Material for Functional Foods Engineering and Technology. *Food Reviews International* 21(1), 139–166.

Shi, J., Yi, C., Xue, S.J., Jiang, Y., Ma, Y. and Li, D. (2009) Effects of modifiers on the profile of lycopene extracted from tomato skins by supercritical CO2. *Journal of Food Engineering*, 93(4), 431–436.

Smith, R.M. (2002) Extractions with superheated water. *Journal of Chromatography A*, 975(1), 31–46.

Soliva-Fortuny, R., Balasa, A., Knorr, D. and Martín-Belloso, O. (2009) Effects of pulsed electric fields on bioactive compounds in foods, a review. *Trends in Food Science and Technology*, 20(11–12), 544–556.

Sovová (2005) Mathematical model for supercritical fluid extraction of natural products and extraction curve evaluation. *The Journal of Supercritical Fluids*, 33(1), 35–52.

Sowbhagya, H.B., Srinivas, P. and Krishnamurthy, N. (2010) Effect of enzymes on extraction of volatiles from celery seeds. *Food Chemistry*, 120(1), 230–234.

Srinivas, K., King, J.W., Monrad, J.K., Howard, L.R. and Hansen, C.M. (2009) Optimization of Subcritical Fluid Extraction of Bioactive Compounds Using Hansen Solubility Parameters. *Journal of Food Science*, 74(6), E342–E354.

Srinivas, K., King, J.W., Howard, L.R. and Monrad, J.K. (2010) Solubility of Gallic Acid, Catechin, and Protocatechuic Acid in Subcritical Water from (298.75 to 415.85) K. *Journal of Chemical and Engineering Data*: null.

Sun, M. and Temelli, F. (2006) Supercritical carbon dioxide extraction of carotenoids from carrot using canola oil as a continuous co-solvent. *The Journal of Supercritical Fluids*, 37(3), 397–408.

Tao, G., He, L., Liu, W.-S., Xu, L., Xiong, W., Wang, T. and Kou, Y. (2006) Preparation, characterization and application of amino acid-based green ionic liquids. *Green Chemistry*, 2006(8), 639–646.

Tao, G., He, L., Sun, N. and Kou, Y. (2005) New generation ionic liquids, cations derived from amino acids. *Chemical Communications*, (28), 3562–3564.

Taylor, L.T. (1996) Properties of supercritical fluids. In: L.T. Taylor (ed.) *Supercritical Fluid Extraction*. New York, John Wiley & Sons Inc., pp. 7–27.

Temelli, F. (2009) Perspectives on supercritical fluid processing of fats and oils. *The Journal of Supercritical Fluids*, 47(3), 583–590.

Tena, M.T., Valcárcel, M., Hidalgo, P.J. and Ubera, J.L. (1997) Supercritical Fluid Extraction of Natural Antioxidants from Rosemary, ГÇë Comparison with Liquid Solvent Sonication. *Analytical Chemistry*, 69(3), 521–526.

Thuy Pham, T.P., Cho, C.W. and Yun, Y.S. (2010) Environmental fate and toxicity of ionic liquids: A review. *Water Research*, 44(2), 352–372.

Töpfl, S. (2006) *Pulsed Electric Fields (PEF) for Permeabilization of Cell Membranes in Food and Bioprocessing Applications, Process and Equipment Design and Cost Analysis*. Berlin, Technical University of Berlin.

Vági, E., Simándi, B., Vásárhelyiné, K/P., Daood, H., Kéry, Á., Doleschall, F. and Nagy, B. (2007) Supercritical carbon dioxide extraction of carotenoids, tocopherols and sitosterols from industrial tomato by-products. *The Journal of Supercritical Fluids*, 40(2), 218–226.

Vasapollo, G., Longo, L., Rescio, L. and Ciurlia, L. (2004) Innovative supercritical CO_2 extraction of lycopene from tomato in the presence of vegetable oil as co-solvent. *The Journal of Supercritical Fluids*, 29(1–2), 87–96.

Vaughn Katherine, L.S., Clausen Edgar, C., King Jerry, W., Howard Luke, R. and Julie, C.D. (2008) Extraction conditions affecting supercritical fluid extraction (SFE) of lycopene from watermelon. *Bioresource Technology*, 99(16), 7835–7841.

Vega, P.J., Balaban, M.O., Sims, C.A., O'Keefe, S.F. and Cornell, J.A. (1996) Supercritical carbon dioxide extraction efficiency for carotenes from carrots by RSM. *Journal of Food Science*, 61(4), 757–759.

Vinatoru, M. (2001) An overview of the ultrasonically assisted extraction of bioactive principles from herbs. *Ultrasonics Sonochemistry*, 8(3), 303–313.

Virot, M., Tomao, V., Le Bourvellec, C., Renard, C.M.C.G. and Chemat, F. (2010) Towards the industrial production of antioxidants from food processing by-products with ultrasound-assisted extraction. *Ultrasonics Sonochemistry*, 17(6), 1066–1074.

Vorobiev, E. and Lebovka, N. (2010) Enhanced Extraction from Solid Foods and Biosuspensions by Pulsed Electrical Energy. *Food Engineering Reviews*, 2(2), 95–108.

Wakelyn, P.J. and Wan, P.J. (2001) Food industry-solvents for exrtacting vegetables oils. In: G. Wypych (ed.) *Handbook of Solvents*, Toronto, William Andrew Publishing; ChemTec Publishing, pp. 923–946.

Wang, L. and Weller, C.L. (2006) Recent advances in extraction of nutraceuticals from plants. *Trends in Food Science and Technology*, 17(6), 300–312.

Wang, W.C. and Sastry, S.K. (2002) Effects of moderate electrothermal treatments on juice yield from cellular tissue. *Innovative Food Science and Emerging Technologies*, 3(4), 371–377.

Wijngaard, H. and Brunton, N. (2009) The Optimization of Extraction of Antioxidants from Apple Pomace by Pressurized Liquids. *Journal of Agricultural and Food Chemistry*, 57(22), 10625–10631.

Xu, H., Zhang, Y. and He, C. (2007) Ultrasonically Assisted Extraction of Isoflavones from Stem of Pueraria lobata (Willd.) Ohwi and Its Mathematical Model. *Chinese Journal of Chemical Engineering*, 15(6), 861–867.

Zimmermann, U., Pilwat, G. and Riemann, F. (1974) Dielectric breakdown of cell membranes. *Biophysical Journal*, 14(11), 881–899.

Zosel, K., inventor. (1981) Process for the decaffeination of coffee. Germany Pat. No. US4260639.

Zuo, Y.B., Zeng, A.W., Yuan, X.G. and Yu, K.T. (2008) Extraction of soybean isoflavones from soybean meal with aqueous methanol modified supercritical carbon dioxide. *Journal of Food Engineering*, 89(4), 384–389.

19 Analytical techniques for phytochemicals

Rong Tsao and Hongyan Li

Guelph Food Research Centre, Agriculture & Agri-Food Canada, Ontario, Canada

19.1 Introduction

Increasing evidence in epidemiology and clinical trials has pointed to the important roles of phytochemicals in the prevention of chronic diseases and promotion of human health. Intakes of dietary phytochemicals have been associated with reduced risks of cancer, cardiovascular diseases, diabetes, chronic inflammation, neural degeneration, and other chronic degeneration and illness. Many phytochemicals are strong antioxidants, which once absorbed into our bodies help counteract excess oxidative stresses from reactive oxygen or nitrogen species (ROS or RNS). The imbalance between the oxidative stress and the body's antioxidant status has been widely recognized as the cause of the aforementioned diseases. Phytochemicals is a general term for plant-originated secondary metabolites that possess various biological activities. These natural products can be categorized into different chemical classes (Liu, 2004; Tsao and Akhtar, 2005; Tsao, 2010). While polyphenols and carotenoids are most frequent targets of studies related to human health, due largely to their antioxidant activities, many other groups including phytosterols, saponins, glucosinolates, and S-containing compounds such as allicin, also contribute significantly to lowered health risks.

Studies in recent years have shown that although direct actions of the antioxidant phytochemicals in neutralizing ROS such as free radicals may be a major mode of actions, other biological activities including the inhibition or enhancement of key enzymes involved in carbohydrate or fat metabolism, and effects on biomarkers of the cell signaling pathways may be equally important (Tsao, 2010).

On the other hand, most activities of the phytochemicals are found during *in vitro* stuies in their original forms in the plants. However, before these bioactive compounds can exert their various physiological activities, they have to survive many steps in the food chain. These compounds may change in their chemical structures and physical chemical properties during post-harvest storage, food processing, or extraction (when used as food supplements), formulation, post-processing storage, the human digestive tract, and cell biochemical processes once absorbed. It is therefore imperative that these factors are considered when

---

Handbook of Plant Food Phytochemicals: Sources, Stability and Extraction, First Edition.
Edited by B.K. Tiwari, Nigel P. Brunton and Charles S. Brennan.
© 2013 John Wiley & Sons, Ltd. Published 2013 by John Wiley & Sons, Ltd.

developing an analytical method for a particular phytochemical or a group of phytochemicals. Detection of phytochemicals in plants or food samples is relatively easier as samples are often available in large quantities. However, analytical chemists are increasingly facing challenges from clinical nutritionists as samples from animal or human trials are normally very limited, the analytes are at minute concentrations, and often with no standard reference materials available, particularly in the case of metabolites, for example, a typical blood sample from a mouse study is <150 µL (Kinniry et al., 2006), and the concentrations can be as low as picomol/L (Gardner et al., 2009; Bolca et al., 2010).

These have brought new challenges to analytical chemists as they can no longer only follow the traditional techniques involved in sample preparation, separation, identification, and quantification of the analytes. These challenges have also necessitated analytical instrumentations that provide fast and sensitive detection, and other new techniques for sample clean up and separation. This chapter will therefore be focused on the latest advances in techniques used in sample preparation and quantitative and qualitative analyses of phytochemicals of food origins with health benefits. It is nearly impossible for this chapter to cover analytical methods for all groups of bioactive phytochemicals,

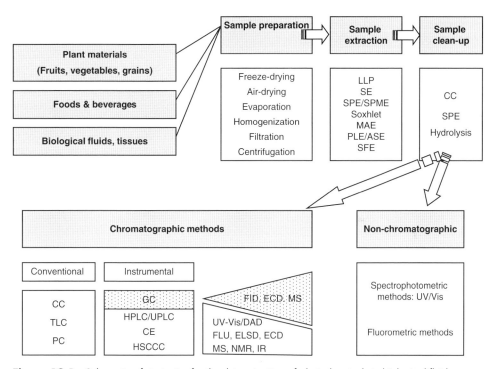

Figure 19.1 Schematic of strategies for the determination of phytochemicals in biological fluids, beverages, plants, and foods.
Abbreviations: SFE: supercritical fluid extraction; SPME: solid-phase microextraction; ASE/PLE: accelerated solvent extraction/pressurized liquid extraction; MAE: microwave-assisted extraction; SFE: supercritical fluid extraction; HSCCC: high-speed counter-current chromatography; TLC: thin layer chromatography; CC: column chromatography; PC: paper chromatography; HPLC: high performance liquid chromatography; UPLC: ultra performance liquid chromatography; CE: capillary electrophoresis; FLU: fluorescence; FID: flame ionization detection; ECD: electron capture detection (GC)/electrochemical detector (LC); MS: mass spectrometry; UV/Vis: ultraviolet/visible; DAD: diode array detector; ELSD: evaporative light scattering detector; NMR: nuclear magnetic resonance; IR: infrared.

therefore for readers who are interested in detailed information and in-depth discussions on the specific phytochemicals of their interest, comprehensive reviews by others are recommended (Tsao and Deng, 2004; Oleszek and Bialy, 2006; de Rijke et al., 2006; Marston, 2007; Marston and Hostettmann, 2009). Figure 19.1 is an illustration that will help discussions in this chapter.

19.2 Sample preparation

Sample preparation is a critical step of a successful analytical method. Again, due to the large variations in sample matrices, for example, plant materials, food formulations, biological fluid and tissue samples, the diverse chemical structures, and physicochemical properties of the phytochemicals, it is unrealistic to develop any definitive procedure or protocol for all types of sample matrices. However, there are important common precautions that must be taken for better preparing samples for the subsequent analyses. The overall purposes of sample preparation are to concentrate or dilute the samples so the analytes can be detected and quantified within the detection limit and liner range; to rid any interference that might affect the detection of the compounds of interest. Therefore, techniques adopted to sample preparation must follow these two principles (Figure 19.1).

19.2.1 Extraction

Many techniques are available for optimized extraction of phytochemicals from various samples (Figure 19.1). The most frequently encountered samples containing phytochemicals are either in a liquid or a solid form. Liquid food and biological samples often have highly complex matrix, other than certain beverages which can be directly applied to spectrophotometric or chromatographic systems and those that can be cleaned up simply by precipitation and centrifugation, such as removal of proteins from soy milk samples. The overwhelming majority of samples are initially extracted by liquid-liquid partitioning (LLP) and solid phase extraction (SPE) (use of adsorbents such as different resins, e.g. C18, LH-20),

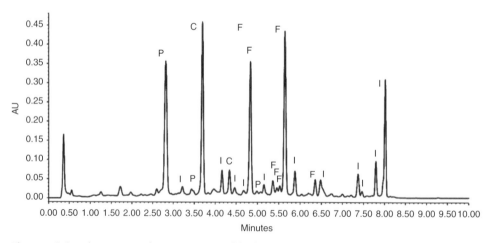

Figure 19.2 Characteristic absorption spectra of the four groups of phenolics identified in *Trifolium* species and UPLC profile of *T. squarrosum* (a); P, phenolic acid; C, clovamid; F, flavanoid; and I, isoflavone. Reproduced with permission from Oleszek, W., Stochmal, A., and Janda, B. (2007). Copyright 2007, American Chemical Society.

with or without pre-concentration. LLP helps move unwanted compounds into one of the two immiscible layers depending the lipophilicity or hydrophilicity of the compounds of interest. For example, polyphenols such as isoflavones in soy milk can be extracted by adding methanol or ethanol and then partitioned against hexane (to remove highly fat-soluble components such as plant sterols and fatty acids) after precipitation of proteins; vise versa, similar approaches can be used for extracting phytosterols with hexane and partitioning against hydrophilic solvents to rid compounds like flavonoids. LLP is not only important for better chromatographic separation, but also critical for removing interference for spectrophotometric analysis of phytochemicals such as the total phenolic content (TPC) assay using the Folin-Ciocalteu method. Some proteins and carbohydrates have been reported to interfere with the reaction, causing false and over estimation of TPC (Singleton *et al.*, 1999; Stevanato *et al.*, 2004).

Some minor modifications can improve the efficiency of the LLP. Acidification with weak acids can keep polyphenols in the neutral form, thus better partitioned into the organic layer. This is particularly important when hydrolysis is involved. Aglycones of flavonoids, for example, are extracted from the aqueous alcohol solution with acetyl acetate after acid hydrolysis. Phytochemicals existing in the free form in food and biological samples can be readily extracted into the solvents, however, some compounds extractable with organic solvents, such as phenolic acids, may exist in a bound form with other soluble contents such as small proteins/peptides and soluble dietary fibers. Acid or alkaline hydrolysis may be performed prior to or during the LLP (Kim *et al.*, 2006).

SPE is both an extraction technique and a sample clean up method. Porous resins have been used as adsorbents of different phytochemicals from liquid samples. Batch soaking using such adsorbents is rare in an analytical procedure, but columns or cartridges pre-packed with these adsorbents are frequently used in extracting phytochemicals from liquid samples. Polyphenols in beverages, for example, are often extracted by passing the liquid through an SPE cartridge, and then eluded with appropriate solvents. Solid phase micro-extraction (SPME) is a special form of SPE, and it is almost exclusively used for sequestering volatile compounds such as essential oil components from herbs and spices, and then analyzed by gas chromatography (GC).

Frequently, however, the samples are in the solid form. This includes those that are originally solid and freeze-dried samples of liquid or fresh plant or animal samples. Solvent extraction (SE) by soaking the sample in a single or mixed solvent is the most simple and efficient method for the majority of phytochemicals in foods. It offers good recovery of phytochemicals from various samples, however, the use of large amount of organic solvents poses health and safety risks, and is environmentally unfriendly. Proper solvents or solvent mixtures are critical; however efficient extraction of targeted phytochemicals may be aided by ultrasound wave, heating or refluxing (e.g. Soxhlet extraction), microwave or high pressure, or combination of these techniques. Under all circumstances, parameters must be chosen to avoid degradation of the bioactive components. High temperature, for example, can lead to degradation of certain polyphenols (Palma *et al.*, 2001). Among different solvent extraction techniques, pressurized liquid extraction (PLE), microwave-assisted extraction (MAE) and supercritical fluid extraction (SFE) are relatively recent technologies, and have been increasingly incorporated into the analytical methods for phytochemicals (Tsao and Deng, 2004).

PLE, also known as accelerated solvent extraction (ASE), is a relatively new technology used in extraction of phytochemicals (Palma *et al.*, 2001; Piñeiro *et al.*, 2004). The only difference between PLE and the conventional solvent extraction is the high pressure in the former. The design allows solid samples to be extracted in a significantly reduced volume of solvent and, applying high pressure and elevated temperature, PLE results in higher recovery rates compared to conventional method (Palma *et al.*, 2001; Piñeiro *et al.*, 2004).

MAE combines the traditional solvent extraction with microwave energy. It is important for the extraction solvent in an MAE to have good polarity, because solvents with high dielectric constants (polar) can absorb more microwave energy, therefore resulting in better extraction efficiency (Hong et al., 2001; Tsao and Deng, 2004). For this reason, polar solvents such as water or alcohol are often added as modifiers in order to achieve an optimal dielectric constant of the extraction solvent. Although, other studies have shown that solvents of low dielectric constants can quickly direct the microwave energy to the moisture inside the cellular structure of the sample, causing the cells to erupt and the walls to break, thus leading to the releasing of the phytochemicals to the surrounding solvent. Microwave has been used to extract phytochemicals in different forms, but solvent-free microwave extraction (SFME), a combination of microwave heating and dry distillation, is a new green technique developed in recent years (Wang et al., 2006). A unique version of SFME, dry-diffusion and gravity (MDG), is worth special noting (Farhat et al., 2010). In MDG, the direct interaction of microwaves with dried plant material first favours the release of essential oils trapped inside the cells of plant tissues, and then the essential oil moves naturally downwards by earth gravity on a spiral condenser outside the microwave cavity where it condenses and is collected. The method was found to be highly efficient and clean as compared to conventional hydrodistillation (Farhat et al., 2010).

SFE is essentially a solvent extraction under high pressure. Certain gases such as carbon dioxide (CO_2) can be liquefied to a state called supercritical fluid when the pressure and temperature are right; and this gas-like characteristic helps the fluid diffuse to the matrix and access to the phytochemical. SFE has been used in recent years in many applications, and supercritical CO_2 is the most widely used solvent for many phytochemicals. A CO_2-based SFE is most suitable for the extraction of phytochemicals as carotenoids, phytosterols, and other relatively lipophilic compounds, owing to the apolar property of CO_2 (Lesellier et al., 1999; Sun et al., 2002; Tsao and Deng, 2004). It is not considered a good method of extraction for polar phytochemicals such as polyphenols despite some reports showing good recovery rate using 95% methanol and 5% CO_2 (Chang et al., 2000), which does not take the full advantage of CO_2 and the SFE. Phytochemicals in herbs and spices such as rosemary have been extracted using SFE with higher recovery than typical organic solvents (Tena et al., 1997). A unique solvent-free SFE has been developed and used for efficient extraction of essential oil of oregano (Bayramoglu et al., 2008).

19.2.2 Sample clean-up

Extracts obtained as described in section 19.2.1 can be directly subjected to quantitative or qualitative analysis, although very often they are further diluted or concentrated prior to direct analysis. Further clean-up or treatment may be required depending on the objectives of the study and the analytical methods employed (Figure 19.1). Additional LLP or SPE can be used to remove unwanted components, for example, methanol extracts containing highly hydrophilic phenolic acids are often re-partitioned into n-butanol, separating these phytochemicals from water-soluble peptides, sugar or polysaccharides (Kim et al., 2006). Similarly, column chromatography or SPE has been used for further fractionating the extracts into different phytochemical groups before instrumental analysis (Tsao and Deng, 2004).

One of the most frequently used clean-up methods is hydrolysis, which is particularly useful in polyphenol analysis. Polyphenols including the many subgroups of flavonoids are often highly glycosylated, not only by association with different sugars, but at different positions of the aglycones. Hydrolysis simplifies the polyphenol profile of the extracts, thus resulting in better separation. It is also necessary because standards of glycosides are difficult to obtain. Hydrolysis of glycosylated compounds can be done under enzymatic, acidic or basic condition (the latter is also called saponification). Different hydrolysis conditions can lead to different hydrolysis products, thus care must be taken in choosing the right method (Muir, 2006; Shao et al., 2011). While β-glucosidase is used for enzymatic hydrolysis of native glycosides in plants and food samples, β-glucronides and sulphatase are used to hydrolyze conjugates in biological samples such as human plasma or urine (Tsao et al., 2004; Shao et al., 2011). Strong acid such as 2–4 N HCl at elevated temperature such as refluxing has been most frequently used for hydrolysis of polyphenol conjugates in fruit and vegetable extracts, while saponification under strong alkaline condition is used for phytochemical esters with fatty acids such as lutein esters, saponins, and phytosterols (Shao et al., 2011). Hydrolysis is also done to release certain types of polyphenols that otherwise exist as complexes. Polyphenols such as seicoisolariciresinol diglucosides (SDG) can only be released from the lignan complex upon acid or alkaline hydrolysis (Muir, 2006). Polymeric procyanidins are another group of phytochemicals that must be hydrolyzed in order to produce bioactive monomer or oligomers (White et al., 2010).

A good clean-up procedure must be able to eliminate efficiently the interfering compounds from the sample, but more importantly it must also have a high rate of recovery of the compounds of interest. This is sometimes a dilemma for analysts as the two requirements cannot always be met together. Additional challenges also exist during the course of sample clean-up. Some phytochemicals can be permanently bound to the packing materials of a column or SPE; some interfering compounds may have too similar physicochemical properties such as polarity and therefore co-elute with the analytes. Hydrolysis may not stop exactly at the cleavage of glycosidic or ester bonds, but could lead to destruction of main structural features of some relatively unstable compounds under acid or alkaline conditions as well (Nuutila et al., 2002).

19.3 Non-chromatographic spectrophotometric methods

Analyses of the crude or cleaned up extracts obtained as described above can be carried out using non-chromatographic spectrophotometric methods or chromatographic methods (Figure 19.1). Spectrophotometric methods are based on the ability of the phytochemicals that absorb light in the ultraviolet (UV) or visible range of the spectrum (e.g. total carotenoid content), or the ability of forming such chromophores after reacting with certain reagents (e.g. total phenolic content), and the quantification is based on Beer-Lambert law. This approach does not require separation of individual compounds in the extract, and often quantification is done as the total amount of similar compounds in the extract. The advantages of the non-chromatographic spectrophotometric methods are simple, fast, and of low cost, but these methods lack the specificity for individual compounds and the results are less accurate. Underestimation or overestimation often occur in these methods because of the interference from large molecules such as proteins and carbohydrates or their monomeric forms (amino acids and sugars) (Singleton et al., 1999; Stevanato et al., 2004), and the lack of appropriate standards. Many of the spectrophotometry based methods use one representative compound for quantification, and the total concentration of the group of the phytochemicals is often expressed in equivalent number to this reference compound, however, in reality, no single compound can truly reflect the real composition of a mixture. For example, the total phenolic content of an extract as measured by the Folin-Chiocalteu method is most often expressed in gallic acid equivalent concentrations.

However, as discussed above, gallic acid by no means can represent all polyphenols. For this reason, some methods use a flavonoid as the reference compound, but for extracts rich in phenolic acids, error will occur. The spectrophotometric methods can be used for quantitative analysis of individual compounds in food supplements if they are formulated with pure compounds. Quantification can be done via a standard curve of a particular reference compound, or be calculated using the molar absorptivity. Many of these methods have now been adopted for microplate readers, making high-throughput analyses possible (Shao et al., 2010; Wang et al., 2010). Typical spectrophotometric methods used for the estimation of phytochemicals are briefly discussed in sections 19.3.1, 19.3.2, 19.3.3, 19.3.4, 19.3.5, and 19.3.6.

19.3.1 Total phenolic content (TPC)

The Folin-Ciocalteu (FC) assay is the most widely used method for the estimation of total phenolic content (TPC) in extracts fruits, vegetables, grains, and other foods. The FC reagent consists of an oxidizing mixture of phosphotungstic acid and phosphomolybdic acid which, when reduced, produce a mixture of blue molybdenum and tungsten oxides (λ_{max} 765 nm). Technically all compounds that can be oxidized by the FC reagent will be measured; therefore the method can potentially produce erroneous results. However, most plants or food extracts contain minimum interfering compounds (Escarpa and Gonzalez, 2001), thus the FC-TPC remains a popular method. Early methods were based on the protocol developed by Slinkard and Singleton (1977), however, it has been adapted to a high throughput using a microplate reader (Wang et al., 2010). Briefly, 25 µL gallic acid standard or a sample was mixed with 125 µL FC reagent in 96-well microplates and allowed to react for 10 min at room temperature. A 125 µL saturated sodium carbonate (Na_2CO_3) solution was then added and allowed to stand for 30 min at room temperature before the absorbance of the reaction mixture was read at 765 nm using a visible-UV microplate kinetic reader. Calibration is achieved with an aqueous gallic acid solution (50–500 µg/mL). The TPC was expressed as mg gallic acid equivalent (GAE) per g or mL of the original sample based on the calibration curve (Wang et al., 2010).

19.3.2 Total flavonoid content (TFC)

The aluminium chloride method has been widely used for the total flavonoids content estimation. This method is based on flavonoids' capability of forming stable complex with Al ions in a solution. The color of the complex depends on the ratio of the Al ions to the flavonoid molecules and the hydroxylation pattern of the latter. For this reason the spectrophotometric readings used in this method can vary from 367 to 510 nm in different experimental procedures. Different modifications of this method have been made, from a simple protocol as used by Bahorun et al. (2004) to a complicated protocol by Dewanto Dewanto et al. (2002). Other methods such as using 2,4-dinitrophenylhydrazine to generate a chromaphore at 495 nm have also been reported (Chang et al., 2002; Meda et al., 2005). The already mentioned disadvantages may be overcome by the new method developed by He et al. (2008). This novel approach, called sodium borohydride/chloranil-based (SBC) assay, is based on a reduction reaction that converts flavonoids with a 4-carbonyl group to flavanols using sodium borohydride catalyzed with aluminum chloride. The flavan-4-ols were then oxidized to anthocyanins by chloranil in an acetic acid solution. The anthocyanins were reacted with vanillin in concentrated hydrochloric acid and then quantified spectrophotometrically at 490 nm (He et al., 2008). This novel SBC TFC assay is specific to flavonoids, thus it eliminates interference from other phenolic compounds, for example, phenolic acids. It is sensitive, and has high accuracy and precision. The method can be widely used for fruits, vegetables, whole grains, and other food or nutraceutical products that contain flavones, flavonols, flavonones, flavononols, isoflavonoids, flavanols (catechins), and anthocyanins (He et al., 2008). The only drawback of the SBC assay is the multiple-step reactions, although in the end a microplate reader was used to analyze a large amount of samples.

19.3.3 Total anthocyanin content (TAC)

The color of anthocyanins depends on the acidity of the medium. At acidic pH = 1–3, anthocyanidins exist predominantly in the form of the red flavylium cation (oxonium), and when the pH increases, the intensity of the color decreases as the flavylium cation becomes the colorless hemketal (pH 4.5). When the pH shifts higher, rapid proton loss occurs and the equilibrium is shifted toward a purple quinoidal anhydrobase at pH < 7 and a deep blue ionized anhydrobase at pH < 8. Analysis of total anthocyanin content (TAC) is therefore based on this pH differential property of the anthocyanins (between pH 1.0 and 4.5). The absorbance of anthocyanins at 520 nm is proportional to the concentration, and the absorbance from the haze (at 700 nm) is deducted during calculation. Results of TAC are expressed on a cyanidin-3-glucoside equivalent basis (AOAC 2005).

19.3.4 Total carotenoid content (TCC)

Carotenoids of food origin exhibit absorption in the visible region of the spectrum typically between 400 and 500 nm. The absorbance can therefore be measured and used to quantify the concentration of a single compound or to estimate the total carotenoid concentration (TCC) in a mixture such as food extract. Quantification of a pure carotenoid is simple, however for a sample with mixed carotenoids, a specific λ_{max} and extinction coefficient cannot be used. For this reason, a λ_{max} of 450 nm or 470 nm and a typical $A^{1\%}$ value of 2500, that is, the extinction coefficient (or the absorbance of a 1% solution), are used for the calculation (Schoefs, 2003; Britton, 1995). In our laboratory, TCC is calculated using a standard curve of β-carotene (0.001–0.005 mg/mL), and the concentration is expressed in β-carotene equivalents. Chlorophylls in fresh produce can be co-extracted with carotenoids, therefore some earlier methods included subtracting the concentration of chlorophylls (which absorb visible light at 662 and 645 nm) in their calculation (Lichtenthaler, 1987). Saponification of the extract can destroy the chlorophylls, thus avoiding over estimation of TCC, however the procedure was also accompanied with 13% loss of carotenoids (Biehler et al., 2010).

19.3.5 Methods based on fluorescence

While the majority of the non-chromatographic analyses of phytochemicals are based on spectrophotometric method; alternative methods have been explored. Recently a fluorescence method has been developed for the determination of TPC in foods (Shanhaghi et al., 2008). This new method was based on the fluorescence sensitization of terbium (Tb^{3+}) by complexation with flavonols at pH 7.0, which fluoresces intensely with an emission maximum at 545 nm when excited at 310 nm. The method was significantly more sensitive than the FC-TPC method. The total concentrations can be expressed in quercetin equivalents (Shanhaghi et al., 2008).

19.3.6 Colorimetric methods for other phytochemicals

Colorimetric methods have also been developed for other phytochemicals. Many of these methods have been developed since decades ago; however, due to the easiness and simplicity of use, many are still used today. Like the spectrophotometric methods for the phenolics, flavonoid, carotenoids, and anthocyanins, colorimetric methods have been used to determine the quantities of phytosterols, saponins, glucosinolates, and other S-containing compounds such as allici. The Liebermann-Burchard test is a method developed for cholesterol analysis, but it is still used in some laboratories to assess the phytosterol content of foods (Kenny, 1952; Okpuzor *et al.*, 2009). A method using vanillin and sulfuric acid was developed for quantitative determination of saponins (and other terpenoids) more than 3 decades ago, but is still used by many for quick and simple analysis (Hiai *et al.*, 1976). Glucosinolates are phytochemicals unique to the mustard family foods. The total glucosinolate content in these samples has been determined by spectrophotometric method as well (Hu *et al.*, 2010). These colorimetric methods are less frequently used currently due to the advancement in instrumentation such as chromatographic techniques.

19.4 Chromatographic methods

Chromatographic methods are powerful analytical tools in phytochemical studies. Owing to the vast number and diverse chemical classes of phytochemicals, and the versatility of chromatography, in-depth discussion on this topic is beyond the scope of this chapter. It is the authors' intention that readers find specific reviews for their specific research subjects (Tsao and Deng, 2004; Oleszek and Bialy, 2006; de Rijke *et al.*, 2006; Marston, 2007; Marston and Hostettmann, 2009). Only the latest advances in both conventional and instrumental chromatographic methods related to phytochemical analysis will be discussed in this chapter (Figure 19.1).

19.4.1 Conventional chromatographic methods

Conventional chromatography includes column and planar chromatography; the latter can be a thin layer chromatography (TLC) or paper chromatography (PC) (Figure 19.1). Open column chromatography or its automated form such as flash chromatography has been mainly used for preparative separation and purification of the various phytochemicals. TLC and PC, although having been used for analytical purposes, are less commonly found in methods for phytochemical analysis, particularly those requiring high sensitivity. Compared to the other conventional chromatographic methods, however, TLC still finds some unique applications in phytochemical analysis and other related research due to the new technological advances in adsorbent materials such as reversed-phase high performance TLC (HPTLC), and in technologies related to methods that provide a constant and optimum mobile phase velocity (forced flow and electroosmotically-driven flow), imaging, and other densitometry (e.g. video densitometry for recording multidimensional chromatograms) and detection technologies such as *in situ* scanning mass spectrometry (Poole, 2003). TLC has attracted more attention as a fast and convenient detection method for various bioactivities of phytochemicals (Marston, 2010). These technologies, in combination with 2D, multiple development, and coupled column–layer separation techniques could dramatically increase the use of TLC for the characterization of complex mixtures such as plant extracts containing bioactive phytochemicals (Poole, 2003). HPTLC has been used for fingerprinting of flavonoids and quantification of tetrahydroamentoflavone, a bioactive flavonoid in *Semecarpus anacardium* plant (Aravind *et al.*, 2008). More recent development and in-depth discussions can be found in the latest reviews and books (Waksmundzka-Hajnos *et al.*, 2008; Marston, 2010).

19.4.2 Instrumental chromatographic methods

Many instrumental chromatographic methods, from GC, HPLC/UPLC to other recent separation technologies such as capillary electrophoresis (CE), supercritical fluid chromatography (SFC), and analytical high speed counter-current chromatography (HSCCC), have been used in phytochemical analysis. Furthermore, various detectors including flame ionization detector (FID), electron capture detector (ECD-GC) and MS for the GC system, and the UV/Vis, DAD, FLU, evaporative light scattering detector (ELSD), electrochemical detector (ECD-LC), nuclear magnetic resonance (NMR), and MS for the LC system, have also been developed and coupled with the advanced chromatographic separation units. In-depth reviews on these separation and detection technologies and their applications can be found in many recent publications (Tsao and Deng, 2004; Liu, 2008; Marston and Hostettmann, 2009). This chapter will only discuss the most widely used methods.

19.4.2.1 *Gas chromatography*

Gas chromatography (GC) is an excellent analytical tool for phytochemicals or their derivatives that are volatile upon heating. GC is a particularly effective separation method and highly sensitive instrument for phytochemicals such as monoterpenoids and other essential oil components in spices and herbs. Compounds containing hydroxyl group(s) but do not have strong UV/Vis absorption such as phytosterols are derivatized to trimethylsilyl (TMS) ethers before being analyzed by GC (Piironen *et al.*, 2002; Iafelice *et al.*, 2009). Similarly, some polyphenols such as isoflavones are also made into TMS derivatives and analyzed by GC (Naim *et al.*, 1974; Campo Fernández *et al.*, 2008; Hsu *et al.*, 2010). Derivatization is time-consuming and might be a source of artifacts.

19.4.2.2 *High performance liquid chromatography/ultra performance liquid chromatography*

While other chromatographic techniques are used in the analysis of phytochemicals, the overwhelming methods use liquid chromatography. It is not an overstatement that high performance liquid chromatography (HPLC) is the most popular and reliable system among all chromatographic separation and detection technique for phytochemicals, particularly the health beneficial food-borne phytochemical components. The versatility of HPLC is also aided by the different separation modes and types of detection methods, among which is the diode array detector (DAD) coupled with mass spectrometer (MS).

Different separation modes (types of column) have been used to separate and analyze different phytochemicals, among them the adsorption/desorption based columns including the normal phase (NP) and reversed phase (RP) columns are most frequently used. Silica-based stationary phase and its interaction with a non-polar mobile phase is the principle of the NP HPLC separation. NP HPLC is preferred for the analysis of phytochemicals with relatively high lipophilicity. Carotenoids, for example, have been analyzed in NP HPLC, for example separation of saponified carotenoids was carried out on a silica column (250×4.6 mm I.D., $5\,\mu$m) using gradient elution from 95% of light petroleum to 95% acetone (Almela,

1990). Polymeric procyanidids from fruits such as blueberry have also been separated and analyzed by NP-HPLC (Gu et al., 2002).

The majority of the food-originated phytochemicals are analyzed by RP HPLC. The most popular RP column is packed with octadecyl carbon chain bonded silica (ODS or C18), while other RP stationary phases (e.g. C4, C8, C30, Phenyl, CN) are commercially available and many have been used for phytochemcial analysis. In addition, these columns come with an array of different particle sizes and end-capping technologies, giving RP HPLC unmatchable versatility in separating thousands of phytochemicals reposted in food.

RP HPLC has been the method of choice for separating the two key bioactive phytochemical groups in foods and biological samples, polyphenols and carotenoids (Tsao and Deng, 2004; Tsao, 2010). RP HPLC separation coupled with different detection technologies have been reviewed extensively (Oliver and Palou, 2000; Tsao and Deng, 2004; de Rijke et al., 2006; Valls et al., 2009). For the separation of polyphenols, the stationary phase is almost exclusively C18, which is coupled with a binary mobile phase system containing acidified water (solvent A) and a polar organic solvent (solvent B), where solvent A usually includes aqueous acids or additives such as phosphate, and solvent B pure or acidified methanol or acetonitrile (Tsao and Deng, 2004). Many methods have been developed for simultaneous detection of multiple polyphenols of different chemical groups. A method using a binary mobile phase consisting of 6% acetic acid in 2 mM sodium acetate aqueous solution (v/v, final pH 2.55) (solvent A) and acetonitrile (solvent B) and a RP-C18 column produced a near baseline separation of 25 polyphenolic compounds commonly found in fruits (Tsao and Yang, 2003). A similar method using a binary system of 50 mM sodium phosphate in 10% methanol (solvent A) and 70% methanol (solvent B) successfully separated 28 polyphenols of different classes in 90 min (Sakakibara et al., 2003).

In terms of carotenoids, while a large number of methods has been developed around the C18 column (Oliver and Palou, 2000; Tsao et al., 2004; Tsao and Yang, 2006; Chandrika, 2010) and good separation has been achieved, for more complex samples, particularly those containing multiple carotenoids and their esters, a C30 column can give more improved separation and selectivity than the conventional C8 and C18 materials. RP C30 column is particularly a good choice for the separation of geometric isomers of carotenoids (Humphries and Khachik, 2003; Aman et al., 2004). A recent report also showed that a C30 column had better separation of carotenoids in corn compared to a C18 column (Burt 2010). Using C30 LC-MS, Breithaupt et al. were able to identify eight regioisomeric monoesters in addition to known lutein mono- and diesters (Breithaupt et al., 2002). Geometric isomers of free carotenoids have been separated using mainly C30 columns, however, we recently developed a method using an RP C18 column in combination with DAD and MS detection, separating several cis-isomers of lutein diesters for the first time (Tsao et al., 2004). Several good review papers have been published in recent years on the separation of carotenoids and readers are referred to those for more detailed discussions (Oliver and Palou, 2000).

In addition to NP and RP HPLC, other separation modes such as size-exclusion chromatography (SEC) and ion exchange chromatography (IEC) have also been found useful in separating and analyzing phytochemicals. For example, using a TSK gel α-2500 column, and a mobile phase consisting of acetone and 8 M urea (pH 2) (6:4), procyanidins with various degrees of polymerization were separated in native forms from apple and other plant extracts (Yanagida et al., 2003). These techniques are often used in combination with conventional RP HPLC (e.g. C18). For instance, ion exchange resins such as Amberlite XAD-7 are often used to separate anthocyanins from other highly water-soluble interference like sugars. Anthocyanins separated by IEC are often further purified on a Sephadex LH-20 column before finally being analyzed on a RP C18 column (Andersen et al., 2004). Recently, we have developed a novel mixed mode HPLC method using a column combining both ion-exchange and reversed-phase separation mechanisms (SiELC PrimeSep B2 column, 250 mm × 4.6 mm i.d.; particle size 5μ), to facilitate analysis of anthocyanins in grapes (McCallum et al., 2007). It was found that chromatographic performance and subsequent analysis of anthocyanidin diglucosides and acylated compounds were significantly improved compared to those associated with conventional C18 RP method. The enhanced chromatographic resolution provides nearly complete separation of 37 anthocyanin types.

HPLC has been used for the analysis of phytochemicals other than polyphenols and carotenoids. Both the analytes and the instrumentations play important roles, because not all phytochemicals contain a chromophore for sensitive on-line detections and, in the mean time, not all detectors are sensitive enough or appropriate for all samples containing phytochemicals. While the discussions here have emphasized polyphenols and carotenoids, one must keep in mind that many HPLC methods have been developed for phytosterols, saponins, glucosinolates, and S-containing compounds such as allicin (Hu et al., 2010; Li et al., 2005a; Rosen et al., 2001; Zarrouk et al., 2010).

A new generation of HPLC, high performance liquid chromatography (UPLC or UHPLC), has become increasingly used in phytochemical analysis in very recent years due to the advancement made in both column technology (sub-2 micron particle size) and instrumental hardware (ultra high pressure pump). UPLC has many significant advantages over the conventional HPLC in the performance of the separation, for example, increased resolution and sensitivity, however, one that stands out above all is the drastically reduced analytical time (<1/10 of the time of a conventional HPLC) and solvent use (Pongsuwan et al., 2008; Oleszek et al., 2007) (Figure 19.2).

Detection methods

Separation in both HPLC and UPLC are coupled with different types of detection devices. Phytochemicals such as polyphenols and carotenoids strongly absorb light in the UV/Vis region of the spectrum, thus are best detected using UV/Vis detector, or a photodiode array detector (PDA or DAD). DAD collects UV/Vis spectral data as the compounds are separated and eluted from the column, therefore it not only provides excellent quantitative capability, but also information for putative structures of unknown compounds by matching the UV/Vis spectrum of the compound and retention time with a standard (Tsao and Yang, 2003). Phytochemicals that have no or very weak UV/Vis absorbance can be detected by other detectors such as ELSD or ECD-LC. Saponins, for example, have been analyzed using HPLC-ELSD (Li et al., 2005a; Chen et al., 2011). Phytochemicals with strong redox potential are best suited for ECD-LC, and those can be excited to emit fluorescence and can be analyzed using fluorescence detector (FLU) with enhanced sensitivity. ECD-LC and FLU detector are far more sensitive than UV/Vis detector or DAD, therefore they are particularly useful for analysis of phytochemicals in biological fluid or tissue samples. Bolarinwa and Linseisen (2005) developed a sensitive RP HPLC method for determination of 23 flavonoids and phenolic acids in plasma samples using ECD-LC with limits of detection between 1.45 and 22.27 nM. RP HPLC-FLU is a particularly sensitive analytical method for phytoestrogens such as lignans and isoflavones and their metabolites in biological samples. A method developed recently used an excitation wavelength of 350 nm and emission wavelength of 472 nm, for the determination of puerarin and daidzein in human serum (Klejdus et al., 2004; Liu et al., 2010). A different excitation and emission wavelengths were set at 277 nm and 617 nm, respectively, for the analysis of flax lignans (Mukker, 2010).

While the above mentioned on-line detection methods for GC or HPLC are sensitive and can be used to analyze various food or biological samples containing phytochemicals, their degradation products during food processing and storage, or metabolites in animals and humans, and their abilities are limited to providing quantitative data. DAD-HPLC can aid the identification of phytochemicals by acquiring UV/Vis absorption spectrum of analyte peaks, however, such spectrophotometric data are often not enough for the positive identification of an unknown compound. Mass spectrometric detector coupled with GC or HPLC (GC-MS or HPLC-MS), on the other hand, provides rich structural information in fragmentation pattern. On-line GC-MS or HPLC-MS, particularly techniques such as tandem MS that employs collision-induced dissociation (CID), can provide sufficient information for the final confirmation of most known polyphenols found in foods (Li et al., 2006). There are two main types of ionization techniques in MS for phytochemicals, the ion-spray techniques such as electro-spray ionization (ESI), thermospray and atmospheric pressure chemical ionization (APCI), and the ion-desorption techniques which include fast atom bombardment (FAB), plasma desorption (PD), and matrix assisted laser desorption ionization (MALDI) (Tsao and Deng, 2004). ESI and APCI are the two most widely used ionization methods for antioxidant phytochemicals, and most commercial chromatography-mass spectrometry (LC-MS) instruments can accommodate both of these techniques. Although there is no clear line, ESI is more often used to ionize molecules such as polyphenols that are polar and exist as ions in aqueous solutions, and APCI is used for less polar and non-ionic antioxidants such as carotenoids (Careri et al., 2002). APCI and ESI can be operated under both positive and negative ion modes (PI and NI). The most frequently used mass analyzers can also be separated into two main groups: analyzers based on ion beam transport such as magnetic field, time-of-flight (TOF), and quadruple mass filter; and those based on ion trapping technology (Tsao and Deng, 2004). These analyzers vary in their capabilities with respect to resolution, accuracy, and mass range. MS detector is highly useful for the identification of phytochemicals because of the complex and diverse structures, and low concentrations of these natural products in plants, foods, and biological systems. Sensitivity and selectivity of detection can be increased using tandem mass spectrometry, that is, two (MS-MS) or more (MS^n) mass analyzers coupled in series. MS-MS and MS^n produce more fragmentation of the precursor and daughter ions, therefore, provide additional structural information for the identification of antioxidant phytochemicals.

Flamini (2003) has summarized the use of LC-MS in studies of polyphenols in grape extracts and wine. The author specifically indicated that LC-MS techniques are the most effective tool in the study of the structure of anthocyanins. The MS/MS approach is a powerful tool that allows great anthocyanin aglycone and sugar moiety characterization. In the same review, other LC-MS techniques such as MALDI-TOF (time-of-flight) was also discussed by the author for the analysis of procyanidin oligomers. In addition, although both positive ion (PI) and negative ion (NI) modes are used for the detection of various phytochemicals, NI-MS methods, both APCI and ESI were found to be excellent for flavonoid analysis, in terms of sensitivity and in providing specific structural information (Pérez-Magariño et al., 1999). The same authors also showed that ESI was the method of choice for the analysis of low-molecular-mass phenols under NI mode, whereas flavan-3-ol compounds were well detected under both PI and NI modes. Negative LC-APCI-MS and low-energy collision induced dissociation (CID) MS-MS were used to provide molecular mass information and product-ion spectra of the other phenolic compounds (Li et al., 2005b). Detection of phytochemicals, particularly the use of the various LC-MS techniques, has been subjected to many recent reviews (Careri et al., 2002; Yang et al., 2009).

Biological fluids and tissues

The rapid increase of interest in the roles of phytochemicals in human health has led to great demands for good analytical methods for detecting the various biologically active compounds beyond the normal plant and food samples. The biological samples generated in animal and human clinical studies such as plasma, urine, and tissue samples, contain extremely low concentrations (high pM to low nM) of the original form of phytochemicals and their metabolites (Gardner *et al.*, 2009; Bolca *et al.*, 2010). Quantitative and qualitative analyses of these compounds in biological samples are therefore highly challenging, and are a new field which has not yet been comprehensively described. Such method will not only require sound techniques during sample collection, preparation and clean up processes, but more importantly the instrumental analysis, particularly using HPLC coupled with various detectors is key. Detectors such as ECD-LC and FLU are considered more sensitive than the UV/Vis or DAD detectors, but they are disadvantaged in only detecting certain groups of phytochemicals, for example, polyphenols. MS detector therefore serves as the best choice. Only a few reviews have been published on the bioanalysis topic, many of which are limited to particular groups of bioactive phytochemicals (Xing *et al.*, 2007; Vacek *et al.*, 2010). Nevertheless, some examples will be briefly discussed here. Using a C30 column, Rajendran *et al.*, (2005) were able to separate a total of 21 carotenoids, including all-*trans* forms of lutein, zeaxanthin, alpha-cryptoxanthin, beta-cryptoxanthin, alpha-carotene, beta-carotene, and lycopene, as well as their 14 *cis*-isomers in human serum samples. The method took place in 51 min at a flow rate of 1.0 mL/min and detection at 476 nm. Mullen *et al.* (2010) developed a HPLC-DAD-FLU-MS method for detection and quantification of 40 polyphenols found in fruit beverages, and identified 13 metabolites in plasma and 20 in urine samples of the subjects who consumed the drinks. The increased separation efficiency of UPLC has also made this technology more favorable for the analysis of phytochemicals and their metabolites in biological samples. Coupled with MS detection, it becomes a powerful tool in that not only is the sensitivity significantly improved, but the analytical time is shortened to one tenth of the conventional HPLC (Serra *et al.*, 2009; Zhang, 2010).

In summary, strategies for good analytical methods for the various phytochemicals in foods, beverages, or biological samples must include good sample preparation, separation, and detection techniques. Although many advanced technologies are available now, application of these technologies in analysis of a specific compound or a group of phytochemicals depends on the physicochemical properties of the analytes or their metabolites, sample matrix, extraction method, and the spectrophotometric or chromatographic techniques. Analysis of phytochemicals and their metabolites in biological samples is a new challenging field in food and nutrition research, and is worthy of special attentions.

References

Almela, L., Lopez-Roca, J.M., Candela, M.E., and Alcazar, M.D. (1990) Separation and determination of individual carotenoids in a Capsicum cultivar by normal-phase high-performance liquid chromatography. *Journal of Chromatography A*, 502, 95–106.

Aman, R., Bayha, S., Carle, R., and Schieber, A. (2004) Determination of carotenoid stereoisomers in commercial dietary supplements by high-performance liquid chromatography. *Journal of Agricultural and Food Chemistry*, 52, 6086–6090.

Andersen, Ø.M., Fossen, T., Torskangerpoll, K., Fossen, A., and Hauge, U. (2004) Anthocyanin from strawberry (*Fragaria ananassa*) with the novel aglycone, 5-carboxypyranopelargonidin. *Phytochemistry*, 65, 405–410.

AOAC Official Method 2005.02, (2005) Total monomeric anthocyanin pigment content of fruit juices, beverages, natural colorants, and wines-ph differential method. *J AOAC International*, 88, 1269–1278.

Aravind, S.G., Arimboor, R., Rangan, M., Madhavan, S.N., and Arumughan, C. (2008) Semi-preparative HPLC preparation and HPTLC quantification of tetrahydroamentoflavone as marker in S*emecarpus anacardium* and its polyherbal formulations. *J Pharm BiomeD Anal.*, 48, 808–813.

Bahorun, T., Luximon-Ramma, A., Crozier, A., and Aruoma, O.I. (2004) Total phenol, flavonoid, proanthocyanidin and vitamin C levels and antioxidant activities of Mauritian vegetables. *J Sci Food Agric.*, 84, 1553–1561.

Bayramoglu, B., Sahin, S., and Sumnu, G. (2008) Solvent-free microwave extraction of essential oil from oregano. *J Food Engineering.*, 88, 535–540.

Biehler, E., Mayer, F., Hoffmann, L., Krause, E., and Bohn, T. (2010) Comparison of 3 spectrophotometric methods for carotenoid determination in frequently consumed fruits and vegetables. *J Food Sci.*, 75, C55–C61.

Bolarinwa, A. and Linseisen, J. (2005) Validated application of a new high-performance liquid chromatographic method for the determination of selected flavonoids and phenolic acids in human plasma using electrochemical detection. *J Chromatogr B*, 823, 143–151.

Bolca, S., Urpi-Sarda, M., Blondeel, P., Roche, N., Vanhaecke, L., Possemiers, S., Al-Maharik, N., Botting, N., De Keukeleire, D., Bracke, M., Heyerick, A., Manach, C., and Depypere, H. (2010) Disposition of soy isoflavones in normal human breast tissue. *Am J Clin Nutr.*, 91, 976–984.

Breithaupt, D.E., Wirt, U., and Bamedi, A. (2002) Differentiation between lutein monoester regioisomers and detection of lutein diesters from marigold flowers (*Tagetes erecta* l.) and several fruits by liquid chromatography–mass spectrometry. *Journal of Agricultural and Food Chemistry*, 50, 66–70.

Britton, G. (1995) UV/Visible spectroscopy. In: G. Britton (ed.) *Carotenoids: Spectroscopy*, Basel, Birkhauser.

Burt, A.J., Grainger, C.M., Young, J.C., Shelp, B.J., and Lee, E.A. (2010) Impact of postharvest handling on carotenoid concentration and composition in high-carotenoid maize (Zea mays L.) kernels. *Journal of Agricultural and Food Chemistry*, 58, 8286–8292.

Campo Fernández, M., Cuesta-Rubio, O., Rosado Perez, A., Montes De Oca Porto, R., Márquez Hernández, I., Piccinelli, A.L., and Rastrelli, L. (2008) GC-MS determination of isoflavonoids in seven red Cuban propolis samples. *Journal of Agricultural and Food Chemistry*, 56, 9927–9932.

Careri, M., Bianchi, F., and Corradini, C. (2002) Recent advances in the application of mass spectrometry in food-related analysis. *Journal of Chromatography* A, 970, 3–64.

Chandrika, U.G., Basnayake, B.M., Athukorala, I., Colombagama, P.W., and Goonetilleke, A. (2010) Carotenoid content and in vitro bioaccessibility of lutein in some leafy vegetables popular in Sri Lanka. *J Nutr Sci Vitaminol.*, 56, 203–207.

Chang, C.C., Yang, M.H., Wen, H.M., and Chern, J.C. (2002) Estimation of Total Flavonoid Content in Propolis by Two Complementary Colorimetric Methods. *J Food Drug Anal.*, 10, 178–182.

Chang, C.J., Wu, S.M., and Yang, P.W. (2000) High-pressure carbon dioxide and co-solvent extractions of crude oils from plant materials. *Innov Food Sci Emerging Technol.*, 1, 187–191.

Chen, X.Q., Zan, K., Yang, J., Liu, X.X., Mao, Q., Zhang, L.Li,, Lai. M.X., and Wang, Q. (2011) Quantitative analysis of triterpenoids in different parts of Ilex hainanensis, Ilex stewardii and Ilex pubescens using HPLC–ELSD and HPLC–MS^n and antibacterial activity. *Food Chem.*, 126, 1454–1459.

de Rijke, E., Out, P., Niessen, W.M., Ariese, F., Gooijer, C., and Brinkman, U.A. (2006) Analytical separation and detection methods for flavonoids. *Journal of Chromatography* A, 1112, 31–63.

Dewanto, V., Wu, X., Adom, K.K., and Liu, R.H. (2002) Thermal processing enhances the nutritional value of tomatoes by increasing total antioxidant activity. *Journal of Agricultural and Food Chemistry*, 50, 3010–3014.

Escarpa, A. and Gonzalez, M.C. (2001) Approach to the content of total extractable phenolic compounds from different food samples by comparison of chromatographic and spectrophotometric methods. *Anal Chim Acta.*, 427, 119–127.

Farhat, A., Fabiano-Tixier, A.S., Visinoni, F., Romdhane, M., and Chemat, F. (2010) A surprising method for green extraction of essential oil from dry spices: Microwave dry-diffusion and gravity. *J Chromatogr.*, 1217, 7345–7350.

Flamini, R. (2003) Mass spectrometry in grape and wine chemistry. Part I: polyphenols. *Mass Spectrom Rev.*, 22, 218–250.

Gardner, C.D., Oelrich, B., Liu, J.P., Feldman, D., Franke, A.A., and Brooks, J.D. (2009) Prostatic soy isoflavone concentrations exceed serum levels after dietary supplementation. *Prostate*, 69, 719–726.

Gu, L., Kelm, M., Hammerstone, J.F., Beecher, G., Cunningham, D., Vannozzi, S., and Prior, R.L. (2002) Fractionation of polymeric procyanidins from lowbush blueberry and quantification of procyanidins in selected foods with an optimized normal-phase hplc–ms fluorescent detection method. *J Agric Food Chem.*, 50, 4852–4860.

He, X., Liu, D., and Liu, R.H. (2008) Sodium borohydride/chloranil-based assay for quantifying total flavonoids. *J Agric Food Chem.*, 56, 9337–9344.

Hiai, S., Oura, H., and Nakajima, T. (1976) Color reaction of some sapogenins and saponins with vanillin and sulfuric acid. *Planta Med*, 29, 116–122.

Hong, N., Yaylayan, V.A., Raghavan, G.S., Pare, J.R., and Belanger, J.M. (2001) Microwave-assisted extraction of phenolic compounds from grape seed. *Nat. Prod. Lett.*, 15, 197–204.

Hsu, B.Y., Inbaraj, B.S., and Chen, B.H. (2010) Analysis of soy isoflavones in foods and biological fluids: an overview. *Journal of Food and Drug Analysis*, 18, 141–154.

Hu, Y., Liang, H., Yuan, Q., and Hong, Y. (2010) Determination of glucosinolates in 19 Chinese medicinal plants with spectrophotometry and high-pressure liquid chromatography. *Nat Prod Res.*, 24, 1195–205.

Humphries, J. and Khachik, F. (2003) Distribution of lutein, zeaxanthin, and related geometrical isomers in fruit, vegetables, wheat, and pasta products. *Journal of Agricultural and Food Chemistry*, 51, 1322–1327.

Iafelice, G., Verardo, V., Marconi, E., and Caboni, M.F. (2009) Characterization of total, free and esterified phytosterols in tetraploid and hexaploid wheats. *Journal of Agricultural and Food Chemistry*, 57, 2267–2273.

Kenny, A.P. (1952) The Determination of Cholesterol by the Liebermann-Burchard Reaction. *Biochem J.*, 52, 611–619.

Kim, K.H., Tsao, R., Yang, R., and Cui, S.W. (2006) Phenolic acid profiles and antioxidant activities of wheat bran extracts and the effect of hydrolysis conditions. *Food Chem.*, 95, 466–473.

Kinniry, P., Amrani, Y., Vachani, A., Solomides, C.C., Arguiri, E., Workman, A., Carter, J., and Christofidou-Solomidou, M. (2006) Dietary flaxseed supplementation ameliorates inflammation and oxidative tissue damage in experimental models of acute lung injury in mice. *J Nutr.*, 136, 1545–1551.

Klejdus, B., Vacek, J., Adam, V., Zehnálek, J., Kizek, R., Trnková, L., and Kubán, V. (2004). Determination of isoflavones in soybean food and human urine using liquid chromatography with electrochemical detection. *J Chromatogr. B*, 806, 101–111.

Lesellier, E., Gurdale, K., and Tchapla, A. (1999) Separation of cis/trans isomers of β-carotene by supercritical fluid chromatography. *J Chromatogr. A*, 844, 307–320.

Li, L., Tsao, R., Dou, J., Song, F., Liu, Z., and sLiu, S. (2005a) Detection of saponins in extract of Panax notoginseng by liquid chromatography–electrospray ionisation-mass spectrometry. *Analytica Chimica Acta.*, 536, 21–28.

Li, L., Tsao, R., Yang, R., Liu, C., Zhu, H., and Young, J.C. (2006) Polyphenolic profiles and antioxidant activities of heartnut (Juglans ailanthifolia Var. cordiformis) and Persian walnut (Juglans regia L.). *Journal of Agricultural and Food Chemistry*, 54, 8033–8040.

Li, L., Tsao, R., Yang, R., Young, J.C., Zhu, H., Deng, Z., Xie, M., and Fu, Z. (2005b) Isolation and purification of acteoside and isoacteoside from Plantago psyllium L. by high-speed counter-current chromatography. *Journal of Chromatography* A, 1063, 161–169.

Lichtenthaler, H.K. (1987) Chlorophylls and carotenoids, the pigments of photosynthetic biomembranes. In: R. Douce and L. Packer (eds) *Method Enzymol*, New York, Academic Press Inc. pp. 350–382.

Liu, E.H., Qi, L.W., Cao, J., Li, P., Li, C.Y., and Peng, Y.B. (2008) *Advances of Modern Chromatographic and Electrophoretic Methods in Separation and Analysis of Flavonoids Molecules*, 13, 2521–2544.

Liu, Y.K., Jia, X.Y., Liu, X., and Zhang, Z.Q. (2010) On-line solid-phase extraction-HPLC-fluorescence detection for simultaneous determination of puerarin and daidzein in human serum. *Talanta*, 82, 1212–1217.

Liu, R.H. (2004) Potential synergy of phytochemicals in cancer prevention: mechanism of action. *Journal of Nutrition*, 134, 3479S–3485S.

Marston, A. and Hostettmann, K. (2009) Natural product analysis over the last decades. *Planta Med.*, 75, 672–682.

Marston, A. (2007) Role of advances in chromatographic techniques in phytochemistry. *Phytochem.*, 68, 2785–2797.

Marston, A. (2010) Thin-layer chromatography with biological detection in phytochemistry. *Journal of Chromatography A*, (in press), doi, 10.1016/j.chroma.2010.12.068.

McCallum, J.L., Yang, R. Young, J.C., Strommer, J.N., and Tsao, R. (2007) Improved high performance liquid chromatographic separation of anthocyanin compounds from grapes using a novel mixed-mode ion-exchange reversed-phase column. *Journal of Chromatography A*, 1148, 38–45.

Meda, A., Lamien, C.E., Romito, M., Millogo, J., and Nacoulma, O.G. (2005) Determination of the total phenolic, flavonoid and proline contents in burkina fasan honey, as well as their radical scavenging activity. *Food Chemistry*, 91, 571–577.

Muir, A.D. (2006) Flax lignans-analytical methods and how they influence our understanding of biological activity. *J AOAC Int.*, 89, 1147–1157.

Mukker, J.K., Kotlyarova, V., Singh, R.S., and Alcorn, J. (2010) HPLC method with fluorescence detection for the quantitative determination of flaxseed lignans. *J Chromatogr B Analyt Technol Biomed Life Sci.*, 878, 3076–3082.

Mullen, W., Borges, G., Lean, M.E., Roberts, S.A., and Crozier, A. (2010) Identification of metabolites in human plasma and urine after consumption of a polyphenol-rich juice drink. *Journal of Agricultural and Food Chemistry*, 58, 2586–2595.

Naim, M., Gestetner, B., Zilkah, S., Birk, Y., and Bondi, A. (1974) Soybean isoflavones. Characterization, determination, and antifungal activity. *Journal of Agricultural and Food Chemistry*, 22, 806–810.

Nuutila, A.M., Kammiovirta, K., and Oksman-Caldentey, K.M. 2002. Comparison of methods for the hydrolysis of flavonoids and phenolic acids from onion and spinach for HPLC analysis. *Food Chemistry*, 76, 519–525.

Okpuzor, J., Okochi, V.I., Ogbunugafor, H.A., Ogbonnia, S., Fagbayi, T., and Obidiegwu, C. (2009) Estimation of Cholesterol Level in Different Brands of Vegetable Oils. *Pakistan J Nutr.*, 8, 57–62.

Oleszek, W. and Bialy, Z. (2006) Chromatographic determination of plant saponins—An update (2002–2005). *Journal of Chromatography A*, 1112, 78–91.

Oleszek, W., Stochmal, A., and Janda, B. (2007) Concentration of Isoflavones and Other Phenolics in the Aerial Parts of Trifolium Species. *Journal of Agricultural and Food Chemistry*, 55, 8095–8100.

Oliver, J. and Palou, A. (2000) Chromatographic determination of carotenoids in foods. *Journal of Chromatography A*, 881, 543–555.

Palma, M., Piñeirom Z., and Barroso, C.G. (2001) Stability of phenolic compounds during extraction with superheated solvents. *Journal of Chromatography A*, 921, 169–174.

Pérez-Magariño, S., Revilla, I., González-SanJosé, M.L., and Beltrán, S. (1999) Various applications of liquid chromatography-mass spectrometry to the analysis of phenolic compounds. *Journal of Chromatography A*, 847, 75–81.

Piironen, V., Toivo, J., and Lampi, A.M. (2002) Plant sterols in cereals and cereal products. *Cereal Chemistry*, 79, 148-154.

Piñeiro, Z., Palma, M., and Barroso, C.G. (2004) Determination of catechins by means of extraction with pressurized liquids. *Journal of Chromatography A*, 1026, 19–23.

Pongsuwan, W., Bamba, T., Harada, K., Yonetani, T., Kobayashi, A., and Fukusaki, E. (2008) High-throughput technique for comprehensive analysis of Japanese green tea quality assessment using ultra-performance liquid chromatography with time-of-flight mass spectrometry (UPLC/TOF MS). *Journal of Agricultural and Food Chemistry*, 56, 10705–10708.

Poole, C.F. (2003) Thin-layer chromatography: challenges and opportunities. *Journal of Chromatography A*, 1000, 963–984.

Rajendran, V., Pu, Y.S., and Chen, B.H. (2005) An improved HPLC method for determination of carotenoids in human serum. *Journal of Chromatography B*, 824, 99–106.

Rosen, R.T., Hiserodt, R.D., Fukuda, E.K., Ruiz, R.J., Zhou, Z., Lech, J., Rosen, S.L., and Hartman, T.G. (2001) Determination of allicin, S-allylcysteine and volatile metabolites of garlic in breath, plasma or simulated gastric fluids. *Journal of Nutrition*, 131, 968S–971S.

Sakakibara, H., Honda, Y., Nakagawa, S., Ashida, H., and Kanazawa, K. (2003) Simultaneous determination of all polyphenols in vegetables, fruits and teas. *Journal of Agricultural and Food Chemistry*, 51, 571–581.

Schoefs, B. (2003) Chlorophyll and carotenoid analysis in food products. A practical case-by-case view. *Trends Anal Chem.*, 22, 335–339.

Serra, A., Macià, A., Romero, M.P., Salvadó, M.J., Bustos, M., Fernández-Larrea, J., and Motilva, M.J. (2009) Determination of procyanidins and their metabolites in plasma samples by improved liquid chromatography-tandem mass spectrometry. *J Chromatogr B Analyt Technol Biomed Life Sci.*, 877, 1169–1176.

Shaghaghi, M., Manzoori, J., and Jouyban, A. (2008) Determination of total phenols in tea infusions, tomato and apple juice by terbium sensitized fluorescence method as an alternative approach to the Folin–Ciocalteu spectrophotometric method. *Food Chemistry*, 108, 695–701.

Shao, S., Hernandez, M., Kramer, J.K.G., Rinker, D.L., and Tsao, R. (2)10. Ergosterol Profiles, Fatty Acid Composition, and Antioxidant Activities of Button Mushrooms as Affected by Tissue Part and Developmental Stage. *Journal of Agricultural and Food Chemistry*, 58, 11616–11625.

Shao, S. and Tsao, R. (2011) Systematic evaluation of pre-HPLC sample processing methods on total and individual isoflavones in soybeans and soy products. *Food Research International*, (in press).

Singleton, V.L., Orthofer, R., and Lamuela-Raventos, R.M. (1999) Analysis of total phenols and other oxidation substrates and antioxidants by means of Folin-Ciocalteu reagent. *Methods in Enzymology*, 299, 152–178.

Slinkard, K. and Singleton, V.L. (1977) Total phenol analysis: automation and comparison with manual methods. *American Journal of Enology and Viticulture*, 28, 49–55.

Stevanato, R., Fabris, S., and Momo, F. (2004) New enzymatic method for the determination of total phenolic content in tea and wine. *Journal of Agricultural and Food Chemistry*, 52, 6287–6293.

Sun, L., Rezaei, K.A., Temelli, F., and Ooraikul, B. (2002) Supercritical Fluid Extraction of Alkylamides from Echinacea angustifolia. *Journal of Agricultural and Food Chemistry*, 50, 3947–3953.

Tena, M.T., Valcarcel, M., Hidalgo, P.J., and Ubera, J.L. (1997) Supercritical fluid extraction of natural antioxidants from rosemary: Comparison with liquid solvent sonication. *Analytical Chemistry*, 69, 521–526.

Tsao, R. and Yang, R. (2006) Lutein in selected Canadian crops and agri-food processing by-products and purification by high-speed counter-current chromatography. *Journal of Chromatography A*, 1112, 202–208.

Tsao, R. and Deng, Z. (2004) Separation Procedures for naturally occurring antioxidant phytochemicals. *Journal of Chromatography B*, 812, 85–99.

Tsao, R., Yang, R., Young, J.C., Zhu, H., and Manolis, T. (2004) Separation of geometric isomers of native lutein diesters in marigold (tagetes erecta l.) by high performance liquid chromatography-mass spectrometry. *Journal of Chromatography A*, 1045, 65–70.

Tsao, R. and Yang, R. (2003) Optimisation of a new mobile phase to know the complex and real polyphenolic composition: towards a total phenolic index using HPLC. *Journal of Chromatography A*, 1018, 29–40.

Tsao, R. (2010) Chemistry and Biochemistry of Dietary Polyphenols. *Nutrients*, 2, 1231–1246.

Tsao, R. and Akhtar, H. (2005) Nutraceuticals and functional foods: I. current trend in phytochemical antioxidants research. *Journal of Food, Agriculture and Environment*, 3, 10–17.

Vacek, J., Ulrichová, J., Klejdus, B., and Šimánek, V. (2010) Analytical methods and strategies in the study of plant polyphenolics in clinical samples. *Analytical Methods*, 2, 604–613.

Valls, J., Millán, S., Martí, M.P., Borràs, E., and Arola, L. (2009) Advanced separation methods of food anthocyanins, isoflavones and flavanols. *Journal of Chromatography A*, 1216, 7143–7172.

Waksmundzka-Hajnos, M., Sherma, J., and Kowalska, T. (eds) (2008) *Thin Layer Chromatography in Phytochemistry*, New York, CRC Press Taylor Francis.

Wang, Z., Ding, L., Li, T., Zhou, X., Wang, L., Zhang, H., Liu, L., Li, Y., Liu, Z., Wang, H., Zeng, H., and He, H. (2006) Improved solvent-free microwave extraction of essential oil from dried Cuminum cyminum L. and Zanthoxylum bungeanum Maxim. *Journal of Chromatography A*, 1102, 11–17.

Wang, S., Meckling, K., Marcone, M., Kakuda, Y., and Tsao, R. (2010) Synergistic, additive and antagonistic effects of food mixtures on total antioxidant capacities. *Journal of Agricultural and Food Chemistry*, (in press).

White, B.L., Howard, L.R., and Prior, R.L. (2010) Release of bound procyanidins from cranberry pomace by alkaline hydrolysis. *Journal of Agricultural and Food Chemistry*, 58, 7572–7579.

Xing, J., Xie, C., and Lou, H. (2007) Recent applications of liquid chromatography-mass spectrometry in natural products bioanalysis. *J Pharm Biomed Anal.*, 44, 368–78.

Yanagida A, Shoji T, Shibusawa Y. 2003. Separation of proanthocyanidins by degree of polymerization by means of size-exclusion chromatography and related techniques. *Journal of Biochemical and Biophysical Methods*, 56, 311–322.

Yang, M., Sun, J., Lu, Z., Chen, G., Guan, S., Liu, X., Jiang, B., Ye, M., and Guo, D.A. (2009) Phytochemical analysis of traditional Chinese medicine using liquid chromatography coupled with mass spectrometry. *Journal of Chromatography A*, 1216, 2045–2062.

Zhang, W., Xu, M., Yu, C., Zhang, G., and Tang, X. (2010) Simultaneous determination of vitexin-4″-O-glucoside, vitexin-2″-O-rhamnoside, rutin and vitexin from hawthorn leaves flavonoids in rat plasma by UPLC-ESI-MS/MS. *Journal of Chromatography B*, 878, 1837–1844.

Zarrouk, W., Carrasco-Pancorbo, A., Segura-Carretero, A., Fernández-Gutiérrez, A., and Zarrouk, M. (2010) Exploratory characterization of the unsaponifiable fraction of tunisian virgin olive oils by a global approach with HPLC-APCI-IT MS/MS analysis. *Journal of Agricultural and Food Chemistry*, 58, 6418–6426.

20 Antioxidant activity of phytochemicals

Ankit Patras,[1] Yvonne V. Yuan,[2] Helena Soares Costa[3] and Ana Sanches-Silva[3]

[1] Department of Food Science, University of Guelph, Guelph, Ontario, Canada
[2] School of Nutrition, Ryerson University, Toronto, Ontario, Canada
[3] National Institute of Health Dr Ricardo Jorge, Food and Nutrition Department, Lisbon, Portugal

20.1 Introduction

Phytochemicals in fruits, vegetables and cereals have attracted a great deal of attention mainly concentrated on their role in preventing diseases caused as a result of oxidative stress. Oxidative stress, which releases free oxygen radicals in the body, has been implicated in a number of disorders including cardiovascular malfunction, cataracts, cancers, rheumatism and many other auto-immune diseases besides ageing.

These phytochemicals act as antioxidants, scavenge free radicals and may inhibit cell death or apoptosis. Epidemiological studies have shown that there may be significant positive associations between intake of fruits and vegetables or cereals and reduced rate of heart disease mortality, common cancers and other degenerative diseases as well as ageing (Steinmetz and Potter, 1996; Joseph et al., 1999; Dillard and German, 2000; Prior and Cao, 2000). The most thoroughly investigated dietary components in fruits, vegetables, cereals or legumes acting as antioxidants are fibre, carotenoids, polyphenols, flavonoids, conjugated isomers of linoleic acid, epigallocatechin, gallate, soya protein, isoflavanones, vitamins A, B, C, E, tocopherols, calcium, selenium, chlorophyl, alipharin, sulphides, catechin, tetrahydrocurecumin, sesaminol, lignans, glutathione, uric acid, indoles, thiocyanates and protease inhibitors (Karakaya and Kavas, 1999). These compounds may act independently or in combination as anti-cancer or cardio-protective agents by a variety of mechanisms. The available scientific data indicates a protective role for fruits and vegetables against certain cancers including those of the pancreas, bladder and breast (American Institute of Cancer Research, 1997). This is attributed to the fact that these foods may provide an optimal mix of phytochemicals such as natural antioxidants, fibres and other bioactive compounds. In contrast, a recent report by the European Food Safety Authority (EFSA, 2010) has issued negative opinions on the actions of antioxidants in human health. The EFSA panel documented that the claimed effects refer to the protection of body cells and molecules (such as DNA, proteins and lipids) from oxidative damage, including UV-induced oxidative damage. The panel considered that the protection of molecules such as DNA, proteins and lipids from oxidative damage may be a beneficial physiological effect (EFSA, 2010). No human studies investigating the effects of the food(s)/food constituent(s) on reliable markers of oxidative damage to body cells or to molecules such as DNA, proteins and lipids have been provided in relation to any of the health claims evaluated in this opinion (EFSA, 2010). Nevertheless, we believe it is too early to make strong judgements about antioxidants and their biological properties.

The concept of antioxidant activity of unprocessed and processed foods is gaining significant momentum and emerging as an important parameter to assess the quality of the product. With the expansion of the global market and fierce competition amongst multinational companies, the parameter of antioxidant activity will soon secure its place in nutritional labelling with accompanying regulatory guidelines. In this context development of a practical method of determining the antioxidant activity for industrial use will become imperative. This will give a further boost to the exploitation of fruits and vegetables and development of nutraceuticals and beverages. Particular focus will be given to mechanisms and measurement of antioxidant activity by different assays.

20.2 Measurement of antioxidant activity

According to its definition, an antioxidant should have a significantly lower concentration than the substrate in the antioxidant activity test. Depending on the type of reactive oxygen species (ROS) and target substrate, a certain antioxidant may play a completely different action or have a completely different role/performance. In line with this, some authors support the use of a selection of methods to measure the antioxidant activity (Yuan *et al.*, 2005a, 2005b).

The choice of substrate is very important in an antioxidant activity test. Depending on the type of substrate and its amount/concentration, different results will be achieved. The application of tests in both aqueous and lypophilic phase systems has also been described as important, in order to study the relative bioactivity of an antioxidant. Because certain stable free radical methods (e.g. ABTS•+, DPPH etc.) generally do not include a substrate other than the stable free radical in question, they are considered artificial because they do not represent the real process in food samples (discussed in section 20.2.1).

The method generally used to determine total antioxidant activity is the Trolox equivalent antioxidant capacity (TEAC) assay, although total oxyradical scavenging capacity (TOSC) assay, oxygen radical absorbance capacity (ORAC) assay, 1,1-diphenyl-2-picrylhydrazyl (DPPH) assay or ferric ion reducing antioxidant parameter (FRAP) assay can also be used.

20.2.1 Assays involving a biological substrate

Assays involving a biological substrate have the advantage of being closer to an *in vivo* situation, where both aqueous and a lipid phase are present and take into account the solubilities and partitioning between different phases. One of these assays measures the inhibition of ascorbate/iron induced lipid peroxidation of cell or liver microsomes (Plumb *et al.*, 1996; Lana and Tijskens, 2006). Other assays that employ biological substrates include the inhibition of human LDL oxidation (Heinonnen *et al.*, 1998a; Meyer *et al.*, 1997, 1998) and the lecithin-liposome oxidation assay (Heinonnen *et al.*, 1998a, 1998b; Frankel and Huang, 1997), both catalysed by copper. These models are important because LDL oxidation is related to coronary disease and liposome oxidation to food oxidation.

20.2.2 Assays involving a non-biological substrate

20.2.2.1 Electron and hydrogen transfer assays

Assays for measurement of antioxidant activity may involve hydrogen atom transfer (HAT) or single electron transfer (SET). These two mechanisms generally occur simultaneously and the prevalence of one of them depends on the structure of the antioxidant and pH. The mechanism and antioxidant efficiency are mainly determined by two factors: the bond dissociation energy (BDE) and the ionisation potential (IP) (Prior et al., 2005; Karadag et al., 2009). HAT methods measure the capacity of an antioxidant (AH, a hydrogen donor) to quench free radicals by hydrogen donation.

$$X^\bullet + AH \rightarrow XH + A^\bullet$$

In HAT based assays, the reactivity is determined by the BDE of the H donating group of the antioxidant and it is higher for compounds with $\Delta BDE \approx 10\,kcal/mol$ and $\Delta IP \leq 36\,kcal/mol$ (Prior et al., 2005).

HAT assays depend on the solvent, pH and are affected by the presence of reducing agents such as metals. HAT reactions are generally quite fast and quantitation is derived from the kinetic curves (Karadag et al., 2009). HAT assays include the oxygen radical absorbance capacity (ORAC), the total peroxyl radical-trapping antioxidant parameter assay (TRAP) and the crocin-bleaching assay.

SET methods measure the capacity of a potential antioxidant to transfer one electron to reduce a compound:

$$X^\bullet + AH \rightarrow X^- + AH^{\bullet+}$$
$$H_2O$$
$$AH^{\bullet+} \leftrightarrow A^\bullet + H_3O^+$$
$$X^- + H_3O^+ \rightarrow XH + H_2O$$
$$M(III) + AH \rightarrow AH^+ + M(II)$$

In the SET based assays, the reactivity is determined by the deprotonation and IP of the functional group. These assays are pH dependent (MacDonald-Wicks et al., 2006). The higher the pH, the lower IP values are and deprotonation increases. In compounds with $\Delta IP \geq 45\,kcal/mol$, the major reaction mechanism is SET (Prior et al., 2005).

SET reactions are usually slow and negatively affected by trace components and contaminants, especially metals (Prior et al., 2005). Generally, these reactions measure the relative percent decrease in product instead of kinetics or total antioxidant capacity (Karadag et al., 2009).

SET assays include the ferric ion reducing antioxidant power (FRAP) and the copper reduction capacity assay. Trolox equivalent antioxidant capacity (TEAC) and 2,2-diphenyl-1-picrylhydrazyl (DPPH) assays are usually classified as SET but both mechanisms may be used (Prior et al., 2005). HAT and SET are competitive reactions but it has been demonstrated that HAT is dominant in biological redox reactions (Karadag et al., 2009).

20.2.2.2 Reduction of the Fremy's radical

The Fremy's radical assay is an indirect method to determine 'chain-breaking antioxidant activity' in food, and is based on the capability of the Fremy's stable free radical to react with H-donors. The Fremy's radical (potassium nitrosodisulfonate) is a specific oxidising salt which converts phenols into quinines (Zimmer et al., 1970). The concentration of the Fremy's radical is monitored by ESR (electron spin resonance) spectroscopy. A low signal indicates the detection of low amounts of radicals and therefore an antioxidant and dominating pro-oxidant effect of the extracts (Summa et al., 2007).

In particular, the method was applied to wine (Burns et al., 2001), extracts made from cherry liqueur pomace (Rødtjer et al., 2006), fruit juices (Gardner et al., 2000), coffee (Summa et al., 2007) and Scotch whiskeys (MacPhail et al., 1999). Gardner et al. (2000) has pointed out the advantages of this assay: it is very sensitive, allowing detection at a sub-micromolar level; analysis can be carried out on turbid or highly coloured solutions and radicals have well-defined spectra, allowing clear resolution from radical intermediates which may be formed during the oxidation process.

20.2.2.3 Copper (II) reduction capacity

This method is a variant of the FRAP assay, using copper (Cu) instead of iron (Fe). It is based on the reduction of Cu (II) to Cu (I) by the action of the reductants (antioxidants) present in a sample (Prior et al., 2005; Huang et al., 2005). This method, however, has not been broadly used (MacDonald-Wicks et al., 2006).

20.2.3 Ferrous oxidation–xylenol orange (FOX) assay

The FOX assay measures the hydroperoxides (ROOHs), which are the initial stable products formed during peroxidation of unsaturated lipids such as fatty acids and cholesterol (Nourooz-Zadeh, 1999). The assay is based on the oxidation of ferrous (Fe^{2+}) to ferric (Fe^{3+}) ions by ROOHs under acidic conditions. FOX is a precise and simple method, but the amount of extract and the incubation time have to be adapted for each sample (Grau et al., 2000). FOX was also reported as being highly specific (Grau et al., 2000). Moreover, the FOX method is sensitive (measures concentrations of $5\,\mu M$ LOOH), inexpensive, rapid, not sensitive to ambient oxygen or light levels and it does not require special reaction conditions (DeLong et al., 2002). FOX assay has been applied, for instance, to lipoprotein and lipossomes (Jiang et al., 1991; Jiang et al., 1992), plasma (Nourooz-Zaheh et al., 1994; Nourooz-Zaheh et al., 1995), vegetable oils (Nourooz-Zaheh et al., 1995), soybean oils (Yildiz et al., 2003), plant tissue (DeLong et al., 2002), fried snacks (Navas et al., 2004) and dark chicken meat (Grau et al., 2000).

20.2.4 Ferric thiocyanate (FTC) assay

The ferric thiocyanate method determines the amount of peroxide at the initial stage of lipid peroxidation. The peroxide reacts with ferrous chloride ($FeCl_2$) to give a ferric chloride dye which has a red colour. Recently, some studies have used this technique to evaluate the antioxidant activity in different matrices such as citrus by-products (Senevirathne et al., 2009); gingers (Ruslay et al., 2007); Malay traditional vegetables (Abas et al., 2006); rosemary extract, blackseed essential oil, carnosic acid, rosmarinic acid and sesamol (Erkan et al., 2008); an edible seaweed (*Kappaphycus alvarezzi*) (Kumar et al., 2008); sugar cane bagasse (Ou et al., 2009); hazelnut skin (Locatelli et al., 2010); wines, grape juices (Sanchez-Moreno et al., 1999), extracts from *Platycodon grandiflorum* A. De Condolle roots (plants used both as a herbal medicine and food in Asia) (Lee et al., 2004) and sweet potatoes (Huang et al., 2006).

20.2.5 Hydroxyl radical scavenging deoxyribose assay

The deoxyribose assay for detection of hydroxyl radical (•OH) scavenging activity described by Halliwell and co-workers (1987) was designed as a relatively simple and cheap spectrophotometric alternative to pulse radiolysis for the determination of the rate constants of •OH scavenging compounds reacting with hydroxyl radicals. The assay relies on the generation of •OH via the Fenton reaction which reacts with deoxyribose under neutral pH conditions, followed by degradation of the sugar molecule and the formation of malondialdehyde (MDA) with heating under acidic conditions. The MDA will yield a pink chromogen upon heating with 2-thiobarbituric acid which can then be detected at 532 nm. The rate of deoxyribose degradation in this assay is enhanced by the inclusion of ascorbic acid which reduces ferric to ferrous ions to facilitate the Fenton reaction. Antioxidant molecules which scavenge •OH will compete with deoxyribose and thereby decrease the final amount of the pink chromogen formed. Interestingly, EDTA itself is an •OH scavenger in this assay (Halliwell et al., 1987), therefore only those •OH which are not scavenged by EDTA go on to degrade the deoxyribose. It is important to note that the deoxyribose assay for •OH scavenging activity is sensitive to contamination by transition metal ions, resulting in high 'blank' or 'control' values, approximately $A532=0.2$ to 0.3 (Aruoma et al., 1987). Common sources of iron contamination may include the phosphate buffer, or other reagents (deoxyribose, ascorbate; Aruoma et al., 1987), thus reagents and water used in analyses must be treated with Chelex resin or otherwise deionised.

The deoxyribose assay can only be used to evaluate the •OH scavenging activity of polar antioxidants (Aruoma, 1994). Aqueous ethanol can be used to solubilise antioxidants under study as needed, keeping in mind that ethanol is an •OH scavenger itself (Halliwell et al., 1987). For example, Yuan and co-workers (2005a) used 0.1% ethanol to solubilise *Palmaria palmata* extracts prior to assessing the •OH scavenging activities of these marine red algal samples; these workers corrected for the ethanol antioxidant activity by using an appropriate solvent 'control'.

20.2.6 1,1-diphenyl-2-picrylhydrazyl (DPPH•) stable free radical scavenging assay

The 1,1-diphenyl-2-picrylhydrazyl (DPPH•) – which is also known as α,α-diphenyl-β-picrylhydrazyl, 2,2-diphenyl-1-picrylhydrazyl or 2, 2-Diphenyl-1-(2,4,6-trinitrophenyl) hydrazyl) – assay originally described by Blois (1958) was designed to take advantage of a common electron spin resonance reagent, a stable free radical with an odd, unpaired valence electron to study antioxidant activity. With its odd electron, DPPH• can be stabilised by accepting an electron or hydrogen radical from an antioxidant molecule such as a sulfhydryl group (Blois, 1958); ascorbic acid as a reducing agent; polyphenols; or more generically by an antioxidant (AH) or free radical (R•; Brand-Williams et al., 1995).

DPPH• is known for its deep violet colour and strong absorbance at 517 nm when dissolved in ethanol at concentrations between 1 mM and 22.5 µM (Blois, 1958; Yen and Chen, 1995; Sharma and Bhat, 2009); this absorbance is decreased with the decolourisation of DPPH• which accompanies the pairing of the lone electron. The A517 of DPPH• is stable between pH 5 and 6.5, but is sensitive to highly alkaline conditions which can be buffered by acetate (Blois, 1958; Sharma and Bhat, 2009). The wide range of DPPH• concentrations used in the literature is no doubt related to the limited solubility of this stable free radical. Moreover, studies using DPPH• have varied widely not only in the solvent used to dissolve the stable free radical, but also the wavelength used to monitor the decolourisation of the stable free radical (Blois, 1958; Brand-Williams et al., 1995; Kitts et al., 2000; Sharma and Bhat, 2009; Shimada et al., 1992; Yan et al., 1998; Yen and Chen, 1995; Yuan et al., 2005a). When Sharma and Bhat (2009) measured the A517 of DPPH• over a wide range of concentrations (approximately 10–250 µM) using different solvent systems, the A517 varied as follows: 60% methanol was slightly > methanol > ethanol, thus, 517 nm may not have been the optimal wavelength to monitor the decolourisation of DPPH• in differing solvents. The peak intensity absorption characteristics of chromophores can be observed to be influenced by not only the structure of the molecule, but also solvent and vibrational effects. The choice of solvent may also have been influenced by the solubility of the antioxidant compounds or extracts under evaluation, since the DPPH• stable free radical scavenging methodology can be used to study both polar and non-polar antioxidants such as ascorbic acid and butylated hydroxyanisole (BHA) or butylated hydroxytoluene (BHT), respectively. On the other hand, many studies of antioxidant molecules or plant extracts with the potential to be functional foods or nutraceuticals will use ethanol as the extraction medium and/or solvent as opposed to the notably toxic methanol. The impact of the choice of solvent is clearly demonstrated with BHT, which exhibited an EC_{50} of 60 µM when methanol was the solvent, but 9.7 µM when the solvent was 60% methanol (Sharma and Bhat, 2009), thus the solubility of the antioxidant molecules in the chosen solvent system plays an important role in these studies.

Perhaps the variable of most interest and debate in attempting to compare and reconcile data from different laboratories and studies from the literature in the evaluation of antioxidant efficacy is the quantitation or expression of DPPH• stable free radical scavenging activity. The length of time that a sample is incubated with DPPH• and monitored by spectrophotometer is highly variable in the literature despite the in depth discussion of the importance of antioxidant kinetic behaviour by Brand-Williams and co-workers (1995). These workers identified three ranges of kinetic behaviour: rapid kinetics exhibited by ascorbic acid which reacts very quickly with DPPH• reaching a steady state plateau in 1 min. or less; intermediate kinetics exhibited by α-tocopherol which reached a steady state plateau between 5 and 30 min; and slow kinetics exhibited by a diversity of antioxidants including phenolic acids and complex marine algae/seaweed-derived extracts which only reached a steady state plateau after 1–6 h incubation (Yuan et al., 2005a).

On the other hand, many investigators have chosen a single time point to quantify the DPPH• stable free radical scavenging efficacy of the antioxidants under study, with the most common choice as 30 min (varies from 20 to 60 or 90 min). However, unless the antioxidants are screened and identified as having rapid or intermediate kinetics, the stable free radical scavenging activity can be underestimated (Brand-Williams et al., 1995; Sharma and Bhat, 2009). For example, Brand-Williams and co-workers (1995) observed EC_{50} values for BHT of 0.943 mol/L after 30 min incubation, but 0.189 mol/L after 240 min, a five-fold difference in antioxidant efficacy. Best practices likely reflect monitoring the decrease in DPPH• absorbance until a steady state plateau is reached, particularly since the majority of antioxidants appear to exhibit intermediate or slow reaction kinetics.

20.2.7 Azo dyes as sources of stable free radicals in antioxidant assays

Azo compounds, or dyes, are distinguished by containing an azo group –N=N- within their structure and comprise a large class of synthetic organic dyes such as Congo red and Tartrazine; approximately 60–70% of dyes used in the food and textile industries are azo dyes. Azo compounds are also regularly used as free radical initiators in the study of antioxidant compounds, and particularly the quantitation of lipid peroxidation *in vitro* and *vivo*, due to the predictable thermal decomposition of these compounds to yield N_2 and two carbon radicals, R• (Niki, 1990). These radicals may then either react with each other to yield a stable non-radical end product (R-R), or react with molecular O_2 to yield peroxyl radicals, ROO• which can then participate in the peroxidation of a polyunsaturated lipid emulsion model system. The structure or composition of R will determine not only the solubility of the azo compound, but also the kinetics of the decomposition. There have been two commonly used hydrophilic radical initiators in the recent literature: 2, 2′-azo-bis-(2-amidipropropane hydrochloride) (ABAP) or 2,2′-azo-bis(2-amidinopropane) dihydrochloride (AAPH). ABAP and AAPH are the same chemical, just differing with one HCl moiety present in the former and two HCl moieties in the latter. Due to its polarity, AAPH generates its radicals in the aqueous region of an oil-in-water emulsion used in studying lipid peroxidation; whereas, 2, 2′-azobis(2,4-dimethylvaleronitrile) (AMVN) is a lipophilic radical initiator which generates its radicals within the lipid regions of emulsion, micelle droplets or membranes (Niki, 1990; Noguchi *et al.*, 1998; Yuan *et al.*, 2005b). More recently, Noguchi and co-workers (1998) described a novel lipophilic azo free radical initiator, 2,2′-azobis (4-methoxy-2, 4-dimethylvaleronitrile) (MeO-AMVN). The decomposition of azo radical initiators is a function of mainly temperature, and to a lesser extent solvent and pH (Niki, 1990). Therefore, the rate of generation of radicals would be constant over the short term in the case of an accelerated lipid oxidation model using a free radical initiator and elevated temperature of incubation (Ng *et al.*, 2000; Zhang and Omaye, 2001; Yuan *et al.*, 2005b; Hu *et al.*, 2007).

The solubility of the azo free radical initiator used is very important with respect to the polarity of the antioxidant(s) under study as well as the composition of the model system (Niki, 1990; Yuan *et al.*, 2005b). For example, Niki (1990) discussed that if AAPH is used to generate free radicals in a liposome system with a lipophilic antioxidant such as α-tocopherol, care should be taken to sonicate multilamellar liposomes to yield unilamellar liposomes to facilitate the interaction of AAPH radicals with the antioxidant, which otherwise would not be possible if the antioxidant was located within the inner membranes of a multilamellar system. Similarly, Yuan and co-workers (2005b) reported a protective effect of red algal *Palmaria palmata* (dulse) extracts on lipid peroxidation in linoleic acid emulsions when AAPH was the free radical initiator, but the absence of a protective effect in the presence of AMVN; the lack of a protective effect of dulse extracts could be associated with the localisation of the free radicals within the lipid phase of the emulsion not in contact with the aqueous dulse extract constituents. Azo compounds are desirable for *in vivo* studies of lipid peroxidation or oxidative stress due to the spontaneous and known breakdown of these compounds under physiological conditions (Niki, 1990). Moreover, because of the ability of azo compounds to generate peroxyl radicals at a known and constant rate, they can also be used in other model systems to good effect, including as a free radical initiator in the scission of supercoiled plasmid pBR322 DNA (Hu *et al.*, 2007) or in the oxygen radical absorbance capacity (ORAC) assay, to be discussed in section 20.2.8.

20.2.8 Oxygen radical absorbance capacity (ORAC) assay

When originally developed, the ORAC assay was designed to evaluate the protective effect of antioxidant compounds against reactive oxygen species-mediated damage to the fluorescent indicator R- or β-phycoerythrin (PE); free radicals were derived from

AAPH or •OH in the presence of Cu1+/2+ and ascorbic acid (Dávalos, Gómez-Cordovés and Bartolomé, 2004). However, despite the linear, zero order kinetics exhibited by PE in the ORAC assay, the natural variability of this reagent and its instability to photobleaching (necessitating making a fresh PE solution daily) and interaction with polyphenols lead researchers to look for an alternate fluorescent indicator. Fluorescein was subsequently demonstrated to not only have excellent photostability within assay conditions, but also not to have any interactions with antioxidant molecules, such as polyphenols (Dávalos *et al.*, 2004). The ORAC assay has gained great acceptance amongst the food science, functional food and nutraceutical research community, and indeed by marketers of such foods, due to its utility in analysing multiple samples quickly using 96-well microplates and the potential for automated (robotic) reagent handling (Dávalos *et al.*, 2004). The ORAC assay is based on quantitation of antioxidant activity from the area under the curve calculated from the decay in fluorescence intensity when fluorescein is degraded by AAPH-derived peroxy radicals (Dávalos *et al.*, 2004). Thus, one of the strengths of the ORAC methodology is that the antioxidant efficacy of compounds is monitored until exhaustion and the fluorescence returns to the baseline. The ORAC methodology was modified from the original to analyse both hydrophilic (H-ORACFL) and hydrophobic antioxidant molecules (L-ORACFL; Wu *et al.*, 2004a, 2004b), with the total ORACFL activity of a food represented by the sum of the two. The H-ORACFL values were roughly ten-fold that of the corresponding L-ORACFL values as expected for fresh and dried fruits, vegetables, nuts, spices, cereals and infant foods.

20.2.9 Total radical-trapping antioxidant parameter (TRAP) assay

The original total peroxyl radical trapping antioxidant parameter (TRAP) methodology described by Wayner and co-workers (1985) was designed to assess the capacity of plasma antioxidant constituents to quench or trap azo compound-derived peroxy radicals from the thermal decomposition of ABAP or AAPH as already discussed. TRAP assay conditions comprised monitoring the uptake of O_2, using an oxygen electrode, by a test sample incubated at 37°C (Wayner *et al.*, 1985). Subsequent modifications to the TRAP assay methodology favoured the use of fluorescent indicators and monitoring the degradation of these molecules, such as PE (Ghiselli *et al.*, 1995), as in the ORAC assay in section 20.2.8. However, as described, the inherent variability of this naturally occurring pigment and its lack of photostability led to its replacement with other indicators such as the nonfluorescent 2,7-dichlorofluorescin-diacetate (DCFH-DA; Valkonen and Kuusi, 1997). In contrast to the ORAC assay, monitoring DCF fluorescence results in an initial lag phase whilst the endogenous antioxidants are depleted, followed by a rapid increase in fluorescence, representing a propagation phase. There is a second lag phase attributed to the effects of the addition of Trolox as an internal standard, and another propagation phase after the Trolox has been depleted. Valkonen and Kuusi (1997) reported lag phases, = 15 min, = 19.5 min and TRAP values of 1292 µM for fresh plasma, whereas after storage at −80°C, 2 mo., the corresponding values were 16.5 min., 23 min. and 1205 µM, respectively.

Alho and Leinonen (1999) modified the TRAP methodology to incorporate chemiluminescence to measure the antioxidant capacities of human plasma and cerebrospinal fluid (CSF). The methodology was further modified to use the lipophilic azo compound AMVN to determine the TRAP activity of low density lipoprotein (LDL) as a measure of LDL oxidisability an indicator of the atherogenicity of these particles (Ahlo and Leinonen, 1999;

Malminiemi et al., 2000). Thus, the TRAP assay has value as a clinical measure of overall antioxidant capacity of biological fluids, but has also been adapted for use in functional food and nutraceutical research.

20.2.10 ABTS•+radical cation scavenging activity

The radical cation form of 2,2′-azino-bis-(3-ethyl-benzthiazoline-6-sulfonic acid) (ABTS) is generated by oxidising ABTS with potassium persulfate to form ABTS•+, a blue-green chromophore with absorbance maxima at 415, 645, 734 and 815 nm (Pellegrini et al., 1999; Re et al., 1999). Similar to the DPPH• stable free radical scavenger in section assay 20.2.6, the evaluation of potential antioxidant activity using ABTS•+involves the decolourisation of the preformed cation radical by the antioxidant molecule donating an electron or hydrogen atom. Free radical scavenging by hydrophilic or lipophilic antioxidants is measured by monitoring the A734 until a steady state plateau is achieved, or by using a single time point measure (Hu and Kitts, 2000). Sample antioxidants can be tested for ABTS•+radical cation scavenging efficacy including phenolics, flavonoids and hydroxycinnamates solubilised in ethanol; anthocyanidins in acidic ethanol, pH 1.3; carotenoids (lycopene and β-carotene) dissolved in dichloromethane; α-tocopherol in ethanol and plasma antioxidants diluted with water (Pellegrini et al., 1999; Re et al., 1999).

Interestingly, the ABTS•+radical cation scavenging EC50 values for L-ascorbic acid, BHA and the red marine alga *Palmaria palmata* observed by Yuan and co-workers (2005a) were relatively similar to the corresponding results obtained for these antioxidants in the DPPH• free radical scavenger assay for both kinetics (rapid versus slow) as well as antioxidant efficacy. For example, the % inhibition of A734 values for L-ascorbic acid reached a steady state plateau within 1–2 min when incubated with ABTS•+, and similar kinetics were observed for % DPPH• quenching; the % inhibition of A734 values for BHA achieved a steady state plateau after 30–50 min with ABTS•+, and similarly with % DPPH• quenching; whereas the % inhibition of A734 values for the marine red alga *Palmaria palmata* extract never reached a steady state plateau even after 180 min incubation with ABTS•+, but did reach a steady state plateau after 50–60 min. for % DPPH• quenching. These differences in kinetic behaviour of antioxidant compounds in the DPPH• and ABTS•+free radical scavenging assay systems are thought to be related to the reaction stoichiometry of the number of electrons available to inactivate the free radicals (Koleva et al., 2002). Slow reacting compounds such as BHT, or the closely related BHA, and P. palmaria extracts herein are hypothesised to have a more complex reaction mechanism involving one or more secondary reactions in the quenching of the DPPH• (Koleva et al., 2002) and thereby also, ABTS•+free radicals.

20.2.11 Ferric reducing ability of plasma (FRAP) assay

The ferric reducing ability of plasma (FRAP) assay was designed as a simple, inexpensive method to quantify the collective non-enzymatic antioxidant capacity of biological fluids such as plasma, saliva, tears, urine and cerebrospinal fluid (Benzie and Strain, 1996, 1999). It was proposed that an assay such as this could evaluate the combined effect of plasma antioxidant constituents and thus directly measure the 'total antioxidant power' of a complex mixture with potential synergistic effects which would not be evident when assayed as single components (Benzie and Strain, 1999). The FRAP assay is based on the single electron transfer by an antioxidant to reduce the ferric to ferrous ion; when the ferric-tripyridyltriazine (Fe^{3+}-TPTZ) complex is reduced to the ferrous counterpart, the complex absorbs at 593 nm with an intense blue colour. The time course of the assay will vary depending on the kinetics of the sample antioxidant being evaluated; for example, L-ascorbic acid and α-tocopherol exhibited rapid kinetics with a steady state plateau reached within 1 min, uric acid reached a steady state plateau after 3 min, whereas, bilirubin did not reach a steady state plateau after 8 min (Benzie and Strain, 1996). The importance of assay temperature is demonstrated in particular with uric acid which exhibits slower reaction kinetics at room temperature versus 37°C (Benzie and Strain, 1999).

A modification to the FRAP assay, named the FRASC assay, allows the dual measurements of ascorbic acid content and FRAP activity in one test system. For the FRASC assay, one sample aliquot is treated with ascorbate oxidase while its pair is left untreated (Benzie and Strain, 1997). A major criticism of the FRAP assay is the lack of physiological relevance of the reaction conditions at pH 3.6; however, the assay does provide a means to determine the potential reducing activity of complex biological fluids, as well as aqueous or ethanolic extracts of potential functional foods and nutraceuticals and solutions of purified antioxidant molecules.

20.2.12 Inhibition of linoleic acid oxidation as a measure of antioxidant activity

The antioxidant activity of molecules is most often attributed to the ability to delay the onset of lipid autoxidation, or peroxidation, by scavenging reactive oxygen species (ROS) or the ability to act as chain-breaking antioxidants to inhibit the propagation phase of lipid autoxidation (Yuan et al., 2005a; Nawar, 1996). In vitro systems designed to study the efficacy of molecules to inhibit lipid peroxidation have often comprised emulsion (Yuan et al., 2005a, 2005b) or liposomal systems (Hu et al., 2007) to model tissue membranes or food systems. The lipids most often incorporated into these systems include linoleic acid (C18:2,ω-6) as well as ethyl linoleate since the majority of foods contain a variety of unsaturated fatty acids and associated esterified forms (Coupland et al., 1996).

The assays chosen to monitor lipid oxidation are of key importance since the data they provide are a function of the stage of the peroxidation process. For example, the alteration of unsaturated fatty acid bond geometry from the native non-conjugated isomer to the formation of conjugated dienes (CD):

$$-CH=CH-CH2-CH=CH- \rightarrow -CH=CH-C=CH-CH- \rightarrow$$
$$-CH=CH-CH=CH-COOH- \text{ or, } -COOH-CH=CH-CH=CH$$

is associated with the resonance stabilisation and shift in double bond position with the formation of isomeric hydroperoxides during the early stages of lipid oxidation (Nawar, 1996). Thus, the UV-absorbance due to CD formation subsequently decreases as the hydroperoxides begin to decompose, prior to increasing once again as decomposition products begin to form (Puhl et al., 1994). On the other hand, 2-thiobarbituric acid reactive substances (TBARS) formation reflects primarily the production of scission and breakdown dialdehyde products such as MDA later in the reaction (Nawar, 1996).

Lipid emulsions and liposome preparations are well suited to the inclusion of pro-oxidants or free radical-generators such as the azo compounds AAPH and AMVN already discussed

(Yuan *et al.*, 2005b). When studying these systems, researchers must also be aware of the 'polar paradox' whereby hydrophilic compounds exhibit weak antioxidant activity in emulsions due to the dilution of these compounds in the aqueous phase; or conversely, lipophilic compounds exhibit strong antioxidant activity due to the concentration of the antioxidant at the lipid-air interface allowing strong protection of an emulsion against oxidation. On the other hand, the opposite antioxidant profile may be observed in bulk lipid or oil systems (Koleva *et al.*, 2002). The solubilities of the free radical-generator and the antioxidant under study are also important variables in study design.

20.2.13 Other assays – methods based on the chemiluminescence (CL) of luminol

The main principle of these methods is based on the ability of luminol and related compounds to luminesce under the flux of free radicals (chemiluminescence, CL) (Roginsky and Lissi, 2005). CL is brought about due to a reaction of a free radical derived from luminol with active free radicals. CL can be easily recorded. The addition of an antioxidant compound, being a scavenger of an active free radical, results in CL quenching, commonly with a pronounced induction period (Roginsky and Lissi, 2005). The quantity of the tested antioxidant can be estimated from the duration of tIND. As a rule, antioxidant activity is given in Trolox equivalents. The attractive feature of CL methods is their productivity; commonly, one run normally takes a few minutes only; in addition, the assay can be easily automated. As for shortcomings of this group of methods, first of all, the mechanism for chemical processes resulting in CL is not known in detail. The latter may create problems with interpreting the data obtained. Different versions of this method differ in the type of active free radical produced and the way of free radical production as well as in details of the protocol. While the majority of assays have been developed for testing biologically relevant samples, they can be easily applied for food testing (Roginsky and Lissi, 2005). Parejo, Codina, Petrakis and Kefalas (2000) suggested inducing CL by reaction of Co^{2+} chelated by EDTA with H_2O_2. Although the authors suggested HO$^{\cdot}$ as an active free radical, which attacks luminol, it is more realistic that O^-_2 plays this role. The method was used to test red wines (Arnous *et al.*, 2001). The method is well-instrumented and computerised and its capability was demonstrated by the example of testing several natural products including wines, tea and medicinal herb extracts. The evident advantage of the method is its very high productivity: the procedure takes commonly a couple of minutes only. At the same time, the kinetic theory of the process underlying the assay is really not suggested (Roginsky and Lissi, 2005).

20.2.14 Comparison of various methods for determining antioxidant activity: general perspectives

The antioxidant content of food samples may be characterised by two independent parameters: antioxidant capacity and reactivity (Roginsky and Lissi, 2005). For individual antioxidants, this corresponds to the stoichiometric coefficient and the rate constant for reaction between antioxidants and highly reactive free radicals. There is no single robust answer to the question of which index of antioxidant activity is more relevant. The main attention is currently paid to determining antioxidant capacity. The absolute majority of the recently developed methods are designed to solve this major problem. Admittedly, the reactivity of food samples may be of interest under certain conditions. Meanwhile, the information on the reactivity of food and individual natural polyphenols is still rather poor and conflicting (Roginsky and Lissi, 2005).

It is well known that indirect methods (DPPH, ABTS•+) are used more frequently than direct methods (competitive crocin bleaching, competitive β-carotene bleaching). The question now arises as to which of the methods, direct or indirect, is better in principle. Each kind of method has both advantages and disadvantages. The direct methods are more adequate in principle, especially those based on the model of the chain controlled reaction. Besides, they are commonly more sensitive. The disadvantage of the direct methods is that most of them are rather time-consuming and their application requires significant experience in chemical kinetics (Roginsky and Lissi, 2005). As a consequence, direct methods are commonly not so suitable for routine testing of natural products.

As a rule, well-developed indirect methods, such as the DPPH and ABTS•+ assays, are more productive and easier in handling (Roginsky and Lissi, 2005). The crucial point concerning the application of indirect assays is their informative capability. The indirect methods commonly provide information on the capability of natural products to scavenge stable free radicals, for example, DPPH and ABTS•+. Undoubtedly, the best indirect methods as well as the Folin–Ciocalteu test allow the estimation of antioxidant activity to the first approximation (Roginsky and Lissi, 2005). However, it is questionable whether the raw data obtained with indirect methods give quantitative information on the capability of natural products to inhibit oxidative processes. To conclude, it should be remembered that the methods described here are intended for the determination of the antioxidant activity of food samples, that is, the antioxidative potential of food. As for the antioxidative action of food substituents in real biological systems, this will mainly depend also on their bioavailability and food antioxidants metabolism *in vivo*.

It should be noted that beneficial influence of many foodstuffs and beverages including fruits, vegetables, tea, red wine, coffee and cacao on human health has been recently recognised to originate from the chain-breaking antioxidant activity of natural polyphenols, a significant constituent of these products. For this reason, the dietary value of such products is determined to a large extent by their antioxidant activity. Although the kinetic approach provides the basis of the majority of these methods, only a few of them have been analysed from the viewpoint of chemical kinetics.

20.2.15 Discrepancies over antioxidant measurement

Different assays have been introduced to measure antioxidant activity of foods and biological samples. The concept of antioxidant activity first originated from chemistry and was later adapted to biology, medicine, epidemiology and nutrition (Pellegrini *et al.*, 2003). It describes the ability of redox molecules in foods and biological systems to scavenge free radicals. This concept provides a broader picture of the antioxidants present in a biological sample as it considers the additive and synergistic effects of all antioxidants rather than the effect of single compounds, and may, therefore, be useful for study of the potential health benefits of antioxidants on oxidative stress-mediated diseases (Brighenti *et al.*, 2005).

Recently, Floegel *et al.* (2011) evaluated the antioxidant activity of various fruits, vegetables and their products by ABTS, DPPH and ORAC assays. Their study showed that relative to DPPH assay, ABTS assay was more strongly correlated with ORAC from USDA database, phenolics and flavonoids content of the 50 most popular antioxidant-rich foods in the US diet. The results suggested that ABTS assay better reflects the antioxidant contents in a variety of foods than DPPH assay. It has been previously reported that antioxidant capacity determined by different *in vitro* assays give different values (Ou *et al.*, 2002). Ou *et al.* (2002) conducted a large scale vegetable analysis using two different *in vitro* assays, FRAP and ORAC, and obtained very different antioxidant capacities from these methods. In their study, antioxidant capacities as determined by FRAP and ORAC assays were only weakly correlated. Pellegrini *et al.* (2003) reported that rankings of several fruits, vegetables and beverages differed based on antioxidant capacity measured by FRAP and ABTS assays suggesting that caution should be exercised when interpreting antioxidant capacities from different assays.

Xu and Chang (2008) studied the effect of soaking, boiling and steaming on antioxidant activities of cool season food legumes by two different methods (FRAP and ORAC). As compared to original unprocessed legumes, all processing steps caused significant ($p<0.05$) decreases in total phenolic content, DPPH and ORAC values in all tested cool season food legumes (green pea, yellow pea, chickpea and lentil). In contrast, oxygen radical absorbance capacities were increased with the increase of pressure in both pressure boiling and pressure steaming treatments. TPC and DPPH were not parallel with ORAC in cases of pressure boiling and pressure steaming treatments. This phenomenon could be attributed to the increases or the formation (after high pressure heat treatments) of specific compounds, which could provide more hydrogen atom during oxidation–reduction reaction.

Gorinstein *et al.* (2010) recently reported a high correlation between polyphenols content in three exotic fruits and antioxidant capacities measured by ABTS, DPPH and FRAP assays. Similarily, Dudonné *et al.* (2009) reported a strong positive correlation between ABTS and DPPH assays with a Pearson correlation coefficient of $r=0.906$ when used for 30 aqueous plant extracts.

By definition, the antioxidant activity is the capability of a compound (composition) to inhibit oxidative degradation, for example, lipid peroxidation. Phenolics are the main antioxidant components of foods (Roginsky and Lissi, 2005). Antioxidant activity of polyphenols is associated with various mechanisms of action, the elevated reactivity of phenolics towards active free radicals is considered as the most common principle mechanism. The authors would like to distinguish between the antioxidant activity and the reactivity. The antioxidant activity gives the information about the duration of antioxidative action; the reactivity characterises the starting dynamics of antioxidation at a certain concentration of an antioxidant or complex antioxidant mixture (Roginsky and Lissi, 2005).

Antioxidant activity may be a key parameter for both food science and technology and nutritional studies, and therefore there is presently a vital need to develop a standardised methodology to measure total antioxidant activity in plant foods. As discussed in the above sections, there are substantial differences in sample preparation, extraction of antioxidants (solvent, temperature etc.), selection of end-points and expression of research results, even for the same antioxidant assay, so that comparison between the values reported by different laboratories can be quite difficult.

Most original works and reviews on antioxidant activity focus mainly on the characteristics of the measurement procedure such as free radical generating system, redox interactions, molecular target, end-point, lipophilic and hydrophilic solubility etc. However, little attention has been paid to critical steps such as sample preparation (Luthria, 2006) or the procedure for extraction of antioxidants (Pellegrini *et al.*, 2007).

It should be remembered that the already mentioned methods are intended for the determination of antioxidant activity of food sample per se, that is, the antioxidative potential of food. As for the antioxidative action of food substituents in real biological systems, this will depend also on their bioavailability and food antioxidants metabolism. The authors also strictly recommend complete standardisation of antioxidant assays as the results in the above studies can be confusing. Discussed here are different processing technologies and their impact on antioxidant activity of fruits, vegetables, juices, cereal, legumes, spices etc. It should be noted that the authors have tried to address all processing technologies and their effects on the antioxidant content. However, not all research studies have carried out two or more antioxidant assays with the aim to estimate their reliability and limitations. It is also very complicated and perplexing to correlate the data on antioxidant activity of natural products reported in various works and measured by various methods. These data are generally poorly repeatable, first of all, because natural products are hardly repeatable in principle (Roginsky and Lissi, 2005).

Some critical points to rememember while assessing antioxidant status of unprocessed and processed foods (Pérez-Jiménez et al., 2008):

- Determination of antioxidant activity of various foods and beverages should include three key steps: sample preparation and extraction of antioxidants, measurement of antioxidant activity and expression of results.
- During sample preparation, the loss of antioxidants in the drying and milling steps must be kept to a minimum.
- In the extraction of antioxidants, at least two extraction cycles with mixtures of different polarity of water and organic solvents must be combined.
- Determination of total antioxidant activity must be performed both in aqueous-organic extracts and in their corresponding residues, which may exhibit higher antioxidant activity than the aqueous-organic extracts, a fact usually ignored in the literature.
- Antioxidant activity values should only be compared where the method, the solvent and the analytical conditions are the same.
- Possible interference from certain food constituents must also be taken into account when determining antioxidant capacity.
- At least two assays should be performed to determine antioxidant activity.
- Expression of kinetic parameters such as EC_{50}, tEC_{50} and AE may also provide a more comprehensive evaluation of antioxidant activity.

20.3 Concluding remarks

The key nutritional role of plant produce is unquestionable, which is in part due to the presence of phytochemicals with various biological activities. These plant foods are a good source of major and minor (polyphenols, vitamins, carotenoids, glucosinolates minerals, etc.) compounds which may have important metabolic and/or physiological effects. More recent evidence provides potential information of their impact on health, so these secondary metabolites are currently marketed as functional foods and nutraceutical ingredients. The authors would like to highlight the fact that there are many methods used to determine total antioxidant activity, and it is important to point out that all of them have some limitations. It has been observed in previous studies that some antioxidant assay methods give different trends. For that reason multiple methods to generate an 'antioxidant profile' might be needed.

References

Abas, F., Lajis, N.H., Israf, D.A., Khozirah, S. and Kalsom, Y.U. (2006) Antioxidant and nitric oxide inhibition activities of selected Malay traditional vegetables. *Food Chemistry*, 95, 566–573.

Alho, H. and Leinonen, J. (1999) Total antioxidant activity measured by chemiluminescence methods. *Methods in Enzymology*, 299, 3–15.

American Institute of Cancer Research. (1997) *Food Nutrition, and the Prevention of Cancer: A Global Perspective*. Washington, DC: American Institute of Cancer Research.

Aramwit, P., Bang, N. and Srichana, T. (2010) The properties and stability of anthocyaninsin mulberry fruits. *Food Research International*, 43, 1093–1097.

Arena, E., Fallico, B. and Maccarone, E. (2001) Evaluation of antioxidant capacity of blood orange juice as influenced by constituents, concentration process and storage. *Food Chemistry*, 74, 423–427.

Arnous, A., Makris, D.P., and Kefalas, P. (2001) Effect of principle polyphenolic components in relation to antioxidant characteristics of aged red wines. *Journal of Agriculture and Food Chemistry*, 49, 5736–5742.

Aruoma, O.I. (1994) Deoxyribose assay for detecting hydroxyl radicals. *Methods in Enzymology* 233, 57–66.

Aruoma, O.I., Grootveld, M. and Halliwell, B. (1987) The role of iron in ascorbate-dependent deoxyribose degradation. Evidence consistent with a site-specific hydroxyl radical generation caused by iron ions bound to the deoxyribose molecule. *Journal of Inorganic Biochemistry*, 29(4), 289–299.

Benkeblia, N. (2000) Phenylalanine ammonia-lyase, peroxidase, pyruvic acid and total phenolics variations in onion bulbs during long-term storage. *Lebensmittel-Wissenschaft und-Technologie*, 33(2), 112–116.

Benzie, I.F.F. and Strain, J.J. (1999) Ferric reducing/antioxidant power assay: direct measure of total antioxidant activity of biological fluids and modified version for simultaneous measurement of total antioxidant power and ascorbic acid concentration. *Methods in Enzymology*, 299, 15–27.

Benzie, I.F.F. and Strain, J.J. (1996) The ferric reducing ability of plasma (FRAP) as a measure of 'antioxidant power': the FRAP assay. *Analytical Biochemistry*, 239(1), 70–76.

Benzie, I.F.F. and Strain, J.J. (1997) Simultaneous automated measurement of total 'antioxidant' (reducing) capacity and ascorbic acid concentration. *Redox Report*, 3(4), 233–238.

Blois, M.S. (1958) Antioxidant determinations by the use of a stable free radical. *Nature* 181 (4617), 1199–1200.

Beyers, M. and Thomas, A.C. (1979) Irradiation of subtropical fruits. 4. Changes in certain nutrients present inmangoes, papayas, and litchis during canning, freezing, and irradiation. *Journal of Agricultural and Food Chemistry*, 27, 48–51.

Brighenti, F., Valtueñ a, S., Pellegrini, N., Ardigo, D., Del Rio, D., Salvatore, S. et al. (2005) Total antioxidant capacity of the diet is inversely and independently related to plasma concentrations of highsensitive C-reactive protein in adult Italian subjects. *British Journal of Nutrition*, 93(5), 619–625.

Brand-Williams, W., Cuvelier, M.E. and Berset, C. (1995) Use of a free radical method to evaluate antioxidant activity. *Lebensmittel-Wissenschaft und-Technologie*, 28, 25–30.

Burns, J., Gardner, P.T., Matthews, D., Duthie, G.G., Lean, M.E.J. and Crozier, A. (2001) Extraction of phenolics and changes in antioxidant activity of red wines during vinification. *Journal of Agricultural and Food Chemistry*, 49, 5797–5808.

Chu, Y.H., Chang, C.L. and Hsu, H.F. (2000) Flavonoid content of several vegetables and their antioxidant activity. *Journal of the Science of Food and Agriculture*, 80, 561–566.

Coupland, J.N., Zhu, Z., Wan, H., McClements, D.J., Nawar, W.W. and Chinachoti, P. (1996) Droplet composition affects the rate of oxidation of emulsified ethyl linoleate. *Journal of the American Oil Chemists Society*, 73(6), 795–801.

Dávalos, A., Gómez-Cordovés, C. and Bartolomé, B. (2004) Extending applicability of the oxygen radical absorbance capacity (ORAC-fluorescein) assay. *Journal of Agricultural and Food Chemistry*, 52(1), 48–54.

DeRuiter, F.E. and Dwyer, J. (2002) Consumer acceptance of irradiated foods: Dawn of a new era? *Food Service Technology*, 2, 47–58.

DeLong, J.M., Prange, R.K., Hodges, D.M., Forney, C.F., Bishop, M.C. and Quilliam, M. (2002) Using a modified ferrous oxidation- xylenol orange (FOX) assay for detection of lipid hydroperoxides in plant tissues. *Journal of Agricultural and Food Chemistry*, 50, 248–254.

Del Pozo-Insfran, D., Balaban, M.O. and Talcott, S.T. (2006) Enhancing the retention of phytochemicals and organoleptic attributes in muscadine grape juice through a combined approach between dense phase CO_2 processing and copigmentation. *Journal of Agricultural and Food Chemistry*, 54, 6705–6712.

Dillard, C.J. and German, J.B. (2000) Phytochemicals: nutraceuticals and human health. *Journal of the Science of Food and Agriculture*, 80, 1744–1756.

Da Porto, C., Decorti, D. and Tubaro, F. (2010) Effects of continuous dense-phase CO_2 system on antioxidant capacity and volatile compounds of apple juice. *International Journal of Food Science and Technology*, DOI: 10.1111/j.1365-2621.2010.02339.x

Eberhardt, M.V., Lee, C.Y. and Liu, R.H. (2000) Antioxidant activity of fresh apples. *Nature*, 405(6789), 903–904.

EFSA (2010) Scientific Opinion on the substantiation of health claims related to various food(s)/food constituent(s) and protection of cells from premature aging, antioxidant activity, antioxidant content and antioxidant properties, and protection of DNA, proteins and lipids from oxidative damage pursuant to Article 13(1) of Regulation (EC) No 1924/2006. *EFSA Journal*, 8(2), 1489.

Erkan, N., Ayranci, G. and Yranci, E. (2008) Antioxidant activities of rosemary (Rosmarinus Officinalis L.) extract, blackseed (Nigella sativa L.) essential oil, carnosic acid, rosmarinic acid and sesamol. *Food Chemistry*, 110, 76–82.

Esterbauer, H., Lang, J., Zadravec, S. and Slater T.F. (1984) Detection of malonaldehyde by high-performance liquid chromatography. *Methods in Enzymology*, 105, 319–328.

FAO/IAEA/WHO (1999) High-dose irradiation: Wholesomeness of food irradiated with doses above 10 kGy. Report of a joint FAO IAEA WHO study group. Rome: Food and Agriculture Organization of the United Nations.

Floegel, A., Kim, D.K., Chung. S.J., Koo, S.I. and Chun, O.K. (2011) Comparison of ABTS/DPPH assays to measure antioxidant capacity in popular antioxidant-rich US foods. *Journal of Food Composition and Analysis*, doi:10.1016/j.jfca.2011.01.008

Frankel, E.N., Huang, S.-W. and Aeschbach, R. (1997) Antioxidant activity of green teas in different lipid systems. *JAOCS*, 74, 1309–1315.

Fraser, D. (1951) Bursting bacteria by release of gas pressure. *Nature*, 167, 33–34.

Garcia-Gonzalez, L., Geeraerd, A.H., Spilimbergo, S., Elst, K., VanGinneken, L. and Debevere, J. (2007) High pressure carbon dioxide inactivation of microorganisms in foods: The past, the present and the future. *International Journal of Food Microbiology*, 117(1), 1–28.

Gardner, P.T., White, T.A.C., McPhail, D.B. and Duthie, G.G. (2000) The relative contributions of vitamine C, carotenoids and phenolics to the antioxidant potential of fruit juices. *Food Chemistry*, 68, 471–474.

Ghiselli, A., Serafini, M., Maiani, G., Azzini, E. and Ferro-Luzzi, A. (1995) A fluorescence-based method for measuring total plasma antioxidant capability. *Free Radical Biology and Medicine*, 18(1), 29–36.

González-Aguilar, G.A., Villegas-Ochoa, M.A., Martínez- Téllez, M.A., Gardea, A.A. and Ayala-Zavala, J.F. (2007) Improving antioxidant capacity of fresh-cut mangoes treated with UV-C. *Journal of Food Science*, 72, S197–S202.

Gorinstein, S., Martín-Belloso, O., Park, Y.-S., Haruenkit, R., Lojek, A., Číž, M., Caspi, A., Libman, I. and Trakhtenberg, S. (2001) Comparison of some biochemical characteristics of different citrus fruits. *Food Chemistry*, 74, 309–315.

Gorinstein, S., Haruenkit, R., Poovarodom, S., Vearasilp, S., Ruamsuke, P., Namiesnik, J., Leontowicz, M., Leontowicz, H., Suhaj, M. and Sheng, G.P. (2010) Some analytical assays for the determination of bioactivity of exotic fruits. *Phytochemical Analysis*, 21, 355–362.

Grau, A., Codony, R., Rafecas, M., Barroeta, A.C. and Guardiola, F. (2000) Lipid hydroperoxide determination in dark chicken meat through a ferrous oxidation-xylenol Orange method. *Journal of Agricultural and Food Chemistry*, 48, 4136–4143.

Granado, F., Olmedilla, B., Blanco, I. and Rojas-Hidalgo, E. (1992) Carotenoid composition in raw and cooked Spanish vegetables. *Journal of Agricultural and Food Chemistry*, 40, 2135–2140.

Gunes, G., Blum, L.K. and Hotchkiss, J.H. (2005) Inactivation of yeasts in grape juice using a continuous dense phase carbon dioxide processing system. *Journal of the Science of Food and Agriculture*, 85(14), 2362–2368.

Guderjan, M., Toepfl, S., Angersbach, A. and Knorr, D. (2005) Impact of pulsed electrifield treatment on the recovery and quality of plant oils. *Journal of Food Engineering*, 67(3), 281–287.

Greefield, H. and Southgate, D.A.T. (1992) *Food Composition Data Production*, Management and Use. Elsevier Applied Science.

Halliwell, B., Gutteridge, J.M.C. and Aruoma, O.I. (1987) The deoxyribose method: a simple 'test-tube' assay for determination of rate constants for reactions of hydroxyl radicals. *Analytical Biochemistry*, 165(1), 215–219.

Heinonnen, I. M., Meyer, A.S. and Frankel, E.N. (1998a) Antioxidant activity of berry phenolics on human low-density lipoprotein and liposome oxidation. *Journal of Agricultural and Food Chemistry*, 46, 4107–4112.

Heinonnen, M., Rein, D., Satué-Gracia, M., Huang, S.-W., German, J.B. and Frankel, E.N. (1998b) Effect of Protein on the antioxidant activity of phenolic compounds in a lecithin-lipossome oxidation system. *Journal of Agricultural and Food Chemistry*, 46, 917–922.

Henríquez, C., Carrasco-Pozo, C., Gómez, M., Brunser, O. and Speisky, H. (2008) Slow and fast-reacting antioxidants from berries: their evaluation through the FRAP (ferric reducing antioxidant power) assay. *Acta Horticulturae*, 777, 531–536.

Hu, C. and Kitts, D.D. (2000) Studies on the antioxidant activity of Echinacea root extract. *Journal of Agricultural and Food Chemistry*, 48, 1466–1472.

Hu, C., Yuan, Y.V. and Kitts, D.D. (2007) Antioxidant activities of the flaxseed lignan secoisolariciresinol diglucoside, its aglycone secoisolariciresinol and the mammalian lignans enterodiol and enterolactone in vitro. *Food and Chemical Toxicology*, 45(11), 2219–2227.

Huang, D., Ou, B. and Prior, R.L. (2005) The chemistry behind antioxidant capacity assays. *Journal of Agricultural and Food Chemistry*, 53, 1841–1856.

Huang, D., Ou, B., Hampsch-Woodill, M., Flanagan, J.A. and Deemer, E.K. (2002) Development and validation of oxygen radical absorbance capacity assay for lipophilic antioxidants using randomly methylated β-cyclodextrin as the solubility enhancer. *Journal of Agricultural and Food Chemistry*, 50(7), 1815–1821.

Huang, D.-J., Chen, H.-J., Hou, W.-C., Lin, C.-D. and Lin, Y.-H. (2006). Sweet potato (Ipomoea batatas [L.] Lam 'Tainong 57") storage root mucilage with antioxidant activities in vitro. *Food Chemistry*, 98, 774–781.

Ismail, A., Marjan, Z.M. and Foong, C.W. (2004) Total antioxidant activity and phenolic content in selected vegetables. *Food Chemistry*, 87, 581–586.

Ismail, A. and Lee, W.Y. (2004) Influence of cooking practice on antioxidant properties and phenolic content of selected vegetables. *Asia Pacific Journal of Clinical Nutrition*, 13, S162–S165, (Suppl).

Jay, J.M., Loessner, M.J. and Golden, D.A. (2005) *Modern Food Microbiology* (7th edition), New York: Springer.

Jiang, Z.-Y., Hunt, J. V. and Wolff, S. P. (1992). Ferrous ion oxidation in the presence of xylenol orange for detection of lipid hydroperoxide in low density lipoprotein. *Analytical biochemistry*, 202, 384–389.

Jiang, Z.-Y., Woollard, A.C.S. and Wolff, P. (1991) Lipid Hydroperoxide measurement by oxidation of Fe^{2+} in the presence of xylenol orange. Comparison with the TBA assay and an iodometric method. *Lipids*, 26, 853–856.

Joseph, J.A., Shukit-Hale, B., Denisova, N.A. et al. (1999) Reversal of age-related declines in neuronal signal transduction, cognitive, and motor behavioural de®cits with blue berry, spinach, or strawberry dietary supplementation. *Journal of Neuroscience*, 19, 8114–8812.

Kabir, H. (1994) Fresh-cut vegetables. In: A.L. Brods and V.A. Herndon (eds), *Modified Atmosphere Food Packaging*, Institute of Packaging Professionals, pp. 155–160.

Karadag, A., Ozcelik, B. and Saner, S. (2009) Review of methods to determine antioxidant capacities. *Food Analysis Methods*, 2, 41–60.

Karakaya, S. and Kavas, A. (1999). Antimutagenic activities of some foods. *Journal of Agricultural Food Chemistry*, 79, 237–242.

Kaur, C. and Kapoor, H. (2001) Review: antioxidants in fruits and vegetables – the millennium's health. *International Journal of Food Science and Technology*, 36, 703–725.

Kitts, D.D., Wijewickreme, A.N. and Hu, C. (2000) Antioxidant properties of a North American ginseng extract. *Molecular and Cellular Biochemistry*, 203, 1–10.

Koleva, I.I., van Beek, T.A., Linssen, J.P.H., de Groot, A. and Evstatieva, L.N. (2002) Screening of plant extracts for antioxidant activity: a comparative study on three testing methods. *Phytochemical Analysis*, 13, 8–17.

Kumar, K.S., Ganesan, K. and Rao, P.V.S. (2008) Antioxidant potential of solvent extracts of Kappaphycus alvarezii (Doty) Doty – an edible seaweed. *Food Chemistry*, 107, 289–295.

Lafuente, M.T., Zacarías, L., Martínez-Téllez, M.A., Sánchez-Ballesta, M.T. and Granell, A. (2003) Phenylalanine ammonia-lyase and ethylene in relation to chilling injury as affected by fruit age in citrus. *Postharvest Biology and Technology*, 29(3), 308–317.

Lee, J.-Y., Hwang, W.-I. and Lim, S.-T. (2004) Antioxidant and anticancer activities of organic extracts from Platycodon grandiflorum A. De Candolle roots. *Journal of Ethnopharmacology*, 93, 409–415.

Lee, S.K. and Kader, A.A. (2000) Preharvest and postharvest factors influencing vitamin C content of horticultural crops. *Postharvest Biology and Technology*, 20, 207–220.

Locatelli, M., Travaglia, F., Coisson, J.D., Martelli, A., Stévigny, C. and Arlorio, M. (2010) Total antioxidant activity of hazelnut skin (Nocciola Piemonte PGI): Impact of different roasting conditions. *Food Chemistry*, 119, 1647–1655.

Luthria, D. (2006) Significance of sample preparation in developing analytical methodologies for accurate estimation of bioactive compounds in functional foods. *Journal of the Science of Food and Agriculture*, 86(14), 2266–2272.

MacDonald-Wicks, L.K., Wood, L.G. and Garg, M.L. (2006) Methodology for the determination of biological antioxidant capacity in vitro: a review. *Journal of the Science of Food and Agriculture*, 86, 2046–2056.

MacPhail, D.B., Gardner, P.T., Duthie, G.G., Steele, G.M. and Reid, K. (1999) Assessement of the antioxidant potential of Scotch whiskeys by electron spin resonance spectroscopy: relationship to hydroxyl-containing aromatic components. *Journal of Agricultural and Food Chemistry*, 47, 1937–1941.

Mahattanatawee, K., Manthey, J. A., Luzio, G., Talcott, S. T., Goodner, K., and Baldwin, E. A. (2006) Total antioxidant activity and fiber content of select Florida- Grown Tropical fruits. *Journal of Agricultural and Food Chemistry*, 54, 7355–7363.

Malminiemi, K., Palomäki, A. and Malminiemi, O. (2000) Comparison of LDL trap assay to other tests of antioxidant capacity; effect of vitamin E and lovastatin treatment. *Free Radical Research*, 33(5), 581–593.

Manvell, C. (1997) Minimal processing of food. *Food Science and Technology Today*, 11, 107–111.

Meyer, A.S., Heinonen, M. and Frankel, E.N. (1998) Antioxidant interactions of catechin, cyaniding, caffeic acid, quercetin, and ellagic acid on human LDL oxidation. *Food Chemistry*, 61, 71–75.

Meyer, A.S., Yi, O.-S., Pearson, D.A., Waterhouse, A.L. and Frankel, E.N. (1997) Inhibition of Human Low-Density lipoprotein oxidation in relation to composition of phenolic antioxidants in grapes (Vitis vinifera). *Journal of Agricultural and Food Chemistr.*, 45, 1638–1643.

Miller, N., Diplock, A. T., and Rice-Evans, C. (1995) Evaluation of the total antioxidant activity as a marker of the deterioration of apple juice on storage. *Journal of Agricultural and Food Chemistry*, 43, 1794–1801.

Musa, K.H., Abdullah, A., Jusoh, K. and Subramaniam, V. (2008) Antioxidant activity of pink-flesh guava (psidium guajava l.): effect of extraction techniques and solvents. *Food Analytical Methods*, 4(1), 100–107, doi: 10.1007/s12161-010-9139-3.

Nawar, W.W. (1996) Lipids. In: Fennema, O.R. (ed.) *Food Chemistry* New York, Marcel Dekker Inc., pp. 225–319.

Navas, J. A., Tres, A., Codony, R., Boatella, J., Bou, R., and Guardiola, F. (2004). Modified ferrous oxidation-xylenol orange method to determine lipid hydroperoxides in fried snacks. *European Journal of Lipid Science and Technology*, 688–696.

Ng, T.B., Lui, F. and Wang, Z.T. (2000) Antioxidatve activity of natural products from plants. *Life Sciences*, 66(8), 709–723.

Niki, E. (1990) Free radical initiators as as source of water- or lipid-soluble peroxyl radicals. *Methods in Enzymology*, 186, 100–108.

Noguchi, N., Yamashita, H., Gotoh, N., Yamamoto, Y., Numano, R., and Niki, E. (1998) 2,2′–Azobis (4-methoxy-2,4-dimethylvaleronitrile), a new lipid-soluble azo initiator: application to oxidations of lipids and low-density lipoprotein in solution and in aqueous dispersions. *Free Radical Biology and Medicine*, 24(2), 259–268.

Nourooz-Zadeh, J. (1999) Ferrous ion oxidation in presence of xylenol orange for detection of lipid hydroperoxides in plasma. In: H. Sies, *Oxidants and Antioxidants*, Methods in Enzymology series, London, Academic Press.

Nourooz-Zadeh, J., Tajaddini-Sarmadi, J., Birlouez-Aragon, I., and Wolff, S. P. (1995). Measurement of hydroperoxides in edible oils using the ferrous oxidation in xylenol orange assay, *Journal of Agricultural and Food Chemistry*, 43, 17–21.

Nourooz-Zadeh, J., Tajaddini-Sarmadi, J., and Wolff, S. P. (1994). Measurement of plasma hydroperoxide concentrations by the ferrous oxidation-xylenol orange assay in conjuction with triphenylphosphine, *Analytical Biochemistry*, 220, 403–409.

Ou, B., Hampsch-Woodill, M. and Prior, R.L. (2001) Development and validation of an improved oxygen radical absorbance capacity assay using fluorescein as the fluorescent probe. *Journal of Agricultural and Food Chemistry*, 49(10), 4619–4626.

Ou, B., Huang, D., Hampsch-Woodill, M., Flanagan, J.A. and Deemer, E.K. (2002) Analysis of antioxidant activities of common vegetables employing oxygen radical absorbance capacity (ORAC) and ferric reducing antioxidant power (FRAP) assays: A comparative study. *Journal of Agricultural and Food Chemistry*, 50, 3122–3128.

Ou, S.Y., Luo, Y.L., Huang, C.H., and Jackson, M. (2009) Production of coumaric acid from sugarcane bagasse. *Innovative Food Science and Emerging Technologies*, 10, 253–259.

O'Beirne, D. and Francis, G.A. (2003) Reducing pathogen risk in MAP-prepared produce. In R. Ahvenainen (ed.) *Novel Food Packaging Techniques*, Cambridge, UK/Boca Raton, FL, Woodhead Publishing Limited/CRC Press LLC, pp. 231–232.

Parejo, I., Codina, C., Petrakis, C. and Kefalas, P. (2000) Evaluation of scavenging activity assessed by Co(II)/EDTA-induced luminal chemiluminescence and DPPH (2,2-diphenyl-1-picrylhydrazyl) freeradical assay. *Journal of Pharmocological and Toxicological Methods*, 44, 3871–3880.

Pellegrini, N., Re, R., Yang, M. and Rice-Evans, C. (1999) Screening of dietary carotenoids and carotenoid-rich fruit extracts for antioxidant activities applying 2,2′-Azinobis(3-ethylene-benzothiazoline-6-sulfonic acid radical cation decolorization assay. *Methods in Enzymology*, 299, 379–389.

Pellegrini, N., Visioli, F., Buratti, S. and Brighenti, F. (2001) Direct analysis of total antioxidant activity of olive oil and studies on the influence of heating. *Journal of Agricultural and Food Chemistry*, 49, 2532–2538.

Pellegrini, N., Colombi, B., Salvatore, S., Brenna, O.V., Galaverna, G., Del Rio, D. *et al.* (2007) Evaluation of antioxidant capacity of some fruit and vegetable foods: Efficiency of extraction of a sequence of solvents. *Journal of the Science of Food and Agriculture*, 87, 103–111.

Pellegrini, N., Serafini, M., Colombi, B., Del Rio, D., Salvatore, S., Bianchi, M. *et al.* (2003) Total antioxidant capacity of plant foods, beverages and oils consumed in Italy assessed by three different in vitro assays. *Journal of Nutrition*, 133, 2812–2819.

Pietta, P., Simonetti, P. and Mauri, P. (1998). Antioxidant activity of selected medicinal plants. *Journal of Agricultural and Food Chemistry*, 46, 4487–4490.

Polydera, A.C., Stoforos, N.G. and Taokis, P.S. (2004) The effect of storage on the antioxidant capacity of reconstituted orange juice which had been pasteurized by high pressure or heat. *International Journal of Food Science and Technology*, 39, 789–791.

Piga, A., Agabbio, M., Gambella, F. and Nicoli, M.C. (2002) Retention of antioxidant activity in minimally processed mandarin and satuma fruits. *Lebensmittel-Wissenschaft und-Technologie-Food Science and Technology*, 35, 344–347.

Pinilla, M.J., Plaza, L., Sánchez-Moreno, C., De Ancos, B. and Cano, M.P. (2005) Hydrophilic and lipophilic antioxidant capacities of commercial mediterranean vegetable soups (gazpachos). *Journal of Food Science*, 70, S60–S65.

Plaza, M.L., Ramirez-Rodrigues, M.M., Balaban, M.O., and Balaban, M.O. (2010) Quality improvement of guava puree by dense phase carbon dioxide (DP-CO2) pasteurization. (236–09). IFT Annual Meeting+Food Expo, Anaheim, CA., 20 July,

Plumb, G.W., Chambers, S.J., Lambert, N., Bartolomé, B., Heaney, R.K., Wanigatunga, S., Aruoma, O.I., Halliwell, B. and Williamson, G. (1996) Antioxidant actions of fruit, herb and spice extracts. *Journal of Food Lipids*, 3, 171–188.

Prior, R.L., Wu, X. and Schaich, K. (2005) Standardized methods for the determination of antioxidant capacity and phenolics in foods and Dietary Supplements. *Journal of Agricultural and Food Chemistry*, 53, 4290–4302.

Prior, R.L. and Cao, G. (2000) Antioxidant phytochemicals in fruits and vegetables: diet and health implications. *Horticulture Science*, 35, 588–592.

Puhl, H., Waeg, G. and Esterbauer, H. (1994) Methods to determine oxidation of low-density lipoproteins. *Methods in Enzymology*, 233, 425–441.

Re, R., Pellegrini, N., Proteggente, A., Pannala, A., Yang, M. and Rice-Evans, C. (1999) Antioxidant activity applying an improved ABTS radical cation decolorization assay. *Free Radical Biology and Medicine*, 26 (9–10), 1231–1237.

Rice-Evans, C.A. (2000) Measurement of total antioxidant activity as a marker of antioxidant status in vivo: Procedures and limitations. *Free Radical Research*, 33 (Suppl.), 559–566.

Rodtjer, A., Skibsted, L.H. and Andersen, M.L. (2006) Antioxidative and prooxidative effects of extracts made from cherry liqueur pomace. *Food Chemistry*, 99, 6–14.

Roginsky, V. and Lissi, E.A. (2005) Review of methods to determine chain-breaking antioxidant activity in food. *Food Chemistry*, 92, 235–254.

Ruslay, S., Abas, F., Shaari, K., Zainal, Z., Maulidiani, Sirat, H., Israf, D.A. and Lajis, N.H. (2007) Characterization of the components present in the active fractions of health gingers (Curcuma xanthrrhiza and Zingiber zerumbet) by HPLC-DAD-ESIMS. *Food Chemistry*, 104, 1183–1191.

Sales, J.M and Resurreccion, A.V.A (2010a) Maximizing phenolics, antioxidants and sensory acceptance of UV and ultrasound-treatbed peanuts. *LWT – Food Science and Technology*, 43, 1058–1066.

Salvini, S., Parpinel, M., Gnagnarella, P., Maisonneuve, P. and Turrini, A. (1998) *in Banca Dati di Composizione degli Alimenti per Studi Epidemiologici in Italia*. Milano, Istituto Europeo di Oncologia.

Salleh-Mack, S.Z. and Roberts, J.S. (2007) Ultrasound pasteurization: The effects of temperature, soluble solids, organic acids and pH on the inactivation of Escherichia coli ATCC 25922. *Ultrasonics Sonochemistry*, 14, 323–329.

Saxena, A., Maity, T., Raju, P.S. and Bawa, A.S. (2010) Degradation kinetics of colour and total carotenoids in jackfruit (Artocarpus heterophyllus) bulb slices during hot air drying. *Food and Bioprocess Technology*, DOI 10.1007/s11947-010-0409-2.

Sharma, O.P. and Bhat, T.K. (2009) DPPH antioxidant assay revisited. *Food Chemistry*, 113, 1202–1205.

Shimada, K., Fujikawa, K., Yahara, K. and Nakamura, T. (1992) Antioxidative properties of xanthan on the autoxidation of soybean oil in cyclodextrin emulsion. *Journal of Agricultural and Food Chemistry*, 40(6), 945–948.

Shui, G. and Leong, L. P. (2006) Residue from star fruit as valuable source for functional food ingredients and antioxidant nutraceuticals. *Food Chemistry*, 97, 277–284.

Steinmetz, K.A. and Potter, J.D. (1996) Vegetable, fruit and cancer epidemiology. *Cancer Causes and Control*, 2, 325–351.

Valkonen, M. and Kuusi, T. (1997) Spectrophotometric assay for total peroxyl radical-trapping antioxidant potential in human serum. *Journal of Lipid Research*, 38(4), 823–833.

Wang, H., Cao, G. and Prior, R.L. (1996) Total antioxidant capacity of fruits. *Journal of Agricultural and Food Chemistry*, 44, 701–705.

Wayner, D.D.M., Burton, G.W., Ingold, K.U. and Locke, S. (1985) Quantitative measurement of the total, peroxyl radical-trapping antioxidant capability of human blood plasma by controlled peroxidation. The important contribution made by plasma proteins. *FEBS Letters*, 187(1), 33–37.

Wood, O.B. and Bruhn, C.M. (2000) Position of the American dietetic association: Food irradiation. *Journal of the American Dietetic Association*, 100, 246–253.

Wu, X., Beecher, G.R., Holden, J.M., Haytowitz, D.B., Gebhardt, S.E. and Prior, R.L. (2004a) Lipophilic and hydrophilic antioxidant capacities of common foods in the United States. *Journal of Agricultural and Food Chemistry*, 52(12), 4026–4037.

Wu, X., Gu, L., Holden, J., Haytowitz, D.B., Gebhardt, S.E., Beecher, G. and Prior, R.L. (2004b) Development of a database for total antioxidant capacity in foods: a preliminary study. *Journal of Food Composition and Analysis*, 17, 407–422.

Xue, J., Chen, L. and Wang, H. (2008) Degradation mechanism of Alizarin Red in hybridgas–liquid phase dielectric barrier discharge plasmas: Experimental and theoreticalexamination. *Chemical Engineering Journal*, 138, 120–127.

Yan, X., Nagata, T. and Fan, X. (1998) Antioxidative activities in some common seaweeds. *Plant Foods for Human Nutrition*, 52, 253–262.

Yen, G.C. and Lin, H.T. (1996) Comparison of high pressure treatment and thermal pasteurisation on the quality and shelf life of guava puree. *International Journal of Food Science and Technology*, 31, 205–213.

Yen, G.-C. and Chen, H.-Y. (1995) Antioxidant activity of various tea extracts in relation to their antimutagenicity. *Journal of Agricultural and Food Chemistry*, 43, 27–32.

Yildiz, G., Wehling, R.L. and Cuppett, S.L. (2003) Comparison of four analytical methods for the determination of peroxide value in oxidized soybean oils. *JAOCS*, 80, 103–107.

Yoshida, Y., Shimakawa, S., Itoh, N. and Niki, E. (2003) Action of DCFH and BODIPY as a probe for radical oxidation in hydrophilic and lipophilic domain. *Free Radical Research*, 37(8), 861–872.

Yuan, Y.V., Bone, D.E. and Carrington, M.F. (2005a) Antioxidant activity of dulse (Palmaria palmata) extract evaluated in vitro. *Food Chemistry*, 91(3), 485–494.

Yuan, Y.V., Carrington, M.F. and Walsh, N.A. (2005b) Extracts from dulse (Palmaria palmata) are effective antioxidants and inhibitors of cell proliferation in vitro. *Food and Chemical Toxicology*, 43(7), 1073–1081.

Yuan, Y.V., Westcott, N.D., Hu, C. and Kitts, D.D. (2009) Mycosporine–like amino acid composition of the edible red alga, Palmaria palmata (dulse) harvested from the west and east coasts of Grand Manan Island, New Brunswick. *Food Chemistry*, 112(2), 321–328.

Zepka, L.Q. and Mercadante, A.Z. (2009) Degradation compounds of carotenoids formed during heating of a simulated cashew apple juice. *Food Chemistry*, 117, 28–34.

Zenker, M., Heinz, V. and Knorr, D. (2003) Application of ultrasound assisted thermalprocessing for preservation and quality retention of liquid foods. *Journal of Food Protection*, 66, 1642–1649.

Zhang, L., Lu, Z.,Yu, Z. and Gao, X. (2005) Preservation of fresh-cut celery by treatment of ozonated water. *Food Control*, 16(3), 279–283.

Zhang, P. and Omaye, S.T. (2001) β-Carotene: interactions with α-tocopherol and ascorbic acid in microsomal lipid peroxidation. *Journal of Nutritional Biochemistry*, 12, 38–45.

Zhang, J., Davis, T.A., Matthews, M.A., Drews, M.J., LaBerge, M. and An, Y.H. (2006) Sterilization using high-pressure carbon dioxide. *Journal of Supercritical Fluids*, 38(3), 354–372.

Zhang, D. and Hamauzu, Y. (2004) Phenolics, ascorbic acid, carotenoids and antioxidant activity of broccoli and their changes during conventional and microwave cooking. *Food Chemistry*, 88(4), 503–509.

Zimmer, H., Lankin, D.C. and Horgan, S.W. (1970) Oxidations with potassium nitrodisulfonate (Fremy's radical). The Teuber reaction. *Chemical Reviews*, 2, 229–246.

21 Industrial applications of phytochemicals

Juan Valverde

Teagasc Food Research Centre Ashtown, Dublin, Ireland

21.1 Introduction

The definition, general chemistry, classification and important sources of phytochemicals, as well as the effects of processing, are extensively outlined in the first part of this book. This chapter is related with the industrial applications of phytochemicals. The economic relevance of phytochemicals in food and other industries resides in their applicability to industry. An industrial application results of the transfer of a scientific knowledge into a technological use in order to control and/or modify an industrial process. The objective of this book chapter is to extensively review the use of phytochemicals in food industry. Phytochemicals are used in the food industry as food ingredients/additives and physico-chemical properties of phytochemicals determine how they are used in industry. For example, if a phytochemical is an antioxidant, this phytochemical could be potentially used to avoid undesirable oxidation of food products (fats or proteins). The use of phytochemicals in food industry is controlled by competent regulating bodies, in individual countries or economic areas. For example, the Food and Drug Administration (FDA) regulates the use of food additives in the United States of America while the European Food Safety Agency (EFSA) does so in all the European Union.

Phytochemicals are naturally occurring in fruits, vegetables and seaweeds and, as shown in previous chapters, the processing involved for their human consumption leads to a loss or degradations of these compounds in the foods, reducing some of their quality properties. Moreover, in most cases the waste or by-products of these processes is particularly rich in phytochemicals such as fruit juice production (apple pomace) or peeling (i.e. carrots and onions). Considerable efforts are carried out by research and development centres and food industry to find innovative ways to reduce this loss without jeopardising in other sensory attributes and/or their cost efficiency. The main trends in this sense have been: (1) the use of new or unique varieties, rich in a specific or family of phytochemicals, so when processed the overall content in phytochemicals is still high; (2) the use of new innovative ways of food processing that improve the contents of phytochemicals when compared with more

Handbook of Plant Food Phytochemicals: Sources, Stability and Extraction, First Edition.
Edited by B.K. Tiwari, Nigel P. Brunton and Charles S. Brennan.
© 2013 John Wiley & Sons, Ltd. Published 2013 by John Wiley & Sons, Ltd.

traditional ways of food processing; and (3) to artificially enhance the food product in order to supplement or fortify the original products. In addition to this last trend, phytochemicals have been used not only to fortify, but also as food ingredients, in order to improve sensory attributes and shelf life of a given food product. As a result phytochemicals are in fact a broad group of compounds that present multiple physico-chemical properties, and consequently can be potentially employed for many industrial applications.

21.2 Phytochemicals as food additives

The use of a substance as a food additive must require that this substance is not-toxic at the recommended levels of use. This requirement also applies for both chronic or acute toxicity. As a consequence the use of food additives is, in most countries, regulated by a food safety authority. Hence additive regulations might differ in part from country to country. Although in some economic areas such as the European Union there has been a compliance of these regulations. Some of these regulations extend to countries that don't belong to this economic area but have strong economic links with it (i.e. Switzerland, Norway, Iceland, Turkey and Australia). At the international level the Food and Agriculture Organization (FAO) of the United Nations and the World Health Organization (WHO) have a joint expert committee in food additives (JECFA). This committee evaluates safety of food additives advices on the standards, guidelines and codes of practice on the use of food additives. The classification of food additives can be done according to their technological use (Figure 21.1). More precisely they can be categorised according to their specific role or their chemical characteristics. Vitamins, minerals, amino acids and other phytochemicals can be used to improve nutritional profile of foodstuffs, but they contribute to this increase in nutritional value in a very different way (due to their chemical properties). Therefore vitamins, amino acids and minerals are

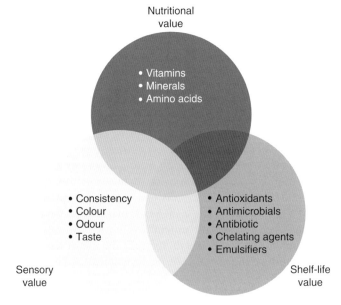

Figure 21.1 Venn diagram representing the technological characteristics that are improved in foods by the use of food additives/ingredients.

normally included in food formulations to balance losses during processing. This is a common practice in fruit juices, canned vegetables, bakery products, snacks and milk. In the case of minerals, fortification is usually done in those that are more fully available such as iron or calcium; this is a common practice in products such as cereal based breakfasts (Belitz *et al.*, 2004; Poletti *et al.*, 2004; Akhtar and Ashgar, 2011). As shown in Figure 21.1 some additives might have multiple functions and they contribute to increase value in more than one way. Ascorbic acid is a strong antioxidant and at the same time can have some beneficial effects as a dough improver in baked products. Some amino acids have important biological value and also can contribute to a protein-rich kind of taste in foods.

21.2.1 Flavourings

Flavourings are substances that are used as food additives to improve taste and/or smell of food. Fruit, vegetables, herbs and spices contain a great variety of volatile compounds many of which have flavouring properties (Figure 21.2). Chemical compounds such as carbonyl compounds, hydrocarbons esters pyranones, furanones, volatile sulphur compounds pyrazines, phenols and terpenes are common volatile compounds extracted from fruits and vegetables for their use as flavouring substances. In particular volatile sulphur compounds are common in vegetables of the *Brassicacea* (broccoli and cabbage) and *Liliaceae* (onion, garlic and leek) family. Pyrazines are contained in some spices and fruits of the *Capsicum* family (black pepper or chilli pepper). Terpenes (mono and sesquiterpenes) are present in fruits, vegetables and spices, for example β-caryophyllene is a volatile with flavouring properties present in many herbs (oregano, rosemary, black pepper). Many non-volatile phytochemicals can easily degrade to volatiles with flavouring properties, such as β-carotene into β-ionone or ferulic acid into vainillin.

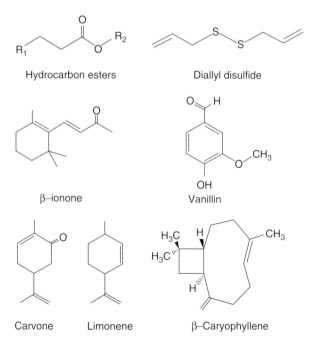

Figure 21.2 Chemical structure of some phytochemicals often used as flavourings in the industry.

21.2.2 Sweeteners and sugar substitutes

Plants and fruits are a great source of phytochemicals with sweetening properties (Figure 21.3). Many of these sweetening compounds are interestingly not sugars, but proteins (Gibbs et al., 1996). This is the case of several sweetening proteins (Monellin, Thaumatin, Brazzein, Pentadin, Miraculin and Mabilin) from plant origin that have been discovered in Africa and/or Asia. The pulp of the serendipity berry (*Dioscorephyllum volkensii*) is rich in Monellin a protein with a molecular weight of 10.5 kDa (Morris et al., 1973; Fan et al., 1993). Monellin is composed of two polypeptide chains that are not covalently bound and is 1500–2000 times sweeter than sucrose (Morris et al., 1973; Belitz et al., 2004).

Thaumatin (E-957) is obtained from *Thaumatococcus danielii*. Thaumatin contains two sweet proteins Thaumatin I and Thaumatin II and is approximately 2000 times sweeter than sucrose. Curculin from *Curculingo latifolia* and Miraculin from the fruit of *Synsepalum dulcificum* are both taste modifiers (Belitz et al., 2004). The mode of action seems to affect the taste buds, making sour or acid foods taste sweet. Brazzein and Pentadin are two sweetening proteins from the climbing plant Oubli (*Pentadiplandra brazzeana*) (Gao et al., 1999). Chemical stability of brazzein makes it a very interesting ingredient for food processing. Brazzein is stable over a broad pH range of 2.5–8 (Caldwell et al., 1998; Hellekant and Danilova, 2005) and heat stable at 98 °C for 2 hours (Ming and Hellekant, 1994; Assadi-Porter et al., 2000; Hellekant and Danilova, 2005).

Mabinlins are sweet-tasting proteins extracted from the seed of Mabinlang (*Capparis masaikai Levl.*), a Chinese plant from the region of Yunnan. The sweetness of mabinlin-2 is unchanged after 48 hours incubation at 80 °C (Kurihara, 1992; Liu et al., 1993) Mabinlin-3 and -4 sweetness stayed unchanged after 1 hour at 80 °C, while mabinlin-1 lost sweetness after 1 hour in the same conditions (Kurihara, 1992). Apart from sweetening proteins there

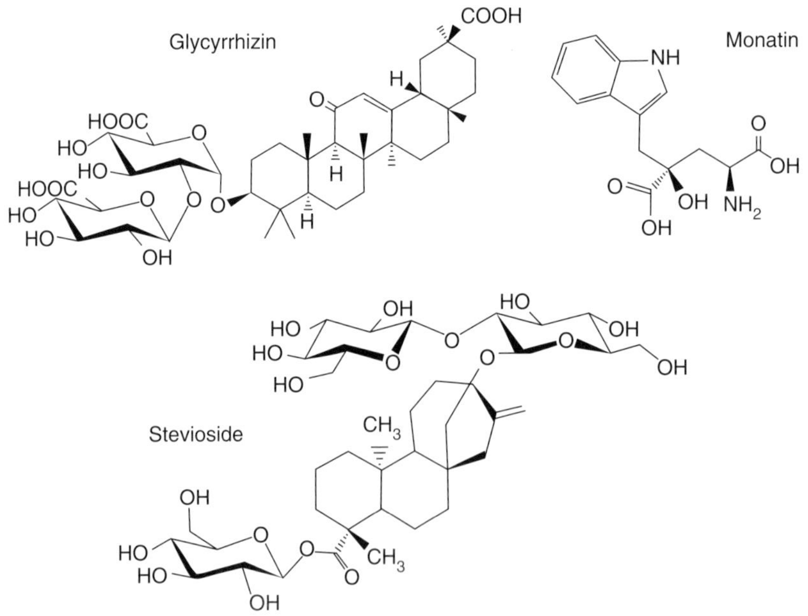

Figure 21.3 Chemical structure of some phytochemicals often used as sweetening agents in the food industry.

are other compounds in plants that can be used as sweeteners. For example, Monatin isolated from the South African plant *Sclerochiton ilicifoliu* (Abraham et al., 2005). Glycyrrhizin is a sweet-tasting compound obtained from liquorice root. It is 30–50 times sweeter than sucrose. Glycyrrhizin is a triterpenoid saponin glycoside of glycyrrhizic acid (Belitz et al., 2004) and it loses its sweetening properties when hydrolysed. The sweet taste of glycyrrhizin is different from sugar but it lasts for longer in the mouth. It is also quite stable upon heating. However the use of glycyrrhizin is limited due to its cortisone (anti-inflammatory) like properties (Akamatsu et al., 1991).

The leaves of stevia (*Stevia rebaudiana*) are rich (ca 6% in mass) in non-protein sugar substitute sweetener substances known as steviosides and rebaudiosides. They are heat-stable, pH-stable and do not ferment. Steviosides are diterpene glycosides of steviol. Its use has been limited due to unclear toxic properties. Stevioside has in addition a bitter taste, the Rebaudiside A, which is steviol β-linked to a glucose, and a glucose-glucose-glucose trisaccharide (also β-linked 2-1 and 3-1 with β-D-glucose) is less bitter and more polar.

A study performed in 1985 reported that steviol was mutagen (Pezzuto et al., 1985). However stevioside as a sweetener was evaluated by the Scientific Committee for Food (SCF) in 1984, 1989 and 1999. JECFA reviewed the safety of steviol glycosides (in 2000, 2005, 2006, 2007 and 2009) and established an ADI for steviol glycosides (expressed as steviol equivalents) of 4 mg/kg bw/day. In 2010 EFSA reviewed the use of stevioside glycosides as food additive establishing similar ADI levels as the JECFA (*EFSA Journal*, 2010, 8(4), 1537).

21.2.3 Colouring substances

Many other phytochemicals are used as colouring substances in food processing (Figure 21.4). These colouring substances are used to adjust or correct food discoloration or colour change during processing or storage (Belitz et al., 2004). There are four main types of phytochemicals used as colouring substances, anthocyanins, betalains, chlorophylls and carotenoids. Anthocyanins, a class of flavonoids derived ultimately from phenylalanine, are water-soluble. Anthocyanins can provide a large range of colours (orange/red to violet/blue)

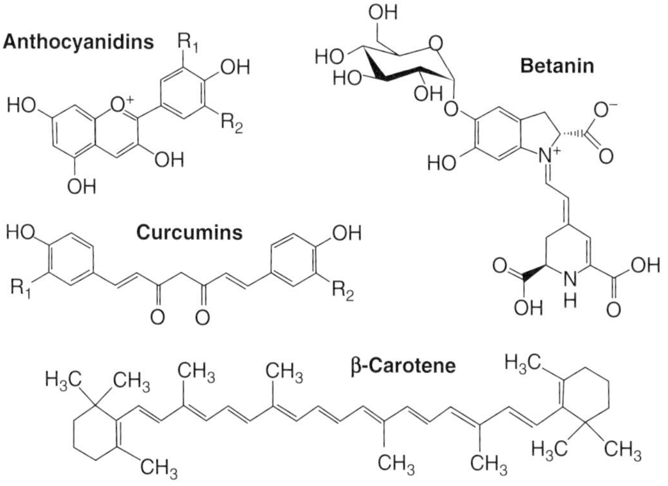

Figure 21.4 Chemical structure of some phytochemicals often used as colouring agents in the food industry.

as a function of their chemical structure and environment. Therefore slight modifications of their chemical structure may lead to colour changes, but their colour also depends on co-pigments, metal ions and pH. They are widely distributed in the plant kingdom and are industrially obtained by aqueous extraction of by-products such as fruit skins and peels (Schieber et al., 2001).

Betalains are nitrogen-containing water-soluble compounds derived from tyrosine that are found only in a limited number of plant lineages and present yellow-to-red colours. Some betalains have a stronger colouring capacity than anthocyanins and their colour exhibits higher pH stability (Stintzing and Carle, 2007; Tanaka et al., 2008; Azeredo, 2009). Betalains are obtained by aqueous extraction of beet roots and are constituted by betacyanins (red) and betaxanthines (yellow). Betanin represent 75–95% of main colouring principle. Interestingly anthocyanins and betalains are mutually exclusive and never have both been found in the same plant. Betalains and anthocyanins are used as colouring agents for fruit preparations, dairy products, ice creams, confectionery, pet-foods, soups, sauces, beverages and drinks.

On the other hand, most of the carotenes and carotenoids are lipid-soluble, yellow-to-red phytochemicals. Carotenes and carotenoids are a subclass of terpenoids, that often are obtained by solvent extraction of carrots (*Daucus carota*), oil palm fruit (*Elaeis guinensis*), sweet potato (*Ipomea batatas*), marigold (*Tagetes erecta*) and microalgae (*Spirulina platensis*). Carotenes and carotenoids are used in multiple industrial applications such as dairy (beverages, cream and dairy desserts), oils, fats and emulsions, fruit based products (spreads, desserts, canned fruits, pastry fillings), mustards, egg based products, soups and broths, edible coatings and cooked fish and fish products.

Chlorophylls and their derivatives are obtained by solvent extraction of grasses, alfalfa (*Mendicago sativa*), nettles (*Urticaceaes*) and other plants or algae materials. During the extraction of chlorophylls and their subsequent solvent removal, the naturally present coordinated magnesium may be wholly or partly removed from the chlorophylls to yield dark olive green pheophytins. In order to keep the bright green colour of chlorophylls the copper complexes of chlorophyll can be synthesised. Copper complexes of chlorophylls can be obtained by addition of an organic salt of copper. However, due to the toxicity of copper salts, its use is limited and maximum levels recommended in Codex Alimentarius of the GSFA (General Standard for Food Additives) are rarely above 500 mg/kg. Curcumin or turmeric yellow is obtained by solvent extraction of turmeric (ground rhizomes of *Curcuma longa* L.) and the extract is purified by crystallisation. This process eliminates the pungent and aromatic essential oil in turmeric, leaving deodorised turmeric which is used in dairy products and baked goods. Curcumin is relatively inexpensive and heat stable, but has poor light stability (Timberlake and Henry, 1986).

21.2.4 Antimicrobial agents/essential oils

Antimicrobials are used for either killing or inhibiting the growth of microorganisms. A large range of antimicrobial substances are used in the food industry (Figure 21.5). Most of them are small organic acids that are chemically synthesised, such as benzoic, sorbic or propoinic acids (Cowan, 1999). However plants present a vast chemical collection of antimicrobial substances that are synthesised to protect themselves and that in recent years have been explored as antimicrobial additives by the food industry as an alternative to chemical synthesised compounds which are perceived by consumers as unhealthy. There are three main groups of antimicrobial compounds in plants, phenolic/polyphenols; terpenoids and alkaloids (Cowan, 1999; Tiwari

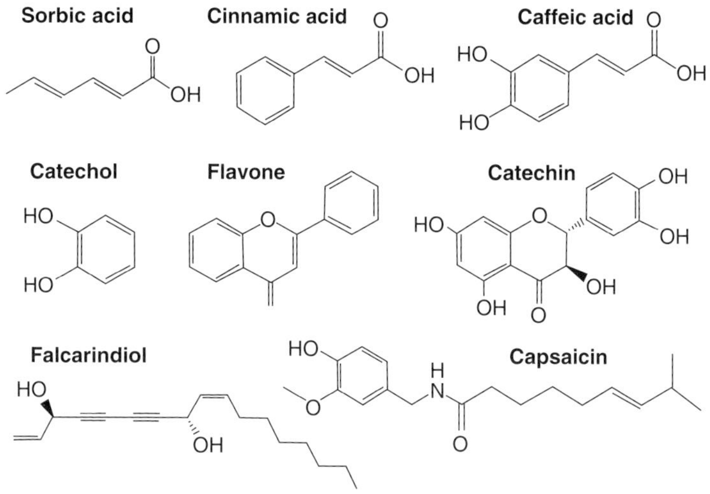

Figure 21.5 Chemical structure of some phytochemicals responsible for antimicrobial properties in some plants.

et al., 2009; Van Vuuren et al., 2009; Tajkarimi et al., 2010). Simple phenols and phenolic acids such as cinnamic or caffeic have shown strong antimicrobial properties against viruses, bacteria and fungi (Cowan, 1999). Catechol and pyrogallol, both hydroxylated phenols, have also shown strong toxicity towards microorganism and the number of hydroxyl groups on the phenol ring has been associated with their level of antimicrobial capacity. Flavones, flavonoids and flavonols are more complex phenolic structures, derived of the 2-phenyl-1,4 benzopyrone polyphenolic structure, and they are known to be synthesised by plants in response to microbial infection (Dixon et al., 2006). Several flavonoids have shown to have strong antimicrobial properties (Proestos et al., 2005).

C-17 polyacetylenes, such as falcarinol and falcarindiol (commonly found in the *Apiaceae* family such as carrots, parsnips, celery and fennel), have antibacterial effects against various micro-organisms such as gram-positive bacteria (*Bacillus* ssp., *Staplylococcus* ssp., *Streptococcus* ssp.) and gram-negative bacteria (*Escherichia* ssp., *Pseudomonas* ssp.) (Christensen et al., 2010). These polyacetylenes also present antimycobacterial effects, of which the most important seems to be the activity against *M. tubercolosis* (Kobaisy et al., 1997). These effects represent pharmacologically useful properties by which falcarinol and related polyacetylenes could have positive effects on human health and may be used to develop antibiotics (Christensen et al., 2010). Recently it was shown that falcarindiol strongly inhibited the growth of *Micrococcus luteus* and *Bacillus cereus*, with a minimum inhibitory concentration (MIC) value of $50\,\mu g\,mL^{-1}$ (Meot-Duros et al., 2010).

Capsaicin, one of nature's most pungent spices from plants from the capsicum genus (peppers), is known to have strong antimicrobial properties and is used traditionally to preserve foods (Belitz et al., 2004). Terpenoids or isoprenoids are the main constituent of essential oils, which are used due to their antimicrobial properties in food preservation (Cowan, 1999; Tiwari et al., 2009).

Essential oils, which are concentrated hydrophobic phases extracted from plants by distillation or solvent extraction, contain high amounts of some of the antimicrobial compounds already described (Van Vuuren et al., 2009). Therefore essential oils do not have any specific chemical or pharmaceutical properties in common, although the individual

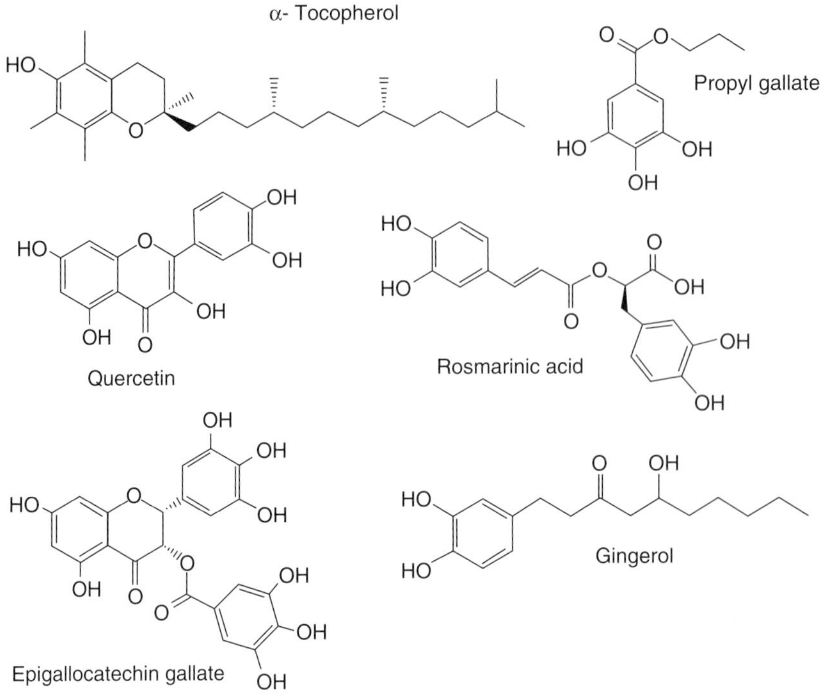

Figure 21.6 Chemical structure of some phytochemicals with antioxidant properties from different plant sources.

properties of essential oils have been extensively studied (Sacchetti et al., 2005). They are used for flavouring food and drinks but also in perfumes, cosmetics, soaps and cleaning products.

21.2.5 Antioxidants

Many phytochemicals have antioxidant properties. Antioxidants are compounds capable of inhibiting oxidation of other molecules (Figure 21.6). Oxidation, from a chemical point of view, is a reaction where the oxidation state of a compound is increased. This often happens by loss of electrons. Consequently oxidation reactions can generate free radicals. Free radicals are known to initialise chain reactions that can interfere with a multitude of biological processes. Antioxidants are able to terminate with chain reactions by removing free radicals, inhibiting further oxidation reactions. This is achieved by oxidising themselves into stable radicals that do not continue reacting. Tocopherols (vitamin D) are obtained by the vacuum steam distillation of edible vegetable oil product. D-α-tocopherol is the most abundant. Tocopherols are used as antioxidants in butter oil (ghee), anhydrous milk fat, fat spreads, dairy fat spreads and blended spreads at a maximum level of 500 mg/kg. Ascorbates (vitamin C) are also used as strong antioxidants, however most of their production nowadays is chemically synthesised. Esters from gallic acid with different alkyl alcohols (propanol, octanol and lauryl-alcohol) although present in nuts and other plants are also chemically synthesised and used in the food industry as antioxidants. Gum guaiacum (E-314) is an extract from the tree species *Guaiacum officinale* that is used as antioxidant. Other plant extracts rich in phenolics and/or flavonoids, such as cathechins from tea (*Camellia sinensis*), gingerol from ginger (*Zingiber officinale*) have shown strong antioxidant capacity (Aruoma *et al.*, 1997; Ho *et al.*, 1997; Moure *et al.*, 2001). Also terpenes and terpenoids such as carnosolic acid and carnosol from herbs like sage (*Salvia officinalis*) and rosemary (*Rosmarinus officinalis*) have shown to be strong participants in antioxidant activity of extracts and essential oils from these plants (Lagouri *et al.*, 1995; Frankel *et al.*, 1996). More recently potato peels and sugar beet pulp extract (Mohdaly *et al.*, 2010), bran and stalks from cereals (Esposito *et al.*, 2005; Lai *et al.*, 2009; Lerma-Garcia *et al.*, 2009) and onion skins (Roldan *et al.*, 2008) have been largely explored as sources of antioxidants that could be used in industry (Shahidi and Wanasundara, 1994; Wanasundara and Shahidi, 1994; Chotimarkorn and Silalai, 2008; Lerma-Garcia *et al.*, 2009).

21.3 Stabilisation of fats, frying oils and fried products

Rancidity or lipid oxidation can take place in three different ways:

1. Hydrolytic rancidity: occurs when water splits fatty acid chains from glycerol backbone in mono-, di- or tri-glicerides.
2. Oxidative rancidity: occurs when double bonds of unsaturated fatty acids react chemically with oxygen.
3. Microbial rancidity: occurs when microorganisms are involved in the lipid oxidation process. This takes places by enzymes called lipases.

These different types of rancidity can yield different products and therefore different sensory appreciation of the food products. For example, microbial rancidity takes place during the ageing of cheeses, developing flavours considered both desirable and/or unpleasant. However, chemical reactions such as hydrolytic or oxidative rancidity yield to mainly undesirable flavour products (Belitz *et al.*, 2004). Lipid oxidation requires oxygen to take place and can occur through enzymatic hydrolysis (lypoxygenases) or autoxidation. Autoxidation involves free-radical mediated reactions, often catalysed by the presence of metals or by the presence of light (UV) and irradiation. The mechanism of lipid autoxidation is complex and involves a vast number of interrelated reactions of intermediates. Model systems have been used in order to try and determine the mechanistic pathways of autoxidation (Shahidi and Zhong, 2005). The rate of autoxidation has been shown to depend on fatty acid composition, degree of insaturation, the presence of and activity of pro-and anti-oxidants, partial pressure of oxygen, the nature of the surface being exposed to oxygen and the storage conditions (Murado and Vazquez, 2010).

The autoxidation process can be explained more easily through a sequential free radical chain reaction mechanism. This chain reaction mechanism is constituted by three main steps: initiation, propagation and termination (Figure 21.7). The initiation step occurs when a hydrogen atom at α methylene group in double bond of the unsaturated fatty acid is removed to form an alkyl radical (R·). The initiation step is followed by the propagation step where the unstable alkyl radical generated in the initiation step reacts with oxygen (in triplet state) generating a peroxy free radical (Figure 21.8). The peroxy free radical continues reacting, propagating the chain reaction. The chain reaction finishes with the termination step, which occurs when the radicals formed during the propagation step react with other radicals generating stable products. The oxidation products generated (alkyl aldehydes) are responsible of the 'off-taste' or rancid flavours. The free radicals generated also react or damage other compounds including vitamins and proteins. The oxidation rate is affected by the

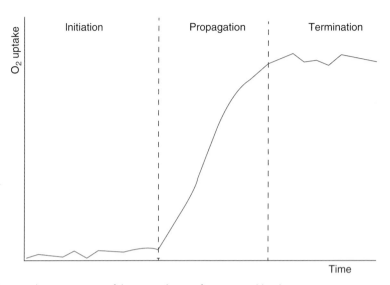

Figure 21.7 Elementary steps of the autoxidation of unsaturated lipids.

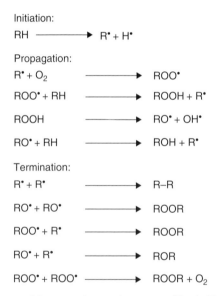

Figure 21.8 Elementary steps of the autoxidation of unsaturated lipids (details on the chain reaction mechanism are given).

number, position and geometry of the double bonds, the position of the fatty acid in the glycerol residue (those in positions 1 and 3 react more easily than those in position 2) and the temperature of the system (at higher temperatures, higher oxidation rates).

The initiation or the rate of lipid oxidation of the propagation can be delayed or slowed down by the presence of antioxidants. Antioxidants work by either inhibiting the formation of free radical lipids in the initiation step or by interrupting propagation of free radical chain. Generally antioxidants are believed to intervene in the chain reaction by donating a hydrogen atom to the peroxy free radicals formed in the propagation step, giving as a result

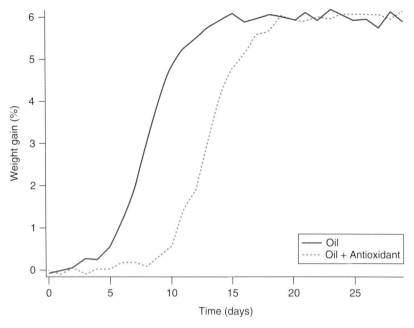

Figure 21.9 Increase of induction time of oxidation (measured as weight gain in %) by the addition of an antioxidant.

a peroxide and a rather stable radical. However, if this radical is able to trap a second peroxide radical; it is known to be an efficient inhibitor of the chain reaction. In the case of phenolics, flavonoids, ascorbic acid and tocopherols the free radical is stabilised through resonance delocalisation. Since the second half of the twentieth century, it is common practice to add synthetic antioxidants to fats and oils in order to stabilise them. Synthetic antioxidants used by industry include butylated hydroxyanisole (BHA, E-320), butylated hydroxytoluene (BHT, E-321), tertiary butylhydroquinone (TBHT, E-319) and propyl-gallate (E-310). The synthetic antioxidants are effective and cheap; however they are not well regarded by consumers who often prefer more natural products. Therefore research in this field has concentrated in the use of natural sources of antioxidants (spices, herbs, teas, seeds, cereals, grains, fruits and vegetables) as alternatives to the synthetic ones (Yanishlieva and Marinova, 2001). Moreover, mixtures of antioxidants can lead to synergism; so when two or more antioxidants are combined together their overall antioxidant capacity is statistically significantly higher than the addition of their individual antioxidants capacity. Substances such as ascorbyl palmitate, phospholipids or organic acids are known to have this reinforcing effect.

The effectiveness of the added antioxidant is estimated on the basis of the induction period (IP), usually this induction period is determined in time units. The comparison between the actions of various inhibitors in different lipid systems can be carried out by comparing the relative stabilisation factor (F), which is the ratio between the induction period with added antioxidants and the induction period of the control sample as shown in Figure 21.9. There have been many studies on the use of natural antioxidants for the stabilisation of edible oils. There is vast material written on the subject and Table 21.1 summarises the use of phytochemicals to the stabilisation of oil.

Table 21.1 Summary of studies using phytochemicals for the stabilisation of edible oils

Type of oil	Phytochemicals used	Comments	References
Corn oil	0.2% sesamol	Stability significantly higher than without sesamol. Sesamol seems to have a strong synergistic effect with tocopherols.	(Fukuda et al., 1986)
	Rosemary extract	In bulk corn oil, the rosemary extract, carnosic acid, rosmarinic acid, and alpha-tocopherol were significantly more active than carnosol.	(Frankel et al., 1996)
Cotton seed oil	0.03% of methanolic extracts from oat.	Pronounced antioxidants effect in oil. In some conditions the extract was more effective than TBHQ	(Tian and White, 1994)
Fish oils	Lecithin (0.02%) and tocopherol concentrates (2–0.02%)	Lecithin and tocopherols increased the oxidative stability of fish oils by synergistic effect.	(Hamilton, Kalu et al., 1998)
	Catechin, morin and quercetin	Effects comparable to that of α-tocopherol	(Nieto, Garrido et al., 1993)
	Rosemary (1% of ground plant)	It was 2.5 times more effective than 0.03% BHT.	(Lagouri, Boskou et al., 1995)
	Green tea extracts	Have shown to exhibit prooxidant effects in menhaden and seal blubber oil. This was due to chlorophyll constituents. When chlorophyll the extracts exhibit excellent antioxidant effect.	(Wanasundara and Shahidi, 1998)
	Oregano	Oregano at 1% (w/w) level had a similar effect to that of 200 ppm TBHQ in mackerel oil	(Tsimidou, Papavergou et al., 1995)
Peanut oil	Catechin	Significant increase of peanut oil stability	(Chu and Hsu, 1999)
	Methanolic extract from Ginger	More effective that BHT.	(Yanishlieva and Marinova, 2001)
	0.02% of essential oils	*Origanum majorana*, *Acanthaolippia seriphioides* and *Tagetes filifolia* exhibited pronounced antioxidant effects	(Maestri, Zygallo et al., 1996)
Rapeseed/ Canola oil	Spice extracts	Rosemary and sage showed strongest effect	(Shahidi and Wanasundara, 1994)
	0.1% Rosemary extract	Showed capacity to retain better tocopherols.	(Gordon and Kourimská, 1995)
	0.10% Ethanol extracts of canola meal	The ethanol extracts of canola meal were more effective than 0.02% of BHT.	(Wanasundara and Shahidi, 1994)

Oil	Antioxidant	Effect	Reference
	Flavonoids	Myricetin, epicatechin, naringin, rutin, morin and quercetin were superior than BHA and BHT in inhibiting oxidation	(Wanasundara and Shahidi, 1994)
	Green tea catechins	Catechin showed strong antioxidant capacity when compared to BHT.	(Chen, Chan et al., 1996)
Sunflower oil	0.001–0.02% β-carotene	Increased oxidative stability of sunflower oil	(Yanishlieva, Marinova et al., 2006)
	0.02% methanolic extracts from rosemary	The F value was 1.6, slightly higher than for BHT.	(Gamel and Kiritsakis, 1999)
	Marine algae extracts	1% Laminaria digitata and himanthalia elongata showed higher antioxidant activity than BHT at 0.05%.	(Le Tutour, 1990)
	Garlic extract	1% garlic extracts gave higher protection against oxidation than BHT	(Iqbal and Bhanger, 2007)
	Oil mill waste	Ethanol extract appeared to be a stronger antioxidant than BHT, ascorbyl palmitate and vitamin E	(Lafka, Lazou et al., 2011)
Soybean oil	Phosphatidylethanolamine (PE) and tocopherols	This mixture has shown synergistic effects that increased the stability of oil.	(Yanishlieva and Marinova, 2001)
	0.03% of methanolic extracts from oat.	The extracts gave better protection to bulk oil than TBHQ, this was even more marked in emulsions.	(Tian and White, 1994)
	Rice bran oil	Different ratios such as 25, 50 and 75% in SBO/RBO retarded significantly the oxidative process and hydrolytic rancidity in fried products during storage	(Chotimarkorn and Silalai, 2008)
Palm oil	Ficus exasperata leaves	The use of leaves from the plant Ficus exasperata decreased free fatty acid content and peroxide values.	(Umerie, Ogbuagu et al., 2004)
Rice bran oil	Naturally rich in antioxidant γ-oryzanol	Rice bran oil has been studied to be used in combination with other oils more prone to oxidation in order to improve their stability	(Lerma-Garcia, Herrero-Martinez et al., 2009)

Lipid oxidation during deep-frying takes place easily due to the high temperatures used (140–200 °C) and because the process is a semi-open system where the presence of oxygen from the atmosphere can be transferred at a high rate. Lipid oxidation has a major impact on the quality of the deep-fired product as well as the oil used for frying. Thus oxidation of fats is characterised by chemical changes such as a decrease of total unsaturated fatty acid content and the formation of polar and polymeric products (Yanishlieva and Marinova, 2001). During the frying process, drastic lipid oxidation takes place. Unpleasant flavours can be developed but also important nutrients from the oils can be degraded rapidly (such as tocopherols). Therefore there is an interest in use of antioxidants capable of resisting the frying process. Again extracts from plants have been used for this purpose. Studies on effect of changes in tocopherols in deep-fat-frying have shown that α-tocopherol is lost much faster than β-, γ- or δ-tocopherols, with a reduction of 50% α-tocopherol after four to five frying operations compared with values of about seven and seven to eight frying operations for β- and γ-tocopherol. The presence of rosemary extract in the frying oil has shown to have a marked reduction in the rate of loss of the tocopherols (Gordon and Kourimská, 1995).

There are potential uses of Pandan (*Pandanus amaryllifolius*) leaf extract in refined, blanched and deodorised palm oil. The extracts (optimum concentration 0.2%) significantly retarded oil oxidation and deterioration ($p < 0.05$), comparable to 0.02% BHT in tests such as peroxide value, anisidine value, iodine value, free fatty acid and oxidative stability index (OSI) (Nor et al., 2008).

Fried products are susceptible to changes over time in their sensory and nutritional quality as they have a layer of fat/oil covering them. This layer can degrade, taking into consideration that fried products often have a large contact surface and therefore are more prone to oxidation. However, some authors have considered the use of frying process to increase nutritional value (phytochemical content) of certain foods (Saguy and Dana, 2003). Frying has been shown to produce less deterioration of water-soluble nutrients and other phytochemicals (Saguy and Dana, 2003). This higher retention has been explained as being due to the fact that the temperature of frying a product is below 100 °C under the crust region, and the fact that water-soluble nutrients will not leach into the frying oil, which does happen when foods are boiled. However it does have an impact in lipid soluble nutrients. Some authors consider that the use of oils naturally or artificially enriched with phytochemicals could help to nutritionally improve some fried foods. Holland and co-workers studied the effect of oil uptake in French fries (Holland et al., 1991). Holland et al. reported that a portion of 100 g of French fries fired in vitamin E rich corn oil could represent 50% of recommended daily allowance (RDA) of vitamin E. Rossi et al. showed that the stability of vitamin E in vegetable oils during deep-fat-frying of French fries depends basically on two factors: (1) the fatty composition of the oil (in particular polyunsaturated fatty acid content) and (2) the type of vitamin E homologues (tocopherols and tocotrienols) present (Rossi et al., 2007). Another possibility for increasing the oxidative stability of fats and oils has been explored by enriching the food product in antioxidative phytochemicals before the frying step. This has been explored for dough-based fried products. Rice flour containing rice bran powder, which is rich in antioxidant γ-oryzanol (Lai et al., 2009), fried in soybean oil showed better PUFA's stability and TBARS values were lower (Chotimarkorn and Silalai, 2008).

Recently the influence of microencapsulation and addition of the phenolic antioxidant caffeic acid on the storage stability of olive oil has been reported (Sun-Waterhouse et al., 2011). Olive oil in the absence or presence of 300 ppm caffeic acid and encapsulated in 1.5% w/w sodium alginate shells was compared. The addition of caffeic acid increased the stability and total phenolic content of the final oil product. Oxidation changes were generally slower in the encapsulated oil samples. Both encapsulation and addition of caffeic acid preserved unsaturated fatty acids including C18:1 (omega-9), C18:2 (omega-6) and C18:3 (omega-3). Oil encapsulation method using alginate microspheres was shown to be a feasible approach to increasing olive oil stability. In addition the presence of caffeic acid in the olive oil provides additional protection to the oil and also improves the nutritional value of the final oil product in terms of elevated total phenolic content and desired unsaturated fatty acids (Sun-Waterhouse *et al.*, 2011).

Microencapsulation is a technique widely used in food manufacturing of powdered edible oil products (Gouin, 2004). Microencapsulation allows a prolonged stability over time by protecting core component oils from oxidation caused by light, moisture and oxygen (Heinzelmann and Franke, 1999; Suh *et al.*, 2007; Velasco *et al.*, 2009; Serfert *et al.*, 2010). However, the outer lipid fraction in the surface of the microcapsules is exposed to oxidation during processing or storage (Velasco *et al.*, 2009). In order to improve the stability of microcapsules two main strategies have been pursued: (1) minimising the lipid content on the surface of microcapsules by improving the microencapsulation efficiency (MEE) and (2) the addition of antioxidants to the lipid fraction to delay oxidative processes in the product. In this regard, the effect of microencapsulated γ-oryzanol as an antioxidant was evaluated during the heat treatment of animal fat lard (Suh *et al.*, 2007).

The stability of microencapsulated fish oil and the effect of various antioxidant phytochemicals and relative humidity has been explored by means of PV and TBARS. Without antioxidants, the encapsulated fat was around ten times more stable against oxidation than non-encapsulated fat. Lipophilic antioxidants (such as tocopherols) seem to be more effective than amphiphilic antioxidants (ascorbyl palmitate). Antioxidants were shown to be more efficient at low relative humidity values (Baik *et al.*, 2004).

Fish and flaxseed oils stability against oxidation has been shown to improve by microencapsulation in conjunction with antioxidants. Flaxseed oils with variable levels of vitamin E, rosmarinic acid in addition to carnosic acid for fish oils, were encapsulated and their oxidative stability was tested over time (Barrett *et al.*, 2010). Stability for both fish oils was improved with encapsulation, most significantly for flaxseed oil rather than for fish oil. Fish oil encapsulated with antioxidants had improved stability, mostly significant with carnosic acid. Results were not so promising for flaxseed oils (Barrett *et al.*, 2010).

Natural product extracts showed great potential for their application as antioxidants in microencapsulation of oil products in the food industry (Ahn *et al.*, 2008). The use of natural plant extracts such as rosemary, broccoli sprout and citrus has been shown to effectively inhibit the lipid oxidation of microencapsulated high oleic sunflower oil. High microencapsulation efficiency was achieved using dextrin-coating method with milk protein isolates, soy lecithins and sodium triphosphate as supplements. Stability was tested by using Rancimat method, peroxide value (PV) p-anisidine value (ASV) and showed that induction period was significantly increased in presence of natural product extracts.

In this section we have detailed the stabilisation of fats, frying oils and fried products by using antioxidant phytochemicals. However, often fats are constituents of more complex foods or used for other uses rather than frying (such as sauces, emulsions, spreads/margarines). Oxidation in these products also occurs but it generally takes places at much lower rate due to structural constraints that lead to lower exposure of fat to oxygen (as happens by microencapsulation effect). Nevertheless some of these products are still sensible to oxidation and much effort has been made to avoid their spoilage during storage. The following section will consider the use of certain phytochemicals for the stabilisation of other food products different from fats and oils for frying.

21.4 Stabilisation and development of other food products

21.4.1 Anti-browning effect of phytochemicals in foods

The discoloration or colour change of some food products (in particular fruits and fruit derived products) is a major concern for the food industry (Wrolstad and Wen, 2001). Most of this discoloration and/or browning is caused by exposure to air and subsequent oxidation of colouring compounds into non-coloured compounds. Oxidation of fruits involves an enzyme-catalysed oxidation of phenolic compounds present in the fruit. Browning of a fruit typically occurs following a mechanical injury to the fruit, such as during the harvesting or processing of such foods. Traditionally sulfites have been used to inhibit the enzymatic oxidation and browning in 'fresh-cut' and processed fruits. Nevertheless, the increase in regulatory attention and consumer awareness of some risks associated with sulfites has drawn some attention to other anti-oxidation and anti-browning alternatives. Since a segment of the population is hypersensitive to sulfites, therefore, food processors prefer to avoid using sulfite compounds and it has been reported that sulfites can have an increased risk for asthmatic patients (Mathison *et al.*, 1985; Bush *et al.*, 1986). As in the case for other food additives, the food industry is particularly interested in the use of browning inhibitors from natural sources rather than synthetic. Other oxidation and browning inhibitors such as citric acid and phosphates in combination with ascorbic acid, however, are not sufficiently effective. The use of inhibitor 4-hexyl resorcinol is limited in the United States, Canada and some Latin American countries to use with shrimp (Montero *et al.*, 2004). Even if 4-hexyl resorcinol was approved for use in other products it is not certain that it would be used by food processors due to be derived from a synthetic chemical rather than from a natural source (Guandalini *et al.*, 1998; Son *et al.*, 2001). The use of other natural inhibitors of polyphenol oxidases (PPOs) has been motivated by the need to replace sulfating agents in order to prevent or minimise the loss of fresh or processed foodstuffs (Billaud *et al.*, 2003).

21.4.1.1 Sulfur-containing compounds

Some sulfur-containing substances such as N-acetylcysteine and reduced glutathione are natural compounds with antioxidant properties, and have been proposed as browning inhibitors to prevent darkening in apple, potato and fresh fruit juices (Friedman and Molnar-Perl, 1990; Molnar-Perl and Friedman, 1990; Friedman *et al.*, 1992; Friedman and Bautista, 1995). Sulfur-containing anti-browning additives seem to react with o-quinones formed during the initial phase of enzymatic browning reactions to yield colourless addition products or to reduce o-quinones to diphenols (Richard-Forget *et al.*, 1992). This means that the sulphur containing compounds are not PPO enzymes inhibitors *per se*, but that they intervene by conjugating with some primary oxidation products formed in this reaction (Richard-Forget *et al.*, 1992; Billaud *et al.*, 2003). Studies have revealed that different PPO enzymes from different plant sources react differently with the same sulphur-containing compound (Sapers and Miller, 1998; Billaud *et al.*, 2003). Fresh-cut apples and pears dipped into an anti-browning solution containing N-acetylcysteine and/or glutathione have been shown to inhibit browning in comparison to non-dipped slices (Rojas-Graü *et al.*, 2006). Pineapple juice has also been shown to inhibit browning and oxidation of fresh fruit (Lozano-De-Gonzalez *et al.*, 1993). The anti-browning/antioxidant effectiveness of pineapple juice is, however, unacceptably variable for use in the food industry. The effectiveness of the pineapple juice as an anti-browning/antioxidising agent varies depending on the type, cultivar and where the pineapple was grown. Specific methods for making natural anti-browning/antioxidant compositions from pineapple juice and/or from pineapple processing plant waste streams have been reported and patented (Wrolstad and Wen, 2001). These compositions are known to comprise S-sinapyl-L-cysteine, N-L-δ-glutamyl-S-sinapyl-L-cysteine and S-sinapyl glutathione as active ingredients.

Onion has been found to have low molecular weight sulphur bioactive compounds capable of reducing enzymatic browning and/or oxidoreductase activity (Eissa et al., 2006). The PPO activities of avocado fruit were significantly reduced by the different onion by-products analysed (Roldan et al., 2008). In addition, some technological and stabilisation processes applied to onion may significantly influence their PPO inhibition capacity. Heated onion extracts have shown to be more effective in prevention of pear and banana browning than fresh onion extracts (Kim et al., 2005). The addition of heated onion extract exhibited stronger inhibitory effect on peach polyphenol oxidase activity than that of the fresh one. The retardation of peach juice browning by onion extract seemed to be caused by inhibition of peach PPO (Kim and Kim, 2007).

The positive effect of a temperature rise in onion extracts towards PPO inhibition of different fruits or vegetables has been widely studied (Ding et al., 2002; Kim and Kim, 2007; Lee, 2007; Roldan et al., 2008). For example, higher anti-browning activity was found in sterilised by-products than in pasteurised and frozen ones. On the other hand, sterilisation of sugar rich products (such as onion) induces the formation of caramelisation and Maillard reaction products with marked sensory properties that could influence final product quality. Therefore milder processes like pasteurisation have been suggested as a more suitable choice in order to develop a food ingredient with an interesting added anti-browning property (Roldan et al., 2008). In addition the safety of the food ingredient could be maintained by the thermal treatment.

21.4.1.2 Phenolic acids

Although PPO normally are able to oxidise phenolic compounds, certain phenolic acids are able to inhibit PPO activity by binding to the active site of the enzyme (Janovitz-Klapp et al., 1990). Kojic acid (5-hydroxy-2-(hydroxymethyl)-4-pyrone) is a phenolic acid from fungal origin produced by many species of Aspergillus and Penicillium that it has been suggested acts as an inhibitor to the PPO enzyme (Chen et al., 1991; Iyidogan and BayIndIrlI, 2004). Chen et al. suggested that the mechanism action of Kojic acid is probably due to its capacity to interfer with the uptake of O_2 required for the enzyme reaction (Chen et al., 1991). This reduces o-quinones to diphenols preventing the formation of melanin via polymerisation and/or by combination of the two previous processes (Chen et al., 1991). Son et al. reported that the minimal concentration of kojic acid for effective anti-browning activity on fresh-cut apples is similar to the concentration for commercial anti-browning additives such as oxalic and cysteine (Son et al., 2001). Other phenolic acids such as p-Coumaric acid, ferulic acid, cinnamic acid and gallic acid showed similar inhibitory activity to ascorbic acid, but chlorogenic acid and caffeic acid were much weaker ($p < 0.05$) (Son et al., 2001).

21.4.1.3 Edible coatings

As mentioned already, the enzymatic browning is often inhibited by a direct immersion of the fruit pieces in an aqueous solution of anti-browning agents. On the other hand, the use of edible coatings as carriers of anti-browning agents for fresh-cut products has also been investigated (Baldwin et al., 1996; Rojas-Graü et al., 2006; Montero-Calderón et al., 2008; Oms-Oliu et al., 2008; Oms-Oliu et al., 2008; Rojas-Graü et al., 2008; Rojas-Graü et al., 2009; Oms-Oliu et al., 2010). Baldwin et al. (1996) reported improved browning inhibition on fresh-cut apples by ascorbic acid incorporated into an edible coating formulation when compared to the apples dipped directly into an aqueous solution of the very same compound (Baldwin et al., 1996). Other authors have reported improved browning inhibition in fresh-cut apples, pears and papayas coated with alginate and gellan based coatings with presence of thiol containing compounds N-acetylcysteine and glutathione (Rojas-Graü et al., 2006; Rojas-Graü et al., 2007; Tapia et al., 2007; Tapia et al., 2008). Some polysaccharide-based coatings (alginate based) are applied in a first step and the anti-browning agents are incorporated afterwards in a dipping solution containing calcium for cross-linking and instant gelation of the coating (Wong et al., 1994; Lee et al., 2003). Increased levels of vitamin C and total phenolic content were observed in pear wedges coated with alginate, gellan and pectin including N-acetylcysteine and glutathione compared with control samples (Oms-Oliu et al., 2008). Tapia and co-workers reported that the addition of ascorbic acid in alginate- and gellan-based coatings helped to preserve the natural ascorbic acid content of fresh-cut papaya (Tapia et al., 2008). The effect of application of a chitosan coated film on enzymatic browning of litchi (*Litch chinensis* Sonn.) fruit was studied by Zhang and Quantick (1997). It was reported that chitosan film coating delayed changes in contents of anthocyanins, flavonoids and total phenolics (Zhang and Quantick, 1997; Jiang et al., 2005). It also delayed the increase in polyphenol oxidase activity and partially inhibited the increase in peroxidase activity. These authors further reported that application of chitosan may form a layer of film on the outer pericarp surface, thus resulting in less browning. The mechanism of action of chitosan is unclear, but may involve adsorption of PPO, its substrates or products, or a combination of such processes.

21.4.2 Colour stabilisation in meat products

The impact of protein oxidation on the quality of meat and meat products is manifested by the presence of a free radical chain similar to those described for lipids. Oxidative degradation of meat proteins involves the modification of amino acid side chains leading to the formation of carbonyl compounds (Xiong, 2000; Stadtman and Levine, 2003). The discolouration of raw burger patties is generally attributed to the oxidation of ferrous heme–iron (Fe^{2+}) into its ferric form (Fe^{3+}) in proteins induced by lipid products (Yin and Faustman, 1993). Oxymyoglobin is transformed into metmyoglobin and consequently the colour changes the characteristic bright red colour of fresh meat to brownish colour, often considered as undesirable (except for aged beef or game). The oxidising proteins could affect the tenderness of fresh pork during chill storage and protein oxidation has been suggested to likely impact certain sensory quality attributes such as colour, texture and flavour of cured products (Ventanas et al., 2007). Other meat products such as burger patties are even more susceptible to oxidation as mincing and salt addition might promote oxidative reactions (Ladikos and Lougovois, 1990). Ganhao et al. (2010) reported significant increases of protein carbonyl during chill storage of burger patties (Ganhão et al., 2010). In parallel, an intense loss of redness and increase in hardness was found to take place throughout the refrigerated storage. The effect of several fruit extracts and quercetin in these trends showed that most phenolic rich wild Mediterranean fruit extracts as well as quercetin reduced the formation of protein carbonyls and inhibited the colour and texture deterioration during refrigerated storage

(Ganhão *et al.*, 2010). Yin and Cheng (2004) showed that isolated sulphur containing compounds from garlic (diallyl sulfide, diallyl disulfide, s-ethyl cysteine and n-acetyl cysteine) inhibited discoloration of ground beef. The exogenous addition of these garlic-derived compounds delayed oxymyoglobin formation significantly (Yin and Cheng, 2003; Sallam *et al.*, 2004; Bozin *et al.*, 2008). In addition, Yin and Cheng showed that these sulphur compounds significantly inhibited the growth of pathogenic bacteria such as *Salmonella typhimurium*, *Escherichia coli* O157:H7, *Listeria monocytogenes*, *Staphyllococcus aureus* and *Campylobacter jejuni*, suggesting that these compounds are interesting for microbiological safety and extending the shelf life of several food products (Yin and Cheng, 2003). The use of phytochemicals as antimicrobials to extend shelf life is considered in the section 21.4.4.2.

21.4.3 Antimicrobials to extend shelf life

Antimicrobial compounds present in essential oils from certain plants can be used to extend the shelf life of foods (Holley and Patel, 2005; Gutierrez *et al.*, 2008; Gutierrez *et al.*, 2009; Tiwari *et al.*, 2009). These antimicrobials extend the shelf life by reducing microbial growth rate or viability (Tiwari *et al.*, 2009). When used as food additives, essential oils might sometimes only be effective in high concentrations (1–3%). These concentrations are often above those that are organoleptic acceptable (Lis-Balchin *et al.*, 1998; Arora and Kaur, 1999; Tajkarimi *et al.*, 2010). The presence of fat, carbohydrate, protein, salt and pH reaction influence the effectiveness of these agents in foods (Holley, 2005). There are many examples of antimicrobial inactivation of essential oils on foods in order to improve their shelf life. The use of essential oils as antimicrobials to improve shelf life were reviewed by Burt (2004) and later by Holley and co-workers (Holley and Patel, 2005) who highlighted two examples where spice/herbal materials have been successfully used as either a dip on poultry carcasses (Dickens and Ingram, 2001) or as a surface coating on salt water fish (Harpaz *et al.*, 2003). In particular some phytochemicals have shown strong antimicrobial properties and consequently their mechanism of action has been studied in more detail. This is the case for allyl isothiocyanate (AIT) commonly found in mustard and horseradish oil; diallyl sufide and diallyl disulfide from garlic and related alliums; eugenol from clove, carvone from spearmint; cinnemaldehyde from cinnamon and carvacrol and thymol from oregano and thyme, respectively.

Due to the great antimicrobial activity that garlic and onion possess, both vegetables could be used as natural preservatives, to control the microbial growth (Pszczola, 2002). Chemical characterisation sulphur compounds contained in garlic have allowed the statement that they are the main active antimicrobial agents (Tsao and Yin, 2001; Rose *et al.*, 2005). However, other compounds such as proteins, saponins and phenolic compounds can also contribute to this activity (Griffiths *et al.*, 2002). Garlic has been proven to inhibit the growth of grampositive, gram-negative and acid-fast bacteria, as well as toxin production. Bacteria against which garlic is effective include strains of *Pseudomonas*, *Proteus*, *Escherichia coli*, *Staphylococcus aureus*, *Klebsiella*, *Salmonella*, *Micrococcus*, *Bacillus subtilis*, *Mycobacterium* and *Clostridium* (Delaha and Garagusi, 1985), some of which are resistant to penicillin, streptomycin, doxycilline and cephalexin, among other antibiotics.

Allyl isothiocyanate is derived from the glucosinolate sinigrin found in plants of the family *Brassicaceae*. It is a well-recognised antimicrobial agent against a variety of organisms, including food-borne pathogens such as *Escherichia coli* O157:H7 (Luciano and Holley, 2009). Interestingly allyl isothiocyanate has been found to be generally more effective against gram-negative bacteria with less or no effect on LAB. AIT possesses strong antimicrobial activity against *E. coli* O157:H7 as well as *V. parahaemolyticus* in ground beef

(200–300 ppm) after 21 days at 4 °C (Nadarajah *et al.*, 2005). AIT, along with other thiocyanates, is known to react with thiols and sulphydryls as well as terminal amino acids, and these reactions may contribute to its loss from products during storage (Ward *et al.*, 1998; Wang, 2003; Wang and Chen, 2010; Wang *et al.*, 2010). In addition allyl isothiocyanate has been reported to have bactericidal effects against *Helicobacter pylori*, which has been investigated due to its association with infections and upper gastrointestinal diseases, such as chronic gastritis, peptic ulcer and gastric cancer (Shin *et al.*, 2004).

21.5 Nutracetical applications

The term nutraceutical was coined as a contraction of nutritional and pharmaceutical by DeFelice and the Foundation for Innovation medicine (Wildman, 2007). At the present time there is no universally accepted definition but the term has evolved since it was first coined and nowadays it is generally accepted that a nutraceutical is any substance that may be considered a food or part of a food and that provides medical or health benefits, including the prevention and treatment of disease. Such products may range from isolated nutrients, dietary, supplements and diets to genetically engineered foods, herbal products and processed foods such as cereals, soups and beverages (Wildman, 2007). On the other hand, functional foods are generally considered to be 'foods or dietary components that may provide a health benefit beyond basic nutrition' (Wildman, 2007).

21.5.1 Phytosterol and phytostanol enriched foods

Phytosterols are plant derived compounds with a similar chemical structure and function to cholesterol. Many clinical trials have reported a cause and effect relationship between the consumption of plant sterols and the reduction of blood cholesterol levels. Phytosterols inhibit the intestinal absorption of cholesterol. Some foods are naturally rich in phytosterols like unrefined vegetable oils, whole grains, nuts and legumes and some of these products have been used as ingredients for processed foods and beverages with added plant sterols or stanols. Many of these products are now available in many countries, and many countries allow health claims for such commercial products. Presently the EU regulatory authorities on food and feed safety have raised and registered 14 questions regarding the use of phytosterols. Ten of these questions have been assed and a scientific opinion has been published; in particular two questions regarding the use of health claims that phytosterols in functional foods reduce blood cholesterol levels have been approved (*EFSA Journal*, 2010, 8(10), 1813–1835). On the other hand, two regarding the effect of phytosterols on prostate cancer have been rejected (*EFSA Journal*, 2010, 8(10), 1813–1835). Three are still under consideration and one, regarding the use of phytosterols in low fat fermented milk product, has been withdrawn by the application (EFSA-Q-2008-3823).

21.5.2 Resveratrol enriched drinks and beverages

Resveratrol is a phenolic compound commonly found in grapes, red wine and berries. Resveratrol has been considered among certain authors to be responsible of the 'French paradox': the fact that French nationals have lower incidence of heart disease than other Westerners despite high red wine consumption. However some other authors claim that resveratrol is not present in sufficient quantities in red wine to explain this paradox.

Resveratrol is well-absorbed when taken orally, but is also rapidly metabolised and eliminated. Cellular and animal models have shown very remarkable results on the capacity of resveratrol to inhibit the growth of cancer cells and to increase lifespan of animal models, but to date little is known about the effect of resveratrol in humans. However this has not stopped the food industry from commercialising beverages and supplements rich in resveratrol and claiming its health benefits. Indeed there are four registered questions by the EU regulatory authorities in food and feed safety on the health claims of resveratrol. Two of them have been assessed and rejected on the basis that the data supplied by applications were insufficient to explain the cause and effect relationship between the consumption of resveratrol and health benefits. Both of the rejected questions assumed that the mechanism of action of resveratrol is based in its antioxidant activity, an assumption that EFSA scientific panel considers insufficient to justify several potential health benefits. The other two questions are still under consideration: one related to cardiovascular health. Nevertheless, in 2009, the Food and Drug Administration of the US considered the status of resveratrol and it was upgraded from GRAS (generally recognised as safe) status to novel food. On the other hand, the Danish Council for strategic research announced in February 2011 that it will commit € 2.47 million to a complete resveratrol study investigating multiple metabolic syndrome endpoints, including obesity, type-2 diabetes and osteoporosis, for the next five years.

21.5.3 Isoflavone enriched dairy-like products

Isoflavones are a class of phytoestrogens (estrogens from plant origin). Soy and soy products are comparatively rich sources of isoflavones in the human diet. It has been claimed that isoflavones are beneficial for the prevention of cardiovascular diseases, hormone associated cancers, osteoporosis, cognitive decline and the treatment of menopausal symptoms. However, in most cases mixed results have been reported. Some of the initial assumptions of the beneficial effects of isoflavones were based on the review of epidemiological studies (Clarkson, 2002; Messina *et al.*, 2004; Cassidy and Hooper, 2006) and these assumptions were not verified with later clinical trials (Lichtenstein *et al.*, 2002; Weggemans and Trautwein, 2003; Dewell *et al.*, 2006; Brink *et al.*, 2008). It has been suggested that differences in results obtained by several studies are due to the significant differences in bioavailability of isoflavones in function of the food matrix used for their delivery (de Pascual-Teresa *et al.*, 2006).

Most of the nutraceutical applications of isoflavones or soy protein (rich in isoflavones) are related to dairy-like products such as milk-soy, soy drinks and beverages, yogurts and dairy desserts. At the present time there are 17 registered health claims applications that have been submitted to EFSA. Three have already been reviewed; two are related to bone health and the third to the antioxidant capacity of isoflavones. All three have been rejected by the scientific panel who considered that in the applications there was insufficient data to explain cause and effect relationship between isoflavones and the claimed health benefits (*EFSA Journal*, 2009, 7(9), 1267–1282; *EFSA Journal*, 2009, 7(9), 1270–1284; *EFSA Journal* 2010, 8(2), 1493–1515).

21.5.4 β-glucans

Polysaccharides of D-glucose, linked by β-glycosidic bonds or β-glucans, are a wide range of molecules of different molecular size. Therefore β-glucan rich fractions have significant differences in solubility and viscosity. β-glucans are found in the bran of many cereal grains and the cell wall of mushrooms. Most nutraceutical applications of β-glucans are oat based. Commonly a soluble fibre rich in β-glucans (above 30%) is introduced into a food formulation, such as smoothies, ready-meals, condiments, dressings and bakery products. Most of these products are low in fat and fine products that claim to be beneficial to heart health. There are 34 registered applications to be reviewed by EFSA which are related to β-glucans. Many scientific opinions have delivered; those relating the mechanism of action to β-glucans to their antioxidant capacity have been rejected. There are many still under revision and there is one, approved in December 2010, which links the consumption of oat β-glucans. The report from the scientific panel states that a cause and effect relationship has been established between the consumption of oat β-glucan and lowering of blood LDL-cholesterol concentrations. Therefore the scientific panel considered that 'Oat beta-glucan has been shown to lower/reduce blood cholesterol. Blood cholesterol lowering may reduce the risk of (coronary) heart disease' (*EFSA Journal* 2010, 8(12), 1885–1900). This confirms earlier health claims from national food and safety authorities, such as the French and Swedish, that approved the claim of cholesterol lowering of oat β-glucan in 2009.

21.5.5 Flavonoids

Nowadays there are a wide range of supplements rich in flavonoids from numerous plant extracts (tea, berries, grape and herbs) that can be easily found in so-called health stores. Also some drink and beverage companies are using flavonoids to target the healthy product market. Large worldwide producers of soft and fizzy drinks have been launching variations of their existing products but enriched with polyphenols. This is the case for Coca Cola company, that launched in October 2007 Diet Coke plus Antioxidants in the UK, a year later in France and has recently launched in Brazil. Some large drink and beverage companies have shown their interest in research on the bioavailability of polyphenols for this kind of drink (Borges *et al.*, 2010).

The fortification or enrichment of polyphenols has been hindered by the chemical properties of some polyphenols, which are prone to interact with proteins (Labuckas *et al.*, 2008; Han *et al.*, 2011). These reactions can have significant effects in nutritional and sensory quality of the enriched products. However the use of polyphenols in cheese has been explored lately. Purified phenolic compounds such as catechin, epigallocatechin gallate (EGCG), tannic acid, homovanillic acid, hesperetin and natural extracts rich in polyphenols like grape, green tea or cranberry extract were added to a prepared cheese. They were shown to have an impact in the retention time of the cheese curds and their gel-formation behaviours. The authors observed that the effects were related to molecular properties and in particular hydrophobicity of phenolic compounds (Han *et al.*, 2011). There are at least 44 health claim applications for flavonoids registered by EFSA. All the claims that related their health benefits to antioxidant capacity have been rejected (as mentioned for many other phytochemicals). There are 18 applications still under revision and at least one application that has been withdrawn.

21.6 Miscellaneous industrial applications

21.6.1 Cosmetic applications

The cosmetic industry uses phytochemicals or extracts from phytochemical rich plants in order to improve the sensory attributes of the product and/or to improve the technical properties of a formulation. For example, polysaccharides are commonly used for stabilising emulsions, foams and gels in many cosmetic products. Phenolic compounds and some carotenoids are used in cosmetics that protect skin from UV-light. These compounds absorb in the UV region, creating the effective radiation filter for sunscreens. Some of the phytochemicals used in the cosmetic industry are triterpene saponins (Balandrin, 1996; Ceppi, 1998) and phenylethanoid glycosides. Triterpene saponins are triterpenes that belong to the group of saponins. The amphiphilic characteristics of these compounds makes them an important surface active compound potentially interesting for stabilising complex dispersed systems such as emulsions and foams (Sarnthein-Graf and La Mesa, 2004; Wang *et al.*, 2005). In addition saponins in general have been shown to possess a vast array of chemical/biological properties and were reviewed by Francis *et al.* (2002).

Phenylethanoid glycosides are a type of phenylpropanoid and therefore they hold a series of antioxidant, antimicrobial and colour stabilising activities (inhibitory properties of tyrosinases) (Kurkin, 2003). In addition to these general properties phenylethanoids have shown great potential for commercial/industrial exploitation in the cosmetics industry (Kurkin, 2003; Fu *et al.*, 2008). Due to their ability to inhibit the 5α reductase enzyme the phenylethanoid glycosides present anti-seborrheic properties and therefore are of great use in the cosmetics industry for development of skin and hair care products (Korkina, 2007).

21.6.2 Bio-pesticides

Some phytochemicals have been shown to present anti-nematode activities. Nematodes represent a serious threat for plants in agronomics. In recent years new methods of pest control using more environmentally friendly (or natural) pesticides have been proposed. Although these are never as effective as the synthetic ones, they could help to reduce the amount of less environmental friendly pesticides used in farming. Integrated pest management, transgenic plant resistance and biological control strategies are being investigated as methods of control (Ghisalberti and Atta, 2002). Glucosinolates, glucosinolate derived compounds, alkaloids, terpenes, phenylpropanoids and sesquiterpenes are some of the natural products from plant origin that have shown biopesticide properties (Ujváry and Robert, 2009; Gonzalez-Coloma *et al.*, 2010; Oka, 2010).

References

Abraham, T.W., Cameron, D.C. *et al.* (2005) Beverage compositions comprising monatin and methods of making same U.S.P. Office. United States, Cargill, Inc.: 81.
Ahn, J.-H., Kim, Y.-P. *et al.* (2008) Antioxidant effect of natural plant extracts on the microencapsulated high oleic sunflower oil. *Journal of Food Engineering*, 84(2), 327–334.
Akamatsu, H., Komura, J. *et al.* (1991) Mechanism of anti-inflammatory action of glycyrrhizin: effect on neutrophil functions including reactive oxygen species generation. *Planta Medica*, 57(2), 119–121.
Akhtar, S. and Ashgar, A. (2011) Mineral Fortification of whole wheat flour: an overview. In: V.R. Preedy, R.R. Watson and V.B. Patel (eds) *Flour and Breads and their Fortification in Health and Disease Prevention*, San Diego, Academic Press, pp. 263–271.
Arora, D.S. and Kaur, J. (1999) Antimicrobial activity of spices. *International Journal of Antimicrobial Agents*, 12(3), 257–262.
Aruoma, O.I. and Spencer, J.P.E. *et al.* (1997) Characterization of food antioxidants, illustrated using commercial garlic and ginger preparations. *Food Chemistry*, 60(2), 149–156.
Assadi-Porter, F.M., Aceti, D.J. *et al.* (2000) Efficient Production of Recombinant Brazzein, a Small, Heat-Stable, Sweet-Tasting Protein of Plant Origin. *Archives of Biochemistry and Biophysics*, 376(2), 252–258.
Azeredo, H.M.C. (2009) Betalains: properties, sources, applications, and stability – a review. *International Journal of Food Science and Technology*, 44(12), 2365–2376.

Baik, M., Suhendro, E. et al. (2004) Effects of antioxidants and humidity on the oxidative stability of microencapsulated fish oil. *Journal of the American Oil Chemists' Society*, 81(4), 355–360.

Balandrin, M.F. (1996) Commercial utilization of plant-derived saponins: an overview of medicinal, pharmaceutical, and industrial applications. *Advances in Experimental Medicine and Biology*, 404, 1–14.

Baldwin, E.A., Nisperos, M.O. et al. (1996) Improving storage life of cut apple and potato with edible coating. *Postharvest Biology and Technology*, 9(2), 151–163.

Barrett, A.H., Porter, W.L. et al. (2010) Effect of various antioxidants, antioxidant levels, and encapsulation on the stability of fish and flaxseed oils: assessment by fluorometric analysis. *Journal of Food Processing and Preservation*, no.

Belitz, H.–D., Grosch, W. et al. (eds) (2004) *Food Chemistry*, Berlin, Springer-Verlag.

Billaud, C., Roux, E. et al. (2003) Inhibitory effect of unheated and heated -glucose, -fructose and -cysteine solutions and Maillard reaction product model systems on polyphenoloxidase from apple. I. Enzymatic browning and enzyme activity inhibition using spectrophotometric and polarographic methods. *Food Chemistry*, 81(1), 35–50.

Borges, G., Mullen, W. et al. (2010) Bioavailability of multiple components following acute ingestion of a polyphenol-rich juice drink. *Molecular Nutrition and Food Research*, 54(S2): S268–S277.

Bozin, B., Mimica-Dukic, N. et al. (2008) Phenolics as antioxidants in garlic (Allium sativum L., Alliaceae). *Food Chemistry*, 111(4), 925–929.

Brink, E., Coxam, V. et al. (2008) Long-term consumption of isoflavone-enriched foods does not affect bone mineral density, bone metabolism, or hormonal status in early postmenopausal women: a randomized, double-blind, placebo controlled study. *The American Journal of Clinical Nutrition*, 87(3), 761–770.

Burt, S. (2004) Essential oils: their antibacterial properties and potential applications in foods: A review. *International Journal of Food Microbiology*, 94(3), 223–253.

Bush, R.K., Taylor, S.L. et al. (1986) Prevalence of sensitivity to sulfiting agents in asthmatic patients. *The American Journal of Medicine*, 81(5), 816–820.

Caldwell, J.E., Abildgaard, F. et al. (1998) Solution structure of the thermostable sweet-tasting protein brazzein. *Nature Structural Biology*, 5(6), 427–431.

Cassidy, A. and L. Hooper (2006) Phytoestrogens and cardiovascular disease. *Journal of the British Menopause Society*, 12(2), 49–56.

Ceppi, P. (1998) Uso de las saponinas en cosemticos. Faculty of Engineering. Santiago, Chile, Catholic University of Chile. B.Sc. thesis.

Chen, J.S., Wei, C.-i. et al. (1991) Inhibitory effect of kojic acid on some plant and crustacean polyphenol oxidases. *Journal of Agricultural and Food Chemistry*, 39(8), 1396–1401.

Chen, J.S., Wei, C.I. et al. (1991) Inhibition mechanism of kojic acid on polyphenol oxidase. *Journal of Agricultural and Food Chemistry*, 39(11), 1897–1901.

Chotimarkorn, C. and Silalai, N. (2008a) Addition of rice bran oil to soybean oil during frying increases the oxidative stability of the fried dough from rice flour during storage. *Food Research International*, 41(3), 308–317.

Chotimarkorn, C. and Silalai, N. (2008b) Oxidative stability of fried dough from rice flour containing rice bran powder during storage. *LWT – Food Science and Technology*, 41(4), 561–568.

Christensen, L.P., Ronald Ross, W. et al. (2010) Bioactivity of Polyacetylenes in food plants. In: V.R. Preedy, R.R. Watson and V.B. Patel (eds) *Bioactive Foods in Promoting Health*, San Diego, Academic Press, pp. 285–306.

Clarkson, T.B. (2002) Soy, soy phytoestrogens and cardiovascular disease. *The Journal of Nutrition*, 132(3), 566S–569S.

Cowan, M.M. (1999) Plant Products as Antimicrobial Agents. *Clinical Microbiology Review*, 12(4), 564–582.

de Pascual-Teresa, S., Hallund, J. et al. (2006) Absorption of isoflavones in humans: effects of food matrix and processing. *The Journal of Nutritional Biochemistry*, 17(4), 257–264.

Delaha, E.C. and Garagusi, V.F. (1985) Inhibition of mycobacteria by garlic extract (Allium sativum). *Antimicrob Agents Chemotherapy*, 27(4), 485–486.

Dewell, A., Hollenbeck, P.L.W. et al. (2006) A critical evaluation of the role of soy protein and isoflavone supplementation in the control of plasma cholesterol concentrations. *Journal of Clinical Endocrinology and Metabolism*, 91(3), 772–780.

Dickens, J.A. and Ingram, K.D. (2001) Efficacy of an Herbal extract, at various concentrations, on the microbiological quality of broiler carcasses after simulated chilling. *Journal of Applied Poultry Research*, 10(2), 194–198.

Ding, C.-K., Chachin, K. et al. (2002) Inhibition of loquat enzymatic browning by sulfhydryl compounds. *Food Chemistry*, 76(2), 213–218.

Dixon, R.A., Dey, P.M. *et al.* (2006) *Phytoalexins: Enzymology and Molecular Biology*, New York, John Wiley & Sons, Inc.

Eissa, H.A., Fadel, H.H.M. *et al.* (2006) Thiol containing compounds as controlling agents of enzymatic browning in some apple products. *Food Research International*, 39(8), 855–863.

Esposito, F., Arlotti, G. *et al.* (2005) Antioxidant activity and dietary fibre in durum wheat bran by-products. *Food Research International*, 38(10), 1167–1173.

Fan, P., Bracken, C. *et al.* (1993) Structural characterization of monellin in the alcohol-denatured state by NMR: Evidence for .beta.-sheet to .alpha.-helix conversion. *Biochemistry*, 32(6), 1573–1582.

Francis, G., Kerem, Z. *et al.* (2002) The biological action of saponins in animal systems: a review. *British Journal of Nutrition*, 88, 587–605.

Frankel, E.N., Huang, S.-W. *et al.* (1996) Antioxidant Activity of a rosemary extract and its constituents, carnosic acid, carnosol, and rosmarinic acid, in bulk oil and oil-in-water emulsion. *Journal of Agricultural and Food Chemistry*, 44(1), 131–135.

Friedman, M. and Bautista, F.F. (1995) Inhibition of Polyphenol Oxidase by Thiols in the Absence and Presence of Potato Tissue Suspensions. *Journal of Agricultural and Food Chemistry*, 43(1), 69–76.

Friedman, M. and Molnar-Perl I. (1990). "Inhibition of browning by sulfur amino acids. 1. Heated amino acid-glucose systems." *Journal of Agricultural and Food Chemistry,* 38(8), 1642–1647.

Friedman, M., I. Molnár-Perl, *et al.* (1992). "Browning prevention in fresh and dehydrated potatoes by SH-containing amino acids." *Food Addit Contam.* 9(5), 499–503.

Fu, G., H. Pang, *et al.* (2008). "Naturally Occurring Phenylethanoid Glycosides, Potential Leads for New Therapeutics" *Curr Med Chem.* 15(2), 2592–2612.

Ganhão, R., D. Morcuende, *et al.* (2010). "Protein oxidation in emulsified cooked burger patties with added fruit extracts: Influence on colour and texture deterioration during chill storage." *Meat Science* 85(3), 402–409.

Gao, G.-H., J.-X. Dai, *et al.* (1999). "Solution conformation of brazzein by 1 H nuclear magnetic resonance: resonance assignment and secondary structure." *International Journal of Biological Macromolecules* 24(4), 351–359.

Ghisalberti, E.L. and R. Atta (2002). Secondary metabolites with antinematodal activity. Studies in Natural Products Chemistry, *Elsevier*. Volume 26, Part 7, 425–506.

Gibbs, B.F., I. Alli, *et al.* (1996). "Sweet and taste-modifying proteins: A review." *Nutrition Research* 16(9), 1619–1630.

Gonzalez-Coloma, A., M. Reina, *et al.* (2010). *Natural Product-Based Biopesticides for Insect Control.* Comprehensive Natural Products II. Oxford, Elsevier: 237–268.

Gordon, M.H. and L. Kourimská (1995). "Effect of antioxidants on losses of tocopherols during deep-fat frying." *Food Chemistry* 52(2), 175–177.

Gouin, S. (2004). "Microencapsulation: industrial appraisal of existing technologies and trends." *Trends in Food Science & Technology NFIF* part 2 15(7–8), 330–347.

Griffiths, G., L. Trueman, *et al.* (2002). "Onions—A global benefit to health." *Phytotherapy Research* 16(7), 603–615.

Guandalini, E., A. Ioppolo, *et al.* (1998). "4-hexylresorcinol as inhibitor of shrimp melanosis: Efficacy and residues studies; evaluation of possible toxic effect in a human intestinal *in vitro* model (caco-2); preliminary safety assessment." *Food Additives and Contaminants* 15(2), 171–180.

Gutierrez, J., C. Barry-Ryan, *et al.* (2008). "The antimicrobial efficacy of plant essential oil combinations and interactions with food ingredients." *International Journal of Food Microbiology* 124(1), 91–97.

Gutierrez, J., C. Barry-Ryan, *et al.* (2009). "Antimicrobial activity of plant essential oils using food model media: Efficacy, synergistic potential and interactions with food components." *Food Microbiology* 26(2), 142–150.

Han, J., M. Britten, *et al.* (2011). "Effect of polyphenolic ingredients on physical characteristics of cheese." *Food Research International* 44(1), 494–497.

Han, J., M. Britten, *et al.* (2011). "Polyphenolic compounds as functional ingredients in cheese." *Food Chemistry* 124(4), 1589–1594.

Harpaz, S., L. Glatman, *et al.* (2003). "Effects of Herbal Essential Oils Used To Extend the Shelf Life of Freshwater-Reared Asian Sea Bass Fish (Lates calcarifer)." *Journal of Food Protection* 66, 410–417.

Heinzelmann, K. and K. Franke (1999). "Using freezing and drying techniques of emulsions for the microencapsulation of fish oil to improve oxidation stability." *Colloids and Surfaces B: Biointerfaces* 12(3–6), 223–229.

Hellekant, G. r. and V. Danilova (2005). "Brazzein a Small, Sweet Protein, Discovery and Physiological Overview." *Chemical Senses* 30(suppl 1), i88–i89.

Ho, C.T., C.W. Chen, *et al.* (1997). Natural Antioxidants from tea. Natural Antioxidants, Chemistry, Health Effects and Applications. F. Shahidi. Champaign, Illinois, AOCS Press, 213–223.

Holland, B., A.A. Welch, *et al.* (1991). *The composition of foods*, Cambridge, UK: Royal Society of Chemistry., 10–21.

Holley, R.A. and D. Patel (2005). "Improvement in shelf-life and safety of perishable foods by plant essential oils and smoke antimicrobials." *Food Microbiology* 22(4), 273–292.

Iyidogan, N.F. and A. BayIndIrlI (2004). "Effect of -cysteine, kojic acid and 4-hexylresorcinol combination on inhibition of enzymatic browning in Amasya apple juice." *Journal of Food Engineering* 62(3), 299–304.

Janovitz-Klapp, A.H., F.C. Richard, *et al.* (1990). "Inhibition studies on apple polyphenol oxidase." *Journal of Agricultural and Food Chemistry* 38(4), 926–931.

Jiang, Y., J. Li, *et al.* (2005). "Effects of chitosan coating on shelf life of cold-stored litchi fruit at ambient temperature." *LWT – Food Science and Technology* 38(7), 757–761.

Kim, C.Y. and M.J. Kim (2007). "Inhibition of Polyphenol Oxidase and Peach Juice Browning by Onion Extract." *Food Science and Biotechnology* 16(3), 421–425.

Kim, M.-J., C.Y. Kim, *et al.* (2005). "Prevention of enzymatic browning of pear by onion extract." *Food Chemistry* 89(2), 181–184.

Kobaisy, M., Z. Abramowski, *et al.* (1997). "Antimycobacterial Polyynes of Devil's Club (Oplopanax horridus), a North American Native Medicinal Plant." *Journal of Natural Products* 60(11), 1210–1213.

Korkina, L.G. (2007). "Phenylpropanoids as naturally occurring antioxidants: from plant defense to human health." *Cellular and Molecular Biology* 53(1), 15–25.

Kurihara, Y. (1992). "Characteristics of antisweet substances, sweet proteins, and sweetness-inducing proteins." *Critical Reviews in Food Science and Nutrition* 32(3), 231–252.

Kurkin, V.A. (2003). "Phenylpropanoids from Medicinal Plants: Distribution, Classification, Structural Analysis, and Biological Activity." *Chemistry of Natural Compounds* 39(2), 123–153.

Labuckas, D.O., D.M. Maestri, *et al.* (2008). "Phenolics from walnut (Juglans regia L.) kernels: Antioxidant activity and interactions with proteins." *Food Chemistry* 107(2), 607–612.

Ladikos, D. and V. Lougovois (1990). "Lipid oxidation in muscle foods: A review." *Food Chemistry* 35(4), 295–314.

Lagouri, V., D. Boskou, *et al.* (1995). Screening for antioxidant activity of essential oils obtained from spices. Developments in Food Science, Elsevier. Volume 37, Part 1, 869–879.

Lai, P., K.Y. Li, *et al.* (2009). "Phytochemicals and antioxidant properties of solvent extracts from Japonica rice bran." *Food Chemistry* 117(3), 538–544.

Lee, J.Y., H.J. Park, *et al.* (2003). "Extending shelf-life of minimally processed apples with edible coatings and antibrowning agents." *Lebensmittel-Wissenschaft und-Technologie* 36(3), 323–329.

Lee, M.-K. (2007). "Inhibitory effect of banana polyphenol oxidase during ripening of banana by onion extract and Maillard reaction products." *Food Chemistry* 102(1), 146–149.

Lerma-Garcia, M.J., J.M. Herrero-Martinez, *et al.* (2009). "Composition, industrial processing and applications of rice bran gamma-oryzanol." *Food Chemistry* 115(2), 389–404.

Lichtenstein, A.H., S.M. Jalbert, *et al.* (2002). "Lipoprotein Response to Diets High in Soy or Animal Protein With and Without Isoflavones in Moderately Hypercholesterolemic Subjects." *Arterioscler Thromb Vasc Biol* 22(11), 1852–1858.

Lis-Balchin, M., G. Buchbauer, *et al.* (1998). "Antimicrobial activity of Pelargonium essential oils added to a quiche-filling as a model food system." *Letters in Applied Microbiology* 27(4), 207–210.

Liu, X., S. Maeda, *et al.* (1993). "Purification, complete amino acid sequence and structural characterization of the heat-stable sweet protein, mabinlin II." *European Journal of Biochemistry* 211(1–2), 281–287.

Lozano-De-Gonzalez, P.G., D.M. Barrett, *et al.* (1993). "Enzymatic Browning Inhibited in Fresh and Dried Apple Rings by Pineapple Juice." *Journal of Food Science* 58(2), 399–404.

Luciano, F.B. and R.A. Holley (2009). "Enzymatic inhibition by allyl isothiocyanate and factors affecting its antimicrobial action against Escherichia coli O157:H7." *International Journal of Food Microbiology* 131(2–3), 240–245.

Mathison, D.A., D.D. Stevenson, *et al.* (1985). "Precipitating Factors in Asthma." *Chest* 87(1 Supplement), 50S–54S.

Meot-Duros, L., S. Cérantola, *et al.* (2010). "New antibacterial and cytotoxic activities of falcarindiol isolated in Crithmum maritimum L. leaf extract." *Food and Chemical Toxicology* 48(2), 553–557.

Messina, M., S. Ho, *et al.* (2004). "Skeletal benefits of soy isoflavones: a review of the clinical trial and epidemiologic data." *Current Opinion in Clinical Nutrition & Metabolic Care* 7(6), 649–658.

Ming, D. and G. Hellekant (1994). "Brazzein, a new high-potency thermostable sweet protein from Pentadiplandra brazzeana B." *FEBS Letters* 355(1), 106–108.

Mohdaly, A.A.A., M.A. Sarhan, et al. (2010). "Antioxidant efficacy of potato peels and sugar beet pulp extracts in vegetable oils protection." *Food Chemistry* 123(4), 1019–1026.

Molnar-Perl, I. and M. Friedman (1990). "Inhibiton of browning by sulfur amino acids. 2. Fruit juices and protein-containing foods." *Journal of Agricultural and Food Chemistry* 38(8), 1648–1651.

Montero-Calderón, M., M.A. Rojas-Graü, et al. (2008). "Effect of packaging conditions on quality and shelf-life of fresh-cut pineapple (Ananas comosus)." *Postharvest Biology and Technology* 50(2-3), 182–189.

Montero, P., O. Martínez-Álvarez, et al. (2004). "Effectiveness of Onboard Application of 4-Hexylresorcinol in Inhibiting Melanosis in Shrimp (Parapenaeus longirostris)." *Journal of Food Science* 69(8), C643–C647.

Morris, J.A., R. Martenson, et al. (1973). "Characterization of Monellin, a Protein That Tastes Sweet." *Journal of Biological Chemistry* 248(2), 534–539.

Moure, A., J.M. Cruz, et al. (2001). "Natural antioxidants from residual sources." *Food Chemistry* 72(2), 145–171.

Murado, M.A. and J.A. Vazquez (2010). "Mathematical Model for the Characterization and Objective Comparison of Antioxidant Activities." *Journal of Agricultural and Food Chemistry* 58(3), 1622–1629.

Nadarajah, D., J.H. Han, et al. (2005). "Inactivation of Escherichia coli O157,H7 in packaged ground beef by allyl isothiocyanate." *International Journal of Food Microbiology* 99(3), 269–279.

Nor, F.M., S. Mohamed, et al. (2008). "Antioxidative properties of Pandanus amaryllifolius leaf extracts in accelerated oxidation and deep frying studies." *Food Chemistry* 110(2), 319–327.

Oka, Y. (2010). "Mechanisms of nematode suppression by organic soil amendments--A review." *Applied Soil Ecology* 44(2), 101–115.

Oms-Oliu, G., M.A. Rojas-Graü, et al. (2010). "Recent approaches using chemical treatments to preserve quality of fresh-cut fruit: A review." *Postharvest Biology and Technology* 57(3), 139–148.

Oms-Oliu, G., R. Soliva-Fortuny, et al. (2008). "Physiological and microbiological changes in fresh-cut pears stored in high oxygen active packages compared with low oxygen active and passive modified atmosphere packaging." *Postharvest Biology and Technology* 48(2), 295–301.

Oms-Oliu, G., R. Soliva-Fortuny, et al. (2008). "Using polysaccharide-based edible coatings to enhance quality and antioxidant properties of fresh-cut melon." *LWT - Food Science and Technology* 41(10), 1862–1870.

Pezzuto, J.M., C.M. Compadre, et al. (1985). "Metabolically activated steviol, the aglycone of stevioside, is mutagenic." *Proceedings of the National Academy of Sciences of the United States of America* 82(8), 2478–2482.

Poletti, S., W. Gruissem, et al. (2004). "The nutritional fortification of cereals." *Current Opinion in Biotechnology* 15(2), 162–165.

Proestos, C., N. Chorianopoulos, et al. (2005). "RP-HPLC Analysis of the Phenolic Compounds of Plant Extracts. Investigation of Their Antioxidant Capacity and Antimicrobial Activity." *Journal of Agricultural and Food Chemistry* 53(4), 1190–1195.

Pszczola, D.E. (2002). "Antimicrobials: setting up additional hurdles to ensure food safety." *Food and Technology* (56), 99–107.

Richard-Forget, F.C., P.M. Goupy, et al. (1992). "Cysteine as an inhibitor of enzymic browning. 2. Kinetic studies." *Journal of Agricultural and Food Chemistry* 40(11), 2108–2113.

Rojas-Graü, M.A., R. Grasa-Guillem, et al. (2007). "Quality Changes in Fresh-Cut Fuji Apple as Affected by Ripeness Stage, Antibrowning Agents, and Storage Atmosphere." *Journal of Food Science* 72(1), S036–S043.

Rojas-Graü, M.A., A. Sobrino-López, et al. (2006). "Browning Inhibition in Fresh-cut 'Fuji' Apple Slices by Natural Antibrowning Agents." *Journal of Food Science* 71(1), S59–S65.

Rojas-Graü, M.A., R. Soliva-Fortuny, et al. (2009). "Edible coatings to incorporate active ingredients to fresh-cut fruits: a review." *Trends in Food Science & Technology* 20(10), 438–447.

Rojas-Graü, M.A., M.S. Tapia, et al. (2008). "Using polysaccharide-based edible coatings to maintain quality of fresh-cut Fuji apples." *LWT - Food Science and Technology* 41(1), 139–147.

Roldan, E., C. Sanchez-Moreno, et al. (2008). "Characterisation of onion (Allium cepa L.) by-products as food ingredients with antioxidant and antibrowning properties." *Food Chemistry* 108(3), 907–916.

Rose, P., M. Whiteman, et al. (2005). "Bioactive S-alk(en)yl cysteine sulfoxide metabolites in the genus Allium: the chemistry of potential therapeutic agents." *Natural Product Reports* 22(3), 351–368.

Rossi, M., C. Alamprese, et al. (2007). "Tocopherols and tocotrienols as free radical-scavengers in refined vegetable oils and their stability during deep-fat frying." *Food Chemistry* 102(3), 812–817.

Sacchetti, G., S. Maietti, et al. (2005). "Comparative evaluation of 11 essential oils of different origin as functional antioxidants, antiradicals and antimicrobials in foods." *Food Chemistry* 91(4), 621–632.

Saguy, I.S. and D. Dana (2003). "Integrated approach to deep fat frying: engineering, nutrition, health and consumer aspects." *Journal of Food Engineering* 56(2–3), 143–152.

Sallam, K.I., M. Ishioroshi, et al. (2004). "Antioxidant and antimicrobial effects of garlic in chicken sausage." *Lebensmittel-Wissenschaft und-Technologie* 37(8), 849–855.

Sapers, G.M. and R.L. Miller (1998). "Browning Inhibition in Fresh-Cut Pears." *Journal of Food Science* 63(2), 342–346.

Sarnthein-Graf, C. and C. La Mesa (2004). "Association of saponins in water and water-gelatine mixtures." *Thermochimica Acta* 418(1-2), 79–84.

Schieber, A., F.C. Stintzing, et al. (2001). "By-products of plant food processing as a source of functional compounds – recent developments." *Trends in Food Science & Technology* 12(11), 401–413.

Serfert, Y., S. Drusch, et al. (2010). "Sensory odour profiling and lipid oxidation status of fish oil and microencapsulated fish oil." *Food Chemistry* 123(4), 968–975.

Shahidi, F. and U. Wanasundara (1994). Stabilization of Canola Oil by Natural Antioxidants. *Lipids in Food Flavors, American Chemical Society.* 558, 301–314.

Shahidi, F. and Y. Zhong (2005). Lipid Oxidation: Measurement Methods. Bailey's Industrial Oil & Fat Products. F. Shahidi. New Jersey, John Wiley & Sons Inc., 357–385.

Shin, I.S., H. Masuda, et al. (2004). "Bactericidal activity of wasabi (Wasabia japonica) against Helicobacter pylori." *International Journal of Food Microbiology* 94(3), 255–261.

Son, S.M., K.D. Moon, et al. (2001). "Inhibitory effects of various antibrowning agents on apple slices." *Food Chemistry* 73(1), 23–30.

Stadtman, E.R. and R.L. Levine (2003). "Free radical-mediated oxidation of free amino acids and amino acid residues in proteins." *Amino Acids* 25(3), 207–218.

Stintzing, F.C. and R. Carle (2007). "Betalains - emerging prospects for food scientists." *Trends in Food Science & Technology* 18(10), 514–525.

Suh, M.-H., S.-H. Yoo, et al. (2007). "Antioxidative activity and structural stability of microencapsulated [gamma]-oryzanol in heat-treated lards." *Food Chemistry* 100(3), 1065–1070.

Sun-Waterhouse, D., J. Zhou, et al. (2011). "Stability of encapsulated olive oil in the presence of caffeic acid." *Food Chemistry* 126(3), 1049–1056.

Tajkarimi, M.M., S.A. Ibrahim, et al. (2010). "Antimicrobial herb and spice compounds in food." *Food Control* 21(9), 1199–1218.

Tanaka, Y., N. Sasaki, et al. (2008). "Biosynthesis of plant pigments: anthocyanins, betalains and carotenoids." *The Plant Journal* 54(4), 733–749.

Tapia, M.S., M.A. Rojas-Graü, et al. (2008). "Use of alginate- and gellan-based coatings for improving barrier, texture and nutritional properties of fresh-cut papaya." *Food Hydrocolloids* 22(8), 1493–1503.

Tapia, M.S., M.A. Rojas-Graü, et al. (2007). "Alginate- and Gellan-Based Edible Films for Probiotic Coatings on Fresh-Cut Fruits." *Journal of Food Science* 72(4), E190-E196.

Timberlake, C.F. and B.S. Henry (1986). "Plant pigments as natural food colours." *Endeavour* 10(1), 31–36.

Tiwari, B.K., V.P. Valdramidis, et al. (2009). "Application of Natural Antimicrobials for Food Preservation." *Journal of Agricultural and Food Chemistry* 57(14), 5987–6000.

Tsao, S.-M. and M.-C. Yin (2001). "In-vitro antimicrobial activity of four diallyl sulphides occurring naturally in garlic and Chinese leek oils." *J Med Microbiol* 50(7), 646–649.

Ujváry, I. and K. Robert (2009). Pest Control Agents from Natural Products. *Hayes' Handbook of Pesticide Toxicology* (Third Edition). New York, Academic Press, 119–229.

Van Vuuren, S.F., S. Suliman, et al. (2009). "The antimicrobial activity of four commercial essential oils in combination with conventional antimicrobials." *Letters in Applied Microbiology* 48(4), 440–446.

Velasco, J., F. Holgado, et al. (2009). "Influence of relative humidity on oxidation of the free and encapsulated oil fractions in freeze-dried microencapsulated oils." *Food Research International* 42(10), 1492–1500.

Ventanas, S., J. Ventanas, et al. (2007). "Extensive feeding versus oleic acid and tocopherol enriched mixed diets for the production of Iberian dry-cured hams: Effect on chemical composition, oxidative status and sensory traits." *Meat Science* 77(2), 246–256.

Wanasundara, U. and F. Shahidi (1994). "Canola extract as an alternative natural antioxidant for canola oil." *Journal of the American Oil Chemists' Society* 71(8), 817–822.

Wang, C.Y. (2003). "Maintaining postharvest quality of raspberries with natural volatile compounds." *International Journal of Food Science & Technology* 38(8), 869–875.

Wang, S.Y. and C.-T. Chen (2010). "Effect of allyl isothiocyanate on antioxidant enzyme activities, flavonoids and post-harvest fruit quality of blueberries (Vaccinium corymbosum L., cv. Duke)." *Food Chemistry* 122(4), 1153–1158.

Wang, S.Y., C.-T. Chen, et al. (2010). "Effect of allyl isothiocyanate on antioxidants and fruit decay of blueberries." *Food Chemistry* 120(1), 199–204.

Wang, Z.W., M.Y. Gu, *et al.* (2005). "Surface Properties of Gleditsia Saponin and Synergisms of Its Binary System " *Journal of Dispersion Science and Technology* 26(3), 341–347.

Ward, S.M., P.J. Delaquis, *et al.* (1998). "Inhibition of spoilage and pathogenic bacteria on agar and pre-cooked roast beef by volatile horseradish distillates." *Food Research International* 31(1), 19–26.

Weggemans, R.M. and E.A. Trautwein (2003). "Relation between soy-associated isoflavones and LDL and HDL cholesterol concentrations in humans: a meta-analysis." *Eur J Clin Nutr* 57(8), 940–946.

Wildman, R.E.C., Ed. (2007). Handbook of Nutraceuticals and Functional Foods Boca Raton, CRC Press.

Wong, D.W.S., S.J. Tillin, *et al.* (1994). "Gas exchange in cut apples with bilayer coatings." *Journal of Agricultural and Food Chemistry* 42(10), 2278–2285.

Wrolstad, R.E. and L. Wen (2001). *Natural antibrowning and antioxidant compositions and methods for making the same*. United States, The State of Oregon Acting By and Through the State Board of Higher Education on Behalf of Oregon State University.

Xiong, X.L. (2000). Protein oxidation and implications for muscle foods quality. *Antioxidants in muscle foods*. C. Faustman and C.J. Lopez-Bote. New York, John Wiley & Sons Ltd., 85–111.

Yanishlieva, N.V. and E.M. Marinova (2001). "Stabilisation of edible oils with natural antioxidants." *European Journal of Lipid Science and Technology* 103(11), 752–767.

Yin, M.-c. and W.-s. Cheng (2003). "Antioxidant and antimicrobial effects of four garlic-derived organosulfur compounds in ground beef." *Meat Science* 63(1), 23–28.

Yin, M.C. and C. Faustman (1993). "Influence of temperature, pH, and phospholipid composition upon the stability of myoglobin and phospholipid: A liposome model." *Journal of Agricultural and Food Chemistry* 41(6), 853–857.

Zhang, D. and P.C. Quantick (1997). "Effects of chitosan coating on enzymatic browning and decay during postharvest storage of litchi (Litchi chinensis Sonn.) fruit." *Postharvest Biology and Technology* 12(2), 195–202.

Index

abiotic stresses, 201, 214
ABTS, 320, 322, 453, 460, 463
accessions, 202, 204–5, 208, 209, 223
achenes, 109, 110
acidification, 338, 342
acylation, 333, 335, 347, 349, 351–2, 359
agglutination, 26, 27, 55
aggregation, 24, 60, 62, 82, 83, 86, 87, 121, 148
aglycones, 30, 81, 341, 346, 360, 361, 418, 422, 439
aleurone, 139, 141, 189, 308, 327
alfalfa, 144, 148, 213, 215, 335, 478
alkaloids, 18, 28, 29, 35–7, 69, 75–6, 80–81, 94, 129, 144, 214, 264, 478, 495
allicin, 30, 75, 434–5
almonds, 163, 165, 167, 168, 171–2, 174–5
amino acids, 18, 35, 40, 117, 126, 189, 428, 441, 474–5, 492
angiogenesis, 30, 73, 88, 91
antagonistic action, 83
anthocyanidins, 13, 71, 75, 88, 148, 341, 342, 356, 357, 441, 460
anthocyanins, 20, 33, 52, 53, 71, 72, 81, 87, 111–13, 138–9, 142, 147–8, 151, 153, 184, 186, 188, 207, 208, 210, 219–20, 223, 237–8, 242, 247, 249, 252–6, 267–8, 270, 274, 278, 280, 283–4, 286, 288–90, 292, 305, 311, 312, 326, 332–3, 335–6, 338, 341–42, 348–50, 351–7, 359, 363, 376–7, 378, 381, 383, 402, 405, 408, 420, 422–3, 426, 440–42, 444–6, 460, 475, 477–8, 490
  stability, 292, 351–4, 356–7, 359, 376–7, 380
  synthesis, 280
anti-cancer, 19, 29–30, 72, 75, 77, 78, 88–94, 141, 146, 265, 332, 452

anti-inflammatory, 33, 34, 53, 60, 77, 78, 81, 84, 90–93, 121, 139, 141, 147, 186, 332, 333, 399, 477
antimetastatic action, 93
antimicrobial activity, 382, 495
antimicrobials, 292, 382, 491
antioxidant activity, 57, 69, 71, 82, 116, 126, 143, 171–4, 183, 191, 210, 211, 216, 237, 239, 241, 249, 255, 261, 266–8, 276, 277, 280, 289, 321–4, 353, 356, 377, 379, 380, 408, 418, 420, 425, 452–65, 473, 481, 485, 493
antioxidant assays, 457–8, 453–65
antioxidant capacity, 57, 58, 73, 83, 141, 145, 150, 154, 174, 181, 207, 210, 213, 220, 221, 235, 238, 240–3, 249, 252, 253, 275, 277–80, 292, 304–5, 313, 318, 320–21, 323, 341, 363, 380, 383, 426, 453, 454, 459, 460, 462–4, 479, 481, 485, 493, 494
antioxidant compounds, 107, 182, 240, 316, 317, 320–23, 454, 456–8
antioxidative properties, 80, 182, 304, 305, 320, 321
antipyretic, 75
antiradical, 333
antitussive, 77
apigenin, 60, 61, 83, 88, 90, 304, 402, 403, 407
apple, 187, 251, 265, 278, 285, 407, 421
apple cider, 292, 407
apple juice, 58, 75, 267, 380, 407, 425
apple peels, 109, 421–23
apple pomace, 187, 407–11, 424
ascorbate, 56, 58, 213, 289, 355, 453, 456, 461
ascorbic acid, 83, 213, 235–7, 238, 240–43, 248, 249, 251–3, 257, 267, 276, 278, 280, 282, 284, 286, 288–92, 332, 342, 344, 348, 352, 354–5, 357, 377–8, 379–81, 380, 381, 383, 385, 456, 457, 461, 483, 488–90
atherosclerosis, 72, 81–4, 86, 164
autoxidation, 358, 461, 481, 482

Handbook of Plant Food Phytochemicals: Sources, Stability and Extraction, First Edition.
Edited by B.K. Tiwari, Nigel P. Brunton and Charles S. Brennan.
© 2013 John Wiley & Sons, Ltd. Published 2013 by John Wiley & Sons, Ltd.

baking, 4, 151, 266–7, 303–4, 307, 312, 315–23, 327, 341, 343
baking time, 266, 315, 322
banana, 109, 114, 182, 188, 252, 280, 288, 489
banana peels, 188
barley, 68, 139–43, 150, 152, 210, 267, 304, 308, 310, 311, 313–15, 323, 324, 326
barley grains, 150, 267
beans, 153
Beer–Lambert law, 441
berries, 10, 52, 57, 72, 79, 88, 109, 110, 116, 117, 126, 188, 202, 208, 254–5, 341, 492, 494
betacyanins, 32, 33, 125, 126, 338, 340–42, 347, 350, 354, 355, 357, 363, 376, 378, 380, 478
betacyanin stability, 338, 340, 355
beta–glucan, 494
betalain stability, 341, 347, 354, 355, 357, 376, 380
betanidin, 126, 338, 350, 361, 376
betanin, 126, 335–6, 341–2, 350, 354, 355, 361, 364, 376, 380
betanin degradation, 342, 355, 361, 364, 380, 381
betanin stability, 335, 361
betaxanthins, 21, 33, 125, 126, 347, 350
bioavailability, 50–54, 57, 63, 93, 95, 96, 257, 261, 267, 303, 312, 315, 324, 332, 356, 364, 463, 493, 494
biodiversity, 7
biomarkers, 57, 58, 63, 143, 153
biosynthesis, 7, 35, 38–5, 222, 266, 280, 376
biosynthetic stages, 40
black rice, 142, 274, 311, 312, 342
black soybeans, 148, 373, 380–81
blackberries, 109–13, 117, 288
blackberry juice, 288, 290, 340
blanching, 247–50, 257, 261, 266, 342–3, 345–8, 381
blueberries, 72, 75, 91, 94, 109, 110, 254, 255, 341, 356
boiling, 150, 152, 248, 251, 266–8, 341, 343, 400, 402, 407, 419, 464
bran, 10, 17, 71, 73, 78, 126, 139, 141, 143, 144, 152, 153, 189, 190, 277, 303, 307–13, 327, 421, 481, 485–6, 493
Brazil nuts, 163–8, 172, 174, 175
bread, 140, 142, 150–153, 174, 189, 304, 312, 315–22, 343
broccoli, 10, 16, 24, 54, 57, 59, 75, 79, 110, 116–18, 126, 201, 203–6, 208, 210–215, 217, 218, 220–24, 238, 242–3, 252, 256, 267, 278, 280, 282, 288, 334, 341, 347, 359–63, 473, 475, 487
broccoli sprouts, 221
browning, 237, 240, 241
buckwheat, 73, 143, 152, 267, 304–5, 310–11, 317–24, 326
buckwheat grains, 304, 305, 318–9
by–products, 187, 413

cabbage, 10, 24–5, 79, 109–10, 119, 126, 129, 202–4, 206, 208, 211, 213, 216, 217, 224, 238, 248–50, 254–6, 332, 334, 357–9, 351, 352, 354, 356, 475
caffeic acid, 60, 87, 145, 150, 171, 237, 239, 254, 326, 351, 352, 486–7, 489
caffeine, 29, 267, 376, 413, 416
cancer cells, 61, 72, 73, 89–93, 123, 493
cancer chemoprevention, 79, 80, 88
canning, 4, 180, 254, 255, 342, 343
capsaicin, 29, 33, 34, 75, 94
capsaicinoids, 21, 33, 419, 423, 424
caramelisation, 489
carcinogenesis, 59, 68, 88, 91–3, 95, 334
cardiovascular diseases, 83, 493
carnosic acid, 92, 93, 418, 420, 455, 484, 487
carotenes, 15, 53, 81, 116, 188–9, 223, 343, 380, 478
carotenoid, 13, 23, 24, 82, 85, 113, 116–17, 142, 151, 153, 182, 188, 207, 209, 242, 254, 255, 266, 281–2, 286–7, 308, 316, 327, 342, 343, 347, 358, 361, 362, 376, 380, 440, 441, 447
carotenoid oxidation, 376–8
carotenoids degradation, 151
carotenoids losses, 151
carrot juice, 58, 188, 285, 338, 358–9
carrots, 19, 80, 109, 113, 116, 121, 123–5, 181, 206–10, 212–15, 217, 218, 221–3, 249–50, 255, 268, 282, 338, 342, 343, 361, 377–80, 383, 405, 416, 419, 476, 478, 479
casein kinase I (CKI), 61
cashew nuts, 163, 165, 167, 171–2, 174
cassava, 189, 218
catechins, 10, 83, 84, 88, 89, 92, 171, 187, 218, 318, 334, 338, 342, 344, 347, 352, 355, 356, 360, 363, 379–80, 381, 383, 406–8, 418, 420, 452
catechol, 82
cauliflower, 16, 24, 109–10, 118–19, 204–6, 208, 213, 223, 226, 256, 334
celery, 121, 124–6, 202, 206, 208, 286, 423, 477, 479

cell permeabilisation, 427
cell proliferation, 55, 60, 61, 72, 90–91, 94
cellulases, 187, 188, 421–3
cellulolytic enzyme, 422
cellulose, 8, 17, 139–40, 144, 145, 182–4, 415, 421
centrifugation, 436
cereals, 2, 49, 73, 78, 80, 85, 126, 138, 139, 141–4, 152, 167, 182, 189, 190, 208, 214, 304, 308, 310–13, 324–5, 338, 421, 452, 459, 481, 483, 492
chalcones, 13, 23, 71, 353
chemoprevention, 68, 74, 78, 88, 95–8
chemopreventive agents, 69, 73, 78, 79, 94, 95
chickpeas, 73, 149, 324, 335
chitosan, 355, 382, 490
chlorogenic acid, 11, 22, 69, 221, 251, 304, 305, 326, 351–2, 378, 381, 407, 420
chlorophyll breakdown, 218
cholesterol absorption, 71, 84, 147, 175
citrus, 10, 51, 79, 124, 181–2, 187, 208, 214, 381, 417, 423, 455, 487
clouding, 182
CO_2 atmospheres, 241, 243
coffee, 11, 28, 75, 202, 218, 413, 416, 455, 463
coffee decaffeination, 413
collision–induced dissociation (CID), 446
color stability, 333, 335, 339–40, 342, 348–54, 358, 376
color tonation, 352
colour stabilising, 495
copigmentation, 335, 342, 349, 351–3, 376
corn, 139, 141–4, 150–153, 265, 267, 308, 324, 326, 340, 341, 421, 444, 484, 486
courmarins, 92
COX enzymes, 61
cranberries, 109, 186, 268, 376, 383
cruciferous, 24, 54, 79, 88, 117, 119, 129, 248, 253, 254, 399
cultivars, 109–21, 124, 126, 129, 141, 143, 146, 175, 183, 189, 202, 204–8, 211, 213–17, 222, 308, 310, 311, 327, 341, 354, 359, 383
cultivation, 25, 123, 215, 222
curcumin, 11, 60, 69, 73, 75, 88, 90–92, 95, 171, 276
cyanidin, 108, 112–13, 142, 147–8, 172, 188, 207, 237, 254–5, 274, 280, 282, 288, 290, 326, 335, 341, 342, 349, 351, 354, 357, 359, 377, 405, 441

cyanidin hexoside, 342
cyanidin–3–glucoside, 148, 274, 280, 288, 290, 335, 342, 354, 357, 377, 441
cytotoxicity, 30, 78, 91, 92, 120, 123

daidzein, 146, 152, 168, 335, 341, 348, 418, 445
decarboxylation, 338, 341–3, 350
degradation, 7, 20, 22, 29, 53, 57, 61, 84, 86, 147, 149–51, 186, 187, 192, 222, 238, 242, 243, 248, 250, 251, 253, 254, 266, 274, 275, 280, 284, 288–91, 304, 313, 315–18, 320–21, 323, 332–4, 336–47, 350–64, 375–81, 380–81, 400, 404–5, 407–9, 422, 424, 437, 446, 456, 459, 464
degradation pathways, 257, 333
dehydrated vegetables, 380
dehydration, 353, 361
demalonylation, 342
detoxification enzymes, 79
dielectric heating, 260, 262
dietary fiber, 17, 26, 138–41, 144, 145, 149, 152–3, 181–4, 437
diogenin, 147
discoloration, 352, 353, 382, 477, 488, 491
diseases, 1, 2, 29, 49, 50, 68–72, 75, 77–9, 81–2, 86, 89, 90, 92, 95, 129, 138, 141, 153, 164, 184, 185, 235–6, 264–5, 303, 318, 332–3, 375, 434, 452, 463, 492
diterpenes, 12, 35, 38, 84, 85, 92
down–regulating cyclin, 91
DPPH assay, 210, 456–7
DPPH scavenging activity, 277, 321, 322
drying, 149, 151, 180–81, 190, 218, 261, 267–8, 307, 327, 340, 342–4, 353, 354, 361, 376, 384–5, 417, 425, 435
durum wheat, 142, 150, 317

edible coatings, 382–3, 478, 490
EDTA, 355, 456, 462
egg–plant, 120
einkorn wheat, 316, 317
ejaculation, 77
electric field strength, 262, 264, 284, 425
electrical conductivity, 264
electrical resistance heating, 264
electron donor, 56, 288
emulsifiers, 385
encapsulation, 355, 362, 384–6, 487
endoplasmic reticulum (ER), 29
enterodiol, 71, 143, 148

enzymatic browning, 149, 187, 327, 340, 361, 488–90
enzyme activity, 218, 241, 281, 313, 334, 422
enzyme inactivation, 250, 281, 344, 348
enzyme–catalyzed reaction, 34
epicatechin, 63, 75, 77, 84, 86, 89, 148, 172, 174, 187, 334, 344, 377–8, 379, 380, 408, 418, 485
epidemiology, 372, 434, 463
epimerization, 344
ethanol, 87, 181, 187, 188, 191, 312, 341–3, 356, 360, 401–03, 405, 407–13, 416–21, 423, 424, 437, 456, 457, 460, 461, 484
exposure, 55, 60, 62, 88, 93, 96, 120, 121, 261, 273, 278, 280, 286, 288, 341, 342, 347, 358, 361, 376, 378, 381, 404, 487, 488
extraction, 4, 7, 126, 144, 171, 181–2, 188, 190–91, 267, 277, 286, 289–91, 306, 314, 318–21, 323, 327, 339, 341, 343, 348, 355, 358–60, 399–401, 403–38, 447, 457, 464, 478
extrudates, 150, 324, 326
extrusion cooking, 149, 324, 326, 343

falcarindiol, 19, 121, 123, 206–7, 209, 213, 214, 222, 249, 250, 479
falcarindiol–3–actetate, 250
falcarinol, 19, 121, 123, 206–7, 209–10, 213–15, 217, 249, 250, 379, 479
fatty acids, 38, 39, 71, 82, 140, 141, 144, 153, 165, 167, 174, 175, 183, 186, 202, 208, 256, 347, 384, 437, 439, 455, 461, 481, 487
fenugreek seeds, 147
fermentation, 4, 16, 17, 113, 140, 181, 182, 218, 312–16, 318–19, 321–3, 327, 347
fertilization, 27, 209–12
ferulic acid, 69, 139, 141, 145, 149, 150, 171, 221, 274, 310, 311, 313, 319, 324, 326, 407, 421, 475, 489
flavanols, 52, 59, 63, 72, 87, 88, 110, 289, 420, 440, 441
flavones, 13, 23, 72, 83, 88, 110, 275, 351, 423, 441
flavonoid, 61, 74, 84, 86, 107, 108, 110, 111, 139, 141, 147, 171, 172, 174, 208, 212–13, 221, 237, 241, 249, 250, 276, 278, 280, 288, 311, 318, 319, 321, 323, 351, 354, 383, 440, 442, 443, 446
flavonols, 13, 23, 52, 71, 72, 88, 110, 111, 171, 188, 201, 202, 206, 207, 210–212, 219, 236, 237, 255, 276–8, 290, 344, 352, 364, 407, 420, 441, 479
flavour, 24, 54, 113, 125, 207, 209, 219, 221, 250, 274, 281, 284, 289, 307, 312, 315, 481, 490
flesh, 24, 109, 110, 112, 113, 120, 268, 278
folates, 307, 312–15
Folin–ciocalteu method, 439
food processors, 3, 4, 488
fractionating, 440
frap (ferric reducing antioxidant power), 454
free radicals, 22, 23, 50, 56, 74, 93, 141, 174, 185, 256, 290, 291, 333, 362–3, 434, 452–4, 456–64, 480–2
fruit extracts, 91, 404, 490
fruit juices, 52, 113, 274, 280, 286, 288, 291–2, 341, 342, 354, 455, 475, 488
frying, 149, 247, 255–7, 267, 268, 340, 341, 481, 486–7
functional foods, 1, 50, 80, 107, 144, 152–3, 222, 380, 457, 459, 492

gallic acid, 22, 35, 77, 82–4, 87–8, 108, 145, 163, 171–2, 305, 418, 421, 441–2, 440, 480
garlic, 19, 20, 30, 61, 75, 79, 80, 88, 202, 207, 208, 211, 212, 353, 382, 475, 485, 491
genistin, 335, 339, 341
germination, 4, 12, 152, 185, 304–6, 307, 313, 314, 327
gingerols, 90
glucobrassicin, 117, 119, 203, 204, 209, 210, 213, 217, 238, 339, 342, 347
gluconapin, 119, 204, 209, 217, 334, 341, 347, 348, 360
glucoraphanin, 117–19, 203, 204, 208–9, 210–215, 217, 220, 223, 238, 242–3, 339, 342, 347, 355, 358, 360–63
glucoside cleavage, 347
glucosides, 52, 75, 108, 188, 207, 254, 255, 335, 341, 348, 359, 361, 407, 418
glucosinolate, 15, 16, 25, 40–41, 60, 62, 117–19, 202–4, 206, 208–9, 211–18, 220, 221, 223, 237–8, 243, 248, 253, 254, 334–5, 341, 344–6, 348, 360, 363, 442, 491, 495
glucuronides, 52–4
glucuronosylation, 349, 350
glutathione, 31, 54, 59, 68, 74, 79, 90, 93, 174, 452, 488–90
glycoalkaloids, 120–22, 129

glycosides, 18, 21, 22, 40, 52, 54, 75–7, 81, 110, 112, 120, 124, 126, 146, 147, 171, 186–7, 189, 204, 208, 210, 237, 241, 304, 318, 319, 349, 383, 404, 407, 418, 421, 424, 441, 477
Gorinstein, 260, 464
grain processing, 303–5, 307, 309, 311, 313, 315, 317, 319, 321, 323, 327, 481
grains, 4, 138, 202, 318, 364
grape juice, 86, 87, 284, 292, 340, 342, 353, 377, 379, 455
grapes, 62, 80, 85, 109, 110, 112, 113, 129, 181, 186, 187, 202, 212, 278, 280, 288, 340, 351, 408, 445, 492

harvest time, 120
harvesting stage, 126
hazelnuts, 163–5, 167–8, 171–2
health benefits, 1, 2, 17, 50, 69, 71, 78, 113, 116, 139, 143, 146, 147, 163–5, 167, 171, 174, 211, 237, 305, 312, 324, 332, 363, 375, 399, 405–7, 435, 492–4
heat stability, 250, 253, 342, 349, 350, 405
hemicelluloses, 144, 145
herbs, 11, 54, 70, 78, 81, 95, 125, 201, 202, 214, 222, 375, 401, 420, 437, 439, 443, 475, 481, 483, 494
hydrodistillation, 400, 401, 407, 440
hydrogen bonding, 27, 349
hydrophilic, 23, 24, 26, 72, 248, 437, 458–62, 464
hydroxyl radicals, 289, 291, 456
hydroxylation, 343, 440
hyperchromic, 335, 350–51
hypertension, 71, 77, 81, 189

inactivation, 248, 254, 266, 274, 280, 284, 286, 289, 292, 315, 343, 344, 347, 348, 402, 460, 491
indicators, 459
indicaxanthin, 21, 126
indoles, 10, 69, 79, 84, 88, 95, 450
inducers, 59, 62
industrial applications, 408, 424, 475–97
inflammation, 30, 56, 60, 62, 87, 185, 236, 402, 434
irradiation, 219, 220, 273, 274, 278, 280
isochorismate, 34, 35
isoflavones, 7, 13, 23, 69, 72, 80, 83, 84, 88, 110, 138, 146, 148, 152, 167–8, 171, 202, 208, 214, 217, 335, 341, 348, 361, 418, 421, 423, 428, 437, 443, 445, 493

isoflavonoids, 8, 23, 37, 69, 426, 439
isomerization, 151, 174, 266, 333, 338, 342, 343, 357–8, 360
isomers, 22, 142, 150, 254, 266, 334, 342, 356–8, 360, 381, 444, 447, 452
isothiocyanates, 25, 54, 61, 62, 79, 88, 117, 204, 224, 248, 253, 254, 334, 346, 348, 360

jams, 340, 352, 356
juice processing, 188, 354, 359
Juices, 407

kaempferol, 61, 75, 92, 110, 171, 204, 206, 207, 210–212, 220, 237–8, 275, 318–19, 326, 352, 380
kernel maturity, 328

lactonization, 35
LDL oxidation, 83, 87, 453
lectins, 18, 26–8, 55, 71, 82
legumes, 2, 85, 143–8, 164, 174, 208, 267, 308, 452, 492
lentil, 18, 144, 145, 148, 152
lettuce, 10, 109, 112, 116, 123, 124, 201, 208, 212, 214, 217, 236–46, 288, 383
light–induced betacyanin, 47
lignan levels, 210
lignans, 13, 22, 35, 71, 72, 138, 139, 141, 143, 148, 167, 171, 202, 208, 303, 305, 313, 314, 319, 445, 452
lignans content, 148
lignin, 16, 17, 35, 71, 144, 182, 238–40, 383
lipid oxidation, 325, 333, 384, 458, 461, 481, 482, 486, 487
lipid peroxidation, 58, 74, 84, 174, 323, 356, 453, 455, 458, 461, 464, 481
lutein, 15, 23, 24, 53, 54, 113–17, 139, 142, 150, 151, 153, 167–8, 188, 206, 207, 209, 213, 236, 238, 242, 255, 281, 288, 308, 327, 342, 343, 353, 358–60, 364, 380, 416–17, 441, 444, 447
lycopene, 23, 24, 53, 54, 80, 88, 113, 116, 129, 188, 208, 210, 214–15, 235, 249, 255, 266, 280–81, 285, 288–9, 342, 343, 358, 362, 380, 381, 416–17, 447, 460
lysosomal catabolism, 29

maceration time, 426
macrophages, 23, 34

Maillard reaction, 149, 261, 291, 304, 322–323, 327, 356, 489
malonyl glucoisdes, 418
malonyldaizin, 339
malvidin, 72, 148, 335, 341, 351, 352, 359
mandarin juice, 426
Mandarin peels, 424
mango, 91, 181–3, 256, 280, 284, 355
maturity, 107, 109–11, 113, 116, 121, 129, 185, 209–10, 219, 375
maturity index, 116
maturity stages, 107, 113, 116, 129, 209–10
mechanical abrasion, 240
mechanical agitation, 386
mechanistic pathways, 483
medioresinol, 143
melatonin, 175
meta–analysis, 56, 217
metabolism, 7, 26, 34, 35, 38, 50, 52–4, 56–60, 62, 79, 140, 189, 212, 219, 222, 264, 348, 358, 383, 434, 463–4
metabolites, 7, 8, 16, 18, 29, 34–9, 50, 53, 54, 59–61, 77, 85, 93, 94, 96, 117, 123, 216, 232, 334, 425, 426, 434, 435, 445–7
methanol, 145, 304, 312, 313, 321, 337, 341, 401, 406, 408, 412, 416, 418, 420, 424, 437, 440, 444, 457
methyl groups, 23
microbial safety, 3
microencapsulation, 384, 486, 487
microorganisms, 29, 35, 38, 261, 273, 274, 280, 284, 291, 344, 353, 382, 425, 478, 481
microwave cavity, 440
microwave treatment, 267
milling, 4, 149, 189, 303, 307–10, 312, 327, 403–4, 425
milling fractions, 308–10, 312
milling process, 149, 307–8, 312
minerals, 10, 58, 71, 82, 95, 139, 144, 149, 185, 189, 267, 315, 356, 382, 385, 474, 475
minimal processing, 4, 235–7, 250
modifiers, 415–19, 440
monoterpenes, 12, 35, 38, 183
myricetin, 74, 110, 171
myrosinase, 25, 54, 117, 118, 202, 204, 213, 215, 218, 238, 243, 248, 253, 334, 339, 341, 347, 348, 355, 358, 360–1

naringin, 91, 182, 424, 485
neutrophils, 23, 86

nitrogen fertilization, 210
non–thermal processing, 257
nutraceuticals, 1, 409, 453, 461
nuts, 2, 10, 68, 70–71, 73, 82, 85, 126, 143, 163–8, 170–172, 174–5, 186, 202, 208, 218, 225, 267, 277, 459, 480, 492

oats, 10, 301, 307, 308, 315, 324
ohmic heating, 260, 261, 264–6
olive oil, 181, 256, 267, 380, 417, 486, 487
onions, 11, 19, 88, 110, 207–8, 212, 218–20, 237, 473
orange juice, 266, 281, 284, 286–9, 291–2, 340, 354, 377, 379
organoleptic, 113, 236, 247, 316, 327, 491
oxidation process, 344, 455
oxidative damage, 16, 56–9, 73, 85, 107, 265, 452, 453
oxidative reactions, 34, 144, 490
oxidative stress, 58, 87, 91, 141, 164, 236, 319, 434, 458, 463
oxygen donors, 353
ozone, 22, 273, 286, 288
ozone applications, 286
ozone processing, 273, 286

packaging conditions, 383
parboiling, 327, 329
parsley, 19, 121, 202, 206, 207, 276, 419
parsnip, 19, 121, 207
pasteurization, 261, 285, 342, 348, 380
peanuts, 73, 80, 85, 144, 164, 168, 172
pearling, 308
peas, 129, 144, 147, 153, 181, 255, 265–6, 324
pecans, 163, 165, 167, 168, 172, 175
pectins, 10, 17, 181, 421
peeling, 4, 120, 123, 240, 344
pelargonidin, 112, 142, 147, 148, 255, 282, 284, 288–90, 335, 341, 349, 351, 357, 359
peppers, 10, 21, 94, 110, 116, 126, 202, 208, 249, 256, 419, 423, 424, 428, 479
permeability, 242, 382, 424
peroxidation, 56, 58, 83, 323, 333, 453, 455, 458, 461
pesticides, 25, 49, 55, 216, 286, 495
petanin, 335
Phase I, 11, 16, 50, 54, 59, 60, 62, 75, 92, 94, 253, 334
Phase II catalyses, 59
phenol rings, 13

phenolics, 23, 35, 36, 71, 72, 89, 107, 109, 138, 139, 141, 144, 149, 150, 167, 171–3, 183, 187, 191, 192, 206–12, 214, 217, 220, 221, 235, 237–2, 250–252, 254, 264, 267, 268, 276, 277, 279, 280, 289, 305, 321, 323, 326, 341, 351, 355, 376, 380, 383, 408, 409, 418, 422, 423, 436, 442, 460, 462, 480, 483, 491
phenylalanine, 34, 35, 40, 238–9, 280, 477
phenylpropanoids, 35, 477
phloridizin, 423
phospholipases, 86
phospholipid, 385
phosphorylation, 34, 60, 73, 82, 89, 94
photodegradation, 358
phototropism, 46
phytic acid, 71, 82, 84, 144, 148, 149, 152
phytoalexins, 86, 125
phytoestrogens, 144, 148, 171, 304, 305, 493
phytonutrients, 7, 96, 152, 174, 248, 267
phytosterols, 71, 84, 168, 174–5, 315, 399, 426, 434, 437–9, 442, 443, 445, 492
pigment stability, 350, 354, 359
pigmentation, 116, 235
pine nuts, 73, 163, 165, 167, 168, 171, 172
pineapple, 109, 181, 182, 280, 284, 286, 288, 488, 489
Pinto Bean, 148, 153
pistachios, 163, 165, 167, 168, 171, 172, 175
plant metabolites, 13, 24, 49, 50, 56, 60, 111, 120, 186
plant oils, 426
plant sterols, 82, 84, 202, 225, 303, 437, 492
plant terpenoids, 69
plant tissues, 111, 208, 222, 235, 239, 425, 440
plant wound, 221
plasma LDL cholesterol, 120
polarity, 400, 416, 419, 420, 440, 441, 458
policosanol, 379
polyacetylenes, 19, 29, 30, 121, 123, 201, 207, 209, 214, 215, 217, 222, 250, 251, 419, 479
polyketides, 35
polymerization, 140, 149, 152, 352, 442
polymorphisms, 54, 79
polyphenol content, 249, 254, 256, 304, 319, 380, 426
polyphenolics, 88, 167, 174, 236, 350, 352, 377, 381, 400
polysaccharides, 16, 26, 184, 421, 422, 426, 440, 494
Pomace, 275
pomegranate accessions, 135

post–harvest, 201, 218–22, 236, 238, 241, 242, 286, 358, 360, 362, 383, 400, 434
postharvest storage, 246
post–harvest treatment, 201, 218, 219, 358
potato peels, 120, 481
potatoes, 110, 112, 120, 121, 222, 255, 256, 265, 266, 268, 340, 359, 403
preservatives, 382, 491
pressurized liquid extraction, 435
proanthocyanins, 113
proapoptotic, 93, 95
procyanidins, 63, 83, 188, 380, 408, 420, 421, 441, 444
proofing, 315, 320, 322
pro–oxidants, 72, 380, 461
propagation phase, 459
protodioscin, 146
protogracillin, 146
protoneodioscin, 146
pulses, 4, 70, 73, 80, 138, 144, 145, 147, 148, 202, 214, 284, 285, 310, 311, 335, 413, 425, 426, 456
pyrazines, 475
pyrolysis, 333

quercetin, 10, 52, 53, 61, 69, 74, 75, 77, 83, 86, 88, 89, 91, 92, 94, 108, 110–112, 148, 171, 187, 204, 206–8, 210–212, 216, 219, 220, 237, 238, 252, 279, 304, 305, 318–20, 326, 352, 354, 404–6, 422, 441, 484, 490
quinoa, 304, 305, 317–19, 321
quinone, 21, 59, 68, 74, 89, 339

radiation processing, 275
radical scavenging assay, 323, 456–7, 460
radio frequency dielectric heating, 264
radish, 15, 112, 113, 202, 203, 211, 213, 215, 334, 352, 353, 359
raspberries, 72, 90, 109, 110, 112, 117, 254, 255, 340, 342, 380
reactivity, 31, 83, 357, 454, 462–4
red beet pigments, 380
redox regulation, 74
re–isomerization, 380
resveratrol, 39, 60, 68, 69, 72, 80, 83, 88–90, 171, 188, 278, 280, 495
roasting, 150, 172, 174, 267, 323, 324
ROS (reactive oxygen species), 141
ROS scavenging enzymes, 221
rosemary, 92, 93, 276, 342, 379, 382, 412, 420, 428, 440, 455, 475, 481, 484, 486, 487, 497

rutin, 52, 82, 86, 91, 110, 207, 275, 304, 305, 319, 320, 351, 422, 485
rye, 10, 139, 140, 143, 152, 167, 305–8, 311–15, 318–22, 324, 326

salt stress treatment, 215
saponification, 441
saponins, 71, 77, 81–3, 139, 146, 147, 152, 432, 441, 442, 445, 491, 495
sensory attributes, 268, 476
sensory quality, 274, 490
sesquiterpenes, 12, 35, 38, 41, 88, 123, 477, 497
sesterterpenes, 12
shelf life, 210, 220, 251, 260, 261, 273, 281, 284, 324, 358, 360, 362, 363, 375, 381, 382, 400, 474, 491
shredding, 237, 243, 347
signal transduction pathways, 364
sinigrin, 62, 119, 206, 334, 347, 348, 360, 491
soaking, 152, 189, 304, 327, 348, 353, 355, 437, 464
solubilization, 361
soluble fiber, 144
soluble phenolic compounds, 237, 239
solvent extraction, 267, 343, 418, 419, 426, 428, 435, 437, 439, 478, 479
sonication, 191, 289–91, 424
sorghum, 71, 141, 144, 150, 152, 310–12, 381
sourdough, 312–315, 318, 319, 321
Sous vide processing, 250, 251
Soxhlet extraction, 398, 412, 418, 437
soy isoflavones, 335
soyasaponins, 84, 147, 428
spectrophotometric, 436–8, 440–42, 446, 447, 456
spectrophotometric methods, 441, 442
spray drying, 340, 384–5
sprouting, 205, 206, 219, 223, 304, 305
stabilisation processes, 489
stability of catechins, 353, 358
stability of isoflavones, 339
starches, 10, 17, 140, 141, 180
steaming, 150, 248, 343, 345, 360
steroids, 35, 77, 120, 144, 175
sterols, 38, 71, 81, 95, 138, 139, 143, 168, 170, 188, 202, 208, 249, 304, 307, 313, 314
storage period, 219, 240, 275–7, 286, 288, 351, 376, 378–83
storage stability, 342, 361, 376, 380, 381, 490
strawberries, 10, 57, 109, 110, 112, 222, 264, 278, 286, 288, 342, 352, 359

strawberry jams, 338
strawberry juice, 288–91
stress compounds, 125
stress responses, 213, 221
subcritical water extraction, 419, 420
sulforaphane, 10, 54, 59, 62, 75, 79, 80, 88, 94, 95, 204, 210–212, 220–221, 223, 224, 258, 339, 348
sulfur fertilization, 211
supercritical fluids, 428
superoxides, 86
sweet potato, 109, 112, 113, 116, 256, 265, 267–8, 340, 341, 347, 352, 405, 409, 421, 455, 478

tannins, 8, 10, 13, 22, 71, 75, 77, 81, 87, 92, 113, 141, 145, 149, 150, 152, 171, 175, 184, 185, 187, 256, 275, 276, 399, 405, 406, 408
taste modifiers, 476
tea polyphenols, 72, 95, 267
terpenes, 7, 8, 12, 13, 21, 38, 40, 51, 69, 81, 88, 167, 214, 417, 423, 477, 483, 495
terpenoids, 12, 35, 37, 38, 51, 53, 69, 88, 123, 125, 144, 202, 399, 442, 478, 481
thermal processing, 53, 149, 150, 247, 248, 255–7, 264, 265, 267–9, 273–4, 281, 289, 291, 304, 320, 324, 341, 342, 423
thermostability, 342
tocols, 126, 139, 142, 143, 175, 308, 312, 317, 318, 322, 324, 327
tocopherols, 71, 72, 126, 129, 139, 142, 171, 172, 175, 308, 313, 315, 318, 324, 426, 452, 483–7
tocotrienols, 82, 126, 129, 139, 142, 303, 306, 308, 313, 318, 324, 486
tomatoes, 10, 53, 80, 88, 110, 113, 116, 120, 121, 188, 201, 208, 222, 255, 280, 288, 342, 380, 417
total anthocyanins, 112, 142, 148, 253–5, 274, 279, 340, 342, 350, 377, 379, 380
total carotenoid content, 441
total flavonoid content, 440
total glucosinolates, 206, 209, 211, 214, 215, 220, 249, 344, 346–7, 360, 363
transcription factors, 61, 72, 75
tree nuts, 2, 3, 163–5, 167–71, 173, 175
triterpenes, 12, 35, 92, 188, 495
tryptophan, 34, 40, 69, 189
turnip, 16, 24, 116, 119, 201–4, 206, 334
turpentine, 8

ultrasonic processing, 290
ultrasound, 4, 191, 250, 273, 289–3, 413, 417, 423, 424, 427, 428, 437
ultrasound–assisted extraction, 437
ultraviolet, 15, 68, 85, 95, 358, 435, 441
UV radiation, 22, 274, 280

vanillin, 90, 146, 150, 421, 440, 442
varietal differences, 317
vegetable oils, 129, 385, 488, 494
vitamins, 49, 58, 69, 72, 82, 95, 129, 139, 189, 217, 260, 265, 266, 281, 315, 381, 382–5, 452, 474, 481
volatile compounds, 19, 475
volatile sulphur compounds, 475

walnuts, 73, 163, 165, 167, 168, 171, 172, 175
water activity, 376, 378
water stress, 201, 214
watercress, 10, 59, 60, 75, 202, 253
water–soluble vitamins, 385
wine production, 426

xanthones, 13, 22, 185, 186

zeaxanthin, 15, 23, 24, 54, 114, 116, 139, 142, 150, 151, 167, 168, 236, 238, 255, 281, 308, 317, 327, 342, 353, 447

Food Science and Technology

GENERAL FOOD SCIENCE & TECHNOLOGY, ENGINEERING AND PROCESSING

Title	Author	ISBN
Organic Production and Food Quality: A Down to Earth Analysis	Blair	9780813812175
Handbook of Vegetables and Vegetable Processing	Sinha	9780813815411
Nonthermal Processing Technologies for Food	Zhang	9780813816685
Thermal Procesing of Foods: Control and Automation	Sandeep	9780813810072
Innovative Food Processing Technologies	Knoerzer	9780813817545
Handbook of Lean Manufacturing in the Food Industry	Dudbridge	9781405183673
Intelligent Agrifood Networks and Chains	Bourlakis	9781405182997
Practical Food Rheology	Norton	9781405199780
Food Flavour Technology, 2nd edition	Taylor	9781405185431
Food Mixing: Principles and Applications	Cullen	9781405177542
Confectionery and Chocolate Engineering	Mohos	9781405194709
Industrial Chocolate Manufacture and Use, 4th edition	Beckett	9781405139496
Chocolate Science and Technology	Afoakwa	9781405199063
Essentials of Thermal Processing	Tucker	9781405190589
Calorimetry in Food Processing: Analysis and Design of Food Systems	Kaletunç	9780813814834
Fruit and Vegetable Phytochemicals	de la Rosa	9780813803203
Water Properties in Food, Health, Pharma and Biological Systems	Reid	9780813812731
Food Science and Technology (textbook)	Campbell-Platt	9780632064212
IFIS Dictionary of Food Science and Technology, 2nd edition	IFIS	9781405187404
Drying Technologies in Food Processing	Chen	9781405157636
Biotechnology in Flavor Production	Havkin-Frenkel	9781405156493
Frozen Food Science and Technology	Evans	9781405154789
Sustainability in the Food Industry	Baldwin	9780813808468
Kosher Food Production, 2nd edition	Blech	9780813820934

FUNCTIONAL FOODS, NUTRACEUTICALS & HEALTH

Title	Author	ISBN
Functional Foods, Nutraceuticals and Degenerative Disease Prevention	Paliyath	9780813824536
Nondigestible Carbohydrates and Digestive Health	Paeschke	9780813817620
Bioactive Proteins and Peptides as Functional Foods and Nutraceuticals	Mine	9780813813110
Probiotics and Health Claims	Kneifel	9781405194914
Functional Food Product Development	Smith	9781405178761
Nutraceuticals, Glycemic Health and Type 2 Diabetes	Pasupuleti	9780813829333
Nutrigenomics and Proteomics in Health and Disease	Mine	9780813800332
Prebiotics and Probiotics Handbook, 2nd edition	Jardine	9781905224524
Whey Processing, Functionality and Health Benefits	Onwulata	9780813809038
Weight Control and Slimming Ingredients in Food Technology	Cho	9780813813233

INGREDIENTS

Title	Author	ISBN
Hydrocolloids in Food Processing	Laaman	9780813820767
Natural Food Flavors and Colorants	Attokaran	9780813821108
Handbook of Vanilla Science and Technology	Havkin-Frenkel	9781405193252
Enzymes in Food Technology, 2nd edition	Whitehurst	9781405183666
Food Stabilisers, Thickeners and Gelling Agents	Imeson	9781405132671
Glucose Syrups – Technology and Applications	Hull	9781405175562
Dictionary of Flavors, 2nd edition	De Rovira	9780813821351
Vegetable Oils in Food Technology, 2nd edition	Gunstone	9781444332681
Oils and Fats in the Food Industry	Gunstone	9781405171212
Fish Oils	Rossell	9781905224630
Food Colours Handbook	Emerton	9781905224449
Sweeteners Handbook	Wilson	9781905224425
Sweeteners and Sugar Alternatives in Food Technology	Mitchell	9781405134347

FOOD SAFETY, QUALITY AND MICROBIOLOGY

Title	Author	ISBN
Food Safety for the 21st Century	Wallace	9781405189118
The Microbiology of Safe Food, 2nd edition	Forsythe	9781405140058
Analysis of Endocrine Disrupting Compounds in Food	Nollet	9780813818160
Microbial Safety of Fresh Produce	Fan	9780813804163
Biotechnology of Lactic Acid Bacteria: Novel Applications	Mozzi	9780813815831
HACCP and ISO 22000 – Application to Foods of Animal Origin	Arvanitoyannis	9781405153669
Food Microbiology: An Introduction, 2nd edition	Montville	9781405189132
Management of Food Allergens	Coutts	9781405167581
Campylobacter	Bell	9781405156288
Bioactive Compounds in Foods	Gilbert	9781405158756
Color Atlas of Postharvest Quality of Fruits and Vegetables	Nunes	9781405158756
Microbiological Safety of Food in Health Care Settings	Lund	9781405122207
Food Biodeterioration and Preservation	Tucker	9781405154178
Phycotoxins	Botana	9780813827001
Advances in Food Diagnostics	Nollet	9780813822211
Advances in Thermal and Non-Thermal Food Preservation	Tewari	9780813829685

For further details and ordering information, please visit www.wiley.com/go/food

Food Science and Technology from Wiley-Blackwell

SENSORY SCIENCE, CONSUMER RESEARCH & NEW PRODUCT DEVELOPMENT

Sensory Evaluation: A Practical Handbook	Kemp	9781405162104
Statistical Methods for Food Science	Bower	9781405167642
Concept Research in Food Product Design and Development	Moskowitz	9780813824246
Sensory and Consumer Research in Food Product Design and Development	Moskowitz	9780813816326
Sensory Discrimination Tests and Measurements	Bi	9780813811116
Accelerating New Food Product Design and Development	Beckley	9780813808093
Handbook of Organic and Fair Trade Food Marketing	Wright	9781405150583
Multivariate and Probabilistic Analyses of Sensory Science Problems	Meullenet	9780813801780

FOOD LAWS & REGULATIONS

The BRC Global Standard for Food Safety: A Guide to a Successful Audit	Kill	9781405157964
Food Labeling Compliance Review, 4th edition	Summers	9780813821818
Guide to Food Laws and Regulations	Curtis	9780813819464
Regulation of Functional Foods and Nutraceuticals	Hasler	9780813811772

DAIRY FOODS

Dairy Ingredients for Food Processing	Chandan	9780813817460
Processed Cheeses and Analogues	Tamime	9781405186421
Technology of Cheesemaking, 2nd edition	Law	9781405182980
Dairy Fats and Related Products	Tamime	9781405150903
Bioactive Components in Milk and Dairy Products	Park	9780813819822
Milk Processing and Quality Management	Tamime	9781405145305
Dairy Powders and Concentrated Products	Tamime	9781405157643
Cleaning-in-Place: Dairy, Food and Beverage Operations	Tamime	9781405155038
Advanced Dairy Science and Technology	Britz	9781405136181
Dairy Processing and Quality Assurance	Chandan	9780813827568
Structure of Dairy Products	Tamime	9781405129756
Brined Cheeses	Tamime	9781405124607
Fermented Milks	Tamime	9780632064588
Manufacturing Yogurt and Fermented Milks	Chandan	9780813823041
Handbook of Milk of Non-Bovine Mammals	Park	9780813820514
Probiotic Dairy Products	Tamime	9781405121248

SEAFOOD, MEAT AND POULTRY

Handbook of Seafood Quality, Safety and Health Applications	Alasalvar	9781405180702
Fish Canning Handbook	Bratt	9781405180993
Fish Processing – Sustainability and New Opportunities	Hall	9781405190473
Fishery Products: Quality, safety and authenticity	Rehbein	9781405141628
Thermal Processing for Ready-to-Eat Meat Products	Knipe	9780813801483
Handbook of Meat Processing	Toldra	9780813821825
Handbook of Meat, Poultry and Seafood Quality	Nollet	9780813824468

BAKERY & CEREALS

Whole Grains and Health	Marquart	9780813807775
Gluten-Free Food Science and Technology	Gallagher	9781405159159
Baked Products – Science, Technology and Practice	Cauvain	9781405127028
Bakery Products: Science and Technology	Hui	9780813801872
Bakery Food Manufacture and Quality, 2nd edition	Cauvain	9781405176132

BEVERAGES & FERMENTED FOODS/BEVERAGES

Technology of Bottled Water, 3rd edition	Dege	9781405199322
Wine Flavour Chemistry, 2nd edition	Bakker	9781444330427
Wine Quality: Tasting and Selection	Grainger	9781405113663
Beverage Industry Microfiltration	Starbard	9780813812717
Handbook of Fermented Meat and Poultry	Toldra	9780813814773
Microbiology and Technology of Fermented Foods	Hutkins	9780813800189
Carbonated Soft Drinks	Steen	9781405134354
Brewing Yeast and Fermentation	Boulton	9781405152686
Food, Fermentation and Micro-organisms	Bamforth	9780632059874
Wine Production	Grainger	9781405113656
Chemistry and Technology of Soft Drinks and Fruit Juices, 2nd edition	Ashurst	9781405122863

PACKAGING

Food and Beverage Packaging Technology, 2nd edition	Coles	9781405189101
Food Packaging Engineering	Morris	9780813814797
Modified Atmosphere Packaging for Fresh-Cut Fruits and Vegetables	Brody	9780813812748
Packaging Research in Food Product Design and Development	Moskowitz	9780813812229
Packaging for Nonthermal Processing of Food	Han	9780813819440
Packaging Closures and Sealing Systems	Theobald	9781841273372
Modified Atmospheric Processing and Packaging of Fish	Otwell	9780813807683
Paper and Paperboard Packaging Technology	Kirwan	9781405125031

For further details and ordering information, please visit www.wiley.com/go/food